Handbook of *Listeria monocytogenes*

Handbook of *Listeria monocytogenes*

Edited by Dongyou Liu

CRC Press
Taylor & Francis Group
Boca Raton London New York

CRC Press is an imprint of the
Taylor & Francis Group, an **informa** business

CRC Press
Taylor & Francis Group
6000 Broken Sound Parkway NW, Suite 300
Boca Raton, FL 33487-2742

First issued in paperback 2019

ISBN-13: 978-1-4200-5140-7 (hbk)
ISBN-13: 978-0-367-38742-6 (pbk)

Library of Congress Cataloging-in-Publication Data

Handbook of Listeria monocytogenes / editor, Dongyou Liu.
p. ; cm.
Includes bibliographical references and index.
ISBN 978-1-4200-5140-7 (hardcover : alk. paper) 1. Listeria monocytogenes. I. Liu, Dongyou.
[DNLM: 1. Listeria monocytogenes. QW 142.5.A8 H236 2008]

QR82.Z9H36 2008
579.3'7--dc22 2007046284

Visit the Taylor & Francis Web site at
http://www.taylorandfrancis.com

and the CRC Press Web site at
http://www.crcpress.com

Contents

SECTION I Biology and Pathogenicity

SECTION II Identification and Detection

SECTION III Genomics and Proteomics

SECTION IV Immunity and Vaccines

Preface

Listeria monocytogenes is a fascinating creature. At a quick glance, it is an innocent, peaceful, and free-living Gram-positive rod no different from many other microbes distributed ubiquitously in the environment; its frequent occurrence and isolation often blunt our vigilance. However, beneath its gentle appearance, it reveals a deadly intracellular bacterium that has the extraordinary capacity to survive under arduous conditions (including many food manufacturing processes) and to succeed as a ferocious food-borne pathogen. *L. monocytogenes* is truly the microbial world's wolf in sheep's clothing or Dr. Jekyll and Mr. Hyde.

Indeed, apart from *L. monocytogenes*, there are few other bacterial species that have the tenacity to withstand such an array of external stresses. Moreover, there are fewer still intracellular pathogens that have the ingenuity to adhere, enter host cells, escape from vacuoles, multiply in cytoplasm, and spread to neighboring cells, thus provoking a cascade of cell-mediated immune responses in their wake without being completely decimated by the host. Not surprisingly, besides offering a useful model for the study of host–pathogen relationships, *L. monocytogenes* has been increasingly recognized as a preferred vector for delivering anti-infective and cancer vaccine molecules, given its unsurpassed ability to elicit both $CD4^+$ and $CD8^+$ T cell responses and to direct protective molecules into the cytoplasm.

Although *L. monocytogenes* was first described in the 1920s, its impact as a food-borne pathogen was not fully appreciated until the 1980s. Extensive research efforts in the preceding decades led to the elucidation of the fundamental aspects of *L. monocytogenes* biology, epidemiology, stress response, pathogenesis, immunology, genetics, and proteomics. A variety of phenotypic and genotypic techniques for identification, detection, strain typing, and virulence determination of this bacterium was also developed.

It is my honor to have as contributors a panel of international scientists with expertise in respective fields of *L. monocytogenes* research; their in-depth knowledge and pertinent technical insights have greatly enriched this book, making it an interesting text to read and a valuable roadmap to possess. In addition, the unwavering support of Senior Editor Judith Spiegel, Ph.D., and dedication of other staff at CRC Press have made the compilation of this book a thoroughly enjoyable experience.

I hope the readers of this book will become as intrigued and fascinated by the multifaceted features of this microbe as I have been during my involvement in *Listeria* research. Due to its ease of maintenance, *L. monocytogenes* often gives rise to a fleeting sense of triumph and a tinge of complacency in a novice investigator. However, the capability of this remarkable bacterium to conceal its true colors and to mislead research endeavors has proved a constant source of frustration for many upcoming as well as experienced microbiologists. If this book is helpful in any way to prepare readers in their study and exploitation of *L. monocytogenes* and other microbes to the advancement of biomedical sciences and intervention strategies, it may have well served its purpose.

Editor

Dongyou Liu, Ph.D., is currently a member of the research faculty in the Department of Basic Sciences, College of Veterinary Medicine, at Mississippi State University in Starkville. Previously, he received a B.V.Sc. degree from Hunan Agricultural University in China. After 1 year of postgraduate training under Dr. Kong Fanyao's supervision at China Agricultural University, he continued his Ph.D. study (on the immunological diagnosis of hydatid disease) in the laboratory of Drs. Michael D. Rickard and Marshall W. Lightowlers at Melbourne University School of Veterinary Science in Australia. During the past 15 years, he has been actively engaged in the analysis of molecular mechanisms of bacterial pathogenesis and in the development of nucleic acid-based assays for improved identification and virulence determination of microbial pathogens (e.g., ovine footrot bacterium, dermatophyte fungi, and listeriae) in several research and clinical laboratories in Australia and the United States. His research efforts have resulted in more than 6 first-authored original and review articles in professional journals and books.

Contributors

A. Jerald Ainsworth
College of Veterinary Medicine
Mississippi State University
Starkville, Mississippi

Frank W. Austin
College of Veterinary Medicine
Mississippi State University
Starkville, Mississippi

Sukhadeo Barbuddhe
Institute for Medical Microbiology
Justus-Liebig-University
Giessen, Germany

Armelle Bigot
Faculté de Médecine René Descartes and Inserm
Unité de Pathogénie des Infections Systémiques
Université Paris Descartes
Paris, France

Trinad Chakraborty
Institute for Medical Microbiology
Justus-Liebig-University
Giessen, Germany

Alain Charbit
Faculté de Médecine René Descartes and Inserm
Unité de Pathogénie des Infections Systémiques
Université Paris Descartes
Paris, France

Yi Chen
Department of Food Science
The Pennsylvania State University
University Park, Pennsylvania

Ying Cheng
Department of Food Science
North Carolina State University
Raleigh, North Carolina

Mickaël Desvaux
INRA Clermont-Ferrand/Theix/Lyon
Microbiology—Food Quality and Safety Team
Saint-Genès Champanelle, France

Cormac G. M. Gahan
Department of Microbiology and Alimentary Pharmabiotic Center
University College, Cork
Cork, Ireland

Gernot Geginat
Institut für Medizinische Mikrobiologie und Hygiene
Medizinische Fakultät Mannheim der Universität Heidelberg
Mannheim, Germany

Lisa Gorski
Produce Safety and Microbiology Research Unit
U.S. Department of Agriculture
Agricultural Research Service
Albany, California

Silke Grauling-Halama
Institut für Medizinische Mikrobiologie und Hygiene
Medizinische Fakultät Mannheim der Universität Heidelberg
Mannheim, Germany

Torsten Hain
Institute for Medical Microbiology
Justus-Liebig-University
Giessen, Germany

Michel Hébraud
INRA Clermont-Ferrand/Theix/Lyon
Microbiology—Food Quality and Safety Team
Saint-Genès Champanelle, France

Colin Hill
Department of Microbiology and Alimentary Pharmabiotic Center
University College, Cork
Cork, Ireland

Sophia Kathariou
Department of Food Science
North Carolina State University
Raleigh, North Carolina

Stephen J. Knabel
Department of Food Science
The Pennsylvania State University
University Park, Pennsylvania

Yukio Koide
Department of Infectious Diseases
Hamamatsu University School of Medicine
Handa-yama, Hamamatsu, Japan

Michael Kuhn
Lehrstuhl für Mikrobiologie
Theodor-Boveri-Institut für Biowissenschaften der Universität Würzburg
Würzburg, Germany

Mark L. Lawrence
College of Veterinary Medicine
Mississippi State University
Starkville, Mississippi

Dongyou Liu
College of Veterinary Medicine
Mississippi State University
Starkville, Mississippi

Jim McLauchlin
Health Protection Agency
London, United Kingdom

Masao Mitsuyama
Department of Microbiology
Kyoto University Graduate School of Medicine
Sakyo-ku, Kyoto, Japan

Toshi Nagata
Department of Health Science
Hamamatsu University School of Medicine
Handa-yama, Hamamatsu, Japan

Yvonne Paterson
Department of Microbiology
University of Pennsylvania
Philadelphia, Pennsylvania

Sylvie M. Roche
INRA
Infectiologie Animale et Santé Publique
Nouzilly, France

Mariela M. Scortti
Centre for Infectious Diseases
School of Biomedical Sciences
University of Edinburgh
Edinburgh, United Kingdom

Matthew M. Seavey
Department of Microbiology
University of Pennsylvania
Philadelphia, Pennsylvania

Robin M. Siletzky
Department of Food Science
North Carolina State University
Raleigh, North Carolina

Helena M. Stack
Department of Microbiology and Alimentary Pharmabiotic Center
University College, Cork
Cork, Ireland

José A. Vázquez-Boland
Centre for Infectious Diseases
School of Biomedical Sciences
University of Edinburgh
Edinburgh, United Kingdom

Philippe Velge
INRA
Infectiologie Animale et Santé Publique
Nouzilly, France

Thorsten Verch
Department of Microbiology
University of Pennsylvania
Philadelphia, Pennsylvania

Martin Wagner
Institute for Milk Hygiene, Milk Technology and Food Science
University of Veterinary Medicine
Vienna, Austria

Section I

Biology and Pathogenicity

1 Biology

Martin Wagner and Jim McLauchlin

CONTENTS

1.1 INTRODUCTION

Listeria monocytogenes (originally named *Bacterium monocytogenes*) is a Gram-positive bacterium first described in 1926 in Cambridge, United Kingdom, as a cause of infection with monocytosis in laboratory rodents.[1] In the following year, Pirie[2] also isolated a Gram-positive bacterium, in this instance from infected wild gerbils in South Africa, and proposed the name *Listerella* for the genus in honor of the surgeon Lord Lister. Murray and Pirie realized that they were dealing with the same species of bacteria and thus combined the names to form *Listerella monocytogenes*. This was later changed for taxonomic reasons to *Listeria monocytogenes*.[3]

In 1929 in Denmark, Nyfeldt isolated *L. monocytogenes* from the blood cultures of patients with a mononucleosis-like infection (a rare manifestation of the disease)[4] and, in 1936, Burn in the United States established listeriosis as a cause of both sepsis among newborn infants and meningitis in adults.[5] Prior to 1926 there were descriptions published that were likely of listeriosis; indeed, a "diphtheroid" isolated from the cerebrospinal fluid of a soldier in Paris in 1919 was later identified as *L. monocytogenes*.[6] However, up to the 1980s, human listeriosis remained a relatively obscure disease attracting limited attention, although large outbreaks of considerable morbidity and mortality but of unknown transmission occurred. For example, 279 and 166 human listeriosis cases occurred in Halle, Germany, during 1966 and in the Anjou region of France between 1975 and 1976, respectively. During the early to mid-1980s, there was a rise in the total numbers of human and animal listeriosis cases in Europe and North America. Furthermore, outbreaks occurred in North America and Europe linked to the consumption of contaminated coleslaw,[7] milk,[8] soft cheese,[9,10] and pâté.[11] The concept that listeriosis is a food-borne disease, however, is not new. In 1926, Pirie[2] produced the prophetic, but somewhat forgotten, statement that "infection can be produced by subcutaneous inoculation or by feeding, and it is thought that by feeding that the disease is spread in nature."

Since the series of outbreaks in the 1980s, it is apparent that *L. monocytogenes* causes an extremely serious, invasive, and often life-threatening food-borne disease with a high economic burden to both public health services and the food industry. Recent estimates ranked listeriosis

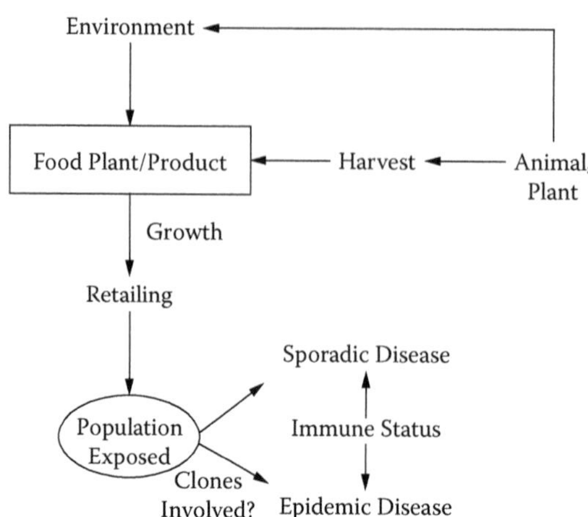

FIGURE 1.1 Ecological interrelationships of *L. monocytogenes*, the environment, food, and cases of listeriosis.

as the second and fourth most common cause of death from food-borne infectious diseases in the United States[12] and in England and Wales,[13] respectively. This disease was estimated as being responsible for 14–15% of all deaths from food poisoning in both studies. A schematic outline of the ecological interrelationships of *L. monocytogenes*, the environment, food, and cases of listeriosis is shown in Figure 1.1.

1.2 LISTERIOSIS

Listeriosis occurs in various animals, including humans, and most often affects the uterus at pregnancy, the central nervous system, or the bloodstream. Serious, and indeed life-threatening, infection is most often recognized. Subclinical infections do occur but are rarely identified, and considering the widespread distribution of *Listeria* spp., the true spectrum of disease and overall disease burden remains to be defined. Almost all cases of human listeriosis are due to *L. monocytogenes*[14]; very rare infections due to *Listeria ivanovii* and *Listeria seeligeri* have been described.[15,16] *L. monocytogenes* is also the major pathogen for other animals, although approximately 10% of septicemia in sheep has been reported as due to *L. ivanovii*.[17] In addition to having a virulence gene cluster (also known as *Listeria* pathogenicity island 1 or LIPI-1) shared with *L. monocytogenes* and *L. seeligeri*, *L. ivanovii* possesses a separate *Listeria* pathogenicity island 2 (LIPI-2), which encodes phosphocholinesterases for efficient utilization of phospholipids in ruminant erythrocytes.[18] This may explain the susceptibility of small ruminants to *L. ivanovii* infection. The factors that increase or decrease the risk of listeriosis associated with food consumption of *L. monocytogenes* in food are outlined in Table 1.1.

In humans, infection is most often recognized in the immunocompromised, the elderly, pregnant women, and unborn or newly delivered babies. Infection can be treated successfully with antibiotics, but despite this, human infection has a mortality of 20–40%.[19] In domestic animals (especially in sheep and goats), listeriosis usually presents as encephalitis, abortion, or septicemia and causes considerable economic loss.[17] Consumption of contaminated food or feed is believed to be the principal route of infection. However, in humans, infection can also be transmitted, albeit rarely, by direct contact with either the environment or infected animals, or by cross-infection between patients during the neonatal period.[20,21]

The incidence of infection increases with age, so the mean age of adult infections is more than 55 years. Men are more commonly infected than women over the age of 40 years, and because women are infected in the child-bearing years, the overall sex distribution is biased toward males

TABLE 1.1
Factors That Affect the Risk of Food-borne Listeriosis

	Increases Risk	Decreases Risk
Factors associated with the bacterium	Contamination of raw product, factory sites Cross-contamination from other foods Growth during the food chain More virulent strains and adaptation to food chain	Exclude *L. monocytogenes* from the food chain and prevent its multiplication by: • adequate temperature and shelf-life control • quality of factory sites and environmental hygiene • quality of raw materials Less virulent or nonpathogenic strains Typing of strains and identification of causally related contamination or infections
Factors outside the host	Poor processing, retailing, and storage of food Incomplete surveillance	HACCP, process controls, guidelines, standards, specification and food regulation Education of all those involved with the food chain Surveillance of clinical cases and foods Withdrawal of contaminated foods
Host factors	Consumption of "risky" foods Increased susceptibility to serious infection by age or immunosuppressive agents, exposures, or conditions	Dietary advice to "at risk" to avoid "risky" foods Immunity

in the elderly immunocompromised population. Immunosuppression is a major risk factor for both the epidemic and sporadic forms of listeriosis and probably accounts for the increasing incidence with age. Human immunodeficiency virus disease is a predisposing factor in some countries. Earlier studies in the United States reported an incidence of 7.1 cases/million persons/year,[22,23] which has now reduced to 3.1.[24] Estimates on the incidence of listeriosis in European countries range from less than one to greater than seven cases/million/year.[25] Although many initiatives have been undertaken to lower the incidence of listeriosis worldwide, recent reports indicate a dramatic increase in its incidence in Germany[26] and England and Wales[27]; in the latter countries, cases are predominantly confined to patients over 60 years of age.

The low incidence of listeriosis, at least in comparison to the leading food-borne pathogens such as *Salmonella* and *Campylobacter,* is in contrast to preventive actions taken when *L. monocytogenes* is detected in foods. More than 60% of all recall actions in the United States between 1996 and 2000 were launched as a result of detecting *L. monocytogenes* contamination.[28] This fact emphasizes that *L. monocytogenes* contamination not only has significance for public health considerations, but also has socioeconomic effects on food production and food enterprises throughout the world, including those involved with international trade.

L. monocytogenes infections usually occur in urban populations and in the absence of specific contact with animals. Human listeriosis has a marked seasonality, with a peak in cases occurring during the late summer and autumn. In contrast, listeriosis in animals has a marked seasonal peak in the spring. The incubation period in humans between exposure (consumption of contaminated foods) and clinical recognition of the disease varies widely between individuals: from 1 to 90 days, with an average for intrauterine infection of about 30 days.[11] The majority of cases appear to be sporadic; however, food-borne outbreaks have been recorded worldwide, particularly in Northern European and North American countries.[11,29–33] Outbreaks can be recognized by field epidemiological observations,[34] although enhanced surveillance (including the application of discriminatory subtyping of isolates) may be needed to identify related clusters of cases.[11,30,35] Rigorous collection of surveillance data and typing of every isolate are therefore invaluable for the identification of clusters of cases and outbreak investigation. A wide range of food types has been associated with transmission in

TABLE 1.2
Food Products Associated with Transmission of Listeriosis

Dairy Products	Meats	Fish	Vegetables	Complex Foods
Soft cheese	Cooked chicken	Fish	Coleslaw salad	Sandwiches
(Raw) milk	Turkey frankfurters	Shellfish	Vegetable rennet	
Ice cream/soft cream	Sausages	Shrimp	Salted mushrooms	
Butter	Pâté and rillettes	Smoked fish	Alfalfa tables	
	Pork tongue in aspic	Shellfish	Raw vegetables	
		Cod roe	Pickled olives	
			Rice salad	
			Cut fruit	

both sporadic cases and outbreaks (Table 1.2). These foods generally have the common features of being capable of supporting the multiplication of *L. monocytogenes,* being processed with extended (usually refrigerated) shelf lives, being consumed without further cooking, and being contaminated with high levels of *L. monocytogenes.*[21] Because of the long incubation period and low attack rate, specific foods associated with transmission are identified in the minority of all cases.

The minimal infective dose for listeriosis is unclear, although this is likely to vary considerably between individuals. However, it is generally accepted that there is some sort of dose–response relationship, and the burden of disease when consuming foods with levels below 100 colony-forming units (CFUs) per gram is likely to be very low. Therefore, EC directive 2073/2005 allows the presence of *L. monocytogenes* in ready-to-eat foods placed on the market, provided that they do not support the growth of this bacterium or that they will not exceed this limit at the end of the shelf life. However, not all food regulators have taken this approach, and the United States has had zero tolerance of any *L. monocytogenes* in a processed food since the 1980s. The contamination level in foods associated with infection has revealed an average contamination level of 10^2–10^6 CFUs/ml/g in the majority of the cases.[36] Data from naturally infected cases and consideration of the rates of contamination of food on sale indicate that the attack rate for serious disease is low.

Listeriae are ubiquitous in the environment and have been isolated from natural environments, including fresh water, wastewater, mud, and soil—especially when decaying vegetable material is present.[19] Studies on the origin of *Listeria* strains from a wide geographical distribution of all listerial species in the environment have confirmed the environment as the natural reservoir for this genus.[37] A prevalence of more then 20% in soil samples was reported almost 30 years ago[38]; however, surprisingly few studies have focused on the lifestyle of *Listeria* in soil and rhizosphere,[39] and the environmental niches of different *Listeria* species are not understood.

Because *Listeria* spp. are distributed so widely, contamination of food may occur before harvest, especially those in contact with soil such as fresh produce (e.g., salads and other vegetables). An extremely wide range of animals (mammals, birds, fish, and invertebrates) has been reported to carry *Listeria* spp. in the feces without apparent disease, and hence environments contaminated by animal feces frequently contain bacteria from this genus. Not surprisingly, *Listeria* spp. are readily recovered from the food–animal environments, and in one study, 42–87% of straw samples collected from the surroundings of cattle in Austria contained bacteria of these species.[40] Depending on the season, 5.3–45% of the cattle were fecal carriers of *Listeria.*[41,42]

The relationship between the feeding of silage to domestic animals and the development of listeriosis has long been realized.[5] This is of particular importance when the pH is greater than 5.5 and the silage is of poor quality or has had prolonged exposure to aerobic conditions.[17] Modern practices of producing silage in large polythene-covered bales ("big bales") favors the growth of *L. monocytogenes* in comparison with production in the more traditional clamps and may, in part, explain the recent apparent increase in the incidence of listeriosis in domestic animals in Britain.[16]

Feeding poor-quality silage has been described as a risk factor for both infection in food animals (especially sheep and goats) and for contamination of raw milk, particularly in the last weeks of winter feeding.[43] Interestingly, corn silage but not grass silage was identified as the major risk factor for *Listeria* transmission in healthy Swiss cattle.[44]

Oral exposure of cattle from contaminated feed can result in localized intramammary infections.[45–47] During listerial mastitis, large numbers of *L. monocytogenes* can be persistently secreted into milk, which has minimal organoleptic changes even when heavily contaminated.[48] With the exception of mastitis, systemic infection in food-producing animals with *L. monocytogenes* is unlikely to result in this bacterium entering the food chain since the diseased animals will be excluded from food production after diagnosis. However, extensive cross-contamination between animals and the environment may occur in on-farm food production systems after cases of abortive listeriosis.[49] Occupationally acquired listeriosis in veterinary and farm workers, particularly presenting as ocular or cutaneous infection following attending bovine abortions, can occur.[20] The most frequent reason for *L. monocytogenes* contamination of raw foods is poor hygiene during milking or slaughter. Sanaa et al.[50] identified the most likely factors leading to the contamination of raw milk as feeding poor-quality silage, lack of cleanliness of cows, the frequency of cleaning the exercise area, and the use of improper towels for cleaning.

After harvest, foods and food components can become contaminated at any stage of food processing, during retailing, and in consumers' homes. However, evidence from outbreaks of listeriosis highlights the importance of contamination from sites within food processing environments. *Listeria* can be readily introduced from the environment into a food processing plant, especially where hygiene barriers are insufficient, and this group of bacteria can persist in specific areas for considerable periods of time. A preliminary factor for contamination is insufficient protection of plant areas where a *Listeria* contamination must be strictly avoided (e.g., ripening area in cheese production; Figure 1.2).

In dairies, the prevalence for *L. monocytogenes* contamination of raw bulk tank milk and dairy milk samples is low.[51] Proper pasteurization is a reliable method to eradicate *L. monocytogenes;* however, hazards can arise when the milk or other dairy products become (re)contaminated at later steps of the processing chain or if unpasteurized or underheated milk is used.

Alternative food production systems such as raw milk processing and specialist cheese making have increased, especially in small food enterprises seeking new economic niches to escape the low margins obtained in industrial food production. The diversification in food production often leads

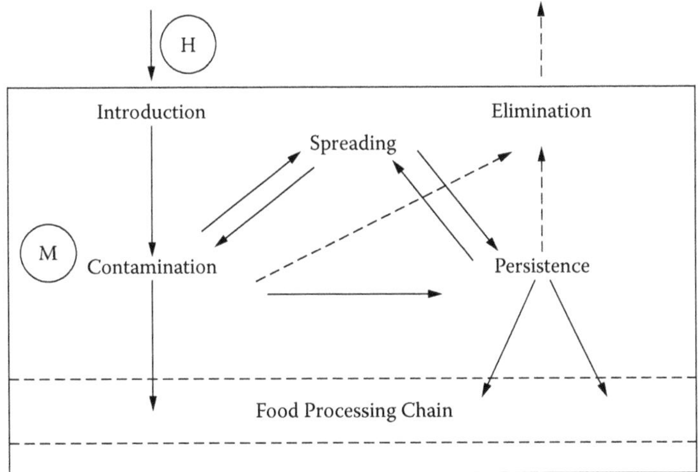

FIGURE 1.2 Schematic outline for *L. monocytogenes* contamination routes within a food production environment (H = hygienic barrier must be intact; M = monitoring to detect contamination periods at early phases needed).

to greater coexistence of highly automated and integrated food production systems with microscale (frequently with a high degree of handling) but innovation-driven food production methods, combined with retailing of products directly from farms or at local markets. The consequence is the establishment of a variety of simple to complex food production and retail management chains, probably with increased inherent risks of transmission of listeriosis. One example is the growing segment of organic food production in many European countries, where small enterprise-driven food production systems are becoming increasingly used in strategies to satisfy consumer demand for these types of products. These food manufacturing and retailing changes are likely to impact a range of food-borne pathogens, including *L. monocytogenes*. The great challenge for current and future food technologists is the implementation of the generally high standards of hygiene now found in the majority of the more "industrial" food producers within the infrastructure used by smaller manufacturers.

Epidemiological and microbiological studies using molecular characterization of *Listeria* isolates from contaminated fish, shrimp, ice cream, and meat plants provide some information on how contamination cycles become established in food production environments.[52–56] Contamination by a single strain of *L. monocytogenes* can affect a single environment for several years.[57] The observation from fish and dairy plants that the proportion of products contaminated increases at specific (usually later) processing steps (e.g., after brining or during ripening) suggests that contamination of food and food contact surfaces in the processing area from sites and machinery within the food plant is of greater importance for the finished product than is the contamination of the raw food components entering the factory. Contamination rates of raw meat and meat products can be similar to or even higher than those found in dairies.[58] Contamination rates of vegetables very much depend on the contamination status of a plant; if a plant is contaminated, more than 50% of the vegetables can contain *L. monocytogenes*, even after heat treatments at 90°C for 4 to 8 minutes (Pappelbaum et al., submitted). Sanitation efforts are usually insufficient to eliminate this pathogen from the environment of the food plant, since listeriae survive in niches within plants where sanitation is not sufficiently effective and this bacterium can persist and remain undetectable for years.[57]

The genetic basis for persistence of *L. monocytogenes* is not understood; however, model systems have demonstrated increased adherence to stainless steel and inorganic materials by "persistent" *L. monocytogenes* isolates, which are more resistant to sanitizers.[59–62] The identification and distribution of the genetic determinants associated with the persistence capability of *L. monocytogenes* would be helpful to predict the probability for long-term food plant contamination.

There is presently a knowledge gap concerning the very important effects of the consumer on listeriosis. Neither the contamination status of household environments nor the contamination level of foods stored at home has been intensively investigated. A case control study in the United States in the early 1990s showed that *L. monocytogenes* was more likely to occur in the foods within the refrigerated cases than in matched controls, even though the strains recovered were not necessarily the same as those causing infection.[63] In Europe, one of the few available studies reported 4.8 and 1.7% of food samples collected from households being contaminated by *L. innocua* and *L. monocytogenes,* respectively.[58] That *L. monocytogenes* may be common in household environments was substantiated by a further study by Beumer et al.[64] The consumer sphere is a neglected area for food hygiene research. More efforts should be directed toward strategies to reduce exposure to food-borne pathogens in patients' domestic environments, together with understanding behavior and consumption patterns, particularly in the more vulnerable segments of the population. The availability of this knowledge would help strengthen the final (and, in some instances, the only) control strategy for listeriosis by the provision of dietary advice to vulnerable groups.

Dietary advice has been given out to vulnerable groups in the United Kingdom and United States to avoid specific food types that have been particularly associated with transmission of listeriosis (Table 1.3). Similar advice has been given in other countries, including France, Australia, and New Zealand. This advice is certainly prudent, although efforts must be made to continually reinforce the advice. It is clearly not possible to advise vulnerable groups against all food types associated with

TABLE 1.3
Dietary Advice for the Prevention of Listeriosis

In the United States:	In the United Kingdom:
Advice to the General Public	
Cook thoroughly raw food from animal sources such as beef, pork, and poultry. Wash raw vegetables thoroughly before eating. Keep uncooked meats separate from vegetables and from cooked foods and ready-to-eat foods. Avoid raw (unpasteurized) milk or foods made from raw milk. Wash hands, knives, and cutting boards after handling uncooked foods.	Keep foods for as short a time as possible, follow the storage instructions carefully, and observe the "best by" and "eat by" dates on the label. Do not eat undercooked poultry or meat products. Make sure you reheat cooked–chilled meals thoroughly and according to the instructions on the label. Wash salad, fruit, and vegetables that will be eaten raw. Make sure that your refrigerator is working properly and keep foods stored in it really cold. When reheating food, make sure that it is piping hot all the way through and do not reheat more than once. When using a microwave oven to cook or reheat food, observe the standing times recommended by the oven manufacturer to ensure that food attains an even temperature before it is eaten. Throw away leftover food. Cooked food that is not eaten straightaway should be cooled as rapidly as possible and stored in the refrigerator.
Advice for At-Risk Groups	
Before eating, cook leftover foods or ready-to-eat foods such as hot dogs until steaming hot before eating. Avoid soft cheese such as feta, brie, Camembert, blue-veined, and Mexican style cheese. Hard cheeses, processed cheese, cottage cheese, or yogurt need not be avoided. Raw vegetables should be thoroughly washed before eating. Although the risk of listeriosis associated with foods from deli counters is relatively low, pregnant women and immunosuppressed persons may choose to avoid these foods or thoroughly reheat cold cuts before serving.	Pregnant women and anyone with low resistance to infection should not eat soft ripened cheeses of the brie, Camembert, or blue veined types, and they should not eat pate. Any bought cooked–chilled meals or ready-to-eat poultry should be reheated until piping hot. Do not eat them cold.

infection, and a balance must be made between providing sensible guidance allowing individuals to make informed choices and fear mongering. Targeting of advice to the general public and attitudes toward risk are outside the scope of this chapter; however, dietary advice is no substitute for controlling the organism in the food chain. To date there has been limited investigation on the assessment of how effective dietary advice is for controlling listeriosis; however, a recent case control study in the United States concluded that reducing the disease burden may require intervention directly in retail and domestic environments.[65] Furthermore, Goulet[66] advocated not only avoiding high-risk foods, but also provided information on how to reduce the risk through cooking, avoiding cross-contamination, and reducing the length of time of refrigerated storage of perishable foods.

1.3 TAXONOMY OF THE GENUS *LISTERIA*

Listeriae are a genus of rod-shaped, nonsporulating, Gram-positive bacteria that, although originally described as monotypic containing only *L. monocytogenes sensu lato,* now comprise six species.

Two of these species are pathogenic (*L. monocytogenes sensu stricto* and *L. ivanovii*)[67] and four are nonpathogenic (*L. innocua*,[68] *L. seeligeri*,[69] *L. welshimeri*,[69] and *L. grayi*[70]). A seventh species (*Listeria murrayi*) was previously recognized within the genus; however, analysis using DNA–DNA hybridization, multilocus enzyme electrophoresis, and rRNA restriction fragment length polymorphism procedures led to the clarification of this as a subspecies within *L. grayi*.[71]

Based on the serological reactions of listerial somatic (O-factor) and flagellar (H-factor) antigens with specific antisera, *Listeria* species are classified into serotypes, or commonly named serovariants or serovars, with *L. monocytogenes* comprising serovars 1/2a, 1/2b, 1/2c, 3a, 3b, 3c, 4a, 4b, 4c, 4d, 4e, and 7 (see chapter 5). Using various genetic subtyping techniques, *L. monocytogenes* is separated into three lineages: Lineage I contains serovars 1/2b, 3b, 4b, 4d, and 4e; lineage II contains serovars 1/2a, 1/2c, 3a, and 3c; and lineage III contains serovars 4a and 4c (see chapter 7). More recent reports indicate that lineage III can be further divided into three subgroups (IIIA, IIIB, and IIIC). Subgroup IIIA is made up of typical rhamnose-positive avirulent serovar 4a and virulent serovar 4c strains; subgroup IIIC is made up of atypical rhamnose-negative virulent serovar 4c strains and subgroup IIIB is made up of atypical rhamnose-negative virulent non-4a and non-4c strains, some of which may be related to serovar 7.

It is possible that subgroup IIIB (including serovar 7) may represent a novel subspecies within *L. monocytogenes*.[72,73] Strains of *L. monocytogenes* within serovars 4b, 1/2a, 1/2b, and 1/2c account for more than 98% of isolations from clinical cases of human listeriosis, and serovar 4b has been associated with the most recent large outbreaks of listerisois. In the experimental mouse models, *L. monocytogenes* serovars 4b, 1/2a, 1/2b, and 1/2c tended to display a higher infectivity than other serotypes through intragastric inoculation,[74] and almost all serovars have the capability to cause mouse mortality via the intraperitoneal route[73,75,76] (see chapter 8). However, the outbreak in Halle, Germany, in 1966 due to a strain of *L. monocytogenes* serovar 1/2a, where 279 cases were recognized, illustrates the potential for strains outside serovar 4b to cause large outbreaks. In addition, natural sporadic infections due to strains other than serovars 1/2a, 1/2b, and 4b, although relatively infrequent, are of similar severity to those due to other serovars. Thus, some caution is needed to apply data from animal or tissue culture models to the naturally occurring infection.

Phenotypic and genotypic studies indicate that all species of the genus *Listeria* form stable and homogeneous groups of bacteria. The most closely related genus is *Brochothrix*. There is a relatedness to other low G+C Gram-positive bacteria such as *Bacillus, Clostridium, Enterococcus, Streptococcus,* and *Staphylococcus*.[77–82] Complete genome sequences are now available for *L. monocytogenes, L. innocua,* and *L. welshimeri*, and work is underway to decipher the genome sequences of *L. ivanovii, L. seeligeri,* and *L. grayi* (see chapter 10). Comparison of *L. monocytogenes* and *L. innocua* genomes suggested a close relationship to that from *Bacillus subtilis* and thus a common origin.[82] The close relatedness was the likely reason that *L. innocua* was not taxonomically distinguished from *L. monocytogenes* earlier than 1977.[67] Genomic analysis, at least among *L. monocytogenes, L. innocua,* and *L. welshimeri,* has provided a remarkably similar phylogenic relationship to that generated by other methods, including 16S rDNA sequence analysis[83] (see chapter 10).

Sequence information is available from many gene regions of *Listeria* spp., especially from within a 9.6-kb gene cluster called *Listeria* pathogenicity island 1 (LIPI-1), which harbors several important virulence genes (see chapter 4) and occurs only in *L. monocytogenes, L. seeligeri,* and *L. ivanovi*.[31] Phylogenetic analyses of sequences within the universal *prs, ldh, vclA, vclB, iap,* 16S, and 23S-rRNA coding genes revealed that *L. grayi* represents the ancestor within the genus. The remaining five species form two groups: one containing *L. monocytogenes* and *L. innocua* and the other *L. welshimeri, L. ivanovii,* and *L. seeligeri*.[84] Schmid et al. postulated that the LIPI-1 virulence gene cluster was present in the common ancestor of *L. monocytogenes, L. innocua, L. ivanovii, L. seeligeri,* and *L. welshimeri* and that the pathogenic capability was lost in two separate events.[84] This hypothesis is supported by comparative examination of the whole genome sequences of *L. monocytogenes, L. innocua,* and *L. welshimeri;* this showed a reduction in the genomes of both

L. innocua and *L. welshimeri* in relation to *L. monocytogenes,* thus rendering an ancestral, formerly pathogenic species into a nonpathogen.[82]

The discussion is still ongoing as to whether all *L. monocytogenes* strains isolated from foods have the same pathogenic potential for humans.[85,86] This discussion is stimulated by the observation that a small number of *L. monocytogenes* types (e.g., 1/2a, 1/2b, and 4b) are responsible for the majority of both sporadic cases and outbreaks worldwide. Furthermore, molecular analysis of strains from some of the larger outbreaks reveals a close relatedness.[87–89] By sequencing parts of the listeriolysin gene, Rasmussen et al.[90] showed that clonal lineages exist within the species *L. monocytogenes* that differ in their biological, epidemiological, and pathological behaviors. Meanwhile, it was demonstrated by several groups that *L. monocytogenes* can be differentiated into three evolutionary lineages. Some authors have suggested an even different degree of pathogenicity among isolates of the different lineages.[86] Limited data on virulence within *in vivo* mouse models together with *in vitro* models, however, provide consistent differences in pathogenicity among various serotypes and strains.[91–93] Given the ongoing need to further establish the relationship between lineages and serotypes and their pathogenic potential, as well as the conspicuous lack of rapid and reliable techniques for laboratory discrimination of pathogenic and nonpathogenic strains, a strong argument has been made for public health, food hygiene, and food regulatory purposes that all isolates of *L. monocytogenes* should be regarded as potentially pathogenic for the present.[14]

1.4 MORPHOLOGY

All members of the genus *Listeria* form regular, short rods, 0.4–0.5 by 1–2 μm with parallel sides and blunt ends and usually occur singly or in short chains; in older or rough cultures, filaments of 6 μm in length may develop. These bacteria are Gram positive with even staining, but some cells, especially in older cultures, lose their ability to retain the Gram stain. Gram-stained preparations of *Brochothrix thermosphacta*, *Lactobacillus casei*, *Bacillus cereus,* and *L. monocytogenes* are shown in Figure 1.3. All species of *Listeria* are motile with peritrichous flagella when cultured

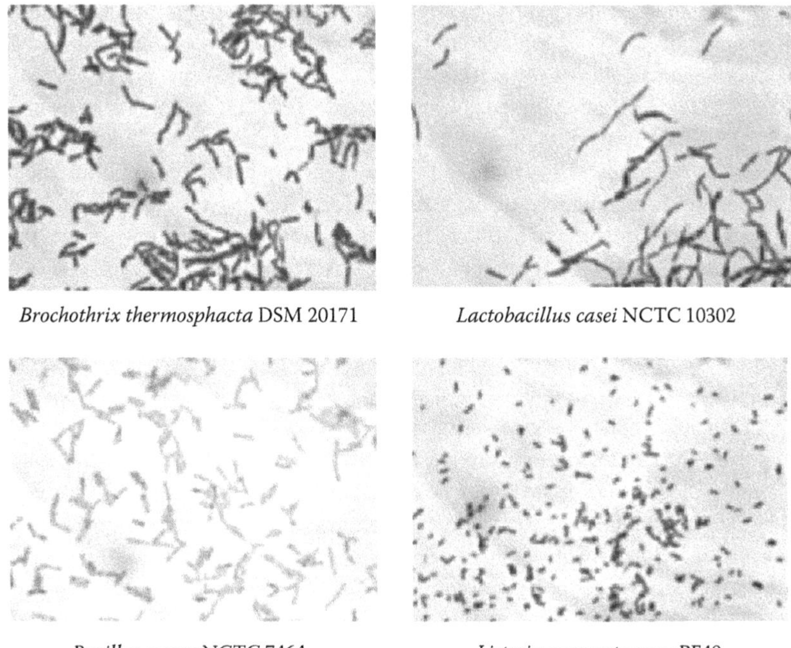

FIGURE 1.3 Gram-stained preparations of *L. monocytogenes* and other low GC Gram-positive bacteria.

between 20 to 25°C but nonmotile at 37°C. The metabolism is aerobic and facultatively anaerobic. Growth occurs between pH 5.2 and 9 and between <0 and 45°C; optimal growth occurs between 30 and 37°C. Cytochromes are produced and homofermentative anaerobic catabolism of glucose results in production of L(+)– lactic acid, acetic acid, and other end products. Acid but no gas is produced from other sugars.

The cell wall composition in *Listeria* shows a multilayer of peptidoglycan containing mesodiaminopimelic acid (meso-DAP) (variation A1γ of Scheifer and Kandler)[94] as the diamino acid typical for Gram-positive bacteria,[95] but arabinose and mycolic acids are not present. In addition to *N*-acetylmuramic acid and *N*-acetylglucosamine, glucosamine occurs as a component of the cell wall polysaccharide.[96,97] Ribitol and lipoteichoic acids (O-factor) present in *Listeria* are responsible, together with flagellal antigens (H-factor), for the 16 different serotypes described in the genus *Listeria*[97–101] (see chapter 5) and serve with some modification as receptors for bacteriophages.[102] The long chain fatty acids consist of predominantly straight chain saturated, anteisomethyl branched chain types. When grown at 37°C, the major fatty acids are 14-methylhexadecanoic (anteiso-$C_{17:0}$) and 12-methyltetradecanoic (anteiso-$C_{15:0}$). Menaquinones are the sole respiratory quinones; the major quinone contains seven isoprene units (MK-7).[103]

All *Listeria* species are morphologically similar on artificial media, and *Listeria* colonies after 24–48 h are 0.5–1.5 mm in diameter, round, translucent, low convex with a smooth surface and entire margin, and nonpigmented with a crystalline central appearance. The bacterial growth may be sticky when removed from agar surfaces, but usually emulsifies easily and may leave a slight impression on the agar surface after removal. Older cultures (3–7 d) are larger, 3–5 mm in diameter, and have a more opaque appearance; sometimes, rough colonial forms may develop with a sunken center.

1.5 BIOCHEMISTRY

Listeria cell surfaces are similar to those of other Gram-positive bacteria and are composed mainly of peptidoglycan (making up 35% of the dry weight of cell wall), teichoic acids (which link to the glycan layers to account for 60–70% of the dry weight of cell wall), and lipoteichoic acids (which show structural and biological resemblance to lipopolysaccharides in Gram-negative bacteria). The teichoic acid component is shown to be about 20% of the cell wall weight of *L. monocytogenes* strain EGD, and it contains *N*-acetylglucosamine, rhamnose, ribitol, and phosphorus in a molar ratio of 0.95:1.0:0.97:0.98. The molecular weight of the teichoic acid chain in *L. monocytogenes* strain EGD is estimated to be about 120 kDa by gel filtration technique. The immunological activity of *Listeria* teichoic acids is attributable to the rhamnose residue that functions as a major antigenic determinant (i.e., O-factor or somatic antigen). Recently, a gene, *lmo2537,* encoding a putative UDP-*N*-acetylglucosamine 2-epimerase, a precursor of the teichoic acid linkage unit, has been identified in the genome of *L. monocytogenes* strain EGD-e. The translated protein of *lmo2537* shares partial homology with the cognate epimerase MnaA of *Bacillus subtilis* (62%) and Cap5P of *Staphylococcus aureus* (55%). By construction and analysis of a conditional knockout mutant, it was found that *lmo2537* is involved in *L. monocytogenes* teichoic acid biogenesis, and the *lmo2537* knockout mutant shows drastically reduced virulence in mice.[104]

Listeria species employ flagella to provide motility *in vitro*, and this property may have some role in enhancing adhesion and invasion of host epithelial cells in an *in vitro* tissue culture invasion and cell spreading model. Five to six peritrichous flagella are produced by *L. monocytogenes.* The expression of flagellar motility genes in *L. monocytogenes* appears to be down-regulated by a regulatory protein (MogR, formerly Lmo0674) as an EGD mutant (EGDeΔ*674*) containing the underlying gene deletion demonstrated altered *flaA* gene expression, leading to abolishment of temperature regulation on *flaA* gene transcription. MogR works by binding to the *flaA* promoter region, thus leading to the decreased expression of *flaA* and other motility genes. Apart from involvement in the temperature-dependent motility gene repression during extracellular growth, MogR also plays an essential role in down-regulation of motility gene expression independent of temperature during intracellular invasion in this tissue culture model.[105] However, the role of this in natural infection is not clear, especially because an *flaA* deleted mutant showed no difference from its isogenic parent in virulence, persistence, or T cell responses in a mouse infection model.[106] *Listeria* species possess distinct combinations of somatic (O) antigens (containing 15 subtypes [I–XV]) and flagellar (H) antigens (consisting of 4 subtypes [A–D]). These can be targeted with specific polyclonal and monoclonal antibodies; they can be effectively separated into serovars by *in vitro* tests (see chapter 5).

Besides generating a number of molecules fundamental to the formation of cell wall and flagella as well as other essential structures, *Listeria* species also produce on demand a large battery of specialized, purpose-made proteins that underpin listerial survival and maintenance under stressful pH, temperature, and salt conditions (e.g., BetL, Bsh, Lmo1421, OpuB, and OpuC) (see chapter 3). These enhance *L. monocytogenes* entry into host cells, escape from vacuoles, replication in cytosol, and spread to neighboring cells (e.g., InlA, InlB, LLO, PLCs, Mpl, and ActA) (see chapter 4). For instance, *L. monocytogenes* synthesizes internalins A and B (InA and InlB) for efficient entry into host epithelial and other cells, respectively. Once inside the host cells, it secretes LLO and phospholipases C (PLCs) for prompt escape from vacuoles, and ActA and Mpl for intra- and intercellular migration. Not surprisingly, these invasion- and virulence-associated proteins have often become the targets of host immune responses that lead to the partial or total elimination of invading *L. monocytogenes* (depending on dosages).

Upon contact with the invading *L. monocytogenes*, an immunocompetent mammalian host launches nonspecific immune attacks that are characterized by activation and expansion of natural killer (NK) cells and dendritic cell-primed CD4 T helper cells. These cell types are potent producers of gamma interferon (IFN-γ), which activates antimicrobial macrophages for secretion of tumor necrosis factor alpha (TNF-α) and interleukin 12 (IL-12). TNF-α and IL-12 in turn prompt NK cells to destroy the infected cells (see chapter 13). The surviving *L. monocytogenes* utilizes a pore-forming toxin LLO to disrupt vacuolar membrane, which further enhances expression of IFN-γ from NK and CD4 T cells and up-regulates molecules in antigen presentation and activation of antimicrobial macrophages. This leads to expansion of antigen-specific, cytotoxic CD8 T cells and subsequent elimination of *L. monocytogenes* bacteria (see chapter 14).

In addition, many *L. monocytogenes* invasion- and virulence-associated proteins (e.g., LLO, InlA, InlD, InlC2, IspA-E) are recognized by host humoral immune systems through production of specific antibodies,[107] although the role of humoral immune response against listeriosis requires further clarification. The fact that *L. monocytogenes* internalins InlA, InlD, and InlC2 and listerial surface proteins IspA-E are recognized by infected rabbits suggests that these proteins are induced or significantly up-regulated *in vivo* during infection and may play critical roles in *Listeria* pathogenesis.[107] Subsequent analysis indicates that IspC is an 86-kDa cell surface protein consisting of 774 amino acids with autolytic (peptidoglycan hydrolase) activity and cell wall binding activity, attributable to its N- and C-terminal regions, respectively.[108] Thus, IspC may constitute a member of *L. monocytogenes* autolysins that include P45, P60, NamA (or MurA), Ami, and Auto, which have been shown previously to contribute to virulence of *L. monocytogenes*.

The characteristic biochemical features of *Listeria* species can be exploited via a number of biochemical reactions for their differentiation from other closely related bacteria, as well as from one another, in spite of their remarkable morphological similarities (Table 1.4). These reactions comprise sugar fermentation, an amino acid peptidase, and lypolytic and hemolytic activities, in addition to the enhancement of hemolysis or CAMP phenomenon,[109] where *Listeria* cultures are grown on sheep blood agar perpendicularly to either *Staphylococcus aureus* or *Rhodococcus equi* (Figure 1.4). Although thousands of clinical food and environmental isolates have been examined to date, only a few unusual *Listeria* phenotypes have been discovered. For instance, *L. monocytogenes* with weak or even negative hemolysis[110] or catalase reaction[111] has been noted. Indeed, the type of strain (ATCC 15313 or NCTC 10357) of *L. monocytogenes* is nonhemolytic, CAMP test negative, and nonmotile.[112] Hemolytic *L. innocua*[113,114] and nonhemolytic *L. seeligeri* isolates[115] have also been

TABLE 1.4
Phenotypic Reactions of *Listeria* Species

	Species[a]					
***In vitro* Character**	***L. monocytogenes***	***L. ivanovii***	***L. innocua***	***L. welshimeri***	***L. seeligeri***	***L. grayi***
Gram staining	+	+	+	+	+	+
Catalase test	+	+	+	+	+	+
β-Hemolysis on blood containing agar	+	++[c]	–	–	±	–
Lipase production	+	+	–	–	+	–
Amino acid peptidase activity	–	+	+	+	+	+
Aesculin hydrolysis	+	+	+	+	+	+
β-D glucosidase	+	+	+	+	+	+
Acid Production from:						
D-mannitol	–	–	–	–	–	+
L-rhamnose	+	–	+	±	–	±
D-xylose	–	+	–	+	+	–
α-Methyl D-mannoside	+	–	+	+	±	±
CAMP Test[b] with:						
Staphylococcus aureus	+	–	–	–	+	–
Rhodococcus equi	–	+	–	–	–	–

[a] +, positive reaction; –, negative reaction; ±, variable or weak reaction.
[b] Enhancement of hemolysis reaction.
[c] ++, strong haemolytic reaction

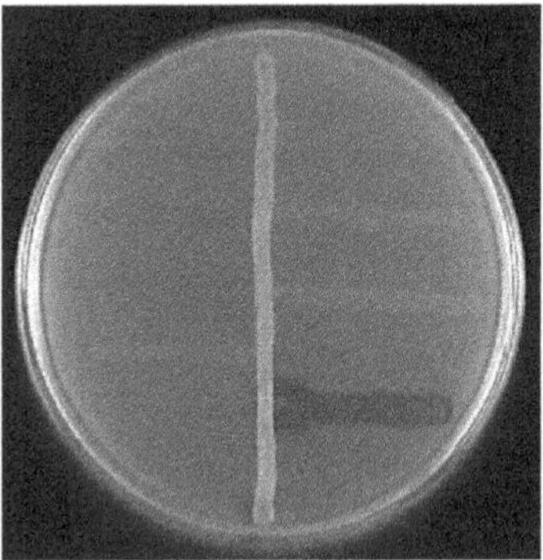

FIGURE 1.4 Enhancement of hemolysis (CAMP) reaction. Organisms are streaked perpendicularly on sheep blood agar with *Rhodococcus equi* (vertical) and (left side) *L. monocytogenes* (four strains) all negative; bottom is the type strain, which is nonhemolytic; right side, *L. seeligeri* (top) negative; *L. innocua* and *L. welshimeri* (middle) negative; and *L. ivanovii* (bottom) positive.

described. The proteins responsible for listerial hemolytic reactions on blood agar plates have been identified in *L. monocytogenes* (i.e., listeriolysin or LLO), *L. ivanovii* (ivanolysin or ILO), and *L. seeligeri* (seeligeriolysin or SLO). As the genes encoding these proteins are located in their respective virulence gene clusters, they are invariably involved in listerial virulence. There is ample evidence that alteration in the LLO-coding gene (*hly*) in *L. monocytogenes* renders it avirulent in mice.

1.6 PHYSIOLOGY

1.6.1 Growth Characteristics

Listeria spp. are relatively nutritionally undemanding and usually require biotin, riboflavin, thiamine, thioctic acid, and amino acids (e.g., cysteine, glutamine, isoleucine, leucine, and valine) for optimal growth. Carbohydrates (e.g., glucose), from which acid is produced as a by-product, are also essential for *Listeria* multiplication. In general, *Listeria* species grow well on a number of nonselective microbiological media designed to support the growth of bacteria. In artificial media, *Listeria* can grow under both aerobic and anaerobic conditions on nonselective media such as tryptone soy broth or brain–heart infusion (BHI) broth; the stationary phase is typically reached after 12 h of growth in BHI broth (incubation at 30°C). Tumbling motility is observed for cultures maintained between 20 and 25°C. Expression and repression of flagella synthesis is temperature- and catabolite regulated and associated with virulence and adherence to abiotic materials.[105,116–118] *Listeria* are psychrophilic organisms highly adapted to low temperatures. pH values of <5.2 result in a loss of growth but do not necessarily eliminate *Listeria* from foods. Survival occurs at temperatures below 0°C.

There is now a range of selective broths and agars available for the isolation of *Listeria* spp. from nonsterile samples such as food, environmental samples, and feces. PALCAM and OXFORD selective agars were recommended by ISO 11290-1, 2 in 1996 and 1998 as primary and secondary plating media for isolation of all *Listeria* spp., which are essentially indistinguishable and grow in a typical form signified by a depressed central area of the colony (so-called bull-eye or fish-eye colony). Selective plating media have now been developed for *L. monocytogenes* based on chromogenic substrates (Figure 1.5). *Listeria* agar as described by Ottaviani and Agosti (ALOA)[119] was the first chromogenic medium proposed in amendments 2004 of ISO 11290-1, 2.[120,121] The agar formulation consists of 5-bromo-4-chloro-3-indolyl-β-D-glucopyranosid, an enzyme substrate utilized by a β-D-glucosidase, which is produced by all *Listeria* spp. Typical blue–turquoise colonies of an average diameter of 1 mm grow on the agar surface.

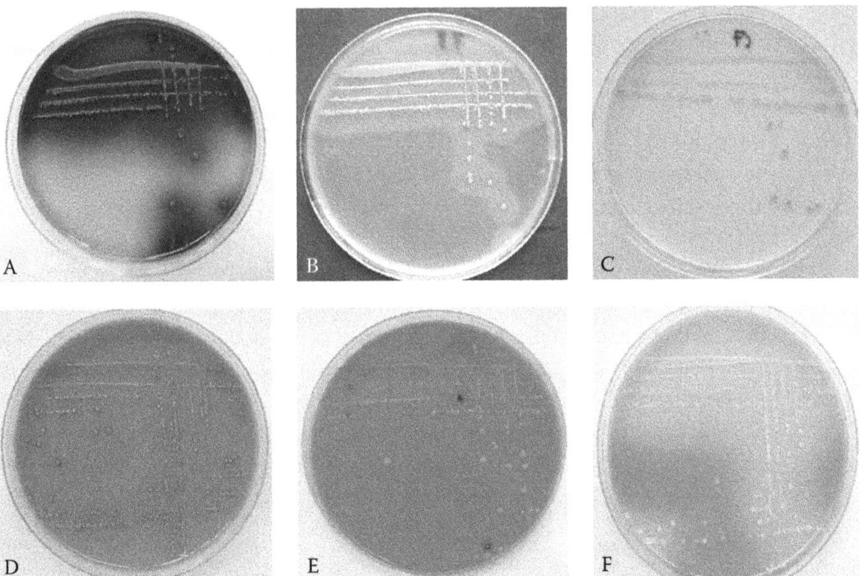

FIGURE 1.5 Selective plating media for the selective isolation of *Listeria* spp. A: Palcam—*Listeria* spp; B: ALOA-like medium (COMPASS)—*L. monocytogenes*; C: ALOA-like medium (CHROMagar)—*L. innocua*; D: Rapid'L.mono—*L. monocytogenes*; E: Rapid'L.mono—*L. monocytogenes* (blue–turqouise colonies) and *L. innocua* (white colonies); F: Rapid'L.mono—*L. welshimeri.*

The second substrate, L-α-phosphatidylinositol, is hydrolyzed by the phospholipase C, a virulence factor produced only by the two pathogenic species *L. monocytogenes* and *L. ivanovii*. Typical colonies are blue–turquoise, surrounded by a white zone of precipitation. Almost all leading manufacturers of microbiological culture media offer plating media similar to the original ALOA formulation: Biolife (ALOA®), AES (ALOA® 1 day), Oxoid (OCLA), CHROMagar (CHROMagar™), Merck (Chromoplate®), Biomerieux (OAA), and Biorad (AL). Other chromogenic media utilize only the phospholipase C detection, producing blue–turquoise colonies of *L. monocytogenes* and white colonies of other *Listeria* spp. Producers of this agar type are Biosynth (BCM® LMPM), Biorad (Rapid'L.mono), and Heipha (LIMONO-Ident). Selective chromogenic *L. monocytogenes* plating media can reduce the presumptive detection of *L. monocytogenes* to within 24 or 48 h of incubation at 37°C; however, further tests to confirm the identity of isolates are required. Examples of the colonial growth of different *Listeria* spp. on selective media are shown in Figure 1.5. Noncultural techniques such as those based upon immunoassays and the polymerase chain reaction are used increasingly for the detection of *Listeria* in enrichment broths for the examination of foods.

L. monocytogenes favors food as an agent in transmission for listeriosis.[122] *Listeria* grow in a wide range of food types that have relatively high water activities ($a_w > 0.92$) and over a wide range of temperatures (–0.15 to 45°C). Growth at refrigeration temperatures is relatively slow, with a maximum doubling time of about 1–2 d at 4°C in dairy products.[122] Similar doubling times were detected in vacuum-packed sliced beef and 100 h at –0.15°C.[123] *Listeria* can survive in the presence of 10% NaCl and 200 ppm $NaNO_2$, as well as in moist and dry environments at specific sites within food manufacturing environments for years. Even when present at high levels in foods, spoilage or taints are not generally produced. Multiplication in food is restricted to the pH range 4.3–9.4.[119] *L. monocytogenes* is not very heat resistant and will not endure pasteurization of milk; its D values (decimal reduction times) in various foods vary from 16.7 to 1.3 min at 60°C and 0.2 to 0.06 min at 70°C.[124]

L. monocytogenes has been isolated from numerous types of raw, processed, cooked, and ready-to-eat foods, usually at levels below 10 organisms per gram. The properties of this bacterium favor transmission through food and a wide variety of food and food matrices will support its growth; especially toward the end of an extended shelf life, this bacterium can become very heavily contaminated. Such "problem" food types that support the growth of *L. monocytogenes* include soft cheese, milk, pâté, frankfurters and other sausages, cooked meat and poultry, smoked fish and shellfish, processed vegetables, and some cut fruit, including melon. Growth can be localized within specific areas of an individual food either because of the source of contamination (i.e., within cut or contact surfaces or where raw herbs and spices have been added) or because of the physicochemical properties of the foods, such as in the areas of higher pH associated with the rind or with mold growth within a soft cheese.

L. innocua is frequently isolated from foods and the physiology of this organism (including its sensitivity to heat and cleaning regimes) is very similar to that of *L. monocytogenes*. Thus, the presence of *L. innocua* can be considered as an indicator organism for the possible presence of a *Listeria* hazard in end products and indicator of inadequate hygiene conditions in food production environments. This holds true also for all other *Listeria* species; these are more rarely isolated from food environments and, thus, soil seems to be their natural habitat. For example, among more than 9300 samples (smears, brine, and cheeses) investigated in Vienna during 20 years of *Listeria* surveillance, *L. grayi*, *L. seeligeri*, *L. welshimeri*, and *L. ivanovii* were identified from 2, 12, 39, and 4 samples, respectively.

1.6.2 Stress Tolerance

Listeria are capable of surviving under harsh environmental conditions such as extreme temperature (–0.1 to 45°C), pH (3.0–9.5), and salt (up to 10%) ranges (see chapter 3). As with other bacterial pathogens, "heat shock" responses occur such that exposure to heat or reduced pH increases the level of survival of *L. monocytogenes* to normally lethal conditions. This can be important in various food processes such as cheese production and has been reported to result in survival, for example, in feta cheese at pH 4.6 for >90 days.[122] It was shown by DeRoin et al.[125] that survival could occur on sand particles for 151 days at 10°C and 73 days at 22°C, thus highlighting the role of dust as a possible vehicle for cross-contamination. The ability of this bacterium to grow (albeit slowly) at refrigeration temperatures makes *L. monocytogenes* of particular concern as a postprocessing contaminant in long-shelf-life refrigerated foods. The widespread distribution of *L. monocytogenes* and its capability to survive on dry and moist surfaces favors postprocessing contamination of foods from both raw product and factory sites.[21]

In addition, *Listeria* spp. also show unusual tolerance to high salt concentrations (up to 10% NaCl and sodium nitrite). The ability of *L. monocytogenes* to withstand severe environmental stresses depends on its efficient stress response mechanisms (see chapter 3). Several salt stress response genes (including *betL, gbuABC, opuC, opuB, lmo1421,* and *bsh*) have been characterized, a majority of which are regulated by an alternative sigma factor, σ^B (encoded by *sigB*)—a protein subunit of RNA polymerase (RNAP). Mutations in *sigB* and related genes result in lower acid and salt tolerance in *L. monocytogenes*. Further, σ^B also influences *L. monocytogenes* virulence gene expressions by co-regulating a pleiotropic virulence regulator gene, *prfA*. The latter encodes a protein (PrfA) that is responsible for regulating itself and other virulence-associated genes (i.e., *plcA, hlyA mpl, actA,* and *plcB*) (see chapter 3). A number of *L. monocytogenes* genes expressed in response to growth at low temperature have also been identified. However, although *Listeria* spp. are known to tolerate alkali and pressure well, the underlying mechanisms against these stresses are poorly understood.

1.6.3 Antimicrobial Susceptibility and Resistance

Listeria spp. are susceptible to a range of antimicrobial agents. After examination of 103 *Listeria* strains (consisting of 21 *L. monocytogenes,* 21 *L. innocua,* 21 *L. seeligeri,* 19 *L. ivanovii,* 11 *L. welshimeri,* and 10 *L. grayi*), Troxler et al.[126] observed that while listeriae are naturally or intermediately sensitive to many antimicrobials, they demonstrate marked resistance to others (Table 1.5). In addition, *L. grayi* appears resistant to trimethoprim, co-trimoxazole, and rifampicin, but it is susceptible to quinolones. Furthermore, *L. ivanovii* is naturally resistant to most quinolones,

TABLE 1.5
Susceptibility and Resistance of *Listeria* Species to Antimicrobial Agents

Natural or Intermediate Sensitivity to	Natural or Intermediate Resistance to
Tetracyclines	Cephalosporins (cefetamet, cefixime, ceftibuten, ceftazidime, cefdinir, cefpodoxime, cefotaxime, ceftriaxone, cefuroxime)
Aminoglycosides	Aztreonam
Penicillins (except oxacillin)	Pipemidic acid
Cephalosporins (loracarbef, cefazoline, cefaclor)	Dalfopristin/quinupristin Sulfamethoxazole
Cefotiam	
Cefoperazone	
Carbapenems	
Macrolides	
Lincosamides	
Glycopeptides	
Chloramphenicol	
Rifampicin (except *L. grayi*)	

but is naturally sensitive to fosfomycin and fusidic acid. However, *L. innocua* and *L. monocytogenes* are naturally tolerant of fusidic acid.[126]

The capability of *L. monocytogenes* to tolerate selective antimicrobial agents was also noted in another recent study involving 38 isolates from four dairy farms.[127] While all 38 *L. monocytogenes* isolates resisted cephalosporin C, streptomycin, and trimethoprim, a majority also tolerated ampicillin, rifampicin, and florfenicol, and about half were resistant to tetracycline, penicillin G, and chloramphenicol. However, these 38 *L. monocytogenes* isolates were notably susceptible to amoxicillin, erythromycin, gentamicin, kanamycin, and vancomycin. Interestingly, antibiotic resistance genes were detected in a significant proportion of these *L. monocytogenes* strains, with *floR* (66%), *penA* (37%), *strA* (34%), *tetA* (32%), and *sulI* (16%) being most prominent. However, antitetracycline (*tetB, tetC, tetD, tetE,* and *tetG*) and other antimicrobial genes (*cmlA, strB, aadA, sulI, vanA, vanB, ampC, ermB, ereA,* and *ereB*) were not identified in the *L. monocytogenes* isolates under investigation.[127]

More recently, through examination of 106 *L. monocytogenes* strains isolated predominantly from human blood cultures and cerebrospinal fluids during the period 1958–2001 in Denmark, Hansen et al. showed that these *L. monocytogenes* strains maintained their susceptibility to penicillin G, ampicillin, meropenem, gentamicin, sulphamethoxazole, trimethoprim, erythromycin, vancomycin, linezolid, chloramphenicol, and tetracycline. Thus, the authors concluded that antimicrobial susceptibility of these *L. monocytogenes* strains had not changed in Denmark from 1958 to 2001, and the multiresistant strains found in human infections elsewhere had not been found in Denmark.[128]

The preceding data provide support for the current use of ampicillin or penicillin, alone or in combination with gentamicin, for effective treatment of listeriosis. Alternative treatment regimens incorporate trimethoprim-sulphamethoxazole, vancomycin plus gentamicin, and possibly erythromycin. Indeed, a combination of ampicillin and aminoglycoside (usually gentamicin) has been widely accepted as the therapy of choice for human listeriosis. Under circumstances where it is not possible to use a beta-lactam antibiotic, second-choice therapy may employ trimethoprim with a sulfonamide (e.g., co-trimoxazole). Other second-line agents for listeriosis are erythromycin and vancomycin. As a new drug against Gram-positive bacteria, linezolid can also be utilized.[129]

1.6.4 Prediction of *Listeria* Growth in Foods

Being a ubiquitous bacterium, *L. monocytogenes* contaminates a wide variety of foods. To ensure that manufactured foods are safe for human consumption, it is essential for the food industry to determine conditions that maintain food freshness while keeping microbial contamination to a minimum level. Toward this goal, a number of models concerning lag times and growth/no growth for any combination of environmental settings (e.g., temperature, pH, and water activity) have been developed.[130] Use of these models helps quantify the combined effects of various hurdles on the probability of growth and determine the chances that microorganisms will grow in specific environmental conditions. By incubating *L. monocytogenes* at 7°C in nutrient broth with different combinations of environmental factors—pH 5.0–6.0, water activity (a_w) 0.960–0.990, and acetic acid concentration 0–0.8% (w/w)—it was found that most *L. monocytogenes* strains were not able to grow at $a_w < 0.930$, pH < 4.3, or a total acetic acid concentration > 0.4% (w/w).[131,132]

In another study, the growth/no growth response of a mixture of five strains of *L. monocytogenes* in tryptic soy broth and on tryptic soy agar was assessed after a combination of temperature (4–30°C), pH (4.24–6.58), and a_w (0.900–0.993) treatments for 30 days. It was noted that at 25°C, the minimum pH values of 4.45 and 5.10 permitted growth in broth and agar, respectively, with respective a_w limits of 0.900 and 0.945.[133] Furthermore, by examining the combined effects of temperature, pH, and organic acids (lactic, acetic, and propionic) on the growth kinetics of *L. innocua* ATCC 33090, Le Marc et al. developed a model that allows an accurate description of the boundary between growth and no growth of *Listeria*.[134]

1.7 CONCLUSIONS AND PERSPECTIVES

Listeriosis is a serious disease. Listeria spp. are a ubiquitous genus of Gram-positive bacteria present in a diversity of environments, including soil, vegetation, and water, where they survive as saprophytes. When consumed via contaminated foods by susceptible mammalian hosts, such as pregnant women, the elderly, or immunosuppressed individuals, *L. monocytogenes* moves into intracellular environments, where it multiplies, evades the host's immunity, and spreads to neighboring cells and other organs. Although some early clinical manifestations of human listeriosis are mild, nonspecific, and influenza-like (e.g., chills, fatigue, headache, muscular and joint pain, as well as gastroenteritis), serious systemic illness occurs, including septicemia, meningitis, encephalitis, abortions, and, in some cases, death[14,31] (see chapter 4). The average mortality rate of human listeriosis is 30%, which is higher than that for most other common food-borne pathogens such as *Salmonella enteritidis* (with a mortality of 0.38%), *Campylobacter* species (0.02–0.1%), and *Vibrio* species (0.005–0.01%).[12,33]

Listeriosis is a preventable disease. Since the majority of human listeriosis cases result from consumption of *L. monocytogenes*-contaminated foods, the primary intervention to prevent disease is to eliminate or reduce this bacterium from food products.[28] Since this bacterium is capable of growth during processing and storage in a wide variety of foods, including at refrigeration temperatures and under conditions of high levels of salt or nitrite, one of the most important critical control points for an integrated food safety approach is food production areas. The contamination pressure on food processing plants is considerable. Knowledge of the plant contamination status is of highest priority for food safety managers and an application of vigorous, proactive hygiene management is the most promising strategy to counteract the risk for plant contamination. These considerations are of particular importance for foods that support the growth of *L. monocytogenes.* The survival of *L. monocytogenes* in foods results from a complex series of interactions influenced by the adaptation of this bacterium to the composition of the food, to the food microflora, and to the conditions of food processing and storage. The increase of the pH value on smeared soft cheeses during ripening due to changes in the microbial composition of the surface flora is an example of an accumulation of risk-enhancing parameters that may enable *L. monocytogenes* to multiply to hazardous numbers during storage.

Although much is now understood by food producers of the significance of *L. monocytogenes,* exchange of knowledge and acting on "lessons learned" are still required to control contamination in production plants. "Out-of-normality" production, such as unpredictable seasonal increases, change in personnel, restructuring of buildings, problems in maintenance and cleaning of plant and equipment, and cost cutting, can precipitate contamination episodes. Therefore, the *Listeria* contamination status of a food plant may also be indicative of the economical and social conditions in food plants (Figure 1.6). The final controlling measure is by dietary advice, particularly to vulnerable groups and behavior in domestic environments.

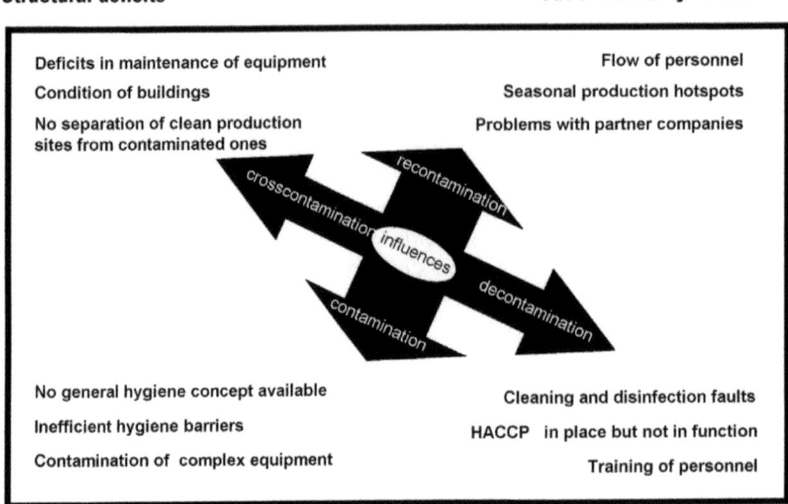

FIGURE 1.6 Problems in food plants and food chain management frequently leading to contamination with listeriae.

REFERENCES

1. Murray, E.G.D., Webb, R.A., and Swann, M.B.R., A disease of rabbits characterized by a large mononuclear leucocytosis, caused by a hitherto undescribed bacillus *Bacterium monocytogenes, J. Path. Bacteriol.,* 28, 407, 1926.
2. Pirie, J.H.H., A new disease of veld rodents, "Tiger River Disease," *Publ. S. Afr. Inst. Med. Res.,* 3, 163, 1927.
3. Pirie, J.H.H., *Listeria*: Change of name for a genus of bacteria, *Nature,* 145, 264, 1940.
4. Nyfeldt, A., Etiologie de la mononucleose infecteuse, *Compt. Rend. Soc. Biol.,* 101, 590, 1929.
5. Gray, M.L., and Killinger, A.H., *Listeria monocytogenes* and listeric infection, *Bacteriol. Rev.,* 30, 309, 1966.
6. Cotoni, L., A propos des bactéries dénommées Listerella rappel d'une observation ancienne de méningite chez l'homme, *Ann. Inst. Pasteur,* 68, 92, 1942.
7. Schlech, W.F. et al., Epidemic listeriosis—Evidence for transmission by food, *N. Engl. J. Med.,* 308, 203, 1983.
8. Fleming, D.W. et al., Pasteurized milk as a vehicle of infection in an outbreak of listeriosis, *N. Engl. J. Med.,* 312, 404, 1985.
9. Bille, J., and Glauser, M.P., Listeriose en Suisse, *Bull. B.A. Gesundheitsw.,* 3, 28, 1988.
10. Linnan, M.J. et al., Epidemic listeriosis associated with Mexican-style cheese, *N. Engl. J. Med.,* 319, 823, 1988.
11. McLauchlin, J. et al., Human listeriosis and paté: A possible association, *Brit. Med. J.,* 303, 773, 1991.
12. Mead, P.S. et al., Food-related illness and death in the United States, *Emerg. Infect. Dis.,* 5, 607, 1999.
13. Adak, G.K., Long, S.M., and O'Brien, S.J., Trends in indigenous food-borne disease and deaths, England and Wales: 1992 to 2000, *Gut,* 51, 832, 2002.
14. McLauchlin, J., The pathogenicity of *Listeria monocytogenes*: A public health perspective, *Rev. Med. Microbiol.,* 8, 1, 1997.
15. Rocourt, J. et al., Meningite purulente aigue à *Listeria seeligeri* chez un adulte immunocompetent, *Schweiz. Med. Wochenschr.,* 116, 248, 1986.
16. Cummins, A.J., Fielding, A.K., and McLauchlin, J., *Listeria ivanovii* infection in a patient with AIDS, *J. Infect.,* 28, 89, 1994.
17. Low, J.C., and Donachie, W., A review of *Listeria monocytogenes* and listeriosis, *Vet. J.,* 153, 9, 1997.

18. Gonzalez-Zorn, B. et al., The smcL gene of *Listeria ivanovii* encodes a sphingomyelinase C that mediates bacterial escape from the phagocytic vacuole, *Mol. Microbiol.*, 33, 510, 1999.
19. Farber, J.M., and Peterkin, P.I., *Listeria monocytogenes*, a food-borne pathogen, *Microbiol. Rev.*, 55, 476, 1991.
20. McLauchlin, J., and Low, C., Primary cutaneous listeriosis in adults: An occupational disease of veterinarians and farmers, *Vet. Rec.*, 135, 615, 1994.
21. McLauchlin, J., The relationship between *Listeria* and listeriosis, *Food Control*, 7, 187, 1996.
22. Gellin, B.C., and Broome, C.V., Listeriosis, *JAMA*, 261, 1313, 1989.
23. Gellin, B.C. et al., The epidemiology of listeriosis in the United States—1986, *Am. J. Epidemiol.*, 133, 392, 1986.
24. Anonymous, Preliminary Food Net data on the incidence of infection with pathogens transmitted commonly through food—10 states, 2006, *MMWR Weekly*, 56, 336, 2007.
25. de Valk, H. et al., Feasibility study for a collaborative surveillance of *Listeria* infections in Europe, Report to the European Commission, 2003, DG SANCO, Paris.
26. Koch, J., and Stark, K., Significant increase of listeriosis in Germany—Epidemiological patterns 2001–2005, *Euro Surveill.*, 11, 85, 2006.
27. Gillespie, I. et al., Changing pattern of human listeriosis in England and Wales, 2001–2005, *Emerg. Infect. Dis.*, 12, 1361, 2006.
28. Wong, S. et al., Recalls of foods and cosmetics due to microbial contamination reported to the U.S. Food and Drug Administration, *J. Food Prot.*, 63, 1113, 2000.
29. Goulet, V. et al., Listeriosis from consumption of raw-milk cheese, *Lancet*, 345, 1581, 1995.
30. Goulet, V. et al., Listeriosis outbreak associated with the consumption of rillettes in France in 1993, *J. Infect. Dis.*, 177, 155, 1998.
31. Vázquez-Boland, J.A. et al., *Listeria* pathogenesis and molecular virulence determinants, *Clin. Microbiol. Rev.*, 14, 584, 2001.
32. Bille, J. et al., Outbreak of human listeriosis associated with tomme cheese in northwest Switzerland, 2005, *Euro Surveill.*, 11, 91, 2006.
33. Mead, P.S. et al., Nationwide outbreak of listeriosis due to contaminated meat, *Epidemiol. Infect.*, 134, 744, 2006.
34. Wiedmann, M., ADSA Foundation Scholar Award—An integrated science-based approach to dairy food safety: *Listeria monocytogenes* as a model system, *J. Dairy Sci.*, 86, 1865, 2003.
35. Anonymous, Empfehlungen zum Nachweis und zur Bewertung von *Listeria monocytogenes* in Lebensmitteln im Rahmen der amtlichen Lebensmittelüberwachung. Anlage 1, Empfehlungen des Bundesinstituts für gesundheitlichen Verbraucherschutz und Veterinärmedizin, 2000, Berlin: BgVV.
36. Dawson, S.J. et al., *Listeria* outbreak associated with sandwich consumption from a hospital retail shop, United Kingdom, *Euro Surveill.*, 11, 89, 2006.
37. Rocourt, J., and Seeliger, H.P., Distribution of species of the genus *Listeria*, *Zentralbl. Bakteriol. Mikrobiol. Hyg.* (A), 259, 317, 1985.
38. Weis, J., and Seeliger, H.P.R., Incidence of *Listeria monocytogenes* in nature, *Appl. Microbiol.*, 30, 29, 1995.
39. Kutter, S., Hartmann, A., and Schmid, M., Colonization of barley (*Hordeum vulgare*) with *Salmonella enterica* and *Listeria* spp., *FEMS Microbiol. Ecol.*, 56, 262, 2006.
40. Pless, P. et al. Influence of feed quality and milking hygiene on *Listeria* prevalence in raw milk, *Proc. 10th Int. Congr. Anim. Hyg.*, 343, 2000.
41. Husu, J.R., Epidemiological studies on the occurrence of *Listeria monocytogenes* in the feces of dairy cattle, *J. Vet. Med.*, 37, 276, 1990.
42. Unnerstad, H. et al., *Listeria monocytogenes* in feces from clinically healthy dairy cows in Sweden, *Acta Vet. Scand.*, 41, 167, 2000.
43. Fenlon, D.R., Rapid quantitative assessment of the distribution of *Listeria* in silage implicated in a suspected outbreak of listeriosis in calves, *Vet. Rec.*, 118, 240, 1986.
44. Boerlin, P., Boerlin-Petzold, F., and Jemmi, T., Use of listeriolysin O and internalin A in a seroepidemiological study of listeriosis in Swiss dairy cows, *J. Clin. Microbiol.*, 41, 1055, 2003.
45. Fedio, W.M. et al., A case of bovine mastitis caused by *Listeria monocytogenes*, *Can. Vet. J.*, 31, 773, 1990.
46. Gitter, M., Bradley, R., and Blampied, P.H., *Listeria monocytogenes* infection in bovine mastitis, *Vet. Rec.*, 25, 390, 1980.

47. Jensen, N.E. et al., *Listeria monocytogenes* in bovine mastitis. Possible implication for human health, *Int. J. Food Microbiol.*, 32, 209, 1996.
48. Schoder, D. et al., A case of sporadic ovine mastitis caused by *Listeria monocytogenes* and its effect on contamination of raw milk and raw-milk cheeses produced in the on-farm dairy, *J. Dairy Res.*, 70, 395, 2003.
49. Winter, P. et al., Clinical and histopathological aspects of mastitis caused by *Listeria monocytogenes* in cattle and ewes, *J. Vet. Med. (B)*, 51, 176, 2004.
50. Sanaa, M. et al., Risk factors associated with contamination of raw milk by *Listeria monocytogenes* in dairy farms, *J. Dairy Sci.*, 76, 2891, 1993.
51. Waak, E., Tham, W., and Danielsson-Tham, M.L., Prevalence and fingerprinting of *Listeria monocytogenes* strains isolated from raw whole milk in farm bulk tanks and in dairy plant receiving tanks, *Appl. Environ. Microbiol.*, 68, 3366, 2002.
52. Dauphin, G., Ragimbeau, C., and Malle, P., Use of PFGE typing for tracing contamination with *Listeria monocytogenes* in three cold-smoked salmon processing plants, *Int. J. Food Microbiol.*, 64, 51, 2001.
53. Destro, M.T., Leitão, M.F., and Farber, J., Use of molecular typing methods to trace the dissemination of *Listeria monocytogenes* in a shrimp processing plant, *Appl. Environ. Microbiol.*, 62, 705, 1996.
54. Fonnesbech Vogel, B. et al., Elucidation of *Listeria* contamination routes in cold-smoked salmon processing plants detected by DNA-based typing methods, *Appl. Environ. Microbiol.*, 67, 2586, 2001.
55. Miettinen, M., Bjorkroth, K.J., and Korkeala, H.J., Characterization of *Listeria monocytogenes* from an ice cream plant by serotyping and pulsed-field gel electrophoresis, *Int. J. Food Microbiol.*, 46, 187, 1999.
56. Suihko, M.L. et al., Characterization of *Listeria monocytogenes* isolates from the meat, poultry, and seafood industries by automated ribotyping, *Int. J. Food Microbiol.*, 72, 137, 2002.
57. Wagner, M., Maderner, A., and Brandl, E., Random amplification of polymorphic DNA for tracing and molecular epidemiology of *Listeria* contamination in a cheese plant, *J. Food Prot.*, 59, 384, 1996.
58. Wagner, M. et al., Survey on the *Listeria* contamination of ready-to-eat food products and household environments in Vienna, Austria, *Zoon. Publ. Health*, 54, 16, 2007.
58a. Pappelbaum, K. et al., Monitoring hygiene on- and at-line is critical for controlling *Listeria monocytogenes* during produce processing, *J. Food Prot.* (in press).
59. Norwood, D.E., and Gilmour, A., Adherence of *Listeria monocytogenes* strains to stainless steel coupons, *J. Appl. Microbiol.*, 86, 576, 1999.
60. Blackmann, I.C., and Frank, J.F., Growth of *Listeria monocytogenes* as a biofilm on various food-processing surfaces, *J. Food Prot.*, 59, 827, 1996.
61. Lunden, J.M. et al., Persistent *Listeria monocytogenes* strains show enhanced adherence to food contact surface after short contact times, *J. Food Prot.*, 63, 1204, 2000.
62. Pan, Y., Breidt, F., Jr., and Kathariou, S., Resistance of *Listeria monocytogenes* biofilms to sanitizing agents in a simulated food processing environment, *Appl. Environ. Microbiol.*, 72, 7711, 2006.
63. Pinner, R.W. et al., Role of foods in sporadic listeriosis: Microbiological and epidemiological investigations, *JAMA*, 267, 2046, 1992.
64. Beumer, R.R. et al., *Listeria* species in domestic environments, *Epidemiol. Infect.*, 117, 437, 1996.
65. Varma, J.K. et al., *Listeria monocytogenes* infection from foods prepared in a commercial establishment: A case-control study of potential sources of sporadic illness in the United States, *Clin. Infect. Dis.*, 44, 521, 2007.
66. Goulet, V., What can we do to prevent listeriosis in 2006? *Clin. Infect. Dis.*, 44, 529, 2007.
67. Seeliger, H.P.R. et al., *Listeria ivanovii*, *Int. J. Syst. Bacteriol.*, 34, 336, 1984.
68. Seeliger, H.P.R, Apathogene Listerien: *Listeria innocua* sp. nov. (Seeliger und Schoofs, 1977), *Zentralbl. Bakteriol. Parasit. Infekt. Hyg.* Abt. 1 Reihe A, 249, 487, 1981.
69. Rocourt, J., and Grimont, F., *Listeria welshimeri* sp. nov. and *Listeria seeligeri* sp. nov., *Int. J. Syst. Bacteriol.*, 3, 866, 1983.
70. Errebo-Larsen, H., and Seeliger, H.P.R., A mannitol fermenting *Listeria*: *Listeria grayi* sp. n. *Proc. 3rd Int. Symp. Listeriosis*, Biltoven, The Netherlands, 1966, 35–39.
71. Rocourt, J. et al., Assignment of *Listeria grayi* and *Listeria murrayi* to a single species, *Listeria grayi*, with a revised description of *Listeria grayi*, *Int. J. Syst. Bacteriol.*, 42, 171, 1992.
72. Roberts, A. et al., Genetic and phenotypic characterization of *Listeria monocytogenes* lineage III, *Microbiology*, 152, 685, 2006.
73. Liu, D. et al., *Listeria monocytogenes* subgroups IIIA, IIIB, and IIIC delineate genetically distinct populations with varied virulence potential, *J. Clin. Microbiol.*, 44, 4229, 2006.
74. Barbour, A.H., Rampling, A., and Hormaeche, C.E., Variation in the infectivity of *Listeria monocytogenes* isolates following intragastric inoculation of mice, *Infect. Immun.*, 69, 4657, 2001.

75. Liu, D. et al., Characterization of virulent and avirulent *Listeria monocytogenes* strains by PCR amplification of putative transcriptional regulator and internalin genes, *J. Med. Microbiol.*, 52, 1066, 2003.
76. Liu, D., *Listeria monocytogenes*: Comparative interpretation of mouse virulence assay, *FEMS Microbiol. Lett.*, 233, 159, 2004.
77. Feresu, S.B., and Jones, D., Taxonomic studies on *Brochothrix, Erysipelothrix, Listeria* and atypical lactobacilli, *J. Gen. Microbiol.*, 134, 1165, 1988.
78. Hartford, T., and Sneath, P.H.A., Optical DNA–DNA homology in the genus *Listeria, Int. J. Syst. Bacteriol.*, 43, 26, 1993.
79. Rocourt, J. et al., DNA relatedness among serovars of *Listeria monocytogenes* sensu lato, *Curr. Microbiol.*, 7, 383, 1982.
80. Wilkinson, B.J., and Jones, D.A., Numerical taxonomic survey of *Listeria* and related bacteria, *J. Gen. Microbiol.*, 98, 399, 1997.
81. Collins, M.D. et al., Phylogenetic analysis of the genus *Listeria* based on reverse transcriptase-sequencing of 16S rRNA, *Int. J. Food Microbiol.*, 41, 240, 1991.
82. Glaser, P. et al., Comparative genomics of *Listeria* species, *Science*, 294, 849, 2001.
83. Hain, T. et al., Whole-genome sequence of *Listeria welshimeri* reveals common steps in genome reduction with *Listeria innocua* as compared to *Listeria monocytogenes, J Bacteriol.*, 188, 7405, 2006.
84. Schmid, M.W. et al., Evolutionary history of the genus *Listeria* and its virulence genes, *Syst. Appl. Microbiol.*, 28, 1, 2005.
85. Hof, H., and Rocourt, J., Is any strain of *Listeria monocytogenes* detected in food a health risk? *Int. J. Food Microbiol.*, 16, 173, 1992.
86. Wiedmann, M. et al., Ribotypes and virulence gene polymorphisms suggest three distinct *Listeria monocytogenes* lineages with differences in pathogenic potential, *Infect. Immun.*, 65, 2707, 1997.
87. Buchrieser, C. et al., Pulsed-field gel electrophoresis applied for comparing *Listeria monocytogenes* strains involved in outbreaks, *Can. J. Microbiol.*, 39, 395, 1993.
88. Czajka, J., and Batt, C.A., Verification of causal relationships between *Listeria monocytogenes* implicated in food-borne outbreaks of listeriosis by randomly amplified polymorphic DNA patterns, *J. Clin. Microbiol.*, 32, 1280, 1994.
89. Kathariou, S. et al., Involvement of closely related strains of a new clonal group of *Listeria monocytogenes* in the 1998–99 and 2002 multistate outbreaks of food-borne listeriosis in the United States, *Foodborne Pathog. Dis.*, 3, 292, 2006.
90. Rasmussen, O.F. et al., *Listeria monocytogenes* isolates can be classified into two major types according to the sequence of the listeriolysin gene, *Infect. Immun.*, 59, 3945, 1992.
91. Brosch. R. et al., Virulence heterogeneity of *Listeria monocytogenes* strains from various sources (food, human, animal) in immunocompetent mice and its association with typing characteristics, *J. Food Prot.*, 56, 297, 1993.
92. Conner, D.E. et al., Pathogenicity of food-borne, environmental and clinical isolates of *Listeria monocytogenes* in mice, *J. Food Sci.*, 54, 1553, 1989.
93. Tabouret, M. et al., Pathogenicity of *Listeria monocytogenes* isolates in immunocompromised mice in relation to listeriolysin production, *J. Med. Microbiol.*, 34, 13, 1991.
94. Schleifer, K.H., and Kandler, O., Peptidoglycan types of bacterial cell walls and their taxonomic implications, *Bacteriol. Rev.*, 36, 407, 1972.
95. Ghosh, B.K., and Murray, R.G., Fine structure of *Listeria monocytogenes* in relation to protoplast formation, *J. Bacteriol.*, 93, 411, 1967.
96. Ullmann, W.W., and Cameron, J.A., Immunochemistry of the cell walls of *Listeria monocytogenes, J. Bacteriol.*, 98, 486, 1969.
97. Hether, N.W., and Jackson, L.L., Lipoteichoic acid from *Listeria monocytogenes, J. Bacteriol.*, 156, 809, 1983.
98. Kamisango, K. et al., Structural and immunochemical studies of teichoic acid of *Listeria monocytogenes, J. Biochem. (Tokyo),* 93, 1401, 1983.
99. Fiedler, F. et al., The biochemistry of murein and cell wall teichoic acids in the genus *Listeria, Syst. Appl. Micobiol.*, 5, 360, 1984.
100. Fiedler, F., Biochemistry of the cell surface of *Listeria* strains: A locating general view, *Infection,* 16, 92, 1988.
101. Seeliger, H.P.R., and Höhne, K., Serotyping of *Listeria monocytogenes* and related species. In *Methods in microbiology,* ed. Bergan, T., and Norris, J., 1–3. Academic Press, New York, 1979.

102. Wendlinger, G., Loessner, M.J., and Scherer, S., Bacteriophage receptors on *Listeria monocytogenes* cells are the N-acetylglucosamine and rhamnose substituents of teichoic acids or the peptidoglycan itself, *Microbiology,* 142, 985, 1996.
103. McLauchlin, J., and Rees, C.E.D., Genus *Listeria.* In *Bergey's manual of systematic bacteriology,* 2nd ed., The low G + C gram-positive bacteria, Vol. 3, eds. DeVos P., Garrity, G., Jones, D., Krieg, N.R., Ludwig, W., Rainey, F.A., Schleifer, K.H., and Whitman, W.B., Williams and Williams, Baltimore, MD, 2008 (in press).
104. Dubail, I. et al., Identification of an essential gene of *Listeria monocytogenes* involved in teichoic acid biogenesis, *J. Bacteriol.,* 188, 6580, 2006.
105. Grundling, A. et al., *Listeria monocytogenes* regulates flagellar motility gene expression through MogR, a transcriptional repressor required for virulence, *Proc. Natl. Acad. Sci. USA,* 101, 12318, 2004.
106. Way, S.S. et al., Characterization of flagellin expression and its role in *Listeria monocytogenes* infection and immunity, *Cell. Microbiol.,* 6, 235, 2004.
107. Yu, W.L., Dan, H., and Lin, M., Novel protein targets of the humoral immune response to *Listeria monocytogenes* infection in rabbits, *J. Med. Microbiol.,* 56, 888, 2007.
108. Wang, L., and Lin, M., Identification of IspC, an 86-kilodalton protein target of humoral immune response to infection with *Listeria monocytogenes* serotype 4b, as a novel surface autolysin, *J. Bacteriol.,* 189, 2046, 2007.
109. Christie, R., Atkins, N.E., and Munch-Peterson, E., A note on a lytic phenomenon shown by group B streptococci, *Aust. J. Exp. Biol. Med. Sci.,* 22, 197, 1944.
110. Allerberger, F. et al., Nonhemolytic strains of *Listeria monocytogenes* detected in milk products using VIDAS immunoassay kit, *Zentralbl. Hyg. Umweltmed.,* 200, 189, 1997.
111. Bubert, A. et al., Isolation of catalase-negative *Listeria monocytogenes* strains from listeriosis patients and their rapid identification by anti-p60 antibodies and/or PCR, *J. Clin. Microbiol.,* 35, 179, 1997.
112. Kathariou, S., and Pine, L., The type strain(s) of *Listeria monocytogenes*: A source of continuing difficulties, *Int. J. Syst. Bacteriol.,* 41, 328, 1991.
113. Johnson, J. et al., Natural atypical *Listeria innocua* strains with *Listeria monocytogenes* pathogenicity island 1 genes, *Appl. Environ. Microbiol.,* 70, 4256, 2004.
114. Volokhov, D.V. et al., The presence of the internalin gene in natural atypically hemolytic *Listeria innocua* strains suggests descent from *L. monocytogenes, Appl. Environ. Microbiol.,* 73, 1928, 2007.
115. Volokhov, D. et al., Discovery of natural atypical nonhemolytic *Listeria seeligeri* isolates, *Appl. Environ. Microbiol.,* 72, 2439, 2006.
116. Dons, L., Rasmussen, O.F., and Olsen, J.E., Cloning and characterization of a gene encoding flagellin of *Listeria monocytogenes, Mol. Microbiol.,* 6, 2919, 1992.
117. Kathariou, S. et al., Repression of motility and flagellin production at 37°C is stronger in *Listeria monocytogenes* than in the nonpathogenic species *Listeria innocua, Can. J. Microbiol.,* 41, 572, 1995.
118. Vatanyoopaisarn. S. et al., Effect of flagella on initial attachment of *Listeria monocytogenes* to stainless steel, *Appl. Environ. Microbiol.,* 66, 860, 2000.
119. Ottaviani, F., Ottaviani, M., and Agosti, M., Differential agar medium for *Listeria monocytogenes, Quinper Froid Symp. Proc.,* June 1997.
120. Anonymous, Horizontal method for the detection of *Listeria monocytogenes* in food- and feeding stuffs (11290:1996, Part 1). International Standardization Organization, 1996, Geneva, Switzerland.
121. Anonymous, Horizontal method for the enumeration of *Listeria monocytogenes* in food- and feeding stuffs (11290:1998, Part 2). International Standardization Organization, 1998, Geneva, Switzerland.
122. Ryser, E.T., and Marth, E.M., *Listeria, listeriosis and food safety.* Marcel Dekker, New York, 1991.
123. Hudson, J.A., and Mott, S.J., Growth of *Listeria monocytogenes, Aeromonas hydrophila* and *Yersinia eneterocolitica* in paté and a comparison with predictive models, *Int. J. Food Microbiol.,* 20, 1, 1993.
124. Bell, C., and Kyriakides, A., *Listeria: A practical approach to the organism and its control in foods,* 2nd ed. Blackwell, Oxford, 2005.
125. DeRoin, M.A. et al., Survival and recovery of *Listeria monocytogenes* from ready-to-eat meals inoculated with a desiccated and nutritionally depleted dust-like vector, *J. Food Prot.,* 66, 962, 2003.
126. Troxler, R. et al., Natural antibiotic susceptibility of *Listeria* species: *L. grayi, L. innocua, L. ivanovii, L. monocytogenes, L. seeligeri,* and *L. welshimeri* strains, *Clin. Microbiol. Infect.,* 6, 525, 2000.
127. Srinivasa, V. et al., Prevalence of antimicrobial resistance genes in *Listeria monocytogenes* isolated from dairy farms, *Foodborne Pathog. Dis.,* 2, 201, 2005.
128. Hansen, J.M., Gerner-Smidt, P., and Bruun, B., Antibiotic susceptibility of *Listeria monocytogenes* in Denmark 1958–2001, *APMIS,* 113, 31, 2005.

129. Poros-Golchowska, J., and Markiewicz, Z., Antimicrobial resistance of *Listeria monocytogenes, Acta Microbiol. Pol.,* 52, 113, 2003.
130. Standaert, A.R. et al., Modeling individual cell lag time distributions for *Listeria monocytogenes, Risk Anal.,* 27, 241, 2007.
131. Gysemans, K.P. et al., Exploring the performance of logistic regression model types on growth/no growth data of *Listeria monocytogenes, Int. J. Food Microbiol.,* 114, 316, 2007.
132. Vermeulen, A. et al., Influence of pH, water activity and acetic acid concentration on *Listeria monocytogenes* at 7°C: Data collection for the development of a growth/no growth model, *Int. J. Food Microbiol.,* 114, 332, 2007.
133. Koutsoumanis, K.P., Kendall, P.A., and Sofos, J.N., A comparative study on growth limits of *Listeria monocytogenes* as affected by temperature, pH and a_w when grown in suspension or on a solid surface, *Food Microbiol.,* 21, 415, 2004.
134. Le Marc, Y. et al., Modeling the growth kinetics of *Listeria* as a function of temperature, pH and organic acid concentration, *Int. J. Food Microbiol.,* 73, 219, 2002.

2 Epidemiology

Dongyou Liu

CONTENTS

2.1 INTRODUCTION

The genus *Listeria* comprises a group of Gram-positive, rod-shaped bacteria, which are classified taxonomically into six species (i.e., *L. monocytogenes, L. ivanovii, L. innocua, L. seeligeri, L. welshimeri,* and *L. grayi*). Whereas *L. monocytogenes* is a facultative intracellular pathogen of both humans and animals, *L. ivanovii* (previously known as *L. monocytogenes* serotype 5) mainly infects ruminant animals (e.g., sheep and cattle). The other four species are free-living saprophytes with no apparent pathogenic inclination. Based on serological reactions, *Listeria* species are divided into over 15 distinct serotypes, with *L. monocytogenes* consisting of serotypes 1/2a, 1/2b, 1/2c, 3a, 3b, 3c, 4a, 4b, 4c, 4d, 4e, and 7; *L. ivanovii* of serotype 5, and other *Listeria* species of serotypes 6a and 6b as well as several shared serotypes with *L. monocytogenes*.[1,2] Using genetic typing methods, *L. monocytogenes* is separated into three lineages: lineage I is composed of serotypes of 1/2b, 3b, 4b, 4d, and 4e; lineage II of serotypes 1/2a, 1/2c, 3a, and 3c; and lineage III of serotypes 4a and 4c.[3,4] In addition, lineage III can be subdivided into subgroups IIIA, IIIB, and IIIC.[5,6]

Listeria spp. demonstrate considerable morphological, biochemical, and molecular resemblances and occupy similar ecological niches in the environment. Given their renowned ability to withstand arduous external conditions such as wide pH, temperature, and salt ranges,[7,8] there is no surprise that *Listeria* spp. are distributed in a diverse range of environments and have been isolated from soil, water, effluents, foods, wildlife, and domestic animals as well as humans. Since *Listeria* spp. are co-present in various environmental, food, and clinical specimens, they often add complexity to the laboratory identification of *L. monocytogenes* and diagnosis of listeriosis. Indeed, it has only become feasible in recent decades to clarify the taxonomical status of *Listeria* spp and to discriminate *L. monocytogenes* from other *Listeria* spp in a reliable and precise manner with the application of molecular techniques. Therefore, any discussion on *L. monocytogenes* epidemiology will be incomplete and impossible without mentioning the other non*monocytogenes Listeria* spp.

Although the association of clinical diseases with *Listeria* bacteria has been noted from the 1920s, *Listeria* infections were frequently regarded as rare and insignificant until the early 1980s, when several high-profile human listeriosis outbreaks occurred. In line with a trend toward increased consumption of ready-to-eat and heat-and-eat foods nowadays, *L. monocytogenes* has emerged as an important human food-borne pathogen. Although it is being infective to all human population groups, *L. monocytogenes* has the propensity to cause particularly serious diseases in pregnant women, neonates, the elderly, and immunosuppressed persons. With its nonspecific, flu-like early symptoms (e.g., chills, fatigue, headache, muscular and joint pain) and gastroenteritis, human listeriosis does not usually attract sufficient attention and receive prompt treatment until it has already evolved into septicemia, meningitis, encephalitis, abortions, and, in some cases, death.[9] In fact, with average mortality rates of about 30%, *L. monocytogenes* is more deadly than other common food-borne pathogens such as *Salmonella enteritidis* (with a mortality of 0.38%), *Campylobacter* species (0.02–0.1%), and *Vibrio* species (0.005–0.01%).[10,11]

Apart from the development and application of rapid, sensitive, and precise diagnostic assays for differentiation of *L. monocytogenes* from other bacterial species (see Chapters 5 and 6), the availability of a detailed knowledge on listerial epidemiology is vital for the control and prevention of listeriosis. Extensive past research has uncovered numerous insights on *Listeria* epidemiology and related aspects. In this chapter, we focus on several key aspects of *Listeria* epidemiology, including the distribution of *Listeria* spp. in the environment (e.g., water, soil, vegetation, food processing facilities, and foods); the mechanisms of *Listeria* survival in the environment, food, and mammalian hosts; the virulence and pathogenicity of *L. monocytogenes* and *L. ivanovii* in humans and/or animals; and the measures to reduce the incidence of *Listeria* bacteria in environments and foods and to prevent the occurrence of listeriosis in animal and human hosts.

2.2 *LISTERIA* IN ENVIRONMENTS

2.2.1 Distribution

2.2.1.1 Water and Sludge

Listeria spp. are rarely isolated from unpolluted seawater, groundwater, and springs[12–14]; however, they are commonly found in river and bay waters, sewage, and industrial and farming effluents. In a bacteriological survey conducted in Northern Italy, Luppi et al.[15] isolated 11 *Listeria* strains (including one *L. monocytogenes*, two *L. seeligeri*, one *L. welshimeri*, and seven *L. innocua*) from 50 river water samples (22.0%), 15 *Listeria* strains (i.e., one *L. monocytogenes*, 11 *L. innocua*) from 80 surface water samples (19%), one *Listeria* strain (*L. innocua*) from 98 groundwater samples (1.0%), and 14 *Listeria* strains (i.e., eight *L. monocytogenes*, five *L. innocua*, and one *L. seeligeri*) from 33 urban sewage samples (42.4%). Colburn et al.[16] examined 37 fresh- or low-salinity water samples from tributaries draining into Humboldt–Arcata Bay in California and identified *Listeria* spp. and *L. monocytogenes* from 30 (81%) and 19 (62%) of these samples, respectively. In the sediment where water samples were taken, *Listeria* spp. and *L. monocytogenes* were recovered from 13 (30%) and 5 (17%) of the 46 samples. MacGowan et al.[17] reported that of the 115 sewage samples, 108 (93.9%) contained *Listeria* spp., 60.0% (i.e., 61/108) of which belonged to *L. monocytogenes*. Although Hansen et al.[14] detected no *L. monocytogenes* in a freshwater stream, they found the bacterium in 2% of the samples from seawater fish farms, 10% from freshwater fish farms, 16% from fish slaughterhouses, and 68% from a fish smokehouse. These results suggest that the incidence of *Listeria* spp. in water increases with animal and human activity, as polluted water and sewage/sludge provide a nutritional medium for *Listeria* growth compared with unpolluted water.

2.2.1.2 Soil

Weis and Seeliger[18] undertook an extensive survey of *Listeria* distribution in soil in Germany during the early 1970s and isolated *L. monocytogenes* and *L. ivanovii* from about 20% of surface soil samples analyzed. These included soil samples from cornfields (13% or 3/23), grain fields (13.9% or 5/36), cultivated fields (18.7% or 3/16), uncultivated fields (51.4% or 20/39), meadows/pastures (8.7% or 4/46), forest (15.2% or 9/59), and wildlife feeding grounds (43.2% or 14/37). Serovars 1/2b and 4b were most predominant in all soil samples, in addition to serovar 1/2a, which was common in soil from uncultivated fields and wildlife feeding grounds. Although Van Renterghem et al.[19] did not find *L. monocytogenes* in stored liquid manure and manured soil samples, they reported that 16% of fresh pig feces, 20% of fresh cattle feces, and 5% of groundwater samples contained the bacterium. MacGowan et al.[17] showed that *Listeria* spp. were present in 20 (14.7%) of the 130 soil samples from the United Kingdom; *L. monocytogenes* and *L. innocua* were common in fecal specimens, whereas *L. ivanovii* and *L. seeligeri* were dominant in soil specimens. More recently, Moshtaghi et al.[20] isolated *Listeria* spp. in 23 (17.7%) of the 130 soil samples collected in India, which consisted of seven (5.4%) *L. monocytogenes*, two (1.5%) *L. ivanovii*, 10 (7.7%) *L. innocua*, and four (3.1%) *L. welshimeri*. Interestingly, *L. ivanovii* was isolated only in the soil from animal-inhabited areas, but not in soil from agricultural fields. These findings highlight that *Listeria* spp. frequently occur in soil (with an average presence of 20%), where they probably live as saprophytes.

2.2.1.3 Plants and Vegetation

In an early study, Welshimer[21] isolated eight *L. monocytogenes* strains in vegetation collected from fields in Virginia. After examining vegetation or soil taken from 12 farms and seven nonagricultural sites, Welshimer and Donker-Voet[22] also obtained 27 *L. monocytogenes* isolates. Weis and Seeliger[18] found that *L. monocytogenes* existed in 20% of plant samples in Germany, which ranged from 9.7% (3/31) of plants from cornfields, 13.3% (4/40) from grain fields, 12.5% (2/16) from cultivated fields, 44% (33/75) from uncultivated fields, 15% (6/38) from meadows/pastures, 21.3% (13/61) from forest, and 23.1% (9/39) from wildlife feeding grounds. The wide distribution of *L. monocytogenes* in soil and plants reinforces the notion that *Listeria* spp. are free-living microbes with a capacity to sustain and grow in plant–soil environments.

2.2.1.4 Food Processing Plants

L. monocytogenes readily adheres to the surfaces of food processing benches, machinery, and floors, and subsequently grows into biofilm matrix with increased resistance to adverse conditions.[23] In particular, *L. monocytogenes* serotypes 1/2a, 1/2b, and 1/2c appear to have better adapted to meat processing plants, as they have more frequently been isolated from these environments.[24–27] Further, Autio et al.[26] reported that *L. monocytogenes* strains were not always plant specific, since certain genetically related strains were recovered in the pork-meat products originating from unrelated sites. Additionally, Chae and Schraft[28] observed that some *L. monocytogenes* strains were more likely to grow into a mature biofilm than others. Norwood and Gilmour[25] showed that *L. monocytogenes* serotype 1/2c had a tendency to attach to stainless steel surfaces in comparison with other serotypes. Chambel et al.[29] isolated 213 *Listeria* isolates from cheese-producing plants/farms (e.g., walls, drains, shelves, and diverse equipment) in Portugal; this included 85 *L. monocytogenes,* 88 *L. innocua*, 39 *L. seeligeri,* and one *L. ivanovii.* Eifert et al.[30] also detected *L. monocytogenes* serotype 4b complex (serotype 4b and the closely related serotypes 4d and 4e) from environmental samples collected in two turkey processing plants in the United States.

The preceding results indicate that, apart from rare appearances in unpolluted seawater, groundwater, and springs, *Listeria* species commonly occur in soil, vegetation, runoff water, sewage, etc. The incidence of *Listeria* spp. generally rises with heightened animal and human activity. Of the six species, *L. monocytogenes* and *L. innocua* are more frequently isolated from the environment, followed by *L. seeligeri, L. ivanovii,* and *L. welshimeri*. However, isolation of *L. grayi* from environmental samples is seldom described in the literature.

2.2.2 Mechanisms of Survival

Although *Listeria* species, especially *L. monocytogenes*, are robust bacteria with relatively low demand for nutrients for maintenance, they require external sources of biotin, riboflavin, thiamine, thioctic acid, amino acids (e.g., cysteine, glutamine, isoleucine, leucine, and valine) and carbohydrates (e.g., glucose) for optimal growth. This is reflected by the fact that *Listeria* spp. are infrequently isolated from unpolluted seawater and groundwater or springs, which may contain insufficient nutrients for listerial maintenance and replication. On the other hand, *Listeria* spp. are abundant in runoff water, sewage, soil, vegetation, and food processing environments, which offer a relatively nutritional medium for listerial expansion.

Being free-living organisms, *Listeria* spp. have to endure various external stresses for their eventual survival in the environment. For example, in river water and sludge, *Listeria* spp. have to withstand near-freezing temperatures during winter and extreme outdoor heat in summer. In food processing environments, *Listeria* spp. are routinely exposed to alkaline detergents and sanitizers that are applied to clean and sterilize food-processing surfaces, machines, and floors. From the studies involving the pathogenic species *L. monocytogenes*, it appears that this bacterium is remarkably agile in its response to pH, temperature, and osmotic extremes. It adroitly synthesizes a range of specialized molecules on demand to enhance its tolerance of heat, acid, and osmotic stresses, and to enable its survival and growth over a diverse range of temperatures (–0.15 to 45°C), pH (3.0–12.0), and osmolarity (up to 40% NaCl)[7,8] (see following discussion and Chapter 3).

2.2.3 Control and Prevention Strategies

As *Listeria* species grow poorly in nutrient-deficient, unpolluted seawater and spring water and thrive on nutrient-rich, contaminated waters, sewage, and sludge, an obvious strategy to lower the occurrence of *Listeria* spp. in the environment is to remove as much residual nutrient in the runoff waters from farms and processing plants as possible, thus leading to a slower the growth of *Listeria* bacteria therein. This can be done by pretreating runoff waters and sewage before release to waterways. In addition, implementation of routine screening procedures to monitor the presence of *Listeria* spp. in various environmental specimens will also offer prompt corrective actions when listerial bacteria reach unacceptable levels and thus prevent the spread of listeriosis in animals and humans.

2.3 *Listeria* in Foods

2.3.1 Distribution

2.3.1.1 Seafood

While *Listeria* spp. are rarely isolated in seafood harvested from unpolluted waters, they often exist in raw fish material from waters with farms and human settlements nearby. For instance, Eklund et al.[31] recovered *L. monocytogenes* in the slime, skin, and belly cavity of salmon, and Draughon et al.[32] detected heavy *L. monocytogenes* contamination in fresh rainbow trout from retail markets in the United States. Miettinen and Wirtanen[33] showed that *L. monocytogenes* was present in 0–75% of farm-raised fish in Finland, with fish gill samples contributing to 95.6% (43/45) of the *L. monocytogenes* isolations and fish skin or viscera accounting for the remaining 4.4% (2/45). These authors also noted that 14.3 and 8.8% of the individual thawed fish contained *Listeria* spp. and *L. monocytogenes*, respectively.[33] Rodas-Suarez et al.[34] examined 66 oyster, 66 fish, and 144 estuarine water samples collected over a 12-month period (June 2001 to May 2002) in Veracruz, Mexico. Even though *Listeria* spp. were not recovered in oyster samples, they were detectable in 22.7 and 30.5% of fish and estuarine water samples, respectively, with *L. monocytogenes* being present in 3 (4.5%) of the 66 fish samples and 12 (8.3%) of the 144 water samples. Chou et al.[35] showed that *L. monocytogenes* dominated *Listeria* isolations from channel catfish fillets, accounting for 25–47% of the strains obtained. Other *Listeria* spp. present included *L. welshimeri, L. innocua, L. ivanovii, L. grayi*, and *L. seeligeri*. It appears that plant-specific and non-plant-specific *L. monocytogenes* coexisted in processed catfish fillets. The fact that some isolates were persistently detected in processed fillets suggested either that inadequate sanitation procedures were implemented by the plants or that these isolates originated from the natural habitats of the catfish; thus, *Listeria* spp. may occur in raw fish material at times. Contamination during processing or postprocessing seems to be a more pressing issue to the fish processing industry. This is further supported by the findings that *L. monocytogenes* strains previously established in the plant were found in finished cold smoked salmon and smoked rainbow trout, and these strains were distinct from those contained in raw fish material coming into the plant.[36,37]

Several reports have linked seafood to listeriosis outbreaks in the past decade. Ericsson et al.[38] described three pregnancy-related and six non-pregnancy-related cases of listeriosis in Sweden, from which two deaths resulted. All of the patients had consumed *L. monocytogenes*-contaminated "gravad" rainbow trout before becoming ill. Similarly, Miettinen et al.[39] reported five cases of febrile gastroenteritis in humans, who had previously eaten cold smoked rainbow trout that contained *L. monocytogenes*. In addition, a small outbreak of listeriosis (with 17 cases) occurred in Auckland, New Zealand, in 1992, which was traced to smoked mussels contaminated with *L. monocytogenes*.[40]

TABLE 2.1
Seafood Products and Their Potential Risks for Listeriosis[a]

High Risk	Low Risk
RTE raw mollusks (e.g., mussels, clams and oysters)	Semipreserved fish (e.g., salted, marinated fish, caviar, and fermented)
RTE raw fish	Heat-processed fish (sterilized and sealed containers)
Lightly preserved fish products (e.g., salted, marinated, fermented, cold-smoked, and gravad)	Dried, dry-salted, and smoke-dried fish
Mildly heat-processed fish products (e.g., pasteurized, cooked, hot smoked) and crustaceans (e.g., precooked, breaded fillets)	Fresh/frozen fish and crustaceans

[a] Adapted from Rocourt, J. et al., *Int. J. Food Microbiol.*, 62, 197, 2000.

Table 2.1 summarizes the potential risk levels of seafood products for *L. monocytogenes*-related illness.

2.3.1.2 Meat

Chicken. Luppi et al.[15] conducted a bacteriological survey in Italy that led to the identification of 13 *Listeria* strains (nine *L. monocytogenes* and four *L. innocua*) in 113 meat samples (11.5%) and 4 *Listeria* strains (two *L. monocytogenes* and two *L. innocua*) in 75 frozen-food samples (5.3%) from retail outlets. MacGowan et al.[17] noted that, in the United Kingdom, *L. monocytogenes* was most likely isolated from poultry products (21/32 or 65.6%), with beef (9/26 or 34.6%), lamb (8/20 or 40%), pork (9/32 or 28.1%), and sausage (8/23 or 34.7%) not far behind. However, it was infrequently identified in pâté (1/40) or soft cheeses (1/251). In a study involving 645 samples collected from plant surfaces, water, and poultry products in a poultry processing plant in Brazil, Reiter et al.[42] detected *L. monocytogenes* in 230 (35.6%) of the samples; a higher proportion of chicken breast and wings than chicken legs was contaminated. Further, Capita et al.[43] found the percentages of store-purchased fresh chicken carcasses contaminated with *Listeria* spp., *L. monocytogenes, L. innocua, L. welshimeri, L. grayi*, and *L. ivanovii* were 95, 32, 66, 7, 4, and 2%, respectively.

Pork. A previous investigation showed that *L. monocytogenes* was detectable in pig fecal samples (ranging from 0 to 47%) and skin of healthy pigs.[44] A number of reports pointed out that pig tonsils (up to 61%) and tongues (up to 14%) were more likely to harbor *L. monocytogenes* than pig feces.[45–47] Autio et al.[48] detected *L. monocytogenes* in up to 64% of pig viscera (e.g., tongue, esophagus, trachea, lungs, heart, diaphragm, kidneys, and liver), which might result from cross-contamination between tonsils/tongues and other viscera during processing. Further, Kanuganti et al.[47] found ground pork samples (45–50%) were more likely to contain *L. monocytogenes* than small intestines (8–9%). Although *L. monocytogenes* serotypes 1/2a, 1/2b, and 1/2c were most commonly identified from processed pig products,[24,49] serotypes 4b and 4e were nonetheless isolated in a number of other cases.[27,50]

Meat products have been associated with several notable listeriosis outbreaks worldwide. A 300-patient listeriosis epidemic occurred in the United Kingdom during 1989 and 1990, which was linked to consumption of pâté.[51] An outbreak of listeriosis involving 279 patients took place in France in 1992, for which *L. monocytogenes* 4b strain-contaminated pork tongue was responsible[52]; another listeriosis epidemic caused illness in 39 patients in France in 1993, and this time contaminated potted pork was the culprit.[53] About 100 cases of human listeriosis, which were attributable to the consumption of hot dogs contaminated with *L. monocytogenes*[54] emerged in the United States in 1998. In the United States, a listeriosis outbreak involving a total of 108 cases occurred in 1998, from which 14 associated deaths and four miscarriages or stillbirths resulted. A subsequent case-control analysis indicated that meat frankfurters were the most likely source of infection.[74]

Beef. Samadpour et al.[55] assessed the prevalence of food-borne pathogens including *L. monocytogenes* in ground beef samples from retail markets in Seattle, Washington. From 1750 ground beef samples, 18 (3.5%) were positive for *L. monocytogenes*. Bosilevac et al.[56] analyzed a total of 1186 beef samples from four countries (i.e., United States [487], Australia [220], New Zealand [223], and Uruguay [256]) and obtained 79 *L. monocytogenes* isolates (including United States [17], Australia [4], New Zealand [5], and Uruguay [53]).

Lamb. Antoniollo et al.[57] found that 23 (34.8%) of lamb carcasses (n = 69) at a packing plant in Brazil contained *L. innocua*, and that 3 (4.3%) had *L. monocytogenes* and 1 (1.5%) had *L. ivanovii.*

Rabbit. Rodriguez-Calleja et al.[58] evaluated rabbit carcasses (n = 24) from two abattoirs and rabbit meat packages (n = 27) from supermarkets, and noted that seven (13.7%) were contaminated with *Listeria*, of which three were *L. monocytogenes*, two *L. seeligeri*, one *L. ivanovii*, and one *L. innocua*.

2.3.1.3 Dairy Products

The presence of *Listeria* spp. and *L. monocytogenes* in raw bulk tank milk and dairy milk samples is generally low,[59,60] with *Listeria* species being detected in 2.2% (11) of the 500 raw milk samples (including eight *L. monocytogenes* and three *L. innocua*) in a recent investigation.[60] Jararao et al.[61] surveyed 248 bulk tank milk samples from Pennsylvania and identified *L. monocytogenes* in 2.2% of the samples. Upon analysis of a total of 76,271 dairy samples during the years 1990–1999 in Switzerland, Pak et al.[62] detected *L. monocytogenes* in 3722 (4.9%) of the samples. It appears that water samples used for cheese washing gave the highest proportion of positive samples (9.5%), followed by cheese-surface swabs (5.0%). Interestingly, out of the 3722 *L. monocytogenes* isolates, 1328 were serologically typable, of which 92.7% belonged to just three serotypes (i.e., 1/2a, 1/2b, and 4b); 1/2b was most prevalent (particularly in hard and semihard cheeses) between 1990 and 1995 and 1/2a most common (notably in soft cheeses) between 1996 and 1999. In another study involving 103 cheese samples, Da Silva et al.[63] detected *L. monocytogenes* in 11 (10.68%), *L. innocua* in 13 (12.62%), *L. grayi* in 6 (5.83%), and *L. welshimeri* in 1 (0.97%) of the samples analyzed. Hofer et al.[64] obtained 228 *Listeria* strains/isolates in Brazil, which included 70 *L. monocytogenes,* 138 *L. innocua*, 19 *L. seeligeri,* and 1 *L. welshimeri*. Interestingly, the 70 *L. monocytogenes* strains/isolates consisted of serovars 1/2a (41), 1/2b (10), 1/2c (1), 4b (14), and 4e (1).

Dairy products have been incriminated in several early cases of listeriosis outbreaks. Pasteurized milk was responsible for 49 cases of listeriosis in Boston in 1983,[65] while chocolate milk was involved in 45 cases of *Listeria*-related gastroenteritis in 1994.[66] In addition, consumption of a Mexican-type cheese led to an epidemic of 142 listeriosis cases in 1985 in California,[67] the Vacherin Mont d'Or cheese caused 122 cases of listeriosis in Switzerland,[68] and soft cheeses were involved in two listeriosis epidemics (with a total of 51 cases) between 1995 and 1997 in France.[69]

2.3.1.4 Vegetables and Fruits

Vegetables and cut fruits constitute an ideal, nutrient-rich medium for *Listeria* maintenance and expansion. After examination of a total of 855 samples (including 425 cabbage, 205 water, and 225 environmental sponge samples) from four cabbage farms in Texas during 1999–2000, Prazak et al.[70] isolated 26 *L monocytogenes* (or 3%) from the 855 samples, with 20 isolates originating from cabbage samples, 3 from water samples, and 3 from environmental sponge samples.

The first documented outbreak of food-borne listeriosis (with 23 patients) occurred in 1979 in a Boston hospital, where *L. monocytogenes* 4b strain-contaminated vegetables prepared within the hospital were responsible.[71] A subsequent food-borne listeriosis epidemic took place in the Maritime Provinces, Canada, in 1981; this time, contaminated coleslaw was involved.[72] In Italy, an *L. monocytogenes*-related outbreak of gastroenteritis (characterized by headache, abdominal pain, and fever, with no death recorded) was detected in 1997, which was traced to the consumption of a *L. monocytogenes* 4b strain-contaminated maize and tuna salad.[73] Consumption of melons and hummus prepared at a commercial establishment was the main cause for a recent outbreak of sporadic listeriosis (with 169 cases) in the United States.[75]

2.3.1.5 Ready-to-Eat Foods

Ready-to-Eat (RTE) foods cover a range of preprocessed fish, meat, and vegetables that can be consumed without further handling (cooking). Many RTE foods (e.g., cold-smoked fish) contain 2–5% NaCl. Although RTE foods are usually kept at low temperatures (4 and –20°C), the survival and growth of *L. monocytogenes* is not drastically hindered during the storage at 4°C. Thus, RTE foods offer an excellent medium supporting *L. monocytogenes* maintenance. Van Coillie et al.[76] undertook a study on the prevalence of *L monocytogenes* in 252 RTE food products in Belgium, and they found the bacterium in 23.4% (29) of the samples, with minced meat and smoked halibut being heavily contaminated. In addition, *L. innocua* and *L. welshimeri* were also detected in prepared minced meat. PCR analysis revealed that a majority of *L. monocytogenes* strains isolated belonged to serotype 1/2a (3a) group.

Angelidis and Koutsoumanis[77] reported that of the 209 RTE meat products (e.g., bacon and cooked ham) from Hellenic retail markets, 17 (8.1%) were positive for *L. monocytogenes*. Zhou and Jiao[78] examined 844 RTE food samples acquired from retail markets in China and isolated *L. monocytogenes* from 21 of these samples. Wagner et al.[79] analyzed 946 RTE food samples from retail markets in Vienna, Austria. Of these samples, 142 (13.1%) and 45 (4.8%) were positive for *Listeria* spp. and *L. monocytogenes*, respectively. Furthermore, 18 (19.4%) of the 93 RTE fish and seafood, 11 (5.5%) of the 200 soft cheese, seven (4.9%) of the 144 RTE raw meat sausages, and five (4.5%) of the 112 cooked meat products/patés were contaminated with *L. monocytogenes*.[79] Goulet et al.[80] documented a listeriosis epidemic between July and October 2002 in the United States that was linked to consumption of deli turkey breast contaminated with *L. monocytogenes*, with 54 cases and eight deaths being recorded.

2.3.2 Mechanisms of Survival

Being ubiquitously distributed in the natural environment, *Listeria* spp. invariably find their way into various food materials. Because of their ability to withstand extreme pH, temperature, and osmotic conditions, these bacteria remain largely unscathed after going through many food manufacturing processes. This is evidenced by the observations that multiple *Listeria* species (particularly *L. monocytogenes, L. innocua,* and *L. welshimeri*) are routinely isolated from manufactured foods such as cheese, milk, and processed meats. For example, with 36 *Listeria* strains isolated from 160 ground beef, pork sausage, ground turkey, and chicken samples, *L. monocytogenes*, *L. innocua*, and *L. welshimeri* accounted for 16 (44%), 12 (33%), and 8 (22%) of the isolations, respectively. In another report, the percentages of 100 store-purchased fresh chicken carcasses that were contaminated with *Listeria* spp., *L. monocytogenes, L. innocua, L. welshimeri, L. grayi*, and *L. ivanovii* were found to be 95, 32, 66, 7, 4, and 2%.[43] Further, upon examination of 103 cheese samples, *L. monocytogenes* was detected in 11 (10.68%), *L. innocua* in 13 (12.62%), *L. grayi* in 6 (5.83%), and *L. welshimeri* in 1 (0.97%).

Although the optimal pH for *L. monocytogenes* growth is at 7.1, this bacterium is known to sustain at pH between 3.0 and 12.0.[8] Thus, use of low pH during processing of cured dried sausages and other fermented specialties is unlikely to completely eradicate this bacterium.[81] Our current knowledge on listerial resistance to pH stress (e.g., acid) has largely come from studies involving the pathogenic species *L. monocytogenes* (see Chapter 3). The identification and characterization of alternative sigma factor (σ^B) (encoded by *sigB*) in *L. monocytogenes*, which is a protein subunit of RNA polymerase (RNAP), have unraveled the molecular mechanisms of listerial responses to acid stress. Sigma^B regulates the glutamate decarboxylases (GAD) system and other acid response networks (e.g., ADI and ATPase) in *L. monocytogenes* to survive under acidic conditions.[82] In addition,

TABLE 2.2
PCR and Southern Blot Analysis of *sigB* gene in *Listeria* spp.

Species	Strain	Origin	*sigB* Gene: PCR (474 bp)[a]	*sigB* Gene: Southern Band (kb)
L. monocytogenes	EGD	Guinea pig	+	2.0, 1.0, 0.9, 0.7
L. grayi	ATCC 19120	Chinchilla feces	–	(0.6)[b]
	ATCC 20402	Corn leaves/stalks	–	(0.6)[b]
	ATCC 20403	Corn leaves/stalks	–	(0.6)[b]
L. innocua	CLIP 11262	Corn	+	0.6
	ATCC 33091	Human feces	+	0.7, 0.5
	ATCC 51742	Cabbage	+	3.0
L. ivanovii	ATCC 19119	Sheep	+	0.7
	SLCC 2379	Unknown	+	0.6
	RM3325	Cheese	+	0.9
L. seeligeri	AT-02	Unknown	+	0.6
	DA-41	Unknown	+	0.9
	ATCC 35967	Soil	+	0.6
L. welshimeri	ATCC 35897	Plant	+	0.6
	ATCC 43550	Soil	+	0.6
	ATCC 43551	Soil	+	0.6

[a] PCR was performed by using primers derived from *L. monocytogenes* EGD-e *sigB* gene (i.e., SigBF: 5′- TATTTGGATTGCCGCTTACC-3′ and SigBR: 5′-CCATCCGAAT-CAGCTTCAAT-3′), which facilitated amplification of a specific 474-bp fragment.

[b] Southern blot was carried out with *Hin*dIII-digested genomic DNA from each *Listeria* strain and probed with chemically labeled *sigB* fragment of 474 bp generated by PCR from *L. monocytogenes* EGD. A faint *Hin*dIII band of 0.6 kb was detected in *L. grayi* in comparison with other *Listeria* species.

σ^B contributes to *L. monocytogenes* invasion and virulence through its regulation of InlA and InlB expression. With prior exposure to sublethal acidic pH, *L. monocytogenes* also has the capacity to develop cross-protection against subsequent osmotic stress.[83,84]

On the other hand, relatively little has been documented on non*monocytogenes Listeria* species' responses to environmental stresses such as wide pH and salt ranges. Recently, we conducted a study to determine if and to what extent non*monocytogenes Listeria* species are able to enable its survival under extreme pH and salt conditions, and to examine the status of the *sigB* gene in non*monocytogenes Listeria* species by PCR and Southern blot (D. Liu, unpublished data). This study involved a collection of 15 non*monocytogenes Listeria* strains (with three each of *L. grayi, L. innocua, L. ivanovii, L. seeligeri*, and *L. welshimeri*) (Table 2.2), along with one control *L. monocytogenes* strain (EGD), whose profile of resistance to acid, alkali, and salt conditions was described previously.[8]

As shown in Figure 2.1, all 15 non*monocytogenes Listeria* strains (representing *L. ivanovii, L. seeligeri, L. innocua, L. welshimeri*, and *L. grayi*) under investigation essentially failed to recover on brain–heart infusion (BHI) agar plates after treatment with 100 m*M* Tris pH 2.0 for 1 h. On the other hand, all nine *L. ivanovii, L. innocua*, and *L. welshimeri* strains as well as two *L. seeligeri* (AT-02 and ATCC 35967) strains tolerated the treatment with 100 m*M* Tris pH 3.0 for 1 h, while the other *L. seeligeri* (DA-41) and three *L. grayi* strains showed poor recovery after such treatment. All 15 non*monocytogenes Listeria* strains recouped well from a 1-h treatment with 100 m*M* Tris pH 4.0; all but one (i.e., *L. grayi* strain ATCC 19120) also survived a 1-h exposure to 100 m*M* Tris

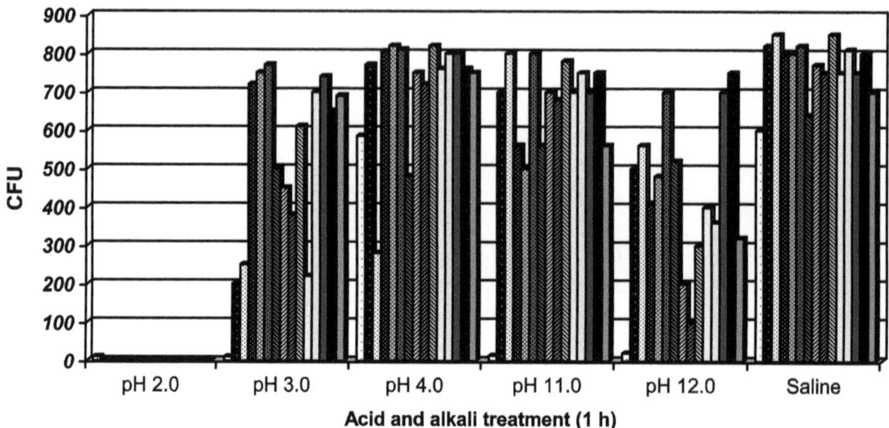

FIGURE 2.1 Recovery of non*monocytogenes Listeria* strains after treatment with 100 m*M* Tris at various pH for 1 h, with saline treatment as a control. The colony forming units (CFUs) shown here represent the averages of duplicate plate counts from 25 μl inoculon at 10^{-8} dilution. As in Figures 2.2, 2.3, and 2.5, the order of *Listeria* strains under each condition is (from left): *L. grayi* ATCC 19120, ATCC 20402, ATCC 20403; *L. innocua* CLIP 11262, ATCC 33091, ATCC 51742; *L. ivanovii* ATCC 19119, SLCC 2379, RM3325; *L. seeligeri* AT-02, DA-41, ATCC 35967; and *L. welshimeri* ATCC 43550, ATCC 43551, ATCC 35897.

pH 11.0 in comparison with the saline controls (Figure 2.1). Although *L. grayi* strain ATCC 19120 was notably affected by Tris pH 11.0, it did show moderate growth upon a 1-h exposure to 100 m*M* Tris pH 10.0 (resulting in a survival rate of 50%) and to 100 m*M* Tris pH 9.0 (with a survival rate of 60%) (data not shown). Further, all nine *L. seeligeri, L. innocua,* and *L. welshimeri* strains, as well as one *L. ivanovii* (ATCC 19119) and two *L. grayi* (ATCC 20402 and ATCC 20403) strains, resisted treatment with 100 m*M* Tris pH 12.0 for 1 h. The other two *L. ivanovii* (i.e., SLCC 2379 and RM3325) and one *L. grayi* (ATCC 19120) were adversely impacted (Figure 2.1).

The ability of non*monocytogenes Listeria* strains to tolerate acid and alkali stresses was confirmed after treating the 15 strains with 10 m*M* HCl and 100 m*M* NaOH. As in the case of 100 m*M* Tris treatments, while other *Listeria* species/strains resisted the 10 m*M* HCl treatment, three *L. ivanovii* (especially RM3325) and one *L. grayi* (ATCC 20402) strains were somewhat susceptible, and one *L. seeligeri* (DA-41) and two *L. grayi* (ATCC 19120 and ATCC 20403) strains were severely affected by this exposure (Figure 2.2). Similar to the 1-h treatment with Tris pH 12.0, while the growth of other *Listeria* species/strains was only marginally affected by a 15-min treatment with 100 m*M* NaOH, the recovery of three *L. ivanovii* and two *L. grayi* (ATCC 20402 and ATCC 20403) strains were notably hindered, and one *L. grayi* strain (ATCC 19120) was significantly impacted by such treatment (Figure 2.3).

These results suggest that *L. innocua, L. welshimeri,* and, to a lesser extent, *L. seeligeri,* demonstrate a higher resistance to acid and alkali than *L. ivanovii* and *L. grayi* and that *L. ivanovii, L. grayi,* and *L. seeligeri* show intraspecies variations in pH tolerance. Through examination of the *sigB* gene in the 15 *Listeria* strains by PCR and Southern blot, it is clear that, apart from *L. grayi,* other non*monocytogenes Listeria* species/strains may possess a similar copy of *sigB* gene to *L. monocytogenes* (Table 2.2 and Figure 2.4). The fact that *L. grayi* strains did not react in PCR with *sigB* primers and reacted poorly in Southern blot with a *sigB* probe derived from *L. monocytogenes* EGD indicates significant nucleotide changes in *L. grayi sigB* gene (Table 2.2). Nonetheless, the alteration in the *L. grayi sigB* gene may contribute only to some but not all of its susceptibility to acid, as one *L. grayi* strain (ATCC 19120) showed poor recovery (2%) after treatment with 100 m*M* Tris pH

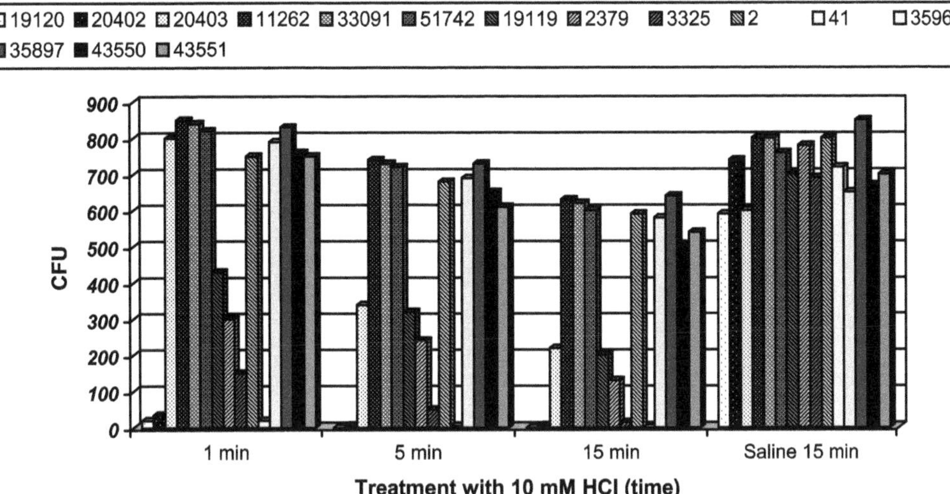

FIGURE 2.2 Recovery of non*monocytogenes Listeria* strains after treatment with 10 m*M* HCl for 1, 5, and 15 min, with saline treatment (for 15 min) as a control. The CFUs shown here represent the averages of duplicate plate counts from 25 μl inoculon at 10^{-8} dilution.

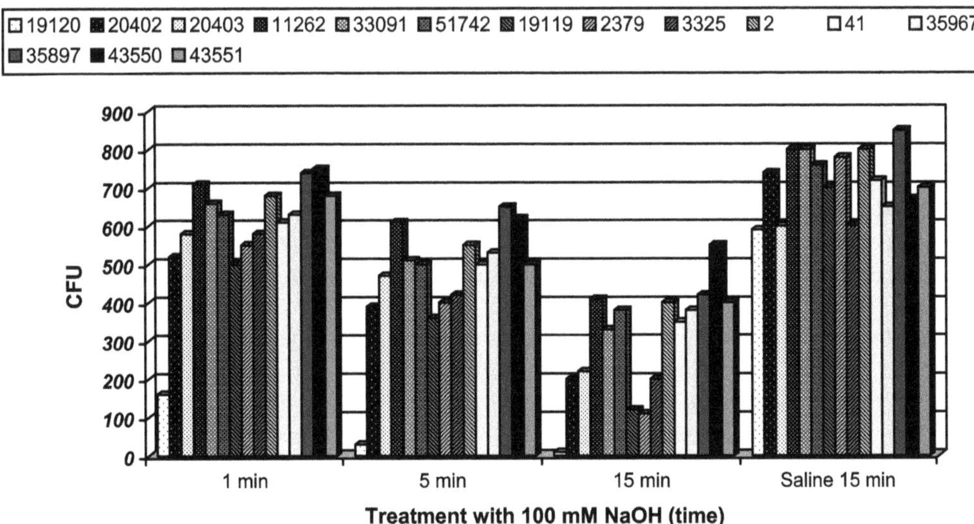

FIGURE 2.3 Recovery of non*monocytogenes Listeria* strains after treatment with 100 m*M* NaOH for 1, 5, and 15 min, with saline treatment (for 15 min) as a control. The CFUs shown here represent the averages of duplicate plate counts from 25 μl inoculon at 10^{-8} dilution.

3.0 for 1 h, while the other two *L. grayi* strains (ATCC 20402 and ATCC 20403) had a recovery of 30%. Moreover, the influence of the *sigB* gene to *Listeria* tolerance of alkali appears to be minimal, as three *L. grayi* strains harboring an altered copy of the *sigB* gene displayed varied vulnerability to treatment with 100 m*M* Tris pH 12.0 (with ATCC 19120 showing marginal recovery of 3%, and ATCC 20402 and ATCC 20403 having a 60–70% recovery). These findings thus imply that other genes/proteins may be in play in listerial response to alkali.

Pasteurization (usually at 68°C) and cold storage (at 4°C) are typical temperature-related measures applied to foods. While pasteurization will destroy a majority of other bacteria, some *L. monocytogenes* strains are known to survive this process. Although heating food at a much

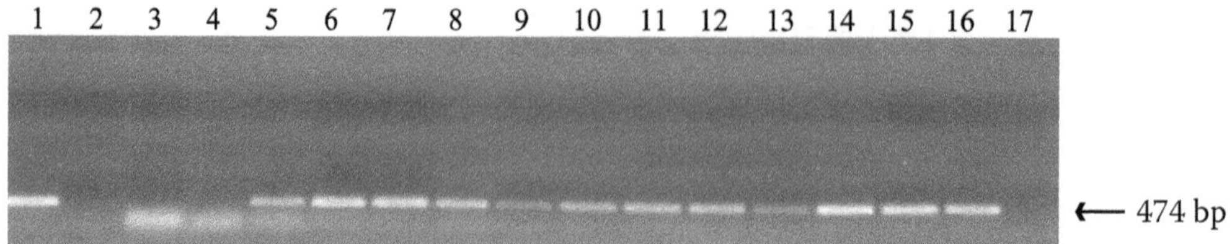

FIGURE 2.4 Electrophoretic analysis of PCR products amplified with *L. monocytogenes sigB* gene primers SigBF/R. Lane 1 contained PCR product from *L. monocytogenes* EGD; lanes 2–4, *L. grayi* ATCC 19120, ATCC 20402, and ATCC 20403; lanes 5–7, *L. innocua* ATCC 11262, ATCC 33091, and ATCC 51742; lanes 8–10, *L. ivanovii* ATCC 19119, SLCC2379, and RM3325; lanes 11–13, *L. seeligeri* AT-02, DA-41, and ATCC 35967; lanes 14–16, *L. welshimeri* ATCC 35897, ATCC 43550, and ATCC 43551; and lane 17, negative control with no template DNA. The expected product of 474 bp is indicated on the right.

higher temperature will result in a complete elimination of all bacteria, it is generally not desirable to do so as this often decreases the appeal of food products in terms of taste and appearance. *L. monocytogenes* readily grows at 4°C, and thus prolonged storage of food in refrigerators favors the replication of this bacterium, which can cause a serious problem if the stored food is consumed without further cooking (as in the case of RTE food products). In addition, there is evidence that *L. monocytogenes* thermotolerance may actually increase after prior exposure to sublethal temperatures.[64,85] The secretion of a number of heat response proteins (e.g., GroESL, DnaKJ, Clp, and HtrA) by *L. monocytogenes* may help this bacterium in mitigating against temperature stress.[82]

Salt is an essential ingredient used for preparation of dried sausage and corned beef. As *L. monocytogenes* can survive at 40% NaCl for a long period[8] and grow at up to 10% NaCl, it is unlikely to be affected terminally by the conventional salt treatment employed in food processing. In fact, the available data suggest that *L. monocytogenes* copes well with cheese brining systems. Apparently, *L. monocytogenes* response to osmotic stress relies on its production of purpose-made proteins (e.g., BetL, Gbu, OpuC, RelA, Ctc, HtrA, KdpE, LisRK, ProBA, and BtlA) to promote its survival and growth at elevated osmolarities[82] (see Chapter 3). Again, there are limited data regarding non*monocytogenes Listeria* responses to salt stress. By subjecting 15 non*monocytogenes Listeria* strains (Table 2.2) to saturated NaCl, it was found that apart from three *L. ivanovii* strains that displayed lowered recovery from a 1-h treatment with saturated NaCl (7 *M*), the 12 other non*monocytogenes Listeria* strains were generally tolerant of such exposure (Figure 2.5). After incubation in saturated NaCl for 20 h, however, one *L. ivanovii* (RM3325) and one *L. grayi* (ATCC 19120) showed relatively poor growth, while the other *Listeria* strains were able to recover moderately in comparison with the saline controls (Figure 2.5) (D. Liu, unpublished data). Overall, *L. innocua* appeared to be more resistant to salt that other *Listeria* species under investigation.

Listeria's capacity to endure harsh pH and salt conditions is advantageous to its survival in the environment and in food processing facilities, since hypochloric acid (bleach), chlorine, potassium hydroxide, sodium hydroxide, and quaternary salts are frequently applied as detergents and sanitizers to remove organic residuals from food processing surfaces, equipment, and plants. One of the commonly used sanitizers is quaternary ammonium compounds (QACs), which cause bacterial membrane damage and inactivate bacterial cellular enzymes, yet are nontoxic to humans and noncorrosive to machines and hard surfaces. While the precise mechanism of *L. monocytogenes* response to alkaline stress has not been fully delineated, this bacterium acquires resistance to QACs through horizontal transfer of an *mdrL* gene encoding an efflux pump.[86] Further, *L. monocytogenes* uses physical adaptation mechanisms (e.g., surface attachment and biofilm formation) to enhance its resistance to disinfectants and sanitizers.[23,87,88]

Perhaps, *Listeria* interspecies variations in pH and salt resistance may partially explain for the more frequent isolations of *L. innocua*, *L. welshimeri*, and *L. seeligeri* than *L. ivanovii* and

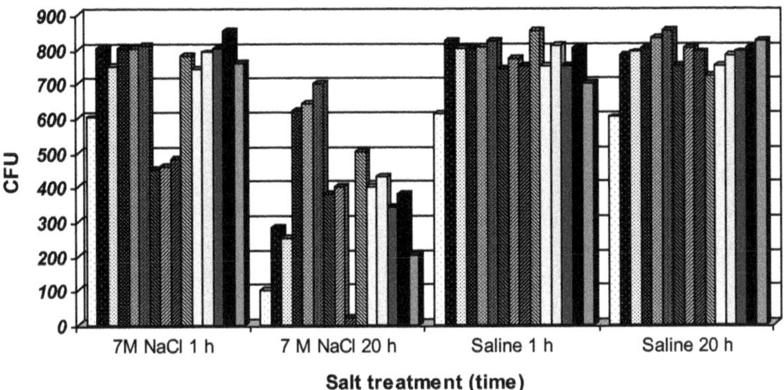

FIGURE 2.5 Recovery of non*monocytogenes Listeria* strains after treatment with saturated salt solution for 1 and 20 h, with saline treatment as a control. The CFUs shown here represent the averages of duplicate plate counts from 25 µl inoculon at 10^{-8} dilution.

L. grayi from food and environmental specimens. While nonmonocytogenes *Listeria* species such as *L. innocua*, *L. welshimeri,* and, to some extent, *L. seeligeri* are more tolerant of acid and alkali, *L. ivanovii* and *L. grayi* are somewhat similar to *L. monocytogenes* in their reduced resistance to acid and alkali conditions. Interestingly, of the two less robust *Listeria* species, *L. ivanovii* is a known animal pathogen, and one of the least resistant *L. grayi* strains, ATCC 19120, came from chinchilla feces. Thus, it is possible that *Listeria* species, which have adapted to the environment, may acquire traits that enable their survival under more stressful conditions, whereas those that have adapted to animal hosts may become less tolerant to external stresses. Nonetheless, as the aforementioned results have been derived mainly from the analysis of *Listeria* pure isolates, the influence of other compounds such as proteins and lipids on the acid, alkali, and salt tolerance in *Listeria* bacteria has not been taken into account. Further investigation is clearly needed to help elucidate the potential roles of these compounds in the modulation of *Listeria* resistance to pH and salt.

2.3.3 Control and Prevention Strategies

Listeria spp., especially *L. monocytogenes*, are tolerant of extreme temperature, pH, and osmotic conditions, which are behind the principal mechanisms of food processing practices. On the whole, food products provide a nutrient-rich medium for listerial maintenance and multiplication, and any viable *Listeria* bacteria that have survived the food processing procedures will likely grow and cause infection if not completely eliminated prior to consumption.

2.3.3.1 Before Processing

It is important to adapt appropriate farming and husbandry practices that reduce and eliminate the occurrence of *L. monocytogenes* in vegetables and farm-raised animals. For instance, use of organic fertilizers free of *L. monocytogenes* contamination in vegetable growing will lead to cleaner raw materials for downstream processing. In addition to maintaining proper hygiene on farms (e.g., boot cleaning and cloth changing), feeding dry feed or silage and rearing in closed houses will decrease the likelihood of *L. monocytogenes* in pigs.[45,89] Use of clean containers and water for preprocessing handling will also ensure a lowered level of *L. monocytogenes* contamination.

2.3.3.2 During Processing

Direct contact with contaminated processing equipment is an important cause of *L. monocytogenes* occurrence in food products. Adoption of several control measures (under a system called the hazard analysis and critical control point [HACCP] plan) by food processing companies during processing (e.g., thorough cleansing of in-process products and food-contact surfaces, clear separation of staff functions, and scrupulous personal hygiene) has led to significant reduction of *L. monocytogenes* contamination in finished food products.[90] In particular, use of heat treatment (e.g., hot steam, hot air, or hot water at 80°C) is effective in cleaning and sanitizing the skinning, slicing, and brining equipment and in eradicating *L. monocytogenes* from food processing plants. In addition, visitors and staff job rotation are potential risk factors for increasing *L. monocytogenes* contamination in food processing plants.[90] Optimization and application of more effective and innovative combinations of detergents and sanitizers may also help eliminate *L. monocytogenes* in food processing environments. For example, Vasseur et al.[84] reported that an alkaline treatment (pH 10.5) followed by an acid treatment (pH 5.4) provided a more efficient means of killing *L. monocytogenes* than other combinations of sanitizers and detergents.

2.3.3.3 After Processing

Implementation of monitoring and quality control procedures represents the most cost-effective postprocessing measures against food-borne pathogens, including *L. monocytogenes.* Besides assisting the formulation of cleaning and sanitizing programs to reduce and eradicate *L. monocytogenes* in food processing environments, these procedures help ensure that the finished food products meet the recommended tolerable levels of *L. monocytogenes* in the specified product categories (e.g., no organisms in 25 g of a food product in the United States, and fewer than 100 colony forming units [CFUs] per gram in certain foodstuffs and zero tolerance in foods with extended shelf-lives in Canada and France) before releasing to the retail markets. Additionally, for many RTE meat products, 1 week of postprocessing storage prior to shipment has been found to inhibit listerial growth and decrease the number of *L. monocytogenes* in these products and can be used as an effective supplemental postprocessing treatment.[91] Finally, there is scope for continued improvement in home food-handling practices, where storage of foods at incorrect refrigeration temperatures often contributes to heightened risk of listeriosis in humans.

2.4 *LISTERIA* IN ANIMALS

Being widespread in the environment, *Listeria* spp. thrive in polluted water, soil, vegetation, and foods. Wildlife and domestic animals often acquire *Listeria* spp. (including pathogenic *L. monocytogenes* and *L. ivanovii*) via contaminated feed (e.g., silage and grasses) and water. Both *L. monocytogenes* and *L. ivanovii* have been frequently isolated from wildlife and domestic animals and have been shown to be responsible for *Listeria*-related animal illness (e.g., abortion and stillbirth).[92] Other *Listeria* spp. (e.g., *L. innocua* and *L. grayi*) may also occur in animals, but their role as either pathogens or pass-by bacteria needs further clarification.

2.4.1 Prevalence

Weis and Seeliger[18] isolated *L. monocytogenes* and *L. ivanovii* from 15.7% (16/102) of the feces from deer and stag and 17.3% (8/46) of the feces from birds in Germany. The 16 strains from deer and stag consisted of two serovar 1/2a, one serovar 1/2b, six serovar 4b, and one serovar 5 (i.e., *L. ivanovii*); the eight strains from birds comprised two serovar 1/2a, three serovar 1/2b, and one serovar 4b. Occasionally, other *Listeria* spp. have been isolated from wildlife. For example, an *L. grayi* strain (ATCC 19120) was retrieved from chinchilla feces, and an *L. ivanovii* strain was acquired from a septicemic chinchilla.[93] These findings suggest that *Listeria* spp. (including *L. monocytogenes* and *L. ivanovii*) commonly occur in wildlife, which may become infected via ingestion of contaminated plants/vegetation and water.

2.4.1.1 Domestic Animals

Hofer et al.[64] conducted a survey on *Listeria* presence in Brazilian healthy cattle, resulting in the isolation of 239 strains. These included 96 *L. monocytogenes*, 141 *L. innocua*, one *L. ivanovii*, and one *L. grayi*. Of the 96 *L. monocytogenes* strains, seven serotypes were detected—that is, 1/2a (5), 1/2b (12), 3a (1), 4a (37), 4ab (5), 4b (34), and 4e (2). Antoniollo et al.[57] examined 35 feces samples from lambs in Brazil and found 7 (20%) with *L. welshimeri* and 3 (8.6%) with *L. innocua*. Hutchison et al.[94] investigated the prevalence of *L. monocytogenes* in livestock wastes in the United Kingdom and noticed that 29.8% (241/810), 19.8% (24/126), 19.4% (13/67), and 29.2% (7/24) of fresh fecal samples from cattle, pig, poultry, and sheep, respectively, contained *L. monocytogenes*. Similarly, Nightingale et al.[95] showed that the presence of *L. monocytogenes* in 52 ruminant (cattle, sheep, and goat) farms was about 20%. Ho et al.[96] reported that among a total of 759 fecal samples from dairy cattle, *L. monocytogenes* and other *Listeria* spp. were isolated from 26 (3.4%) and 112 (14.8%), respectively. Of the 674 samples from the processing facilities on these farms, similar percentages of *L. monocytogenes* (2.7%) and *Listeria* spp. (13.9%) were detected. These observations underscore the common occurrence of *Listeria* spp. in domestic animals.

L. monocytogenes has been shown to cause a spectrum of clinical diseases in domestic animals (e.g., cattle, sheep, horses, and chickens). Winter et al.[97] described *L. monocytogenes*-related subclinical mastitis in two cows and two sheep, which demonstrated no overt clinical signs apart from having an elevated somatic cell count and a persistent shedding of *L. monocytogenes*. An investigation of bovine listeriosis that occurred on a Pacific Northwest dairy farm revealed that an *L. monocytogenes* clinical strain (of serotype 1/2a) was closely related genetically to fecal strains from asymptomatic cows.[98] After analysis of a large number of clinical samples (i.e., 239 blood, 243 milk, and 243 fecal swabs) from cattle with mastitis, Rawool et al.[99] isolated 12 *Listeria* strains including four *L. monocytogenes* and one *L. ivanovii*. Wagner et al.[100] reported a listeriosis outbreak in a flock of 55 sheep caused by feeding *L. monocytogenes*-contaminated grass silage. During the outbreak phase, abortive (nine ewes), encephalitic (one ewe), and septicemic (four ewes) forms of listeriosis were observed clinically, with the septicemic cases showing *Listeria* accumulation in visceral organs but not in the brain, and the encephalitic ewe developing central nervous system symptoms and rhombencephalitis. Gudmundsdottir et al.[101] identified 20 *L. monocytogenes* isolates from five confirmed and four suspected incidents of listeriosis in horses in Iceland, with large numbers of the bacterium being detected in the feces of horses with severe signs of disease. One *L. monocytogenes* serovar 1/2a genotype was isolated from the confirmed listeriosis cases. Kurazono et al.[102] also reported *L. monocytogenes*-related encephalitis in Japanese chickens with neural signs (torticollis and drowsiness) and mortality; *L. monocytogenes* was detected immunohistochemically in the medulla oblongata, cerebellum, and spinal cord.

On the other hand, *L. ivanovii* mainly causes septicemia, enteritis, and abortion in sheep and cattle, unlike *L. monocytogenes*, which tends to also produce meningoencephalitis in these animals. Sheep abortions due to infection with *L. ivanovii* were reported in Australia and India,[103,104] and bovine abortions attributable to *L. ivanovii* were described in the United States and Australia.[105,106] Pathologic features of *L. ivanovii*-related bovine abortions often resembled those seen in abortions due to *L monocytogenes*.[105]

Furthermore, Peters et al.[107] isolated five *L. innocua* strains from brain stems of ruminants with central nervous system (CNS) disturbances and/or pathoanatomical CNS alterations. However, the precise role of *L. innocua* in the pathogenesis of ruminant CNS diseases was poorly understood.

2.4.1.2 Laboratory Animals

Because of its broad host specificity, *L. monocytogenes* infection can be evaluated in a variety of laboratory animal models with the aim to elucidate the immunology, invasion mechanisms, and pathological characteristics of this intracellular bacterium. The most commonly applied animal model is inbred mice, although other animal models have also been utilized for investigation of specific aspects of listeriosis.

Murine model—The experimental murine model (including inbred mice and rats) has been used extensively for studies of *L. monocytogenes* virulence and pathogenicity owing to its ease of handling and relatively low cost. By inoculating mice via oral (intragastric or i.g.), intranasal (i.n.), intraperitoneal (i.p.), intravenous (i.v.), or subcutaneous (s.q.) route, the virulence of a given *L. monocytogenes* strain is determined by the resulting mouse mortality in relation to the dosages, or by the number of the bacteria that arrive at the spleen and liver of mice 2–3 days postinfection. The pathogenicity of an *L. monocytogenes* strain can be assessed by the resulting pathological characteristics.

From the data available, it appears that the infectivity of *L. monocytogenes* in mice varies with the routes of inoculation. *L. monocytogenes* is highly potent when administered intravenously, which is followed by intraperitoneal, intranasal, subcutaneous, and intragastral inoculations. In contrast to humans, who become readily infected with *L. monocytogenes* through oral ingestion of contaminated foods, mice are somewhat resistant to listeriosis via oral or intragastric inoculation, and thus often require a much larger oral dose to establish the infection than humans do. It is apparent that wild-type mice generally lack a human-like E-cadherin receptor, which interacts with *L. monocytogenes* species-specific surface protein internalin A (InlA) to facilitate its breach of the host intestinal barrier. A transgenic mouse strain harboring human-like E-cadherin has thus been developed recently[108] to enable closer simulation of human listeriosis in a murine model.

In addition, different inbred mouse strains also display varied susceptibility to listeriosis. Cheers and McKenzie[109] showed that, among several inbred mouse strains, C57BL/6J and SJL/WEHI mice were relatively resistant, while CBH/HeJ, A/J, DBA/iJ, BALB/cJ and CBA/H mice were susceptible to listeriosis through intravenous inoculation. It is most notable that the susceptibility of the A/J mouse strain to *L. monocytogenes* can be traced to a defect in its *c5* gene.[110] This gene encodes for the C5 complement, whose chief function is to help garner macrophages and polymorphonuclear cells to the site of infection. With a 2-bp deletion in its *c5* gene, the A/J mouse strain loses its capacity to produce C5 protein, and thus becomes enfeebled in its inflammatory response to *L. monocytogenes* invasion.

Further, *L. monocytogenes* serotypes also have a bearing on listerial infectivity and pathogenicity in murine hosts. Among the 12 common *L. monocytogenes* serotypes (i.e., 1/2a, 1/2b, 1/2c, 3a, 3b, 3c, 4a, 4b, 4c, 4d, 4e, and 7), serotypes 1/2a, 1/2b, 1/2c, and 4b often reached the spleen and liver in mice in larger quantity than other serotypes via intragastric inoculation.[111] However, serotype 1/2a (EGD), 4c (e.g., ATCC 19116 and 874), and 7 (e.g., R2-142) strains produced more mortalities than other serotypes via intraperitoneal injection, whereas serotype 4a strains (e.g., ATCC 19114, HCC23, and HCC25) caused no mortality.[6,112,113] On the other hand, *L.ivanovii* appears to be less virulent than *L. monocytogenes* in the murine model and grows mainly in the liver.

Other animal models—Gerbils, guinea pigs, rabbits, sheep, and nonhuman primates, as well as zebrafish, have been utilized occasionally in listeriosis research.[108,114–118] These animal models sometimes disclose valuable insights on *L. monocytogenes* pathogenicity that are not available with the mouse modeL. For example, besides offering an appropriate model for *L. monocytogenes*-related rhombencephalitis, gerbils are especially useful for investigation on the molecular mechanisms of listerial crossing of the blood–brain barrier.[115] Inoculating gerbils in the middle ear with a low infective dose of *L. monocytogenes* helped create prolonged otitis media with persistent bacteremia. The infected gerbils developed a severe rhombencephalitis with circling syndrome, paresia, ataxia, and rolling movements; histological lesions were mainly located in the brainstem (e.g., coalescent, necrotic abscesses with perivascular sheaths), and mimicked those observed in human rhombencephalitis.[115]

As guinea pigs possess a human-like E-cadherin receptor for InlA-mediated internalization, they are suited for oral infection with *L. monocytogenes*. Listeriosis in pregnant guinea pigs often results in stillbirths, with the overall disease being similar to that observed in humans. Further, with a close genetic relation to humans, nonhuman primates provide an ideal alternative to the study of listerial pathogenesis and immune response in humans.[117] Specifically, by administering *L. monocytogenes* orally, 4 of 10 nonhuman primates delivered stillborn infants; *L. monocytogenes* was isolated from fetal tissue, and the pathology was consistent with listeriosis as the cause of pregnancy loss.[117]

2.4.2 Mechanisms of Survival and Virulence

Listeria spp., in particular *L. monocytogenes,* have the ability to sustain external pH, temperature, and salt stresses, which enables their survival under the acidic conditions within the mammalian stomach without being totally destroyed. Specifically, *L. monocytogenes* produces alternative sigma factor σ^B for regulation of several stress-response genes (e.g., *opuCA, lmo1421,* and *bsh*) and related proteins.[7] Then, *L. monocytogenes* synthesizes surface proteins called internalins (e.g., InA and InlB) to enable its entry into epithelial and other host cells. Through its binding to an adhesion protein E-cadherin in host enterocytes leading to local cytoskeletal rearrangements, InlA facilitates *L. monocytogenes* entry into epithelial cells. By interacting with a hepatocyte growth factor receptor Met, InlB paves the way for *L. monocytogenes* entry into hepatocytes and other host cells. Inside the vacuoles, *L. monocytogenes* secretes listeriolysin (LLO) and phosphatidylinositol-phospholipase C (PlcA) to lyse the vacuolar membrane and escape to the cytosol, where it undergoes rapid intracellular expansion. Afterwards, *L. monocytogenes* generates another surface protein ActA and phosphatidylcholine-phospholipase C (PlcB) to aid its spread to neighboring cells. With the help of PlcB and a metalloprotease (Mpl), *L. monocytogenes* disrupts the secondary double-layer membrane vacuoles and initiates a new cycle of infection.

The virulence-associated proteins PlcA, LLO (encoded by *hly*), Mpl, ActA, and PlcB are encoded by five adjacent genes, which are located in a 9.6-kb virulence gene cluster named *Listeria* pathogenicity island 1 (or LIPI-1). LIPI-1 is regulated overall by a pleiotropic virulence regulator PrfA, whose corresponding gene *prfA* lies downstream of *plcA*. The *prfA* gene possesses two promoters, one of which is recognized by PrfA (i.e., a mechanism of self-regulation) and the other by the stress-response regulator σ^B. Besides its presence in *L. monocytogenes*, LIPI-1 is also found in pathogenic *L. ivanovii* and nonpathogenic *L. seeligeri.* Interestingly, the virulence regulator PrfA from *L. seeligeri* shows a weak binding to the PrfA-dependent promoters Phly and PactA, due possibly to the amino acid exchanges in its C-terminal domain, while PrfA from both *L. monocytogenes* and *L. ivanovii* demonstrates similarly high efficiency in transcriptional initiation at Phly and PactA. This may contribute partly to the lowered virulence of nonpathogenic *L. seeligeri* in comparison with pathogenic *L. monocytogenes* and *L. ivanovii.*[119] The fact that *L. innocua, L. welshimeri,* and *L. grayi* do not possess this important virulence gene cluster may partially explain their apparent lack of virulence and pathogenicity. However, given that *L. innocua* has been isolated from brainstems of ruminants showing CNS disturbances,[107] it may play an unspecified role in the disease process.

L. monocytogenes' crossing over mammalian intestine relies on the interaction of its InlA with the host's E-cadherin receptor protein. In the permissive or susceptible mammalian species (e.g., humans, guinea pigs, ovine and bovine animals), the E-cadherin receptor protein contains a proline at position 16. In the nonpermissive or resistant species (e.g., mice and rats), on the other hand, the proline in the E-cadherin receptor protein is replaced by glutamic acid. This amino acid substitution in the E-cadherin protein results in a reduced capacity of murine enterocytes to react with *L. monocytogenes* InlA and hampers *L. monocytogenes*' entry into murine enterocytes. As a consequence, *L. monocytogenes* does not usually cause gastroenteritis in mice, as it does in humans.

Further, *L. monocytogenes* does not show tropism toward the murine brainstem and fetoplacental unit, which possess nonhuman-like E-cadherin. Thus, wild-type mice do not offer an ideal model for the study of *L. monocytogenes*-related gastroenteritis, encephalitis, and fetoplacental disease.

By putting human E-cadherin cDNA under the control of the promoter of the intestinal fatty-acid-binding protein (iFABP) gene in mice, a transgenic murine model that permits direct *L. monocytogenes* interaction with murine enterocyte E-cadherin was produced.[108] Thus, the entry of *L. monocytogenes* into murine enterocytes becomes feasible, facilitating its efficient passage through the intestinal barrier, subsequent growth in the small intestine lamina propria, and migration to mesenteric lymph nodes, liver, and spleen. In other words, this transgenic mouse model offers a more appropriate means to the investigation of *L. monocytogenes* gastric, CNS, and fetoplacental pathogenesis without resorting to the alternative, more expensive guinea pig model.

Recently, after Southern blot analysis of the *inlA* and *inlB* gene structures in various *L. monocytogenes* serotypes, it became clear that the *inlA* gene in the 11 common serotypes was largely unchanged with the formation of *Hin*dIII bands of 1.3–1.5 kb, whereas the *inlB* gene in serotypes 1/2a, 1/2b, 1/2c, 3a, 3b, and 3c demonstrated a larger *Hin*dIII band (4.0 kb) than that in serotypes 4a, 4b, 4c, 4d, and 4e (1.5 kb) (D. Liu, unpublished data). These observations suggest that serotypes 4a, 4b, 4c, 4d, and 4e may possess an altered *inlB* gene in relation to serotypes 1/2a, 1/2b, 1/2c, 3a, 3b, and 3c, which may result in their relative inefficiency in crossing the intestinal and other barriers in wild-type murine hosts.

This hypothesis was supported by the findings that serotype 4b strain (i.e., ATCC 19115, which was originated from a human listeriosis case) produced a somewhat lower mouse mortality than serotype 1/2a strain (i.e., EGD), in contrast to the human listeriosis situation where serotype 4b is a disease-inducing agent as potent as (if not more potent than) serotype 1/2a. Therefore, a fully functional InlB may be more important than InlA for *L. monocytogenes'* entry into murine host cells, since wild-type mice are known to harbor an inappropriate E-cadherin receptor for InlA. As serotype 4b strain ATCC 19115 has an intact *inlA* gene but an imperfect *inlB* gene, it may exhibit a higher pathogenicity in humans than in mice. The recent observation that, while InlB was vital for *L. monocytogenes'* colonization of murine liver and spleen, it was not required for crossing the murine intestinal barrier[120] provided further backing for this notion.

In addition to InlA and InlB, another *L. monocytogenes* internalin InlJ has been shown to play a part in listerial virulence by contributing partially to its successful passage through the intestinal barrier as well as to the successive phases of infection.[121] As the corresponding *inlJ* (*lmo2821*) gene exists only in *L. monocytogenes* strains/serotypes that are capable of causing human listerial outbreaks and mouse mortality and is absent in avirulent, nonpathogenic serotype 4a strains, it offers an excellent target for rapid discrimination of virulent from avirulent *L. monocytogenes* strains. Further, *L. monocytogenes* along with *L. ivanovii* secretes hexose phosphate transporter (Hpt), under the control of the central virulence regulator PrfA, to transport glucose-6-phosphate from the cytosol into the endoplasmic reticulum for its intracellular growth.[122]

Since InlA, InlB and InlJ are limited to *L. monocytogenes, L. ivanovii's* entry to host enterocytes may have to depend on other internalins or internalin-like proteins. Being present in both *L. monocytogenes* and *L. ivanovii,* but absent in *L. innocua, L. seeligeri, L. welshimeri,* and *L. grayi,* InlC has been shown to be involved in *L. monocytogenes'* postintestinal invasion of other host cells.[123] Therefore, InlC may also play a role in *L. ivanovii* cell invasion. Further, a 4.25-kb internalin locus unique to *L. ivanovii*, termed *i-inlFE,* has been identified.[124] It appears that, apart from LIPI-1, *L. ivanovii* harbors a second pathogenicity island (i.e., LIPI-2), which is composed of a 22-kb chromosomal locus encoding 10 internalin, with i-InlB1-B2 being surface-associated internalins similar to *L. monocytogenes* InlB and i-InlE-L being excreted internalins. Except i-*inlB*1, all LIPI-2 internalins genes are controlled by the virulence regulator PrfA,[124] as characteristic palindromic sequences ("PrfA-boxes") are found in the promoter regions of the tandemly arranged *i-inlF* and *i-inlE*.

Given that mutant strains containing nonpolar *i-inlE* and/or *i-inlF* deletion failed to cause mouse mortality at high doses, the *i-InlE* and *i-InlF* genes may be critical for *L. ivanovii*'s entry into host cells and subsequent virulence.[125] *L. ivanovii* also harbors an *smcL* gene encoding a sphingomyelinase C (SMase), which shares more than 50% identity with the SMases from *Staphylococcus aureus* (beta-toxin), *Bacillus cereus,* and *Leptospira interrogans.* Being independent of PrfA regulation, SmcL mediates *L. ivanovii*'s escape from the phagocytic vacuole and migration into the cytosoL. Thus, together with PlcA and PlcB, *L. ivanovii* utilizes a third phospholipase with membrane-damaging activity, which may function in synergy with the pore-forming toxin ILO (ivanolysin O) to enhance disruption of the vacuolar compartment. As the 5′ end of *smcL* is contiguous with the internalin locus *i-inlFE,* which is also specific to *L. ivanovii* and is required for full virulence in mice, *smcL* may constitute part of the *L. ivanovii* virulence gene cluster essential for its virulence and pathogenicity.[126]

Interestingly, of the three lineages within *L. monocytogenes* species (with lineage I consisting of serotypes of 1/2b, 3b, 4b, 4d, and 4e; lineage II of serotypes 1/2a, 1/2c, 3a, and 3c; and lineage III of serotypes 4a and 4c), lineage I strains are frequently detected in encephalitic cases and lineage II strains are found in a much wider spectrum of clinical cases (including encephalitis, septicemia, and fetal infections) in cattle.[127] These findings imply that lineage I strains may have acquired the genetic traits to specifically target host brain issue, while lineage II strains have developed a much broader tissue specificity. Indeed, there is evidence that *L. monocytogenes* lineages possess distinct gene sets that facilitate their adaptation to their unique environmental niches. For example, lineage I and II strains appear to have more transcriptional regulator genes than lineage III strains. This was reflected by the fact that several transcriptional regulator genes (e.g., *lmo0833, lmo1116, lmo1134,* and *lmo2672*) are present in lineages I and II strains but are largely absent in lineage III strains.[112] In addition, lineage I and II strains harbor a full-length *inlJ* gene, whereas lineage III strains either have a much shortened *inlJ* gene (as in serotype 4c) or no *inlJ* gene at all (as in serotype 4a). As examined by Southern blot, genomic DNA from the more pathogenic *L. monocytogenes* lineage I and II serotypes (e.g., 1/2a, 1/2b, 1/2c, and 4b) displayed a 5.0-kb *Hin*dIII band, while DNA from the less pathogenic lineage III serotypes (e.g., 4a and 4c) showed a 1.5- or 2.0-kb *Hin*dIII fragment using an *inlJ* gene probe.[6,126] Moreover, with regard to the species-specific gene *lmo0733,* lineage I and II strains differed from lineage III strains in that the former tended to form a *Hin*dIII band of 5.0 or 6.0 kb while the latter gave a *Hin*dIII band of 1.0 or 1.5 kb in the Southern blot analysis[6,128] Thus, apart from maintaining an uninterrupted *inlJ* gene, the more pathogenic lineage I and II strains may require a complete copy of the *lmo0733* gene for successful establishment in the hosts, although the precise role of *lmo0733* in listeriosis still needs verification.

As facultative intracellular pathogens, *L. monocytogenes* and *L. ivanovii* are capable not only of producing specialized molecules (e.g., LLO, ILO, PlcA, PlcB, Mpl, and ActA) for their invasion of host cells, but also of utilizing various mechanisms for their evasion of host immune assault. One mechanism that is utilized by *L. monocytogenes* to evade host immune surveillance is through N-deacetylation of peptidoglycan (PG), which is a large polymer providing strength and rigidity to the bacterial cell wall. The modification of PG renders *L. monocytogenes* resistant to host lysozyme, thus preventing its degradation and subsequent release of immunostimulants.[127] As a direct outcome, *L. monocytogenes* remains undetected by host pattern-recognition receptors (PRRs), which specifically recognize pathogen-associated molecular patterns (PAMPs) on the outer surface of bacteria (see Chapter 13), thus avoiding host immune attack. Indeed, an *L. monocytogenes* mutant lacking PG N-deacetylase gene (i.e., *pgdA* or *lmo0415*) was highly sensitive to killing by lysozyme and has impaired ability to survive and/or replicate within macrophages.[129]

Recently, the status of *lmo0415* (*pgdA*) in *Listeria* species was examined by PCR and Southern blot (D. Liu, unpublished data). From the genome sequence data at GenBank, it is clear that *L. monocytogenes* F2365 (serotype 4b), *L. innocua* CLIP11262, and *L. welshimeri* SLCC5334 all contain a gene with similarity (at 97% or 1359/1401, 85% or 1191/1401, and 74% or 1038/1401, respectively) to

L. monocytogenes EGD-e (serotype 1/2a) *lmo0415*. Using primer pair derived from *L. monocytogenes* EGD-e (serotype 1/2a) *lmo0415* (i.e., Lmo0415F-1/2a: 5′-CAAAGTCGCGCAACAAAGTA-3′ and Lmo0415R-1/2a: 5′-GGCAATTCGTTTGTTCGTTT-3′) in PCR, only *L. monocytogenes* serotypes 1/2a, 1/2c, 3a, and 3c strains formed a specific band of 713 bp, while other *L. monocytogenes* serotypes and *Listeria* species were negative (Table 2.3).

Interestingly, whereas the forward primer Lmo0415F-1/2a: 5′-CAAAGTCGCGCAACAAAGTA-3′ finds a perfcet match (20/20) in *L. monocytogenes* F2365 (4b) and *L. innocua* CLIP11262 as well as a slightly imperfect match (at 95% or 19/20) in *L. welshimeri* SLCC5334, the reverse primer Lmo0415R-1/2a: 5′-GGCAATTCGTTTGTTCGTTT-3′ demonstrates only a partial identity with *L. monocytogenes* F2365 (4b) (at 85% or 17/20), *L. innocua* CLIP11262 (at 85% or 17/20), and *L. welshimeri* SLCC5334 (at 50% or 10/20). The latter may be responsible for the negative PCR results obtained with *L. monocytogenes* serotypes 1/2b, 3b, 4a-e, *L. innocua, L. welshimeri,* and other *Listeria* strains. Therefore, a second reverse primer Lmo415R-4b (5′-CGCAATTCGCTTGTTGGTTT-3′) was designed from *L. monocytogenes* 4b strain F2365. Application of primer pair Lmo0415F-1/2a and Lmo415R-4b facilitated amplification of a 713-bp band in PCR from all *L. monocytogenes* serotypes, but not from non*monocytogenes Listeria* species (Table 2.3). When examined in Southern hybridization using a PCR-amplified *lmo0415* fragment as probe, *L. monocytogenes* serotypes 4a and 4c showed a smaller band of 1.5 kb, while other serotypes had a larger band of 4.0 or 5.0 kb. Additionally, *L. innocua* and *L. welshimeri* displayed a band of 4.0 kb, *L. seeligeri* a band of 3.0 kb, and *L. ivanovii* a band of 1.5 kb; *L. grayi* formed no band (Table 2.3). In the presence of 100 μg/ml lysozyme, the growth of all *L. monocytogenes* serotypes together with *L. innocua* and *L. grayi* was not significantly affected. However, *L. ivanovii, L. seeligeri,* and *L. welshimeri* were somewhat vulnerable to the lysozyme treatment (Table 2.3).

These results suggest that considerable variations exist in the PG N-deacetylase gene (i.e., *pgdA* or *lmo0415*) among *Listeria* species, as well as within *L. monocytogenes.* Despite possessing a somewhat modified *pgdA* gene, all *L. monocytogenes* serotypes and non-monocytogenes *Listeria* species (with the exception of *L. grayi*) were largely resistant to lysozyme as assessed by an agar diffusion assay. Namely, *L. monocytogenes* serotypes 1/2b, 3a, 3b, 3c, 4b, 4d, and 4e as well as *L. innocua* CLIP 11262 and *L. ivanovii* ATCC 19119 displayed a high level of tolerance to lysozyme (with the formation of 6.0–7.0 mm clearing zones on BHI agar). This was followed by *L. monocytogenes* serotypes 1/2a, 1/2c, 4a, and 4c (with clearning zones of 9.0–10.0 mm). *L. seeligeri* ATCC 35967 and *L. welshimeri* ATCC 43550 were slightly vulnerable to lysozyme (with a clearing zone of 12.0 mm) while *L. grayi* ATCC19120 was extremely susceptible to lysozyme (with a clearing zone of 20 mm) (Table 2.3 and data not shown). Thus, the involvement of *pgdA* appeared to be crucial to *Listeria* tolerance of lysozyme as *L. grayi* ATCC 19120 lacking the *pgdA* gene demonstrated a heightened vulnerability to *in vitro* lysozyme treatment. In addition, *Listeria* species may also utilize other unidentified mechanisms to enable their resistance to the distructive effects of lysozyme, at least in vitro. These mechanisms may be overcome easily by the host defense system since an *L. monocytogenes pgdA* mutant was sensitive to lysozyme killing and had impaired ability to survive within macrophages.[129] Moreover, it is possible that N-deacetylation of PG may represent only one of the mechanisms employed by *Listeria* to evade host immune onslaughter. Indeed, it was reported recently that *L. monocytogenes* expresses phospholipases (e.g., PlcA and PlcB) to facilitate its escape from autophagic degradation, which also forms an important aspect of the host innate immune cellular surveillance network.[130]

2.4.3 Control and Prevention Strategies

In order to reduce the incidence of listeriosis in animals, there are several issues that need to be addressed. First of all, as animals invariably become infected with listerial bacteria via contaminated feed and water, it is vital to ensure that only dry silage and water free of *Listeria* bacteria are

TABLE 2.3
Examination of *Listeria* Strains by PCR and Southern Blot Targeting *lmo0415* (*pgdA*) Gene and Lysozyme Agar Diffusion Assay

Strain	Source	Serotype	PCR for *lmo0415* (713 bp)[a] Primer Pair 1	Primer Pair 2	Southern Blot (kb)[b]	Lysozyme Agar Diffusion Assay (Zone Diameter in mm)[c]
L. monocytogenes EGD	Guinea pig	1/2a	+	+	5.0	10.0
F6854	Turkey frank	1/2a	+	+	ND	ND
RM2991	Sheep brain	1/2b	–	+	4.0	7.0
RM3368	Environment	1/2b	–	+	ND	ND
RM3017	Blood	1/2c	+	+	5.0	10.0
RM3367	Environment	1/2c	+	+	ND	ND
RM3162	Human	3a	+	+	5.0	7.0
RM3026	Food	3a	+	+	ND	ND
RM3121	Chicken	3b	–	+	5.0	7.0
RM3845	Hot dog	3b	–	+	ND	ND
RM3027	Chicken	3c	+	+	5.0	8.0
RM3159	Human	3c	+	+	ND	ND
ATCC 19114	Ruminant brain	4a	–	+	1.5	10.0
HCC23	Catfish	4a	–	+	ND	ND
F2365	Jalisco cheese	4b	–	+	5.0	6.0
RM3177	Human	4b	–	+	ND	ND
ATCC 19116	Chicken	4c	–	+	1.5	9.0
874	Cow brain	4c	–	+	ND	ND
RM3390	Human	4d	–	+	4.0	6.0
RM3108	Chicken	4d	–	+	ND	ND
RM3821	Jalisco cheese	4e	–	+	4.0	6.0
RM2218	Oyster	4e	–	+	ND	ND
L. grayi ATCC 19120	Corn leaves	Unknown	–	–	–	20.0
L. innocua CLIP 11262	Cow brain	6a	–	–	4.0	6.0
L. ivanovii ATCC 19119	Sheep	5	–	–	1.5	6.0
L. seeligeri ATCC 35967	Soil	Unknown	–	–	3.0	12.0
L. welshimeri ATCC 43550	Soil	1/2b	–	–	4.0	12.0

[a] PCR was done with primers derived from the *lmo0415* (or *pgdA*) gene of *L. monocytogenes* strains EGD-e (serotype 1/2a) and F2365 (serotype 4b). Primer pair 1 consisted of Lmo0415F-1/2a: 5′-CAAAGTCGCGCAACAAAGTA-3′ and Lmo0415R-1/2a: 5′-GGCAATTCGTTTGTTCGTTT-3; and primer pair 2 consisted of Lmo0415F-1/2a: 5′-CAAAGTCGCGCAACAAAGTA-3′ and Lmo0415R-4b: 5′-CGCAATTCGCTTGTTGGTTT-3′. Both primer pairs facilitated amplification of a specific product of 713 bp. +, positive; –, negative.

[b] Southern blot was conducted with *Hind*III-digested *Listeria* genomic DNA and probed with a PCR-amplified *lmo0415* fragment using primer pair 1. –, negative. ND, not done.

[c] *Listeria* resistance to lysozyme was evaluated in agar diffusion assay. *Listeria* strains were grown in 5 ml BHI broth at 37°C overnight in an orbital shaker, and the $OD_{600\ mm}$ value for each strain was adjusted to 1.000 with BHI broth. On the BHI agar plates (11 cm in diameter), four wells (2 mm in diameter) were punched out. In each well, 1 mg of lysozyme (i.e., 5 µl of a lysozyme stock at 200 mg/ml, suspended in water, sterile filtered) was added. Then, 25 µl of each *Listeria* suspension was spread over half of a BHI agar plate containing 2 wells. After an 18-h incubation at 37°C, the clearing zones (indicative of growth inhibition or cell lysis) around punched wells were recorded. The values shown here represent the averages from two separate experiments performed in duplicate. ND, not done.

fed to the animals. Poor-quality silage has long been recognized as a risk factor for contributing to listeriosis in food and dairy animals. In addition to feeding good-quality silage, frequent cleaning of animal housing and exercise areas may help decrease the incidence of listeriosis in animals.

Second, it is necessary to implement a monitoring system to test any animals showing clinical manifestations of listeriosis. Once diagnosed, infected animals should be separated from other healthy ones and treated properly. With asymptomatic animals, procedures should be put in place to diagnose them from uninfected, healthy animals. Cows suffering from listerial mastitis are known to secrete large numbers of *L. monocytogenes* into milk. Cross-contamination between animals and the environment may occur in farms after cases of abortive listeriosis. Prompt diagnosis and treatment will help limit the spread of the bacteria and subsequent infections.

Third, future development and application of a vaccine against *L. monocytogenes* and *L. ivanovii* will also help prevent listeriosis in animals. Based on a study in mouse models, it has been shown that naturally avirulent *L. monocytogenes* serotype 4a strains (e.g., HCC23) are capable of eliciting a strong, long-lasting immunity against the *L. monocytogenes* virulent strain challenge.[131] Given its low hemolytic activity accompanied with its induction of negligible amounts of IFN-γ, naturally avirulent strains such as HCC23 may have advantages over intact or attenuated virulent strains as a candidate vaccine. In fact, induction of excessive amounts of IFN-γ, such as is the case with virulent strain EGD, may be not only unnecessary, but also detrimental to the host. As IFN-γ enhances secretion of TNF-α, it may bring about so potent an immune response that the host may succumb as a consequence. Thus, naturally avirulent *L. monocytogenes* serotype 4a strains may provide a potential candidate vaccine against listeriosis in farm animals such as sheep and cattle. In addition, while the existing data suggest that vaccination with *L. ivanovii* failed to protect experimental mice against *L. monocytogenes* virulent strain challenge, we noted recently that immunization of mice with naturally avirulent *L. monocytogenes* serotype 4a strain HCC23 conferred sufficient protection against *L. ivanovii* (D. Liu, unpublished data). Although the mechanism of *Listeria* cross-protection is yet to be elucidated, the ability to apply naturally avirulent *L. monocytogenes* serotype 4a strains as a vaccine may offer an opportunity to protect animals (e.g., sheep and cattle) against both *L. monocytogenes* and *L. ivanovii* infections.

2.5 *LISTERIA* IN HUMANS

Since *Listeria* spp. are robust bacteria with the ability to endure many food processing treatments, such as pH, temperature, and salt conditions, they frequently end up in the various foods that humans consume and then become infected with. Once inside the human stomach, *L. monocytogenes* overcomes the acidic environment, moves across the intestine, enters the blood circulation, and migrates to other parts of the body (e.g., liver, spleen, brain, and fetoplacental tissues), where it grows and stimulates a range of host reactions and clinical diseases. Consequently, while child-bearing women with listeriosis may present mild, flu-like symptoms (e.g., fever and headache) or remain asymptomatic, the infected fetus often aborts or becomes an infant with neonatal listeriosis. Nonpregnant and immunocompromised individuals may develop meningitis or meningoencephalitis. The high mortality rates (about 30%) associated with listeriosis make *L. monocytogenes* one of the most deadly human food-borne pathogens. While human *L. monocytogenes* infection is usually sporadic, outbreaks of listeriosis epidemics do sometimes occur. Interestingly, human listeriosis demonstrates certain seasonal patterns, with the peak of infection in late summer and early autumn. Although no seasonal patterns of *L. monocytogenes* isolation in RTE foods or hot dogs were observed,[132] it was reported recently that *L. monocytogenes* was more frequently isolated in processed catfish during the winter, presumably due to its ability to outcompete other bacterial species at low temperatures.[35]

2.5.1 Prevalence

The overall prevalence of listeriosis in humans is low and is trending lower in more recent years in countries where listeriosis has become a notifiable disease and where appropriate and effective surveillance and prevention measures against listeriosis have been implemented. For example, in the United States, the annual incidence of listeriosis declined from 7.7 cases per million population in 1990 to 4.2 in 1996, and further down to 3.1 in 2003. In France, the incidence of listeriosis decreased from 4.5 cases per million population in 1999–2000 to approximately 3.4 in 2002–2003. Even though *L. monocytogenes* is infective to all population groups, it has a tendency to cause severe, invasive disease in pregnant women, infants, the elderly, and immunocompromised individuals. On the other hand, *L. monocytogenes*-infected healthy individuals often develop febrile gastroenteritis rather than invasive disease.

2.5.1.1 Pregnant Women

Pregnant women account for about 17–24% of clinical cases of listeriosis, and a significant proportion (up to 28%) of pregnant women with listeriosis may experience spontaneous abortion or stillbirth (with fetal loss occurring at 16–35 weeks of gestation).[75] The overall infant mortality for both epidemic and sporadic cases is around 20–30%. The infected mothers may be asymptomatic or only have a flu-like illness and rarely develop CNS infections (Table 2.4). *L. monocytogenes* serotype 4b has been shown to be the most common cause of listeriosis in pregnant women, based on a study undertaken in the United States.[75]

2.5.1.2 Infants

Infants with listeriosis who survive the first 48 hours after birth often develop severe neonatal septicemia and meningitis—a condition known as granulomatosis infantiseptica, which is characterized by the appearance of disseminated granulomatous microabscesses and high mortality rates (about 30%).

TABLE 2.4
Listeriosis and Its Common Symptoms in Humans[a]

Population Group	No. of Cases (%)	Common Symptoms
Pregnant women and infants	141/603 (23%)	Bacteremia (e.g., fever, headache, myalgia, arthralgia, and malaise) Amnionitis Abortion and stillbirth Septicemia and meningitis in infants
Elderly and immunocompromised individuals	414/603 (65%)	Bacteremia Septicemia
Healthy individuals	78/603 (12%)	Asymptomatic fecal carriers or bacteremia (e.g., fever and headache) Gastroenteritis (e.g., nausea, vomiting, and diarrhea) CNS listeriosis (e.g., meningitis, meningoencephalitis, and brain abscess)

[a] Based on a survey of 603 human listeriosis cases that occurred in France during 2001–2003.
Source: Goulet, V. et al., *Euro Surveill.*, 11, 79, 2006.

2.5.1.3 The Elderly

In Denmark, a total of 299 invasive cases of listeriosis were reported during 1994–2003, of which 50% of the patients were aged >70 years and 21% died of the disease.[133] In Finland, of all patients with listeriosis between 1995 and 2004, 57% were aged >65.[134] In the Czech Republic, 75 cases of listeriosis were reported in 2006, with 52% of the patients aged >60.[135] In the United States, a survey revealed that the median age of nonpregnancy-associated listeriosis cases (n = 141) was 71 years (range 1–100 years).[75] Most elderly patients often had an underlying immunocompromising condition.[136]

2.5.1.4 Immunocompromised Individuals

Immunocompromised individuals are persons with underlying diseases (e.g., cancer, organ transplantation, AIDS, chronic hepatic disorder, and diabetes), who are taking medications (e.g., systemic steroids or cyclosporine) or receiving radiation treatment or chemotherapy. In France, the incidence of listeriosis among organ-transplantation recipients was 200 cases per 100,000; among patients with cancer, it was 13 cases per 100,000; and among patients aged >65 years without underlying diseases, it was 1.4 cases per 100,000.[137] In the United States, the incidence of listeriosis among patients with HIV infection and those with AIDS was 52 and 115 cases per 100,000, respectively.[138] Thus, immunocompromised populations with decreasing order of risk appear to be organ-transplantation recipients, AIDS patients, patients with HIV infection, cancer patients, and the elderly without underlying disease. Mortalities (at 38–45%) among immunocompromised or elderly patients were higher than those among other population groups.[136] Because of their subdued immune status, immunocompromised persons often run the risk of becoming infected with pathogens that do not normally occur in humans, such as *L. ivanovvii*.[139]

2.5.1.5 Healthy (Asymptomatic) Individuals

Healthy individuals are individuals who are not pregnant, aged <60 years, and have no predisposing illness (i.e., nonimmunocompromised). The prevalence of *Listeria* spp. in asymptomatic healthy adults ranges from 1 to 5%.[140] Luppi et al.[15] conducted a survey in Italy and isolated seven (1.4%) *L. monocytogenes* and three (0.6%) *L. innocua* strains in 513 fecal specimens from asymptomatic humans. Given that *L. innocua* is generally considered a nonpathogenic species (with its conspicuous lack of virulence gene cluster), its occasional isolation in asymptomatic humans suggests its ability to remain intact within the human gastrointestinal tract (given its exceptional tolerance of pH and salt stresses), either as a way of colonization or as a carrier state. Further investigation is required to determine the precise rate of *L. innocua* in the human disease process. In healthy individuals, the predominant symptom of listeriosis is febrile gastroenteritis (e.g., nausea, vomiting, and diarrhea) rather than invasive disease. These patients often become ill within 24–48 hours of exposure to the contaminated foods (e.g., salad, shrimp, and chocolate milk).[66]

2.5.2 Serotype/Subtype Specificity

Hofer et al.[64] isolated 247 *Listeria* strains in Brazilian human subjects (including listeriosis patients and healthy individuals), which comprised 237 *L. monocytogenes*, six *L. innocua*, and four *L. grayi*. The most common *L. monocytogenes* serotypes were 4b (146), 1/2a (72), and 1/2b (6). Leite et al.[141] characterized 15 *L. monocytogenes* isolates from humans in Portugal, with 13 (86.7%) belonging to serotype 4b and 2 belonging to serovar 1/2b (13.3%). Voetsch et al.[132] examined 530 *L. monocytogenes* isolates from human cases in the United States and showed that serotypes 1/2a (201 or 38%), 4b (191 or 36%), and 1/2b (122 or 23%) were the most common. In France, serotypes 4b, 1/2a, and 1/2b accounted for 96% of the 603 human listeriosis cases that occurred during 2001–2003 (Table 2.5). These data highlight the virulence potential of *L. monocytogenes* serotypes 4b, 1/2a, and 1/2b for human hosts in relation to other serotypes.

TABLE 2.5
***L. monocytogenes* Serotypes Involved in Human Listeriosis**[a]

Serotype	No. of Isolates (%)	Tendency to Cause
4b	294/603 (49%)	CNS infections > M/N diseases > bacteremia
1/2a	163/603 (27%)	Bacteremia > M/N diseases > CNS infections
1/2b	120/603 (20%)	M/N diseases > bacteremia > CNS infections
1/2c	22/603 (4%)	Bacteremia > CNS infections > M/N diseases
3a/3b	4/603 (<1%)	Bacteremia

[a] Based on the analysis of 603 *L. monocytogenes* isolates from 603 French patients during 2001–2003. M/N diseases: maternal-neonatal diseases; CNS infections: central nerve system infections.

Source: Goulet, V. et al., *Euro Surveill.*, 11, 79, 2006.

2.5.3 Mechanisms of Survival and Virulence

2.5.3.1 Gastrointestinal Tract

In a typical mode of infection, *L. monocytogenes* enters the stomach via contaminated foods. With a pH down to 2.5, the stomach's highly acidic environment proves insurmountable for many ordinary microbial pathogens. However, *L. monocytogenes* is by no means an average microbe, as it has been well endowed with the capability to cope with a diversity of external stresses, including acid. Inside the stomach, *L. monocytogenes* produces, under the control of stress-response regulator σ^B, several molecules (e.g., glutamate decarboxylases [GAD] and antiporter proteins BetL, Gbu, and OpuC) to thwart the adverse effects of extracellular acids. Specifically, GAD enzymes convert glutamate γ aminobutyrate (GABA) via decarboxylation. GABA is then transported out of the cell and a new source of glutamate is moved in. This process consumes cytoplasmic protons, resulting in a notable increase in cytoplasmic pH in the bacterium and thus enhancing its survival under the acidic condition.[142]

Having survived this stage, *L. monocytogenes* migrates to the small intestine along with food content, where bile salts create an environment of high salinity (equaling to 0.3 *M* NaCl). In response, *L. monocytogenes* synthesizes transporters BetL, Gbu, and OpuC for uptaking osmolytes glycine betaine and carnitine, which prevent water loss from the bacterial cytoplasm. *L. monocytogenes* also secretes two proteins, Bsh (for bile salt hydrolase) and BtlB (for bile tolerance locus B with bile acid dehydrotase activity), to decrease the potency of bile salts. Remarkably, the corresponding *bsh* gene in *L. monocytogenes* is also regulated by the virulence regulator PrfA. In addition, *L. monocytogenes* utilizes its bile exclusion system (bilE) to rid the bacterial cytoplasm of bile salts.[142]

As mentioned earlier, *L. monocytogenes* employs two key surface proteins, InlA and InlB, to enter epithelial and other host cells, with InlA interacting with E-cadherin present in enterocytes and some other tissues (e.g., placenta), and InlB binding to Met (hepatocyte growth factor receptor) and gC1qR (receptor for complement protein C1q). It is apparent that InlA is essential and sufficient for *L. monocytogenes* entry into the enterocytes in humans and guinea pigs, while InlB may be redundant for entering other cells in these hosts. This is evidenced by the observations that 96% of the 300 *L. monocytogenes* isolates from human patients contained a full-length, functional copy of InlA[143] and that an *L. monocytogenes* Δ*inlB* mutant strain retained its virulence in guinea pigs.[120] It is likely that other internalins (InlC, InlJ, and InlE-H) are able to compensate for the lack of InlB and to assist *L. monocytogenes*' entry into other cells. By contrast, *L. monocytogenes* relies more on InlB than InlA for entry into murine host cells, as the E-cadherin on murine enterocytes is mutated and inefficient for interaction with InlA.

Besides these specific nonphagocytic mechanisms for *L. monocytogenes'* entry into host cells, this bacterium is also capable of exploiting a host's nonspecific phagocytic transport system (e.g., M cells and other phagocytic cells) for translocation of gastrointestinal mucosa and entering the blood circulation.

2.5.3.2 Liver

The liver represents an important initial port of call for *L. monocytogenes.* In experimental mice, 60% of recoverable *L. monocytogenes* came from the liver within 10 min of intravenous inoculation, and this increased to >93% within 6 h of infection. There was an obvious reduction in the number of *L. monocytogenes* in the liver between 10 min and 6 h postinfection, which was accompanied by a sevenfold increase of neutrophils.[144]

InB is known to play a critical role in *L. monocytogenes'* entry into hepatocytes through its interaction with Met. However, given *L. monocytogenes* Δ*inlB* mutant's ability to maintain virulence in guinea pigs, other internalins such as InlC and InlJ may fill the vacuum left by the removal of InlB. Indeed, InlC and InlJ have been shown to become involved in postintestinal invasion of host cells.[121,123] Other internalins, such as InlE-H, may also aid *L. monocytogenes'* entry into hepatocytes at varying capacities.

2.5.3.3 Central Nervous System

L. monocytogenes demonstrates tropism for the mammalian CNS, as it often causes clinical diseases with CNS manifestations (e.g., meningitis, encephalitis, and brain abscess). Meningitis is characterized by high fever, nuchal rigidity, and movement disorders (e.g., tremor, ataxia, and seizures). Through involvement of the brainstem, encephalitis (or rhombencephalitis) is a subacute illness characterized by ataxia and multiple cranial nerve abnormalities; in about 10% of CNS listeriosis, brain abscess is also observed, especially in immunocompromised individuals.[143] *L. monocytogenes'* crossing the blood–brain barrier is ably assisted by the combined actions of InlA, InlB, and possibly other internalins (e.g., InlC and InlJ).

2.5.3.4 Placenta

Pregnancy in mammals creates a distinct immune status that tilts toward a humoral (Th2) response with a subdued cellular (Th1) response. While this immune imbalance favors the protection of the fetus, it inevitably weakens the ability of the mother to launch an effective cell-mediated response against certain microbial pathogens (e.g., *L. monocytogenes*) and allergens. Thus, there is no surprise that pregnant women are 20 times more likely than the general population to develop listeriosis, leading to spontaneous abortions, stillbirths, and neonatal meningitis.

Considering that placenta is made up of multiple layers of tissues that form a significant barrier to many pathogens, the ability of *L. monocytogenes* to get through this hurdle and infect the fetus inside is truly remarkable. It appears that *L. monocytogenes* also exploits the E-cadherin present in the placental–fetal interface to target and cross the human placental barrier via InlA-E-cadherin interaction. Inside the placenta, *L. monocytogenes* is protected by the placental tissues from further maternal immune attacks, where it can grow and release to other maternal organs. *L. monocytogenes* can only be dispelled with the expulsion of infected placental tissues along with the fetus.[145–147] Thus, spontaneous abortion in listeriosis is the host's ultimate attempt to rid itself of the invading *L. monocytogenes.*

2.5.3.5 Gallbladder

Although *L. monocytogenes* is a known intracellular bacterium, it can also grow extracellularly inside the host, such as the lumen of the gallbladder, where the host's antimicrobial immune response is relatively inefficient. *L. monocytogenes* hiding and replicating in gallbladder can be moved to the intestinal tract via the bile duct when the gallbladder contracts to release bile fluid in response to food or cholecystokinin during its normal physiological function. Thus, the gallbladder offers a refuge and also a reservoir for *L. monocytogenes* reinfection of the intestinal tract.[148]

2.5.4 Control and Prevention Strategies

Humans usually become infected with *L. monocytogenes* via contaminated foods and milk. Therefore, the most important step in preventing listeriosis in humans is to reduce and eliminate *Listeria* bacteria in food products. This will require implementation of monitoring procedures at food processing facilities to ensure that only *Listeria*-free products are released to the retail markets for consumers. In addition, regular checking of food products from retail markets will also ensure that any products that have since become contaminated or in which existing low levels of *Listeria* have grown into detectable quantities are removed from the shelves.

Next, it is important to educate consumers to ensure storage of food products at required refrigerator temperatures and proper handling and cooking of the food products before consumption. In addition, it is advisable to use perishable or RTE foods as soon as possible, to clean out refrigerators regularly, and to ensure that refrigerator temperature stays at $<4°C$.

Finally, as mentioned earlier, naturally avirulent *L. monocytogenes* serotype 4a strains (e.g., HCC23) are capable of stimulating a durable, protective immunity in experimental mice against listeriosis.[131] Since naturally avirulent serotype 4a strains induce negligible amounts of IFN-γ, they are a much safer candidate vaccine than intact or attenuated virulent strains. Indeed, although deletion or modification of one or two key genes in *L. monocytogenes* virulent strains reduces their pathogenicity, the ability of these attenuated strains to elicit an unnecessarily potent immune response through high-level IFN-γ production may be unchanged. Further research will help determine if naturally avirulent *L. monocytogenes* serotype 4a strains are equally effective in initiating a protective immune response in human hosts against listeriosis.

2.6 CONCLUSIONS AND PERSPECTIVES

Listeria spp. are robust bacteria with wide distribution. Despite their poor survival in nutrient-deficient, unpolluted seawater and spring water, *Listeria* spp. readily replicate in nutrient-rich, contaminated waters, sewage, sludge, soil, foods, and animal hosts. As a means to reduce the occurrence of *Listeria* spp. in the environment, it is helpful to pretreat runoff waters and sewage and remove residual nutrients before release to waterways. Similarly, in order to lower the incidence of *Listeria* spp. in raw food materials and farm-raised animals, it is important to adopt appropriate farming and husbandry practices that lessen *Listeria* contamination. Next, to eliminate *L. monocytogenes* in manufactured food products, it is critical to implement effective control measures before, during, and after processing stages. Furthermore, to decrease the incidence of listeriosis in humans, it is essential to improve home food-handling practices. However, the successful implementation of the preceding strategies is dependent on the availability of sensitive, rapid, and specific tests for monitoring the presence and levels of *Listeria* spp. in various environmental specimens and for taking corrective measures when required (see Chapters 5 and 6).

We have come a long way toward the understanding of the fundamental aspects of *Listeria* epidemiology, mechanisms of survival, and virulence during the past two decades. The identification of σ^B in *L. monocytogenes* has uncovered the molecular mechanism of listerial response to acid stress. In addition to its regulation of the listerial glutamate decarboxylases (GAD) system and other acid response networks, σ^B also plays a key role in *L. monocytogenes* invasion and virulence through its regulation of InlA and InlB expression. Further, *L. monocytogenes* secretes a large number of proteins (e.g., BetL, Gbu, OpuC, RelA, Ctc, HtrA, KdpE, LisRK, ProBA, and BtlA) to promote its survival and growth at elevated osmolarities. The characterization of several heat response proteins (e.g., GroESL, DnaKJ, Clp, and HtrA) in *L. monocytogenes* highlights the innate ability of this bacterium to counter temperature stress (see Chapter 3). However, *Listeria* responses to stressful alkali and pressure conditions are inadequately understood at this stage, and continued investigation in these areas is critical to the future design of more effective measures to reduce listerial contamination in the environment and in food.

The detailed examination of *Listeria* virulence gene cluster and other virulence-associated genes has yielded valuable insights on the molecular mechanisms of listerial virulence and pathogenicity. Besides producing specialized molecules (e.g., LLO, PlcA, PlcB, Mpl, and ActA) for its invasion of host cells (see Chapter 4), *L. monocytogenes* also deploys some ingenious ways (e.g., N-deacetylation of peptidoglycan and expression of phospholipases) to evade host immune assault. There is no doubt that ongoing research will pinpoint additional mechanisms of *L. monocytogenes* virulence and immune evasion. An in-depth knowledge of host–pathogen immune interaction is vital for devising improved immunological prophylactics against listeriosis. Given that naturally avirulent *L. monocytogenes* serotype 4a strains are capable of inducing protective immunity in mice against listeriosis, further experimentation will help validate if these strains are also useful for immunization in humans against *L. monocytogenes* and in animals against both *L. monocytogenes* and *L. ivanovii* infections.

ACKNOWLEDGMENTS

The author thanks Kara Grubb and Jackie Burns (with financial support from Merck–Merial Summer Research Scholars Program) for assistance in the analysis of *Listeria* resistance to acid, alkali, salt, and lysozyme.

REFERENCES

1. Seeliger, H.P.R., and Langer, B., Serological analysis of the genus *Listeria.* Its values and limitations, *Int. J. Food Microbiol.,* 8, 245, 1989.
2. Liu, D., Identification, subtyping and virulence determination of *Listeria monocytogenes,* an important foodborne pathogen, *J. Med. Microbiol.,* 55, 645, 2006.
3. Brosch, R., Chen, J., and Luchansky, J.B., Pulsed field fingerprinting of Listeriae: Identification of genomic divisions for *Listeria monocytogenes* and their correlation with serovar, *Appl. Environ. Microbiol.,* 60, 2584, 1994.
4. Wiedmann, M. et al., Ribotypes and virulence gene polymorphisms suggest three distinct *Listeria monocytogenes* lineages with differences in pathogenic potential, *Infect. Immun.,* 65, 2707, 1997.
5. Roberts, A. et al., Genetic and phenotypic characterization of *Listeria monocytogenes* lineage III, *Microbiology,* 152, 685, 2006.
6. Liu, D. et al., *Listeria monocytogenes* subgroups IIIA, IIIB and IIIC delineate genetically distinct populations with varied virulence potential, *J. Clin. Microbiol.,* 44, 4229, 2006.
7. Sleator, R.D., Gahan, C.G.M., and Hill, C., Minireview: A postgenomic appraisal of osmotolerance in *Listeria monocytogenes, Appl. Environ. Microbiol.,* 69, 1, 2003.
8. Liu, D. et al., Comparative assessment of acid, alkali and salt tolerance in *Listeria monocytogenes* virulent and avirulent strains, *FEMS Microbiol. Lett.,* 243, 373, 2005.
9. Vázquez-Boland, J.A. et al., *Listeria* pathogenesis and molecular virulence determinants, *Clin. Microbiol. Rev.,* 14, 584, 2001.
10. Altekruse, S.F., Cohen, M.L., and Swerdlow, D.L., Emerging foodborne diseases, *Emerg. Infect. Dis.,* 3, 285, 1997.
11. Mead, P.S. et al., Food-related illness and death in the United States, *Emerg. Infect. Dis.,* 5, 607, 1999.
12. Huss, H.H., Jorgensen, L.V., and Vogel, B.F., Control options for *Listeria monocytogenes* in seafoods, *Int. J. Food Microbiol.,* 62, 267, 2000.
13. Schaffter, N., and Parriaux, A., Pathogenic-bacterial water contamination in mountainous catchments, *Water Res.,* 36, 131, 2002.
14. Hansen, C.H., Vogel, B.F., and Gram, L., Prevalence and survival of *Listeria monocytogenes* in Danish aquatic and fish-processing environments, *J. Food Prot.,* 69, 2113, 2006.

15. Luppi, A. et al., Ecological survey of *Listeria* in the Ferrara area (northern Italy), *Zentralbl. Bakteriol. Mikrobiol. Hyg. A.*, 269, 266, 1988.
16. Colburn, K.G. et al., *Listeria* species in a California coast estuarine environment, *Appl. Environ. Microbiol.*, 56, 2007, 1990.
17. MacGowan, A.P. et al., The occurrence and seasonal changes in the isolation of *Listeria* spp. in shop bought food stuffs, human feces, sewage and soil from urban sources, *Int. J. Food Microbiol.*, 21, 325, 1994.
18. Weis, J., and Seeliger, H.P.R., Incidence of *Listeria monocytogenes* in nature, *Appl. Microbiol.*, 30, 29, 1995.
19. Van Renterghem, B. et al., Detection and prevalence of *Listeria monocytogenes* in the agricultural ecosystem, *J. Appl. Bacteriol.*, 71, 211, 1991.
20. Moshtaghi, H., Garg, S.R., and Mandokhot, U.V., Prevalence of *Listeria* in soil, *Indian J. Exp. Biol.*, 41, 1466, 2003.
21. Welshimer, H.J., Isolation of *Listeria monocytogenes* from vegetation, *J. Bacteriol.*, 95, 300, 1968.
22. Welshimer, H.J., and Donker-Voet, J., *Listeria monocytogenes* in nature, *Appl. Microbiol.*, 21, 516, 1971.
23. Blackmann, I.C., and Frank, J.F., Growth of *Listeria monocytogenes* as a biofilm on various food-processing surfaces, *J. Food Prot.*, 59, 827, 1996.
24. Jay, J.M., Prevalence of *Listeria* spp. in meat and poultry products, *Food Control*, 7, 209, 1996.
25. Norwood, D.E., and Gilmour, A., The growth and resistance to sodium hypochlorite of *Listeria monocytogenes* in a steady-state multispecies biofilm, *J. Appl. Microbiol.*, 88, 512, 2000.
26. Autio, T. et al., Similar *Listeria monocytogenes* pulsotypes detected in several foods originating from different sources, *Int. J. Food. Microbiol.*, 77, 83, 2002.
27. Thevenot, D. et al., Prevalence of *Listeria monocytogenes* in 13 dried sausage processing plants and their products, *Int. J. Food Microbiol.*, 102, 85, 2005.
28. Chae, M.S., and Schraft, H., Comparative evaluation of adhesion and biofilm formation of different *Listeria monocytogenes* strains, *Int. J. Food Microbiol.*, 62, 103, 2000.
29. Chambel, L. et al., Occurrence and persistence of *Listeria* spp. in the environment of ewe and cow's milk cheese dairies in Portugal unveiled by an integrated analysis of identification, typing and spatial-temporal mapping along production cycle, *Int. J. Food Microbiol.*, 116, 52, 2007.
30. Eifert, J.D. et al., Molecular characterization of *Listeria monocytogenes* of the serotype 4b complex (4b, 4d, 4e) from two turkey processing plants, *Foodborne Pathog. Dis.*, 2, 192, 2005.
31. Eklund, M.W. et al., Incidence and sources of *Listeria monocytogenes* in cold smoking fishery products and processing plants, *J. Food Prot.*, 58, 502, 1995.
32. Draughon, F.A., Anthony, B.A., and Denton, M.E., *Listeria* species in fresh rainbow trout purchased from retail markets, *Dairy Food Environ. Sanit.*, 19, 90, 1999.
33. Miettinen, H., and Wirtanen, G., Prevalence and location of *Listeria monocytogenes* in farmed rainbow trout, *Int. J. Food Microbiol.*, 104, 135, 2005.
34. Rodas-Suarez, O.R. et al., Occurrence and antibiotic sensitivity of *Listeria monocytogenes* strains isolated from oysters, fish, and estuarine water, *Appl. Environ. Microbiol.*, 72, 7410, 2006.
35. Chou, C.H., Silva, J.L., and Wang, C., Prevalence and typing of *Listeria monocytogenes* in raw catfish fillets, *J. Food Prot.*, 69, 815, 2006.
36. Autio, T. et al. Sources of *Listeria monocytogenes* contamination in a cold-smoked rainbow trout processing plant detected by pulsed-field gel electrophoresis typing, *Appl. Environ. Microbiol.*, 65, 150, 1999.
37. Rorvik, L.M., Caugant, D.A., and Yndestad, M., Contamination pattern of *Listeria monocytogenes* and other *Listeria* spp. in a salmon slaughterhouse and smoked salmon processing plant, *Int. J. Food Microbiol.*, 25, 19, 1995.
38. Ericsson, H. et al., An outbreak of listeriosis suspected to have been caused by rainbow trout, *J. Clin. Microbiol.*, 35, 2904, 1997.
39. Miettinen, M., Bjorkroth, K.J., and Korkeala, H.J., Characterization of *Listeria monocytogenes* from an ice cream plant by serotyping and pulsed-field gel electrophoresis, *Int. J. Food Microbiol.*, 46, 187, 1999.
40. Brett, M.S., Short, P., and McLauchline, J., A small outbreak of listeriosis associated with smoked mussels, *Int. J. Food Microbiol.*, 43, 223, 1998.
41. Rocourt, J., Jacquet, C., and Reilly, A., Epidemiology of human listeriosis and seafoods, *Int. J. Food Microbiol.*, 62, 197, 2000.
42. Reiter, M.G. et al., Occurrence of *Campylobacter* and *Listeria monocytogenes* in a poultry processing plant, *J. Food Prot.*, 68, 1903, 2005.
43. Capita, R. et al., Occurrence of *Listeria* species in retail poultry meat and comparison of a cultural/immunoassay for their detection, *Int. J. Food Microbiol.*, 65, 75, 2001.

44. Skovgaard, N., and Norrung, B., The incidence of *Listeria* spp in feces of Danish pigs and in minced pork meat, *Int. J. Food Microbiol.,* 8, 59, 1989.
45. Bunčić, S., Paunovic, L., and Radisic, D., The fate of *Listeria monocytogenes* in fermented sausages and in vacuum-packaged frankfurthers, *J. Food Prot.,* 54, 413, 1991.
46. Fenlon, D.R., Wilson, J., and Donachie, W., The incidence and level of *Listeria monocytogenes* contamination of food sources at primary production and initial processing, *J. Appl. Microbiol.,* 81, 641, 1996.
47. Kanuganti, S.R. et al., Detection of *Listeria monocytogenes* in pigs and pork, *J. Food Prot.,* 65, 1470, 2002.
48. Autio, T. et al., *Listeria monocytogenes* contamination pattern in pig slaughterhouses, *J. Food Prot.,* 63, 1438, 2000.
49. Hof, H., and Rocourt, J., Is any strain of *Listeria monocytogenes* detected in food a health risk? *Int. J. Food Microbiol.,* 16, 173, 1992.
50. Thevenot, D., Dernburg, A., and Vernozy-Rozand, C., An updated review of *Listeria monocytogenes* in the pork meat industry and its products, *J. Appl. Microbiol.,* 101, 7, 2006.
51. McLauchlin, J. et al., Human listeriosis and paté: A possible association, *Brit. Med. J.,* 303, 773, 1991.
52. Jacquet, C. et al., Investigations related to the epidemic strain involved in the French listeriosis outbreak in 1992, *Appl. Environ. Microbiol.,* 61, 2242, 1995.
53. Goulet V, et al. Listeriosis outbreak associated with the consumption of rillettes in France in 1993, *J. Infect. Dis.,* 177, 155, 1998.
54. Evans, M.R. et al., Genetic markers unique to *Listeria monocytogenes* serotype 4b differentiate epidemic clone II (hot dog outbreak strains) from other lineages, *Appl. Environ. Microbiol.,* 70, 2383, 2004.
55. Samadpour, M. et al., Incidence of enterohemorrhagic *Escherichia coli, Escherichia coli* O157, *Salmonella,* and *Listeria monocytogenes* in retail fresh ground beef, sprouts, and mushrooms, *J. Food Prot.,* 69, 441, 2006.
56. Bosilevac, J.M. et al., Microbiological characterization of imported and domestic boneless beef trim used for ground beef, *J. Food Prot.,* 70, 440, 2007.
57. Antoniollo, P.C. et al., Prevalence of *Listeria* spp. in feces and carcasses at a lamb packing plant in Brazil, *J. Food Prot.,* 66, 328, 2003.
58. Rodriguez-Calleja, J.M. et al., Rabbit meat as a source of bacterial foodborne pathogens, *J. Food Prot.,* May, 69(5), 1106–1112, 2006.
59. Waak, E., Tham, W., and Danielsson-Tham, M.L., Prevalence and fingerprinting of *Listeria monocytogenes* strains isolated from raw whole milk in farm bulk tanks and in dairy plant receiving tanks, *Appl. Environ. Microbiol.,* 68, 3366, 2002.
60. Moshtaghi, H., and Mohamadpour, A.A., Incidence of *Listeria* spp. in raw milk in Shahrekord, Iran, *Foodborne Pathog. Dis.,* 4, 107, 2007.
61. Jararao M.B. et al., A survey of foodborne pathogens in bulk tank milk and raw milk consumption among farm families in Pennsylvania, *J. Dairy Sci.,* 89, 2451, 2006.
62. Pak, S.L. et al., Risk factors for *L. monocytogenes* contamination of dairy products in Switzerland, 1990–1999, *Prev. Vet. Med.,* 53, 55, 2002.
63. Da Silva, M.C., Hofer, E., and Tibana, A., Incidence of *Listeria monocytogenes* in cheese produced in Rio de Janeiro, Brazil, *J. Food Prot.,* 61, 354, 1998.
64. Hofer, E., Ribeiro, R., and Feitosa, D.P., Species and serovars of the genus *Listeria* isolated from different sources in Brazil from 1971 to 1997, *Mem. Inst. Oswaldo Cruz.,* 95, 615, 2000.
65. Fleming, D.W. et al., Pasteurized milk as a vehicle of infection in an outbreak of listeriosis, *N. Engl. J. Med.,* 312, 404, 1985.
66. Dalton, C.B. et al., An outbreak of gastroenteritis and fever due to *Listeria monocytogenes* in milk, *N. Engl. J. Med.,* 336, 100, 1997.
67. Linnan. M.J. et al., Epidemic listeriosis associated with Mexican-style cheese, *N. Engl. J. Med.,* 319, 823, 1988.
68. Bille, J. et al., Outbreak of human listeriosis associated with tomme cheese in northwest Switzerland, 2005, *Euro Surveill.,* 11, 91, 2006.
69. Goulet, V. et al., Listeriosis from consumption of raw-milk cheese, *Lancet,* 345, 1581, 1995.
70. Prazak, A.M. et al., Prevalence of *Listeria monocytogenes* during production and postharvest processing of cabbage, *J. Food Prot.,* 65, 1728, 2003.
71. Ho, J.L. et al., An outbreak of type 4b *Listeria monocytogenes* infection involving patients from eight Boston hospitals, *Arch. Intern. Med.,* 146, 520, 1986.
72. Schlech, W.F. et al., Epidemic listeriosis—Evidence for transmission by food, *N. Engl. J. Med.,* 308, 203, 1983.

73. Aureli, P. et al., An outbreak of febrile gastroenteritis associated with corn contaminated by *Listeria monocytogenes, N. Engl. J. Med.,* 342, 1236, 2000.
74. Mead, P.S. et al., Nationwide outbreak of listeriosis due to contaminated meat, *Epid. Infect.,* 134, 744, 2006.
75. Varma, J.K. et al., *Listeria monocytogenes* infection from foods prepared in a commercial establishment: A case-control study of potential sources of sporadic illness in the United States, *Clin. Infect. Dis.,* 44, 521, 2007.
76. Van Coillie, E. et al., Prevalence and typing of *Listeria monocytogenes* in ready-to-eat food products on the Belgian market, *J. Food Prot.,* 67, 2480, 2004.
77. Angelidis, A.S., and Koutsoumanis, K., Prevalence and concentration of *Listeria monocytogenes* in sliced ready-to-eat meat products in the Hellenic retail market, *J. Food Prot.,* 69, 938, 2006.
78. Zhou, X., and Jiao, X., Prevalence and lineages of *Listeria monocytogenes* in Chinese food products, *Lett. Appl. Microbiol.,* 43, 554, 2006.
79. Wagner, M. et al., Survey on the *Listeria* contamination of ready-to-eat food products and household environments in Vienna, Austria, *Zoon. Publ. Health,* 54, 16, 2007.
80. Goulet, V. et al., Surveillance of human listeriosis in France, 2001–2003, *Euro Surveill.,* 11, 79, 2006.
81. Foegeding, P.M. et al., Enhanced control of *Listeria monocytogenes* by in situ-produced pediocin during dry fermented sausage production, *Appl. Environ. Microbiol.,* 58, 884, 1992.
82. Cotter, P.D., and Hill, C., Surviving the acid test: Responses of Gram-positive bacteria to low pH, *Microbiol. Mol. Biol. Rev.,* 67, 429, 2003.
83. O'Driscoll, B., Gahan, C.G.M., and Hill, C., Adaptive acid tolerance response in *Listeria monocytogenes*: Isolation of an acid-tolerant mutant which demonstrates increased virulence, *Appl. Environ. Microbiol.,* 62, 1693, 1996.
84. Vasseur, C. et al., Combined effects of NaCl, NaOH, and biocides (monolairin or lauric acid) on inactivation of *Listeria monocytogenes* and *Pseudomonas* spp., *J. Food Prot.,* 64, 1442, 2001.
85. Linton, R.H. et al., The effect of sublethal heat shock and growth atmosphere on the heat resistance of *Listeria monocytogenes* Scott A, *J. Food Prot.,* 55, 84, 1992.
86. Romanova, N., Favrin, S., and Griffiths, M.W., Sensitivity of *Listeria monocytogenes* to sanitizers used in the meat processing industry, *Appl. Environ. Microbiol.,* 68, 6405, 2002.
87. Bremer, P.J., Monk, I., and Butler, R., Inactivation of *Listeria monocytogenes/Flaviobacterium* spp. biofilms using chlorine: Impact of substrate, pH, time and concentration, *Lett. Appl. Microbiol.,* 35, 321, 2002.
88. Holah, J.T. et al., Biocide use in the food industry and the disinfectant resistance of persistent strains of *Listeria monocytogenes* and *Escherichia coli, J. Appl. Microbiol.,* 92, 111, 2002.
89. Beloeil, P.A. et al., *Listeria monocytogenes* contamination of finishing pigs: An exploratory epidemiological survey in France, *Vet. Res.,* 34, 737, 2003.
90. Rorvik, L.M. et al., Risk factors for contamination of smoked salmon with *Listeria monocytogenes, Int. J. Food Microbiol.,* 37, 215, 1997.
91. Ingham, S.C. et al., Survival of *Listeria monocytogenes* during storage of ready-to-eat meat products processed by drying, fermentation, and/or smoking, *J. Food Prot.,* 67, 2698, 2004.
92. Low, J.C., and Donachie, W., A review of *Listeria monocytogenes* and listeriosis, *Vet. J.,* 153, 9, 1997.
93. Kimpe, A. et al., Isolation of *Listeria ivanovii* from a septicemic chinchilla (*Chinchilla lanigera*), *Vet. Rec.,* 154, 791, 2004.
94. Hutchison, M.L. et al., Levels of zoonotic agents in British livestock manures, *Lett. Appl. Microbiol.,* 39, 207, 2004.
95. Nightingale, K.K. et al., Ecology and transmission of *Listeria monocytogenes* infecting ruminants and in the farm environment, *Appl. Environ. Microbiol.,* 70, 4458, 2004.
96. Ho, A.J., Lappi, V.R., and Wiedmann, M., Longitudinal monitoring of *Listeria monocytogenes* contamination patterns in a farmstead dairy processing facility, *J. Dairy Sci.,* 90, 2517, 2007.
97. Winter, P. et al., Clinical and histopathological aspects of mastitis caused by *Listeria monocytogenes* in cattle and ewes, *J. Vet. Med. B.,* 51, 176, 2004.
98. Borucki, M.K. et al., Genetic diversity of *Listeria monocytogenes* strains from a high-prevalence dairy farm, *Appl. Environ. Microbiol.,* 71, 5893, 2005.
99. Rawool, D.B. et al., Detection of multiple virulence-associated genes in *Listeria monocytogenes* isolated from bovine mastitis cases, *Int. J. Food Microbiol.,* 113, 201, 2007.
100. Wagner M. et al., Outbreak of clinical listeriosis in sheep: Evaluation from possible contamination routes from feed to raw produce and humans, *J. Vet. Med. B. Infect. Dis. Vet. Public Health,* 52, 278, 2005.
101. Gudmundsdottir, K.B. et al., *Listeria monocytogenes* in horses in Iceland, *Vet. Rec.,* 155, 456, 2004.
102. Kurazono, M. et al., Pathology of listerial encephalitis in chickens in Japan, *Avian Dis.,* 47, 1496, 2003.

103. Sergeant, E.S., Love, S.C., and McInnes, A., Abortions in sheep due to *Listeria ivanovii*, *Aust. Vet. J.*, 68, 39, 1991.
104. Chand, P., and Sadana, J.R., Outbreak of *Listeria ivanovii* abortion in sheep in India, *Vet. Rec.*, 145, 83, 1999.
105. Alexander, A.V. et al., Bovine abortions attributable to *Listeria ivanovii*: Four cases (1988–1990), *J. Am. Vet. Med. Assoc.*, 200, 711, 1992.
106. Gill, P.A. et al., Bovine abortion caused by *Listeria ivanovii*, *Aust. Vet. J.*, 75, 214, 1997.
107. Peters, M., Amtsberg, G., and Beckmann, G.T., The diagnosis of *Listeria* encephalitis in ruminants using cultural and immunohistological techniques. I. Comparison of different selective media and culture techniques for the detection of *Listeria* from ruminant brains, *Zentralbl. Veterinarmed. B.*, 39, 410, 1992.
108. Lecuit, M. et al., A transgenic model for listeriosis: Role of internalin in crossing the intestinal barrier, *Science*, 292, 1722, 2001.
109. Cheers, C., and McKenzie, I.F., Resistance and susceptibility of mice to bacterial infection: Genetics of listeriosis, *Infect. Immun.*, 19, 755, 1978.
110. Gervais, F., Stevenson, M., and Skamene, E., Genetic control of resistance to *Listeria monocytogenes*: Regulation of leukocyte inflammatory responses by the Hc locus, *J. Immunol.*, 132, 2078, 1984.
111. Barbour, A.H., Rampling, A., and Hormaeche, C.E., Variation in the infectivity of *Listeria monocytogenes* isolates following intragastric inoculation of mice, *Infect. Immun.*, 69, 4657, 2001.
112. Liu, D. et al., Characterization of virulent and avirulent *Listeria monocytogenes* strains by PCR amplification of putative transcriptional regulator and internalin genes, *J. Med. Microbiol.*, 52, 1066, 2003.
113. Liu, D., *Listeria monocytogenes*: Comparative interpretation of mouse virulence assay, *FEMS Microbiol. Lett.*, 233, 159, 2004.
114. Schlech, W.F., III, An animal model of foodborne *Listeria monocytogenes* virulence: Effect of alterations in local and systemic immunity on invasive infection, *Clin. Invest. Med.*, 16, 219, 1993.
115. Blanot, S. et al., A gerbil model for rhombencephalitis due to *Listeria monocytogenes*, *Microbial Pathogen.*, 23, 39, 1997.
116. Czuprynski, C.J., Faith, N.G., and Steinberg, H., A/J mice are susceptible and C57BL/6 mice are resistant to *Listeria monocytogenes* infection by intragastric inoculation, *Infect. Immun.*, 71, 682, 2003.
117. Smith, M.A. et al, Nonhuman primate model for *Listeria monocytogenes*-induced stillbirths, *Infect. Immun.*, 71, 1574, 2003.
118. Bakardjiev, A.I. et al., Listeriosis in the pregnant guinea pig: A model of vertical transmission, *Infect. Immun.*, 72, 489, 2004.
119. Mauder, N. et al., Species-specific differences in the activity of PrfA, the key regulator of listerial virulence genes, *J. Bacteriol.*, 188, 7941, 2006.
119a. Liu, D. et al., A multiplex PCR for species- and virulence-specific determination of *Listeria monocytogenes*, *J. Microbiol. Methods*, 71, 133, 2007.
120. Khelef, N. et al., Species specificity of the *Listeria monocytogenes* InlB protein, *Cell. Microbiol.*, 8, 457, 2006.
121. Sabet, C. et al., LPXTG protein InlJ, a newly identified internalin involved in *Listeria monocytogenes* virulence, *Infect. Immun.*, 73, 6912, 2005.
122. Chico-Calero, L. et al., Hpt, a bacterial homolog of the microsomal glucose-6-phosphate translocase, mediates rapid intracellular proliferation in *Listeria*, *Proc. Natl. Acad. Sci. USA*, 99, 431, 2002.
123. Engelbrecht, F. et al., A new PrfA-regulated gene of *Listeria monocytogenes* encoding a small, secreted protein which belongs to the family of internalins, *Mol. Microbiol.*, 21, 823, 1996.
124. Dominguez-Bernal, G. et al., A spontaneous genomic deletion in *Listeria ivanovii* identifies LIPI-2, a species-specific pathogenicity island encoding sphingomyelinase and numerous internalins, *Mol. Microbiol.*, 59, 415, 2006.
125. Engelbrecht, F. et al., A novel PrfA-regulated chromosomal locus, which is specific for *Listeria ivanovii*, encodes two small, secreted internalins and contributes to virulence in mice, *Mol. Microbiol.*, 30, 405, 1998.
126. Gonzalez-Zorn, B. et al., The smcL gene of *Listeria ivanovii* encodes a sphingomyelinase C that mediates bacterial escape from the phagocytic vacuole, *Mol. Microbiol.*, 33, 510, 1999.
127. Pohl, M.A., Wiedmann, M., and Nightingale, K.K., Associations among *Listeria monocytogenes* genotypes and distinct clinical manifestations of listeriosis in cattle, *Am. J. Vet, Res.*, 67, 616, 2006.
128. Liu, D. et al., *Listeria monocytogenes* serotype 4b strains belonging to lineages I and III possess distinct molecular features, *J. Clin. Microbiol.*, 44, 214, 2006.

129. Boneca, I.G. et al., A critical role for peptidoglycan N-deacetylation in *Listeria* evasion from the host innate immune system, *Proc. Natl. Acad. Sci. USA,* 104, 997, 2007.
130. Py, B.F., Lipinski, M.M., and Yuan, J., Autophagy limits *Listeria monocytogenes* intracellular growth in the early phase of primary infection, *Autophagy,* 3, 117, 2007.
131. Liu, D. et al., Characteristics of cell-mediated, antilisterial immunity induced by a naturally avirulent *Listeria monocytogenes* serotype 4a strain HCC23, *Arch. Microbiol.,* 188, 251, 2007.
132. Voetsch, A.C. et al., Reduction in the incidence of invasive listeriosis in foodborne diseases active surveillance network sites, 1996–2003, *Clin. Infect. Dis.,* 44, 513, 2007.
133. Gerner-Smidt, P. et al., Invasive listeriosis in Denmark 1994–2003: A review of 299 cases with special emphasis on risk factors for mortality, *Clin. Microbiol. Infect.,* 11, 618, 2005.
134. Lyytikainen, O., Surveillance of listeriosis in Finland during 1995–2004, *Euro Surveill.,* 11, 82, 2006.
135. Vit, M. et al., Outbreak of listeriosis in the Czech Republic, late 2006—Preliminary report, *Euro Surveill.,* 12, E070208.1, 2007.
136. Gottlieb, S.L. et al., Multistate outbreak of listeriosis linked to turkey deli meat and subsequent changes in U.S. regulatory policy, *Clin. Infect. Dis.,* 42, 29, 2006.
137. Rebiere, I., and Goulet, V., La listériose: Revue générale et référence à l'épidémie française de 1992, *Lett. Infectiol.,* 8, 130, 1993.
138. Jurado, R.L. et al., Increased risk of meningitis and bacteremia due to *Listeria monocytogenes* in patients with human immunodeficiency virus infection, *Clin. Infect. Dis.,* 17, 224, 1993.
139. Cummins, A.J., Fielding, A.K., and McLauchlin, J., *Listeria ivanovii* infection in a patient with AIDS, *J. Infect.,* 28, 89, 1994.
140. Hof, H., *Listeria monocytogenes*: A causative agent of gastroenteritis? *Euro. J. Clin. Microbiol. Infect. Dis.,* 20, 369, 2001.
141. Leite, P. et al., Comparative characterization of *Listeria monocytogenes* isolated from Portuguese farmhouse ewe's cheese and from humans, *Int. J. Food Microbiol.,* 106, 111, 2006.
142. Sleator, R.D. et al., A PrfA-regulated bile exclusion system (BilE) is a novel virulence factor in *Listeria monocytogenes, Mol, Microbiol.,* 55, 1183, 2005.
143. Jacquet, C. et al., A molecular marker for evaluating the pathogenic potential of foodborne *Listeria monocytogenes, J. Infect. Dis.,* 189, 2094, 2004.
144. Gregory, S.H., Sagnimeni, A.J., and Wing, E.J., Bacteria in the bloodstream are trapped in the liver and killed by immigrating neutrophils, *J. Immunol.,* 157, 2514, 1996.
145. Abram, M. et al., Murine model of pregnancy-associated *Listeria monocytogenes* infection, *FEMS Immunol. Med. Microbiol.,* 35, 177, 2003.
146. Bakardjiev, A.L., Stacey, B.A., and Portnoy, D.A., Growth of *Listeria monocytogenes* in the guinea pig placenta and role of cell-to-cell spread in fetal infection, *J. Infect. Dis.,* 191, 1889, 2005.
147. Bakardjiev, A.L., Theriot, J.A., and Portnoy, D.A., *Listeria monocytogenes* traffics from maternal organs to the placenta and back, *PLoS Pathog.,* 2, e66, 2006.
148. Hardy, J. et al., Extracellular replication of *Listeria monocytogenes* in the murine gall bladder, *Science,* 303, 851, 2004.

3 Stress Responses

Helena M. Stack, Colin Hill, and Cormac G. M. Gahan

CONTENTS

3.1 INTRODUCTION

3.1.1 *Listeria monocytogenes* and Its Environment

L. monocytogenes, the causative agent of listeriosis in humans and animals, is found ubiquitously in the environment. It has been isolated from soil, decaying vegetation,[1–3] river water, sewage sludge,[4] and animal feed.[5] The frequency with which *L. monocytogenes* can be found during environmental surveys has raised doubts that the food processing industry can effectively eliminate *L. monocytogenes* contamination. Food-borne transmission is considered the most common means of contracting both epidemic and sporadic listeriosis, with 99% of all human cases caused by consumption of contaminated food products.[6] Since 1981, several large outbreaks of listeriosis have been described and sources of food have been determined for each outbreak.[7] The frequent occurrence of *L. monocytogenes* in foods coupled with a high mortality rate of 20–30% in infected individuals makes this pathogen a serious public health issue and of great concern to the food industry.[6,8] In addition to *L. monocytogenes* present in foods, listeriosis in animals also represents a link between this pathogen in the environment and human disease. While the connections between the presence of pathogenic *Listeria* in the environment, foods, animals, and humans are not clear, the transmission and pathogenesis of listeriosis is dependent on the ability of *L. monocytogenes* to survive each of these harsh environments.

3.1.2 Voyage from the Natural Environment to the Mammalian Host

3.1.2.1 Outside the Host

Considering that *L. monocytogenes* appears to be a normal resident of the human and animal intestinal tract, with 8–10% of the general population and 1–5% of animals being carriers,[8] it is not surprising that *L. monocytogenes* is found so routinely in the environment, being shed in the feces of its mammalian hosts.[1–4,9] Any successful pathogen must be able to surmount the numerous stresses it encounters during its life cycle—from the natural environment to the food processing plant and, finally, within the mammalian host. After evacuating the host, *L. monocytogenes* experiences a number of arduous conditions, such as a temperature downshift and possible fluctuations in pH and osmolarity, followed by a variety of progressively serious nutrient limitations. Conversely, *L. monocytogenes* may survive and multiply within protozoans, a normal inhabitant of the natural environment, providing a more favorable milieu while allowing the pathogen to adapt for survival within the mammalian host.[10] The high frequency with which *L. monocytogenes* is found in the environment, coupled with its ability to survive and multiply under conditions frequently used for food preservation (such as dehydration, freezing, freeze-thawing, refrigeration, heat, low pH, and salt[11]), makes this pathogen particularly problematic to the food industry and a great risk to human and animal health.

3.1.2.2 Inside the Host

As contaminated food is the major source of infections in both epidemic and sporadic cases of listeriosis,[8,12] the gastrointestinal tract is thought to be the primary site of entry of pathogenic *L. monocytogenes* into the host (Figure 3.1). Upon consumption, *L. monocytogenes* is normally subjected to an increase in temperature and must also survive the low pH of the stomach. Following this, the organism must circumvent the intestinal stresses, such as high osmolarity, low oxygen levels, intestinal bile salts, weak acids, competition with resident microflora for nutrients, and resistance to the cationic peptides of the innate immune system. Bacteria that survive gastric transit access intestinal lymph and blood vessels by invading the intestinal mucosa via the M cells of Peyer's patches and intestinal epithelial cells.[13–15] Once in the blood and lymph, most bacteria are captured and killed by resident macrophages of the liver cells,[16] but those that survive spread into the liver parenchymal cells and travel to the phagocytic cells of the spleen, a highly permissive environment for bacterial

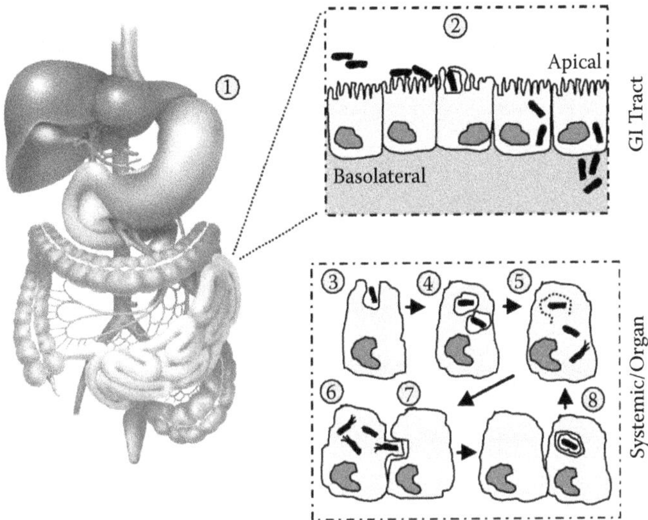

FIGURE 3.1 Pathogenesis of *L. monocytogenes*. (1) Ingestion of the bacterium. (2) Passage through gut into the circulation. (3) Internalization by macrophages, polymorphonuclear cells, and other cells. (4) *L. monocytogenes* in intracellular vacuoles. (5) *L. monocytogenes* escapes from vacuoles, multiplies, and forms actin filament tail that enables movement. (6) By pushing across the membrane, the bacterium forms pseudopod-like structures. (7) The pseudopod is phagocytosed by a neighboring cell. (8) The phagocytosed cell forms a double-membrane internalized vacuole.

growth during the first 24 h of infection.[17,18] Only a small proportion of *L. monocytogenes* cells survive the hostile conditions, such as oxidative stress, imposed by the phagocytic vacuole and reach the cytoplasm, where they proliferate[19,20] and where listerial rearrangement of the macrophage actin facilitates spread of infection.[21,22]

3.1.3 Stress Survival Strategies

Given that the virulence potential of *Listeria* is linked to its ability to survive harsh environmental conditions encountered both in its natural environment and subsequently within the host, understanding the molecular mechanisms by which it overcomes these seemingly insurmountable odds is fundamental to understanding and ultimately controlling this pathogen. This review aims to summarize three important stresses (heat, acid, and osmotic stress) encountered by *L. monocytogenes* during its life cycle and discuss the molecular mechanisms the organism has developed in order to arrive at its site of infection and cause disease. Tables 3.1–3.3 present an overview of the stress response genes discussed in this review and their roles in pathogenicity.

3.2 HEAT STRESS RESPONSE

Exposure to temperatures above the range for normal cell growth leads to progressive loss of bacterial viability.[23] *L. monocytogenes* experiences heat stress at a number of stages during its infectious cycle: the high temperatures employed in the preservation of food (i.e., blanching, pasteurization, and sterilization) and the subtle but significant increase in temperature once *L. monocytogenes* enters the host at the outset of infection. Therefore, it is essential that the bacterium is able to sustain this heat stress in order to arrive at its site of infection. The fundamental heat resistance of microorganisms is owing to the inherent stability of its cell membrane and consequently its intrinsic macromolecules (i.e., ribosomes, DNA, enzymes, and intracellular proteins).[24,25]

In order to combat both lethal and sublethal heat stress, many organisms, including *Listeria*, have evolved sophisticated mechanisms collectively known as the heat shock response. The heat

TABLE 3.1
Overview of Heat-Stress Loci and Their Roles in Pathogenesis

Category	Genetic Locus	Role in Virulence	Ref.
Heat stress	*groESL*	Yes	29, 67, 68
	dnaK	Yes	67, 68, 69
	clpB	Yes	67, 68, 77
	clpC	Yes	67, 78, 79, 105
	clpE	Yes	67, 80
	clpP	Yes	67, 81
	htrA	Yes	31, 119

shock response is a highly conserved cellular defense mechanism characterized by the increased expression and accumulation of heat shock proteins (Hsps), whose expression enhances the survival of microorganisms at elevated temperatures.[26] Most Hsps are expressed at low levels under nonstress conditions but are transiently induced upon exposure to elevated temperatures.[27] Indeed, several Hsps are themselves virulence factors,[28] while others affect pathogenesis indirectly, by increasing bacterial resistance to host defenses or regulating virulence genes.[29–31]

Most Hsps belong to either of two classes: namely, molecular chaperones or adenosine triphosphate (ATP)-dependent proteases (ATPases). In general, proteins with well-folded structures are stable; however, under certain stress conditions proteins become damaged and hydrophobic residues that are normally buried are exposed on the surface. Chaperones bind to these hydrophobic residues and attempt to refold these proteins. If refolding is not possible, these misfolded proteins may bind to regulatory components of proteases and undergo degradation. In many instances, chaperones and proteases work together, with chaperones refolding and preventing aggregation of denatured or misfolded proteins and proteases degrading those denatured proteins unable to adopt their native conformation.[32,33] To date, a number of Hsps have been identified in *L. monocytogenes*, including GroES, GroEL, DnaK, DnaJ, HtrA, and Clp proteins (Table 3.1).

3.2.1 GroESL and DnaKJ

GroES, GroEL, DnaK, and DnaJ all belong to the chaperone family of heat shock proteins whose function is to maintain the integrity of cellular proteins under particular stress conditions.[32,34] In *L. monocytogenes, groES* and *groEL* form a bicistronic *groE* operon driven from a promoter upstream of *groES*, while *dnaK* and *dnaJ* form part of a six-gene operon consisting of *hrcA–grpE–dnaK–dnaJ–orf35–orf29,* termed the *dnaK* operon, which functions from three promoters giving rise to at least four distinct transcripts. Upstream of *hrcA* is the transcriptional start site P1, which is heat shock inducible, while upstream of *dnaJ* and *grpE* are transcriptional start sites P2 and P3, respectively. P2 is predicted to be regulated by the vegetative σ^A subunit, while the transcriptional start site P3, whose sequence does not resemble any known promoter, is thought to be initiated by an alternative sigma factor.[35] GroEL is among the most highly conserved proteins in nature[36] and together with GroES it maintains protein stability under adverse environmental conditions.[37] Fayet et al.[38] showed that the *groE* operon is essential for growth at all temperatures. The chaperone function of DnaK has been well characterized[39] and, in addition, DnaK has been shown to be involved in the transcriptional regulation of *flaA* and *lmaB* in *L. monocytogenes.*[30] Earlier studies in *L. monocytogenes* demonstrated a role for both GroEL and DnaK in the heat shock response following their induction after sublethal heat shock.[29,40,41] Studies in a variety of bacteria have demonstrated increased GroEL expression after exposure to low pH, osmolarity, ethanol, and bile salts, suggesting a role for this protein in the general stress as well as the heat stress response.[29,42–45]

3.2.1.1 Regulation of the Heat Stress Response

The Hrc/CIRCE system represents one of the most conserved heat shock repressor systems described to date, not only in Gram-positive but also in some Gram-negative bacteria.[46–48] The chromosomal organization of the six genes (encompassing the *dnaK* operon) at a single locus and the presence of the *hrcA* homologue along with a novel heat shock CIRCE (controlling inverted repeat of chaperone expression) element[49] in *L. monocytogenes* closely resembles the arrangement seen in other Gram-positive bacteria such as *Clostridium acetobutylicum,*[50] *Bacillus subtilis,*[51] and *Staphylococcus aureus,*[52] though slight disparities are observed. Interestingly, in *B. subtilis* the *hrcA* gene and the regulatory CIRCE sequence play a role in the regulation of expression of the *dnaK* locus in response to heat shock.[53] Given the significant homology between the *dnaK* operon in *L. monocytogenes* and *B. subtilis,*[35] it is not surprising that the listerial *dnaK* operon is regulated in a manner similar to its counterpart in *B. subtilis.*[54]

In *B. subtilis* it has been demonstrated that HrcA acts as a repressor of the heat shock response by binding to the CIRCE element in the *dnaK* operon.[55] In addition, Mogk et al.[56] established that the activity of the repressor (HrcA) is modulated by the GroE chaperonin system. The HrcA repressor is released from the ribosome in an inactive form; to become active, it must interact with the GroE chaperonin system. Once active, HrcA binds to its operators, resulting in repression of transcription of both operons (*groE* and *dnaK*). Upon heat stress, HrcA dissociates from its DNA-binding sites and is once again present in an inactive form, allowing de-repression of both operons. Immediately after heat shock (5–10 min), GroE proteins are involved in the refolding of proteins and preventing aggregation and, as a consequence (lack of available GroE proteins), prevent repression of the operators by HrcA. Under normal conditions, there are sufficient amounts of the GroE proteins available, allowing for interaction with HrcA and once again converting it to its active state, thus permitting repression of both *dnaK* and *groE* operons (Figure 3.2).[47,53]

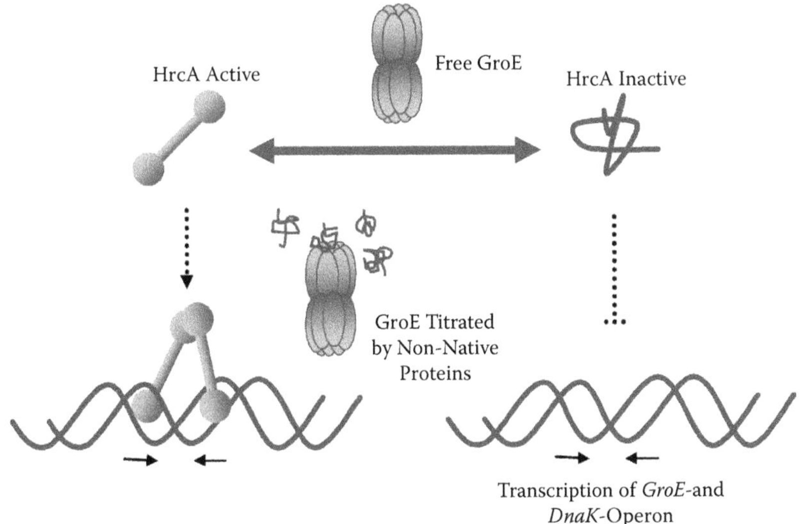

FIGURE 3.2 The Hrc–GroE–CIRCE regulation system. Within the cells, the HrcA repressor is present in an active and an inactive form, where the latter is unable to bind to its operator. The CIRCE element is symbolized by the two inverted arrows (→ ←). The equilibrium between these two forms on HrcA is influenced by the GroE chaperonin system. In the absence of heat stress, the GroE system converts most of the HrcA molecules into the active form. After a sudden temperature upshift, non-native proteins titrate the GroE chaperonins, leading to an increase in the amount of inactive HrcA repressor and resulting in enhanced transcription of the *groE* and *dnaK* operons.

3.2.1.2 Contribution of *gro* Genes to Virulence Potential

In order to survive the infectious process, pathogenic bacteria require an increased production of molecular chaperones. These molecular chaperones appear to be multifunctional, with a number of bacteria expressing specific molecular chaperones on their cell surface that function as adhesins. In addition, the secretion of these chaperones into the extracellular milieu may act as signaling virulence factors.[57–59] The classic chaperones GroESL and DnaK have been shown to be induced during *Staphylococcus aureus* and *Streptococcus pyogenes* infection as GroESL and DnaK specific antibodies and have been detected in the sera from infected patients.[60–62] In *Clostridium difficile* and *Legionella pneumophila,* GroESL has been shown to play a role in cell adherence and cell invasion, respectively.[63–65] In addition, GroEL of *Bartonella bacilliformis* has been shown to be located both on the cell surface and secreted into the extracellular milieu, where it plays a role in the mitogenic effect of *B. bacilliformis* on human vascular endothelial cells.[58] Earlier studies in *L. monocytogenes* by Hevin et al.[66] demonstrated that synthesis of *groEL* and *dnaK* decrease during the early stages of phagocytosis within mouse phagocytes. In addition, Hanawa et al.[41] showed that these heat shock proteins were not induced within the macrophage phagosome.

It has been suggested that the mechanism by which *L. monocytogenes* rapidly escapes from the phagosome prevents the expression of stress proteins during macrophage infection.[41] Conversely, studies by Gahan et al.[29] demonstrated an increase in the transcription of listerial *groESL* following internalization by J774 macrophage cells, indicating that the expression of heat shock/stress proteins may occur upon phagocytosis. In support of this, recent studies using microarray-based approaches showed that *groES* and *groEL,* as well as *dnaK, dnaJ, grpE*, and *hrcA,* are induced intracellularly.[67,68] Further, Hanawa et al.[69] showed that an *L. monocytogenes dnaK* deletion mutant displayed reduced macrophage phagocytosis due to inefficient binding, suggesting a possible role for this protein in adhesion.

3.2.2 Clp Proteins

Clp (caseinolytic protease) protein complexes play a critical role in energy-dependent proteolysis, a common mechanism in prokaryotic and eukaryotic cells for intracellular homeostasis and regulation, particularly under stress conditions.[70,71] The Clp complex is composed of a proteolytic subunit, ClpP, which associates with a Clp ATPase. Although designated a Clp protein to reflect its caseinolytic properties, ClpP itself is unrelated to the Clp ATPase family as it only has peptidase activity; however, when ClpP associates with members of the Clp ATPase family, it acts as a serine protease and prevents the accumulation of altered proteins that might be toxic for the bacteria under stress-inducing conditions.[72] Clp ATPases are ubiquitous among prokaryotes and eukaryotes and are members of the highly conserved Clp/Hsp100 family of proteases, whose function is to regulate ATP-dependent proteolysis and also play a role as molecular chaperones involved in protein folding and assembly.[73,74] Clp ATPases are composed of a catalytic and a regulatory component and are classified on the basis of the presence of either one or two ATP binding domains, on the length of the spacer region separating the two conserved nucleotide-binding regions, and on the occurrence of specific signature sequences.[75]

Analysis of the complete genome sequence of *L. monocytogenes* EGDe reveals several uncharacterized genes encoding proteins belonging to the Clp family.[76] To date, four *L. monocytogenes clp* genes have been characterized, and each displays a role in virulence: *clpB*,[77] *clpC*,[78,79] *clpE*,[80] and *clpP*.[81] ClpB has recently been characterized and no obvious role in stress tolerance was observed, although it is required for induced thermotolerance. In addition, ClpB function in *L. monocytogenes* is restricted to chaperone activity without interaction with ClpP as previously seen in *E. coli*[77,82]; while it has been annotated as being part of a three-gene operon (Listilist: http://genolist.pasteur.fr/ListiList/), it is driven from its own σ^A type promoter.[77] ClpC ATPase is a general stress protein encoded by *clpC*, which is the last gene in a four-gene operon, *ctsR–orf2–orf3–clpC*.[78] CtsR, which encodes a repressor of *clp* gene expression,[83] has been shown to play a role in piezotolerance, stress resistance, motility, and virulence in *L. monocytogenes*.[84]

Interestingly, *orf2* and *orf3* exhibit significant homology to *mcsA* and *mcsB* (51 and 70% similarity, respectively), which form part of the *clpC* operon in *B. subtilis*[85] (Listilist: http://genolist.pasteur.fr/ListiList/; Subtilist: http://genolist.pasteur.fr/SubtiList/). This *clpC* operon is governed by two promoters: a σ^A type promoter upstream of the *ctsR*-like gene, which is thermoregulated and essential to the entire operon in stress conditions, and a second immediately upstream of *clpC*, which is constitutively expressed at low levels and is not thermoregulated.[78] Unlike ClpC, ClpE ATPase encoded by *clpE* forms a monocistronic operon driven from a σ^A type promoter, and its expression is not stimulated by various stressors, including elevated temperature; however, interestingly, it is required for prolonged survival at high temperature.[80] ClpP seine protease, encoded by *clpP*, also forms a monocistronic operon driven from a σ^A type promoter and has been characterized as a general stress protein involved in growth at high temperatures.[81]

3.2.2.1 Regulation of *clp* Genes

Expression of the *clp* genes is regulated by the negative regulator CtsR in a variety of Gram-positive bacteria such as *Lactococcus lactis, Streptococcus mutans,* and *Bacillus subtilis*.[53,86–89] In contrast to the *L. monocytogenes* σ^B regulon, the CtsR-like regulon coding for the Class III heat shock gene repressor is composed of only eight members to date: the negative regulator of this regulon, CtsR, the ClpB, ClpC, and ClpE ATPases, as well as the ClpP serine protease.[77,78,80,81,83] The listerial CtsR regulon seems to differ from that of the closely related *B. subtilis* CtsR regulon, which appears to lack the clpB operon.[77] In short, the listerial CtsR repressor regulates the expression of seven genes located in four transcriptional units. The mechanism of CtsR regulation is not clearly understood, but it has been suggested that *L. monocytogenes* CtsR regulation acts by competing or interfering with RNA polymerase σ^A binding sites.[83]

A recent study by Karatzas et al.[84] demonstrated that a deletion in a single glycine residue in the glycine rich repeat that is essential for CtsR to adopt its proper conformation results in loss of repressor function of this regulator. Given the significant homology observed between the *clpC* operon in *L. monocytogenes* and its counterpart in *B. subtilis*, it seems reasonable to assume that CtsR regulation might function in a similar manner in both these Gram-positive organisms. It has been suggested that at 37°C, *B. subtilis* CtsR regulon repression is possibly due to the low steady-state level of CtsR present in the cell, which negatively autoregulates its own transcription.[83] In *L. monocytogenes, clpC* is constitutively expressed at low levels, allowing CtsR repression of the entire operon under nonstress conditions.[78] Studies in both *B. subtilis* and *L. monocytogenes* have shown that increased levels of ClpC following heat stress modulate CtsR activity, resulting in repression of the CtsR regulon.[80,86,87] Under heat stress conditions, *B. subtilis* decisively induces Clp proteins in order to ensure stress tolerance and degradation of heat-damaged proteins, and the resulting outcome of this induction is the inactivation and removal of the CtsR repressor.[90]

Recently, Kirstein et al.[85,91] demonstrated that the regulation of the CtsR regulon in *B. subtilis* involves all the genes of the *clpC* operon: *ctsR, mcsA, mcsB,* and *clpC*. McsA and McsB are both modulators of CtsR.[90,92] McsB, a putative tyrosine kinase, has been shown to repress the DNA-binding ability of CtsR and modify CtsR to target it for degradation by ClpPC. The tyrosine kinase activity of McsB requires activation by McsA, which results in the phosphorylation of both McsA and McsB and immediate and concurrent phosphorylation of CtsR. CtsR is degraded *in vivo* upon heat shock, and this heat-shock-induced degradation depends on the presence of McsA.[85] ClpC, on the other hand, inhibits the kinase activity of McsB. Therefore, the inhibition of the MscB kinase activity by ClpC supports a model where this interaction would serve as a sensor for heat shock or general stress. The model of regulation suggested for *L. monocytogenes* is that under nonstress conditions a constitutive level of ClpC is available, therefore preventing degradation of CtsR and allowing repression of the entire regulon. Conversely, under heat stress conditions, the *clpC* operon

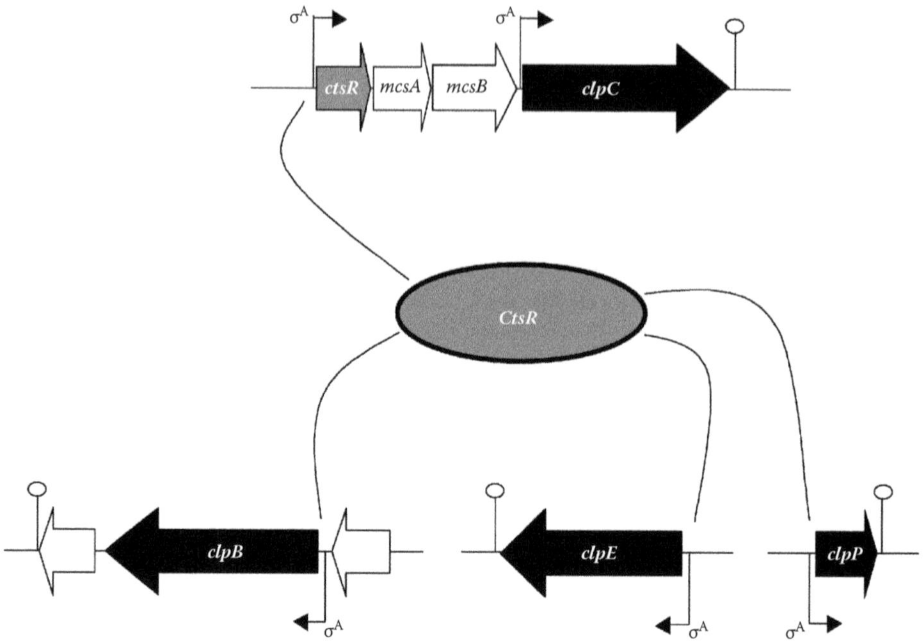

FIGURE 3.3 The CtsR regulon. This regulon consists of four transcriptional units: the tetracistronic *clpC* operon, the tricistronic *clpB,* and the two monocistronic *clpE* and *clpP* operons. All four operons are under the negative control of the CtsR repressor.

is induced, but ClpC is involved in the refolding of unfolded or misfolded proteins, therefore relieving the repression it affords on McsB. McsA phosphorylates McsB; this results in the rapid phosphorylation of CtsR and finally in the degradation of CtsR (Figure 3.3).[85,93]

A physiologically economic principle used by pathogens, including *L. monocytogenes,* is the efficient expression of virulence genes at the temperature of the infected host (i.e., 37°C) with a barely detectable expression level at lower temperatures. Expressional interplay between true virulence genes and general stress response networks has been observed in *L. monocytogenes.* PrfA, the thermoregulated global virulence gene regulator, has been shown to negatively regulate *clpC* expression at 37°C.[94] This is a highly sophisticated mechanism of regulation, which *L. monocytogenes* appears to use in conjunction with CtsR, allowing it to distinguish between the *in vitro* and *in vivo* environments.[95,96] The down-regulation of *clpC* at 37°C by PrfA may alleviate the inhibition of CtsR degradation, thereby permitting expression of this stress regulon *in vivo.* In addition, Chaturongakul and Boor[97] demonstrated σ^B-dependent regulation of ClpC under certain stress conditions in *L. monocytogenes,* and ClpE is overexpressed in the absence of ClpC, indicating that certain Clp ATPases of this pathogen are cross-regulated.[80] Wempkamp-Kamphuis et al.[98] identified ClpP as a putative σ^B-regulated gene in addition to being CtsR regulated. Further, ClpC and ClpP have been shown to be involved in the negative regulation of virulence genes in *L. monocytogenes.*[99] Therefore, it is obvious that *L. monocytogenes* thermoregulation is a very complicated network involving cross-talk between many different regulators.

3.2.2.2 Role of Clp Proteins in Virulence Potential

Clp proteins have been shown to play a role in the pathogenesis of several Gram-positive pathogens, including *Salmonella enterica* serovar Typhimurium,[100] *Staphylococcus aureus,*[101] and *Streptococcus pneumoniae.*[102–104] In *L. monocytogenes* several Clp proteins have also been implicated in its virulence potential. ClpC ATPase has been shown to be required for its survival inside host cells by promoting early escape from the phagosome of macrophages.[78] In addition, Nair et al.[105] pinpointed a role for ClpC in adhesion to and invasion of hepatocytes. In contrast to ClpC, ClpE ATPase, although required for virulence and growth in the liver of *L. monocytogenes,* is not required for invasion of hepatocytes.[80,105] Both ClpP serine protease and ClpB ATPase also play a role in *L. monocytogenes* pathogenesis; ClpP modulates the expression of listeriolysin O, a major factor required to escape from the phagosomes of macrophages[81] and a clpB mutant displaying a 100-fold decrease in virulence in the murine model of infection.[77] Nair et al.[83] showed that the absence of CtsR does not alter the virulence phenotype of *L. monocytogenes,* but overexpression of this protein does confer a virulence defect as expected. More recent studies demonstrated that a single codon deletion in a region encoding a highly conserved glycine repeat in CtsR displayed significantly attenuated virulence when compared to the wild type.[84] In addition, *ctsR, clpB, clpC, clpE,* and *clpP* have all been shown to be induced intracellularly using microarray-based approaches.[67,68] Clearly, many members of the Clp family, in addition to their CtsR repressor, aid in the pathogenesis of *L. monocytogenes.*

3.2.3 HtrA: High-Temperature Requirement

The conserved family of high temperature requirement (*htr*) genes has been identified in several bacteria. Of these, *htrA* is the best characterized, encoding a serine protease (HtrA).[106–108] These proteins are characterized by an amino-terminal domain that determines subcellular localization, a catalytic domain, and a carboxy terminal PDZ (PDZ: PSD95/DLG/ZO-1 proteins[109]) that functions to assemble protein monomers into the functional hexamic complex.[110] Initially characterized in *E. coli,* HtrA functions as a protease in the degradation of misfolded or aggregated proteins produced as a consequence of adverse environmental conditions such as elevated temperatures.[107] Moreover, HtrA can also function as a molecular chaperone in the refolding of misfolded or unfolded proteins and prevent aggregation. The switch between these activities is temperature dependent, with the chaperone activity predominating at lower temperatures and the protease activity predominating at higher temperatures.[111] Orthologs of HtrA have been found in many Gram-positive bacteria, and several have been implicated in bacterial stress response and virulence.[112–114] A significant number of bacterial genomes encode more than one HtrA-like serine protease, such as *B. subtilis,* which encodes three HtrA homologues: *htrA*, *htrB* (*yvtB*), and *yycK*.[115]

A single HtrA homologue has been identified in *L. monocytogenes* and the immediate genomic organization of the region encoding the HtrA-like serine protease in *L. monocytogenes* corresponds to the *B. subtilis* six-gene operon (*yycF–yycK*).[31] Many Gram-positive orthologs of HtrA are predicted to be peripheral membrane proteins anchored to the membrane by a single transmembrane domain located near their N-terminal.[108,116] Interestingly, the *B. subtilis* HtrA can exist both in membrane-associated and extracellular forms, suggesting dual localization and possibly a dual function for HtrA allowing it to function in protein quality control at this site.[117] HtrA in *L. monocytogenes* has been shown to encode two potential start codons. Like *B. subtilis* it is predicted that listerial HtrA has dual localization, both anchored and membrane bound.[31,117] Virulence studies demonstrated a role for the listerial HtrA signal sequence in pathogenesis, suggesting that stresses associated with the *in vivo* environment may signal HtrA secretion.[31] HtrA has been shown to play a role in stress resistance in *L. monocytogenes*—that is, acid, antibiotics, heat, and oxidative and osmotic stress (see section 3.4.2), as well as in biofilm formation.[31,118–120] Although involved in growth at elevated temperatures, listerial *htrA* is not transcriptionally induced upon heat shock, a phenomenon also observed with *yycK* of *B. subtilis*.[31,108] This observation was further supported by an *L. monocytogenes* heat shock microarray.[121]

3.2.3.1 Regulation of HtrA

Regulation of HtrA has been the focus of study in many pathogenic bacteria. In *E. coli* and *Y. enterocolitica, htrA* regulation has been shown to be mediated by both CpxRA and RpoE, respectively.[122,123]

In *S. pneumoniae*, the CiaRH two-component regulatory system has been shown to positively regulate *htrA* and this regulation is directly due to the presence of a CiaR binding site upstream of *htrA*.[124,125] In addition, the CssRS two-component regulatory system has recently been shown to regulate HtrA and HtrB in *B. subtilis*.[126] To date, very little is known about the regulation of HtrA in *L. monocytogenes*. However, recent studies in our laboratory revealed a role for the *L. monocytogenes* two-component system, *lisRK*, in positively regulating *htrA*.[31] The two-component regulatory system LisRK consists of a sensor histidine kinase, *lisK*, anchored to the cell membrane and a cytoplasmic response regulator, *lisR*, involved in modulating a number of stress responses—namely, acid, osmotic, antibiotic, and heat stress, as well as contributing to listerial virulence.[118,127,128] It has been suggested that LisRK may sense perturbations in the external environment and in turn transcriptionally activate *htrA* either directly or indirectly through the regulation of another system. In addition, the presence of a putative σ^B-dependent promoter-binding site upstream of the putative HtrA signal sequence suggests that, under certain circumstances, secretion of HtrA may occur under the transcriptional control of σ^B.

3.2.3.2 Contribution of HtrA to Virulence Potential

HtrA has been shown to play a role in pathogenicity in a number of Gram-positive and Gram-negative bacteria, such as *Salmonella enterica* serovar typhimurium,[129] *Klebsiella pneumoniae*,[130] *Legionella pneumophila*,[131] *Streptococcus pyogenes*,[113] *Streptococcus pneumoniae*,[112] and *Streptococcus mutans*.[132] HtrA has also been shown to play a role in *L. monocytogenes* virulence. A study carried out in our laboratory revealed a role for listerial HtrA in pathogenesis using the murine model of infection.[31] A separate study by Wilson et al.[119] also demonstrated that HtrA is essential for full virulence of *L. monocytogenes*. In addition, mutation of the putative signal sequence affected virulence, suggesting that stresses associated with the *in vivo* environment may signal HtrA secretion.[31]

3.3 ACID STRESS RESPONSE

Acid stress can be defined as the combined biological effect of low pH and weak (organic) acids present in the environment. Weak acids in their uncharged, protonated forms can diffuse freely across the cell membrane, where they dissociate due to a higher cytoplasmic pH (pH_i) than that of the external medium (pH_o), releasing a proton and leading to acidification of the cytoplasm. The lower the external pH (pH_o) is, the more undissociated weak acid is available to cross the membrane and affect cytoplasmic pH (pH_i). Further, intracellular accumulation of weak acids is thought to have harmful effects on the cell beyond simply acidifying the cytoplasm.[133] Acid is likely to be one of the most frequently encountered hostile conditions (i.e., in low pH foods, during gastric transit, following exposure to fatty acids in the intestine, in the phagosome of macrophages during systemic infection, and even upon exiting the host, due to fluctuations in environmental pH).[134] Like most food-borne pathogens, *L. monocytogenes* is a neutrophile (optimum pH 6 or 7)[27,135]; therefore, keeping the cytoplasmic pH (pH_i) at a value close to neutrality, despite fluctuations in external pH is imperative to its survival and a prerequisite for infection.

Thus, *L. monocytogenes* has evolved a number of cellular mechanisms for the adaptation to and survival in acidic environments in order to preserve pH homeostasis. The adaptive response, termed the acid tolerance response (ATR), involves the acquisition of acid tolerance following brief exposure to mild acidic conditions resulting in an altered pattern of protein synthesis. This creates a situation in which the organism can produce acid shock proteins (ASPs) when subjected to a more challenging or even normally lethal pH.[136–139] The ATR in *L. monocytogenes* has been the focus of intense study over the past decade with many studies correlating the ability to mount an ATR with the pathogenesis of the organism.[140–144] In addition, *L. monocytogenes* not only survives acid stress but also requires a drop in pH in order to activate hemolysin, a potent virulence factor that permits its escape from the macrophage phagosome.[19] In addition to the ATR, *L. monocytogenes* has a

TABLE 3.2
Overview of Acid Stress Loci and Their Roles in Pathogenesis

Category	Genetic Loci	Role in Virulence	Ref.
Acid stress	*gadD1T1*	Yes	150, 151, 153, 154
	gadD2T2	n/t[a]	150, 151, 153
	gadD3	n/t[a]	153
	arcB	n/t[a]	154
	arcD	n/t[a]	154
	arcC	Yes	154
	lmo0038	n/t[a]	154
	lmo0040	n/t[a]	154
	lmo0041	n/t[a]	154
	lmo0042	n/t[a]	154
	arcA	n/t[a]	154
	argG	n/t[a]	154
	argR	Yes	154
	atpI	n/t[a]	181
	atpB	n/t[a]	181
	atpE	n/t[a]	181
	atpF	n/t[a]	181
	atpH	n/t[a]	181
	atpA	n/t[a]	181
	atpG	n/t[a]	180
	atpD	Yes	181, 203
	atpC	n/t[a]	181

[a] Not tested.

number of other acid resistance mechanisms such as glutamate decarboxylase, arginine deiminase, and F_0F_1-ATPase systems (Table 3.2).

3.3.1 Glutamate Decarboxylase

The glutamate decarboxylase (GAD) system has been associated with acid resistance in many bacteria. Not surprisingly, it has been found primarily in bacterial species that need to transit the stomach en route to invading or colonizing the gut epithelium.[145–148] In the GAD system, the cytoplasmic glutamate decarboxylase enzyme irreversibly decarboxylates a molecule of extracellularly sourced glutamate, producing γ-aminobutyrate (GABA). During this reaction, an intracellular proton is consumed, thereby alleviating acidification of the cytoplasm and contributing to pH homeostasis. The intracellular molecule of GABA produced via the decarboxylation reaction is subsequently exchanged at the cell membrane for a molecule of glutamate by a glutamate:GABA antiporter (Figure 3.4). In addition to reducing the proton concentration within the cell, GABA is less acidic than glutamate, which contributes to an alkanization of the environment.[149] In order for this decarboxylation reaction to take place, free glutamate must be present in the environment. Interestingly, glutamate is readily available in foods, and its presence in food ingredients and food flavor enhancers, in addition to its use as a food preservation method to adjust the acidity of the food, contributes significantly to the survival of *L. monocytogenes* strains with GAD activity, even at low levels.[150] Thus, the significance of the GAD system is linked to the presence of glutamate in foods, whether it is present naturally or as an additive.

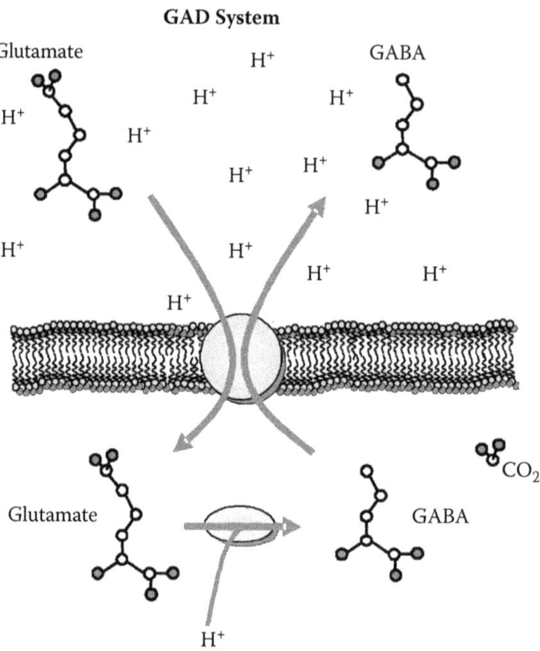

FIGURE 3.4 Schematic representation of the glutamate decarboxylase (GAD) system. Under low pH conditions, glutamate is decarboxylased to γ-aminobutyrate (GABA) and CO_2, consuming a proton in the reaction. The GABA produced is transported out of the cell in an energy-independent manner via a glutamate:GABA antiporter in exchange for another molecule of glutamate.

Cotter et al.[151] showed that expression of GAD by *L. monocytogenes* is an absolute requirement for survival in the acidic environment of the stomach (gastric fluid), further supported by the observation that expression of GAD activity in *L. monocytogenes* strains correlates directly with acid tolerance. Therefore, GAD represents the key mechanism for maintenance of pH homeostasis in *L. monocytogenes* in this environment (gastric fluid). In addition, the GAD system (*gadT1*) has been shown to play a role in survival in bile, possibly by combating the lowered cytoplasmic pH arising from the intracellular dissociation of bile salts.[152]

The GAD system classically involves two proteins: a cytoplasmic glutamate decarboxylase and a glutamate:GABA antiporter at the cell membrane. To date, three glutamate decarboxylase homologs (*gadD1:lmo0447, gadD2:lmo2363,* and *gadD3:lmo2434*) and two antiporters (*gadT1:lmo0448* and *gadT2:lmo2362*), located in three distinct loci, have been identified in *L. monocytogenes.*[141,151,153] The *gadD1T1* genes form part of a five-gene operon[154] and are involved in growth in mildly acidic environments, while the bicistronic *gadD2T2* operon[151] has been shown to play a role in survival in severe acid stress. The *gadD3* gene has recently been mutated in our laboratory but its role in acid stress has not yet been investigated.[155]

Recent work in our laboratory has shown that the *gadD1T1* five-gene operon is absent from serotype 4b strains, which are more frequently associated with disease than with foods. A mutant lacking the entire five-gene islet is less able to survive in a model food system (frankfurters) and is impaired in initial survival within the murine gastrointestinal tract 24 h after oral infection.[154] However, the absence of the Gad islet in serotype 4b *L. monocytogenes* indicates that its presence is not required to cause epidemic listeriosis. While this Gad islet has been shown to benefit *L. monocytogenes* strains grown in mildly acidic conditions, whether its absence from a number of strains may under certain circumstances enhance the pathogenic potenital of the organism remains to be determined.[150,151,153,156] In addition, the percentage of strains with high GAD activity at pH 4, which is approximately the gastric pH after feeding[157] is greater among human carriage isolates than among food and environmental isolates.[158] The generation of metabolic energy (ATP) has been suggested as an alternative function for the GAD system. The ATP generated during the decarboxylation reaction may contribute to pH homeostasis mediated by the F_0F_1-ATPase.[159]

3.3.1.1 Regulation of GAD

Although the GAD system has been identified in a number of bacteria, most of the research into its regulation has been carried out in *E. coli*.[148,160,161] To date very little is known about the regulation of the GAD system in *L. monocytogenes*. However, the *gad* genes are subject to acid-induced transcriptional regulation,[141,151] suggesting that regulators of the acid stress response may in turn regulate the GAD system. The complex transcriptional profiling of the *gad* genes in *L. monocytogenes* revealed the existence of multiple promoters and putative transcriptional termination sites, indicating that multiple regulatory factors are involved in the regulation of the GAD system.[141]

The role of the alternative sigma factor, σ^B, in the stress responses of Gram-positive bacteria has received considerable attention over recent years. In *L. monocytogenes*, σ^B encoded by *sigB* has been shown to play a role in acid, oxidative, and osmotic stress resistance, as well as in low temperature and carbon starvation stress response. In addition, σ^B has been shown to be involved in the stationary phase of growth.[162–166] The association of σ^B with core RNA polymerase (RNAp) provides a mechanism by which this holoenzyme targets specific recognition sequences and controls the transcription of specialized regulons that are activated in response to cellular signals.[167,168]

Earlier studies in *L. monocytogenes* LO28 demonstrated an absolute role for GAD activity in survival in gastric juice (pH 2.5), and a role for σ^B in survival in gastric juice (pH 2.5) in *L. monocytogenes* 10403S was also observed, suggesting that σ^B could possibly regulate the GAD system in *L. monocytogenes*. Interestingly, in *L. monocytogenes* 10403S, GAD activity in both the σ^B mutant and wild type, as well as their net proton movement across the cell membrane, did not differ, indicating that GAD activity is not σ^B dependent during stationary phase in gastric juice, at least in this strain.[151,164] However, recent studies in *L. monocytogenes* strains LO28 and EGDe have shown that the *gad* genes (*gadD1T1*, *gadD2T2,* and *gadD3*) are regulated by σ^B under acidic conditions.[98,154] Moreover, a microarray- and RT-PCR-based approach revealed that stationary phase and NaCl afforded σ^B-dependent regulation of *gadD1* and *gadD3* in *L. monocytogenes* 10403S.[169,170] Thus, *L. monocytogenes* σ^B regulation of GAD activity appears to depend upon the strain, in addition to phase of growth and pH of the medium.

3.3.1.2 Contribution of the GAD System to Virulence Potential

Although the glutamate decarboxylase system has been characterized in a number of bacteria,[145–148] very little is known about its contribution to pathogenesis. In *E. coli*, H-NS, a DNA binding protein involved in the hierarchical regulation of *gad* expression,[171] has been shown to be required to survive murine passage.[172] In *L. monocytogenes*, σ^B has been shown to regulate *gad* genes during acid stress.[98,154] Virulence studies showed σ^B to have only a very minor effect on the spread of *L. monocytogenes* to mouse liver or spleen following intragastric or intraperitioneal infection.[166] However, a more recent study, using the guinea pig model of infection, demonstrated a critical role for σ^B during the gastrointestinal stage of listeriosis. A significant reduction in bacterial numbers of a σ^B mutant in comparison to the wild type was observed in the liver, spleen, lymph nodes, and small intestine after intragastric inoculation. In addition, σ^B was shown to play a significant role in invasion of human host cells. Conversely, σ^B did not contribute to systemic spread of *L. monocytogenes* using this model of infection.[173] Given the absolute requirement for the GAD system in the gastrointestinal phase of infection,[151] it is likely that this system, like σ^B, plays a critical role in *L. monocytogenes* pathogenesis. A recent study by Joseph et al.,[68] using a microarray-based approach, identified *gadD3* as being induced intracellularly within Caco-2 cells. Furthermore, an observed role for the five-gene

Gad islet in murine gastrointestinal survival has been demonstrated, showing the importance of this GAD system in listerial pathogenesis.[154]

3.3.2 Arginine Deiminase

The arginine deiminase (ADI) system has been characterized in a number of bacteria[174–177] and has frequently been implicated in bacterial resistance to acidic environments.[178–180] Three enzymes—arginine deiminase (ADI), catabolic ornithine carbamoyltransferase (cOTCase), and carbamate kinase (CK)—constitute the deiminase pathway that catalyzes the conversion of arginine into ornithine, ammonia, and carbon dioxide. The ornithine produced intracellularly as a result of arginine catabolism is transported out of the cell in exchange for a molecule of arginine in an energy-independent manner by a membrane-bound antiporter. For each mole of arginine catabolized via the ADI pathway, two moles of ammonia (NH_3) and one mole of ATP are produced. This ammonia produced as a result of arginine catabolism can combine with intracellular cytoplasmic protons to produce ammonium ions (NH_4^+), thereby increasing intracellular pH and maintaining pH homeostasis (Figure 3.5). The provision of ATP can be used for microbial growth under a variety of environmental conditions or, alternatively, to extrude protons via the F_0F_1ATPase,[159] which has been shown to play a role in pH homeostasis in *L. monocytogenes*.[181] The ADI system also supplies carbamoyl phosphate, which is required for the biosynthesis of citrulline and pyrimidines.[182]

Ryan[154] recently characterized the ADI system in *L. monocytogenes* and demonstrated a role for this system in growth and survival in acidic conditions. Although the ADI system of *L. monocytogenes* is involved in growth and survival at low pH, it is not as effective in combating the deleterious effects of acidic conditions as the GAD system. The listerial ADI system appears to be more complex than its counterparts found in other genera. Unusually, extracellular arginine is not the primary source of substrate for the reaction, but intracellularly synthesized arginine appears vital. Therefore, it is not surprising that *Listeria* contains genes that are required for both a fully functional deiminase and an arginine biosynthetic pathway (Figure 3.5).

The genetic organization of the listerial arginine deiminase (*arc*) genes is unique in that the gene arrangement is *arcB–arcD–arcC–arcA*. In addition, a number of ancillary genes of unknown function flank the *arc* genes at this locus, a phenomenon also seen in other genera.[177,179,183] In

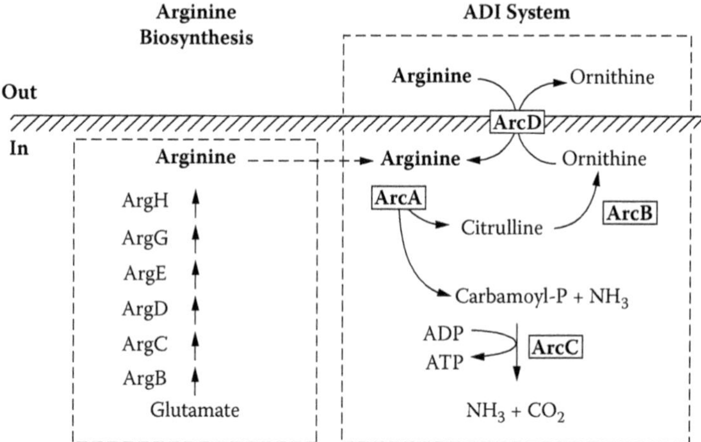

FIGURE 3.5 Schematic representation of the arginine deiminase pathway. This pathway involves three enzymes: arginine deiminase, catabolic ornithine carbamoyltransferase, and carbamate kinase, encoded by *arcA*, *arcB*, and *arcC*, respectively. Arginine is transported into the cell via an arginine:ornithine antiporter encoded by *arcD*. Two moles of ammonia are generated for every mole of arginine consumed by the pathway and one mole of ATP is also produced. In *Listeria*, intracellular arginine produced from glutamate also seems to be an important source (see text for details).

Listeria, the catabolic ornithine carbamoyltransferase (cOTCase) encoded by *arcB* (*lmo0036*), the arginine:ornithine antiporter encoded by *arcD* (*lmo0037*), and the carbamate kinase encoded by *arcC* (*lmo0039*) are part of an operon that also contains an ancillary gene (*lmo0038*) encoding a hypothetical protein of the peptidylarginine deiminase family. The arginine deiminase gene encoded by *arcA* (*lmo0043*) is physically separated from the other genes involved in the ADI pathway by an uncharacterized three-gene operon (*lmo0040–lmo0042*). Further, a gene encoding a regulator of the pathway, *argR* (*lmo1367*), was also identified at some distance from the other genes. Cotranscription of *lmo0042* and *arcA* has been shown, indicating that these uncharacterized genes may play a role in the *Listeria* ADI pathway. In support of this, transcriptional and physiological analysis revealed a role for the listerial *arcA, B C, D,* and accessory genes (*lmo0038, lmo0040, lmo0041,* and *lmo0042*), as well as the regulator *argR* in the acid and anaerobic stress response. Interestingly, *argG*, a gene involved in arginine biosynthesis, was also transcriptionally up-regulated in these conditions—an apparent paradox where genes involved in the catabolism and anabolism of arginine are expressed concurrently.

In support of this, an observed increase in survival of an arginine auxotroph (*argG* mutant) upon supplementation of the low pH medium with arginine suggested that arginine biosynthetic genes might function in the provision of the amino acid as a substrate for the reaction. However, loss of the arginine–ornithine antiporter (*arcD*) resulted in decreased listerial acid resistance, indicating that arginine uptake can also be used to supply the system; the simultaneous mutation of this antiporter and the arginine biosynthetic gene resulted in a further increase in acid sensitivity. Therefore, the theory proposed by Ryan[154] is that both intracellular arginine derived from *de novo* amino acid synthesis and extracellular arginine imported via ArcD or other transporters[184] are required for optimal operation of the ADI system under these conditions, but the intracellular synthesis of arginine is the predominant source of substrate. Significantly, *argG* (anabolic pathway) is repressed in the presence of arginine while the *arc* genes (catabolic pathway) are induced, indicating that simultaneous expression of the two pathways only occurs when intracellular arginine stores are required for provision of sufficient substrate for maximal functioning of the ADI system. Thus far, this has been shown to be the case in conditions of low pH or anaerobicity.

3.3.2.1 Regulation of the ADI System

A recent comprehensive study of the listerial ADI pathway revealed that this system is highly regulated and mediated by a complex interplay among ArgR, Lmo0041, PrfA, and σ^B.[154] ArgR (*lmo1367*) has been identified as dual regulator—activating expression of the ADI pathway while simultaneously repressing expression of the arginine biosynthesis pathway. In an *argR* negative background, reduced expression of the *arc* and accessory genes and increased expression of the *argG* gene were observed. It has been proposed that ArgR mediates its effects as a result of conformational change in the ArgR molecule[185] and not as a result of increased levels of ArgR. This study also revealed a repressor of the ADI system, *lmo0041*, showing significant homology to the RpiR family of transcriptional regulators. Interestingly, a double mutant lacking both the activator (*argR*) and repressor (*lmo0041*) of the ADI pathway demonstrated increased expression of the *arc* genes, although not to the same degree as that generated by a mutation in *lmo0041* alone. This suggests that the repressive function of *lmo0041* takes precedence over the activation of the *arc* genes mediated by *argR*.

A similar situation has been observed in *L. lactis* with two ArgR-type regulators identified: AhrC appears to be involved in activation and ArgR involved in repression of ADI pathway. Like the situation in *Listeria*, the function of AhrC (activator) seems to be overruled by the removal of ArgR (repressor).[186] In addition to ArgR, PrfA and σ^B have also been shown to positively regulate both the ADI and arginine biosynthetic pathways with transcriptional analysis showing reduced transcript of the catabolic *arcA* and *arcD* genes, apart from the anabolic *argG* gene in PrfA and σ^B negative strains. Moreover, loss of σ^B resulted in significant decrease in expression of *argR*, a positive regulator of the ADI pathway, while loss of PrfA showed only a slight reduction in the transcription of this gene relative to the wild type. It has been suggested that regulation of the ADI system could be mediated by a cascade effect, in which induction of PrfA and σ^B would transcriptionally activate *argR*, which in turn induces expression of the genes of the ADI pathway.[154] Regulation of arginine metabolism in *Listeria* is a very complex network, involving at least four regulators: ArgR, Lmo0041, PrfA, and σ^B. It is plausible that *Listeria* exhibits a hierarchical regulatory system whereby σ^B and PrfA control the expression of ArgR, and possibly Lmo0041, as well as other yet unidentified regulators. This results in a cascade effect on the transcription of the arginine metabolic genes.

3.3.2.2 Role of the ADI System in Virulence Potential

The ADI pathway has been implicated in the pathogenesis of *Streptococcus pyogenes*[178,187] and *Streptococcus suis*.[188] Recent studies in *Listeria* revealed a definite role for the ADI pathway in pathogenesis and these represent the first documentation of the contribution of an ADI system to bacterial virulence *in vivo*.[154] Both ArgR and ArcC have been demonstrated to play a significant role in murine systemic infection, with a 10-fold decrease in survival observed for both *argR* and *arcC* null mutants. Further, ArgR, but not ArcC, has been shown to play a significant role in intracellular growth within the macrophage phagosome, but neither ArgR nor ArcC is required for host cell invasion. It has been proposed that *Listeria* may import arginine that is present at low concentrations in several animal tissues and cell lines[189] to reduce its concentration in the host cell.

A precursor of nitric oxide (NO) is arginine. Nitric oxide is a reactive nitrogen species, which plays an important role in host defense against intracellular pathogens including *L. monocytogenes*.[190–192] Therefore, reduced arginine in the host cell leads to a reduced capacity to generate NO, and hence a diminished ability to mount an immune response to *L. monocytogenes* infection. Interestingly, a recent study by Joseph et al.[68] demonstrated that the *L. monocytogenes arg* genes of the arginine anabolic pathway were significantly up-regulated in Caco-2 cells. Furthermore, while the eight genes comprising the ADI pathway (*lmo0036–lmo0043*) are arranged in an identical manner in each of the sequenced *L. monocytogenes* genomes (EGDe, 1/2aF6845, 4bF2365, and 4bH7858), a six-gene cluster (*lmo0036–lmo0041*) is absent from the sequenced nonpathogenic strain *L. innocua* Clip11262 and is specifically absent from *L. monocytogenes* strains of lineage III, which appear to be exclusively associated with animal morbidity.[193]

3.3.3 F_0F_1ATPase

The F_0F_1ATPase is a multisubunit enzyme complex that couples ATP synthesis/hydrolysis with a transmembrane proton translocation. Functionally, the enzyme is organized into two distinct, but physically linked, domains that are easily and reversibly separated into two portions, termed F_0 and F_1. F_0, the integral membrane bound portion, incorporates a, b, and c subunits, which function as a membranous channel for proton translocation; F_1, the ATPase portion which is cytoplasmic bound, consists of α, β, γ, δ, and ε subunits, which may synthesize or hydrolyse ATP.[181,194,195] This enzyme complex provides a well-conserved mechanism by which bacteria can maintain intracellular pH homeostasis. In actively respiring bacteria, such as *E. coli* and *B. subtilis*, the primary role of the F_0F_1ATPase is to synthesize ATP aerobically, as a result of protons passing into the cell, and generate a proton motive force (PMF) anaerobically via the expulsion of protons. When the PMF drops below the thermodynamic threshold of ATP synthesis, the reaction is reversed and the enzyme operates as an H^+-pumping ATPase.

The generation of the PMF allows the F_0F_1ATPase to increase the intracellular pH in situations where it becomes acidified.[181] However, in those bacteria that lack a respiratory chain, such as *Enterococcus hirae*, the role of the F_0F_1ATPase is to create a proton gradient driven by ATP hydrolysis that functions in the extrusion of hydrogen (H^+) ions and consequently the establishment of pH homeostasis.[196,197] To date, very little work has been carried out on the F_0F_1ATPase of

L. monocytogenes. Cotter et al.[181] have characterized the F_0F_1ATPase operon (nine genes) located in a 7-kb region in *L. monocytogenes* and suggested that this organism lacks a respiratory chain but may have an alternative method of ATP synthesis, using the decarboxylases of amino acids. The F_0 subunit consists of *atpI* (unknown), *atpB* (a subunit), *atpE* (c subunit), *atpF* (b subunit), and *atpH* (δ subunit); the F_1 subunit comprises *atpA* (α subunit), *atpG* (γ subunit), *atpD* (β subunit), and *atpC* (ε subunit). Interestingly, *atpG* is expressed threefold less than *atpA* or *atpD*, which is thought to be at least partially due to an alternative start codon, TTG, in combination with a poor ribosomal binding site that has also been identified in other bacteria.[181]

A role for the listerial F_0F_1ATPase has been demonstrated in the acid-tolerance response; however, the listerial acid-tolerance response is not solely dependent on the activity of this complex. This is supported by the observation that at least 17 proteins are induced as a result of mild acid treatment,[139] which includes *atpF* (b subunit).[44] Attempts to create a deletion mutation in *atpA* proved unsuccessful; however, a strategy of insertional mutagenesis aimed at reducing rather than eliminating expression of the downstream genes was achieved. This mutant exhibited enhanced resistance to neomycin and a reduced growth rate, and retained some ATPase activity, as expected. Interestingly, it was capable of eliciting an ATR. However, this could be due to the fact that its relative acid sensitivity was difficult to assess as a consequence of its slow growth.[181]

3.3.3.1 Regulation of ATPase

A common regulatory mechanism found in bacterial ATP synthases is "ADP inhibition." When ADP is bound at the high-affinity catalytic site, ATP synthase is inactivated, allowing ATP hydrolysis. During membrane energization (PMF activation), ADP is released, therefore not altering the catalytic site of ATP synthase allowing the production of ATP. The affinity of ATP synthase for phosphate dramatically increases in the presence of PMF, which is considered to be necessary to prevent the competitive inhibition by ATP during its synthesis. In addition, this increase in affinity for phosphate is thought to diminish ADP inhibition in the presence of a sufficiently high PMF. When the PMF drops below the thermodynamic threshold of ATP synthesis, the reaction is reversed and the enzyme operates as an H^+-pumping ATPase.[198]

Another mechanism of ATP synthase regulation is associated with subunit ε. Inhibition by subunit ε is clearly distinct from the ADP inhibition.[199–201] Feniouk and Junge[198] proposed that regulation results from large conformational changes of the ε α-helical C-terminal domain in response to membrane energization, change in ATP/ADP ratio, or addition of inhibitors. Its conformation can change from a contracted to extended form, and this is determined by the γ subunit. It is when it is in its extended form that ATP hydrolysis is inhibited.[198,202] In *L. monocytogenes*, subunits ε and γ are encoded by *atpC* and *atpG*,[181] but their exact role in the regulation of the listerial F_0F_1ATPase remains to be elucidated. Interestingly, *atpC* (ε subunit) and *atpD* (β subunit) are preceded by PrfA consensus binding sites, the positive regulator of virulence gene expression in *L. monocytogenes*; therefore, PrfA could be contributing to F_0F_1ATPase regulation *in vivo*.[134]

3.3.3.2 Contribution of ATPase to Virulence Potential

To date, the exact role of F_0F_1ATPase in bacterial pathogenesis remains to be determined. The F_0F_{1}-ATPase plays a crucial role in the *L. monocytogenes* acid tolerance response; therefore, it is thought that it would contribute significantly to *in vivo* survival of this organism. Interestingly, a recent study carried out in our laboratory revealed the induction of *atpD*, the β subunit (F_1) during murine gastrointestinal infection.[203] This finding, coupled with the presence of a consensus PrfA promoter binding site,[134] suggests that this F_0F_1ATPase gene may play a significant role *in vivo*. It is interesting to note that comparative genomic analysis between *L. welshimeri* (nonpathogenic) and *L. monocytogenes* revealed the loss of all genes required for "fitness" and virulence of the organism including the F_0F_1ATPase,[204] indicating the important role of this system in *L. monocytogenes* pathogenesis.

3.4 OSMOTIC STRESS RESPONSE

Osmotic stress is the increase or decrease in the osmotic strength of the environment of an organism.[205,206] This occurs as a result of desiccation or the presence of high amounts of osmotically active compounds such as salt or sugars in the environment, thus lowering its water activity.[27] Bacterial maintenance of intracellular osmotic pressure is critical to survival in osmotic stress conditions. Because the bacterial cytoplasmic membrane is permeable to water but not to most other metabolites, hyper- or hypo-osmotic shock causes an instantaneous efflux or influx of water, which is accompanied by a concomitant decrease or increase in intracellular volume, respectively; this results in alteration of the cytoplasmic osmotic pressure. In general, internal osmotic pressure is higher than that of the surrounding medium, generating turgor, which is the pressure exerted outward on the cell wall. It is this turgor pressure that is the driving force for cell extension, growth, and division.[206] Therefore, bacteria must be able to maintain turgor despite variations in the osmolarity of the surrounding medium. A universal response to loss of turgor following hyperosmotic shock is the cytoplasmic accumulation of compatible solutes/osmolytes.[206–208] *L. monocytogenes* encounters elevated osmolarity at a number of stages during its infectious cycle—in the food processing industry as a preservation method and in the gastrointestinal tract of its host; therefore, its ability to sense and appropriately respond to conditions of elevated osmolarity remains crucial to its survival and initiation of infection.[209] Not surprisingly, *Listeria* utilizes compatible solutes for osmoadaptation, as well as a number of other noncompatible solute associated systems (Table 3.3; Figure 3.6).

3.4.1 Compatible Solutes

The accumulation of intracellular compatible solutes offers osmo-, cryo-, and barotolerance in a number of bacteria.[210–213] Compatible solute osmoadaptation is a biphasic response in which elevated levels of K^+ (and its counter ion, glutamate) represent the primary response, followed by a dramatic increase in the cytoplasmic concentrations of a number of compatible solutes.[214] These compounds are small organic molecules that do not alter cellular functions and have a number of common features: they are soluble to high concentrations and can be accumulated to very high levels in the cytoplasm of osmotically stressed cells; they are usually neutral or zwitterionic molecules; and they are unable to cross the cell membrane rapidly without the aid of transport systems.[27,206]

A number of compatible solutes have been identified in *L. monocytogenes,* such as glycine betaine (trimethylammonium compound), proline, proline betaine, acetylcarnithine, carnitine, γ-butyrobetaine, and 3-dimethylsulphoniopropionate, that function as osmoprotectants, with glycine

TABLE 3.3
Overview of Osmotic Stress Loci and Their Roles in Pathogenesis

Category	Genetic Locus	Role in Virulence	Ref.
Osmotic stress	*betL*	No	226
	gbu	No	227
	opuC	Yes	68, 227, 234, 236
	relA	Yes	265
	ctc	n/t[a]	254, 255, 258, 259
	htrA	Yes	31, 119
	kdpE	Yes	261
	lisRK	Yes	127, 261
	proB	No	262, 263
	btlA	No	264

[a] Not tested.

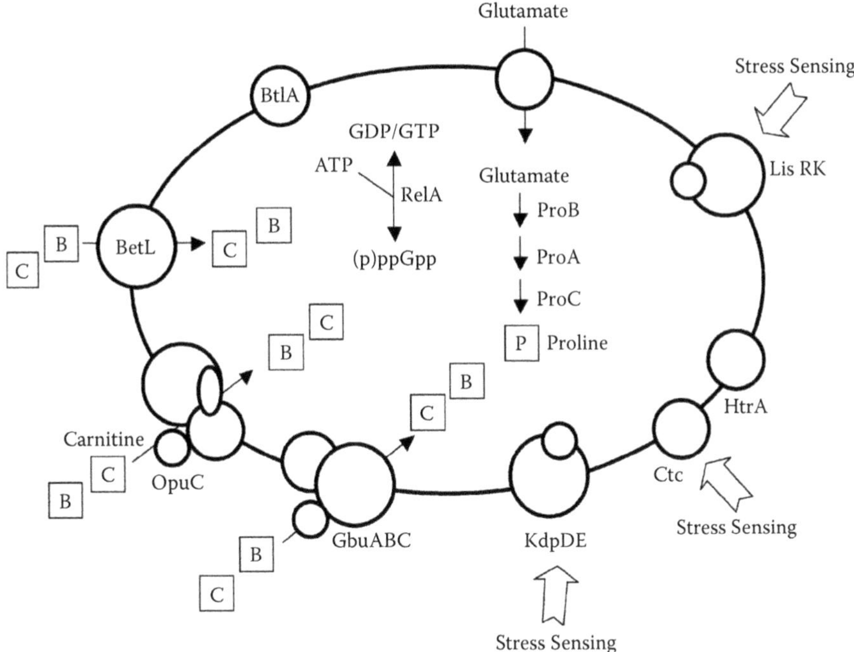

FIGURE 3.6 Overview of the listerial salt stress response discussed in detail in this section. B, C, and P denote betaine, carnitine, and praline, respectively.

betaine and carnitine being the most effective and extensively studied.[214–216] For the majority of bacteria, these osmolytes can be taken up from the environment from which they are readily available or, alternatively, synthesized *de novo*. However, it is thought that in *L. monocytogenes, de novo* synthesis does not occur despite the presence of genes encoding enzymes involved in the production of these solutes, and thus accumulation must occur via a transport mechanism.[217,218] Betaine is predominantly found in plant tissue,[219] while carnitine is found mainly in mammalian tissue.[220] Verheul et al.[221] demonstrated that the duration of betaine and carnitine transport into the cell is directly related to the osmotic strength of the environment.

In *L. monocytogenes* the uptake of glycine betaine and carnitine is mediated by three osmolyte transporters: BetL, Gbu and OpuC.[222–227] Glycine betaine has two specific permeases (BetL and Gbu) devoted to its transport, which have been described as being highly specific, constitutive, and energy-dependent systems. Glycine betaine porter I, a Na^+-glycine betaine symporter encoded by *betL*, is activated by increased osmolarity. Glycine betaine porter II is an ATP-dependent transporter, encoded by the *gbuABC* operon, which is activated by increased osmotic pressure or decreased temperature.[214,223,225,226,228–232] OpuC, an ATP-dependent carnitine transporter encoded by the *opuC* operon (*opuCABCD*), transports carnitine into the cell in response to osmotic and cold stress.[222,233–235]

The characteristics of each transporter have been analyzed by the generation of mutations in each transporter singly and in tandem with other transporters. Sleator et al.[225,226] demonstrated the reduced ability of a *betL* mutant to withstand elevated osmolarity in comparison to the parent, with significantly attenuated ability to accumulate glycine betaine in the presence or absence of salt.

Like BetL, glycine betaine transport in a *gbu* mutant is markedly reduced in comparison to the wild type in the presence of salt, in addition to chill stress. However, in the absence of salt, a more severe reduction of glycine betaine transport is observed for a *gbu* mutant (1%), indicating that all residual uptake observed for a *gbu* mutant is due to BetL.[218,223,236] It has been shown that Gbu is the primary mechanism by which glycine betaine enters *Listeria* at elevated osmolarity[231,236] as well as the major contributor to osmolyte transport during listerial growth and survival in high-risk foods.[227] Following osmotic shock, BetL appears to provide immediate protection in low salt stress (≤4% NaCl), but it is Gbu that plays the major role in long-term osmoadaptation, particularly at higher concentrations (≥4% NaCl).[230,231,237] Interestingly, a role for Gbu and BetL in carnitine transport was observed with Gbu, the primary alternate carnitine porter for *Listeria*.[238] This is particularly interesting, given that earlier studies from the same group proposed that carnitine is, in fact, an inhibitor of ATP-dependent glycine betaine transportation (Gbu).[231]

OpuC is the principal carnitine transporter[222–233]; however, radio-labeled osmolyte studies, coupled with growth experiments, linked OpuC with glycine betaine uptake in *Listeria*. A separate study[238] also demonstrated a role for OpuC in glycine betaine transport, albeit a minor one. In a *gbu*, *betL* double mutant, no betaine was transported into the cytoplasm, and the cell was completely reliant on OpuC for osmolyte uptake. This mutant was severely impaired in growth at elevated osmolarity even though high levels of carnitine were present, indicating the ineffectiveness of carnitine in comparison to betaine in listerial osmotolerance[236,239]—possibly due to its longer carbon backbone. Furthermore, OpuC was observed as being the least-favored osmolyte uptake system in listerial survival in high-risk foods.[227]

3.4.1.1 Regulation of Osmolyte Uptake

In *L. monocytogenes*, compatible solute regulation is achieved at the transcriptional, translational, and post-translational levels. *betL*, *gbu*, and *opuC* are all transcriptionally up-regulated in response to osmotic shock.[226,240–242] In addition, osmotic upshift is one of the most potent stimulators of σ^B activity,[162,242,243] and the presence of a putative σ^B promoter binding site upstream of *betL*, *gbu*, and *opuC* suggests that induction of these genes in response to elevated osmolarity is at least in part σ^B dependent. *betL* is constitutively expressed at low NaCl concentrations (1–2%), possibly from the vegetative σ^A type promoter, and is transcriptionally up-regulated in response osmotic shock (≥2% NaCl), perhaps due to σ^B.[226] However, recent studies indicate that *betL* expression is constitutive and, independent of σ^B, *gbu* is transcribed from dual promoters, a σ^A type promoter and one of which is σ^B dependent, whereas *opuC* is transcribed exclusively from a σ^B-dependent promoter.[169,170,240,241,244]

Promoter fusions and reverse transcriptase PCR have confirmed the dependence of *opuC* and *gbuA* expression on σ^B under conditions of elevated osmolarity. Moreover, kinetic analysis of transcript accumulation after osmotic upshift suggested that σ^B activity is maintained for long periods of time after osmotic stress and may therefore be the mechanism used for expression of *gbuA* during prolonged periods of osmotic stress.[240,241] The σ^B mutant phenotype of impaired betaine and carnitine transport has been shown to be mediated through its effects on *gbu* and *opuC*. This is consistent with *betL* and *gbu* being independent of and dependent on σ^B, respectively, and the absolute requirement for *opuC* expression on σ^B during osmotic stress.[97,240,241,243,245] Moreover, σ^B-dependent regulation of *betL* and *gbu* appears to vary between *Listeria* strains, which is not unusual given that the compatible solute uptake systems also vary between strains.[242]

It has been proposed that BetL is controlled by the activation of a preexisting enzyme[230,231] regulated by a novel osmolyte-sensing mechanism.[221] Sleator et al.[237] demonstrated that, in addition to *betL* induction at elevated osmolarity, the BetL protein itself is activated in response to changes in salinity. The situation in *Listeria* appears to be similar to that of the OpuD system in *B. subtilis*, where maximal uptake activity by OpuD results from a combination of *de novo* synthesis of OpuD and activation of a preexisting OpuD protein.[246] This rapid activation of preexisting BetL protein in response to relatively low concentrations of NaCl suggests that BetL is one of the primary respondents for transport of betaine during the early stages of osmotic upshift.

Only a few genes are known to solely rely upon σ^B for their transcription[247–249]; therefore, the absolute requirement[242] of *opuC* for σ^B provides a mechanism for the preferential use of betaine in the maintenance of osmotic balance under normal conditions, since σ^B activity is minimal under those conditions. However, under conditions of osmotic stress, σ^B is induced, which in turn transcriptionally activates *gbu* and *opuC,* thus providing the cell with an enhanced array of transporters for optimal osmoprotection. Besides transcriptional regulators, *Listeria* may possess proteins that merely modulate transcriptional activity, such as FlaR, which modulates the superhelicity of DNA and is itself induced under conditions of hyperosmotic stress.[250] The presence of alternative initiation codons upstream of *betL, gbuA,* and *opuCA* suggest that these genes are regulated to some extent at the level of translation, as the use of non-ATG initiation codons serves to alter the expression at the translational level.[251,252] Earlier studies by Verheul et al.[221] demonstrated the existence of a novel osmolyte sensing system in the regulation of betaine and carnitine uptake, in which these solutes are subjected to inhibition by preaccumulated betaine and carnitine. Cytoplasmic betaine/carnitine inhibits transport of external betaine and carnitine. This inhibition is alleviated by osmotic upshift, thus allowing an increased uptake of these compatible solutes to combat the stress encountered, and subsequently released as a consequence of osmotic downshift by the activation of a channel-like activity.

3.4.1.2 Contribution of Compatible Solutes to Virulence Potential

For food-borne pathogens, the ability to sense and appropriately respond to conditions of elevated osmolarity in the gastrointestinal lumen is a prerequisite for infection. The osmolarity of the intestinal lumen is equivalent to 0.3 *M* NaCl,[253] the concentration at which maximum carnitine uptake occurs in *Listeria.*[239] The relative abundance of carnitine in mammalian tissues[220] makes it readily available and therefore it was thought that carnitine may play an important role in listerial infection. As predicted, altering the carnitine uptake system (OpuC) resulted in a significant reduction in the ability of *Listeria* to colonize the upper small intestine and cause subsequent systemic infection, thus suggesting that carnitine (produced from the desquamation of the gastrointestinal epithelial layer) represents a key osmoprotectant facilitating growth in this environment.[227,234]

In support of this, a separate study[236] demonstrated that an *opuC* mutant showed a similar phenotype to that of a triple mutant, defective in all three compatible uptake systems (BetL, Gbu, and OpuC), displaying a significant reduction in its ability to cause systemic infection in comparison to the parent. A recent study by Joseph et al.[68] also identified OpuCA and OpuCB as being induced intracellularly. In addition, *betL* and *gbu* have been shown not to play a significant role in *L. monocytogenes* pathogenesis.[226,227] Taken together, these data suggest that it is the carnitine uptake system that plays the major role in listerial pathogenesis. Thus, the individual contribution of each transporter is dependent on the external environment, creating a situation whereby each system is tailored for optimal effects within a specific environmental niche.

3.4.2 Osmotic Stress Proteins

Another mechanism used by *Listeria* to combat osmotic stress, in the absence of or in tandem with compatible solutes, is the alteration of gene expression leading to increased or decreased synthesis of various proteins. Recent studies on the global expression pattern of proteins expressed in response to elevated osmolarity[254–256] as well as transposon mutagenesis studies have facilitated the identification of a number of osmotic stress proteins in *L. monocytogenes* such as RelA,[257] Ctc,[258,259] HtrA,[118,120] KdpE,[260,261] LisRK,[118] ProBA,[262,263] and BtlA.[264]

A recent study identified *relA*, a gene encoding a (p)ppGpp synthetase as a gene whose product is involved in listerial osmotolerance and biofilm formation.[257,265] In many organisms, (p)ppGpp is accumulated under nutrient-limited conditions and is considered a stress-response-related factor because accumulation induces the so-called stringent stress response.[266] In *L. monocytogenes* this enzyme has been proposed to be bifunctional, having both synthetic and degradative properties, as seen in other bacteria.[257,267–269] A role for (p)ppGpp in the growth of *L. monocytogenes* in elevated osmolarity has been observed. In addition, intracellular accumulation of (p)ppGpp is independent of the mechanisms involved in compatible solute accumulation.[257]

Ctc has been identified as an osmotic stress protein in *L. monocytogenes*[254,255,258] belonging to the L25 family of ribosomal proteins and has been shown to facilitate growth in minimal media under conditions of high osmolarity. Further, microscopic analysis showed impaired morphology of the *ctc* mutant grown under osmotic stress conditions in minimal medium in comparison to the wild type. Addition of compatible solutes to the medium completely restored salt resistance of the *ctc* mutant, suggesting that like RelA, Ctc is involved in osmotic stress tolerance in the absence of any osmoprotectant in the medium. It has been proposed that the ribosome of this protein serves as a sensor of osmotic stress through the activity of Ctc.[259]

HtrA has been shown to play a role in the osmotolerance of *L. monocytogenes* (10403S)[118,120]; however, there appears to be some strain variation with respect to the role of *htrA* in the listerial osmotic stress response. In strain EGDe, *htrA* does not appear to be involved in osmotolerance, and it is not transcriptionally up-regulated in response to elevated osmolarity. In addition, growth in minimal medium supplemented with 4% NaCl and 1 m*M* betaine did not result in any significant difference observed between the wild type and the *htrA* mutant,[270,271] indicating that HtrA does not seem to play a role in osmotic stress in EGDe, either in the presence or absence of osmoprotectants in minimal media.

KdpE forms part of a three-gene operon, *kdpD*, *kdpE,* and *orfX,* and encodes a transcriptional response regulator.[260,261] A role for KdpE in listerial osmotolerance was first described by Kallipolitis and Ingmer,[261] where growth of a *kpdE* mutant was significantly impaired in comparison to the parent at a salt concentration of 7.5%; this sensitivity became more prominent at higher salt concentrations. A separate study also revealed a role for this *kdpE* locus in listerial osmotolerance. Interestingly, in this study *kdpE* and its downstream gene *orfX* were both required for optimal growth in salt stress conditions, as a mutation in either gene alone did not confer osmosensitivity. Further, this observed role for the *kdpE* locus is dependent on the potassium concentration of the medium.[260] It has been suggested that *kdpE* may represent part of a two-component system as observed in *E. coli*, where *kdpE* is the response regulator of a two-component sensor–regulator pair that is phosphorylated by its cognate sensor histidine kinase during high osmolarity to prevent plasmolysis and restore turgor pressure, or when potassium concentrations become low.[209,260,261,270,271] Thus, in *L. monocytogenes* the expression of the high-affinity potassium uptake system is dependent on the *kdpE* locus; consequently, this locus influences the ability of potassium to serve as an osmoprotectant for *L. monocytogenes*.[260] Interestingly, *orfX* has been shown to have considerable synteny with RsbQ of *B. subtilis*, which takes part in the activation cascade of the alternative sigma factor, σ^B, during energy stress.[272] Similar phenotypes have been observed upon temperature shift in *L. monocytogenes* σ^B and *orfX* mutants in comparison to the wild type, further suggesting that this gene may encode an RsbQ homologue.[260]

The two-component LisRK system has been shown to play a role in the listerial stress response,[127,128] including elevated osmolarity.[118] Disruption of LisK resulted in a significant reduction in the ability of *L. monocytogenes* to tolerate environments of osmotic stress (8%). LisRK-mediated osmotolerance, like RelA and Ctc, appears to function independently of compatible solutes.[118] Recent work in our lab[31] demonstrated that *lisRK* regulates HtrA in *L. monocytogenes* strain LO28. In addition, a role in osmotolerance for *htrA*, albeit a minor one, has been established in this strain, and it has been proposed that apart from HtrA, LisRK probably controls one or more additional systems necessary for optimal listerial osmotolerance.[118] It has been suggested that LisRK may function at the initial stages of the listerial osmotic stress response, with LisK functioning as the primary sensor and LisR as the primary respondent to elevated osmolarity.[118]

ProBA, encoding enzymes involved in proline biosynthesis, have been linked to listerial osmotolerance. The role of proline as an effective osmolyte has previously been described for a variety of bacteria, including *Listeria*.[215,216] Disruption of the *proBA* operon revealed a significant role for proline synthesis in contributing to the growth and survival of *L. monocytogenes* in environments of elevated osmolarity.[262] Interestingly, while mutations in the listerial *proB* gene leading to overproduction of proline had no obvious effect on listerial osmotolerance, expression of the mutated operon in *E. coli proBA* null background resulted in significant increase in growth rate in conditions of elevated osmolarity. This could be due to the fact that Gram-positive organisms have more extreme turgor requirements than those of its Gram-negative counterparts. In addition, *Listeria* requires more proline than *E. coli* in order to promote maximal growth at elevated osmolarities.[263,273] *Listeria* is thought to possess only one proline biosynthesis pathway in contrast to that observed for *B. subtilis*, as a listerial *proBA* double mutant was unable to grow in praline-deficient medium either at normal or elevated osmolarity.[262] Finally, BtlA, a bile tolerance locus in *L. monocytogenes* has been shown to play a role in the listerial stress response, including growth in elevated osmolarity (7%).[264]

3.4.2.1 Regulation of Osmotic Stress Proteins

To date, very little is known about the regulation of the so-called osmotic stress proteins in *L. monocytogenes*. However, from the studies carried out thus far we know that these genes are osmoregulated, with increases in osmolarity resulting in a requirement for these proteins to combat the stress encountered. But whether this osmoregulation is mediated via σ^B remains to be determined.

Little is known about the regulation of RelA in *Listeria*. However from studies carried out in both Gram-negative and Gram-positive bacteria it appears that RelA is itself a regulatory component of the alternative sigma factors, σ^S and σ^B, respectively, during the stringent stress response through the production of (p)ppGpp.[274–276] Perhaps the same is true for *Listeria*; under conditions of elevated osmolarity or stringent stress, *relA* catalyses the production of (p)ppGpp, which in turn induces the σ^B regulon.

Given the significant homology observed between *ctc* from *L. monocytogenes* and its counterpart in *B. subtilis*, the fact that σ^B regulates *ctc* expression in *B. subtilis* and the presence of a putative σ^B-dependent promoter upstream of the listerial *ctc* gene, it is likely that σ^B regulates *ctc* expression in *L. monocytogenes*. Moreover, using a microarray-based approach, Kazmierczak et al.[169] revealed *ctc* as being part of the σ^B regulon in *L. monocytogenes*. Under normal conditions, a constitutive level of *ctc* is transcribed; however, under conditions of elevated osmolarity, there is an increase in *ctc* transcript, possibly regulated by σ^B.[259]

HtrA regulation is discussed in more detail in section 3.2.3. HtrA is transcriptionally regulated by LisRK. Furthermore, a putative σ^B-dependent promoter-binding site is situated upstream of its putative signal sequence, suggesting that under certain conditions HtrA may be secreted, possibly under the control of σ^B.[31]

ProBA forms a bicistronic operon and its transcription is initiated from a single vegetative σ^A type promoter upstream of *proB*.[262] It is thought that proline biosynthesis is regulated by feedback inhibition of the ProB protein, and mutations in this locus lead to decreased sensitivity of the enzyme for its allosteric effector proline, resulting in overproduction.[263]

BtlA has been shown to be constitutively expressed but is not transcriptionally up-regulated in response to stress conditions. This protein may be regulated at the translational level as an alternative start codon (UUG) is also present.[264,277]

LisRK and the Kpd locus are characterized and predicted as two-component regulatory systems, respectively.[127,128,261] Two-component systems allow target gene expression to be adjusted in response to changes in the environment in order to ensure optimal growth. A novel role has been observed for LisRK in osmosensing and osmoregulation in *L. monocytogenes*. In addition to osmotic stress, LisRK is also activated in response to numerous other stress conditions,[31,118,127,128] which in turn transcriptionally activates another subset of genes to deal with the change in environmental conditions.[31] Interestingly, the final gene in the *kpd* operon (*orfX*) shows homology to RsbQ in *B. subtilis*, which has been shown to be a positive regulator of the alternative sigma factor σ^B during energy stress in *B. subtilis*.[260]

3.4.2.2 Contribution of Osmotic Stress Proteins to Virulence Potential

In *L. monocytogenes*, RelA, the enzyme responsible for the production of (p)ppGpp, has been shown to be absolutely required for virulence using the murine model of infection. A mutant in this locus was avirulent, but hemolysin levels remained unchanged, indicating that the ability to synthesize (p) ppGpp is an essential physiological adaptation that is required for virulence, as seen in other bacteria.[265,278] To date, the role of Ctc in listerial pathogenesis remains to be determined, but its putative regulation by σ^B suggests that it may play some role in the virulence potential of this organism. The role of HtrA in virulence has been discussed in section 3.2.3. Sleator et al.[262,263] demonstrated that proline anabolism does not contribute to listerial virulence. Neither an in-frame *proBA* deletion mutant, abolishing proline biosynthesis, nor a *proB* mutant, which led to hyperproduction of proline, had any measurable effect on *Listeria* pathogenesis. In addition, *btlA* appears not to play a role *in vivo*.[264] Conversely, a role for the both LisR and KdpE response regulators has been demonstrated *in vivo*, with an *lisR* mutant being significantly attenuated after both intragastric and intraperitoneal infection, while the *kdpE* mutant was only significantly attenuated after intragastric infection. Therefore, KdpE may be important in the early steps of the intragastric infection process including translocation from the intestinal lumen, whereas LisR may affect a step common to both infectious routes.[261] Moreover, a separate study revealed a role for the sensor histidine kinase, LisK in listerial virulence.[127]

3.5 STRESS CROSS-PROTECTION

Cross-protection, or stress hardening, can be defined as the ability of one stress condition to provide protection against other stresses. It is dependent on the type of stress encountered and the lethality of the stress.[279] This phenomenon has been described in many bacteria[280–283] and should be considered when current food processing technologies are modified and new ones introduced, as it may counteract the effectiveness of food preservation hurdles and compromise food safety. Not surprisingly, this cross-protection phenomenon also exists in *L. monocytogenes*. It seems that the cross-protection system induced in *L. monocytogenes* by osmotic, hydrogen peroxide, heat, ethanol, and glucose adaptation is more limited than cross-protection by acid adaptation; that is, exposure of *L. monocytogenes* to pH 4.5–5 for 1 h improved the survivability of cells when subsequently exposed to pH 3.5, 17.5% ethanol, or 0.1% hydrogen peroxide.[139,144,152,279,284–287] This suggests that acid adaptation is perceived by bacteria as a more general stress indicator, whereas salt, heat, ethanol, hydrogen peroxide, and glucose may be more specific stress signals. Consequently, exposure to acid in the food processing industry will ensure that the organism is well prepared for *in vivo* stresses such as those encountered in the stomach and intestine and, finally, during systemic infection. Therefore, adaptive cross-resistance can enhance the ability of *L. monocytogenes* to cause disease by contributing to bacterial survival in a variety of challenges imposed by both the food industry and the host.

3.6 CONCLUSIONS AND PERSPECTIVES

The importance of *L. monocytogenes* as the causative agent of potentially fatal food-borne diseases has led to significant research devoted to its ability to withstand both *in vitro* and *in vivo* insults. These studies have confirmed that the virulence potential of this pathogen is directly linked to its ability to survive the hostile environments encountered throughout its life cycle. In particular, copious data have been accumulated over the past decades on the molecular mechanisms of *L. monocytogenes'* tolerance of heat, acid, and osmotic stresses, which are invariably encountered by this bacterium during its various stages of infection within the mammalian host. The idenfication and characterization of a large number of key proteins involved in *L. monocytogenes* heat- (e.g., GroESL, DnaKJ, Clp, and HtrA), acid- (e.g., GAD, ADI, and F_0F_1ATPase), and osmotic- (e.g., BetL, Gbu, OpuC, RelA, Ctc, HtrA, KdpE, LisRK, ProBA, and BtlA) responses have undoubtedly contributed to our appreciation of the ingenuity of this remarkbale bacterium to ably confront overwhelmingly adverse reactions from the mammalian host and to successfully establish itself there.

Being a facultative intracellular bacterium, *L. monocytogenes* also spends a significant part of its life cycle in the natural environment, in which it continuously experiences stressful conditions (e.g., alkali, salt, and pressure) in addition to heat, acid, and osmolarity (discussed previously). The ability of *L. monocytogenes* to sense and appropriately respond to these extra, potentially lethal changes in environmental milieu is crucial to its eventual success as an intracellular pathogen, considering that alkaline detergents are routinely applied in food manufacturing industries to clean and sterilize food processing surfaces and surroundings. Unfortunately, not a great deal is known about *L. monocytogenes*' ability to survive in alkaline conditions. In *Bacillus subtilis,* a 5.7-kb operon (containing seven genes, i.e., *mrpA-G*—*m*ulti*r*esistance *p*rotein—flanked by *yuxB* at the 3′-end and *yuxO* at the 5′-end) has been identified and shown to be involved in pH adaptation and salt resistance, of which *mrpA* encodes an Na+/H+ antiporter or a K+/H+ antiporter. Disruption of *mrpA* results in an impairment of cytoplasmic pH regulation upon a sudden shift in external pH from 7.5 to 8.5.[288,289] Interestingly, the genomes of *L. monocytogenes* EGD-e and *L. innocua* CLIP 11262 also contain a stretch of nucleotides that displays partial homology to *B. subtilis mrpA-G.* The 6.0-kb operon in *L. monocytogenes* EGD-e is also made up of seven genes (i.e., *lmo2378–2384*) flanked by a multidrug resistance efflux pump (*lmo2377*) at the 5′-end and *yuxO* at the 3′-end.[76] Collectively, these genes encode putative proteins that are involved in resistance to cholate/Na(+) and in pH homeostasis. Therefore, future investigations on the *lmo2378–2384* operon will likely yield new insights on the mechanism of *L. monocytogenes*' resistance to alkali.

In order to maintain an upper hand in our fight against this deadly pathogen, we must first unravel the molecular mechanisms by which it survives, grows, adapts, and subsequently cross-adapts in these adverse conditions. Fundamental to understanding these molecular mechanisms is the identification of "fitness" loci that permit growth and survival in such severe environments. Methods to functionally evaluate these numerous genetic loci that are expressed during growth in specific and composite ecological niches, such as microarrays, selective capture of transcribed sequences (SCOTS), and signature tagged mutagenesis (STM), are invaluable. While many of these techniques allow the determination of gene expression in different environments, they subsequently fail to address the expression patterns of individual loci in complex and dynamic ecological niches. One such strategy, which overcomes this problem and allows us to functionally evaluate each locus in diverse environmental niches, is *in vivo* expression technology (IVET) (see chapter 9). With an increasing demand for minimally processed and ready-to-eat foods, these methods are indispensable not only in identifying genetic loci induced under a variety of different stresses such as heat, acid, and salt, but also in identifying genes expressed *in vivo* during the gastrointestinal and systemic phase of infection. These genetic loci may represent attractive targets of the future development of novel therapeutic agents in the ongoing battle against this disease.

REFERENCES

1. Weis, J., and Seeliger, H.P., Incidence of *Listeria monocytogenes* in nature, *Appl. Microbiol.*, 30, 29, 1975.
2. Welshimer, H.J., Survival of *Listeria monocytogenes* in soil, *J. Bacteriol.*, 80, 316, 1960.
3. Welshimer, H.J., Isolation of *Listeria monocytogenes* from vegetation, *J. Bacteriol.*, 95, 300, 1968.
4. Watkins, J., and Sleath, K.P., Isolation and enumeration of *Listeria monocytogenes* from sewage, sewage sludge and river water, *J. Appl. Bacteriol.*, 50, 1, 1981.

5. Smith, M.C., Nervous system. In *Goat medicine,* Smith M.C., and Sherman, D.M., eds. Lea and Febiger, Malvern, PA, 1994, p. 141.
6. Mead, P.S. et al., Food-related illness and death in the United States, *Emerg. Infect. Dis.*, 5, 607, 1999.
7. Schlech, W.F., Foodborne listeriosis, *Clin. Infect. Dis.*, 31, 770, 2000.
8. Farber, J.M., and Peterkin, P.I., *Listeria monocytogenes*, a food-borne pathogen, *Microbiol. Rev.*, 55, 476, 1991.
9. McCarthy, S.A., *Listeria* in the environment. In *Foodborne listeriosis*, Miller, A.J., Smith, J.L., and Somkuti, G.A., eds. Society for Industrial Microbiology. Elsevier Science Publishing, Inc., New York, 1990.
10. Ly, T.M., and Muller, H.E., Ingested *Listeria monocytogenes* survive and multiply in protozoa, *J. Med. Microbiol.*, 33, 51, 1990.
11. Kathariou, S., *Listeria monocytogenes* virulence and pathogenicity, a food safety perspective, *J. Food Prot.*, 65, 1811, 2002.
12. Pinner, R.W. et al., Role of foods in sporadic listeriosis. II. Microbiologic and epidemiologic investigation. The *Listeria* Study Group, *JAMA*, 267, 2046, 1992.
13. Marco, A.J. et al., Penetration of *Listeria monocytogenes* in mice infected by the oral route, *Microb. Pathog.*, 23, 255, 1997.
14. Racz, P., Tenner, K., and Mero, E., Experimental *Listeria* enteritis. I. An electron microscopic study of the epithelial phase in experimental *Listeria* infection, *Lab. Invest.*, 26, 694, 1972.
15. Vazquez-Boland, J.A. et al., *Listeria* pathogenesis and molecular virulence determinants, *Clin. Microbiol. Rev.*, 14, 584, 2001.
16. Mackaness, G.B., Cellular resistance to infection, *J. Exp. Med.*, 116, 381, 1962.
17. Conlan, J.W., Early pathogenesis of *Listeria monocytogenes* infection in the mouse spleen, *J. Med. Microbiol.*, 44, 295, 1996.
18. North, R.J., and Conlan, J.W., Immunity to *Listeria monocytogenes*, *Chem. Immunol.*, 70, 1, 1998.
19. Portnoy, D.A., Jacks, P.S., and Hinrichs, D.J., Role of hemolysin for the intracellular growth of *Listeria monocytogenes, J. Exp. Med.*, 167, 1459, 1988.
20. Tilney, L.G., and Portnoy, D.A., Actin filaments and the growth, movement, and spread of the intracellular bacterial parasite, *Listeria monocytogenes, J. Cell Biol.*, 109, 1597, 1989.
21. Portnoy, D.A. et al., Molecular determinants of *Listeria monocytogenes* pathogenesis, *Infect. Immun.*, 60, 1263, 1992.
22. Tilney, L.G., Connelly, P.S., and Portnoy, D.A., Actin filament nucleation by the bacterial pathogen, *Listeria monocytogenes*, *J. Cell Biol.*, 111, 2979, 1990.
23. Ingraham, J., Effect of temperature, pH, water activity and pressure on growth. In *Escherichia coli and Salmonella typhimurium*: *Cellular and molecular biology,* 2nd ed., Neidhardt, F.C. et al., eds. American Society for Microbiology, Washington, D.C., 1987, p. 1543.
24. Earnshaw, R.G., Appleyard, J., and Hurst, R.M., Understanding physical inactivation processes: combined preservation opportunities using heat, ultrasound and pressure, *Int. J. Food Microbiol.*, 28, 197, 1995.
25. Miller, L.L., and Ordal, Z.J., Thermal injury and recovery of *Bacillus subtilis*, *Appl. Microbiol.*, 24, 878, 1972.
26. Yura, T., and Nakahigashi, K., Regulation of the heat-shock response, *Curr. Opin. Microbiol.*, 2, 153, 1999.
27. Abee, T., and Wouters, J.A., Microbial stress response in minimal processing, *Int. J. Food Microbiol.*, 50, 65, 1999.
28. Sokolovic, Z., and Goebel, W., Synthesis of listeriolysin in *Listeria monocytogenes* under heat shock conditions, *Infect. Immun.*, 57, 295, 1989.
29. Gahan, C.G., O'Mahony, J., and Hill, C., Characterization of the *groESL* operon in *Listeria monocytogenes*: Utilization of two reporter systems (*gfp* and *hly*) for evaluating *in vivo* expression, *Infect. Immun.*, 69, 3924, 2001.
30. Hanawa, T. et al., Participation of DnaK in expression of genes involved in virulence of *Listeria monocytogenes, FEMS Microbiol. Lett.*, 214, 69, 2002.
31. Stack, H.M. et al., Role for HtrA in stress induction and virulence potential in *Listeria monocytogenes, Appl. Environ. Microbiol.*, 71, 4241, 2005.
32. Georgopoulos, C., and Welch, W.J., Role of the major heat shock proteins as molecular chaperones, *Annu. Rev. Cell Biol.*, 9, 601, 1993.
33. Gottesman, S., Proteases and their targets in *Escherichia coli*, *Annu. Rev. Genet.*, 30, 465, 1996.

34. Hartl, F.U., Molecular chaperones in cellular protein folding, *Nature*, 381, 571, 1996.
35. Hanawa, T. et al., Cloning, sequencing, and transcriptional analysis of the *dnaK* heat shock operon of *Listeria monocytogenes, Cell Stress Chaperones*, 5, 21, 2000.
36. Zeilstra-Ryalls, J., Fayet, O., and Georgopoulos, C., The universally conserved GroE (Hsp60) chaperonins, *Annu. Rev. Microbiol.*, 45, 301, 1991.
37. Hendrick, J.P., and Hartl, F.U., Molecular chaperone functions of heat shock proteins. *Annu. Rev. Biochem.*, 62, 349, 1993.
38. Fayet, O., Ziegelhoffer, T., and Georgopoulos, C., The *groES* and *groEL* heat shock products of *Escherichia coli* are essential for bacterial growth at all temperatures, *J. Bacteriol.*, 171, 1379, 1989.
39. Craig, E.A., Gambill, B.D., and Nelson, R J., Heat shock proteins: Molecular chaperones of protein biogenesis, *Microbiol. Rev.*, 57, 402, 1993.
40. Bunning, V.K. et al., Thermotolerance of *Listeria monocytogenes* and *Salmonella typhimurium* after sublethal heat shock, *Appl. Environ. Microbiol.*, 56, 3216, 1990.
41. Hanawa, T., Yamamoto, T., and Kamiya, S., *Listeria monocytogenes* can grow in macrophages without the aid of proteins induced by environmental stresses, *Infect. Immun.*, 63, 4595, 1995.
42. Flahaut, S.A. et al., Relationship between stress response toward bile salts, acid and heat treatment in *Enterococcus faecalis, FEMS Microbiol. Lett.*, 138, 49, 1996.
43. Kilstrup, M. et al., Induction of heat shock proteins DnaK, GroEL, and GroES by salt stress in *Lactococcus lactis, Appl. Environ. Microbiol.*, 63, 1826, 1997.
44. Phan-Thanh, I., and Mahouin, F., A proteomic approach to study the acid response in *Listeria monocytogenes, Electrophoresis*, 20, 2214, 1999.
45. Salotra, P. et al., Expression of DnaK and GroEL homologs in *Leuconostoc esenteroides* [*sic*] in response to heat shock, cold shock or chemical stress, *FEMS Microbiol. Lett.*, 131, 57, 1995.
46. Roncarati, D. et al., Expression, purification and characterization of the membrane-associated HrcA repressor protein of *Helicobacter pylori, Protein. Expr. Purif.*, 51, 267, 2006.
47. Schulz, A., and Schumann, W., *hrcA*, the first gene of the *Bacillus subtilis dnaK* operon encodes a negative regulator of class I heat shock genes, *J. Bacteriol.*, 178, 1088, 1996.
48. Yuan, G., and Wong, S.L., Isolation and characterization of *Bacillus subtilis groE* regulatory mutants: evidence for *orf39* in the *dnaK* operon as a repressor gene in regulating the expression of both *groE* and *dnaK, J. Bacteriol.*, 177, 6462, 1995.
49. Zuber, U., and Schumann, W., CIRCE, a novel heat shock element involved in regulation of heat shock operon *dnaK* of *Bacillus subtilis, J. Bacteriol.*, 176, 1359, 1994.
50. Narberhaus, F., Giebeler, K., and Bahl H., Molecular characterization of the *dnaK* gene region of *Clostridium acetobutylicum*, including *grpE, dnaJ* and a new heat shock gene, *J. Bacteriol.*, 174, 3290, 1992.
51. Wetzstein, M. et al., Cloning, sequencing, and molecular analysis of the *dnaK* locus from *Bacillus subtilis, J. Bacteriol.*, 174, 3300, 1992.
52. Kuroda, M. et al., The hsp operons are repressed by the hrc37 of the hsp70 operon in *Staphylococcus aureus, Microbiol. Immunol.*, 43, 19, 1999.
53. Schumann, W., The *Bacillus subtilis* heat shock stimulon, *Cell Stress Chaperones*, 8, 207, 2003.
54. Hecker, M., Schumann, W., and Volker, U., Heat-shock and general stress response in *Bacillus subtilis, Mol. Microbiol.*, 19, 417, 1996.
55. Schulz, A., Tzschaschel, B., and Schumann, W., Isolation and analysis of mutants of the *dnaK* operon of *Bacillus subtilis, Mol. Microbiol.*, 15, 421, 1995.
56. Mogk, A. et al., The GroE chaperonin machine is a major modulator of the CIRCE heat shock regulon of *Bacillus subtilis, J. EMBO*, 16, 4579, 1997.
57. Ensgraber, M., and Loos, M., A 66-kilodalton heat shock protein of *Salmonella typhimurium* is responsible for binding of the bacterium to intestinal mucus, *Infect. Immun.*, 60, 3072, 1992.
58. Minnick, M.F., Smitherman, L.S., and Samuels, D.S., Mitogenic effect of *Bartonella bacilliformis* on human vascular endothelial cells and involvement of GroEL, *Infect. Immun.*, 71, 6933, 2003.
59. Henderson, B., Allan, E., and Coates, A.R.M., Stress wars: The direct role of host and bacterial molecular chaperones in bacterial infection, *Infect. Immun.*, 74, 3693, 2006.
60. Lemos, J.A., Giambiagi-Demarval, M., and Castro, A.C., Expression of heat-shock proteins in *Streptococcus pyogenes* and their immunoreactivity with sera from patients with streptococcal disease, *J. Med. Microbiol.*, 47, 711, 1998.

61. Qoronfleh, M.W., Weraarchakul, W., and Wilkinson, B.J., Antibodies to a range of *Staphylococcus aureus* and *Escherichia coli* heat shock proteins in sera from patients with *S. aureus* endocarditis, *Infect. Immun.*, 61, 1567, 1993.
62. Qoronfleh, M.W., Gustafson, J.E., and Wilkinson, B.J., Conditions that induce *Staphylococcus aureus* heat shock proteins also inhibit autolysis, *FEMS Microbiol. Lett.*, 166, 103, 1998.
63. Garduno, R.A., Garduno, E., and Hoffman, P.S., Surface-associated hsp60 chaperonin of *Legionella pneumophila* mediates invasion in a HeLa cell model, *Infect. Immun.*, 66, 4602, 1998.
64. Hennequin, C. et al., GroEL (Hsp60) of *Clostridium difficile* is involved in cell adherence, *Microbiology*, 147, 87, 2001.
65. Waligora, A.J. et al., Characterization of a cell surface protein of *Clostridium difficile* with adhesive properties, *Infect. Immun.*, 69, 2144, 2001.
66. Hevin, B., Morange, M., and Fauve, R. M. Absence of an early detectable increase in heat-shock protein synthesis by *Listeria monocytogenes* within mouse mononuclear phagocytes, *Res. Immunol.*, 144, 679, 1993.
67. Chatterjee, S.S. et al., Intracellular gene expression profile of *Listeria monocytogenes*, *Infect. Immun.*, 74, 1323, 2006.
68. Joseph, B. et al., Identification of *Listeria monocytogenes* genes contributing to intracellular replication by expression profiling and mutant screening, *J. Bacteriol.*, 188, 556, 2006.
69. Hanawa, T. et al., The *Listeria monocytogenes* DnaK chaperone is required for stress tolerance and efficient phagocytosis with macrophages, *Cell Stress Chaperones*, 4, 118, 1999.
70. Gottesman, S., Regulation by proteolysis: Developmental switches, *Curr. Opin. Microbiol.*, 2, 142, 1999.
71. Katayama, Y. et al., The two-component, ATP-dependent Clp protease of *Escherichia coli*. Purification, cloning, and mutational analysis of the ATP-binding component, *J. Biol. Chem.*, 263, 15226, 1988.
72. Maurizi, M.R. et al., Clp P represents a unique family of serine proteases, *J. Biol. Chem.*, 265, 12546, 1990.
73. Gottesman, S., Wickner, S., and Maurizi, M.R., Protein quality control: triage by chaperones and proteases, *Genes Dev.*, 11, 815, 1997.
74. Wawrzynow, A., Banecki, B., and Zylicz, M., The Clp ATPases define a novel class of molecular chaperones, *Mol. Microbiol.*, 21, 895, 1996.
75. Schrimer, E.C. et al., HSP100/Clp proteins: A common mechanism explains diverse functions, *Trends Biochem. Sci.*, 21, 289, 1996.
76. Glaser, P. et al., Comparative genomics of *Listeria* species, *Science*, 294, 849, 2001.
77. Chastanet, A. et al., *clpB*, a novel member of the *Listeria monocytogenes* CtsR regulon, is involved in virulence but not in general stress tolerance, *J. Bacteriol.*, 186, 1165, 2004.
78. Rouquette, C. et al., Identification of a ClpC ATPase required for stress tolerance and *in vivo* survival of *Listeria monocytogenes*, *Mol. Microbiol.*, 21, 977, 1996.
79. Rouquette, C. et al., The ClpC ATPase of *Listeria monocytogenes* is a general stress protein for virulence and promoting early bacterial escape from the phagosome of macrophages, *Mol. Microbiol.*, 27, 1235, 1998.
80. Nair, S. et al., ClpE, a novel member of the HSP100 family, is involved in cell division and virulence of *Listeria monocytogenes*, *Mol. Microbiol.*, 31, 185, 1999.
81. Gaillot, P. et al., The ClpP serine protease is essential for the intracellular parasitism and virulence of *Listeria monocytogenes*, *Mol. Microbiol.*, 35, 1286, 2000.
82. Woo, K.M. et al., The heat-shock protein ClpB in *Escherichia coli* is a protein-activated ATPase, *J. Biol. Chem.*, 267, 20429, 1992.
83. Nair, S. et al., CtsR controls class III heat shock gene expression in the human pathogen *Listeria monocytogenes*, *Mol. Microbiol.*, 35, 800, 2000.
84. Karatzas, A. et al., The CtsR regulator of *Listeria monocytogenes* contains a variant glycine repeat region that affects piezotolerance, stress resistance, motility and virulence, *Mol. Microbiol.*, 49, 1227, 2003.
85. Kirstein, J. et al., A tyrosine kinase and its activator control the activity of the CtsR heat shock repressor in *B. subtilis*, *J. EMBO*, 24, 3435, 2005.
86. Derre, I. et al., ClpE, a novel type of HSP100 ATPase, is part of the CtsR heat shock regulon of *Bacillus subtilis*, *Mol. Microbiol.*, 32, 581, 1999.

87. Derre, I., Rapport, G., and Msadek, T., CtsR, a novel regulator of stress and heat shock response, controls *clp* and molecular chaperone gene expression in Gram-positive bacteria, *Mol. Microbiol.*, 31, 117, 1999.
88. Lemos, J.A.C., and Burne, R.A., Regulation and physiological significance of ClpC and ClpP in *Streptococcus mutans, J. Bacteriol.*, 184, 6357, 2002.
89. Varmanen, P. et al., ClpE from *Lactococcus lactis* promotes repression of CtsR-dependent gene expression, *J. Bacteriol.*, 185, 5117, 2003.
90. Kruger, E. et al., Clp-mediated proteolysis in Gram-positive bacteria is autoregulated by the stability of a repressor, *J. EMBO*, 20, 852, 2001.
91. Kirstein, J. et al., Adaptor protein controlled oligomerization activates the AAA$^+$ protein ClpC, *J. EMBO*, 25, 1481, 2006.
92. Kruger, E. et al., The *clp* proteases of *Bacillus subtilis* are directly involved in degradation of misfolded proteins, *J. Bacteriol.*, 182, 3259, 2000.
93. Kirstein, J., and Turgay, K., A new tyrosine phosphorylation mechanism involved in signal transduction in *Bacillus subtilis, J. Mol. Microbiol. Biotechnol.*, 9, 182, 2005.
94. Ripio, M.T. et al., Evidence for expressional crosstalk between the central virulence regulator PrfA and the stress response mediator ClpC in *Listeria monocytogenes, FEMS Microbiol.*, 158, 45, 1998.
95. Johansson, J. et al., An RNA thermosensor controls expression of virulence genes in *Listeria monocytogenes, Cell*, 110, 551, 2002.
96. Renzoni, A., Cossart, P., and Dramsi, S., PrfA, the transcriptional activator of virulence genes, is upregulated during interaction of *Listeria monocytogenes* with mammalian cells and in eukaryotic cell extracts, *Mol. Microbiol.*, 34, 552, 1999.
97. Chaturongakul, S., and Boor, K. J., σ^B activation under environmental and energy stress conditions in *Listeria monocytogenes, Appl. Environ. Microbiol.*, 72, 5197, 2006.
98. Wempkamp-Kamphuis, H.H. et al., Identification of sigma factor σ^B-controlled genes and their impact on acid stress, high hydrostatic pressure and freeze survival in *Listeria monocytogenes* EGD-e, *Appl. Environ. Microbiol.*, 70, 3457, 2004.
99. Raynaud, C., and Charbit, A., Regulation of expression of type 1 signal peptidases in *Listeria monocytogenes, Microbiology*, 151, 3769, 2005.
100. Hensel, M. et al., Simultaneous identification of bacterial virulence genes by negative selection, *Science*, 269, 400, 1995.
101. Mei, J.M. et al., Identification of *Staphylococcus aureus* virulence genes in a murine model of bacteremia using signature-tagged mutagenesis, *Mol. Microbiol.*, 26, 399, 1997.
102. Charpentier, E., Novak, R., and Tuomanen, E., Regulation of growth inhibition at high temperatures, autolysis, transformation and adherence in *Streptococcus pneumoniae* by ClpC, *Mol. Microbiol.*, 37, 717, 2000.
103. Ibrahim, Y.M. et al., Contribution of the ATP-dependent protease ClpCP to the autolysis and virulence of *Streptococcus pneumoniae, Infect. Immun.*, 73, 730, 2005.
104. Kwon, H.Y. et al., Effect of heat shock and mutations in ClpL and ClpP on virulence gene expression in *Streptococcus pneumoniae, Infect. Immun.*, 71, 3757, 2003.
105. Nair, S., Milohanic, E., and Berche, P., ClpC ATPase is required for cell adhesion and invasion of *Listeria monocytogenes, Infect. Immun.*, 68, 7061, 2000.
106. Foucaud-Scheunemann, C., and Poquet, I., HtrA is a key factor in the response to specific stress conditions in *Lactococcus lactis, FEMS Microbiol. Lett.*, 224, 53, 2003.
107. Lipinska, B. et al., Identification, characterization, and mapping of the *Escherichia coli htrA* gene, whose product is essential for bacterial growth only at elevated temperatures, *J. Bacteriol.*, 171, 1574, 1989.
108. Noone, D., Howell, A., and Devine, K.M., Expression of *ykdA*, encoding a *Bacillus subtilis* homologue of HtrA, is heat shock inducible and negatively autoregulated, *J. Bacteriol.*, 182, 1592, 2000.
109. Fanning, A.S., and Anderson, M., PDZ domains: Fundamental building blocks in the organization of protein complexes at the plasma membrane, *J. Clin. Invest.*, 103, 767, 1999.
110. Sassoon, N., Arie, J.P., and Betton, J.M., PDZ domains determine the native oligomeric structure of the DegP (HtrA) protease, *Mol. Microbiol.*, 33, 583, 1999.
111. Spiess, C., Beil, A., and Ehrmann, M., A temperature-dependent switch from chaperone to protease in a widely conserved heat shock protein, *Cell*, 97, 339, 1999.
112. Ibrahim, Y.M. et al, Role of HtrA in the virulence and competence of *Streptococcus pneumoniae, Infect. Immun.*, 72, 3584, 2004.

113. Jones, C.H. et al., Conserved DegP protease in Gram-positive bacteria is essential for thermal and oxidative tolerance and full virulence in *Streptococcus pyogenes, Infect. Immun.*, 69, 5538, 2001.
114. Utaida, S. et al., Genome-wide transcriptional profiling of the response of *Staphylococcus aureus* to cell-wall-active antibiotics reveals a cell-wall-stress stimulon, *Microbiol.*, 149, 2719, 2003.
115. Kunst, F. et al., The complete genome sequence of the Gram-positive bacterium *Bacillus subtilis, Nature*, 390, 249, 1997.
116. Poquet, I. et al., HtrA is the unique surface housekeeping protease in *Lactococcus lactis* and is required for natural protein processing, *Mol. Microbiol.*, 35, 1042, 2000.
117. Antelmann, H. et al., The extracellular proteome of *Bacillus subtilis* under secretion stress conditions, *Mol. Microbiol.*, 49, 143, 2003.
118. Sleator, R.D., and Hill, C., A novel role for the LisRK two-component regulatory system in listerial osmotolerance, *Clin. Micobiol. Infect.*, 11, 599, 2005.
119. Wilson, R.L. et al., *Listeria monocytogenes* 10403S HtrA is necessary for resistance to cellular stress and virulence, *Infect. Immun.*, 74, 765, 2006.
120. Wonderling, L.D., Wilkinson, B.J., and Bayles, D.O., The *htrA* (*degP*) gene of *Listeria monocytogenes* 10403S is essential for optimal growth under stress conditions, *Appl. Environ. Microbiol.*, 70, 1935, 2004.
121. O'Mahoney, J., unpublished data, 2005.
122. Danese, P.N. et al., The Cpx two-component signal transduction pathway of *Escherichia coli* regulates transcription of the gene specifying the stress-inducible periplasmic protease, DegP, *Genes Dev.*, 9, 387, 1995.
123. Heusipp, G. et al., Regulation of *htrA* expression in *Yersinia enterocolitica, FEMS Microbiol. Lett.*, 231, 227, 2004.
124. Mascher, T. et al., The *Streptococcus pneumoniae* cia regulon: CiaR target sites and transcription profile anlaysis, *J. Bacteriol.*, 185, 60, 2003.
125. Sebert, M.E. et al., Microarray-based identification of *htrA*, a *Streptococcus pneumoniae* gene that is regulated by the CiaRH two-component system and contributes to nasopharyngeal colonization, *Infect. Immun.*, 70, 4059, 2002.
126. Westers, H. et al., The CssRS two-component regulatory system controls a general secretion stress response in *Bacillus subtilis, J. FEBS*, 273, 3816, 2006.
127. Cotter, P.D. et al., Identification and disruption of *lisRK*, a genetic locus encoding a two-component signal transduction system involved in stress tolerance and virulence in *Listeria monocytogenes, J. Bacteriol.*, 181, 6840, 1999.
128. Cotter, P.D., Guinane, C.M., and Hill, C., The LisRK signal transduction system determines the sensitivity of *Listeria monocytogenes* to nisin and cephalosporins, *Antimicrob. Agents Chemother.*, 46, 2784, 2002.
129. Johnson, K. et al., The role of a stress-response protein in *Salmonella typhimurium* virulence, *Mol. Microbiol.*, 5, 401, 1991.
130. Cortes, G. et al., Role of the *htrA* gene in *Klebsiella pneumoniae* virulence, *Infect. Immun.*, 70, 4772, 2002.
131. Pedersen, L.L. et al., HtrA homologue of *Legionella pneumophila*: An indispensable element for intracellular infection of mammalian but not protozoan cells, *Infect. Immun.*, 69, 2569, 2001.
132. Biswas, S., and Biswas, I., Role of HtrA in surface protein expression and biofilm formation by *Streptococcus mutans, Infect. Immun.*, 73, 6923, 2005.
133. Bearson, S., Bearson, B., and Foster, W., Acid stress responses in enterobacteria, *FEMS Microbiol. Lett.*, 147, 173, 1997.
134. Cotter, P.D., and Hill, C., Surviving the acid test: responses of Gram-positive bacteria to low pH, *Microbiol. Mol. Microbiol. Rev.*, 67, 429, 2003.
135. Hill, C., O'Driscoll, B., and Booth, I., Acid adaptation and food poisoning microorganisms, *Int. J. Food Microbiol.*, 28, 245, 1995.
136. Foster, J.W., *Salmonella* acid shock proteins are required for the acid tolerance response, *J. Bacteriol.*, 173, 6896, 1991.
137. Foster, J.W. et al., Regulatory circuits involved with pH-regulated gene expression in *Salmonella typhimurium, Microbiology*, 140, 341, 1994.
138. Kroll, R.G., and Patchett, R.A., Induced acid tolerance in *Listeria monocytogenes, Lett. Appl. Microbiol.*, 14, 224, 1992.

139. O'Driscoll, B., Gahan, C.G.M., and Hill, C., Two-dimensional polyacrylamide gel electrophoresis analysis of the acid tolerance response in *Listeria monocytogenes* LO28, *Appl. Environ. Microbiol.*, 63, 2679, 1997.
140. Conte, M.P. et al., Acid tolerance in *Listeria monocytogenes* influences invasiveness of enterocyte-like cells and macrophage cells, *Microbial Pathog.*, 29, 137, 2000.
141. Conte, M.P. et al., Effect of acid adaptation on the fate of *Listeria monocytogenes* in THP-1 human macrophages activated by gamma interferon, *Infect. Immun.*, 70, 4369, 2002.
142. Gahan, C.G.M., and Hill, C., The relationship between acid stress responses and virulence in *Salmonella typhimurium* and *Listeria monocytogenes*, *Int. J. Food Microbiol.*, 50, 93, 1999.
143. Marron, L. et al., A mutant of *Listeria monocytogenes* LO28 unable to induce an acid tolerance response displays diminished virulence in a murine model, *Appl. Environ. Microbiol.*, 63, 4945, 1997.
144. O'Driscoll, B., Gahan, C.G.M., and Hill, C., Adaptive acid tolerance response in *Listeria monocytogenes*: Isolation of an acid tolerant mutant which displays increased virulence, *Appl. Environ. Microbiol.*, 62, 1693, 1996.
145. Banks, E.R. et al., Characterization of anaerobic bacteria by using a commercially available rapid tube test for glutamic acid decarboxylase, *J. Clin. Microbiol.*, 27, 361, 1989.
146. Cozzani, I., Misuri, A., and Santoni, C., Purification and general properties of glutamate decarboxylase from *Clostridium perfringens*, *J. Biochem.*, 118, 135, 1970.
147. Smith, K. et al., *Escherichia coli* has two homologous glutamate decarboxylase genes that map to distinct loci, *J. Bacteriol.*, 174, 5820, 1992.
148. Waterman, S.R., and Small, P.L., Identification of sigma S-dependent genes associated with the stationary-phase acid-resistance phenotype of *Shigella flexneri*, *Mol. Microbiol.*, 21, 925, 1996.
149. Small, P.L., and Waterman, S.R., Acid stress, anaerobiosis and *gadCB*: Lessons from *Lactococcus lactis* and *Escherichia coli*, *Trends Microbiol.*, 6, 214, 1998.
150. Cotter, P.D., O'Reilly, K., and Hill, C., Role of the glutamate decarboxylase acid resistance system in the survival of *Listeria monocytogenes* LO28 in low pH foods, *J. Food. Prot.*, 64, 1362, 2001.
151. Cotter, P., Gahan, C.G.M., and Hill, C., A glutamate decarboxylase system protects *Listeria monocytogenes* in gastric fluid, *Mol. Microbiol.*, 40, 465, 2001.
152. Begley, M., Gahan, C.G.M., and Hill, C., Bile stress response in *Listeria monocytogenes* LO28: adaptation, cross-protection and identification of genetic loci involved in bile resistance, *Appl. Environ. Microbiol.*, 68, 6005, 2002.
153. Cotter, P.D. et al., Presence of GadD1 glutamate decarboxylase in selected *Listeria monocytogenes* strains is associated with an ability to grow at low pH, *Appl. Environ. Microbiol.*, 71, 2832, 2005.
154. Ryan, S., Ph.D. thesis, Molecular and phenotypic characterisation of selected acid resistance systems in *Listeria monocytogenes*, University College Cork, Cork, Ireland, 2006.
155. Cotter, P., Personal communication.
156. Maurelli, A.T. et al., "Black holes" and bacterial pathogenicity: A large genomic deletion that enhances the virulence of *Shigella* spp. and enteroinvasive *Escherichia coli*, *Proc. Natl. Acad. Sci. USA*, 95, 3943, 1998.
157. Hill, M.J., Factors controlling the microflora of the healthy upper gastrointestinal tract, In *Human microbial ecology,* Hill, M.J., and Marsh, P.D., eds. CRC Press, Boca Raton, FL, 2002, p. 57.
158. Olier, M. et al., Screening of glutamate decarboxylase activity and bile salt resistance of human asymptomatic carriage, clinical, food and environmental isolates of *Listeria monocytogenes*, *Int. J. Food Microbiol.*, 93, 87, 2004.
159. Higuchi, T., Hayashi, H., and Abe, K., Exchange of glutamate and γ-aminobutyrate in a *Lactobacillus* strain, *J. Bacteriol.*, 179, 3362, 1997.
160. Bhagwat, A.A., Regulation of the glutamate-dependent acid-resistance system of diarrheagenic *Escherichia coli* strains, *FEMS Microbiol. Lett.*, 227, 39, 2003.
161. Ma, Z. et al., GadE (YhiE) activates glutamate decarboxylase-dependent acid resistance in *Escherichia coli* K-12, *Mol. Microbiol.*, 49, 1309, 2003.
162. Becker, L.A. et al., Role of σ^B in adaptation of *Listeria monocytogenes* to growth at low temperature, *J. Bacteriol.*, 182, 7083, 2000.
163. Ferreira, A., O'Byrne, C.P., and Boor, K.J., Role of σ^B in heat, ethanol, acid and oxidative stress resistance and during carbon starvation in *Listeria monocytogenes*, *Appl. Environ. Microbiol.*, 67, 4454, 2001.
164. Ferreira, A. et al., Role of *Listeria monocytogenes* σ^B in survival of lethal acidic conditions and in the acquired acid tolerance response, *Appl. Environ. Microbiol.*, 69, 2692, 2003.

165. Sue, D. et al., σ^B-dependent induction and expression in *Listeria monocytogenes* during osmotic and acid stress conditions simulating the intestinal environment, *Microbiology*, 150, 3843, 2004.
166. Wiedmann, M. et al., General stress transcription factor δ^B and its role in acid tolerance and virulence of *Listeria monocytogenes*, *J. Bacteriol.*, 180, 3650, 1998.
167. Helmann, J.D., and Chamberlin, M.J., Structure and function of bacterial sigma factors, *Annu. Rev. Biochem.*, 57, 839, 1988.
168. Van Schaik, W., and Abee, T., The role of σ^B in the stress response of Gram-positive bacteria-targets for food preservation and safety, *Curr. Opin. Biotechnol.*, 16, 218, 2005.
169. Kazmierczak, M.J. et al., *Listeria monocytogenes* σ^B regulates stress response and virulence functions, *J. Bacteriol.*, 185, 5722, 2003.
170. Kazmierczak, M., Wiedmann, M., and Boor, K.J., Contributions of *Listeria monocytogenes* σ^B and PrfA to expression of virulence and stress response genes during extra- and intracellular growth, *Microbiol.*, 152, 1827, 2006.
171. De Biase, D. et al., The response to stationary-phase stress conditions in *Escherichia coli*: Role and regulation of the glutamic acid decarboxylase system, *Mol. Microbiol.*, 32, 1198, 1999.
172. Erol, I. et al., H-NS controls metabolism and stress tolerance in *Escherichia coli* O157:H7 that influence mouse passage, *BCM Microbiol.*, 6, 72, 2006.
173. Garner, M.R. et al., σ^B contributes to *Listeria monocytogenes* gastrointestinal infection but not to systemic spread in the guinea pig infection model, *Infect. Immun.*, 74, 876, 2006.
174. Barcelona-Andres, B., Marina, A., and Rubio, V., Gene structure, organization, expression and potential regulatory mechanisms of arginine catabolism in *Enterococcus faecalis*, *J. Bacteriol.*, 184, 6289, 2002.
175. Bourdineaud, J.P. et al., Characterization of the *arcD* arginine:ornithine exchanger of *Pseudomonas aeruginosa*. Localization in the cytoplasmic membrane and a topological model, *J. Biol. Chem.*, 268, 5417, 1993.
176. Budin-Verneuil, A. et al., Genetic structure and transcriptional analysis of the arginine deiminase (ADI) cluster in *Lactococcus lactis* MG1363, *Can. J. Microbiol.*, 52, 617, 2006.
177. Dong, Y. et al., Isolation and molecular analysis of the gene cluster for the arginine deiminase system from *Streptococcus gordonii* DL1, *Appl. Environ. Microbiol.*, 68, 5549, 2002.
178. Degnan, B.A. et al., Characterization of an isogenic mutant of *Streptococcus pyogenes* Manfredo lacking the ability to make streptococcal acid glycoprotein, *Infect. Immun.*, 68, 2441, 2000.
179. Gruening, P. et al., Structure, regulation and putative function of the arginine deiminase system of *Streptococcus suis*, *J. Bacteriol.*, 188, 361, 2006.
180. Marquis, R.E. et al., Arginine deiminase system and bacterial adaptation to acid environments, *Appl. Environ. Microbiol.*, 53, 198, 1987.
181. Cotter, P.D., Gahan, C.G.M., and Hill, C., Analysis of the role of the *Listeria monocytogenes* F_0F_1AT-Pase operon in the acid tolerance response, *Int. J. Food Microbiol.*, 60, 137, 2000.
182. Casiano-Colon, A., and Marquis, R., Role of the arginine deiminase system in protecting oral bacteria and an enzymatic basis for acid tolerance, *Appl. Environ. Microbiol.*, 54, 1318, 1988.
183. Zuniga, M. et al., Structural and functional analysis of the gene cluster encoding the enzymes of the arginine deiminase pathway of *Lactobacillus sake*, *J. Bacteriol.*, 180, 4154, 1998.
184. Klarsfeld, A.D., Goossens, P.L., and Cossart, P., Five *Listeria monocytogenes* genes preferentially expressed in infected mammalian cells: *plcA*, *purH*, *purD*, *pyrE* and an arginine ABC transporter gene, *arpJ*, *Mol. Microbiol.*, 13, 585, 1994.
185. Ni, J. et al., Structure of the arginine repressor from *Bacillus stearothermophilus*, *Nat. Struct. Biol.*, 6, 427, 1999.
186. Larsen, R. et al., ArgR and AhrC are both required for regulation of arginine metabolism in *Lactococcus lactis*, *J. Bacteriol.*, 186, 1147, 2004.
187. Degnan, B.A. et al., Inhibition of human peripherial blood mononuclear cell proliferation by *Streptococcus pyogenes* cell extract is associated with arginine deiminase activity, *Infect. Immun.*, 66, 3050, 1998.
188. Benga, L. et al., Nonencapsulated strains reveal novel insights in invasion and survival of *Streptococcus suis* in epithelial cells, *Cell. Microbiol.*, 92, 867, 2004.
189. Klarsfeld, A.D., and Cossart, P., Response from Klarsfeld and Cossart, *Trends Microbiol.*, 3, 85, 1995.
190. Bogdan, C., Rollinghoff, M., and Diefenbach, A., The role of nitric oxide in innate immunity, *Immunol. Rev.*, 173, 17, 2000.

191. MacMicking, J. et al., Altered responses to bacterial infection and endotoxic shock in mice lacking inducible nitric oxide synthase, *Cell*, 81, 641, 1995.
192. MacMicking, J., Xie, Q.W., and Nathan, C., Nitric oxide and macrophage function, *Annu. Rev. Immunol.*, 15, 323, 1997.
193. Doumith, M. et al., New aspects regarding evolution and virulence of *Listeria monocytogenes* revealed by comparative genomics and DNA arrays, *Infect. Immun.*, 72, 1072, 2004.
194. Penefsky, H.S., and Cross, R.L., Structure and mechanism of F_0F_1-type ATP synthases and ATPases, *Adv. Enzymol. Relat. Areas Mol. Biol.*, 64, 173, 1991.
195. Sebald, W. et al., Structure and genetics of the H^+-conducting F_0 portion of the ATP synthase, *Ann. NY Acad. Sci.*, 402, 28, 1982.
196. Harold, F.M., Pavlasova, E., and Baarda, J.R., A transmembrane pH gradient in *Streptococcus faecalis*: Origin and dissipation by proton conductors and *N, N^1-dicyclohexylcarbodimide, Biochem. Biophys. Acta*, 196, 235, 1970.
197. Shibata, C. et al., Gene structure of *Enterococcus hirae* (*Streptococcus faecalis*) F_0F_1-ATPase, which functions as a regulator of cytoplasmic pH, *J. Bacteriol.*, 174, 6117, 1992.
198. Feniouk, B.A., and Junge, W., Regulation of the F_0F_1-ATPase synthase: The conformation of subunit ε might be determined by directionality of subunit γ rotation, *FEBS Lett.*, 579, 5114, 2005.
199. Dunn, S.D., Tozer, R.G., and Zadorozny, V.D., Activation of *Escherichia coli* F_1-ATPase by lauryldimethylamine oxide and ethylene glycol: relationship of ATPase activity to the interaction of the epsilon and beta subunits, *Biochemistry*, 29, 4435, 1990.
200. Laget, P.P., and Smith, J.B., Inhibitory properties of endogenous subunit epsilon in the *Escherichia coli* F_1 ATPase, *Arch. Biochem. Biophys.*, 197, 83, 1979.
201. Peskova, Y.B., and Nokamoto, R.K., Catalytic control and coupling efficiency of the *Escherichia coli* F_0F_1 ATP synthase: influence of the F_0 sector and epsilon subunit on the catalytic transition state, *Biochemistry*, 39, 11830, 2000.
202. Feniouk, B.A., Suzuki, T., and Yoshida, M., The role of subunit epsilon in the catalysis and regulation of F_0F_1-ATP synthase, *Biochem. Biophys. Acta*, 1757, 326, 2006.
203. Stack, H., Hill, C., and Gahan, C.G.M., unpublished data.
204. Hain, T. et al., Whole genome sequence of *Listeria welshimeri* reveals common steps in genome reduction with *Listeria innocua* as compared to *Listeria monocytogenes, J. Bacteriol.*, 188, 7405, 2006.
205. Conska, L.N., and Hanson, A.D., Prokaryotic osmoregulation: Genetics and physiology, *Annu. Rev. Microbiol.*, 45, 569, 1991.
206. Conska, L.N., Physiological and genetic responses of bacteria to osmotic stress, *Microbiol. Rev.*, 53, 121, 1989.
207. Booth, I.R. et al., Mechanisms controlling compatible solute accumulation: A consideration of the genetics and physiology of bacterial osmoregulation, *J. Food Eng.*, 22, 381, 1994.
208. Conska, L.N., and Epstein, W., *Osmoregulation.* In Escherichia coli *and* Salmonella: *Cellular and molecular biology,* 2nd ed., Neidhardt, F.C. et al., eds. American Society for Microbiology, Washington, D.C., 1996, p. 1210.
209. Sleator, R.D., and Hill, C., Bacterial osmoadaptation: The role of osmolytes in bacterial stress and virulence, *FEMS Microbiol. Rev.*, 26, 49, 2001.
210. Horn, C. et al., Biochemical and structural analysis of the *Bacillus subtilis* ABC transporter OpuA and its isolated subunits, *J. Mol. Microbiol. Biotechnol.*, 10, 76, 2005.
211. Morbach, S., and Kramer, R., Structure and function of the betaine uptake system BetP of *corynebacterium glutamicum*: Strategies to sense osmotic and chill stress, *J. Mol. Microbiol. Biotechnol.*, 10, 143, 2005.
212. Nagata, S. et al., Effect of compatible solutes on the respiratory activity and growth of *Escherichia coli* K-12 under NaCl stress, *J. Biosci. Bioeng.*, 94, 384, 2002.
213. Smiddy, M. et al., Role for compatible solutes glycine betaine and L-carnitine in listerial barotolerance, *Appl. Environ. Microbiol.*, 70, 7555, 2004.
214. Patchett, R.A., Kelly, A.F., and Kroll, R.G., Effect of sodium chloride on the intracellular solute pools of *Listeria monocytogenes, Appl. Environ. Microbiol.*, 58, 3959, 1992.
215. Bayles, D.O., and Wilkinson, B.J., Osmoprotectants and cryoprotectants for *Listeria monocytogenes, Lett. Appl. Microbiol.*, 30, 23, 2000.
216. Beumer, R.R. et al., Effect of exogenous proline, betaine and carnitine on growth of *Listeria monocytogenes* in a minimal medium, *Appl. Environ. Microbiol.*, 60, 1359, 1994.

217. Amezaga, M.R., Ph.D. thesis, University of Aberdeen, Scotland, 1996.
218. Sleator, R.D., Gahan, C.G.M., and Hill, C., A postgenomic appraisal of osmotolerance in *Listeria monocytogenes, Appl. Environ. Microbiol.*, 69, 1, 2003.
219. Rhodes, D., and Hanson, A.D., Quaternary ammonium and tertiary sulfonium compounds in higher plants, *Annu. Rev. Plant Physiol.*, 44, 357, 1993.
220. Bieber, L.L., Carnitine, *Annu. Rev. Biochem.*, 57, 261, 1988.
221. Verheul, A. et al., Betaine and L-carnitine transport by *Listeria monocytogenes* Scott A in response to osmotic signals, *J. Bacteriol.*, 179, 6979, 1997.
222. Fraser, K.R. et al., Identification and characterization of an ATP binding cassette L-carnitine transporter in *Listeria monocytogenes, Appl. Environ. Microbiol.*, 66, 4696, 2000.
223. Ko, R., and Smith, L.T., Identification of an ATP-driven, osmoregulated glycine betaine transport system in *Listeria monocytogenes, Appl. Environ. Microbiol.*, 65, 4040, 1999.
224. Patchett, R.A., Kelly, A.F., and Kroll, R.G., Transport of glycine betaine by *Listeria monocytogenes, Arch. Microbiol.*, 162, 205, 1994.
225. Sleator, R.D. et al., Identification and disruption of BetL, a secondary glycine betaine transport system linked to the salt tolerance of *Listeria monocytogenes* LO28, *Appl. Environ. Microbiol.*, 67, 2078, 1999.
226. Sleator, R.D. et al., Analysis of the role of *betL* in contributing to the growth and survival of *Listeria monocytogenes* LO28, *Int. J. Food Microbiol.*, 60, 261, 2000.
227. Sleator, R.D. et al., Betaine and carnitine uptake systems in *Listeria monocytogenes* affect growth and survival in foods and during infection, *J. Appl. Microbiol.*, 95, 839, 2003.
228. Gerhardt, P.N.M., Smith, L.T., and Smith, G.M., Sodium-driven, osmotically acitivated glycine betaine transport in *Listeria monocytogenes* membrane vesicles, *J. Bacteriol.*, 178, 6105, 1996.
229. Gerhardt, P N.M., Smith, L.T., and Smith, G.M., Osmotic and chill activation of glycine betaine porter II in *Listeria monocytogenes* membrane vesicles, *J. Bacteriol.*, 182, 2544, 2000.
230. Ko, R., Smith, L.T., and Smith, G.M., Glycine betaine confers enhanced osmotolerance and cryotolerance on *Listeria monocytogenes, J. Bacteriol.*, 176, 426, 1994.
231. Mendum, M.L., and Smith, L.T., Characterization of glycine betaine porter I from *Listeria monocytogenes* and its roles in salt and chill tolerance, *Appl. Environ. Microbiol.*, 68, 813, 2002.
232. Sheehan, V.M. et al., Heterologous expression of BetL, a betaine uptake system, enhances the stress tolerance of *Lactobacillus salivarius* UCC118, *Appl. Environ. Microbiol.*, 72, 2170, 2006.
233. Angelidis, A.S. et al., Identification of OpuC as a chill-activated and osmotically activated carnitine transporter in *Listeria monocytogenes, Appl. Environ. Microbiol.*, 68, 2644, 2002.
234. Sleator, R.D. et al., Analysis of the role of OpuC, an osmolyte transport system, in salt tolerance and virulence potential of *Listeria monocytogenes, Appl. Environ. Microbiol.*, 67, 2692, 2001.
235. Verheul, A. et al., An ATP-dependent L-carnitine transporter in *Listeria monocytogenes*, Scott A is involved in osmoprotection, *J. Bacteriol.*, 177, 3205, 1995.
236. Wemekamp-Kamphuis, H.H. et al., Multiple deletions of the osmolyte transporters BetL, Gbu and OpuC of *Listeria monocytogenes* affect virulence and growth at high osmolarity, *Appl. Environ. Microbiol.*, 68, 4710, 2002.
237. Sleator, R.D., Wood, J.M., and Hill, C., Transcriptional regulation and posttranslational activity of the betaine transporter BetL in *Listeria monocytogenes* are controlled by environmental salinity, *J. Bacteriol.*, 185, 7140, 2003.
238. Angelidis, A.S., and Smith, G.M., Three transporters mediate uptake of glycine betaine and carnitine by *Listeria monocytogenes* in response to hyperosmotic stress, *Appl. Environ. Microbiol.*, 69, 1013, 2003.
239. Smith, L.T., Role of osmolytes in adaptation of osmotically stressed and chill-stressed *Listeria monocytogenes* grown in liquid media and on processed meat surfaces, *Appl. Environ. Microbiol.*, 62, 3088, 1996.
240. Peddie, B.A. et al., Relationship between osmoprotection and the structure and intracellular accumulation of betaines by *Escherichia coli, FEMS Microbiol. Lett.*, 120, 125, 1994.
241. Cetin, M.S. et al., Regulation of transcription of compatible solute transporters by the general stress sigma factor, σ^B, in *Listeria monocytogenes, J. Bacteriol.*, 186, 794, 2004.
242. Fraser, K.R. et al., Role of σ^B in regulating the compatible solute uptake systems of *Listeria monocytogenes* osmotic inductions of *opuC* is σ^B-dependent, *Appl. Environ. Microbiol.*, 69, 2015, 2003.
243. Becker, L.A. et al., Identification of the gene encoding the alternative sigma factor σ^B from *Listeria monocytogenes* and its role in osmotolerance, *J. Bacteriol.*, 180, 4547, 1998.

244. Milohanic, E. et al., Transcriptome anlaysis of *Listeria monocytogenes* identifies three groups of genes differently regulated by PrfA, *Mol. Microbiol.*, 47, 1613, 2003.
245. Sue, D., Boor, K.J., and Wiedmann, M., σ^B-dependent expression patterns of compatible solute transporter genes *opuCA* and *lmo1421* and the conjugated bile salt hydrolase gene *bsh* in *Listeria monocytogenes*, *Microbiology*, 149, 3247, 2003.
246. Kappes, R.M., Kempf, B., and Bremer, E., Three transport systems for the osmoprotectant glycine betaine operate in *Bacillus subtilis*: Characterization of OpuD, *J. Bacteriol.*, 178, 5071, 1996.
247. Akbar, S.S. et al., Two genes from *Bacillus subtilis* under the sole control of the general stress transcription factor σ^B, *Microbiology*, 145, 1069, 1999.
248. Maul, B. et al., σ^B-dependent regulation of *gsiB* in response to multiple stimuli in *Bacillus subtilis*, *Mol. Gen. Genet.*, 248, 114, 1995.
249. Petershon, A. et al., Identification and transcriptional analysis of new members of the σ^B regulon in *Bacillus subtilis*, *Microbiology*, 145, 869, 1999.
250. Sanchez-Campillo, M. et al., Modulation of DNA topology by *flaR*, a new gene from *Listeria monocytogenes*, *Mol. Microbiol.*, 18, 801, 1995.
251. Peterson, J.A., Lorence, M.C., and Amarneh, B., Putidaredoxin reductase and putidaredoxin. Cloning, sequence determination and heterologous expression of the proteins, *J. Biol. Chem.*, 265, 6066, 1990.
252. Reddy, P., Peterkofsky, A., and McKenney, K., Translational efficiency of the *Escherichia coli* adenylate cyclase gene: Mutating the UUG initiation codon to CUG or AUG results in increased gene expression, *Proc. Natl. Acad. Sci. USA*, 82, 5656–5660, 1985.
253. Chowdhury, R., Sahu, G. K., and Das, J., Stress response in pathogenic bacteria. *J. Biosci.*, 21, 149–160, 1996.
254. Duche, O. et al., Salt stress proteins induced in *Listeria monocytogenes*, *Appl. Environ. Microbiol.*, 68, 1491–1498, 2002.
255. Duche, O. et al., A proteomic analysis of the salt stress response of *Listeria monocytogenes*, *FEMS Microbiol. Lett.*, 215, 183–188, 2002.
256. Esvan, H. et al., Protein variations in *Listeria monocytogenes* exposed to high salinities, *Int. J. Food Microbiol.*, 55, 151, 2000.
257. Okada, Y. et al., Cloning of *rel* from *Listeria monocytogenes* as an osmotolerance involvement gene, *Appl. Environ. Microbiol.*, 68, 1541, 2002.
258. Gardan, R. et al., Identification of *Listeria monocytogenes* genes involved in salt and alkaline-pH tolerance, *Appl. Environ. Microbiol.*, 69, 3137, 2003.
259. Gardan, R. et al., Role of *ctc* from *Listeria monocytogenes* in osmotolerance, *Appl. Environ. Microbiol.*, 69, 154, 2003.
260. Brondsted, L. et al., *kdpE* and a putative RsbQ homologue contribute to growth of *Listeria monocytogenes* at high osmolarity and low temperature, *FEMS Microbiol. Lett.*, 219, 233, 2003.
261. Kallipolitis, B.H., and Ingmer, H., *Listeria monocytogenes* response regulators important for stress tolerance and pathogenesis, *FEMS Microbiol. Lett.*, 204, 111, 2001.
262. Sleator, R.D., Gahan, C.G.M., and Hill, C., Identification and disruption of the *proBA* locus in *Listeria monocytogenes*: Role of proline biosythesis in salt tolerance and murine infection, *Appl. Environ. Microbiol.*, 67, 2571, 2001.
263. Sleator, R.D., Gahan, C.G.M., and Hill, C., Mutations in the listerial *proB* gene leading to proline overproduction: Effects on salt tolerance and murine infection, *Appl. Environ. Microbiol.*, 67, 4560, 2001.
264. Begley, M., Hill, C., and Gahan, C.G.M., Identification and disruption of *btlA*, a locus involved in bile tolerance and general stress resistance in *Listeria monocytogenes*, *FEMS Microbiol. Lett.*, 218, 31, 2003.
265. Taylor, C.M. et al., *Listeria monocytogenes relA* and *hpt* mutants are impaired in surface-attached growth and virulence, *J. Bacteriol.*, 184, 621, 2002.
266. Cashel, M. et al., The stringent response. In Escherichia coli *and* Salmonella: *Cellular and molecular biology,* 2nd ed., Neidhardt, F.C. et al., eds. American Society for Microbiology, Washington, D.C., 1996, p. 1458.
267. Mechold, U. et al., Functional analysis of a *relA/spoT* homolog from *Streptococcus equisimilis*, *J. Bacteriol.*, 178, 1401, 1996.
268. Sarubbi, E. et al., Characterization of the *spoT* gene of *Escherichia coli*, *J. Biol. Chem.*, 264, 15074, 1989.
269. Wendrich, T.M., and Marahiel, M.A., Cloning and characterization of a *relA/spoT* homologue from *Bacillus subtilis*, *Mol. Microbiol.*, 26, 65, 1997.

270. Stack, H., unpublished data.
271. Epstein, W., Osmoregulation by potassium transport in *Escherichia coli*, *FEMS Microbiol. Rev.*, 39, 73, 1986.
272. Brody, M.S., Vijay, K., and Price, C.W., Catalytic function of an alpha/beta hydrolase is required for energy stress activation of the σ^B transcription factor in *Bacillus subtilis*, *J. Bacteriol.*, 183, 6422, 2001.
273. Kempf, B., and Bremer, E., Uptake and synthesis of compatible solutes as microbial stress responses to high osmolarity environments, *Arch. Microbiol.*, 170, 319, 1998.
274. Gentry, D.R. et al., Synthesis of the stationary-phase sigma factor sigma s is positively regulated by ppGpp, *J. Bacteriol.*, 175, 7982, 1993.
275. Kvint, K., Farewell, A., and Nystrom, T., RpoS-dependent promoters require guanosine tetraphosphate for induction even in the presence of high levels of sigma(s), *J. Biol. Chem.*, 275, 14795, 2000.
276. Zhang, S., and Haldenwang, W.G., RelA is a component of the nutritional stress activation pathway of the *Bacillus subtilis* transcription factor δ^B, *J. Bacteriol.*, 185, 5714, 2003.
277. Vellanoweth, R.L., Translation and its regulation. In *Bacillus subtilis and other Gram-positive bacteria: Biochemistry, physiology, and genetics,* Sonenshein, A.L., Hoch, J.A., and Losick, R., eds. American Society for Microbiology, Washington, D.C., 1993, p. 699.
278. Hammer, B.K., and Swanson, M.S., Co-ordination of *Legionella pneumophila* virulence with entry into stationary phase by ppGpp, *Mol. Microbiol.*, 33, 721, 1999.
279. Lou, Y., and Yousef, A.E., Adaptation to sublethal environmental stresses protects *Listeria monocytogenes* against lethal preservation factors, *Appl. Environ. Microbiol.*, 63, 1252, 1997.
280. Browne, N., and Dowds, B.C.A., Heat and salt stress in the food pathogen *Bacillus cereus*, *J. Appl. Microbiol.*, 91, 1085, 2001.
281. Browne, N., and Dowds, B.C.A., Acid stress in the food pathogen *Bacillus cereus*, *J. Appl. Microbiol.*, 92, 404, 2002.
282. Casey, P.G., and Condon, S., Sodium chloride decreases the bacteriocidal effect of acid pH on *Escherichia coli* 0157:H45, *Int. J. Food Microbiol.*, 76, 199, 2002.
283. Flahaut, S. et al., Alkaline stress response in *Enterococcus faecalis*: Adaptation, cross-protection, and changes in protein synthesis, *Appl. Environ. Microbiol.*, 63, 812, 1997.
284. Faleiro, M.L., Andrew, P.W., and Power, D., Stress response of *Listeria monocytogenes* isolated from cheese and other foods, *Int. J. Food Microbiol.*, 84, 207, 2003.
285. Gahan, C.G.M., O'Driscoll, B., and Hill, C., Acid adaptation of *Listeria monocytogenes* can enhance survival in acidic foods during milk fermentation, *Appl. Environ. Microbiol.*, 62, 3128, 1996.
286. Koutsoumanis, K.P., Kendall, P.A., and Sofos, J.N., Effect of food processing-related stresses on acid tolerance of *Listeria monocytogenes*, *Appl. Environ. Microbiol.*, 69, 7514, 2003.
287. Van Schaik, W., Gahan, C.G.M., and Hill, C., Acid-adapted *Listeria monocytogenes* displays enhanced tolerance against the lantibiotics nisin and lacticin 3147, *J. Food Prot.*, 62, 536, 1999.
288. Ito, M. et al., *mrp*, a multigene, multifunctional locus in *Bacillus subtilis* with roles in resistance to cholate and to Na+ and in pH homeostasis, *J. Bacteriol.*, 181, 2394, 1999.
289. Krulwich, T.A., Ito, M., and Guffanti, A.A., The Na(+)-dependence of alkaliphily in *Bacillus*, *Biochim. Biophys. Acta,* 1505, 158, 2001.

4 Pathogenesis

Michael Kuhn, Mariela Scortti, and José A. Vázquez-Boland

CONTENTS

4.1 INTRODUCTION

Listeria monocytogenes is a multisystemic invasive pathogen capable of colonizing multiple host tissues, causing a range of clinical conditions. Some of these conditions, namely meningoencephalitis and abortion, are much more common than others, reflecting the specific tropism *L. monocytogenes* has for the brain and the placenta. In most cases, the clinical manifestations of listeriosis occur in debilitated, immunocompromised individuals and therefore *L. monocytogenes* can be considered as an opportunistic pathogen. *L. monocytogenes* is also a multihost pathogen capable of infecting and causing disease in a wide variety of animal species, including birds and mammals. The clinical and pathological manifestations of listeriosis are very similar in humans and animals, indicating that the major underlying mechanisms of pathogenesis are essentially alike in all susceptible hosts. Listeriosis has one of the highest hospitalization and mortality rates (90% and 20–30%, respectively) of all food-borne infections.[1,2–4]

The pathogenesis, virulence determinants, and regulation have been the major topics of research on *L. monocytogenes* for the past few decades. Several in-depth reviews have been published in the past few years covering different aspects of recent progress in the field.[1,5–7] This chapter summarizes the essence of our current understanding on the pathogenic features of *L. monocytogenes*, presents an updated overview of the knowledge available on the genetic and molecular virulence determinants of this model intracellular parasite, and reviews recently gathered insights resulting from the current genomic approaches in *Listeria* research.

4.1.1 Clinical Manifestations

Listeriosis can present as a generalized infection with sepsis and bacteremia, or as a localized infection in different parts of the body. Two main forms of listeriosis can be distinguished according to the age band of patients: fetomaternal/neonatal listeriosis, and listeriosis in adults.[1,3,8] The former is thought to result from the blood-borne colonization of the placenta, subsequently leading to listerial invasion of the fetus. The outcome is either abortion—typically within the last third of the gestation period—or the birth of a baby or stillborn fetus with a severe, most often fatal septicemic syndrome known as "granulomatosis infantiseptica" (Figure 4.1). The infection is generally asymptomatic in the mother, although women having suffered a listerial miscarriage often refer to having experienced flu-like episodes within the previous 15 days. Late-onset perinatal listeriosis, affecting newborns between weeks 1 and 8 postpartum, is less frequent and can result from low-level transplacental infection, aspiration of contaminated maternal exudates, or horizontal transmission at neonatology wards via contaminated fomites or inadequate hygiene of the attending personnel. The babies affected by late-onset listeriosis show a febrile syndrome usually associated with meningitis, although gastroenteritis and pneumonia can also be observed. The mortality is generally lower than in early-onset listeriosis.[1,2,9,10]

Listeriosis in nonpregnant adults is typically associated (50–70% of cases) with central nervous system (CNS) infection manifesting as meningitis or, most often, meningoencephalitis (Figure 4.1). Purely encephalitic forms with no meningeal involvement are sometimes observed but these are more common in animals (ruminants). Brain abscess accounts for about 10% of listerial CNS infections. Mortality of brain infection due to *Listeria* ranges between 20 and 60% depending on the severity of the process and the underlying condition of the patient. *L. monocytogenes* accounts for up to 10% of community-acquired bacterial meningitis and is the third most common cause of bacterial meningitis in the adult population after *Streptococcus pneumoniae* and *Neisseria meningitidis*. In some risk groups, such as cancer patients, *L. monocytogenes* ranks first among the causes of bacterial meningitis.[3,11,12] The second most common manifestation of *L. monocytogenes* infection in the adult patient is bacteremia. This condition generally responds well to treatment but again the prognosis is largely dependent on the circumstances of the patient and can be associated with up to 70% mortality in severely debilitated individuals, who may develop acute septicemic disease. *L. monocytogenes* infections can also localize to the cardiovascular system (endocarditis—third most common form of listeriosis, myocarditis, arteritis), the respiratory system (pneumonia, pleuritis, sinusitis), the musculoskeletal system (arthritis, osteomyelitis), the eye (ophthalmitis, conjunctivitis), the mammary gland (mastitis in cows) and the skin (primary pyogranulomatous dermatitis).[10,13–16] *L. monocytogenes* has been recently associated with enteric illness characterized by fever, diarrhea, and vomiting, without concomitant invasive infection.[17]

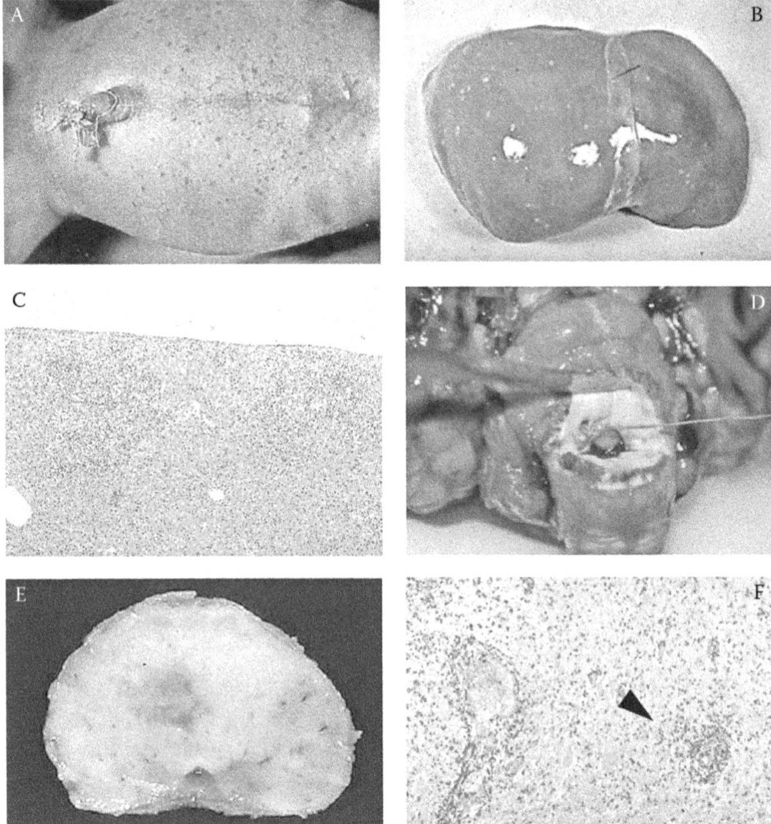

FIGURE 4.1 Clinical and pathological features of *Listeria* infection. (A) Fetomaternal listeriosis. Stillborn fetus with septicemic invasion ("granulomatosis infantiseptica"). (B) Liver from the stillborn fetus in (A) showing typical disseminated pyogranulomatous necrotic foci. (C) Histopathological image of the liver from an experimentally infected sheep with milliary listerial pyogranulomatous hepatitis (hematoxilin/eosin-stained section, 60×). (D) Meningoencephalitis due to *L. monocytogenes* in a cow. (E) Section of the brainstem of a sheep with listerial rhombencephalitis showing inflammatory lesions in the nerve tissue. (F) Parenchymal inflammatory infiltration of the brainstem in (E) showing typical perivascular cuffing (arrow) indicative of blood-borne invasion of the brain tissue by *L. monocytogenes*. Clinical and pathological manifestations of listeriosis are essentially similar in humans and animals. (Panels A–C, E, and F taken from Vazquez-Boland, J.A. et al., *Clin. Microbiol. Rev.*, 14, 584, 2001. With permission.)

4.1.2 Physiopathogenesis

The primary source of infection for both sporadic and epidemic listeriosis is almost invariably contaminated food and hence the gastrointestinal tract is the major portal of entry of *L. monocytogenes* into the host (Figure 4.2).[1,2] Other possible infection routes include

direct transmission via the skin, characterized by a pyogranulomatous rash—generally on hands and arms—that is sporadically seen among farmers and veterinarians exposed to genital secretions or aborted fetuses from cases of listerial miscarriage in ruminants[16,18]

a particular form of food-borne listeriosis that occurs in ruminants, in which *L. monocytogenes*, vehiculated by contaminated silage, is believed to gain access directly to the rhombencephalon by ascending through the cranial nerves by cell-to-cell spread (see later

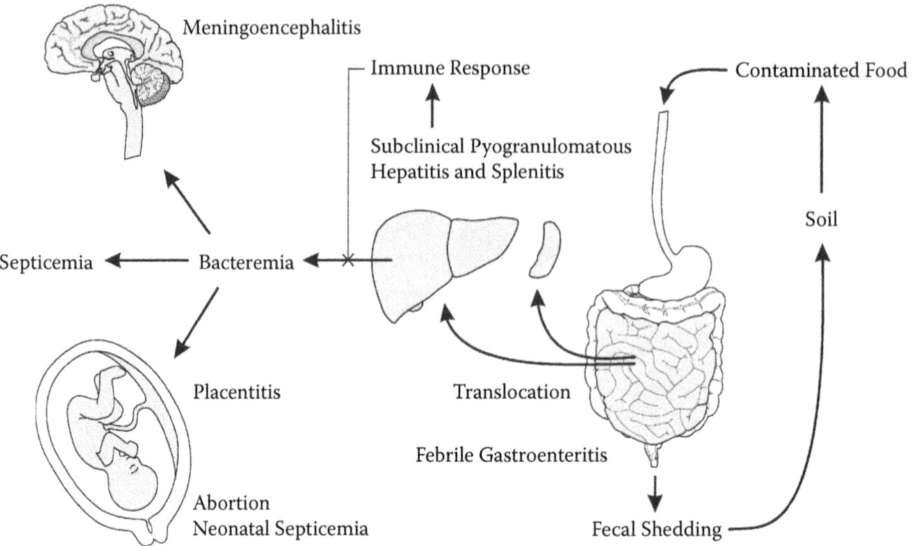

FIGURE 4.2 Schematic representation of the physiopathology of *L. monocytogenes* infection. (Modified from Vázquez-Boland, J.A. et al., *Clin. Microbiol. Rev.*, 14, 584, 2001. With permission.)

discussion) after local invasion of the nerve terminals in the mouth, nasopharynx, or even the eye[1]

the transplacental vertical transmission that occurs in fetomaternal listeriosis, although the infection in the mother is primarily acquired via contaminated food

There are two mechanisms by which *L. monocytogenes* can enter into the host through the intestinal mucosa. One is the direct invasion of enterocytes lining the absorptive epithelium of the microvilli, leading to infection of the intestinal cells. This "enteropathogenic" mechanism involves specific ligand–receptor interactions and is thought to occur only in those animal hosts in which the intestinal cells express "susceptible" isoforms of the receptors to the listerial invasins (e.g., humans or guinea pigs; see later discussion).[1,19] The other entry pathway involves phagocytosis by the M cells of the Peyer's patches.[20,21] This second intestinal translocation mechanism is unspecific (i.e., the nonpathogenic species *L. innocua* is translocated equally well as *L. monocytogenes*), occurs in host systems that do not express the susceptible isoform of the receptors for the listerial invasins (e.g., mice and rats), and is thought to be less efficient than that involving the invasion of enterocytes.[19,22]

Whatever the initial mechanism of entry, the bacteria subsequently localize within professional phagocytes and antigen presenting cells in inflammatory foci in the lamina propria, in particular in the subepithelial follicular tissue, where bacterial replication takes place.[22,23] *In vivo* experimental data using a rat ileal loop model show that *Listeria* bacteria disseminate very rapidly (within 6 h) from the gut to the mesenteric lymph nodes (MLN), presumably vehiculated within dendritic cells.[24] Antigen-presentation events during the intestinal phase of infection by *L. monocytogenes* may be crucial for the early mounting (or boosting) of an antilisterial protective immune response important for preventing the further spread of the pathogen within/from the primary target organs. (The immunological aspects of *L. monocytogenes* infection are extensively discussed in chapters 13 and 14).

The primary target organs are the spleen and liver, which *L. monocytogenes* reaches by lymphohematogenous dissemination from the MLN (Figure 4.2).[1] Animal models of infection show that this happens within the first 24 h after the delivery of an oral inoculum.[25–27] Most of the bacterial burden (≈90%) localizes in the liver parenchyma. Hepatocytes are highly permissive to *L. monocytogenes* both in terms of receptor-mediated internalization and intracellular replication and are believed to be the principal site of listerial multiplication after intestinal translocation. Discrete infectious foci are formed presumably by direct cell-to-cell spread. These foci become immediately surrounded by neutrophils and later by macrophages, producing typical pyogranulomatous lesions.[28–34] The rest of the listerial inoculum localizes in the spleen. In the mouse, *L. monocytogenes* induces apoptosis of the splenic cells, in particular in type I IFN-activated lymphocytes, causing a dramatic regression of the white pulp due to depletion of both $CD4^+$ and $CD8^+$ lymphocytes. This massive lymphocyte death may down-regulate early innate immune responses, creating a permissive environment for bacterial colonization.[34–37] It is unclear whether splenic lymphocyte apoptosis also occurs in humans or other *Listeria*-susceptible animals.

In an immunocompetent host, the listerial replication foci in the primary target organs are efficiently contained by cytotoxic ($CD8^+$) T lymphocytes, leading to complete resolution of the pyogranulomes after day 6 or 7 after infection.[38] These early stages of *L. monocytogenes* infection are essentially subclinical. Given the relatively high frequency with which the pathogen is found in food, such subclinical infections may occur on a regular basis, contributing to the "natural" antilisterial immunity and the relatively low incidence of listeriosis among the normal, healthy population.[39] However, in debilitated hosts with an impaired cellular immune response, the primary infectious foci are inefficiently contained, resulting in the release of the bacteria to the bloodstream (Figure 4.2). If this low-level bacteremia is sufficiently prolonged, *L. monocytogenes* infection may progress toward clinical listeriosis in the form of septicemic disease or localized infection in the brain or the fetoplacental unit.[1,40–42] There is evidence that *L. monocytogenes* disseminates throughout the body using a Trojan horse mechanism, carried inside infected phagocytes.[43,44] *L. monocytogenes* infection is therefore a multistage process in which the pathogen has first to cross the intestinal barrier, then establish and multiply in primary target organs, and finally cross two further barriers (the endothelial barrier—particularly at the level of the brain microcapillaries and the maternofetal barrier) to establish and multiply in secondary target organs where it causes clinical illness (Figure 4.2). A prolonged initial silent phase of infection is consistent with the long incubation period of invasive listeriosis, typically 20–30 (even up to 70) days after consumption of contaminated food.[45]

4.1.3 Factors Affecting the Outcome of *Listeria* Infection

The consensus view, based on the numbers of *L. monocytogenes* bacteria typically found in foods incriminated in listeriosis cases, dose–response risk assessments, and data from animal models of infection, is that high numbers of *L. monocytogenes*, in excess of 10^6 colony forming units (CFUs), are required to cause invasive listeriosis upon oral exposure. Foods contaminated with $\leq 10^2$ CFUs/g, which corresponds to the numbers normally found in *Listeria*-contaminated retail food, are not considered to pose a risk to humans, even for individuals with increased susceptibility.[26,46–48] Indeed, if the minimum infective dose (MID) were not high, the incidence of listeriosis would be substantially greater than the normally reported figures (between 0.5 and 5 cases per million inhabitants per year), considering the normal rates of food contamination by *L. monocytogenes.*

Another factor that may obviously influence the MID is the virulence of the *L. monocytogenes* strain in question. Although all bacteria belonging to this species are assumed to be pathogenic, epidemiological evidence indicates that strain-to-strain differences in virulence exist. Thus, only 4 of the 13 *Listeria* serotypes—namely, 1/2a, 1/2b, 1/2c, and 4b—account for 96–98% of all cases of human and animal listeriosis worldwide. Of these, serovar 1/2c is found in a minority of clinical isolates (2–4%) but predominates among food isolates. Similarly, serovar 4b, belonging to one of the two major genetic lineages of *L. monocytogenes*, predominates among clinical isolates (>50% of listeriosis cases), whereas it is much less frequently found among food isolates than serogroup 1/2 (i.e., serovars 1/2a, 1/2b, and 1/2c) strains. Moreover, a restricted number of 4b strains, representing distinct genotypes, are responsible for most food-borne outbreaks of human listeriosis worldwide, suggesting that certain clones of *L. monocytogenes* may be more pathogenic for humans.[1–3,49,50]

Virulence heterogeneity among *L. monocytogenes* isolates—often associated to natural attenuating mutations in key virulence loci—is also supported by experimental evidence.[52–55] However, the most critical factor is the underlying condition and immunological status of the host as this determines the susceptibility to a given strain of *L. monocytogenes*. The vast majority of listeriosis patients have a physiological or pathological condition that impairs the capacity to mount an effective cellular immune response.

Two major population groups at risk for invasive listeriosis are the neonates and the elderly (>60 years), in which the immune system is immature or declining, respectively. Pregnant women, another major risk group, are assumed to be more susceptible to listeriosis due to the pregnancy-associated depression of cell-mediated immunity that prevents rejection of the fetoplacental allograft. In nonpregnant adults, almost all cases of listeriosis are seen in individuals with chronic, debilitating illnesses or subjected to immunosuppressive therapy. Specific risk groups in the intermediate-age band include cancer and organ transplant patients, HIV-infected and AIDS patients, and individuals with chronic liver disease (alcoholism and cirrhosis), diabetes, and lupus.[1,56,57]

In summary, the severity and clinical outcome of *L. monocytogenes* infection depend on three principal variables: (1) the number of bacteria ingested with food; (2) the pathogenic properties and virulence of the infecting strain; and (3) the underlying condition and immunological status of the host, as follows. In immunocompetent individuals with no predisposing condition, ingestion of low to moderate doses of *L. monocytogenes* ($\leq 10^5$ CFUs) has no effect other than boosting antilisterial protective immunity, whereas ingestion of large doses of the bacteria ($\geq 10^6$ CFUs, sometimes doses as high as 10^{11}) may cause acute febrile gastroenteritis within 24 h of consumption of the contaminated food due to massive invasion of the intestinal mucosa. Depending on the pathogenicity/virulence of the strain, some healthy nonpregnant adults exposed to a large *L. monocytogenes* inoculum may develop invasive listeriosis. In immunocompromised individuals, however, invasive disease is facilitated by the inefficient mobilization of the host defenses and the blood-borne dissemination of *L. monocytogenes* from the primary infectious foci in the liver and spleen (silent phase of infection). Bacteremia may lead to meningoencephalitis if bacteria traverse the brain microcapillaries, to abortion or perinatal septicemia if they traverse the placental barrier, or to septicemic disease in cases of severe immunosuppression (Figure 4.2).[1]

4.2 MOLECULAR DETERMINANTS OF VIRULENCE INVOLVED IN THE INTRACELLULAR LIFE CYCLE

4.2.1 The Cell Biology of Infection

Macrophages and epithelial cells are widely used to study the interaction of *L. monocytogenes* with mammalian host cells. However, it was shown that also neutrophils, dendritic cells, hepatocytes, fibroblasts, endothelial cells, or glial cells may become infected with, and serve as host cells for *L. monocytogenes in vitro* and *in vivo*. Macrophages actively engulf *L. monocytogenes* spontaneously, but internalization of the bacterium by normally nonphagocytic cells is triggered by *L. monocytogenes*-specific factors. Aside from the internalization step, the intracellular life cycle of the bacterium in phagocytes or normally nonphagocytic mammalian cells is, however, essentially identical (Figure 4.3). *L. monocytogenes* induces its own internalization without an extensive remodeling of the host cell surface. Entry occurs via zipper-like phagocytosis, characterized by the emission of small pseudopods that firmly entrap the bacteria and the intimate contact of the bacterial surface with the host cell plasma membrane. Upon uptake, the pathogen appears in a membrane-bound vacuole, which is subsequently lysed by the combined action of the pore-forming hemolysin, listeriolysin (LLO), and two phospholipases (see below). The bacteria that are released into the cytoplasm begin to replicate while making use of specific transporters to gain carbohydrates from the host cell, whereas those remaining in the phagosome are killed and digested.

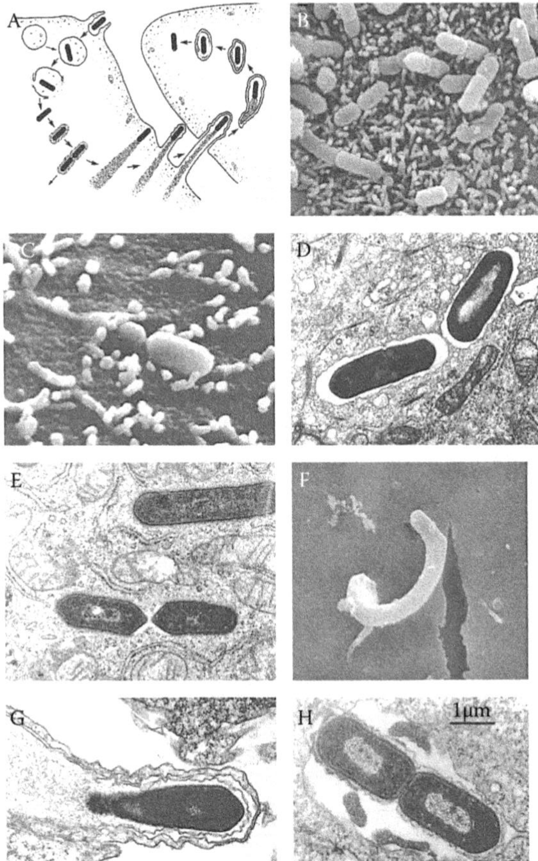

FIGURE 4.3 Stages of listerial intracellular parasitism. (A) Scheme of the intracellular life cycle of pathogenic *Listeria* spp. (B–H) Scanning and transmission electron micrographs of cell monolayers infected with *L. monocytogenes.* (B) Numerous bacteria adhering to the microvilli of a Caco-2 cell (30 min after infection). (C) Two bacteria in the process of invasion (Caco-2 cell, 30 min postinfection). (D) Two intracellular bacteria soon after phagocytosis, still surrounded by the membranes of the phagocytic vacuole (Caco-2 cell, 1 h postinfection). (E) Intracellular *Listeria* cells free in the host cell cytoplasm after escape from the phagosome (Caco-2 cell, 2 h postinfection). (F) Pseudopod-like membrane protrusion induced by moving *Listeria* cells, with the bacterium being evident at the tip (brain microvascular endothelial cell, 4 h postinfection. (G) Section of a pseudopod-like structure in which a thin cytoplasmic extension of an infected cell is protruding into a neighboring noninfected cell, with the protrusion being covered by two membrane layers (Caco-2 cell, 4 h postinfection). (H) Bacteria in a double membrane vacuole formed during cell-to-cell spread (Caco-2 cell, 4 h postinfection). (Taken from Vázquez-Boland, J.A. et al., *Clin. Microbiol. Rev.*, 14, 584, 2001. With permission.)

Concomitant with the onset of intracellular replication, *L. monocytogenes* induces the expression of the surface protein ActA which, through the activation of the cellular Arp2/3 complex, induces the nucleation of host actin filaments. The formation of a polar tail and the permanent polymerization of F-actin at the interface between the bacteria and the actin tails produce a propulsive force, which moves the bacteria through the cytoplasm. Those bacteria that in their random movement reach the plasma membrane push outwards inducing the formation of pseudopod-like structures with the bacterium at the tip. These invading pseudopods or “listeriopods” are taken up by the neighboring cells, in which the bacteria become entrapped within a double membrane. This vacuole is again lysed by LLO and the phospholipases, a broad-specificity phospholipase, releasing the bacteria into the cytoplasm of the newly infected host cell where they initiate a new cycle of replication and actin-based motility (Figure 4.3). This direct cell-to-cell invasion mechanism allows the bacteria to spread through host tissues without leaving the host cytosolic compartment, protected from the humoral effectors of the immune system and phagocytosis (reviewed in references 1 and 5–7).

4.2.2 Surface Proteins of the Internalin Family and the Invasion of Nonprofessional Phagocytic Cells

Several genes encoding internalin proteins were identified before the genome of *L. monocytogenes* became available.[58–61] Internalins belong to the superfamily of LLR (leucine-rich repeat) proteins, characterized by an N-terminal domain containing several successive repeats of 22 amino acids and a signal peptide allowing their export to the cell surface. The internalins can be divided into three distinct classes. The first class, exemplified by InlA, comprises internalins relatively large in size and bacterial cell wall associated due to a LPXTG motif in their C-terminal region. The second class, with only one member in *L. monocytogenes* strain EGD-e (InlB), lacks the LPXTG motif and has instead a C-terminal region with GW-repeat modules, which mediate loose attachment to the bacterial cell surface. The third class, again with only one member in *L. monocytogenes* (InlC), corresponds to considerably smaller LRR proteins, which are not surface attached and are therefore released to the supernatant.

The first internalin to be discovered was InlA, an acidic protein of 800 amino acids that possesses two extended repeat domains. The first consists of 15 LLRs, whereas the second designated domain B region consists of 2.5 repeats of about 70 amino acids each.[58] The InlA protein has a typical N-terminal transport signal sequence and a cell wall anchor in the C-terminal part comprising the sorting motif LPXTG followed by a hydrophobic membrane-spanning region. This distal LPXTG motif has been shown to be responsible for the firm attachment of InlA to the bacterial peptidoglycan envelope in a process mediated by the enzyme sortase A.[63] Sortase A catalyzes the covalent linkage of the LPXTG motif to the peptidoglycan after cleavage of the T–G bond of this motif. The *inlA* gene encoding InlA forms an operon together with the *inlB* gene encoding InlB, although the two genes can be transcribed individually.[60,64] Transcription of the *inlA* gene can occur from up to four promoters, one of which is controlled by the central regulator PrfA (see later discussion for details). Post-transcriptional and -translational control of InlA expression has been also reported.[65] The PrfA-dependent *inlA* promoter is also responsible for *inlB* gene expression.[60,64,66] InlB, a protein of 630 amino acids, also carries an N-terminal transport signal sequence, eight LRRs, and, as a distinctive feature, three C-terminal modules each beginning with the amino acids glycin (G) and tryptophan (W) (GW modules) in lieu of the LPXTG motif and cell-wall-spanning region.[58] InlB is targeted to the bacterial surface via the noncovalent and hence loose association of the GW modules with lipoteichoic acid in the listerial cell wall.[67]

LRR motifs are present in several virulence factors of Gram-positive and Gram-negative bacteria. In eukaryotes, LLRs are often involved in protein–protein interactions. The three-dimensional structures of the LRRs of four internalins, InlA, InlB, InlC, and InlH, have been solved at the atomic level.[68–71] In all four proteins, the N-terminal parts are combined to form a contiguous internalin domain with the LRR region as the central part flanked by a truncated EF-hand-like cap and an immunoglobulin-like fold. The extended beta-sheet resulting from the LRRs constitutes an adaptable concave surface proposed (and shown for InlA) to interact with the respective mammalian receptor molecules during infection.

The internalins InlA and InlB are critically involved in the internalization of *L. monocytogenes* by various nonphagocytic mammalian cells, as demonstrated by experiments in which they were expressed in noninvasive bacteria or coated onto latex beads, which in all cases resulted in internalization by different normally nonphagocytic cell types.[58,72–74] InlA mediates entry predominantly into human enterocyte-like epithelial cells and certain hepatocytes,[30,73,75] whereas InlB promotes invasion into a broader range of cells, including epithelial cells, endothelial cells, hepatocytes, and

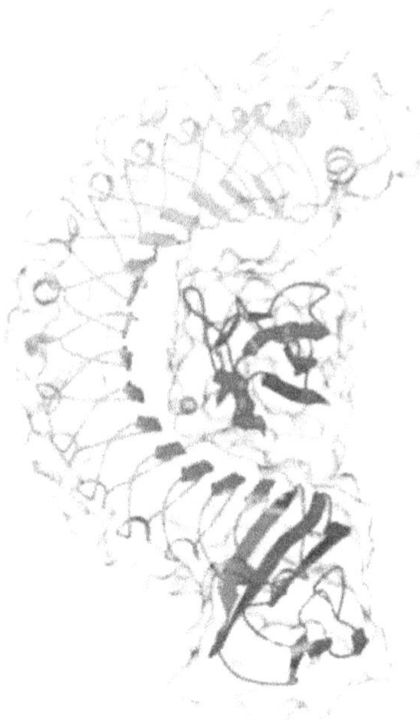

FIGURE 4.4 The complex of the functional domains of InlA (denoted InlA) (left) from *L. monocytogenes* and human E-cadherin (right). The protein backbones are depicted schematically, while the translucent surfaces give an impression of the extended surface of interaction between the two molecules. Note the bridging position of Ca^{2+} (open arrowhead) and Cl^- (filled arrowhead) between the two. (Taken from Schubert, W.D. and Heinz, D.W., *Chembiochemistry,* 4, 1285, 2003. With permission.)

fibroblasts.[30,76–78] These differences in cell specificity of the two internalins are probably due to the differential presence of their respective receptors on the different cell types.

Human E-cadherin was identified as the InlA receptor by affinity chromatography.[79] E-cadherin is an intercellular adhesion protein highly expressed at the basolateral membrane of polarized epithelial cells but also present on other cell types like endothelial cells, hepatocytes, and dendritic cells. The interaction of InlA with human E-cadherin is highly specific since the closely related mouse homolog of human E-cadherin is not recognized as a receptor.[80] This specificity is due to a single amino acid, proline at position 16 in human E-cadherin, which is glutamic acid in mouse E-cadherin and which explains the failure of InlA to promote passage through the interstinal barrier in the mouse model of infection.[80,81] Transgenic mice expressing human E-cadherin in their enterocytes finally allowed demonstration of the role of InlA during the crossing of the intestinal barrier.[19,82] The determination of the crystal structure of the N-terminal part of InlA bound to human E-cadherin showed that the InlA-E-cadherin interaction is based on the concave binding domain, which surrounds and specifically recognizes human E-cadherin (Figure 4.4).[70] The cytoplasmic domain of E-cadherin interacts with α- and β-catenins, which, on the other hand, directly bind to actin filaments and hence complete the link between the listerial surface protein and the host cell cytoskeleton.[83]

InlB interacts with several cellular receptors that may cooperate to promote bacterial uptake by the respective mammalian cells. The hepatocyte growth factor (HGF) receptor c-Met tyrosine kinase was identified as the key InlB receptor.[72] Met is a transmembrane protein with a tyrosine kinase activity in its cytoplasmic domain, which activates several signaling pathways triggered either by InlB or HGF. Upon binding of InlB, rapid tyrosine phosphorylation of the cytoplasmic domain of Met is induced, followed by a complex series of intracellular signal transduction events, with PI-3 kinase at the center of the signaling pathway leading to activation of the Arp2/3 complex, cortical actin cytoskeleton rearrangement, and phagocytosis.[72,84,85] A detailed review about InlB signaling and phagocytosis can be found in Hamon, Bierne, and Cossart.[86] It was long believed that clathrin-based uptake processes were actin independent and could not support entry of vesicles larger than 120 nm. However, Veiga and Cossart recently demonstrated that *L. monocytogenes* uptake induced by InlB-Met-interaction is accompanied by, and dependent on, the recruitment of clathrin and dynamin, the major proteins involved in clathrin-dependent endocytosis.[87] Our previous view on the mechanisms of bacteria-induced phagocytosis through the zipper mechanism needs hence to be corrected to emphasize the crucial role of clathrin in the uptake process.[88]

The complement receptor for the globular part of the C1q-fragment (gC1q-R) is another InlB ligand. gC1qR is a ubiquitously expressed protein present in different subcellular locations and involved in various cellular processes. Direct interaction of gC1q-R and InlB was demonstrated but the role of the gC1q-R in InlB-mediated entry of *L. monocytogenes* is unclear.[89] This role was further questioned by recent data demonstrating that gC1q-R specifically antagonizes, rather than enhances, InlB signaling and that interaction between InlB and gC1q-R is unnecessary for bacterial invasion.[90] Finally, glucosaminoglycans (GAGs) were identified as a third type of InlB ligand.[91] GAGs are present on the surface of mammalian cells where they decorate the proteoglycans. InlB binds to GAGs directly through its C-terminal GW repeats, which normally anchor the protein to the bacterial cell surface. GAGs are hence believed to detach InlB from the bacterial surface, allowing the clustering of the protein at the cellular surface during binding to Met via the InlB-LRR domain. This proposed GAG-dependent InlB clustering might favor the local activation of signal transduction pathways, finally leading to the uptake of the bacteria.[92]

Lipid rafts have been known for many years to be involved in different membrane-based processes.[93] Recent work has documented for the first time a role of lipid rafts in *L. monocytogenes* invasion, since cholesterol depletion, a method to destroy lipid-raft function, reversibly inhibited *Listeria* entry.[94] It was found that the presence of E-cadherin in lipid rafts was critical for initial interaction with InlA to promote bacterial entry. In contrast, the interaction of InlB with Met did not require membrane cholesterol, whereas downstream signaling leading to F-actin polymerization and engulfment was cholesterol dependent.[94,95]

A large number of internalin-related proteins in addition to the well-studied InlA and InlB proteins are encoded in the *L. monocytogenes* genome.[96,97] Of these, InlC, InlE, InlF, InlG, InlH, and InlJ have been studied in some details and it was shown that none of them is able to induce phagocytosis in mammalian cells on its own.[59–61,98] However, examination of double mutants in different internalin genes, including *inlA, inlB, inlC, inlG, inlH,* and *inlE,* showed that InlA-mediated bacterial entry is facilitated by the presence of other internalins, suggesting a cooperation between various internalins during the entry process.[62] The recently identified internalin InlJ is involved in virulence since a mutant lacking the *inlJ* gene is significantly attenuated in virulence after intravenous or oral infection of mice. However, unlike other internalins, InlJ plays no role in adhesion to or invasion into different mammalian cells.[98] The role of the small, secreted, and strictly PrfA-dependent internalin InlC is also unclear. Deletion of the *inlC* gene reduces virulence in a mouse sepsis model and participation of InlC in InlA-mediated internalization has been suggested,[59,62] but the details of its function and mode of action remain unknown.

4.2.3 Other Surface Proteins Involved in Host Cell Invasion

4.2.3.1 p60 and Related Proteins

L. monocytogenes expresses a large number of secreted or cell-wall-associated proteins with autolytic activities. Most of these proteins have not been thoroughly studied yet, but for several, an involvement in host cell invasion was shown experimentally. Protein p60 is a major secreted protein found in all *L. monocytogenes* isolates.[99,100] p60 is also present on the cell surface of the bacteria and possesses murein hydrolase activity, which appears to be involved in a late step of cell division.[101] The gene for this protein, initially called *iap* (*i*nvasion *a*ssociated *p*rotein), codes for an extremely basic protein of 484 amino acids, with a typical signal sequence and an extended repeat domain consisting of 19 threonine-asparagine units.[101,102] p60 belongs to a protein family with two other members in *L. monocytogenes*: p45, a peptidoglycan lytic protein encoded by the *spl* gene,[103] and the hypothetical protein encoded by *lmo394*.[96]

At least one type of rough mutants of *L. monocytogenes*, characterized by the expression of reduced amounts of p60, shows significantly reduced uptake by 3T6 fibroblast cells.[99] Other rough mutants of *L. monocytogenes* were isolated that show normal or even increased levels of p60 expression.[104] These mutants are adherent and invasive like wild-type bacteria, despite the formation of long filaments. The genetic basis of these phenotypes remains unknown.

p60 has been long regarded as an essential protein.[105] However, viable mutants with in-frame deletions in the *iap* gene have been recently described, disputing this assumption.[106,107] The characterization of one of the *iap* deletion mutants showed that the mutant forms cell chains, in spite of which it retains wild-type invasivity. However, the mutant grows in microcolonies inside host cells and does not form F-actin tails but only induces actin clouds around the bacteria. A defect in polar ActA-distribution (see later discussion) caused by impaired cell division was shown to be the reason for the lack of intracellular motility in this p60-defective strain. According to these findings, p60 seems not to be directly linked to cell invasion, but rather to indirectly modify bacterial behavior via its impact on cell division.[107] Recent work shed further light on the genetic and biochemical basis of the reduced p60 expression and the rough phenotype by the description of an additional *secA* gene in *L. monocytogenes,* called *secA2*, located immediately upstream of the *iap* gene, which is involved in the smooth–rough transition.[106,108] Rough mutants with reduced p60 expression synthesize no or nonfunctional SecA2 proteins and the deletion of the *secA2* gene results in reduced p60 expression and the conversion to the rough phenotype.

Another peptidoglycan hydrolase, called MurA or p66, was recently identified as a major surface protein of *L. monocytogenes*. The deletion of the MurA-encoding gene, which shows homology with p60 in its C-terminal domain, also results in the formation of long cell chains. It is currently believed that both proteins, p60 and p66, may act in concert to control proper cell separation during the last step of bacterial division.[109] Whether MurA has a role in cell infection *in vivo* or *in vitro* is currently unknown.

By in silico comparison of the surface protein repertories of *L. monocytogenes* and *L. innocua*, a gene encoding an *L. monocytogenes* surface protein absent in *L. innocua* was identified.[110] The gene, called *aut*, encodes Auto, a surface-associated protein of 572 amino acids containing a signal sequence, an N-terminal autolysin domain, and a C-terminal cell-wall-anchoring domain composed of by four GW modules. The *aut* gene is, like p60 and the other surface-associated listerial autolysins, expressed independently of the central virulence gene regulator, PrfA. Interestingly, *aut* mutants do not show any defect in septation and cell division, suggesting no role of Auto in these functions. Auto is required for entry of *L. monocytogenes* into nonphagocytic mammalian cells but is not involved in adhesion. Since mutants lacking Auto show reduced in virulence in animal models, the autolytic protein may thus represent a novel type of virulence factor.[110]

Up to now, the only listerial protein clearly shown to mediate *L. monocytogenes* adhesion without promoting uptake is the surface molecule Ami.[111,112] Ami is a 102-kD autolysin with the catalytic activity in the N-terminal part of the molecule. The C-terminal region is anchored to the cell wall via GW-dipeptide repeat modules as found in InlB.[67] Ami mediates binding to epithelial cells and hepatocytes in a Δ*inlAB* background, and the C-terminal GW modules were shown to be responsible for conferring the cell-adhesive properties to the protein.[111]

4.2.3.2 The Surface Proteins ActA, LpeA, FbpA, LAP, and Vip

The listerial actin-polymerizing protein ActA, primarily involved in actin-based intracellular motility (see later discussion),[113,114] was suggested to play also a role in internalin-independent uptake of *L. monocytogenes* by epithelial cells.[115] Analysis of the invasive capacity of strains either lacking or overexpressing ActA suggests that ActA may function as an invasion-mediating protein—at least when overexpressed.[116] ActA-mediated attachment and invasion of CHO epithelial-like cells as well as IC-21 murine macrophages could involve electrostatic interactions between positively charged residues in the N-terminal domain of the listerial surface protein and the negatively charged heparan-sulfate (HS) moiety of cell surface HS proteoglycans (HSPG), which are widely distributed in mammalian cells.[115] Whether low-stringency binding of *L. monocytogenes* ActA to HSPG directly triggers uptake or results in adequate presentation of other bacterial factors to the host cell membrane, subsequently leading to phagocytosis, remains to be clarified.

L. monocytogenes encodes a large number of putative lipoproteins whose functions remain largely unknown.[96,97] A lipoprotein of 35 kD, called LpeA (*l*ipoprotein *p*romoting *e*ntry) identified in silico in the *L. monocytogenes* sequence,[117] shows homology to an *S. pneumoniae* adherence factor and is implicated in the invasion of hepatocytes and to a lesser extent of epithelial cells. The corresponding gene mutant was not impaired in cell adhesion or intracellular growth. To date, LpeA is the first listerial lipoprotein identified to be potentially involved in virulence.

Signature-tagged mutagenesis allowed the identification of an *L. monocytogenes* gene called *fbpA*, required for efficient liver colonization.[118] *fbpA* encodes a protein of 60 kD that has strong homologies to atypical fibronectin-binding proteins. FbpA binds human fibronectin and increases adherence of *L. monocytogenes* to HEp-2 cells in the presence of fibronectin. FbpA is present on the bacterial surface and, interestingly, co-immunoprecipitates with LLO and InlB, but not with other known virulence factors. FbpA hence seems to act like a chaperone for two listerial virulence factors and may be a novel multifunctional virulence factor of *L. monocytogenes*.[118]

The 104-kD *Listeria* adhesion protein LAP, determined to be alcohol acetaldehyde dehydrogenase,[119] is expressed by all *Listeria* spp. except *L. grayi* and mediates binding of *L. monocytogenes* exclusively to intestinal epithelial cells.[120] This specificity suggests that LAP may be involved in the intestinal phase of *L. monocytogenes* infection. The heat shock protein Hsp60 was recently identified as a receptor for LAP in the human epithelial cell line Caco-2.[121]

Vip, a 43-kD listerial surface protein anchored to the cell wall through an LPXTG motif was identified by comparative genomics in *L. monocytogenes*. This protein, which appears to be at least partially regulated by PrfA despite the *vip* genes not having a recognizable PrfA box (see later discussion), is required for entry into some mammalian cells and binds to these cells via its interaction with Gp96, a protein of the endoplasmic reticulum and the cell surface. Infection studies using different animal models suggested that Vip plays a role during the intestinal and late stages of the infection.[122]

4.2.4 Secreted Membrane-damaging Proteins Allowing Escape from the Phagocytic Vacuole

4.2.4.1 Listeriolysin O

The hemolytic activity of *L. monocytogenes* is due to the action of listeriolysin O (LLO), which is a major listerial virulence determinant. Nonhemolytic mutants are always avirulent and nonpathogenic in the mouse model, and virulence and pathogenicity are restored in hemolytic revertants or by the introduction of the cloned *hly* gene into a nonhemolytic *L. monocytogenes* mutant.[123–126] LLO is a secreted protein of 58 kDa belonging to the family of thiol-activated, cholesterol-dependent, pore-forming toxins (CDTX) for which streptolysin O is the prototype.[127] All members of this toxin family are irreversibly inhibited by low concentrations of cholesterol, reversibly inactivated by oxidation, and reactivated by reducing agents such as DTT. Cholesterol is believed to be the receptor for these cytolysins since only membranes containing cholesterol are susceptible to LLO-mediated pore formation.[127] Upon addition to erythrocytes, toxin monomers oligomerize in the target cell membrane to form stable pores that can be visualized by electron microscopy (Figure 4.5).[128] During oligomerization, several raft-associated molecules aggregate in a process independent of the

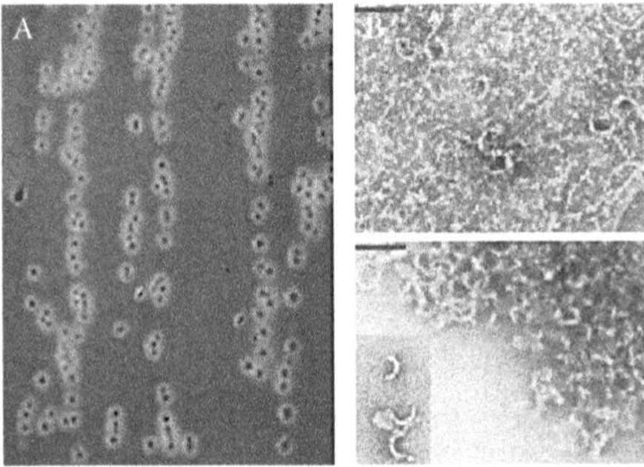

FIGURE 4.5 The hemolysin, a key virulence factor of *L. monocytogenes*. (A) Hemolytic activity of *L. monocytogenes* on sheep blood agar. (B) Theoretical three-dimensional structure of the soluble LLO monomer based on the crystal structure of PFO.[134] Domains are indicated by numbers (the arrow points to the Trp-rich conserved undecapeptide). Ribbon model generated using MacPyMol 0.99. (C) Electron micrographs of rabbit eryrthrocyte membrane after treatment with listeriolysin; large pores in the membrane formed by LLO oligomers are clearly visible. (Panels A and B taken from Vázquez-Boland, J.A. et al., *Clin. Microbiol. Rev.*, 14, 584, 2001, and Parrisius, J. et al., *Infect. Immun.*, 51, 314, 1986, respectively. With permission.)

hemolytic activity of LLO.[129] Optimal hemolytic activity occurs at pH 5.5, a pH value much lower than that determined for the other CDTX,[130] a property in agreement with the compartmentalized function of LLO in the acidified phagosome (see later discussion).

The LLO determinant *hly* codes for a protein of 529 residues including an N-terminal signal sequence.[131–133] As expected, the LLO sequence shows extended homologies with the protein sequences of other CDTX. The three-dimensional structure of LLO is not known. However, given the extensive overall similarity of CDTXs in terms of amino acid sequence, it can be assumed that the listerial toxin has the same spatial configuration as that determined for the closely related toxin perfringolysin O (PFO) from *Clostridium perfringenes*.[134] Thus, LLO is most likely composed of four domains, which form an L-shaped molecule with domain 4 involved in initial membrane binding, located at the end of the longer arm. This domain contains the conserved undecapeptide ECTGLAWEWWR, a hallmark of all members of this toxin family that carries the single Cys residue responsible for the distinctive thiolactivation property. Domains 1–3 are involved in toxin oligomerization, pore formation, and membrane disruption (reviewed in Vázquez-Boland[127]) (Figure 4.5).

The role of LLO in virulence was determined by experimental infection in mice using wild type and *hly* mutants of *L. monocytogenes*. In these experiments, the wild-type strain survived and proliferated, whereas the nonhemolytic mutants were rapidly eliminated.[124–126,135] The role of LLO in the intracellular survival was determined using different cell lines. The nonhemolytic mutants were as invasive as the isogenic wild-type strains; however, they were incapable of intracellular survival and growth within host cells. Electron microscopy of the infected cells revealed that *L. monocytogenes hly* mutants were unable to escape from the phagocytic vacuole to the cytoplasm.[135,136] Additional evidence for LLO being essential for lysis of the phagosomal membrane was obtained by infection of macrophages with a *Bacillus subtilis* strain expressing LLO.[137] This recombinant strain escaped into the cytoplasm, whereas the nonhemolytic *B. subtilis* parental strain remained in the phagosome as the nonhemolytic *L. monocytogenes* mutants did.

The low pH optimum of LLO, which is controlled by a rapid and irreversible denaturation of its structure at neutral pH at temperatures above 30°C,[138] is in agreement with its function in the acidified phagosome (reviewed in Vázquez-Boland[127]). Bafilomycin or ammonium chloride treatment inhibits vacuolar acidification and *L. monocytogenes* escape from phagosomes. These findings underpin the importance of the low pH optimum for LLO activity and its role as a phagocytic vacuole opener.[139,140] The molecular features responsible for the adaptation of LLO to function as an "intracellular" virulence factor were explored using *L. monocytogenes* mutant strains expressing PFO, which is active at a broader pH range, instead of LLO. Such recombinant bacteria escaped from the vacuole but damaged the host cells.[141,142] Using an elegant selection procedure, PFO mutants were isolated that did not damage the host cell. The mutant PFO proteins expressed by these noncytotoxic *L. monocytogenes* recombinants had either a generally reduced hemolytic activity, a reduced activity at neutral pH, or a shorter half-life in the host cell cytoplasm. Together with the finding that *L. monocytogenes* mutants that fail to compartmentalize LLO activity are cytotoxic and avirulent,[143] these findings indicated that the low activity at neutral pH and the short half-life in the cytoplasm are critical to avoid toxin-mediated damage on the host cell and thus to maintain the integrity of the within-host listerial replication niche.

The importance of the restriction of LLO activity to the phagosomal compartment is further supported by other lines of evidence. A PEST-like sequence was identified in the N-terminal region of LLO.[144,145] PEST sequences target proteins to the proteasome degradation pathway, hence dramatically reducing their cytoplasmic half-life.[146] When expressed in *L. monocytogenes*, LLO variants lacking the PEST sequence conferred normal hemolytic activity and allowed vacuolar escape, but the bacteria again were cytotoxic and showed reduced virulence in mice.[144] Lety et al.[147] reported, however, that it was not the PEST-like sequence but instead amino acids immediately downstream of it that were responsible for the correct function of the toxin. A recent report has shown that, although LLO is indeed degraded in a proteasome-dependent manner with ubiquitination and phosphorylation within the PEST-like sequence, both wild-type LLO and PEST-region mutants have similarly short intracellular half-lives.[148] Together with the observation that PEST region mutants exhibit higher intracellular LLO levels than wild-type bacteria, these data suggest that the LLO's PEST-like region, rather than mediating proteasomal degradation by the host, controls LLO production in the cytosol.[149]

Fusion of *L. monocytogenes*-containing phagosomes with endosomes has been observed in electron microscopic studies.[136] However, it is not known whether such an event is necessary for *L. monocytogenes* to progress through its intracellular life cycle.[150] The recent description of Rab5-regulated fusion of *L. monocytogenes*-containing phagosomes with endosomes and the results, which indicate that Rab5a controls early phagosome–endosome interactions and governs the maturation of the early phagosome leading to phagosome–lysosome fusion, show that phagosome maturation events take place upon ingestion of *L. monocytogenes* into macrophages.[151,152] On the other hand, it was shown that an LLO-dependent delay of *L. monocytogenes* phagosome maturation results from disruption of ion gradients across the vacuolar membrane.[153,154] It is believed that this allows the bacteria to prolong their survival inside the phagosome–endosome, ensuring their viability as a prelude to escape into the cytoplasm.[155] Prolonged intraphagosomal survival of *L. monocytogenes* in macrophages was recently demonstrated,[156] which fits nicely with the postulated *L. monocytogenes*-induced delay in phagosome maturation.

4.2.4.2 Phospholipases and Metalloprotease

Two phospholipases and a metalloprotease cooperate with LLO in the disruption of the phagosomal membrane.[157,158] The LLO-independent escape of *L. monocytogenes* from primary vacuoles, shown to occur in some human epithelial cells,[125] is mediated by the additional membrane-damaging capacity afforded by these enzymes. One of the listerial phospholipases is encoded by the *plcA* gene, which is located immediately upstream from the *hly* gene in the listerial central pathogenicity island, LIPI-1[159]. The secreted, mature enzyme is a 33-kDa phospholipase C highly specific for phosphatidylinositol (PI-PLC).[160–162] The crystal structure of this PI-PLC consists of a single $(\beta\alpha)_8$-barrel domain with the active site located at the C-terminal side of the β-barrel.[163] Two histidine residues (His38 and His86) that are not present in the catalytic site are, however, important for enzyme function since their mutagenesis results in PI-PLC variants with a 100-fold reduced activity.[164] The *L. monocytogenes* PI-PLC has been shown to be required for the efficient escape from primary phagosomes in mouse bone-marrow-derived macrophages.[165] The broad pH optimum of the PI-PLC enzyme, ranging from pH 5.5 to 7.0, is consistent with its postulated role in the disruption of acidified phagocytic vacuoles.[167]

The second phospholipase C produced by *L. monocytogenes* hydrolyzes phosphatidylcholine, the major phospholipid component of lecithin, and is thus known as PC-PLC or "lecithinase." The purified PC-PLC enzyme was shown to have a broad substrate range, hydrolyzing phosphatidylethanolamine, phosphatidylserine, and sphingomyelin, but not phosphatidylinositol. PC-PLC is a zinc-dependent phospholipase C of 29 kDa with a pH optimum between 6 and 7.[167,168] The gene encoding PC-PLC, *plcB*, is the third gene of the lecithinase operon together with *mpl* and *actA* and the downstream small *orfs* X and Z (Figure 4.6).[169] PC-PLC is synthesized as a 289-amino-acids polypeptide with an N-terminal transport signal of 25 residues followed by a putative propeptide of 26 residues. Processing of the 30-kDa mature proenzyme to the 29-kDa active PC-PLC enzyme is operated by Mpl and occurs most likely during secretion.[170,171] The contribution of PC-PLC to virulence is primarily associated with its role in the lysis of the double-membrane phagosomes formed after cell-to-cell spread (see later discussion). The role of PC-PLC in escape from primary phagosomes (those formed after phagocytosis of extracellular bacteria) is less clear and differs from cell type to cell type. In the human epithelial-like cell lines Henle 407, HEP-2 and HeLa, where escape of *L. monocytogenes* occurs at low efficiency independently from LLO,[125,158,172] PC-PLC is required for lysis of the phagocytic vacuole together with the metalloprotease.

The metalloprotease Mpl of *L. monocytogenes* contributes to intracellular proliferation and virulence via its role in the proteolytic processing of pro-PC-PLC into its active form.[169,170] *mpl* mutants only produce the pro-PC-PLC form.[170,171] The *mpl* gene is located immediately downstream

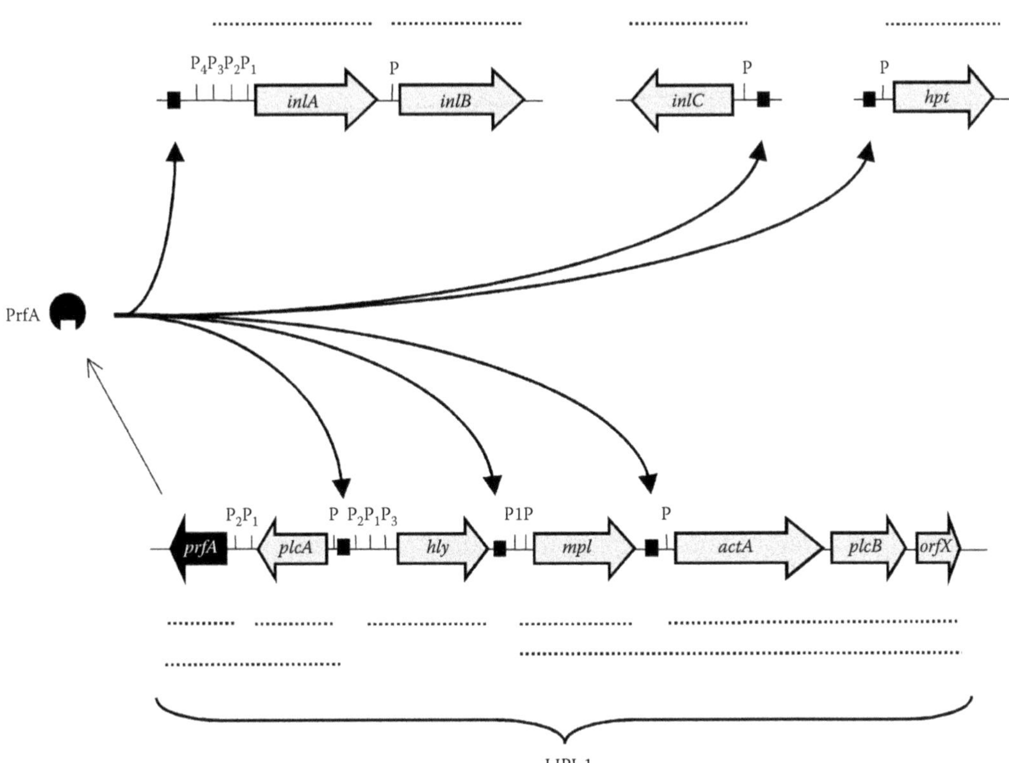

FIGURE 4.6 Physical and transcriptional organization of the the central listerial pathogenicity island LIPI-1 and of the core regulon controlled by the LIPI-1-encoded PrfA transcriptional activator. PrfA boxes are indicated by black squares, known promoters and transcripts by "P" and dotted lines, respectively. (Taken from Scortti, M. et al., *Microb. Infect.*, in press. With permission.)

of the *hly* gene in LIPI-1 and is the first gene of the 5.7-kb lecithinase operon (Figure 4.6).[169,173] The 510-residue amino acid sequence of Mpl shows high homology to several zinc-dependent metalloproteases from *Bacillus* species and possesses a typical N-terminal signal sequence and a putative internal cleavage site. Like other metalloproteases, the enzyme is activated by proteolytic maturation resulting in a mature 35-kDa protein.[173,174] The Mpl protease was purified and biochemically characterized and shown to be active at a wide range of temperatures and pH values; it exhibited high thermal stability and narrow substrate specificity.[175] The translocation through the membrane and the release of PC-PLC from the bacterial cell wall occur most efficiently upon a decrease in pH as regularly encountered in the primary and secondary phagosomes. This release coincides with the proteolytic activation of PC-PLC and is controlled by Mpl, which co-localizes with PC-PLC at the cell wall-membrane interface.[176,177]

4.2.4.3 Other Factors

In addition to LLO and the phospholipases, other factors contributing directly or indirectly to phagosomal escape were recently identified. As mentioned previously, *L. monocytogenes* encodes a large number of putative lipoproteins.[96,97] The deletion of a gene encoding a putative lipoprotein-specific signal peptidase provided the first insights into the possible role in virulence of at least some members of this protein family, as the mutant failed to process several lipoproteins and showed a reduced virulence in the mouse model. The expression of the signal peptidase was strongly induced while the bacteria resided in the phagosome and the mutant bacteria were clearly impaired in phagosomal escape.[178] Whether or not and how listerial lipoproteins contribute to the lysis of the phagosomal membrane require further investigation.

SvpA (*s*urface *v*irulence-associated *p*rotein) is a secreted, surface-exposed 64-kDa protein; a mutant lacking its coding gene, *svpA*, showed impaired growth *in vivo* in mouse organs as well as *in vitro* in macrophages. Growth restriction was most likely due to a reduced capability of the mutant to escape from the phagosome into the host cell cytoplasm. It remains to be determined if SvpA promotes bacterial escape from phagosomes by protecting intraphagosomal bacteria from killing or by acting synergistically with LLO and the phospholipases.[179]

A recent genome-wide RNA interference (RNAi) screen conducted in *Drosophila* cells to identify host factors involved in listerial intracellular pathogenesis led to the identification of more than 300 RNAi targeting a wide range of cellular functions.[180] Further characterization of some of the host factors identified that affected the access of *L. monocytogenes* to the cytosol may provide clues as to the bacterial systems with which they interact or interfere.

4.2.5 Nutritional and Metabolic Determinants of Cytosolic Replication

L. monocytogenes starts multiplying intracellularly shortly after escape from the vacuole with intracellular generation time of 40–60 min, which is comparable to those observed in rich broth culture.[125] The host cell cytoplasm hence allows listerial growth with high efficiency. However, the cytosol is poorly characterized as a substrate supporting bacterial growth and the relative abundance of the nutrients that it may provide is largely unknown. Whereas auxotrophic mutants of *L. monocytogenes* are able to grow intracellularly,[181] the expression of several metabolic genes is intracellularly increased as shown by an early study by Klarsfeld et al.,[182] suggesting that at least some metabolites may be present at limiting concentrations in the host cytosol. The availability of whole-genome microarrays (based on PCR products or oligonucleotides) allowed exploration of the *L. monocytogenes* determinants of intracellular growth within macrophages and epithelial cells by gene expression profiling.[183,184]

These investigations showed that up to 19% of the listerial genes are differentially expressed during the transition from growth in rich medium to intracellular growth. Intracellularly expressed genes showed responses possibly indicative of glucose limitation within host cells, with a decrease in the amount of mRNAs encoding enzymes of central carbon metabolism and a temporal induction of genes involved in alternative carbon source utilization pathways and their regulation. Adaptive intracellular gene expression responses involved genes that are associated with virulence, the general stress response, cell division, and changes in cell wall structure, together with many genes of unknown function.[183] Furthermore, transporter proteins essential for the uptake of carbon and nitrogen sources, factors involved in anabolic pathways, and transcriptional regulators were differentially expressed with respect to broth-grown bacteria.[184] These studies provided evidence that *L. monocytogenes* can use alternative carbon and nitrogen sources during intracellular replication and that the pentose phosphate cycle, but not glycolysis, is the predominant pathway of carbohydrate metabolism in the host environment.

Early studies addressing the question of whether nonpathogenic bacteria not naturally adapted to an intracytoplasmic lifestyle can grow in the host cell cytoplasm used either *B. subtilis* or *L. innocua* strains heterologously expressing LLO to allow phagosomal escape after uptake.[137,185] The results of these studies showed that both engineered nonpathogenic bacteria were able to multiply—at least to some extent—in the host cell cytoplasm, implying that this compartment is generally permissive for bacterial growth. However, a recent study in which the bacteria were directly microinjected into the cytoplasm of mammalian, thus bypassing the normal infection pathway involving internalization/phagosomal escape, showed that only bacteria naturally capable of intracytoplasmic growth, like *L. monocytogenes* and *Shigella flexneri*, replicated efficiently within cells, while bacteria not naturally adapted to life in the cytosol, like *L. innocua* and *B. subtilis*, did not.[186] Furthermore, a Δ*prfA* mutant of *L. monocytogenes* lacking the central virulence regulator PrfA (see section 4.4) multiplied poorly upon microinjection, pointing to the need of specific virulence determinants for efficient intracytoplasmic multiplication. The discrepancies in the studies concerning the ability of bacteria to use the host cell cytosol for growth have been discussed in detail elsewhere, [187–189] and the bottom line conclusion is that while the cytosol may provide all necessary nutrients for bacterial growth, efficient replication in this compartment requires specific adaptation in the microbe.

A clear example of such adaptation is provided by the hexose phosphate permease Hpt, which allows *L. monocytogenes* to take up sugar phosphates from the cytosolic compartment and to use these host-derived metabolites as a carbon and energy source to fuel rapid intracellular growth.[190] Hpt is a member of the organophosphate–inorganic phosphate antiporter family of transporters like the related UhpT permease from *E. coli*. Expression of the *hpt* gene is tightly controlled by the central virulence regulator PrfA (see section 4.4) and is selectively induced intracellularly.[191] Hpt is dispensable for optimal bacterial growth in rich medium but its loss results in impaired intracytosolic proliferation in different cell types and in attenuated virulence in mice, clearly indicating it is a bona fide virulence factor.[190] Interestingly, Hpt alone seems not to be sufficient for efficient intracellular growth since the cytosolic replication of *L. innocua* expressing Hpt together with LLO was not significantly improved within infected macrophages.[192]

Recently, *L. monocytogenes* gene *lplA1* encoding a lipoate protein ligase (LplA1) was also shown to be necessary for efficient intracellular proliferation and full virulence in mice. Mutants lacking the *lplA1* gene showed a phenotype different from that of Hpt mutants, characterized by abortive intracellular growth after about five rounds of replication. A major target for LplA is the E2 subunit of the pyruvate dehydrogenase enzyme (PDH) complex, which is activated by lipoylation. PDH was no longer lipoylated in intracellularly grown *lplA1* mutants, indicating that PDH function is required for listerial replication in the cytosol,[193] consistent with the importance of fermentative carbohydrate metabolism for listerial intracellular growth.

4.2.6 The Surface Protein ActA and Its Role in Actin-Based Motility and Cell-to-Cell Spread

Upon escape from the phagosome and concomitant with the onset of intracellular multiplication, *L. monocytogenes* becomes surrounded by a network of cellular actin filaments and actin-binding proteins. Soon after, the actin filaments form a comet-like tail at one pole of the bacterium, with actin filaments constantly growing toward the bacterial surface, thereby generating a motive force that pushes the bacteria through the host cell cytoplasm at speeds of up to 1.5 μm/sec.[194,195] Interestingly, *L. monocytogenes* rotates around its long axis like a screw as it is propelled by actin polymerization.[196]

Mutants defective in actin-based intracellular motility were initially obtained by transposon mutagenesis. In these mutants the transposon insertion mapped to a gene called *actA* located downstream of *mpl* in the lecithinase operon (Figure 4.6) and encoding a proline-rich protein (ActA) of 639 amino acids.[169] ActA is a surface protein consisting of three domains: the N-terminal domain with the transport signal sequence, the central proline-rich repeat region, and the C-terminal part, which includes a membrane anchor. *actA* mutants are no longer capable of polymerizing actin and of actin-based intracellular movement, do not spread from cell to cell and thus form microcolonies near the host cell nucleus and are totally avirulent/nonpathogenic in mice.[113,114]

ActA alone is sufficient to stimulate F-actin assembly and to promote intracellular movement, as shown by several lines of evidence. First, the nonmotile species *L. innocua* was engineered to express ActA, and the recombinant bacteria induced the formation of actin tails and moved in cytoplasmic extracts as wild-type *L. monocytogenes.*[197] Second, *actA* was transfected into mammalian cells where the ActA protein was targeted to mitochondria, which subsequently assembled F-actin on their surfaces.[198–200] Third, polystyrene beads were coated with purified ActA and these beads induced F-actin polymerization and motility *in vitro* in cell extracts.[201]

The ActA protein is distributed asymmetrically on the surface of *L. monocytogenes.* After cell division, it is concentrated at the old pole and is absent, or present only at low concentrations, at the new bacterial pole.[202–204] This asymmetric distribution of the ActA protein was shown to be required and sufficient for unidirectional actin-based motility by coating streptococci asymmetrically with ActA protein. In a cell-free system, asymmetrically ActA-coated streptococci, but not uniformly coated ones, induced F-actin tails and moved efficiently in cytoplasmic extracts.[205] Additionally, only polystyrene beads asymmetrically coated with ActA induced F-actin tails and moved in the extracts.[201] A structure for ActA is not available, as the protein has resisted all crystallization efforts. It has been reported that ActA, which is believed to have an elongated shape, may be present on the bacterial surface as a dimer,[206] but these data were questioned by the demonstration of functional ActA's being a monomeric protein.[207]

The elucidation of the precise mechanisms by which ActA promotes actin recruitment/assembly and intracellular movement is at the center of the research interests of several laboratories and has been reviewed in detail elswhere.[86,208–210] The expression of mutated forms of ActA either in mammalian cells or in *L. monocytogenes* made it possible to define regions of the ActA protein with specific functions in the recruitment of cellular proteins and hence in actin polymerization and movement. Deletion of the whole N-terminal domain of ActA was followed by a total abolishment of actin polymerization and intracellular movement in both systems, showing the absolute necessity of this domain in ActA function.[194] Within the N-terminal part of ActA, two smaller regions were identified that are required either for filament elongation (aa 117 to 121) or for the continuity of actin polymerization process (aa 21 to 97), because their deletion led to discontinuous actin tail formation.[212–214] In contrast, deletion of the C-terminal domain did not inhibit actin assembly.[211] The actin tails produced by *L. monocytogenes* strains expressing ActA without the central proline-rich repeats were significantly shorter and the speed of the movement was drastically reduced.[215]

The F-actin tails contain several actin-binding proteins and proteins regulating the actin dynamics (Figure 4.7). Profilin, Mena, the Arp2/3 complex, and the *va*sodilator-*s*timulated *p*hosphoprotein (VASP) are associated with the surface of the moving bacteria and co-localize with ActA. Of these, only VASP, Mena, and the Arp2/3 complex were shown to directly bind to ActA.[194,213,216–218] The Arp2/3 complex consists of seven host cell proteins and was initially characterized as a profilin-binding protein complex.[219] Arp2/3 initiates F-actin polymerization due to its nucleation activity,

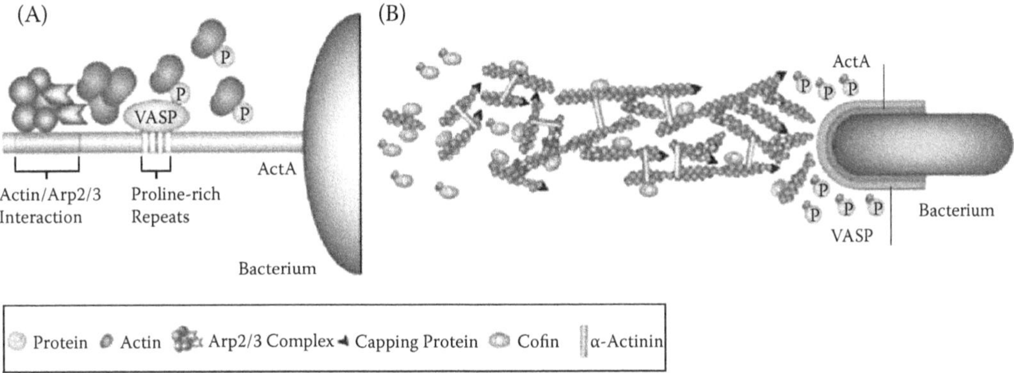

FIGURE 4.7 Model of the cellular and bacterial components required for actin-based motility of *L. monocytogenes.* (A) Interactions between host–cell proteins and ActA at the bacterial surface. The amino-terminal domain activates actin filament nucleation through Arp2/3. The central proline-rich domain binds VASP and profilin interacts with VASP, enhancing filament elongation. (B) Host protein functions throughout the comet tail. In addition to the factors that act at the bacterial surface, capping protein binds to the barbed end of actin filaments to prevent elongation of older filaments, α-actinin cross-links filaments to stabilize the tail structure, and ADF/cofilin disassembles old filaments. (Taken from Cameron, L.A. et al., *Nat. Rev. Mol. Cell Biol.*, 1, 110, 2000. With permission.)

which is significantly activated by the presence of ActA.[220,221] Studies with ActA mutants clearly demonstrated that the N-terminal region of ActA is sufficient for this activation.[222] The regions of the ActA protein that directly interact with proteins from the Arp2/3 complex were mapped to two regions spanning the acidic aa 41 to 46 and the basic aa 146 to 150.[213,218,223] In this interaction ActA is thought to mimic the activity of proteins of the WASP family, which are natural ligands and activators of Arp2/3.[223] Arp2/3 can also bind to the side of preexisting actin filaments and initiate a new filament at this point, creating branched structures found throughout the comet tails associated with moving *L. monocytogenes.*[220]

In summary, the current data point to a central role of the Arp2/3 complex in initiating actin-based listerial intracellular movement. However, a recent study showed that ActA can also generate new actin structures in an Arp2/3-independent, but VASP-dependent, manner,[224] indicating that a complex protein like ActA can interfere with the host cell actin polymerization machinery in multiple ways. In addition to the Arp2/3-dependent nucleation step, the process of *L. monocytogenes* actin-based motility involves an Arp2/3-independent elongation step that requires the host cell protein fascin.[225]

The phosphoprotein VASP binds directly to the proline-rich repeats of ActA via its Ena/VASP-homology domain.[216,226] On the other hand, VASP is a natural ligand of profilin and could hence stimulate actin assembly by binding to ActA and enhancing the profilin concentration in the vicinity of the bacterium. Mena, which is closely related to VASP, also binds ActA and profilin directly and might function in concert with VASP to recruit profilin–actin complexes to the site of actin polymerization.[217] However, profilin is dispensable, at least *in vitro*, because profilin-depleted cytoplasmic extracts still support actin assembly and bacterial movement.[195] ActA itself can bind G-actin within a region in its N-terminal part, but deletion of this region does not interfere with actin tail formation in infected cells.[222] It is believed that inside cells the VASP-mediated profilin/G-actin recruitment can bypass defects in actin binding of ActA.[227] Another function recently attributed to Ena/VASP proteins is the control of the temporal and spatial persistence of bacterial actin-based motility.[228] Furthermore, purified VASP binds to F-actin and enhances the actin-nucleating activity of wild-type ActA and the Arp2/3 complex while also reducing the frequency of actin branch formation.

The ability of VASP to contribute to actin filament nucleation and to regulate actin filament architecture highlights the central role of VASP in actin-based motility.[227,229] The actin tails that rapidly grow at the rear of the moving bacterium are simultaneously disassembled at the opposite end in order to refill the cellular pool of monomeric actin. Disassembly is controlled by actin depolymerizing factor (ADF)/cofilin facilitated by coronin and Aip1[230,231] and by gelsolin in a calcium-dependent process.[232] ADF/cofilin and gelsolin thus play a complementary function in *L. monocytogenes* actin-based movement by mediating the recycling of the actin filaments that build up the comet tail.

The 92-kDa ActA protein on the bacterial surface is cleaved by the listerial metalloprotease Mpl, resulting in a major 72-kDa degradation product and, depending on the *Listeria* strains tested, additional smaller degradation products.[232,234] These products are either found on the bacterial surface or in the supernatant as 65- and 30-kDa fragments. ActA is also degraded inside the host cell cytoplasm by the action of the proteasome.[235] Additionally, ActA becomes phosphorylated inside the host cell,[236] but phosphorylation may not be necessary for movement because a genetically engineered ActA variant lacking the C-terminal region is no longer phosphorylated inside host cells but is fully functional.[211] The roles of Mpl- or proteasome-mediated ActA degradation as well as the phosphorylation inside the host cell are currently not understood. However, it is conceivable that different variants of ActA may interact differently with host cell proteins interfering with the dynamics of the actin cytoskeleton. Another cellular protein, called LaXp180, was found to directly bind to ActA-expressing intracellular *L. monocytogenes*, resulting in the recruitment of the cellular phosphoprotein stathmin.[237] It is unknown at present whether LaXP180, which binds to the central proline-rich repeats of ActA,[238] interacts with one of the ActA variants mentioned before. The role of the recruitment of LaXp180 and stathmin in intracellular motility is still unknown. However, recent data showed that the cytosolically replicating pathogen *Shigella flexneri,* which also displays actin-based intracellular motility, disrupts the host cell microtubule network through the action of its secreted microtubule-severing protein VirA.[239] One could hence speculate that the recruited microtubule-depolymerizing molecule stathmin could be of similar function during *L. monocytogenes* movement.[237]

L. monocytogenes spreads from cell to cell by forming microvilli- or pseudopod-like protrusions, which are phagocytosed by neighboring cells. The mechanisms involved in the formation of such microvilli and of induction of phagocytosis by the neighboring cell are largely unknown except for the notion that members of the ERM-family of host cell proteins, which link the membrane to the actin cytoskeleton, are exploited by *L. monocytogenes* during spreading.[240] Phagocytosis during cell-to-cell spread leads to the formation of a double-membrane vacuole. This "secondary phagosome" rapidly acidifies, and its disruption requires the action of LLO and the listerial phospholipase PC-PLC. Upon lysis of the membrane, it takes one to two bacterial generations before motility recovers.[241] Electron micrographs of *plcB* mutants inside mammalian cells show numerous bacteria followed by actin tails and being trapped within vacuoles surrounded by double membranes, indicating that the *plcB* mutants are impaired in lysis of the double membrane of the vacuole.[169] Careful examination of the plaque formation capacity, which is thought to be a good correlate for intercellular spread of different mutants, revealed that besides the broad spectrum phospholipase PC-PLC, PI-PLC and metalloprotease also contribute to plaque formation, most likely by supporting lysis of the double-membrane vacuole.[205,242] There is evidence that host cell proteases may also cleave and hence activate PC-PLC in the absence of Mpl.[242]

The pivotal role of LLO in escape from the secondary vacuole was shown in a study using an *L. monocytogenes hly* mutant coated with recombinant LLO, which allowed the bacteria to escape from the primary vacuole and to spread into neighboring cells once. However, due to the dilution of the recombinant LLO through bacterial replication, the further spreading of the bacteria was inhibited.[243] The predominant role of LLO in the lysis of the secondary vacuole and hence in cell-to-cell spread, was confirmed by the microinjection of *hly* mutants into the cytosol[186] and by the use of strains allowing the temporal control of LLO expression.[156] Listerial factors other than ActA, LLO, Mpl, and PC-PLC are not known to contribute to intercellular spread.

4.3 MOLECULAR DETERMINANTS INVOLVED IN COLONIZATION OF THE HOST

4.3.1 Bile Salt Hydrolase

The ability to colonize the gallbladder together with extracellular multiplication was recently shown to be an important feature of virulent *L. monocytogenes*.[244] The significance of growth in the gallbladder with respect to the pathogenesis and spread of listeriosis depends on the ability of the bacterium to live in this organ and be disseminated to other tissues and into the environment. Using whole-body imaging, induced biliary excretion of *L. monocytogenes* from the murine gallbladder through the bile duct into the intestine was demonstrated (Figure 4.8).[245] The bacteria have hence to cope with the bactericidal effects of bile when residing in the gall bladder and also in the small intestine. The comparison of the *L. monocytogenes* EGDe and *L. innocua* genomes revealed the presence of an *L. monocytogenes*-specific gene, termed *bsh*, encoding a bile salt hydrolase (BSH).[96,246] Bile salts are the end products of cholesterol metabolism in the liver and are stored in the gall bladder and released into the duodenum, helping fat digestion. In addition, bile salts are known to have antimicrobial activity because they are amphipathic molecules that can attack and degrade lipid membranes.

Some intestinal microorganisms have hence evolved mechanisms to resist the detergent action of bile, including the synthesis of porins, efflux pumps, and transport proteins.[246] Others produce bile salt hydrolases that transform and inactivate the bile salts. The deletion of the *bsh* gene from the *L. monocytogenes* chromosome results in an increase in susceptibility to bile, in reduced virulence, and in reduced liver colonization after infection of mice, thus demonstrating that Bsh is a virulence factor involved in the intestinal and hepatic phases of listeriosis. Analysis of *L. monocytogenes* deletion mutants in two other bile-tolerance loci (*pva* and *btlB*) revealed a role at least for the *btlB* gene product in resisting the acute toxicity of bile and bile salts, particularly glycoconjugated bile salts at low pH. However, the *btlB* gene product has no obvious bile salt hydrolase activity.[247] Finally,

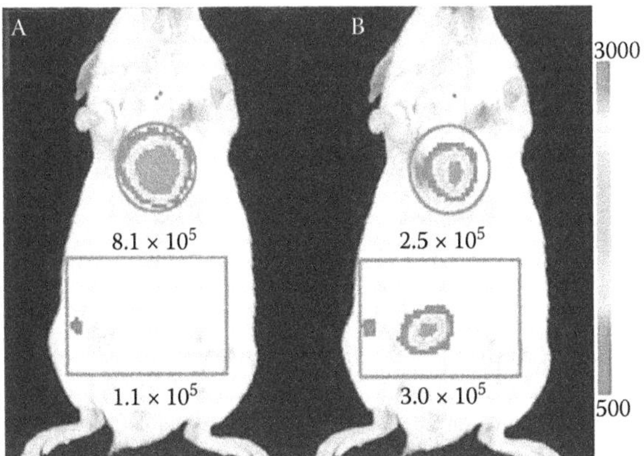

FIGURE 4.8 Transfer of *L. monocytogenes* from the gall bladder to the intestine. A mouse infected intravenously with 4×10^4 CFUs of the virulent luciferase-expressing *L. monocytogenes* strain was starved on day 4 postinfection. The animal was then fed whole milk and imaged 5 min (A) and 50 min (B) after feeding. Passage of *L. monocytogenes* from the gall bladder to the intestine is indicated by the shift of photon detection from the gall bladder (circle) to the intestine (square). Bars indicate photon counts per pixel registered during the 5-min exposures. (Taken from Hardy, J. et al., *Infect. Immun.*, 74, 1819, 2006. With permission.)

in silico analysis of the *L. monocytogenes* EGDe genome revealed a putative bile exclusion system encoded by the *bilEA* and *bilEB* genes.[248] The *bilE* system mediates resistance to bile through the active export of bile from the bacterial cell, especially under conditions that mimic the situation of the upper intestinal tract. Furthermore, an *L. monocytogenes* Δ*bilE* mutant shows much higher levels of intracellular bile than the wild-type strain, and the mutant is severely impaired in virulence upon oral administration to mice.

4.3.2 Vitamin B_{12} Biosynthesis and Anaerobic Use of Ethanolamine

The ability to synthesize vitamin B_{12} is unevenly distributed in living organisms, and about one-third of the bacteria sequenced carry vitamin B_{12} biosynthesis genes. A gene cluster (*cbi* and *cob* genes), identified in the genome sequences of *L. monocytogenes* and *L. innocua*,[96,249,250] shares high homology with vitamin B_{12} biosynthesis genes from *Salmonella* spp., suggesting that the two *Listeria* species use the oxygen-independent pathway like *Salmonella enterica*. Close to the cobalamin biosynthesis genes, both *Listeria* species contain orthologs of genes necessary in *S. enterica* for the coenzyme B_{12}-dependent degradation of ethanolamine and propanediol (*eut* and *pdu* genes, respectively). All three gene clusters (from *cbiP* to *pduS*) may have been horizontally acquired en bloc by a *Listeria* ancestor.[250] *L. monocytogenes* is a microaerophilic organism that thrives best at reduced oxygen tension. Vitamin B_{12}–dependent anaerobic degradation of ethanolamine and propanediol could enable *L. monocytogenes* to use ethanolamine and 1,2-propanediol as carbon and energy sources for growth under the anaerobic conditions encountered in the mammalian gut, where both substances are believed to be abundant. It is tempting to speculate that the vitamin B_{12} synthesis genes, together with the *pdu* and *eut* operons, play a role during a listerial infection and hence may represent a novel type of virulence determinants of *L. monocytogenes*.

4.3.3 Stp Serine-Threonine Phosphatase

The genome sequence of *L. monocytogenes* revealed the presence of genes encoding putative tyrosine phosphatases, serine-threonine kinases, and serine-threonine phosphatases.[96] One of the latter genes, *stp*, encodes a functional Mn^{2+}-dependent serine-threonine phosphatase with similarity to eukaryotic phosphatases. Mutational analysis showed that Stp is required for growth in mouse tissues and hence may represent a new type of listerial virulence factor. As a cellular target for Stp, the elongation factor EF-Tu was identified. Phosphorylation of EF-Tu is known to interfere with amino-acylated transfer RNA-binding, and Stp may play a role in regulating EF-Tu, thereby controlling bacterial survival in the infected host.[251]

4.3.4 The Peptidoglycan N-deacetylase PgdA

The evasion of early innate immune defenses may represent an important virulence mechanism of pathogenic bacteria. Recently, a gene encoding a peptidoglycan N-deacetylase, called *pdgA*, was identified in *L. monocytogenes* and demonstrated to be necessary for the deacetylation of N-acetylglucosamine residues in the listerial peptidoglycan.[252] Inactivation of *pgdA* by insertion mutagenesis suggested a role for peptidoglycan modification in listerial virulence because the mutant was extremely sensitive to the bacteriolytic activity of lysozyme. Furthermore, the mutant was severely impaired in growth in the livers and spleens of mice after oral and i.v. inoculations. This growth defect may be explained by a dramatic reduction in the mutant's ability to escape from the vacuoles of infected macrophages, where the bacteria were rapidly destroyed. The inability to escape from the hostile environment of the macrophage phagosome resulted in a massive IFN-β response triggering early innate immune defenses. Peptidoglycan N-acetylation therefore might represent a mechanism allowing *L. monocytogenes* to evade innate host defenses.[252]

4.4 REGULATION OF VIRULENCE

For facultative pathogens like *L. monocytogenes*, the environment in which they dwell as saprophytes is certainly not less competitive than the infected host as a survival/replication niche. Either of these two habitats poses specific challenges to which the pathogen must adapt in order to ensure its transmissibility and perpetuation. This is generally achieved in pathogenic bacteria by placing the virulence determinants under tight regulation by DNA-binding proteins that coordinately induce or repress virulence genes in response to physicochemical signals that indicate the presence in a host system or in the environment, respectively. The goal is to optimize bacterial fitness by selectively activating the synthesis of virulence factors during infection and by down-regulating it when these factors are not needed, thus avoiding wasteful expenditure of energy and resources.

In *L. monocytogenes*, this key regulatory role is fulfilled by the PrfA protein, a transcription factor of the Crp/Fnr family.[253,254] In this section we will focus on PrfA, as it is the essential regulator identified to date that is directly and specifically involved in the environmental control of listerial virulence determinants. Other regulatory systems are only briefly discussed, as in most cases they are also present in the nonpathogenic *Listeria* spp. and, consequently, it is unlikely that their reported functions in virulence are specific. *L. monocytogenes prfA* (PrfA⁻) mutants are nonpathogenic, despite all virulence determinants being present in the bacteria,[255,256] reflecting the central role that PrfA plays in pathogenesis. PrfA is a homodimer of two 237-residue monomers of 27 kDa with a modular architecture comprising (1) an N-terminal domain with a β(jelly)-roll and a long α-helix that provides most of the dimer interface, (2) a hinge/interdomain region, (3) a C-terminal domain that carries the DNA-binding helix-turn-helix (HTH) motif, and (4) a C-terminal extension that stabilizes the HTH motif (Figure 4.9).[257,258]

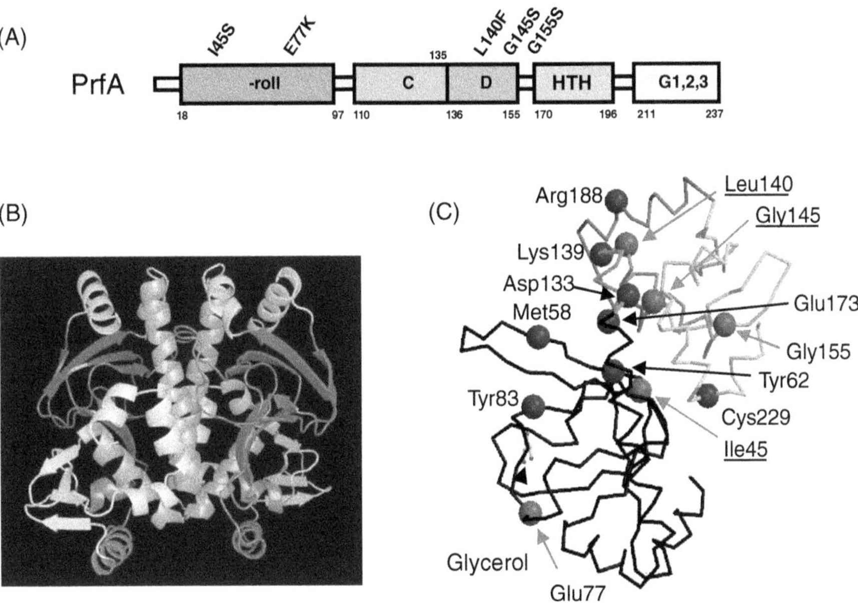

FIGURE 4.9 Structure of the central virulence gene regulator PrfA. (A) Schematic representation of the domains of PrfA with indication of the position of known PrfA* mutations. (B) Crystal structure of the PrfA dimer (constructed with MacPyMol 0.99). (C) Cα trace of the PrfA monomer showing the positions of single amino acid mutations that affect the function of the protein (bullets with respective amono acids; amino acids underlined: PrfA* mutations; others: PrfA⁻ minus mutations). (Panels A and B modified from Scortti, M. et al., *Microb. Infect.*, 9, 1196, 2007; panel C taken from Vega, Y. et al., *Mol. Microbiol.*, 52, 1553, 2004. With permission.)

4.4.1 The PrfA Regulon

Every step of the intracellular life cycle of *L. monocytogenes* (host cell invasion, phagosomal escape, cytosolic replication, and cell-to-cell spread) is mediated by products of the PrfA regulon. Only 10 of the 2853 genes of the *L. monocytogenes* EGD-e genome have been confirmed as bona fide (directly regulated) members of this regulon[59,64,190,256,259]—a number surprisingly small, given the apparent complexity of the listerial intracellular parasitic lifestyle.[254] These genes are organized in seven transcriptional units. Four of them—namely, the *hly* monocistron, the *mpl–actA–plcB(–orfXZ)* ("lecithinase") operon, and the *plcA–prfA* bicistron—are clustered together in the central listerial pathogenicity island LIPI-1 (Figure 4.6)[1,159]; the three others (i.e., the *inlAB* locus encoding InlA and InlB[58]; *inlC* encoding the small, secreted internalin InlC[59]; and *hpt* encoding the hexose phosphate transporter required for rapid bacterial growth in the host cell cytosol[190]) are inserted at different places of the *L. monocytogenes* chromosome (Figure 4.6). The 10 members of the "core" PrfA regulon are all tightly regulated by PrfA, with strong induction during intracellular infection and almost complete down-regulation during extracellular growth in rich medium.[183,184,191,260–263] The PrfA dimer activates transcription of these genes by binding to a palindromic promoter element with canonical sequence tTAACanntGTtAa, called "PrfA box," centered at position –41 with respect to the transcription start point.[264] A functional PrfA box contains seven invariant nucleotides (in capitals) with one-, exceptionally (*inlA* promoter) two-mismatch tolerance.[254] Binding of PrfA to its target sequence recruits RNA polymerase (RNAP) to the promoter, initiating transcription.[257,265]

Up to 145 additional genes from the *L. monocytogenes* EGD-e genome appear to be differentially regulated by PrfA according to transcriptomic profiling studies.[266,267] As only a few of these are preceded by putative PrfA boxes (13 up-regulated, 9 down-regulated), the effect exerted by PrfA is probably indirect. The PrfA boxes from these few genes either are intragenic, are located at a very short distance from the start codon (≤51 bp), or do contain two mismatches, of which one at least affects an invariant nucleotide of the canonical sequence.[254] Moreover, none of these putative PrfA box-associated genes has been independently found as PrfA-regulated in the two independent genome-wide PrfA transcriptome studies carried out to date.[266,267] One such gene is *bsh*, encoding the bile salt hydrolase that facilitates the colonization of the intestinal tract by *L. monocytogenes*.[246] The indirectly regulated genes encode transporters, metabolic enzymes, stress response mediators, regulators, and many proteins of unknown function, suggesting that PrfA plays a more global regulatory role in *L. monocytogenes* physiology.[266,267] In contrast to the core PrfA regulon members, only a fraction of these genes are induced during intracellular infection (17–56%, depending on the study),[183,184] consistent with a marginal involvement of PrfA in controlling their expression.

4.4.2 PrfA Regulation Mechanisms and Modulatory Signals

PrfA regulation operates primarily through changes in the transcriptional activity of the protein. PrfA is normally weakly active in native form, and the available evidence suggests that the strong induction in expression of the PrfA regulon observed during intracellular infection (between ≈10- and >300-fold, depending on the gene) is due to the activation of PrfA function when *L. monocytogenes* senses signals from the host cytosolic compartment (reviewed in Scortti et al.[254]) (Figure 4.10). This activation presumably results from a conformational shift in PrfA caused by the binding of a low-molecular-weight cofactor to the N-terminal β-roll domain. The conformational change would then be allosterically transmitted to the C-terminal HTH motif by a mechanism analogous to that leading to the activation of the structurally related enterobacterial regulator Crp (also known as CAP) by cAMP.[257,258] The nature of the putative PrfA-activating cofactor remains to be identified, but structural and functional data clearly rule out cAMP.[254] *L. monocytogenes* mutants have been characterized that produce a constitutively active PrfA form, called PrfA*, which is locked in the conformation presumably adopted by the regulator *in vivo* during infection.[259,260] PrfA* mutant

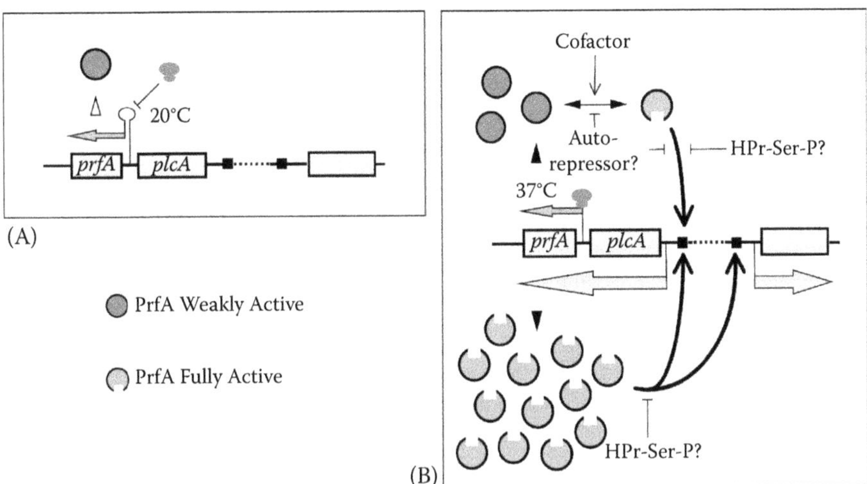

FIGURE 4.10 PrfA regulation model with representation of control levels and underlying mechanisms. Central to this model is the capacity of PrfA to undergo allosteric transition from weakly active to highly active conformations upon interaction with a putative low-molecular-weight cofactor (see text for details). (Modified from Scortti, M. et al., *Microb. Infect.*, in press. With permission.)

proteins carry single amino acid substitutions that increase the specific DNA-binding activity of the regulator.[256,264] Five PrfA* mutations have been identified to date (Figure 4.9): Gly145Ser, Ile45Ser, Glu77Lys, Leu140Phe, and Gly155Ser,[257,259,268,269] of which the first is the most common and best characterized one.[257,258]

PrfA-dependent expression levels also depend on the concentration of the regulatory protein in the bacterial cytoplasm. The amount of PrfA is controlled at the transcriptional and translational levels. Any change in PrfA activity (or PrfA amount) is automatically amplified about tenfold by a transcriptional autoregulatory loop mediated by the PrfA-dependent promoter of the *plcA–prfA* bicistron (Figure 4.6 and Figure 4.10).[165,256,270] P*plcA* promoter-driven *plcA–prfA* read-through transcription is essential for ensuring that sufficient amounts of PrfA are synthesized, as indicated by the avirulence of *L. monocytogenes* mutants, in which the positive feedback loop has been disrupted by insertional mutagenesis.[165,256] The *prfA* gene can also be expressed at low levels in the form of monocistronic transcripts generated from a vegetative promoter, P1*prfA*, situated in the *plcA–prfA* intergenic region.[256,270] The 5′ untranslated region (UTR) of these constitutive transcripts is responsible for the thermoregulated expression of PrfA-dependent genes. At temperatures ≤ 30°C, the *prfA* 5′ UTR forms a secondary structure that masks the ribosome binding site, whereas at 37°C this structure melts, allowing the translation of the *prfA* message (Figure 4.10).[270] This mechanism is thought to play a key role in preventing the production of PrfA-regulated factors when *L. monocytogenes* lives in the environment and to allow the rapid synthesis of PrfA from preformed transcripts when the bacteria enter a warm-blooded animal host (reviewed in Scortti et al.[254]).

PrfA-dependent expression is also controlled by negative regulatory pathways. Indeed, the thermoregulation mechanism on its own is not sufficient for reaching elevated levels of virulence gene activation, as demonstrated by the fact that only very weak PrfA-dependent expression is observed in rich medium at 37°C. In these conditions, however, substantial induction of the PrfA regulon is achieved when the medium is supplemented with an adsorbent such as activated charcoal.[260] This effect is due to the removal by the adsorbent of a small hydrophobic diffusible autorepressor that is released into the medium by exponentially growing *L. monocytogenes*.[272] A second repressor pathway is mediated by PTS-transported sugars such as glucose, fructose, manose, or, particularly, cellobiose and other plant-derived β-glucosides.[273,274] Virulence gene down-regulation occurs only when the amount of carbohydrate is sufficient to promote bacterial growth, excluding that the sugars themselves act as repressor signal molecules and suggesting that the underlying mechanism is somehow related to carbon catabolite repression (CCR).[273,274]

However, growth on hexose phosphates, which are taken by the non-PTS permease Hpt, do not cause virulence gene down-regulation despite obviously generating the same glycolytic intermediates as PTS sugars that are responsible for the CCR response, such as fructose 1,6-bisphosphate.[275] It seems, therefore, that central pathways of CCR are not involved in sugar-mediated PrfA-dependent gene repression; rather, upstream elements are closely linked to the transport itself of the sugar by the PTS permease. This notion is supported by experimental data showing that CcpA, PtsK (HPrK), or Hpr-Ser-P do not play a major role in the repression of the PrfA system caused by sugars.[276–278] The PrfA-dependent gene down-regulation caused by the diffusible autorepressor and PTS-transported sugars is not accompanied by a concomitant reduction in the amount of PrfA protein in the bacterial cytosol, indicating that the underlying mechanism reduces or interferes with PrfA activity (Figure 4.10).[272,279] Both PrfA repressor pathways may cooperate with the thermoregulation mechanism to prevent the production of virulence factors by *L. monocytogenes* colonies when growing in the natural habitat (soil rich in decaying vegetation), where β-glucosides and other PTS-transported sugars are abundant. It could also be that the autorepressor and the glucose/hexose phosphate balance help fine-tune the expression levels of the PrfA regulon during intracellular infection (reviewed in Scortti et al.[254]).

Finally, a third level of control is conferred by *cis*-acting elements in PrfA-regulated transcriptional units. Promoters having perfectly palindromic PrfA boxes with no deviations from the canonical sequence, such as P*plcA* and P*hly*, respond more efficiently to PrfA than those having PrfA boxes with sequence deviations, such as P*actA* (one mismatch) and P*inlAB* (two mismatches). The latter type of PrfA box has less affinity for PrfA, requires a larger activation input by the regulator, and therefore displays a less linear response.[257,259,270,280–282] The affinity of the promoters for RNAP may also influence their expression pattern. Thus, promoters with low affinity for PrfA and RNAP, like P*actA*, display an "all or nothing" response, whereas those binding efficiently both PrfA and RNAP, like P*plcA*, are of the "sensitive" response type characterized by a lower activation threshold and net inducibility by PrfA.[255] Other features that endow PrfA-regulated genes with differential expression characteristics include the presence of additional, PrfA-independent promoters (e.g., the case of *hly*, *mpl*, and the *inlAB* operon),[283,284] the architecture of the –10 region,[284] and the configuration of the 5′ UTR region, which may affect translation efficiency.[65,286,287]

4.4.3 Other Regulatory Systems

The *L. monocytogenes* EGDe genome encodes an unusually high proportion of transcriptional regulators (7.3%, corresponding to 209 regulatory genes), consistent with the dual saprophytic/pathogenic lifestyle of these ubiquitous bacteria and their ability to adapt to a variety of environmental conditions and habitats.[96] Mutational inactivation of a number of these regulators has been reported to result in alterations in virulence-associated phenotypes (e.g., invasion, intracellular proliferation) and/or in impaired *in vivo* survival in animal infection models. Examples include

- the Clp stress protein regulator CtsR[288]
- the transcriptional repressor MogR, involved in regulation of flagellar motility[289]
- PerR and FurR that regulate iron uptake and storage[290]
- the RelA (p)ppGpp synthase and CodY proteins involved in adaptation to postexponential phase growth[291]
- a number of two-component regulatory systems, such as LisrR/K involved in stress tolerance,[292] CesR/K involved in tolerance to ethanol and cell wall–acting antibiotics,[293] CheA/Y involved in chemotaxis,[294] the response regulators AgrA encoded by a quorum-sensing

locus similar to staphylococcal *agr*,[295] DegU involved in the control of motility,[296,297] the respiration regulator ResD,[298] and VirR controlling a regulon of 12 genes, some of which encode products involved in surface components' modifications[299]

However, these regulators are also present in the nonpathogenic species *Listeria innocua*, indicating that they fulfill general regulatory functions in listerial physiology. One has, therefore, to be very cautious on interpreting their involvement in virulence because, clearly, any alteration in housekeeping functions may cause unspecific attenuation due to impairment of the vegetative growth capacity, loss of bacterial fitness/viability, or perturbation of the general regulatory network. In addition, a recent study in which 15 of the 16 response regulators encoded in the *L. monocytogenes* genome were systematically deleted could not confirm, in most cases, previously reported effects on virulence.[297]

It cannot be excluded, of course, that by adaptive evolution some of the listerial "housekeeping" regulators could have become more specialized in virulence control functions. Such seems to be the case, at least, for the stress-response alternative sigma factor σ^B, encoded by *sigB*. A second promoter in the *plcA–prfA* intergenic region, P2*prfA*, situated downstream from P1*prfA* and directing the synthesis of a monocistronic *prfA* message preferentially during stationary phase, is partially regulated by both the vegetative sigma factor, α^A, and σ^B.[300] The partial σ^B-dependence of P2*prfA* may account for the observed overproduction of the PrfA-regulated virulence factors LLO and ActA in stress conditions such as high temperature (>42°C), exposure to oxidants, or nutrient starvation.[301,302]

However, the contribution of σ^B to stress-responsive PrfA regulon activation via P2*prfA* remains unclear, as similar levels of PrfA-dependent gene expression were observed in wild-type and Δ*sigB* *L. monocytogenes* during environmental stress or intracellular infection.[303] σ^B appears also to directly regulate the *inlAB* locus (see section 3.2), a core PrfA regulon member, via one of the four promoters in front of *inlA*.[304,305] σ^B-dependent induction of the synthesis of the internalins InlA and InlB by exposure to stress conditions, such as those presumably encountered by *L. monocytogenes* in the gastrointestinal tract, may play a role in the invasion of the intestinal epithelium during the entry phase of host infection.[306] Additional evidence for a possible interplay between the PrfA (virulence regulation) and the σ^B (stress tolerance) systems was provided by a transcriptomic study carried out on strain EGDe by Milohanic et al., in which many of the 51 genes identified as indirectly regulated by PrfA were found to belong to the σ^B regulon.[266] However, the existence of expressional cross-talk between the PrfA and σ^B regulons could not be independently confirmed by a recent PrfA transcriptome study carried out by another group based on the same *L. monocytogenes* strain.[267]

Post-translational mechanisms may also contribute to regulate *L. monocytogenes* virulence. There is evidence that the serine kinase RsbT plays a role in listerial intracellular survival through its involvement in the regulation of σ^B activity. RbsT forms part of the signal transduction machinery that removes the blockade of σ^B by the antisigma factor RbsV under environmental stress conditions.[307] Also related to the stress tolerance pathways of *L. monocytogenes*, class III heat shock proteins, such as the Clp-HSP1000 ATPase/chaperone subunits B, C, and E, as well as the ClpP proteolytic subunit of the ATP-dependent Clp proteolytic complex, have been shown to be required for listerial intracellular and/or *in vivo* survival in mice.[308–311] Again, however, these stress proteins are also present in the nonpathogenic *Listeria* spp. and therefore their involvement in virulence is likely to just reflect their role in the tolerance to hostile environmental conditions, including those found during infection. Another post-translational regulation mediator with an impact on *L. monocytogenes* virulence is the aforementioned Stp protein (section 4.3), one of the two putative serie-threonine phosphatases encoded in the *L. monocytogenes* EGDe genome. Stp is likely to influence the infectious process indirectly, through regulation of protein synthesis via its target, the elongation factor EF-Tu.[251]

4.5 CONCLUSIONS AND PERSPECTIVES

The emergence of listeriosis as an important and deadly food-borne disease in human populations in the 1980s has provided impetus for undertaking detailed experimental studies on the virulence mechanisms and pathogenesis of *L. monocytogenes.* As a consequence, a large number of key virulence- and invasion-associated proteins (e.g., LLO, PlcA, PlcB, ActA, Mpl, InlA, and InlB) in *L. monocytogenes* have been identified over the preceding two decades and have contributed to our knowledge of *L. monocytogenes* virulence and pathogenicity and to the improved diagnosis and prevention of listeriosis. The deciphering of the complete genome sequence of *L. monocytogenes* EGD-e serovar 1/2a and of the nonpathogenic species *L. innocua* serovar 6a, in 2001,[96] represents a quantum leap in the field of *L. monocytogenes* research. Today, the genomic sequences, partial or complete, of additional *L. monocytogenes* strains, including serovar 4b predominantly associated with human listeriosis, are also available,[50,312] and the sequence of an *L. monocytogenes* serovar 4a strain is complete and will be published soon.[313] The key genomic information from at least one representative strain of all major lineages of *L. monocytogenes*[50] will be thus accessible to the international scientific community soon. The availability in the near future of the complete genome sequence of at least one strain each of the six *Listeria* species[96,313–315] will dramatically expand our genomic information resource from this group of bacteria and facilitate our understanding of their biology and evolution.

The documentation of the *L. monocytogenes* EGD-e genomic sequence makes it possible to approach this important pathogen and model intracellular parasite from a global, genome-wide perspective using microarray-based transcriptome profiling and proteomic technologies (reviewed in Kuhn and Goebel[316]). These novel approaches have not only provided new critical insight into the mechanisms of listerial intracellular parasitism[183,184] but also shed light on the evolutionary mechanisms and dynamics of the genus *Listeria* and the genetic basis of diversity within the pathogenic species *L. monocytogenes.*[50,250,317] Thanks to this freely accessible genomic information, many new hitherto unknown virulence-associated factors and pathways have been discovered at an astonishing pace in the last few years.[98,110,117,122,190,193,246,251,56] The integration of different cell biological methods into the study of the *Listeria*–host cell interaction allows drawing a comprehensive picture of the complex *L. monocytogenes* life cycle within its eukaryotic host cells. Efforts must now also be invested in developing our understanding of the infection process *in vivo*, at the organismal level; exploiting the new possibilities offered by genetically modified, humanized animal models of infection[23,82]; surrogate models[318,319]; and novel *in vivo* imaging techniques.[244,245,320] These new approaches, which have already proven to be extremely valuable in helping interpret the behavior of *L. monocytogenes* within its mammalian host,[82,245] will certainly lead to new ground-breaking details on the still poorly understood aspects of *L. monocytogenes* physiopathogenesis like the dissemination phase and the crossing of the blood–brain and fetoplacental barriers, processes ultimately responsible for the high lethality still today associated with listerial infection.

ACKNOWLEDGMENTS

Supported by the EU ERA-NET Pathogenomics program (project SPATELIS to MK and MS) and by The Wellcome Trust (program grant 04702 to JAVB).

REFERENCES

1. Vazquez-Boland, J.A. et al., *Listeria* pathogenesis and molecular virulence determinants, *Clin. Microbiol. Rev.*, 14, 584, 2001.
2. Farber, J.M. and Peterkin, P.I., *Listeria monocytogenes*, a food-borne pathogen, *Microbiol. Rev.,* 55, 476, 1991.
3. Schuchat, A., Swaminathan, B., and. Broome C.V., Epidemiology of human listeriosis, *Clin. Microbiol. Rev.*, 4, 169, 1991.

4. Adak, G.K., Long, S.M., and O'Brien, S.J., Trends in indigenous food-borne disease and deaths, England and Wales: 1992–2000, *Gut,* 51, 832, 2002.
5. Dussurget, O., Pizarro-Cerda, J., and Cossart, P., Molecular determinants *of Listeria monocytogenes* virulence, *Annu. Rev. Microbiol.*, 58, 587, 2004.
6. Khelef, N. et al., *Listeria monocytogenes* and the genus *Listeria, Prokaryotes*, 4, 404, 2006.
7. Kuhn, M. and Goebel, W., Molecular virulence determinants of *Listeria monocytogenes.* In *Listeria, listeriosis and food safety*, 3rd ed., E. T. Ryser and E. H. Marth, eds. Marcel Dekker, Inc., New York, 2007, 111.
8. Lorber, B., Listeriosis, *Clin. Infect. Dis.,* 24, 1, 1996.
9. Farber, J.M. et al., Neonatal listeriosis due to cross-infection confirmed by isoenzyme typing and DNA fingerprinting, *J. Infect. Dis.,* 163, 927, 1991.
10. Slutsker, L. and Schuchat, A., Listeriosis in humans. In *Listeria, listeriosis, and food safety*, 2nd ed., E.T. Ryser and E.H. Marth, eds. Marcel Dekker, Inc., New York, 1999, 75.
11. Schuchat, A. et al., Bacterial meningitis in the United States in 1995. Active surveillance team, *N. Engl. J. Med.*, 337, 970, 1997.
12. Brouwer, M.C. et al., Community-acquired *Listeria monocytogenes* meningitis in adults. *Clin. Infect. Dis.*, 43, 1233, 2006.
13. Gallaguer, P.G. and Watakunakorn, C., *Listeria monocytogenes* endocarditis: A review of the literature 1950–1986, *Scand. J. Infect. Dis.*, 20, 359, 1998.
14. Gauto, A.R. et al., Arterial infections due to *Listeria monocytogenes*: Report of four cases and review of the world literatura, *Clin. Infect. Dis.*, 14, 23, 1992.
15. Doganay, M., Listeriosis: Clinical presentation, *FEMS Immunol. Med. Microbiol.,* 35, 173, 2003.
16. Allcock, J., Cutaneous listeriosis, *Vet. Rec.*, 130, 18, 1992.
17. Ooi, S.T. and Lorber, B., Gastroenteritis due to *Listeria monocytogenes, Clin. Infect. Dis.*, 40, 1327, 2005.
18. McLauchlin, J. and Low, J.C., Primary cutaneous listeriosis in adults: An occupational disease of veterinarians and farmers, *Vet. Rec.*, 135, 615, 1994.
19. Lecuit, M., Understanding how *Listeria monocytogenes* targets and crosses host barriers, *Clin. Microbiol. Infect.*, 11, 430, 2005.
20. Marco, A.J. et al., Penetration of *Listeria monocytogenes* in mice infected by the oral route, *Microb. Pathogen.*, 23, 255, 1997.
21. Corr, S., Hill, C., and Gahan, C.G.M., An *in vitro* cell-culture model demonstrates internalin- and hemolysin-independent translocation of *Listeria monocytogenes* across M cells, *Microb. Pathogen*, 41, 241, 2006.
22. Pron, B. et al., Comprehensive study of the intestinal stage of listeriosis in a rat ligated ileal loop system, *Infect. Immun.*, 66, 747, 1998.
23. Lecuit, M. et al., Functional genomic studies of the intestinal response to a food-borne enteropathogen in a humanized gnotobiotic mouse model, *J. Biol. Chem.* 282, 15065, 2007.
24. Pron, B. et al., Dendritic cells are early cellular targets of *Listeria monocytogenes* after intestinal delivery and are involved in bacterial spread in the host, *Cell. Microbiol.,* 3, 331, 2001.
25. Czuprynski, C.J., Faith, N.G., and Steinberg, H., Ability of the *Listeria monocytogenes* strain Scott A to cause systemic infection in mice infected by the intragastric route, *Appl. Environ. Microbiol.,* 68, 2893, 2002.
26. Bakardjiew, A.I. et al., Listeriosis in the pregnant guinea pig: A model of vertical transmission, *Infect. Immun.,* 72, 489, 2004.
27. Chico-Calero, I. and Vázquez-Boland, J.A., unpublished data.
28. Conlan, J.C. and North, R.J., Neutrophil-mediated dissolution of infected host cells as a defense strategy against a facultative intracellular bacterium, *J. Exp. Med.*, 174, 741, 1991.
29. Portnoy, D.A., Innate immunity to a facultative intracellular bacterial pathogen, *Curr. Opin. Immunol.,* 4, 20, 1992.
30. Dramsi, S. et al., Entry of *Listeria monocytogenes* into hepatocytes requires the expression of InlB, a surface protein of the internalin multigene family, *Mol. Microbiol.*, 16, 251, 1995.
31. Gaillard, J.-L., Jaubert, F., and Berche P., The *inlAB* locus mediates the entry of *Listeria monocytogenes* into hepatocytes *in vivo, J. Exp. Med.*, 183, 359, 1996.
32. Cousens, L.P. and Wing, E.J., Innate defenses in the liver during *Listeria* infection, *Immunol. Rev.,* 174, 150, 2000.
33. Gregory, S.H. and C.C. Liu., $CD8^+$ T-cell-mediated response to *Listeria monocytogenes* taken up in the liver and replicating within hepatocytes, *Immunol. Rev.*, 174, 112, 2000.

34. Conlan, J.W., Early pathogenesis of *Listeria monocytogenes* infection in the mouse spleen, *J. Med. Microbiol.*, 44, 295, 1996.
35. Marco, A.J. et al., Pathogenesis of lymphoid lesions in murine experimental listeriosis, *J. Comp. Path.*, 105, 1, 1991.
36. Carrero, J.A., Calderon, B. and Unanue, E.R., Listeriolysin O from *Listeria monocytogenes* is a lymphocyte apoptogenic molecule, *J. Immunol.*, 172, 4866, 2004.
37. Carrero, J.A., Calderon, B. and Unanue, E.R., Type I interferon sensitizes lymphocytes to apoptosis and reduces resistance to *Listeria* infection, *J. Exp. Med.*, 200, 535, 2004.
38. Pamer, E.G., Immune response to *Listeria monocytogenes, Nat. Rev. Immunol.*, 4, 812, 2004.
39. Munk, M.E. and Kaufmann, S.H.E., *Listeria monocytogenes* reactive T lymphocytes in healthy individuals, *Microb. Pathogen.*, 5, 49, 1988.
40. Berche, P., Bacteriemia is required for invasion of the murine central nervous system by *Listeria monocytogenes, Microb. Pathogen.*, 18, 323, 1995.
41. Blanot, S. et al., A gerbil model for rhombencephalitis due to *Listeria monocytogenes*, *Microb. Pathogen.*, 23, 39, 1997.
42. Bakardjiev, A.I., Theriot, J.A., and Portnoy, D.A., *Listeria monocytogenes* traffics from maternal organs to the placenta and back, *PLoS Pathog.*, 2, 623, 2006.
43. Drevets, D.A., Dissemination of *Listeria monocytogenes* by infected phagocytes, *Infect. Immun.*, 67, 3512, 1999.
44. Drevets, D.A. et al., *Listeria monocytogenes* infects human endothelial cells by two distinct mechanisms, *Infect. Immun.*, 63, 4268, 1995.
45. Riedo, F.X. et al., A point-source food-borne listeriosis outbreak: Documented incubation period and possible mild illness, *J. Infect. Dis.*, 170, 693, 1994.
46. Smith, M.A. et al., Nonhuman primate model for *Listeria monocytogenes*-induced stillbirths, *Infect. Immunity.*, 71, 1574, 2003.
47. Rocourt, J. et al., Quantitative risk assessment of *Listeria monocytogenes* in ready-to-eat foods; the FAO/WHO approach, *FEMS Immunol. Med. Microbiol.*, 1, 263, 2003.
48. Chen, Y. et al., *Listeria monocytogenes.* Low levels equal low risk, *J. Food Protect.*, 66, 570, 2003.
49. Jacques, C. et al., Investigations related to the epidemic strain involved in the French listeriosis outbreak in 1992, *Appl. Environ. Microbiol.*, 61, 2242, 1995.
50. Doumith, M. et al., New aspects regarding evolution and virulence of *Listeria monocytogenes* revealed by comparative genomics and DNA arrays, *Infect. Immun.*, 72, 1072, 2004.
51. Goulet, V. et al., Surveillance of human listeriosis in France, 2001–2003, *Euro Surveill.*, 11, 79, 2006.
52. Brosch, R. et al., Virulence heterogeneity of *Listeria monocytogenes* strains from various sources (food, human, animal) in immunocompetent mice and its association with typing characteristics, *J. Food Protect.*, 56, 296, 1993.
53. Norrung, B. and Andersen, J.K., Variations in virulence between different electrophoretic types of *Listeria monocytogenes, Lett. Appl. Microbiol.*, 30, 228, 2000.
54. Olier, M. et al., Expression of truncated internalin A is involved in impaired internalization of some *Listeria monocytogenes* isolates carried asymptomatically by humans, *Infect. Immun.*, 71, 1217, 2003.
55. Roberts, A., Chan, Y., and Wiedmann, M., Definition of genetically distinct attenuation mechanisms in naturally virulence-attenuated *Listeria monocytogenes* by comparative cell culture and molecular characterization, *Appl. Environ. Microbiol.*, 71, 3900, 2005.
56. Rocourt, J., Risk factors for listeriosis, *Food Control*, 7, 195, 1996.
57. Orndorff, P.E. et al., Host and bacterial factors in listeriosis pathogenesis, *Vet. Microbiol.*, 114, 1, 2006.
58. Gaillard, J.L. et al., Entry of *Listeria monocytogenes* into cells is mediated by internalin, a repeat protein reminiscent of surface antigens from Gram-positive cocci, *Cell*, 65, 1127, 1991.
59. Engelbrecht, F. et al., A new PrfA-regulated gene of *Listeria monocytogenes* encoding a small, secreted protein which belongs to the family of internalins, *Mol. Microbiol.*, 21, 823, 1996.
60. Dramsi, S. et al., Identification of four new members of the internalin multigene family of *Listeria monocytogenes* EGD, *Infect. Immun.*, 65, 1615, 1997.
61. Raffelsbauer, D. et al., The gene cluster *inlC2DE* of *Listeria monocytogenes* contains additional new internalin genes and is important for virulence in mice, *Mol. Gen. Genet.*, 260, 144, 1998.
62. Bergmann, B. et al., InlA- but not InlB-mediated internalization of *Listeria monocytogenes* by non-phagocytic mammalian cells needs the support of other internalins, *Mol. Microbiol.*, 43, 557, 2002.
63. Bierne, H. et al., Inactivation of the *srtA* gene in *Listeria monocytogenes* inhibits anchoring of surface proteins and affects virulence, *Mol. Microbiol.*, 43, 869, 2002.

64. Lingnau, A. et al., Expression of the *Listeria monocytogenes* EGD *inlA* and *inlB* genes, whose products mediate bacterial entry into tissue culture cell lines, by PrfA-dependent and -independent mechanisms, *Infect. Immun.,* 63, 3896, 1995.
65. Stritzker, J., Schoen, C., and Goebel, W., Enhanced synthesis of internalin A in aro mutants of *Listeria monocytogenes* indicates posttranscriptional control of the *inlA*B mRNA, *J. Bacteriol.,* 187, 2836, 2005.
66. Sheehan, B. et al., Differential activation of virulence gene expression by PrfA, the *Listeria monocytogenes* virulence regulator, *J. Bacteriol.,* 177, 6469, 1995.
67. Braun, L. et al., InlB: An invasion protein of *Listeria monocytogenes* with a novel type of surface association, *Mol. Microbiol.,* 25, 285, 1997.
68. Marino, M. et al., Structure of the InlB leucine-rich repeats, a domain that triggers host cell invasion by the bacterial pathogen *L. monocytogenes*, *Mol. Cell.,* 4, 1063, 1999.
69. Schubert, W.D. et al., Internalins from the human pathogen *Listeria monocytogenes* combine three distinct folds into a contiguous internalin domain, *J. Mol. Biol.* 312, 783, 2001.
70. Schubert, W.D. et al., Structure of internalin, a major invasion protein of *Listeria monocytogenes*, in complex with its human receptor E-cadherin, *Cell,* 111, 825, 2002.
71. Ooi, A. et al., Structure of internalin C from *Listeria monocytogenes*, *Acta Crystallogr. D. Biol. Crystallogr.,* 62, 1287, 2006.
72. Shen, Y. et al., InlB-dependent internalization of *Listeria* is mediated by the Met receptor tyrosine kinase, *Cell,* 103, 501, 2000.
73. Lecuit, M. et al., Internalin of *Listeria monocytogenes* with an intact leucine-rich repeat region is sufficient to promote internalization, *Infect. Immun.,* 65, 5309, 1997.
74. Braun, L., Ohayon, H., and Cossart, P. The InlB protein of *Listeria monocytogenes* is sufficient to promote entry into mammalian cells, *Mol. Microbiol.,* 27, 1077, 1998.
75. Lecuit, M. et al., Targeting and crossing of the human maternofetal barrier by *Listeria monocytogenes*: Role of internalin interaction with trophoblast E-cadherin, *Proc. Natl. Acad. Sci. USA,* 101, 6152, 2004.
76. Gregory, S.H., Sagnimeni, A.J., and Wing, E.J., Internalin B promotes the replication of *Listeria monocytogenes* in mouse hepatocytes, *Infect. Immun.,* 65, 5137, 1997.
77. Greiffenberg, L. et al., Interaction of *Listeria monocytogenes* with human brain microvascular endothelial cells: InlB-dependent invasion, long-term intracellular growth and spread from macrophages to endothelial cells, *Infect. Immun.,* 66, 5260, 1998.
78. Parida, S.K. et al., Internalin B is essential for adhesion and mediates the invasion of *Listeria monocytogenes* into human endothelial cells, *Mol. Microbiol.,* 28, 81, 1998.
79. Mengaud, J. et al., E-cadherin is the receptor for internalin, a surface protein required for entry of *Listeria monocytogenes* into epithelial cells, *Cell,* 84, 923, 1996.
80. Lecuit, M. et al., A single amino acid in E-cadherin responsible for host specificity towards the human pathogen *Listeria monocytogenes*, *EMBO J.,* 18, 3956, 1999.
81. Schubert, W.D. and Heinz, D.W., Structural aspects of adhesion to and invasion of host cells by the human pathogen *Listeria monocytogenes*, *Chembiochemistry,* 4, 1285, 2003.
82. Lecuit, M. et al., A transgenic model for listeriosis: Role of internalin in crossing the intestinal barrier, *Science,* 292, 1722, 2001.
83. Lecuit, M. et al., A role for α- and β-catenins in bacterial uptake, *Proc. Natl. Acad. Sci. USA,* 97, 10008, 2000.
84. Ireton, K. et al., A role for phosphoinositide 3-kinase in bacterial invasion, *Science,* 274, 780, 1996.
85. Ireton, K., Payrastre, B., and Cossart, P., The *Listeria monocytogenes* protein InlB is an agonist of mammalian phosphoinositide 3-kinase, *J. Biol. Chem.,* 274, 17025, 1999.
86. Hamon, M., Bierne, H., and Cossart, P., *Listeria monocytogenes*: A multifaceted model, *Nat. Rev. Microbiol.,* 4, 423, 2006.
87. Veiga, E. and Cossart, P., *Listeria* hijacks the clathrin-dependent endocytic machinery to invade mammalian cells, *Nat. Cell Biol.,* 7, 894, 2005.
88. Veiga, E. and Cossart, P., The role of clathrin-dependent endocytosis in bacterial internalization, *Trends Cell Biol.,* 16, 499, 2006.
89. Braun, L., Ghebrehiwet, B., and Cossart, P., gC1q-R/p32, a C1q-binding protein, is a receptor for the InlB invasion protein of *Listeria monocytogenes*, *EMBO J.,* 19, 1458, 2000.
90. Banerjee, M. et al., GW domains of the *Listeria monocytogenes* invasion protein InlB are required for potentiation of Met activation, *Mol. Microbiol.,* 52, 257, 2004.
91. Jonquieres, R., Pizarro-Cerda, J., and Cossart, P., Synergy between the N- and C-terminal domains of InlB for efficient invasion of nonphagocytic cells by *Listeria monocytogenes*, *Mol. Microbiol.,* 42, 955, 2001.

92. Marino, M. et al., GW domains of the *Listeria monocytogenes* invasion protein InlB are SH3-like and mediate binding to host ligands, *EMBO J.*, 21, 5623, 2002.
93. Simons, K. and Ikonen, E., Functional rafts in cell membranes, *Nature,* 387, 569, 1997.
94. Seveau, S. et al., The role of lipid rafts in E-cadherin- and HGF-R/Met-mediated entry of *Listeria monocytogenes* into host cells, *J. Cell. Biol.,* 166, 743, 2004.
95. Seveau, S. et al., A FRET analysis to unravel the role of cholesterol in Racl and PI 3-kinase activation in the InlB/Met signaling pathway, *Cell. Microbiol.,* 9, 790, 2007.
96. Glaser, P. et al., Comparative genomics of *Listeria* species, *Science,* 294, 849, 2001.
97. Cabanes, D., Surface proteins and the pathogenic potential of *Listeria monocytogenes, Trends Microbiol.,* 10, 238, 2002.
98. Sabet, C. et al., LPXTG protein InlJ, a newly identified internalin involved in *Listeria monocytogenes* virulence, *Infect. Immun.,* 73, 6912, 2005.
99. Kuhn, M. and Goebel, W., Identification of an extracellular protein of *Listeria monocytogenes* possibly involved in the intracellular uptake by mammalian cells, *Infect. Immun.,* 57, 55, 1989.
100. Bubert, A. et al., Structural and functional properties of the p60 proteins from different *Listeria* species, *J. Bacteriol.,* 174, 8166, 1992.
101. Wood, S., Maroushek, N., and Czuprynski, C.J., Multiplication of *Listeria monocytogenes* in a murine hepatocyte cell line, *Infect. Immun.,* 61, 3068, 1993.
102. Köhler, S. et al., The gene coding for protein p60 of *Listeria monocytogenes* and its use as a species specific probe for *Listeria monocytogenes, Infect. Immun.,* 58, 1943, 1990.
103. Schubert, K. et al., P45, an extracellular 45-kDa protein of *Listeria monocytogenes* with similarity to protein p60 and exhibiting peptidoglycan lytic activity, *Arch. Microbiol.,* 173, 21, 2000.
104. Rowan, N.J. et al., Virulent rough filaments of *Listeria monocytogenes* from clinical and food samples secreting wild-type levels of cell-free p60 protein, *J. Clin. Microbiol.,* 38, 2643, 2000.
105. Wuenscher, M.D. et al., The *iap* gene of *Listeria monocytogenes* is essential for cell viability and its gene product, p60, has bacteriolytic activity, *J. Bacteriol.,* 175, 3491, 1993.
106. Lenz, L.L. et al., SecA2-dependent secretion of autolytic enzymes promotes *Listeria monocytogenes* pathogenesis, *Proc. Natl. Acad. Sci. USA,* 100, 12432, 2003.
107. Pilgrim, S. et al., Deletion of the gene encoding p60 in *Listeria monocytogenes* leads to abnormal cell division and to loss of actin-based motility, *Infect. Immun.,* 71, 3473, 2003.
108. Lenz, L.L. and Portnoy, D.A., Identification of a second *Listeria secA* gene associated with protein secretion and the rough phenotype, *Mol. Microbiol.,* 45, 1043, 2002.
109. Carroll, S.A. et al., Identification and characterization of a peptidoglycan hydrolase, MurA, of *Listeria monocytogenes*, a muramidase needed for cell separation, *J. Bacteriol.,* 185, 6801, 2003.
110. Cabanes, D. et al., Auto, a surface associated autolysin of *Listeria monocytogenes* required for entry into eukaryotic cells and virulence, *Mol. Microbiol.,* 51, 1601, 2004.
111. Milohanic, E. et al., The autolysin Ami contributes to the adhesion of *Listeria monocytogenes* to eukaryotic cells via its cell wall anchor, *Mol. Microbiol.,* 39, 1212, 2001.
112. Milohanic, E. et al., Identification of new loci involved in adhesion of *Listeria monocytogenes* to eukaryotic cells. *Microbiol.,* 146, 731, 2000.
113. Domann, E. et al., A novel bacterial virulence gene in *Listeria monocytogenes* required for host cell microfilament interaction with homology to the proline-rich region of vinculin, *EMBO J.,* 11, 1981, 1992.
114. Kocks, C. et al., *Listeria monocytogenes*-induced actin assembly requires the *actA* gene product, a surface protein, *Cell,* 68, 521, 1992.
115. Alvarez-Dominguez, C. et al., Host cell heparan sulfate proteoglycans mediate attachment and entry of *Listeria monocytogenes*, and the listerial surface protein ActA is involved in heparan sulfate receptor recognition, *Infect. Immun.,* 65, 78, 1997.
116. Suarez, M. et al., A role for ActA in epithelial cell invasion by *Listeria monocytogenes*, *Cell. Microbiol.,* 3, 853, 2001.
117. Reglier-Poupet, H. et al., Identification of LpeA, a PsaA-like membrane protein that promotes cell entry by *Listeria monocytogenes*, *Infect. Immun.,* 71, 474, 2003.
118. Dramsi, S. et al., FbpA, a novel multifunctional *Listeria monocytogenes* virulence factor, *Mol. Microbiol.,* 53, 639, 2004.
119. Kim, K.P. et al., Adhesion characteristics of *Listeria* adhesion protein (LAP)-expressing *Escherichia coli* to Caco-2 cells and of recombinant LAP to eukaryotic receptor Hsp60 as examined in a surface plasmon resonance sensor, *FEMS Microbiol. Lett.,* 256, 324, 2006.

120. Pandiripally, V.K. et al., Surface protein p104 is involved in adhesion of *Listeria monocytogenes* to human intestinal cell line, Caco-2, *J. Med. Microbiol.,* 48, 117, 1999.
121. Wampler, J.L. et al., Heat shock protein 60 acts as a receptor for the *Listeria* adhesion protein in Caco-2 cells, *Infect Immun.,* 72, 931, 2004.
122. Cabanes, D. et al., Gp96 is a receptor for a novel *Listeria monocytogenes* virulence factor, Vip, a surface protein, *EMBO J.,* 24, 2827, 2005.
123. Gaillard, J.L., Berche, P., and Sansonetti, P.J., Transposon mutagenesis as a tool to study the role of hemolysin in the virulence of *Listeria monocytogenes*, *Infect. Immun.,* 52, 50, 1986.
124. Kathariou, S. et al., Tn*916*-induced mutations in the hemolysin determinant affecting virulence of *Listeria monocytogenes, J. Bacteriol.,* 169, 1291, 1987.
125. Portnoy, D.A., Jacks, P.S., and Hinrichs, D.J. Role of hemolysin for the intracellular growth of *Listeria monocytogenes*, *J. Exp. Med.,* 167, 1459, 1988.
126. Cossart, P. et al., Listeriolysin O is essential for virulence of *Listeria monocytogenes*: Direct evidence obtained by gene complementation, *Infect. Immun.,* 57, 3629, 1989.
127. Vázquez-Boland, J.A. et al., Listeriolysin. In *Sourcebook of bacterial protein toxins,* J.E. Alouf and M. Popoff, eds. Academic Press, San Diego, CA, 2006, 698.
128. Parrisius, J. et al., Production of listeriolysin by β-hemolytic strains of *Listeria monocytogenes*, *Infect. Immun.*, 51, 314, 1986.
129. Gekara, N.O. et al., The cholesterol-dependent cytolysin listeriolysin O aggregates rafts via oligomerization, *Cell. Microbiol.,* 7, 1345, 2005.
130. Geoffroy, C. et al., Purification, characterization, and toxicity of the sulfhydyl-activated hemolysin listeriolysin O from *Listeria monocytogenes, Infect. Immun.,* 55, 1641, 1987.
131. Domann, E. and Chakraborty, T., Nucleotide sequence of the listeriolysin gene from a *Listeria monocytogenes* serotype 1/2a strain, *Nuc. Acids Res.,* 17, 6406, 1989.
132. Mengaud, J. et al., Identification of the structural gene encoding the SH-activated hemolysin in *Listeria monocytogenes*: Listeriolysin O is homologous with streptolysin O and pneumolysin, *Infect. Immun.,* 55, 3225, 1987.
133. Mengaud, J. et al., Expression in *Escherichia coli* and sequence analysis of the listeriolysin O determinant of *Listeria monocytogenes*, *Infect. Immun.,* 56, 766, 1988.
134. Rossjohn, J. et al., Structure of a cholesterol-binding, thiol-activated cytolysin and a model of its membrane form, *Cell,* 89, 685, 1997.
135. Gaillard, J.L. et al., *In vitro* model of penetration and intracellular growth of *Listeria monocytogenes* in the human enterocyte-like cell line Caco-2, *Infect. Immun.,* 55, 2822, 1987.
136. Tilney, L.G. and Portnoy, D.A., Actin filaments and the growth, movement, and spread of the intracellular bacterial parasite, *Listeria monocytogenes*, *J. Cell Biol.,* 109, 1597, 1989.
137. Bielecki, J. et al., *Bacillus subtilis* expressing a hemolysin gene from *Listeria monocytogenes* can grow in mammalian cells, *Nature,* 345, 175, 1990.
138. Schuerch, D.W. et al., Molecular basis of listeriolysin O pH dependence, *Proc. Natl. Acad. Sci. USA,* 102, 12537, 2005.
139. Beauregard, K.E. et al., pH-dependent perforation of macrophage phagosomes by listeriolysin O from *Listeria monocytogenes*, *J. Exp. Med.,* 186, 1159, 1997.
140. Conte, M.P. et al., The effects of inhibitors of vacuolar acidification on the release of *Listeria monocytogenes* from phagosomes of Caco-2 cells, *J. Med. Microbiol.,* 44, 418, 1996.
141. Jones, S. and Portnoy, D.A., Characterization of *Listeria monocytogenes* pathogenesis in a strain expressing perfringolysin O instead of listeriolysin O, *Infect. Immun.,* 62, 5608, 1994.
142. Jones, S., Preiter, K., and Portnoy, D.A., Conversion of an extracellular cytolysin into a phagosome-specific lysin which supports the growth of an intracellular pathogen, *Mol. Microbiol.,* 21, 1219, 1996.
143. Glomski, I.J., Decatur, A.L., and Portnoy, D.A., *Listeria monocytogenes* mutants that fail to compartmentalize listerolysin O activity are cytotoxic, avirulent, and unable to evade host extracellular defenses, *Infect. Immun.,* 71, 6754, 2003.
144. Decatur, A.L. and Portnoy, D.A., A PEST-like sequence in listeriolysin O essential for *Listeria monocytogenes* pathogenicity, *Science,* 290, 992, 2000.
145. Lety, M.A. et al., Identification of a PEST-like motif in listeriolysin O required for phagosomal escape and for virulence in *Listeria monocytogenes*, *Mol. Microbiol.,* 39, 1124, 2001.
146. Rechsteiner, M. and Rogers, S.W., PEST sequences and regulation by proteolysis, *Trends Biochem. Sci.,* 21, 267, 1996.

147. Lety, M.A. et al., Critical role of the N-terminal residues of listeriolysin O in phagosomal escape and virulence of *Listeria monocytogenes*, *Mol. Microbiol.*, 46, 367, 2002.
148. Schnupf, P., Portnoy, D.A., and Decatur, A.L., Phosphorylation, ubiquitination and degradation of listeriolysin O in mammalian cells: Role of the PEST-like sequence, *Cell. Microbiol.*, 8, 353, 2006.
149. Schnupf, P. et al., Regulated translation of listeriolysin O controls virulence of *Listeria monocytogenes*, *Mol. Microbiol.*, 61, 999, 2006.
150. Webster, P., Early intracellular events during internalization of *Listeria monocytogenes* by J774 cells, *J. Histochem. Cytochem.*, 50, 503, 2002.
151. Alvarez-Dominguez, C. and Stahl, P.D., Increased expression of Rab5a correlates directly with accelerated maturation of *Listeria monocytogenes* phagosomes, *J. Biol. Chem.*, 274, 11459, 1999.
152. Alvarez-Dominguez, C. et al., Phagocytosed live *Listeria monocytogenes* influences Rab5-regulated *in vitro* phagosome-endosome fusion, *J. Biol. Chem.*, 271, 13834, 1996.
153. Henry, R. et al., Cytolysin-dependent delay of vacuole maturation in macrophages infected with *Listeria monocytogenes*, *Cell. Microbiol.*, 8, 107, 2006.
154. Shaughnessy, L.M. et al., Membrane perforations inhibit lysosome fusion by altering pH and calcium in *Listeria monocytogenes* vacuoles, *Cell. Microbiol.*, 8, 781, 2006.
155. Alvarez-Dominguez, C., Roberts, R., and Stahl, P.D., Internalized *Listeria monocytogenes* modulates intracellular trafficking and delays maturation of the phagosome, *J. Cell Sci.*, 110, 731, 1997.
156. Dancz, C.E. et al., Inducible control of virulence gene expression in *Listeria monocytogenes*: Temporal requirement of listeriolysin O during intracellular infection, *J. Bacteriol.*, 184, 5935, 2002.
157. Smith, G.A. et al., The two distinct phospholipases C of *Listeria monocytogenes* have overlapping roles in escape from a vacuole and cell-to-cell spread, *Infect. Immun.*, 63, 4231, 1995.
158. Marquis, H., Doshi, V., and Portnoy, D.A., The broad-range phospholipase C and a metalloprotease mediate listeriolysin O-independent escape of *Listeria monocytogenes* from a primary vacuole in human epithelial cells, *Infect. Immun.*, 63, 4531, 1995.
159. Vázquez-Boland, J.A. et al., Pathogenicity islands and virulence evolution in *Listeria*, *Microb. Infect.*, 3, 571, 2001.
160. Leimeister-Wächter, M., Domann, E., and Chakraborty, T., Detection of a gene encoding a phosphatidylinositol-specific phospholipase C that is coordinately expressed with listeriolysin in *Listeria monocytogenes*, *Mol. Microbiol.*, 5, 361, 1991.
161. Mengaud, J., Braun-Breton, C., and Cossart, P., Identification of phosphatidylinsitol-specific phospholipase C activity in *Listeria monocytogenes*: A novel type of virulence factor? *Mol. Microbiol.*, 5, 367, 1991.
162. Gandhi, A.J., Perussia, B., and Goldfine, H., *Listeria monocytogenes* phosphatidylinositol (PI)-specific phospholipase C has low activity on glycosyl-PI-anchored proteins, *J. Bacteriol.*, 175, 8014, 1993.
163. Moser, J. et al., Crystal structure of the phosphatidylinositol-specific phospholipase C from the human pathogen *Listeria monocytogenes*, *J. Mol. Biol.*, 273, 269, 1997.
164. Bannam, T. and Goldfine, H., Mutagenesis of active-site histidines of *Listeria monocytogenes* phosphatidylinositol-specific phospholipase C: Effects on enzyme activity and biological function, *Infect. Immun.*, 67, 182, 1999.
165. Camilli, A., Tilney, L.G., and Portnoy, D.A., Dual roles of PlcA in *Listeria monocytogenes* pathogenesis, *Mol. Microbiol.*, 8, 143, 1993.
166. Goldfine, H. and Knob, C., Purification and characterization of *Listeria monocytogenes* phosphatidylinositol-specific phospholipase C, *Infect. Immun.*, 60, 4059, 1992.
167. Geoffroy, C. et al., Purification and characterization of an extracellular 29-kilodalton phospholipase C from *Listeria monocytogenes*, *Infect. Immun.*, 59, 2382, 1991.
168. Goldfine, H., Johnston, N.C., and Knob, C., Nonspecific phospholipase C of *Listeria monocytogenes*: activity on phospholipids in Triton X-100-mixes micelles and in biological membranes, *J. Bacteriol.*, 175, 4298, 1993.
169. Vazquez-Boland, J.A. et al., Nucleotide sequence of the lecithinase operon in *Listeria monocytogenes* and possible role of lecithinase in cell-to-cell spread, *Infect. Immun.*, 60, 219, 1992.
170. Poyart, C. et al., The zinc metalloprotease of *Listeria monocytogenes* is required for maturation of the phosphatidylcholine phospholipase C: Direct evidence obtained by gene complementation, *Infect. Immun.*, 61, 1576, 1993.
171. Raveneau, J. et al., Reduced virulence of a *Listeria monocytogenes* phospholipase-deficient mutant obtained by transposon insertion into the zinc metalloprotease gene, *Infect. Immun.*, 60, 916, 1992.
172. Gründling, A., Gonzalez, M.D., and Higgins, D.E., Requirement of the *Listeria monocytogenes* broad-range phospholipase PC-PLC during infection of human epithelial cells, *J. Bacteriol.*, 185, 6295, 2003.

173. Mengaud, J., Geoffroy, C., and Cossart, P., Identification of a new operon involved in *Listeria monocytogenes* virulence: Its first gene encodes a protein homologous to bacterial metalloproteases, *Infect. Immun.*, 59, 1043, 1991.
174. Domann, E. et al., Molecular cloning, sequencing, and identification of a metalloprotease gene from *Listeria monocytogenes* that is species specific and physically linked to the listeriolysin gene, *Infect. Immun.*, 59, 65, 1991.
175. Coffey, A. et al., Characteristics of the biologically active 35-kDa metalloprotease virulence factor from *Listeria monocytogenes, J. Appl. Microbiol.*, 88, 132, 2000.
176. Snyder, A. and Marquis, H., Restricted translocation across the cell wall regulates secretion of the broad-range phospholipase C of *Listeria monocytogenes, J. Bacteriol.*, 185, 5953, 2003.
177. Yeung, P.S., Zagorski, N., and Marquis, H., The metalloprotease of *Listeria monocytogenes* controls cell wall translocation of the broad-range phospholipase C, *J. Bacteriol.*, 187, 2601, 2005.
178. Reglier-Poupet, H. et al., Maturation of lipoproteins by type II signal peptidase is required for phagosomal escape of *Listeria monocytogenes, J. Biol. Chem.*, 278, 49469, 2003.
179. Borezee, E. et al., Identification in *Listeria monocytogenes* of MecA, a homologue of the *Bacillus subtilis* competence regulatory protein, *J. Bacteriol.*, 182, 5931, 2000.
180. Agaisse, H. et al., Genome-wide RNAi screen for host factors required for intracellular bacterial infection, *Science,* 309, 1248, 2005.
181. Marquis, H. et al., Intracytoplasmic growth and virulence of *Listeria monocytogenes* auxotrophic mutants, *Infect. Immun.*, 61, 3756, 1993.
182. Klarsfeld, A.D., Goossens, P.L., and Cossart, P., Five *Listeria monocytogenes* genes preferentially expressed in infected mammalian cells: *plcA, purH, purD, pyrE* and an arginine ABC transporter gene *arpJ, Mol. Microbiol.*, 13, 585, 1994.
183. Chatterjee, S.S. et al., Intracellular gene expression profile of *Listeria monocytogenes, Infect. Immun.*, 74, 1323, 2006.
184. Joseph, B. et al., Identification of *Listeria monocytogenes* genes contributing to intracellular replication by expression profiling and mutant screening, *J. Bacteriol.*, 188, 556, 2006.
185. Falzano, L. et al., Induction of phagocytic behavior in human epithelial cells by *Escherichia coli* cytotoxic necrotizing factor type 1, *Mol. Microbiol.*, 9, 1247, 1993.
186. Goetz, M. et al., Microinjection and growth of bacteria in the cytosol of mammalian host cells, *Proc. Natl. Acad. Sci. USA,* 98, 12221, 2001.
187. Goebel, W. and Kuhn, M., Bacterial replication in the host cell cytosol, *Curr. Opin. Microbiol.*, 3, 49, 2000.
188. O'Riordan, M. and Portnoy, D.A., The host cytosol: front-line or home front? *Trends Microbiol.*, 10, 361, 2002.
189. Vazquez-Boland, J.A., Bacterial growth in the cytosol: Lessons from *Listeria, Trends Microbiol.*, 10, 493, 2002.
190. Chico-Calero, I. et al., Hpt, a bacterial homolog of the microsomal glucose-6-phosphate translocase, mediates rapid intracellular proliferation in *Listeria, Proc. Natl. Acad. Sci. USA,* 99, 431, 2002.
191. Scortti, M. et al., Coexpression of virulence and fosfomycin susceptibility in *Listeria*: Molecular basis of an antimicrobial *in vitro–in vivo* paradox, *Nat. Med.*, 12, 515, 2006.
192. Slaghuis, J. et al., Inefficient replication of *Listeria innocua* in the cytosol of mammalian cells, *J. Infect. Dis.,* 189, 393, 2004.
193. O'Riordan, M., Moors, M.A., and Portnoy, D.A., *Listeria* intracellular growth and virulence require host-derived lipoic acid, *Science,* 302, 462, 2003.
194. Theriot, J.A. et al., Involvement of profilin in the actin-based motility of *Listeria monocytogenes* in cells and cell free extracts, *Cell,* 76, 505, 1994.
195. Marchand, J.B. et al., Actin-based movement of *Listeria monocytogenes*: actin assembly results from the local maintenance of uncapped filament barbed ends at the bacterium surface, *J. Cell. Biol.,* 130, 331, 1995.
196. Robbins, J.R. and Theriot, J.A., *Listeria monocytogenes* rotates around its long axis during actin-based motility, *Curr. Biol.,* 13, 754, 2003.
197. Kocks, C. et al., The unrelated surface proteins ActA of *Listeria monocytogenes* and IcsA of *Shigella flexneri* are sufficient to confer actin-based motility on *Listeria innocua* and *Escherichia coli*, respectively, *Mol. Microbiol.*, 18, 413, 1995.
198. Friederich, E. et al., Targeting of *Listeria monocytogenes* ActA protein to the plasma membrane as a tool to dissect both actin-based cell morphogenesis and ActA function, *EMBO J.*, 14, 2731, 1995.

199. Pistor, S. et al., The ActA protein of *Listeria monocytogenes* acts as nucleator inducing reorganization of the actin cytoskeleton, *EMBO J.*, 13, 758, 1994.
200. Pistor, S. et al., The bacterial actin nucleator protein ActA of *Listeria monocytogenes* contains multiple binding sites for host microfilament proteins, *Curr. Biol.*, 5, 517, 1995.
201. Cameron, L.A. et al., Motility of ActA protein-coated microspheres driven by actin polymerization, *Proc. Natl. Acad. Sci. USA*, 96, 4908, 1999.
202. Kocks, C. et al., Polarized distribution of *Listeria monocytogenes* surface protein ActA at the site of directional actin assembly, *J. Cell Sci.*, 3, 699, 1993.
203. Temm-Grove, C.T. et al., Exploitation of microfilament proteins by *Listeria monocytogenes*: Microvillus-like composition of the comet tails and vectorial spreading in polarized epithelial sheets, *J. Cell. Sci.*, 107, 2951, 1994.
204. Rafelski, S.M. and Theriot, J.A., Bacterial shape and ActA distribution affect initiation of *Listeria monocytogenes* actin-based motility, *Biophys. J.*, 89, 2146, 2005.
205. Smith, G.A., Portnoy, D.A., and Theriot, J.A., Asymmetric distribution of the *Listeria monocytogenes* ActA protein is required and sufficient to direct actin-based motility, *Mol. Microbiol.*, 17, 945, 1995.
206. Mourrain, P. et al., ActA is a dimer, *Proc. Natl. Acad. Sci. USA*, 94, 10034, 1997.
207. Machner, M.P. et al., ActA from *Listeria monocytogenes* can interact with up to four Ena/VASP homology domains simultaneously, *J. Biol. Chem.*, 276, 40096, 2001.
208. Bear, J.E., Krause, M., and Gertler, F.B., Regulating cellular actin assembly, *Curr. Opin. Cell Biol.*, 13, 158, 2001.
209. Cameron, L.A. et al., Secrets of actin-based motility revealed by a bacterial pathogen, *Nat. Rev. Mol. Cell Biol.*, 1, 110, 2000.
210. Cossart, P., Actin-based motility of pathogens: The Arp2/3 complex is a central player, *Cell. Microbiol.*, 2, 195, 2000.
211. Lasa, I. et al., The amino-terminal part of ActA is critical for the actin-based motility of *Listeria monocytogenes*; the central proline-rich region acts as a stimulator, *Mol. Microbiol.*, 18, 425, 1995.
212. Lasa, I. et al., Identification of two regions in the N-terminal domain of ActA involved in the actin comet tail formation by *Listeria monocytogenes*, *EMBO J.*, 16, 1531, 1997.
213. Pistor. S. et al., Mutations of arginine residues within the 146-KKRRK-150 motif of the ActA protein of *Listeria monocytogenes* abolish intracellular motility by interfering with the recruitment of the Arp2/3 complex, *J. Cell Sci.*, 113, 3277, 2000.
214. Lauer, P. et al., Systematic mutational analysis of the amino-terminal domain of the *Listeria monocytogenes* ActA protein reveals novel functions in actin-based motility, *Mol. Microbiol.*, 42, 1163, 2001.
215. Smith, G. A., Theriot, J.A., and Portnoy, D.A., The tandem repeat domain in the *Listeria monocytogenes* ActA protein controls the rate of actin-based motility, the percentage of moving bacteria, and the localization of vasodilator-stimulated phosphoprotein and profiling, *J. Cell Biol.*, 135, 647, 1996.
216. Chakraborty, T. et al., A focal adhesion factor directly linking intracellularly motile *Listeria monocytogenes* and *Listeria ivanovii* to the actin-based cytoskeleton of mammalian cells, *EMBO J.*, 14, 1314, 1995.
217. Gertler, F.B. et al., Mena, a relative of VASP and *Drosophila* enabled, is implicated in the control of microfilament dynamics, *Cell*, 87, 227, 1996.
218. Zalevsky, J., Grigorova, I., and Mullins, R.D., Activation of the Arp2/3 complex by the *Listeria* ActA protein. ActA binds two actin monomers and three subunits of the Arp2/3 complex, *J. Biol. Chem.*, 276, 3468, 2001.
219. Machesky, L.M. et al., Purification of a cortical complex containing two unconventional actins from *Acanthamoeba* by affinity chromatography on profilin-agarose, *J. Cell Biol.*, 127, 107, 1994.
220. Welch, M.D., Iwamatsu, A., and Mitchison, T.J., Actin polymerization is induced by Arp2/3 protein complex at the surface of *Listeria monocytogenes*, *Nature*, 385, 265, 1997.
221. Welch, M.D. et al., Interaction of human Arp2/3 complex and the *Listeria monocytogenes* ActA protein in actin filament nucleation, *Science*, 281,105, 1998.
222. Skoble, J., Portnoy, D.A., and Welch, M.D., Three regions within ActA promote Arp2/3 complex-mediated actin nucleation and *Listeria monocytogenes* motility, *J. Cell Biol.*, 150, 527, 2000.
223. Boujemaa-Paterski, R. et al., *Listeria* protein ActA mimics WASp family proteins: It activates filament barbed end branching by Arp2/3 complex, *Biochemistry*, 40, 11390, 2001.
224. Fradelizi, J. et al., ActA and human zyxin harbor Arp2/3-independent actin-polymerization activity, *Nat. Cell Biol.*, 3, 699, 2001.
225. Brieher, W.M., Coughlin, M., and Mitchison, T.J., Fascin-mediated propulsion of *Listeria monocytogenes* independent of frequent nucleation by the Arp2/3 complex, *J. Cell. Biol.*, 165, 233, 2004.

226. Niebuhr, K. et al., A novel proline-rich motif present in ActA of *Listeria monocytogenes* and cytoskeletal proteins is the ligand for the EVH1 domain, a protein module present in the Ena/VASP family, *EMBO J.*, 16, 5433, 1997.
227. Skoble, J. et al., Pivotal role of VASP in Arp2/3 complex-mediated actin nucleation, actin branch-formation, and *Listeria monocytogenes* motility, *J. Cell Biol.*, 155, 89, 2001.
228. Auerbuch, V. et al., Ena/VASP proteins contribute to *Listeria monocytogenes* pathogenesis by controlling temporal and spatial persistence of bacterial actin-based motility, *Mol. Microbiol.*, 49, 1361, 2003.
229. Samarin, S. et al., How VASP enhances actin-based motility, *J. Cell. Biol.*, 163, 131, 2003.
230. Rosenblatt, J. et al., *Xenopus* actin depolymerizing factor/cofilin (XAC) is responsible for the turnover of actin filaments in *Listeria monocytogenes* tails, *J. Cell. Biol.*, 136, 1323, 1997.
231. Brieher, W.M. et al., Rapid actin monomer-insensitive depolymerization of *Listeria* actin comet tails by cofilin, coronin, and Aip1, *J. Cell. Biol.*, 175, 315, 2006.
232. Larson, L. et al., Gelsolin mediates calcium-dependent disassembly of *Listeria* actin tails, *Proc. Natl. Acad. Sci. USA*, 102, 1921, 2005.
233. Goebel, W. et al., *Listeria monocytogenes*—A model system for studying the pathomechanisms of an intracellular microorganism, *Zbl. Bakt.*, 278, 334, 1993.
234. Niebuhr, K. et al., Localization of the ActA polypeptide of *Listeria monocytogenes* in infected tissue culture cell lines: ActA is not associated with actin "comets," *Infect. Immun.*, 61, 2793, 1993.
235. Moors, M.A., Auerbuch, V., and Portnoy, D.A., Stability of the *Listeria monocytogenes* ActA protein in mammalian cells is regulated by the N-end rule pathway, *Cell. Microbiol.*, 1, 249, 1999.
236. Brundage, R.A. et al., Expression and phosphorylation of the *Listeria monocytogenes* ActA protein in mammalian cells, *Proc. Natl. Acad. Sci. USA*, 90, 11890, 1993.
237. Pfeuffer, T. et al., LaXp180, a mammalian ActA-binding protein, identified with the yeast two-hybrid system, colocalizes with intracellular *Listeria monocytogenes*, *Cell. Microbiol.*, 2, 101, 2000.
238. Bauer, S., Pfeuffer, T., and Kuhn, M., Identification and characterization of regions in the cellular protein LaXp180 and the *Listeria monocytogenes* protein ActA necessary for the interaction of the two proteins, *Mol. Genet. Genom.*, 268, 607, 2003.
239. Yoshida, S. et al., Microtubule-severing activity of *Shigella* is pivotal for intercellular spreading, *Science*, 314, 985, 2006.
240. Pust, S. et al., *Listeria monocytogenes* exploits ERM protein functions to efficiently spread from cell to cell, *EMBO J.*, 24, 1287, 2005.
241. Robbins, J.R. et al., *Listeria monocytogenes* exploits normal host cell processes to spread from cell to cell, *J. Cell Biol.*, 146, 1333, 1999.
242. Marquis, H., Goldfine, H., and Portnoy, D.A., Proteolytic pathways of activation and degradation of a bacterial phospholipase C during intracellular infection by *Listeria monocytogenes*, *J. Cell Biol.*, 137, 1381, 1997.
243. Gedde, M.M. et al., Role of listeriolysin O in cell-to-cell spread of *Listeria monocytogenes*, *Infect. Immun.*, 68, 999, 2000.
244. Hardy, J. et al., Extracellular replication of *Listeria monocytogenes* in the murine gall bladder, *Science*, 303, 851, 2004.
245. Hardy, J., Margolis, J.J., and Contag, C.H., Induced biliary excretion of *Listeria monocytogenes*, *Infect. Immun.*, 74, 1819, 2006.
246. Dussurget, O. et al., *Listeria monocytogenes* bile salt hydrolase is a PrfA-regulated virulence factor involved in the intestinal and hepatic phases of listeriosis, *Mol. Microbiol.*, 45, 1095, 2002.
247. Begley, M. et al., Contribution of three bile-associated loci, *bsh*, *pva*, and *btlB*, to gastrointestinal persistence and bile tolerance of *Listeria monocytogenes*, *Infect. Immun.*, 73, 894, 2005.
248. Sleator, R.D. et al., A PrfA-regulated bile exclusion system (BilE) is a novel virulence factor in *Listeria monocytogenes*, *Mol. Microbiol.*, 55, 1183, 2005.
249. Raux, E., Schubert, H.L., and Warren, M.J., Biosynthesis of cobalamin (vitamin B12): A bacterial conundrum, *Cell. Mol. Life Sci.*, 57, 1880, 2000.
250. Buchrieser, C. et al., Comparison of the genome sequences of *Listeria monocytogenes* and *Listeria innocua*: Clues for evolution and pathogenicity, *FEMS Immunol. Med. Microbiol.*, 35, 207, 2003.
251. Archambaud, C. et al., Translation elongation factor EF-Tu is a target for Stp, a serine-threonine phosphatase involved in virulence of *Listeria monocytogenes*, *Mol. Microbiol.*, 56, 383, 2005.
252. Boneca, I.G. et al., A critical role for peptidoglycan N-deacetylation in *Listeria* evasion from the host innate immune system, *Proc. Natl. Acad. Sci. USA*, 104, 997, 2007.

253. Kreft, J. and Vázquez-Boland, J.A., Regulation of virulence genes in *Listeria*, *Int. J. Med. Microbiol.*, 291, 145, 2001.
254. Scortti, M. et al., The PrfA virulence regulon of pathogenic *Listeria*, *Microb. Infect*, 9, 1196, 2007.
255. Chakraborty, T. et al., Coordinate regulation of virulence genes in *Listeria monocytogenes* requires the product of the *prfA* gene, *J. Bacteriol.*, 174, 568, 1992.
256. Mengaud, J. et al., Pleiotropic control of *Listeria monocytogenes* virulence factors by a gene that is autoregulated, *Mol. Microbiol.*, 5, 2273, 1991.
257. Vega, Y. et al., New *Listeria monocytogenes prfA** mutants, transcriptional properties of PrfA* proteins, and structure-function of the virulence regulator PrfA, *Mol. Microbiol.*, 52, 1553, 2004.
258. Eiting, M. et al., The mutation G145S in PrfA, a key virulence regulator of *Listeria monocytogenes*, increases DNA-binding affinity by stabilizing the HTH motif, *Mol. Microbiol.*, 56, 433, 2005.
259. Ripio, M.T. et al., A Gly145Ser substitution in the transcriptional activator PrfA causes constitutive overexpression of virulence factors in *Listeria monocytogenes*, *J. Bacteriol.*, 179, 1533, 1997.
260. Ripio, M.T. et al., Transcriptional activation of virulence genes in wild-type strains of *Listeria monocytogenes* in response to a change in the extracellular medium composition, *Res. Microbiol.*, 147, 311, 1996.
261. Renzoni, A. et al., PrfA, the transcriptional activator of virulence genes, is up-regulated during interaction of *Listeria monocytogenes* with mammalian cells and in eukaryotic cell extracts, *Mol. Microbiol.*, 34, 552, 1999.
262. Moors, M.A. et al., Expression of listeriolysin O and ActA by intracellular and extracellular *Listeria monocytogenes*, *Infect. Immun.*, 67, 131, 1999.
263. Shetron-Rama, L. M. et al., Intracellular induction of *Listeria monocytogenes actA* expression, *Infect. Immun.*, 70, 1087, 2002.
264. Vega, Y. et al., Functional similarities between the *Listeria monocytogenes* virulence regulator PrfA and cyclic AMP receptor protein: The PrfA* (Gly145Ser) mutation increases binding affinity for target DNA, *J. Bacteriol.*, 180, 6655, 1998.
265. Böckmann, R. et al., PrfA mediates specific binding of RNA polymerase of *Listeria monocytogenes* to PrfA-dependent virulence gene promoters resulting in transcriptionally active complex, *Mol. Microbiol.*, 36, 487, 2000.
266. Milohanic, E. et al., Transcriptome analysis of *Listeria monocytogenes* EGDe identifies three groups of genes differently regulated by PrfA, *Mol. Microbiol.*, 47, 1613, 2003.
267. Marr, A.K. et al., Overexpression of PrfA leads to growth inhibition of *Listeria monocytogenes* in glucose-containing culture media by interfering with glucose uptake, *J. Bacteriol.*, 188, 3887, 2006.
268. Shetron-Rama, L.M. et al., Isolation of *Listeria monocytogenes* mutants with high-level *in vitro* expression of host cytosol-induced gene products, *Mol. Microbiol.*, 48, 1537, 2003.
269. Wong, K.K.Y. and Freitag N.E., A novel mutation within the central *Listeria monocytogenes* regulator PrfA that results in constitutive expression of virulence gene products, *J. Bacteriol.*, 186, 6265, 2004.
270. Freitag, N.E., Rong, L., and Portnoy, D.A., Regulation of the *prfA* transcriptional activator of *Listeria monocytogenes*: Multiple promoter elements contribute to intracellular growth and cell-to-cell spread, *Infect. Immun.*, 61, 2537, 1993.
271. Johansson, J. et al., An RNA thermosensor controls the expression of virulence genes in *Listeria monocytogenes*, *Cell*, 110, 551, 2002.
272. Ermolaeva, E. et al., Negative control of *Listeria monocytogenes* virulence genes by a diffusible autorepressor, *Mol. Microbiol.*, 52, 601, 2004.
273. Milenbachs, A.A. et al., Carbon-source regulation of virulence gene expression in *Listeria monocytogenes*, *Mol. Microbiol.*, 23, 1075, 1997.
274. Brehm, K. et al., The *bvr* locus of *Listeria monocytogenes* mediates virulence gene repression by β-glucosides, *J. Bacteriol.*, 181, 5024, 1999.
275. Ripio, M.T. et al., Glucose-1-phosphate utilization by *Listeria monocytogenes* is PrfA dependent and coordinately expressed with virulence factors, *J. Bacteriol.*, 179, 7174, 1997.
276. Behari, J. and Youngman, P., A homolog of CcpA mediates catabolite control in *Listeria monocytogenes* but not carbon source regulation of virulence genes, *J. Bacteriol.*, 180, 6316, 1998.
277. Herro, R. et al., How seryl-phosphorylated HPr inhibits PrfA, a transcription activator of *Listeria monocytogenes* virulence genes, *J. Mol. Microbiol. Biotechnol.*, 9, 224, 2005.
278. Mertins, S. et al., Interference of components of the phosphohenolpyruvate system with the central virulence gene regulator PrfA of *Listeria monocytogenes*, *J. Bacteriol.*, 189, 473, 2007.
279. Renzoni, A. et al., Evidence that PrfA, the pleiotropic activator of virulence genes in *Listeria monocytogenes*, can be present but inactive, *Infect. Immun.*, 65, 1515, 1997.

280. Sheehan, B. et al., Differential activation of virulence gene expression by PrfA, the *Listeria monocytogenes* virulence regulator, *J. Bacteriol.*, 177, 6469, 1995.
281. Bohne, J. et al., Differential regulation of the virulence genes of *Listeria monocytogenes* by the transcriptional activator PrfA, *Mol. Microbiol.*, 20, 1189, 1996.
282. Williams, J.R., Thayyullathil, C., and Freitag, N.E., Sequence variations within PrfA DNA binding sites and effects on *Listeria monocytogenes* virulence gene expression, *J. Bacteriol.*, 182, 837, 2000.
283. Domann, E. et al., Detection of a *prfA*-independent promoter responsible for listeriolysin gene expression in mutant *Listeria monocytogenes* strains lacking the PrfA regulator, *Infect. Immun.*, 61, 3073, 1993.
284. Luo, Q. et al., *In vitro* transcription of the *Listeria monocytogenes* virulence genes *inlC* and *mpl* reveals overlapping PrfA-dependent and -independent promoters that are differentially activated by GTP, *Mol. Microbiol.*, 52, 39, 2004.
285. Intentionally omitted.
286. Wong, K.K., Bouwer, H.G., and Freitag, N.E., Evidence implicating the 5′ untranslated region of *Listeria monocytogenes actA* in the regulation of bacterial actin-based motility, *Cell Microbiol.*, 6, 155, 2004.
287. Shen, A. and Higgins, D., The 5′ untranslated region-mediated enhancement of intracellular listeriolysin O production is required for *Listeria monocytogenes* pathogenicity, *Mol. Microbiol.*, 57, 1460, 2005.
288. Nair, S. et al., CtsR controls class III heat shock gene expression in the human pathogen *Listeria monocytogenes, Mol. Microbiol.*, 35, 800, 2000.
289. Gründling, A. et al., *Listeria monocytogenes* regulates flagellar motility gene expression through MogR, a transcriptional repressor required for virulence, *Proc. Natl. Acad. Sci. USA.*, 101, 12318, 2004.
290. Rea, R.B., Gahan, C.G., and Hill, C., Disruption of putative regulatory loci in *Listeria monocytogenes* demonstrates a significant role for Fur and PerR in virulence, *Infect. Immun.*, 72, 717, 2004.
291. Bennett, H.J. et al., Characterization of *relA* and *codY* mutants of *Listeria monocytogenes:* Identification of the CodY regulon and its role in virulence, *Mol. Microbiol.*, 63, 1453, 2007.
292. Cotter, P.D. et al., Identification and disruption of *lisRK,* a genetic locus encoding a two-component signal transduction system involved in stress tolerance and virulence in *Listeria monocytogenes, J. Bacteriol.*, 181, 6840, 1999.
293. Kallipolities, B.H. et al., CesRK, a two-component signal transduction system in *Listeria monocytogenes*, responds to the presence of cell wall-acting antibiotics and affects beta-lactam resistence, *Antimicrob. Agents Chemother.*, 47, 3421, 2003.
294. Dons, L. et al., Role of flagellin and the two-component CheA/CheY system of *Listeria monocytogenes* in the host cell invasion and virulence, *Infect. Immun.*, 72, 3237, 2004.
295. Autret, N. et al., Identification of the agr locus of *Listeria monocytogenes:* Role in bacterial virulence, *Infect. Immun.*, 71, 4463, 2003.
296. Knudsen, G.M., Olsen, J.E., and Dons, L., Characterization of DegU, a response regulator in *Listeria monocytogenes,* involved in regulation of motility and contributes to virulence, *FEMS Microbiol. Lett.*, 240, 171, 2004.
297. Williams, T. et al., Construction and characterization of *Listeria monocytogenes* mutants with in-frame deletions in the response regulator genes identified in the genome sequence, *Infect Immun.*, 73, 3152, 2005.
298. Larsen, M.H. et al., The response regulator ResD modulates virulence gene expression in response to carbohydrates in *Listeria monocytogenes, Mol. Microbiol.*, 61, 1622, 2006.
299. Mandin, P. et al., VirR, a response regulator critical for *Listeria monocytogenes* virulence, *Mol. Microbiol.*, 57, 1367, 2005.
300. Rauch, M. et al., SigB-dependent *in vitro* transcription of *prfA* and some newly identified genes of *Listeria monocytogenes* whose expression is affected by PrfA *in vivo*, *J. Bacteriol.*, 187, 800, 2005.
301. Sokolovic, Z. et al., Synthesis of species-specific stress proteins by virulent strains of *Listeria monocytogenes*, *Infect. Immun.*, 58, 3582, 1990.
302. Sokolovic, Z. et al., Surface-associated, PrfA-regulated proteins of *Listeria monocytogenes* synthesized under stress conditions, *Mol. Microbiol.*, 8, 219, 1993.
303. Kazmierczak, M.J. et al., *Listeria monocytogenes* σ^B regulates stress response and virulence functions, *J. Bacteriol.*, 185, 5722, 2003.
304. Kim, H., Marquis, H., and Boor, K., σ^B contributes to *Listeria monocytogenes* invasion by controlling expression of *inlA* and *inlB*, *Microbiology*, 151, 3215, 2005.
305. Kazmierczak, M.J., Wiedmann, M., and Boor, K.J., Contributions of *Listeria monocytogenes* σ^B and PrfA to expression of virulence and stress response genes during extra- and intracellular growth, *Microbiology*, 152, 1827, 2006.

306. Gray, M.J., Freitag, N.E., and Boor, K., How the bacterial pathogen *Listeria monocytogenes* mediates the switch from environmental Dr. Jekyll to pathogenic Mr. Hyde, *Infect. Immun.*, 74, 2505, 2006.
307. Chaturohngakul, S. and Boor, K., RsbT and RsbV contribute to σ^B-dependent survival under environmental, energy and intracellular stress conditions in *Listeria monocytogenes, Appl. Environ. Microbiol.*, 70, 5349, 2004.
308. Rouquette, C. et al., Identification of a ClpC ATPase required for stress tolerance and *in vivo* survival of *Listeria monocytogenes, Mol. Microbiol.*, 21, 977, 1996.
309. Nair, S. et al., ClpE, a novel member of the HSP100 family, is involved in cell division and virulence of *Listeria monocytogenes, Mol. Microbiol.*, 31, 185, 1999.
310. Gaillot, O. et al., The ClpP serine protease is essential for the intracellular parasitism and virulence of *Listeria monocytogenes, Mol. Microbiol.*, 35, 1286, 2000.
311. Chastanet, A. et al., *clpB*, a novel member of the *Listeria monocytogenes* CtsR regulon, is involved in virulence but not in general stress tolerance, *J. Bacteriol.*, 186, 1165, 2004.
312. Nelson, K.E. et al., Whole genome comparisons of serotype 4b and 1/2a strains of the food-borne pathogen *Listeria monocytogenes* reveal new insights into the core genome components of this species, *Nucleic Acids Res.*, 32, 2386, 2004.
313. Hain, T., personal communication, 2007.
314. Hain, T. et al., Whole-genome sequence of *Listeria welshimeri* reveals common steps in genome reduction with *Listeria innocua* as compared to *Listeria monocytogenes, J. Bacteriol.*, 188, 7405, 2006.
315. Glaser, P. and Buchrieser, C., personal communication, 2007.
316. Kuhn, M. and Goebel, W., Genomics of *Listeria monocytogenes*. In *Pathogenomics: Genome analysis of pathogenoc microbes*, J. Hacker and U. Dobrindt, eds. Wiley–VCH, Weinheim, Germany, 2006, chap. 16.
317. Schmid, M.W. et al., Evolutionary history of the genus *Listeria* and its virulence genes, *Syst. Appl. Microbiol.*, 28, 1, 2005.
318. Mansfield, B.E. et al., Exploration of host–pathogen interactions using *Listeria monocytogenes* and *Drosophila melanogaster, Cell. Microbiol.*, 5, 901, 2003.
319. Thomsen, L.E. et al., *Caenorhabditis elegans* is a model host for *Listeria monocytogenes, Appl. Environ. Microbiol.*, 72, 1700, 2006.
320. Yu, Y.A. et al., Visualization of tumors and metastases in live animals with bacteria and vaccinia virus encoding light-emitting proteins, *Nat. Biotechnol.*, 22, 313, 2004.

Section II

Identification and Detection

5 Phenotypic Identification

Lisa Gorski

CONTENTS

5.1 INTRODUCTION

The recognition of *Listeria monocytogenes* as a causative agent of food-borne illness in the early 1980s stimulated the search for simple, accurate enrichments to isolate the organism from food sources. The first description of food-borne listeriosis was reported in 1983. It documented a 1981 outbreak in Canada resulting most likely from contaminated cabbage used to make coleslaw.[1] In the following decade several high-profile outbreaks, including one caused by contaminated, pasteurized milk in 1983[2] and another from contaminated cheese in 1985,[3] showed the need for efficient detection protocols.

L. monocytogenes is a Gram-positive, facultatively anaerobic, non-spore-forming, short, rod-shaped bacterium. It is one of six species in the genus *Listeria*, and the vast majority of human illnesses are caused by the species *L. monocytogenes*. Relatively few cases of human listeriosis are caused by *L. ivanovii*, and at least one case attributed to *L. seeligeri* has been described.[4] Most strains of *L. monocytogenes*, *L. ivanovii*, and *L. seeligeri* demonstrate β-hemolysis—the lysing of blood cells visualized by a zone of clearing on blood agar. Hemolysis is not seen, except in rare cases, in the nonpathogenic *L. innocua, L. welshimeri*, and *L. grayi*. All of the *Listeria* spp. are easy to grow in the laboratory on any common complex medium. Sugars are the primary carbon sources; however, one can use amino acids and peptides as carbon sources as well. Generation times for

L. monocytogenes in rich media at generally used growth temperatures (22–37°C) are in the range of 1–2 h. It is a very hardy bacterium with a growth temperature range reported from –0.4 to 50°C. The pH range for growth is from pH 4.3 to pH 9; however, that range is dependent on the growth temperature.[5] The organism can also withstand high osmotic pressure and has been demonstrated to grow at almost 2 *M* NaCl. For these reasons, it is adapted to grow well in foods, particularly processed foods, which might use salt, pH, and/or temperature to control the growth of spoilage bacteria. Pasteurization is effective in killing the organism, but foods can become contaminated after pasteurization.

Previous to its recognition as a food-borne pathogen, the enrichments for *L. monocytogenes* often took the form of cold temperature enrichment. The organism has the ability to survive and grow at refrigeration temperatures. Its ability to be more readily isolated from samples that were refrigerated was noted by Gray et al.[6] in 1948, about 20 years after the organism was first described. With the refrigeration method of isolation, researchers and clinicians simply kept suspect samples refrigerated, sometimes in rich medium, and periodically transferred samples onto agar plates. Occasionally, it took several months or even years before a positive sample was detected. After the recognition of *L. monocytogenes* as a food-borne pathogen, this was hardly an allowable time frame for rapid detection of contaminated food samples.

On occasion, *L. monocytogenes* can be isolated from contaminated foods by direct plating (e.g., unpasteurized milk samples)[7,8]; however, the levels of the organism must be high, which is most often not the case. Usually, the high levels of other fast-growing bacteria in the foods that *L. monocytogenes* contaminates make it difficult for the pathogen to compete on nutrient media without prior enrichment. Enrichment works on a Darwinian concept to shift the balance in the environment to favor the organism of interest over the other bacteria present in the sample. For this purpose, the environment is the medium provided for the bacterial growth. Knowledge of the basic physiology of the target organism is essential in order to select medium components that favor that organism. If simple nutritional sources or electron acceptors are not unique enough to allow for enrichment of the target bacterium on the basis of nutrients alone, inhibitors of competing bacteria are usually added so the target bacterium will be more fit than those competitors in the medium. Antibiotics and dyes are some of the commonly used compounds that inhibit competing bacteria. The nutrients and inhibitory compounds combine to give an enrichment medium its selectivity.

Generally, the first step in an enrichment is growth in at least one selective, liquid medium before plating onto selective and differential agar. Growth in enrichment broth will allow amplification of the numbers of *L. monocytogenes*, if any are present in the food sample. An understanding of what types of bacteria are likely to compete with the target is important in designing enrichment media. Sometimes the media best suited for isolation of *L. monocytogenes* from dairy, for example, are not the most appropriate choice of media for isolation from ground meat, seafood, or vegetables.[9,10] This is because different types of background microbiota are found in the two very different food sources, and the components of one enrichment medium may not be as efficient in dealing with competitors of *L. monocytogenes* in one food source as the other.

Some protocols utilize two enrichment broths, with the secondary enrichment broth containing higher concentrations of antibiotics and dyes to allow for more robust selection of *L. monocytogenes* after enriching in a primary broth. Regardless of the enrichment broth used, for isolation of colonies and strain identification, portions of the enrichment must be plated onto selective and/or differential agar media. Many of these agars contain components similar to the enrichment broths as well as other compounds that allow for differentiation of the colonies. They have been designed to take advantage of *L. monocytogenes* physiology such that its colonies are somehow distinctive from colonies of other bacteria on the plate, either by their color or by precipitates surrounding them. Older plating media, unless they contained blood, had no such distinctions, and identification of presumptive *L. monocytogenes* colonies on these media is done through the observation of bluish-gray/bluish-green colonies under oblique illumination, a process originally described by Henry.[11]

Some have reported, however, that background microbiota resulting from food enrichments could also appear bluish upon oblique illumination, which made identification of *L. monocytogenes* more difficult.[10]

Potassium tellurite was the first compound found to work as a selective agent for *Listeria* enrichment media[12]; however, a review from the same laboratory from 1966 states that later experiments showed it was inhibitory to some strains.[13] McBride and Girard[14] described McBride *Listeria* agar for the isolation of *L. monocytogenes* in 1960. Subsequent modifications of McBride *Listeria* agar led to the description of several formulations that varied by the presence or absence of blood, the substitution of glycine anhydride for glycine, changes in LiCl concentration, and the addition of cycloheximide.[14–17] McBride *Listeria* agar is no longer used in official protocols for isolation of *L. monocytogenes*. In 1986, Lee and McClain[16] described lithium chloride–phenylethanol–moxalactam (LPM) agar, a modification of McBride *Listeria* agar. LPM agar was a marked improvement and is still listed in some standard protocols as a recommended plating medium; however, it does not contain chemicals that allow for differential appearance of *Listeria* colonies.

The several different liquid and agar media used to enrich *L. monocytogenes* have been refined over the years. Many of the components were originally incorporated into *L. monocytogenes* selective media before the recognition of the bacterium as a food-borne pathogen, and these components were used in media to isolate *L. monocytogenes* from fecal samples. The components still work to inhibit competitive microbiota from food samples. Broad selections of antibiotics are used in the various plating media to inhibit potential competitors.

All of the commonly used enrichment techniques for *L. monocytogenes* will also enrich any other *Listeria* spp. (most of which are not pathogens) present in the food sample. Therefore, species identification of potential colonies is essential for confirmation of *L. monocytogenes* contamination. Many identification schemes exist to specifically differentiate *L. monocytogenes* from its nonpathogenic cousin *L. innocua*, which has many physiological traits in common. The classic plating media also will not distinguish *L. monocytogenes* colonies from colonies of other *Listeria* spp.; however, chromogenic media were developed recently that make it easier to differentiate *L. monocytogenes* from other *Listeria* spp. This differentiation is based on the detection of products of virulence genes resulting in a different colony color or a precipitate visualized by a halo around the colony. With many of these media, further tests must be done to distinguish between *L. monocytogenes* and other virulent *Listeria* spp. such as *L. ivanovii*. Identification of *L. monocytogenes* and speciation of *Listeria* spp. can be done with physiological tests such as carbohydrate utilization and hemolysis patterns that differentiate the species. These tests can be done individually on each isolate, or biochemical kits can be used to run multiple tests at a time. Immunologically based kits are commonly used to specifically detect either *Listeria* spp. in general or *L. monocytogenes* in particular, depending on the type of antisera in the kit, from enrichment cultures or isolated colonies.

Once isolation and identification of the *L. monocytogenes* strain is complete, further characterization can be done by phenotypic or genetic methods. In the case of *L. monocytogenes*, the most utilized form of subtyping is determination of serotype of the strain. Because most cases of clinical listeriosis result from three of the thirteen known serotypes (1/2a, 1/2b, or 4b), the assigning of serotype is an important first step in subtyping of isolates when tracking outbreaks, treating sporadic cases, or simply characterizing food isolates.

In this chapter, targets for the enrichment, detection, identification, and enumeration of *L. monocytogenes* will be reviewed. The most common enrichment protocols and identification methods currently in use will be described. Aspects of enrichment from foods, such as the detection of injured or stressed cells and the possible outgrowth of *L. monocytogenes* by the nonpathogenic *L. innocua*, will be discussed. Biochemical and immunological methods for species identification also will be covered. Phenotypic methods for subtyping, such as serotype identification, will be explained, but discussion about breakdown into genetic lineages and genotyping will be described in chapter 7.

5.2 DETECTION TARGETS

The design of enrichment media for *L. monocytogenes* has developed over the years to take advantage of characteristics of the bacterium that favor its propagation in mixed cultures with competing bacteria, and then to differentiate *L. monocytogenes* from other organisms and, finally, other *Listeria* spp. The methods currently in use for the enrichment and detection of *L. monocytogenes* exploit both specific and nonspecific aspects of the organism. The nonspecific features are those shared by all *Listeria* spp., such as a Gram-positive cell wall or use of specific carbon sources. They can be used to inhibit many non-*Listeria* organisms and make *Listeria* colonies distinctive. Specific aspects for detection relate to targets sufficiently unique to *L. monocytogenes*, such as virulence proteins, that other species of *Listeria*, especially *L. innocua*, appear differently on the medium.

5.2.1 Shared Targets

All of the current standardized enrichment protocols for *L. monocytogenes* will also enrich other species of *Listeria*. After selective enrichment, potential *L. monocytogenes* colonies are identified on selective and differential agars. Nonspecific aspects of *Listeria* physiology that are used in the design of enrichment and isolation protocols are the Gram-positive cell wall, the ability to hydrolyze esculin for a carbon and energy source, and its temperature growth range.

5.2.1.1 Inhibitor Resistance

Table 5.1 lists the common enrichment and plating media for *Listeria*, and the concentrations of selective compounds used in each. Not included in Table 5.1 are media that have proprietary formulations. First among selective compounds for *L. monocytogenes* enrichment are compounds inhibitory to Gram-negative bacteria. Among the common enrichment broths, the most common antibiotic to select for Gram-positive bacteria is nalidixic acid, which inhibits many species of Gram-negative bacteria. Nalidixic acid blocks DNA replication in susceptible bacteria by acting on DNA gyrase.[18] It was recognized in 1966 to have an inhibitory effect on many Gram-negative species, especially the Enterobacteriaceae, but not pseudomonads or *Proteus*. It was then used as a common ingredient in enrichments for *Listeria* as well as streptococci.[19,20] While the Gram-negatives are inhibited, studies show that nalidixic acid has no effect on *L. monocytogenes* growth in culture,[21–23] making it an effective additive for enrichment. Its broad inhibitory nature makes it a good addition to primary and secondary enrichment broths to reduce the growth of many of the Gram-negative bacteria that might be present on a food sample before plating onto selective agar. Therefore, it is present in the common primary and secondary enrichment broths currently in use, including University of Vermont *Listeria* enrichment broth I (UVM I), Fraser broth, half Fraser broth, and buffered *Listeria* enrichment broth (BLEB).

Acriflavine is a toxic dye that was originally added to *L. monocytogenes* enrichment media to inhibit the growth of fecal streptococci. This was especially important at the time when clinicians were attempting to isolate *L. monocytogenes* from patient fecal samples. It works well at inhibiting many Gram-positive cocci, including *Staphylococcus aureus* and some lactic acid bacteria. However, acriflavine has been shown to have strain-specific effects on the time in lag phase,[22] and higher concentrations decrease the growth rate of *L. monocytogenes.*[21] Some strains may be slightly inhibited by its use, but it is still a common component in *L. monocytogenes* enrichment media. It is used in enrichment broths and in the selective-differential agars as well, including UVM, Fraser broth, half Fraser broth, BLEB, polymyxin–acriflavine–LiCl–ceftazidime–esculine–mannitol (PALCAM) agar, and Oxford (OXA) agar. Care must be practiced in its use, as acriflavine binds to all DNA and is harmful to humans.

Cycloheximide is used in some enrichment protocols to reduce the amount of fungal growth in enrichment. This is significant, especially for plant or cheese samples where fungi may be plentiful. It is used in the BLEB enrichment broth and OXA agar. Cycloheximide is harmful to humans, and

TABLE 5.1
Some Common Media Used for *L. monocytogenes* Enrichment and Isolation, the Concentrations (g/L) of Inhibitory Compounds They Contain, and Whether They Contain Esculin

Medium Components	University of Vermont I Broth (UVM I)	Buffered Listeria Enrichment Broth (BLEB)	Fraser Broth	Half Fraser Broth	Oxford Agar (OXA)	Polymyxin–acriflavine–LiCl–ceftazidime–aesculin–mannitol (PALCAM) Agar	Modified Oxford Agar (MOX)	Lithium–phenylethanol–moxalactam (LPM) Agar	Agar Listeria Ottavani and Agosti (ALOA)
Acriflavine	0.012	0.01	0.024	0.012	0.005	0.005	—	—	—
Nalidixic acid	0.02	0.04	0.02	0.01	—	—	—	—	0.02
Cycloheximide	—[a]	0.05[c]	—	—	0.4	—	—	—	0.05 (or 0.01 amphotericin B)
Cefotetan	—	—	—	—	0.002	—	—	—	—
Fosfomycin	—	—	—	—	0.01	—	—	—	—
Colistin	—	—	—	—	0.02	—	0.01	—	—
Ceftazidime	—	—	—	—	—	0.04	—	—	0.02
Polymyxin B	—	—	—	—	—	0.01	—	—	76,700 Units
Moxalactam	—	—	—	—	—	—	0.02	0.0002	—
NaCl	20	5	20	20	—	5	—	—	5
LiCl	—	—	3	3	15	15	15	5	—
Esculin + Fe^{3+}	±[b]	—	+	+	+	+	+	—[d]	—

[a] "—," not present.
[b] Contains esculin, but not Fe^{3+}.
[c] Pimaricin can be used in place of cycloheximide.
[d] FDA protocol calls for the addition of esculin and Fe^{3+} to LPM, if used.

care should be taken in its use. Pimaricin (natamycin) is an antifungal that is less toxic to humans that has been recommended as an alternative to cycloheximide in BLEB enrichment medium.[24,25]

Sodium chloride and lithium chloride are present in many *L. monocytogenes* enrichment and plating media. Both take advantage of the high salt tolerance of *L. monocytogenes*. Strains of *L. monocytogenes* have grown in media containing 2 *M* NaCl. UVM I, Fraser broth, and half Fraser broth contain 2% (0.34 *M*) NaCl, and BLEB contains 0.5% (0.08 *M*) NaCl. It is also included in LPM agar.[26] McBride and Girard[14] incorporated LiCl into McBride *Listeria* agar, the first selective plating medium for *Listeria*. In liquid enrichment medium, it first appeared in 1988 in Fraser broth,[27] which is the only broth in wide use today that includes it. The use of LiCl also takes advantage of the high salt tolerance of *L. monocytogenes*, and it also serves to inhibit entercocci and some Gram-negative bacteria.[28] *L. monocytogenes* can grow in concentrations of up to 2% LiCl,[28] although some strains show slower growth at those concentrations.[21] *Staphylococcus* can withstand higher NaCl concentrations than *Listeria*, and *Staphylococcus* and *Bacillus* are not inhibited by LiCl, so other components must be added to the media to inhibit those organisms. LiCl is more commonly used in agar media including OXA, modified Oxford (MOX) agar, PALCAM agar, and LPM agar.[26,29]

Colistin (also known as polymyxin E) and polymyxin B disrupt the cell membrane of Gram-negative organisms. They are particularly effective against *Pseudomonas* spp., which are often encountered in produce samples.[18,30] Colistin is used in both OX and MOX agars. Polymyxin B is used in PALCAM agar.

Moxalactam is a broad-spectrum, cephalosporin, β-lactam antibiotic with activity against *Staphylococcus*, *Pseudomonas* spp. and several of the Enterbacteriaceae.[31] Larsson et al.[32] found 175 clinical isolates of *L. monocytogenes* to be resistant to moxalactam in concentrations of 128 μg/ml or higher. It is used in both LPM and MOX agars.

Other antimicrobials used in *Listeria* plating media are ceftazidime (PALCAM agar) and cefotetan (OXA agar). They are both broad-spectrum cephalosporins active against many Gram-negative bacteria, including many pseudomonads. Fosfomycin (OXA agar) is an organic phosphonate that interferes with bacterial cell wall synthesis by inhibiting phosphoenolpyruvate transferase and exhibits activity against a broad range of Gram-negative bacteria.

Phenylethanol, available in phenylethyl alcohol agar, causes various Gram-negative bacteria to become elongated as a result of inhibition of DNA synthesis.[30] Many Gram-positive bacteria, including *Listeria,* are resistant, although some have reported phenyl ethyl alcohol to be slightly inhibitory to some *L. monocytogenes* strains.[21,33] It is used only in LPM agar.

5.2.1.2 Esculinase

Esculin is a fluorescent dye that is extracted from the bark of the horse chestnut tree. All *Listeria* spp. hydrolyze esculin using esculinase (a β-D-glucosidase) to liberate a glucose molecule with the production of 6,7-dihydroxycoumarin. The latter compound will react with ferric iron and form a black precipitate. This reaction has become the most common diagnostic tool for the presence of *Listeria* spp. in media. It is one way to distinguish *Listeria* from its morphologically similar cousin *Erysipelothrix*, an animal pathogen that inhabits some of the same agricultural environments as *Listeria.*[34] The reaction is not exclusive to *Listeria*, however, and other bacteria that can hydrolyze esculin include some species of *Pseudomonas, Bacillus*, *Enterococcus*, and some members of the Enterbacteriaceae. The esculinase reaction was originally used to characterize isolates of *Streptococcus.*[35] Dominguez Rodriguez et al.[36] described the first use of esculin and ferric ammonium citrate in media to isolate *Listeria* from sheep feces. Fraser and Sperber[27] used the combination when they developed Fraser broth, which was adopted by both the U.S. Department of Agriculture-Food Safety Inspection Service (USDA-FSIS) and the International Standards Organization (ISO) as a secondary enrichment medium in the standard USDA and ISO protocols.[37,38] Van Netten et al.[29] were the first to describe agar media with the combination when they developed PALCAM medium, which was later adopted by the U.S. Food and Drug Administration (FDA) and ISO as a preferred plating medium. The ferric iron and esculin combination was later included in OXA and MOX agars, also accepted plating media by the FDA. The only non-esculin-based agar currently sanctioned by an official enrichment protocol is LPM; however, the FDA requires that esculin and ferric iron be added to the formulation if LPM is used.[24]

Esculinase is a secreted enzyme, so hydrolysis takes place in the surrounding medium. In the case of plating media, a positive reaction for esculin hydrolysis is a black precipitate that surrounds the colony. If the plates are incubated too long, or if there are many esculinase positive bacteria, the entire medium can turn black, making potential *Listeria* colonies difficult to locate. Typical *Listeria* colonies appear bluish-white or bluish-gray on these media with zones of black in the agar milieu surrounding the colonies. The bluish aspect to the colonies is readily visible, so oblique illumination is not necessary to see it. Older colonies on PALCAM, OXA, and MOX may have an indentation in the center, possibly indicative of autolysis in the oldest part of the colony. Depending on the medium and the strain, full growth of the *Listeria* colonies can take from 24 to 72 h at 35–37°C.[29,39] For liquid medium, such as Fraser broth, a positive reaction is the entire tube or flask of medium turning black.

It is important, to remember that esculin hydrolysis is not confirmative of *Listeria*. Many species of bacteria will hydrolyze it and form the black precipitate. While the combination of inhibitors employed in the various media serves to restrain or slow the growth of these competitors, often they will survive the enrichment and plating protocols. Presumptive *Listeria* colonies must be selected for further analysis and speciation, and if diagnosed as *L. monocytogenes*, serological analysis.

5.2.1.3 Temperature Tolerance

For many years, enrichment at 4°C was the only way to selectively enhance the growth of *L. monocytogenes* in samples. Later it was found that, when trying to isolate *L. monocytogenes* from human clinical samples, growth at 30°C was successful because it was optimum for *L. monocytogenes*, and other competing microbiota might be better suited to 37°C.[13,20,23] Many present *L. monocytogenes* enrichment protocols utilize 30°C and 37°C for different steps, partly because some *L. monocytogenes* strains show an increased susceptibility to the antibiotics used in the selective media at 37°C as opposed to 30°C.[40] *L. monocytogenes* can grow very well at either temperature in rich media, while some of the competitive microbiota in the food samples cannot. Many of the lactic acid bacteria that are present in dairy samples have their growth optima at warmer temperatures, and have slower growth at 30°C; whereas many of the epiphytic plant bacteria are better suited to cooler temperatures and do not grow well, or at all, at 35–37°C. Therefore, if the combination of antibiotics and dyes does not manage to inhibit the growth of a competitor, temperature can serve as a selective agent.

5.2.1.4 Somatic and Flagellar Antigens

Listeria species possess multiple surface markers such as somatic (O) and flagellar (H) antigens that demonstrate group specificity and can be utilized as targets for immunological identification and serotyping with corresponding monoclonal and polyclonal antibodies.[41] *Listeria* somatic (O) antigens have been separated into 15 subtypes (I–XV), and flagellar (H) antigens into four subtypes (A–D). The serotypes of individual *Listeria* strains are ascertained through their unique combinations of O and H antigens. Upon analysis of group-specific *Listeria* O and H antigens via slide agglutination, at least 13 serotypes (i.e., 1/2a, 1/2b, 1/2c, 3a, 3b, 3c, 4a, 4ab, 4b, 4c, 4d, 4e, and 7) are recognized in *L. monocytogenes,* several (e.g., 1/2a, 1/2b, 3b, 4a, 4b, 4c, and 6b) in *L. seeligeri*, one (i.e., 5) in *L. ivanovii*, and a few (e.g., 1/2b, 6a, and 6b) in *L. innocua, L. welshimeri,* and *L. grayi* (Table 5.2). The H-antigens are useful for identification of *Listeria* spp. because the antisera generated against them do not cross-react with other bacterial species.[42] The H-antigenic factors A, B, and C can be used to detect all species of *Listeria* except for *L. grayi*, which can be identified by

TABLE 5.2
Antigen Complements of *Listeria* Serotypes

Serotype	O-antigen[a]	H-antigen
1/2a	I, II	A, B
1/2b	I, II	A, B, C
1/2c	I, II	B, D
3a	II, IV	A, B
3b	II, IV	A, B, C
3c	II, IV	B, D
4a	(V), VII, IX	A, B, C
4ab	V, VI, VII, IX	A, B, C
4b	V, VI	A, B, C
4c	V, VII	A, B, C
4d	(V), VI, VIII	A, B, C
4e	V, VI (VIII), (IX)	A, B, C
7	XII, XIII	A, B, C
5	(V), VI, (VIII), X	A, B, C
6a	V, (VI), (VII), (IX), XV	A, B, C
6b	(V), (VI) (VII), IX, X, XI	A, B, C

[a] Antigens in parentheses may not be present in all isolates.

detection of the H-antigenic factor E. In addition to the somatic and flagellar antigens, a plethora of other shared or distinct protein and teichoic acid components have also been characterized and exploited in various in-house and commercial immunological tests for differentiation of *L. monocytogenes* from non*monocytogenes Listeria* species and contaminating bacteria.

5.2.2 Specific Targets

All of the nonspecific targets act to enhance the growth of *Listeria* spp.; however, what is really needed is the capacity to quickly differentiate *L. monocytogenes* from *L. innocua*. The nonspecific targets for enrichment do not distinguish between these two species, which are similar physiologically except for virulence. Since *L. innocua* is more often detected in food than *L. monocytogenes*[43,44] and has a faster growth rate in enrichments,[45–47] this distinction is important to make. Standard protocols call for the analysis of five suspect colonies from any of the esculin-based agars. If the majority of *Listeria* colonies on one of the esculin-based agars are *L. innocua*, then *L. monocytogenes* could be missed. Specific and fast detection of *L. monocytogenes* is desirable in order to quickly recall a product if necessary, or to save the difficulty and cost of product recalls if not.

The appearance of several chromogenic media formulations, starting in the late 1990s, and one medium based on hemolysis of blood cells allow the specific differentiation of *L. monocytogenes* from other species of *Listeria*. Because the colonies of *L. monocytogenes* look different from those of nonpathogenic *Listeria* spp., these new media can permit quick verification without biochemical tests of individual colonies. One benefit of the chromogenic media is that they considerably shorten the time necessary for diagnosis of *L. monocytogenes* in food samples. With quick differentiation between *L. monocytogenes* and *L. innocua* in particular, confirmation time can be shortened from 5 or more days with traditional enrichment and plating media to 2–3 days. Because of this specificity, chromogenic media are being accepted in standardized enrichment protocols and are used more widely in general. These media take advantage of the expression of virulence genes in *L. monocytogenes* to differentiate it from nonvirulent *Listeria* spp. However, *L. monocytogenes* must still be differentiated from *L. ivanovii*, which possesses similar virulence genes. Importantly, these media will distinguish between typical *L. monocytogenes* and *L. innocua* strains, since both are esculinase positive, and will turn esculin agar and broths black. The formulations of many of these media are proprietary, but the manufacturers do report on the targets for differentiation.

5.2.2.1 Phosphatidylinositol Phospholipase C

All of the described chromogenic media differentiate *L. monocytogenes* from nonpathogenic *Listeria* spp. by virtue of the activity of secreted phosphatidylinositol phospholipase C (PI-PLC), which is associated with virulence in *L. monocytogenes* and *L. ivanovii.*[4,48] The phospholipase C (PLC) activity is encoded by two genes, *plcA* and *plcB*, located in the pathogenicity island that is regulated by the global virulence gene regulator in *L. monocytogenes*, PrfA. Incubation of media that detect virulence factors must be done at 37°C, since the *prfA* gene has higher expression at 37°C.[49] The PLC encoded by *plcB* is a broad range enzyme that cleaves many different types of phospholipids and is thought to be involved in cell-to-cell spread during pathogenesis; the PLC encoded by *plcA* is specific for phosphatidyl inositol and is thought to be involved in escape from the phagosome.[50] Nonpathogenic species of *Listeria* do not contain this PLC activity,[51] and strains in which either of the phospholipases have been deleted have decreased virulence.[50] In addition to *plcA* and *plcB*, *L. ivanovii* produces a third PLC that is specific for sphingomyelin.[52] The media designed to take advantage of PI-PLC activity also contain chromogens.

Depending on the media formulation, the colonies of all *Listeria* spp. may be pigmented, or only the colonies of *L. monocytogenes* and *L. ivanovii* may be pigmented. The chromogenic traits of these media have an advantage over the classical, esculin-based media, since the chromogen is nondiffusible, so with the chromogenic media, only the colony, not the agar, is pigmented. This is a benefit if the sample is rich in esculinase-positive bacteria. In some of the formulations, the PLC activity results in the formation of a halo surrounding the colony resulting from hydrolysis of phosphatidylinositol in the medium, and in others PLC activity results in a pigmented colony.

Many of these media, unless specifically stated, will not differentiate between *L. monocytogenes* and *L. ivanovii*. Some companies market separate media designed only to differentiate between these two pathogens, and they should be used in concert with chromogenic media that detect PLC activity. Media that distinguish between the two pathogens take advantage of carbon source utilization differences that make the colonies appear different. However, since *L. ivanovii* is not seen frequently in foods, a recommendation by the FDA, if using chromogenic media that do not differentiate between the two, is to presume the colony is *L. monocytogenes.*[24] The FDA, ISO, and other standardized procedures allow the use of chromogenic media in addition to esculin-based agars for confirmation of the presence of *L. monocytogenes* in food.[24,53] A major benefit of the chromogenic media is the time saved in *L. monocytogenes* identification. The chromogenic media usually cost more than the classic media; however, the time saved can be a cost benefit. From primary enrichment to confirmation using nonchromogenic media can take 5–7 days. Using chromogenic media to confirm *L. monocytogenes* can cut 3–5 days off that timetable. The greatest benefit of these media is their ability to distinguish *L. monocytogenes* from *L. innocua*. Some of the chromogenic media that are available are described next and are summarized in Table 5.3.

The R&F® *Listeria monocytogenes* chromogenic detection system consists of two selective and differential plating media—the chromogenic plating medium and the confirmatory medium—that will differentiate *L. monocytogenes* and *L. ivanovii* from *L. innocua* or other nonpathogenic *Listeria* by colony color.[54] On the R&F *Listeria monocytogenes* chromogenic plating medium (licensed from Biosynth), *L. monocytogenes* and *L. ivanovii* form blue colonies, while the nonpathogenic species form white colonies when the plates are incubated at 37°C. This differentiation is based on secreted phosphatidylinositol phospholipase C (PI-PLC) activity, associated with virulence in *L. monocytogenes* and *L. ivanovii*. Other bacteria that possess PLC activity can form blue colonies on the chromogenic plating medium, such as certain *Bacillus* spp.; however, the large, rough colony

TABLE 5.3
Appearance of *L. monocytogenes, L. ivanovii,* and *L. innocua* Colonies on Various Chromogenic Media

Medium	Appearance of		
	L. monocytogenes	*L. ivanovii*	*L. innocua*
Agar *Listeria* Ottavani and Agosti (ALOA)	Blue colony + white halo	Blue colony + white halo	Blue colony
Oxoid chromogenic *Listeria* agar (OCLA)	Blue colony + white halo	Blue colony + white halo	Blue colony, no halo
CHROMagar *Listeria*	Blue colony + white halo	Blue colony + white halo	Blue colony, no halo
CHROMagar identification *Listeria*	Mauve colonies	White colonies	N.A.[a]
R&F chromogenic plating medium (Biosynth)	Blue colony	Blue colony	White colony
R&F *L. monocytogenes* confirmatory medium	Fluorescent (366 nm) colony + visible yellow halo	No fluorescence, no yellow halo	N.A.
RAPID'*L. mono.*	Blue colony, no halo	Blue colony + yellow halo	White colony

[a] N.A. = not applicable, not designed to differentiate *L. innocua*.

morphology of *Bacillus* is quite different from that of *Listeria*, and a trained eye can easily discern them.

Blue colonies from the chromogenic plating medium are plated on the R&F *L. monocytogenes* confirmatory medium to differentiate between *L. monocytogenes* and *L. ivanovii*. The confirmatory medium, which contains a pH indicator, a fluorescent dye, and rhamnose, distinguishes between *L. monocytogenes* and *L. ivanovii* based on rhamnose fermentation. Most *L. monocytogenes* strains ferment rhamnose, and *L. ivanovii* strains do not. A positive reaction for *L. monocytogenes* on the confirmatory medium is a colony that is fluorescent under UV light (366 nm), with a visible yellow color, indicative of acid production, surrounding the colony. *L. ivanovii* will not fluoresce or produce acid on the confirmatory medium. The confirmatory medium is recommended by the FDA to differentiate *L. monocytogenes* and *L. ivanovii*, and it has been collaboratively validated by the FDA.[24,55] The components of both the chromogenic and confirmatory media that give the differential reactions are proprietary.

Agar *Listeria* according to Ottaviani and Agosti (ALOA) is another chromogenic medium that is also based on detection of PI-PLC activity. It is a selective and differential medium, and the formulation is known. Its selectivity for *Listeria* is determined by NaCl, LiCl, nalidixic acid, ceftazidime, polymyxin B, cycloheximide, and amphotericin or cycloheximide (Table 5.1 and Table 5.3). This medium detects the PI-PLC activity at 37°C of pathogenic *Listeria* by the addition of phosphatidylinositol to the medium. A positive response is an opaque halo surrounding the colony. The halo is insoluble diacylglycerol that results from the hydrolysis of the lipid in the agar. In addition to detection of PI-PLC activity, the chromogenic compound 5-bromo-4-chloro-3-indolyl-β-D-glucopyranoside (X-glc) in ALOA reacts to β-glucosidase activity present in all *Listeria* spp., resulting in blue colonies. Therefore, pathogenic *Listeria* will form blue colonies with a turbid zone surrounding them on ALOA agar, and nonpathogenic *Listeria* will form blue colonies with no halo.

Most non-*Listeria* bacteria that grow on ALOA will form white colonies; however, just as with the R&F chromogenic agar, some *Bacillus* spp. can grow and form blue colonies on ALOA. *L. ivanovii* colonies will appear just as *L. monocytogenes* colonies on ALOA, so for conclusive identification, suspect colonies must be tested further by biochemical tests to distinguish between *L. monocytogenes* and *L. ivanovii*. Evaluation by Vlaemynck et al.[56] showed that ALOA was superior to esculin-based PALCAM and Oxford agars with 4.3% more positives detected in naturally contaminated samples. Becker et al.[57] also found ALOA to be favorable in comparison with OXA and PALCAM. ALOA or an equivalent has been included in the ISO protocol for detection of *L. monocytogenes*.[38,53]

Two other media also give colony morphologies similar to that seen on ALOA. On Oxoid chromogenic *Listeria* agar (OCLA) and CHROMagar™ *Listeria*, *L. monocytogenes* and *L. ivanovii* will form blue colonies with an opaque halo surrounding them. Once again, the white halo is a result of the activity of PI-PLC causing the hydrolysis of phospholipid incorporated into the medium. The blue color is a result of a chromogen that is taken up by *Listeria* spp. Inhibitory substances in the medium will stop the growth of Gram-negative bacteria, yeast, and fungi. With these media *L. monocytogenes* detection can occur within 24 h at 37°C. CHROMagar *Listeria* will not support the growth of *L. seeligeri*. Both OCLA and CHROMagar *Listeria* have been validated by the Association Française de Normalisation (AFNOR) in the standard ISO enrichment protocol. Another medium, CHROMagar Identification *Listeria,* will differentiate between *L. monocytogenes* and *L. ivanovii* by color with *L. monocytogenes* forming mauve colonies and *L. ivanovii* forming colorless colonies.

RAPID'*L. mono* agar also differentiates pathogenic from nonpathogenic *Listeria* spp. based on PI-PLC activity. Both *L .monocytogenes* and *L. ivanovii* produce blue colonies on this agar, and nonpathogenic *Listeria* spp. form white colonies. Furthermore, RAPID'*L. mono* distinguishes between *L. monocytogenes* and *L. ivanovii* based on the inability of *L. monocytogenes* to metabolize xylose. Xylose is also incorporated into the medium, and the ability to ferment it is reflected in a yellow halo surrounding the colony. On this medium after 24 h of growth at 37°C, *L. monocytogenes* will form blue colonies with no yellow surrounding them. *L. ivanovii* colonies appear similar to *L. monocytogenes* colonies with the addition of a yellow halo around the colony reflecting the ability to metabolize xylose. The selective components of the medium will inhibit many competing Gram-positive and Gram-negative bacteria as well as yeasts and molds. Other non-*Listeria* bacteria, such as *Bacillus* spp. and streptococci will form blue colonies without halos on this medium.[58] Modifications by the manufacturer since 2005 have led to a reduction in the amount of background flora on RAPID'*L. mono.* A recent study using the new formulation of RAPID'*L. mono* with 310 naturally contaminated food samples found it to compare favorably with OXA, PALCAM, and ALOA, with RAPID'*L. mono* finding four more positive samples than OXA or PALCAM.[57] This medium has been validated by AFNOR in the ISO protocol.

5.2.2.2 Listeriolysin O

Contained within the virulence gene cluster in the *L. monocytogenes* genome is the *hly* gene, which encodes a hemolytic protein called listeriolysin O (LLO). During the infection process, LLO allows the bacterium to escape from the phagosome. Strains that have lost LLO function are avirulent.[4] The cytotoxic properties of LLO make it a useful indicator because it will cause the lysis of red blood cells on a blood agar plate (i.e., β-hemolysis), and it is considered an indicator of virulence gene activity. Similar proteins are also produced by *L. ivanovii* (i.e., ivanolysin O or ILO) and *L. seeligeri* (seeligerilysin O, SLO), which render them β-hemolytic as well. The hemolysis by *L. seeligeri* strains is weaker than by both *L. monocytogenes* and *L. ivanovii* because of lower expression of the hemolysin gene.[4] In addition, nonhemolytic *L. seeligeri* strains have been reported.

Since only virulent species of *Listeria* show hemolytic reactions, hemolysis is frequently included as an important step for *Listeria* identification. *Listeria monocytogenes* blood agar (LMBA) was described in 1998 to differentiate *L. monocytogenes* from the nonhemolytic *L. innocua* on a solid medium based on β-hemolysis by *L. monocytogenes*.[59] The base of LMBA is trypticase soy agar to which 5% sheep blood is added. Selectivity is achieved with the addition of LiCl, polymyxin B, and ceftazidime. Previous attempts to design media that used hemolysis as an indicator for *L. monocytogenes* incorporated blood in an overlay, but making the media was laborious. LMBA was an improvement upon hemolysis-based identification media because it incorporated the blood directly into the agar rather than in an overlay. It also reduced the number of selective agents that had been shown to inhibit *L. monocytogenes* hemolytic activity.[59] In plating of enrichments, LMBA was found to be superior to PALCAM and OXA for specific detection of *L. monocytogenes* in naturally contaminated foods after 48 h of incubation.[59–61] Because of the inability to control for variations in blood quality from different batches, LMBA is not widely used.

5.3 IDENTIFICATION METHODS

All *Listeria* spp. possess many traits in common, so the enrichment protocols for *L. monocytogenes* will also enrich other *Listeria* spp. After plating on selective and differential media, most of the standard protocols dictate that at least five potential *L. monocytogenes* colonies are picked for identification. If chromogenic media yield no potential *L. monocytogenes* colonies, then any esculinase-positive colonies can be tested. Presumptive colonies must be characterized as to whether they are, in fact, *L. monocytogenes* or another *Listeria* spp. Specific identification of *L. monocytogenes* can be done with a collection of biochemical-based tests or by using commercially available kits that run several of these tests concomitantly. Immunologic based tests are also used, and kits are commercially available for them. The level and type of identification depend on the needs and the mission of the lab. Labs that routinely screen for *L. monocytogenes* and other pathogens may invest in equipment that will allow for rapid screening of isolates via immuno-based kits, such as a microplate reader or VIDAS®. Depending on the charge of the laboratory, strains may need to be subtyped after identification. Serotype assignment is the most common subtyping used for *L. monocytogenes*, and it is especially important for investigations of food-borne outbreaks.

5.3.1 Nonspecific Methods

The first steps in the identification of any bacterium will start with common and quick tests such as Gram stains, wet mounts, and motility, catalase, and oxidase tests. Gram stains of *Listeria* on young cultures will show short, Gram-positive rods. Wet mounts of young cultures grown at 30°C or less will show tumbling motility. Motility tests must be done on cultures grown at 22–30°C, because the gene encoding the flagellin subunit, *flaA*, is down-regulated at higher temperatures. *Listeria* motility may be difficult to assess in wet mounts, so a tube of motility agar can be used in which motile *Listeria* demonstrate umbrella type growth. Occasional nonmotile isolates of *L. monocytogenes* are found, so a negative motility test alone should not be used to discount the isolate as *Listeria.*[62] *Listeria* are catalase positive and oxidase negative. Catalase tests must be done using colonies that have not been grown on blood agar because blood itself contains catalase. Colonies on non-esculin-based and nonchromogenic agar can be assessed using Henry oblique illumination for typical bluish-gray/bluish-green color.

5.3.2 Biochemical Methods

Biochemical testing must be done on isolated colonies from either direct plating or after enrichment and plating. Tests may be done individually for each isolate, but that can be time consuming. Several commercial kits that speciate strains based on biochemical reactions are available and include API-*Listeria*, *Listeria* ID, Microbact 12L, and MICRO-ID *Listeria*. These kits consist of test strips that contain the media and reagents to conduct several physiological tests concurrently. Many of these kits are accepted for use in standardized protocols. Whether done individually or with a kit, specific tests are required to first confirm *Listeria* spp, and then to determine if the isolate is *L. monocytogenes.* Beyond the simple tests outlined above, the most important biochemical tests are assessments of hemolytic activity and carbohydrate fermentation patterns. There is a colorimetric assay produced commercially that distinguishes *L. monocytogenes* from other *Listeria* spp. based on the enzyme activity of D-alanyl aminopeptidase. This enzyme is absent in *L. monocytogenes* and present in all other species of *Listeria*. However, one should not confirm *L. monocytogenes* based solely on one negative test.

5.3.2.1 Hemolysis

L. monocytogenes, *L. ivanovii*, and *L. seeligeri* demonstrate β-hemolysis due to the production of LLO, ILO, and SLO, as described previously. *L. ivanovii* strains demonstrate the most robust hemolytic activity due to a PLC, a sphingomyelinase, not produced in *L. monocytogenes*,[63] while *L. seeligeri* strains are weakly hemolytic due to low levels of the hemolysin SLO. However, non-hemolytic *L. seeligeri* strains have been described.[64] The protein is encoded by the *hly* gene located in the virulence gene cluster, and is therefore indicative of virulence gene activity. Because only virulent species of *Listeria* are capable of the hemolytic reaction, it is an important identification step for the differentiation of *L. monocytogenes* from *L. innocua*. It should be noted, however, that while *L. innocua* is defined as a nonpathogenic, nonhemolytic strain, there have been descriptions of *L. innocua* strains that are β-hemolytic.[65] Some commercial identification systems include hemolysis of blood cells as part of the strip itself. If those strips are not used, then hemolysis can be assessed on blood agar. Horse or sheep blood agar plates are used for testing hemolysis of individual strains.

Assessing β-hemolysis with *L. monocytogenes* can be problematic because the zone of clearing produced is very small, and some strains are more difficult to judge than others. Sometimes it is easier to see clearing of the blood if the colony is gently removed from the surface of the plate with a loop. Visualization of hemolysis in all three hemolytic *Listeria* species can be enhanced by using the Christie–Atkins–Munch–Peterson (CAMP) test on problematic isolates. The CAMP test, first designed to assess weakly β-hemolytic streptococci, takes advantage of the observations that the hemolytic activity of *L. monocytogenes* is enhanced in the presence of a weakly hemolytic *S. aureus* strain (SA^+) but, in almost all cases, not in the presence of *Rhodococcus equi* (RE^-).[66] Both *S. aureus* and *R. equi* secrete PLCs (known as β-lysin) into the medium, weakening but not lysing the red blood cells, which can then be attacked by the CAMP factors released by virulent *Listeria* spp.[66] A positive CAMP reaction for *L. monocytogenes* is associated with pathogenicity.[67]

Because of strain variability, the *S. aureus* and *R. equi* strains appropriate for the *Listeria* CAMP test have been standardized. Official protocols such as FDA and ISO give strain numbers of *S. aureus* and *R. equi* strains to use, and they can be bought from culture collections such as the American Type Culture Collection and the National Collection of Type Cultures. The test is done by inoculating the *S. aureus* and *R. equi* strains onto a sheep blood agar plate, each in a single line parallel to each other on opposite ends of the plate. The *Listeria* isolates for testing are streaked perpendicular to and between the *S. aureus* and *R. equi* streak lines without touching them. After incubation at 35°C for 24–48 h, the plates are examined for zones of enhanced hemolysis, similar to a small arrowhead, where the streak lines nearly intersect (Figure 5.1). In the CAMP test *L. ivanovii* is SA^-/RE^+, a reaction opposite of *L. monocytogenes*, which makes the CAMP test an important test for distinguishing between *L. monocytogenes* and *L. ivanovii*. *L. seeligeri* strains are weakly SA^+ in the CAMP test.

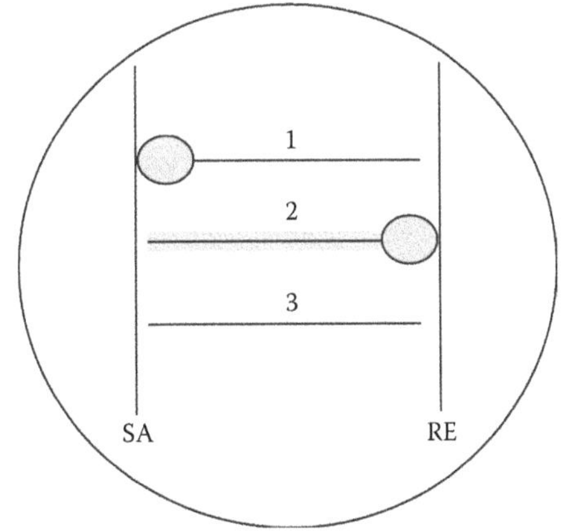

FIGURE 5.1 CAMP test. This representation of a CAMP test shows the proper orientations for inoculation of *S. aureus* (SA), *R. equi* (RE), and *Listeria* isolates 1 (*L. monocytogenes*), 2 (*L. ivanovii*) and 3 (CAMP-negative strain). Hemolysis is indicated by the gray-shaded areas. The ovals represent areas of enhanced hemolysis.

An alternative to the CAMP test with live cultures of *S. aureus* and *R. equi* is to use commercially available β-lysin discs. The β-lysin from *S. aureus* is impregnated into the disc, which is placed onto the surface of a sheep blood agar plate. *Listeria* isolates are streaked outward in a radial fashion from the disc, the plates incubated, and the hemolytic activity observed. *L. ivanovii* strains, which are SA^-, will show no synergistic hemolysis with these discs but will show robust hemolysis along the entire streak line.

5.3.2.2 Carbohydrate Utilization

Although the utilization of up to a dozen sugars can be tested in some of the available commercial kits, the sugar utilization patterns recommended by various official methods for distinguishing *Listeria* and for differentiating between *Listeria* spp. are dextrose, esculin, maltose, rhamnose, mannitol, xylose, and α-methyl-D-mannoside. If sugar utilization is measured individually, isolates should be inoculated into carbohydrate broths, each containing a sugar with a pH indicator, and incubated up to 7 days at 35°C. A positive reaction is acid production with no gas. Isolates can be speciated by their utilization patterns (Table 5.4 and Figure 5.2). With regard to differentiating *L. monocytogenes* from *L. ivanovii*, the most important sugars are rhamnose and xylose. *L. monocytogenes* is rhamnose positive and xylose negative, whereas the opposite is true for *L. ivanovii.* However, rhamnose-negative and slow rhamnose utilization isolates of *L. monocytogenes* have been described and seem to cluster in genetic lineage subgroups IIIB and IIIC (discussed in further detail in a later chapter).[68–70]

Differentiating *L. monocytogenes* from *L. innocua* solely by carbon utilization is difficult. The sugar fermentation patterns of *L. monocytogenes* and *L. innocua* are very similar (Table 5.4), with rhamnose use in *L. innocua* also being strain dependent. Therefore, hemolysis is the important biochemical test to distinguish between *L. monocytogenes* and *L. innocua*. However, as mentioned before, rare β-hemolytic *L. innocua* strains have been described.[65] Because of the possibility of rhamnose-negative *L. monocytogenes* being identified as *L. innocua* in a commercial biochemical test kit, the USDA identification protocol states that all β-hemolytic, rhamnose-negative isolates must be tested further with a ribosomal RNA-based identification system (discussed in further detail in chapter 6).[71]

TABLE 5.4
Carbohydrate Utilization Patterns for *Listeria* spp.

Acid Production from	*L. monocytogenes*	*L. ivanovii*	*L. innocua*	*L. seeligeri*	*L. welshimeri*	*L. grayi*
Dextrose	+	+	+	+	+	+
Esculin	+	+	+	+	+	+
Maltose	+	+	+	+	+	+
Mannitol	–	–	–	–	–	+
Rhamnose	+[b]	–	V[a]	–	V	V
Xylose	–	+	–	+	+	–
α-methyl-D-mannoside	+	–	+	–	+	

[a] Variable.
[b] Rhamnose-negative isolates of *L. monocytogenes* have been described.

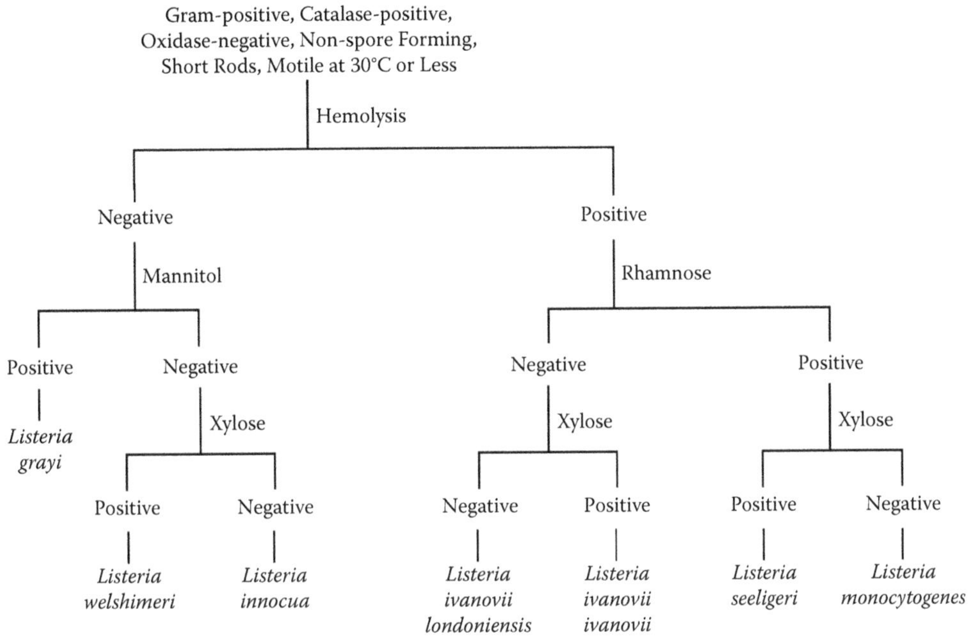

FIGURE 5.2 Schematic of biochemical identification for *Listeria* spp. based on carbohydrate fermentation tests and hemolysis.

5.3.3 Immunological Methods

The most commercially available kit for the rapid phenotypic identification of *Listeria* is the immuno-based kit. Many of these kits are available to detect either *Listeria* spp. or *L. monocytogenes* by immunoassay or enzyme-based immunoassay (EIA). Although the field changes quickly and new products are introduced frequently, some examples are Assurance *Listeria* EIA, Dynabeads anti-*Listeria*, EIA Foss *Listeria*, ListerTest *Listeria* Rapid Test, *Listeria*-Tek, TECRA Listeria Visual Immuno Assay, Transia Plate *Listeria*, Vidas LIS, Vidas LMO, and VIP for *Listeria*. Most detect *Listeria* spp. rather than *L. monocytogenes*. The methods for immunological detection fall into four basic categories: visual immunoprecipitation (based on immunochromatography), latex agglutination, immunomagnetic separation (IMS), and EIAs. All work on the same principle, in which specific anti-*Listeria* or anti-*L. monocytogenes* antisera provided in a matrix bind to antigens on target bacteria presented by the user.

Depending on the kit and/or the needs of the user, the antigen can be presented in a specially prepared food sample, an enrichment culture, a pure liquid culture, or an isolated colony. The kit determines how a positive or negative reaction is detected. Depending on the type of kit, results can be determined by visual detection of agglutination, a color reaction, screening of colonies made by captured cells, or spectrophotometric assay. Some kits will identify *L. monocytogenes* specifically, but most will detect only *Listeria* spp. One benefit of testing only for *Listeria* spp. is that, if the result is negative, further characterization of an enrichment may not be necessary. Immunodetection kits are simple to use and give results rapidly. Nearly all of them require enrichment or pre-enrichment prior to detection, and some are packaged with ready-made enrichment broths. Sometimes the choice of enrichment media affects the reliability of the test, and product brochures will usually specify the best media for testing. The needs of the laboratory will dictate the type of immunoassay chosen.

Lateral flow cassettes utilize visual immunoprecipitation and give a result within 10 min of sample application. The sample is pipetted onto a nitrocellulose matrix contained within a chamber, where it flows to a specific antibody conjugated to colloidal gold or some other visual detection chemistry located on the nitrocellulose membrane. Depending on the kit, a positive reaction can be the appearance of a black line or some other indicator on the membrane in a designated area where the antisera are located. A negative reaction would be the absence of the colored line in the designated area. The cassettes contain positive control areas to ensure the proper function of the antisera, membrane, and visualization chemistry.

Latex agglutination kits utilize latex particles that are coated with antisera. Colony material, aliquots of enrichment cultures, or pure cultures are mixed with the latex particles on a dark colored card. If the antigens are present they will bind to the antibodies, which will cause the white latex particles to agglutinate and form visible, white clumps. This reaction takes only a few minutes.

IMS also utilizes beads that are coated with antisera, but the beads are magnetic. For IMS, the beads are mixed with enrichments, pre-enrichments, or prepared food samples, and incubated for a short time. The beads are then collected with a magnet, washed, and spread onto any of the selective and differential *Listeria* plating media. Some kits package IMS with another immunological procedure to identify *Listeria* by a colony immunoblot after plating rather than by visual inspection of colonies. One benefit of IMS is that it can act as a selective enrichment without selective compounds that might damage or inhibit the growth of some strains.

EIAs include enzyme-linked immunosorbant assay (ELISA) and enzyme-linked fluorescent assays (ELFA), and use a "sandwich" type assay to amplify the signal from antigen binding to specific antisera. Specific primary antibody is attached to a matrix—for example, the well of a microtiter plate—and sample is added. The sample can be an enrichment culture or a pure culture. After a series of incubations and washes, a secondary antibody conjugated to an enzyme used for detection is added. The secondary antibody only binds if the primary antibody has bound to antigen first. After more incubations and washes a compound is added for colorimetric or fluorimetric detection of the secondary antibody. High-throughput laboratories may choose to use EIA-based kits, because they allow screening of multiple samples in multiwell plate formats or self-contained individual strips. Depending on the type of EIA, results can be read visually, in a microplate reader, or by the use of specific equipment designed for the matrix format.

There may be problems with incompatibility of certain immuno-based kits with components of enrichment media. EIAs can be affected by higher salt concentrations in some standard enrichment broths. The instructions that come with the kit will usually mention the best media to use for growing samples for testing.

Most of the immunologically based kits do not report on the targets for identification of *L. monocytogenes* or *Listeria* spp. Some of the visual immunoprecipitation kits report that detection is based on the *Listeria* flagellar antigens by the use of a polyvalent antiserum that recognizes the H-antigens A, B, C, and E. This polyvalent antiserum will react with all *Listeria* spp. that are actively motile. The H-antigens are good targets for *Listeria* detection because the antisera will not cross-react with other species.[42] Samples for use in kits that detect flagellar antigens must be grown at temperatures that will permit expression of *flaA* and, thus, motility.

Once *L. monocytogenes* is identified, the traditional way to subtype isolates is serotyping. This is especially important for investigations of outbreaks or sporadic cases of listeriosis. At least 13 serotypes have been isolated. Although avirulent and reduced pathogenicity strains have been described, all strains of *L. monocytogenes* by definition contain virulence genes and have the ability to cause disease.[72–76] Still, more than 95% of listeriosis cases are caused by only three serotypes: 1/2a, 1/2b, and 4b.[77,78] The serotype subgrouping scheme was first developed by Seeliger and Höhne[41] as groupings based on agglutination reactions of somatic (O) and flagellar (H) antigens with a series of polyvalent antisera. It is therefore based on cell surface antigenic characteristics of the strains. How the cells react to the 15 O-antisera and 4 H-antisera determines their serotype grouping (Table 5.2). Note that serotype 4 strains all possess the same H antigens, and assignment is based entirely on reactions with O-antisera, and that serotype 1/2 strains and serotype 3 strains each possess the same O-antigens, and final determination is based on the reactions with H-antisera. As a result, occasional nonmotile isolates of serotype 1/2 or 3 may not be fully identified as to serotype.[79] While molecular typing methods have been investigated and are being developed that are more distinctive and provide a more detailed picture of *L. monocytogenes* genomic groupings, these classifications are still compared with serotypes.

The O antigens are thermostable, while the H antigens are not.[41] Teichoic acids and lipoteichoic acids, which are covalently linked to the peptidoglycan, are the O antigens.[80] Various sugar substitutions on the repeating polyribitol phosphate units of the teichoic acids determine O antigen specificity. Uchikawa et al.[80] determined the teichoic acid structures of several *L. monocytogenes* serotypes. Strains of serotype 1/2 contain teichoic acids with polyribitol phosphate substituted with N-acetylglucosamine and rhamnose at C-2 and C-4. Serotype 3 strains contained only N-acetylglucosamine substitutions joined by phosphodiester bonds at C-1 of the ribitol residues. Serotype 4 strains differed in their glycosidic substitutions depending on the specific serotype, but all contain N-acetylglucosamine in the teichoic acid chains with glycosidic substitutions on the N-acetylglucosamine residues. Serotype 4b strains are unique in that they contain both galactose and glucose substitutions on the N-acetylglucosamine residues, while other serotype 4 strains contain either galactose or glucose substitutions. Serotype 4b strains lacking galactose in their teichoic acids have been described, although these strains still serotyped as 4b.[81]

The classical method for serotype analysis is the observation of agglutination reactions with killed cells and specific antisera. The cells must be prepared in different ways for the O and H antisera testing. O antigen testing uses autoclaved cells, and H antigen testing is done with formalin treated cells. Prepared cells are mixed with the individual antisera on glass slides or in tubes and visually observed for agglutination, the formation of visible clumps. Calling agglutination reactions can be tricky for some strains. Difficult-to-interpret reactions are observed under a low-power objective on a microscope. The entire process can be subjective, more so when results are compared between labs and between procedures in the same lab. People can read the agglutination reactions in diverse ways, and strains may react differently when retested. A World Health Organization study[82] of 80 strains serotyped independently by six different laboratories showed complete agreement on serotype assignment for only 61.3% of the strains. Reproducibility within each of the labs on 11 strains ranged from 82 to 100%. Because of ambiguous results some food and environmental isolates are very difficult to type. They may not fall into the traditional typing scheme in Table 5.2, or they might fall into more than one serotype group. However, clinical and outbreak isolates are usually very easy to serotype.

An alternative to the agglutination serotype method has been described that uses the same battery of antisera as the agglutination reaction tests. This is an ELISA-based method, which uses a fraction of the antisera of the agglutination method.[79] Whereas a few hundred isolates might be serotyped using the agglutination method with the current antisera kits, up to 10,000 can be typed with the ELISA method. The ELISA method is semiquantitative because absorbance readings are measured on reaction levels to all the antisera, and normalized to the highest reacting O and H antiserum reactions in the antisera collection. Positive and negative reactions are scored relative to these maxima, and the results can be displayed on a bar graph. The patterns on the bar graph relate to the standard serotype scheme. In a study of 101 *L. monocytogenes* isolates conducted between two independent laboratories, agglutination and ELISA serotypes agreed for 88% of the strains.[79]

Most labs will not need to discern serotype. The kits containing all 12 antisera are expensive and not always available, and, depending on the method used, results can be subjective. In cases where it is needed, isolates can be sent to reference laboratories that have more experience at serotyping. For a quick assessment of isolates, however, O-antiserum type 1 and O-antiserum type 4 are readily available to test for *L. monocytogenes* of serotypes 1 and 4, which cause the vast majority of clinical illness.

Methods to assign serotype based on polymerase chain reaction (PCR) and ribotyping have been described,[83–85] but they cannot yet discern all of the serotypes designated in the Seeliger and Höhne scheme.[41] Still, these methods hold promise for a faster, less expensive way to determine serotype or possibly to replace serotyping with more refined genetic-based subtyping.

5.4 PRACTICAL CONSIDERATIONS

In selecting an isolation and identification scheme for *L. monocytogenes* there are several issues to consider, including cost, time, and the number and types of foods or samples that will be tested. Another consideration is that because of differences among *L. monocytogenes* strains, they may not all behave identically in the enrichment and isolation protocols. The low numbers most likely present in any adulterated food make enrichment a necessity; however, the stressful nature of enrichment may impede the growth of some strains in favor of other bacteria that might not be as stressed, or might grow faster than *L. monocytogenes*. Foods might contain *L. innocua*, which some studies show has a faster growth rate, and may produce bacteriocins inhibitory to *L. monocytogenes*, which would make *L. innocua* compete favorably in enrichment cultures and possibly mask *L. monocytogenes*.[45–47,86–89] In addition to competition from other *Listeria* spp., lactic acid bacteria, which are commonly found in dairy products and processed meats, make bacteriocins that are inhibitory to *L. monocytogenes*.[90,91] Foods may also be contaminated with more than one strain of *L. monocytogenes* of different pathogenic potential, and enrichment and isolation may not equally detect each of them. In fact, some strains might be completely outcompeted.[39,92] While detection of any *L. monocytogenes* in a food sample can lead tov a product recall in the United States, inability to isolate the same strain from patients and food sources could hinder trace-back investigations to identify the source of an outbreak.

5.4.1 Selecting an Enrichment and Isolation Protocol

There are several official, standardized protocols designed by regulatory agencies and organizations for *L. monocytogenes* enrichment and isolation, as well as commercial enrichment protocols using media with proprietary ingredients. Three of the official protocols are outlined in Figure 5.3. These are not the only standardized protocols, but many others are based on these, or are very similar. The FDA protocol is the most commonly used method in the United States and is designed for use with any food product. As part of its regulatory mission, the USDA designed its method for isolations from meat, poultry, and egg products. The ISO method is used most commonly in European countries and can also be used with any food product. The natural microbial flora present in a given food source would dictate the type of enrichment to be used. All the protocols call for homogenization of 25 g of food into 225 ml of culture medium.

The USDA and ISO protocols use a primary and secondary enrichment before plating onto selective media. In both cases the secondary enrichment medium is Fraser broth. The primary enrichment media for both (UVM I in the USDA method and half Fraser broth in the ISO method) contain smaller concentrations of inhibitors than the secondary enrichment broth (Table 5.1). Secondary enrichment is therefore more selective, giving the process more discriminatory power for *L. monocytogenes* as the enrichment moves along. The FDA protocol uses a single enrichment culture with a pre-enrichment step with no selective compounds for 4 h. Both steps are in BLEB medium, and actual selective enrichment starts with the addition of inhibitory compounds to the pre-enrichment culture. BLEB enrichment broth contains low amounts of acriflavine, similar to UVM I and half Fraser broth, but higher amounts of nalidixic acid than either of the others. Therefore, BLEB may allow more inhibition of Gram-negative bacteria than UVM I and half Fraser broth. Because of the inclusion of cycloheximide, BLEB may be the enrichment medium of choice for foods containing background fungal flora such as cheeses and plants.

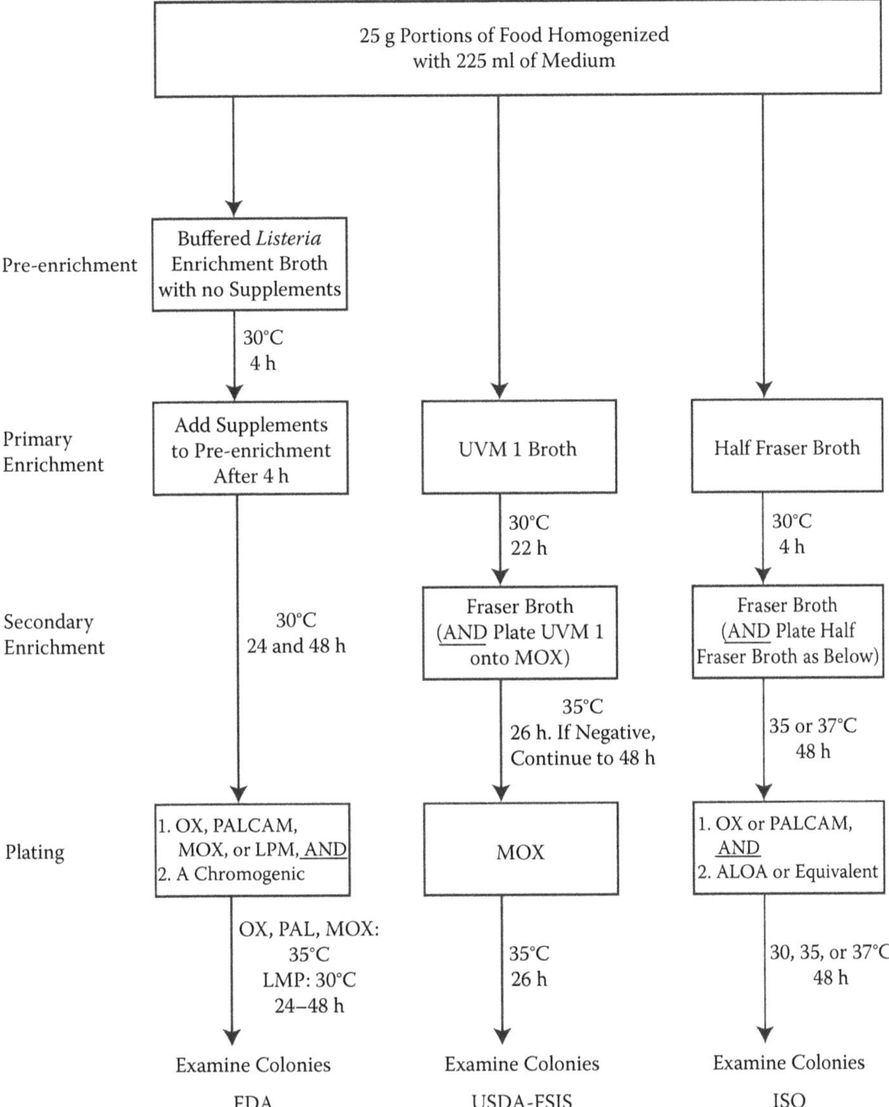

FIGURE 5.3 Schematics of some common standardized enrichment and isolation protocols for *L. monocytogenes*. (FDA protocol from Hitchins, A. D. In *Bacterial analytical manual* online (http://www.cfsan.fda.gov/~ebam/bam-10.html) 2003; USDA-FSIS protocol from (http://www.fsis.usda.gov/ophs/Microlab/Mlg8.03.pdf); ISO protocol according to the 11290-1 reference method, International Organization for Standardization, Geneva, 1996.)

The commonly used *L. monocytogenes* enrichment broths in the preceding protocols are buffered. Buffered broths are more efficient for recovery in some foods, especially dairy products, since lactic acid bacteria can reduce the pH of the culture to the point where growth of *L. monocytogenes* is inhibited.[25] BLEB medium (used in the FDA protocol), UVM I (in the USDA method), and half Fraser broth (in the ISO method) are all buffered with phosphate. The enrichment broths and plating media are complex media that use various protein digests and, in some cases, glucose to provide nutrients. The nutritional components of the three media are as follows: UVM I contains proteose peptone, tryptone, Lab-Lemco powder, yeast extract, and esculin; half Fraser broth contains pancreatic digest of casein, proteose peptone, beef extract, yeast extract, and esculin; and

BLEB contains pancreatic digest of casein, soytone, dextrose, yeast extract, and sodium pyruvate. All pre-enrichments, primary enrichments, and secondary enrichments are incubated at 30°C. As mentioned earlier, this temperature is beneficial to reduce the growth of some bacteria that might compete in the enrichment, and it is the optimum temperature for *L. monocytogenes* growth. After enrichment in broth, incubations for agar media are carried out at higher temperatures unless LPM agar is used, which is incubated at 30°C.

Another concern for detecting *L. monocytogenes* contamination is the possibility that the bacteria on the food surface are stressed since foods are routinely exposed to conditions that injure and kill the bacteria on their surface. Many of these stressful conditions are used during food processing and storage to reduce or slow the growth of the bacterial load to increase the shelf life of the product. Exposure to high or freezing temperatures, acids, food preservatives, and sanitizers can cause sublethal damage to *L. monocytogenes*.[93–96] Damaged cells are less likely to withstand the stress of enrichment culture.[95,97] The antibiotics used in the *L. monocytogenes* enrichments were selected because they inhibit competing microbiota without affecting *L. monocytogenes*. However, if the original population is sufficiently weakened, then the added stress of dealing with an immediate antibiotic challenge may kill it. Over time, the enrichment protocols and the media have evolved to permit stressed cells to recover before full exposure to antibiotics. These methods include a pre-enrichment for a short time in the absence of inhibitory compounds, or enrichment in reduced concentrations of the compounds.

Busch and Donnelly[97] developed *Listeria* repair broth to aid the recovery of heat-injured *L. monocytogenes* cells and found that 5 h of growth prior to addition of the inhibitors were necessary for near full recovery of the injured cells. Osborne and Bremer[98] found that the best medium for recovering *L. monocytogenes* from seafood products exposed to a variety of processing techniques known to cause stress depended on the type of stress that caused the damage. Testing Fraser broth, UVM I, BLEB, and *Listeria* repair broth as enrichment media, they found that after exposure of foods to alcohol-based preservation, full retrieval of *L. monocytogenes* from affected samples was inhibited if they were allowed a nonselective pre-enrichment. They also found that all four media were statistically equivalent in the retrieval of *L. monocytogenes* from all samples treated with acetic acid or osmotic stress. Silk et al.[99] measured the growth kinetics of heat-injured *L. monocytogenes* in UVM I, BLEB, Fraser broth, *Listeria* repair broth, and several nonselective media, and found that Fraser broth was the only selective enrichment broth to negatively affect the time required for the repair of damaged cells. This may be due to the higher concentration of acriflavine in Fraser broth. Half Fraser broth, which contains reduced levels of acriflavine, is the primary enrichment broth for the USDA and ISO protocols. Sodium pyruvate, a component of *Listeria* repair broth found to have an effect on the repair of heat-stressed cells,[97] is a component of BLEB.

Because of many reported advantages of pre-enrichment,[100–102] the FDA protocol includes a pre-enrichment step in BLEB that does not contain nalidixic acid, cycloheximide, and acriflavine. The three inhibitors are added after a 4-h incubation at 30°C after stressed cells have recovered and are better able to withstand antibiotic stress. The FDA also allows using the antifungal pimaricin as an alternative to the toxic cycloheximide.[24] Other protocols, including USDA and ISO, do not include a pre-enrichment step, but rather employ a two-step enrichment protocol with two different broths. In contrast to the single enrichment broth method, many different studies show that the use of both a primary and secondary enrichment increases the probability of isolating *L. monocytogenes* from contaminated samples.[10,27,58,103–105]

In the case of the USDA and ISO methods, the secondary enrichment broth is Fraser broth, which contains esculin and ferric iron and will turn black if *Listeria* spp. is present.[27] This is a useful indicator if the food sample is not contaminated, because if the Fraser broth has not turned black after a 48-h incubation at 35–37°C, then the food sample may be considered negative for *L. monocytogenes*, and further analysis is unnecessary.[71] However, potential false negative results with Fraser broth can occur, so all cultures should be plated for further analysis. Both UVM I and half Fraser broth have lower concentrations of nalidixic acid and acriflavine than Fraser broth. Various studies concluded that nalidixic acid has no effect on *L. monocytogenes* growth in culture[21,22]; however, acriflavine has been shown to have strain-specific effects on the time in lag phase,[22] and negative effects on growth at higher concentrations.[21,36]

Another concern for enrichment of *L. monocytogenes* is the presence of *L. innocua* in the enrichment. Several researchers report that *L. innocua* strains have faster growth rates than *L. monocytogenes*, and that *L. monocytogenes* is more difficult to detect in samples that also contain *L. innocua*.[45–47,92] On the other hand, there have been isolated reports that *L. innocua* in foods did not affect the recovery of *L. monocytogenes*.[106] It is likely that strain variations affect the amount of competition between the two species with some strains of *L. monocytogenes* better adapted than others to competition with *L. innocua*. The USDA and ISO protocols call for streaking plates from both the primary and secondary enrichment cultures, while the FDA calls for streaking plates after both 24 and 48 h of enrichment specifically to detect *L. monocytogenes* in samples that might also contain *L. innocua*. The reasoning behind the measure is that *L. monocytogenes* may be detected early in the enrichment process before *L. innocua* has the opportunity to outcompete it.

Different strains of *L. monocytogenes* can also compete with each other in enrichment broths, resulting in differential detection of the strains. These variations may be related to serotype. While strains of serotypes 1/2a, 1/2b, and 4b cause over 95% of listeriosis cases, the most common type found in outbreaks is serotype 4b. Yet serotype 1/2a is more commonly isolated from foods.[107] Loncarevic et al.[108] found a greater variety of strains of *L. monocytogenes* from food samples after direct plating of foods than after enrichment of the same samples, indicating that strains do not survive equally in the enrichment procedure. Physiological and genetic differences have been described for serotype 1/2a and 4b strains. For example, serotype 1/2a strains survive exposure to bacteriocins at 4°C better than serotype 4b strains, and serotype 4b strains survive heat treatment after a cold storage better than serotype 1/2a strains.[109] Another study with 25 isolates of different serotypes found that serotype 4b isolates had lower heat resistance than strains of the other serotypes.[110] Comparisons of genomes between a serotype 1/2a and several 4b strains showed that the genome of the serotype 1/2a strain is slightly larger.[111] It is possible that a combination of physiological and ecological differences allows for better survival of serotype 1/2a strains in foods; however, it is also possible that serotype 1/2a strains simply enrich better than serotype 4b strains. It is certainly likely that, regardless of serotype, some strains are better at survival in enrichment than others.

There have been reports of different strains of *L. monocytogenes* not enriching or growing as well as others in standard enrichment broths. If the sample is contaminated with more than one strain of *L. monocytogenes* or even with *L. innocua*, then some strains might not be as readily recovered as others. Bruhn et al.[92] found a bias in UVM I medium, used in the USDA protocol, when it was co-inoculated with genetic lineage I (including serotypes 4b and 1/2b) and lineage II (including serotype 1/2a) strains of *L. monocytogenes*. The study of eight strains found that lineage II strains were consistently recovered at higher ratios than lineage I strains when the two were co-inoculated into UVM I at equivalent levels. They also report that an *L. innocua* strain outcompeted the *L. monocytogenes* lineage I strains, but not the lineage II strains when co-inoculated with them into UVM I. In monoculture, the eight strains also showed differences in growth rates in UVM I medium, but not in the nonselective brain heart infusion broth.

In a different study[39] using the FDA enrichment protocol and plating onto MOX, it was shown that some strains were more fit in the enrichment protocol than others. When strains of serotypes 1/2a and 4b were co-inoculated into BLEB with and without added food, some strains were consistently recovered at higher levels than others; yet when each of the strains was put through the FDA protocol individually, it was recovered at equivalent levels in the presence or absence of added food. While strains of both serotypes were among the most fit of the eight in the study, the strains with the highest fitness were of serotype 1/2a. The addition of cheese and lettuce to the coculture enrichments resulted in slightly better competition and more equivalent recovery of both strains in some cases. In both of these studies, among the poor competitors were clinical isolates of *L. monocytogenes*.

Regarding esculin-based plating media, the FDA protocol allows for the largest variety of these media to be used, and the order of preference among them is PALCAM, OXA, MOX, LPM. If LPM is used, however, esculin and ferric iron must be added, since they are not part of its regular composition. The ISO protocol calls for plating onto PALCAM or OXA. The USDA protocol calls for plating of the primary and secondary enrichments onto only MOX agar. Both the FDA and ISO methods call for plating on two different media, which increases the chances of finding *L. monocytogenes* contamination. When selecting an esculin-based agar to use, many studies comparing them find similar levels of detection for OXA, PALCAM, and MOX after enrichment.[55,57,61,112] PALCAM is more selective than OXA, with fewer non-*Listeria* colonies arising on PALCAM after an enrichment with blue-veined cheese[113] and with meat products.[114] PALCAM agar allows the growth of some *Staphylococcus* strains; however, acid production from mannitol fermentation by *Staphylococcus* and the resulting formation of a yellow halo on PALCAM allow it to be differentiated from *Listeria* strains.[29] Of the four esculin-based agars, PALCAM is the only one that has mannitol and will make this distinction. The presence of moxalactam in MOX inhibits most *Staphylococcus* strains. MOX is more selective than OXA, but *L. monocytogenes* colonies might take slightly longer to arise on MOX.

Studies with injured cells show that plating directly onto a selective agar significantly decreases the recovery levels of those cells in comparison to nonselective agars such as trypticase soy agar.[115,116] Therefore, plating of food samples directly onto selective agar might decrease the detection of *L. monocytogenes*. However, after a 1- to 2-day enrichment process, injured cells should have recovered sufficiently to increase plating efficiencies. Studies and observations with healthy cells show equivalent plating efficiencies with MOX, OXA, and PALCAM as compared to nonselective agars.[117,118] There have been reports of enrichment in nonselective media prior to plating on selective agar, and, in some cases, the instances of recovery are better with a nonselective enrichment.[106,119] However, nonselective enrichment might not be suitable for samples containing high levels of resident microbiota.

The FDA and ISO protocols each allow for chromogenic agars to be used and require one esculin-based agar in addition to a chromogenic agar. Using a combination of chromogenic and esculin-based agars as the plating media led to better recovery of *L. monocytogenes* isolates than use of two esculin-based agars alone.[55,57] Chromogenic agars, unlike esculin-based agars, allow simple differentiation between *L. monocytogenes* and *L. innocua*. There are still benefits to esculin-based agars, however, since not all strains are recovered equally on both types of media.[61,120] The chromogenic agars are specific for virulent *Listeria* spp., and nonpathogenic *Listeria* spp. can be easily eliminated from analysis. Since enrichment and identification protocols call for at least five presumptive colonies from esculin-based agars to be picked for identification, using a chromogenic agar will increase the chances of making the correct colony selection. Chromogenic agars are not foolproof, however, and there have been cases of *L. monocytogenes*-like colonies identifying as other *Listeria* spp., and vice versa.[58] There have also been reports of *L. monocytogenes* strains, including hypovirulent isolates, having atypical colony morphologies or not being detected on Rapid'*L. mono*, ALOA, OX, and PALCAM. Because of strain and media differences, using a combination of chromogenic and esculin-based media gives a better chance of *L. monocytogenes* detection than use of either one on its own.[120,121]

If the resources are available, using more than one enrichment protocol for the same sample would increase the chance of finding *L. monocytogenes* contaminants if they are present. Many studies comparing enrichment and isolation protocols have concluded that no one protocol finds every *L. monocytogenes* isolate, and using more than one increases the chances of finding contamination.[104,105,122–124] Most studies directly comparing the USDA and FDA protocols were not done with the enrichment broth compositions currently in use. Comparisons of growth rates of stressed and healthy *L. monocytogenes* cells in the current versions of UVM I and BLEB found no statistical difference.[99] The ISO protocol was developed in 1996, and there have not been comparative studies among the three protocols. In schematic the ISO and USDA protocols are similar, and the concentration of acriflavine in half Fraser broth and UVM I are the same, but UVM I has twice the amount of nalidixic acid. Theoretically, since nalidixic acid reportedly has no effect on the growth kinetics of *L. monocytogenes* strains,[21,22] the ISO and USDA protocols should perform similarly in the liquid enrichment process, but this has not been tested. The nonselective components of the broths are different, and at least one study has suggested that these nonselective media ingredients might affect the growth kinetics of strains.[99] The USDA method still does not recommend a chromogenic agar for isolation purposes, although considering the ease of use of such media, and the increasing numbers of products available, future revisions of the protocol may call for them. The USDA protocol is designed for isolation and detection of *L. monocytogenes* from meat and poultry products, which should be taken into consideration when selecting a protocol to use. Therefore, if choosing two of the three official protocols to use, then selecting the FDA protocol along with either ISO or USDA might provide for the most diversity in enrichment and isolation schemes.

The three enrichment protocols listed in Figure 5.3 are not the only methods available. There are proprietary media designed for enrichment and isolation of *L. monocytogenes*, such as Oxoid Novel Enrichment Broth—*Listeria* (ONE Broth) designed for the recovery of *L. monocytogenes* in 24 h without the need for secondary enrichment. After 24 h the enrichment can be plated onto any suitable medium, including a chromogenic agar, or can be used in a kit for specific detection of *Listeria* spp. or *L. monocytogenes*. Another *L. monocytogenes* enrichment and detection system is the BCM® *Listeria monocytogenes* Detection System (LMDS) by Biosynth,[54] which uses a nonselective pre-enrichment followed by a selective enrichment and plating onto chromogenic plating media. The BCM selective enrichment broth contains 4-methylumbelliferyl-myo-inositol-1-phosphate, a fluorogenic substrate that recognizes PI-PLC activity, so only fluorescent enrichment broths need to be plated onto chromogenic agar. However, in order to find low levels of *L. monocytogenes* that might be masked by high levels of *L. innocua*, all enrichments should be plated. In a study comparing the LMDS to the USDA method, 26.5% of environmental sponge samples analyzed by the USDA method gave false positives, while no false positives were found using LMDS.[54]

Using one of the official methods outlined in Figure 5.3 takes up to 4 days after sample processing before colonies are present to analyze. Using a chromogenic agar shortens that time by a day. ONE broth from Oxoid produces colonies to analyze after 2–3 days, and the BCM LMDS provides colonies to analyze in 3–4 days.

5.4.2 Selecting an Identification Protocol

If a large volume of sample testing is done, using a kit to identify isolates is probably the best course of action. The type of kit selected will depend on the needs of the laboratory. If a simple yes or no answer for *Listeria* presence with no colony isolation is needed, then any of the immunologic based kits could be used with primary and/or secondary enrichment cultures. If the lab is testing many samples on a regular basis, then an ELISA- or ELFA-based kit using a plate reader would provide fast, high-throughput results of enrichment cultures. One benefit of the immunologically based kit is that isolated colonies are not necessary to do the analysis. The specific antisera in the kits will bind with *Listeria* or *L. monocytogenes* antigen present in the enrichments and give a positive or negative reaction. If isolated colonies are necessary, the immuno-based kits could still be used, and the enrichments plated after the immunological kit give a positive *L. monocytogenes* result. Using an IMS type of kit can serve as an enrichment since the antibodies on the latex beads will capture the target bacteria, and the beads can be collected with a magnet, washed, and plated onto selective media. This type of isolation was used to detect *Listeria* spp. in feces of zoo animals using the two-stage ISO enrichment method followed by IMS and plating on ALOA.[125] In this case, the authors report that use of a secondary enrichment increased the number of positive samples over a single enrichment alone.

Most immuno-based kits detect for *Listeria* spp., not *L. monocytogenes* specifically. The antibody formulations used in most of these detection kits are proprietary, so the detection targets are unknown. Some report that the basis for *Listeria* distinction are antisera based on detection of flagellar antigens. Most strains of *L. monocytogenes* are motile, but occasional nonmotile isolates are found.[62] If these strains are nonmotile because of the lack of flagella, then use of antiflagella antisera will not detect them. Some immuno-based detection systems are affected by components of the enrichment media. EIAs in particular are negatively affected by high salt concentrations. There are reports of ELISA assays giving poor results when done with *L. monocytogenes* grown in the higher salt enrichment media UVM I and Fraser broth.[126,127] The immuno-based kits usually state which medium is best to use for optimum detection, and it is sometimes included in the kit. Therefore, if an immuno-based test gives a negative result for *L. monocytogenes*, it would be wise to follow up with a biochemical- or genetic-based test.

Use of individual tests or a biochemical-based kit requires isolated colonies. Figure 5.2 shows the scheme of tests that are needed for species identification of *Listeria*. The data from these individual biochemical tests can be used with hemolysis data from single cultures on blood plates or the CAMP test (Figure 5.1), and tests of Gram reaction, catalase, oxidase, and motility to confirm the presence of *L. monocytogenes*. Once again, use of a biochemical-based kit that runs multiple tests concurrently will save time since each individual medium will not have to be prepared by lab personnel.

There also exist colorimetric, biochemical kits to differentiate *L. monocytogenes* from other *Listeria* spp. based on the detection of the enzyme D-alanyl aminopeptidase, which is present in all species except *L. monocytogenes*.[128,129] While a positive reaction in this system is indicative that the organism is not *L. monocytogenes*, further characterization of negative testing isolates should be done to confirm *L. monocytogenes*.

With biochemical tests, one should always be aware of the potential for strain vagaries. The majority of *Listeria* isolates show the physiology patterns displayed in Figure 5.2; however, there are descriptions of strains that are atypical and stray from the described physiology. Two important biochemical tests for *L. monocytogenes* are rhamnose utilization and β-hemolysis, yet strains have been described that are negative or slow for one of these traits.[68–71] These strains can test negative for *L. monocytogenes* in biochemical kits. Sado et al.[70] described slow rhamnose utilizing *L. monocytogenes* isolates that tested negative with the micro-ID test strip, but positive with the API *Listeria* strip. Strain differences are not limited to *L. monocytogenes*. *L. innocua* is defined as a nonhemolytic species, but at least one β-hemolytic *L. innocua* strain has been described.[65] Similar to the potential problems with immuno-based tests mentioned before, the lesson from these unusual isolates is that *L. monocytogenes* should not be dismissed based solely on one negative result. A genetic-based test would be a rapid and easy secondary test.

5.4.3 Enumeration

The FDA, ISO, and USDA protocols call for enumeration of *L. monocytogenes* present in any food sample. In the United States, detection of *L. monocytogenes* at any level is reason for recall of some contaminated products as part of the zero-tolerance policy of the FDA. Still, even in the published FDA and USDA protocols, enumeration is necessary. All three official protocols use a most probable number (MPN) technique where the sample to be tested is homogenized and placed into enrichment medium. Direct plating is done of this suspension. Aliquots of the suspension are also dispensed into tubes at full strength, and in a series of dilutions in enrichment broth in replicates, such that the tubes represent diluted portions of the food sample. The tubes are put through the enrichment regimen and plated onto appropriate media, and colonies are identified to determine the smallest dilution that tests positive for *L. monocytogenes*. The number of positive tubes is compared to a table that uses statistical analysis to determine the level of contamination of the original sample. The table is dependent on the number of replicates made of each dilution.

Use of a multiple tube protocol for MPN analysis can be time consuming. Variations have been devised using a colony lift immunoassay on a filter,[130,131] by passing food homogenates through a filter, and by the use of specific media to detect *L. monocytogenes*.[132] If a simple yes or no answer is all that is required, MPN analysis may not need to be performed.

5.5 CONCLUSIONS AND PERSPECTIVES

L. monocytogenes is an important food-borne pathogen that shares many morphological, biochemical and molecular features with other *Listeria* species as well as other common bacteria. Phenotypic identification methods exploit *L. monocytogenes*' ability to sustain and grow at low temperatures, to tolerate selective antibiotics, and to display distinct biochemical and immunological reactions. Although use of one or two such techniques may only result in a partial characterization, a combinational application of several phenotypic procedures will often lead to a correct identification of *L. monocytogenes* in a majority of cases.

L. monocytogenes is known to encompass a diversity of strains within the species, which frequently demonstrate enormous variations in biological, biochemical, and immunological characteristics. As with any microbiological method described to assess an entire genus and species, there are occasional atypical strains that stray from the norm. Therefore, one must be careful if using only phenotypic tests to identify an *L. monocytogenes* isolate. To be sure of the identity of an isolate, a combination of biochemical, immunological, and genetic tests should be used. The ease of use of the various rapid diagnostic kits makes them practical and ideal for many different types of laboratories.

Despite continuing improvement in phenotypic methodologies, the existing standardized phenotypic tests for the presence of *L. monocytogenes* in foods still require an enrichment step prior to detection. The state of enrichment culture to date is such that only one proprietary formulation, the BCM *Listeria monocytogenes* Detection System, will detect *L. monocytogenes* specifically during the enrichment culture process. As this system is effective with only limited types of processed specimens, more research needs to be done with naturally contaminated samples to compare the three official protocols in their current formulations (i.e., ISO 11290, FDA BAM, and USDA) to determine which method, if any, is best at enriching *L. monocytogenes* strains from all sample types.

Past efforts to further enhance the performance of phenotypic methods for *L. monocytogenes* have been somewhat hampered by the lack of a well-defined strain collection. Recently, a collection of 44 isolates was described that includes 25 genetically diverse *L. monocytogenes* strains, a hemolytic *L. innocua* strain, and a set of outbreak *L. monocytogenes* strains from nine major human listeriosis outbreaks.[133] Such a standardized strain collection would be beneficial in assessing current and future enrichment protocols for strain differences. To be sure of detecting contamination levels, variety is the best weapon. Use of multiple enrichment protocols with an assortment of plating media will strengthen the chances of finding the most *L. monocytogenes* contamination in any group of samples.

REFERENCES

1. Schlech, W. F. I. et al., Epidemic listeriosis—Evidence for transmission by food, *New Engl. J. Med.* 308, 203, 1983.
2. Fleming, D. W. et al., Pasteurized milk as a vehicle of infection in an outbreak of listeriosis, *New Engl. J. Med.* 312, 404, 1985.
3. Linnan, M. J. et al., Epidemic listeriosis associated with Mexican-style cheese, *New Engl. J. Med.* 319, 823, 1988.
4. Vázquez-Boland, J. A. et al., *Listeria* pathogenesis and molecular virulence determinants, *Clin. Microbiol. Rev.* 14, 584, 2001.
5. Cole, M., Jones, M., and Holyoak, C., The effect of pH, salt concentration and temperature on the survival and growth of *Listeria monocytogenes*, *J. Appl. Microbiol.* 69, 63, 1990.
6. Gray, M. L. et al., A new technique for isolating Listerellae from the bovine brain, *J. Bacteriol.* 55, 471, 1948.

7. Van Kessel, J. S. et al., Incidence of *Salmonella, Listeria monocytogenes* and fecal coliforms in bulk tank milk, NAHMS dairy 2002 survey, 87, 2822, 2004.
8. Buchanan, R. L. et al., Comparison of lithium chloride–phenylethanol–moxalactam and modified Vogel Johnson agars for detection of *Listeria* spp. in retail-level meats, poultry, and seafood, *Appl. Environ. Microbiol.* 55, 599, 1989.
9. Loessner, M. J. et al., Comparison of seven plating media for enumeration of *Listeria* spp., *Appl. Environ. Microbiol.* 54, 2003.
10. Hao, D. Y.-Y., Beuchat, L. R., and Brackett, R. E., Comparison of media and methods for detecting and enumerating *Listeria monocytogenes* in refrigerated cabbage, *Appl. Environ. Microbiol.* 53, 955, 1987.
11. Henry, B. S., Dissociation in the genus *Brucella, J. Infect. Dis.* 52, 374, 1933.
12. Gray, M. L., Stafseth, H. J., and Thorp, F., Jr., The use of potassium tellurite, sodium azide, and acetic acid in a selective medium for the isolation of *Listeria monocytogenes, J. Bacteriol.* 59, 443, 1950.
13. Gray, M. L. and Killinger, A. H., *Listeria monocytogenes* and listeric infections, *Bacteriol. Rev.* 30, 309, 1966.
14. McBride, M. E. and Girard, K. F., A selective method for the isolation of *Listeria monocytogenes* from mixed bacterial populations, *J. Lab. Clin. Med.* 55, 153, 1960.
15. Lovett, J., Francis, D. W., and Hunt, J. M., *Listeria monocytogenes* in raw milk: Detection, incidence, and pathogenicity, *J. Food Prot.* 50, 188, 1987.
16. Lee, W. H. and McClain, D., Improved *Listeria monocytogenes* selective agar, *Appl. Environ. Microbiol.* 52, 1215, 1986.
17. Vera, H. D. and Dumoff, M., Culture media, in *Manual of clinical microbiology*, Blair, J. E., Lennette, E. H., and Truant, J. P., eds. American Society for Microbiology, Washington, D.C., 1970, p. 633.
18. Estavez, E. G., Bacteriologic plate media: Review of mechanisms of action, *Lab Med.* 15, 258, 1984.
19. Beerens, H. and Tahon-Castel, M. M., Milieu à l'acide nalidixique pour l'isolement des streptocoques, *D. pneumoniae, Listeria, Erysipelothrix, Ann. de Inst. Pasteur* 111, 90, 1966.
20. Mavrothalassitis, P., A method for the rapid isolation of *Listeria monocytogenes* from infected material, *J. Appl. Bacteriol.* 43, 47, 1977.
21. Jacobsen, C. N., The influence of commonly used selective agents on the growth of *Listeria monocytogenes, Int. J. Food Microbiol.* 50, 221, 1999.
22. Beumer, R. R. et al., The effect of acriflavine and nalidixic acid on the growth of *Listeria* spp. in enrichment media, *Food Microbiol.* 13, 137, 1996.
23. Kramer, P. A. and Jones, D., Media selective for *Listeria monocytogenes, J. Appl. Bacteriol.* 32, 381, 1969.
24. Hitchins, A. D., Detection and enumeration of *Listeria monocytogenes* in foods. In *Bacterial analytical manual* online (http://www.cfsan.fda.gov/~ebam/bam-10.html), 2003.
25. Asperger, H. et al., A contribution of *Listeria* enrichment methodology—Growth of *Listeria monocytogenes* under varying conditions concerning enrichment broth composition, cheese matrices and competing microflora, *Microbiology* 16, 419, 1999.
26. Curtis, G. D. W. et al., A selective differential medium for the isolation of *Listeria monocytogenes, Lett. Appl. Microbiol.* 8, 95, 1989.
27. Fraser, J. A. and Sperber, W. H., Rapid detection of *Listeria* spp. in food and environmental samples by esculin hydrolysis, *J. Food Prot.* 51, 762, 1988.
28. Cox, L. J., Dooley, D., and Beumer, R., Effect of lithium chloride and other inhibitors on the growth of *Listeria* spp., *Food Microbiol.* 7, 311, 1990.
29. van Netten, P. et al., Liquid and solid selective differential media for the detection and enumeration of *L. monocytogenes* and other *Listeria* spp., *Int. J. Food Microbiol.* 8, 299, 1989.
30. Estevez, E. G., Bacteriologic plate media: Review of mechanisms of action, *Lab Med.* 15, 258, 1984.
31. Yoshida, T. et al., Moxalactam (6059-S), a novel 1-oxa-β-lactam with an expanded antibacterial spectrum: Laboratory evaluation, *Antimicrob. Agents Chemother.* 17, 302, 1980.
32. Larsson, S. et al., Antimicrobial susceptibilities of *Listeria monocytogenes* strains isolation from 1958 to 1982 in Sweden, *Antimicrob. Agents Chemother.* 28, 12, 1985.
33. van Netten, P. et al., A selective and diagnostic medium for use in the enumeration of *Listeria* spp. in foods, *Int. J. Food Microbiol.* 6, 187, 1988.
34. Seeliger, H. P. R. and Jones, D., *Listeria*, in *Bergey's manual of systematic bacteriology* (vol. 2), Sneath, P. H. A., ed. Williams & Wilkins, Baltimore, 1986, p. 1235.
35. Facklam, R. R. and Moody, M. D., Presumptive identification of group D streptococci: The bile-esculin test, *Appl. Microbiol.* 20, 245, 1970.

36. Dominguez Rodriguez, L. D. et al., New methodology for the isolation of *Listeria* microorganisms from heavily contaminated environments, *Appl. Environ. Microbiol.* 47, 1188, 1984.
37. Scotter, S. L. et al., Validation of ISO method 11290 part 1—Detection of *Listeria monocytogenes* in foods, *Int. J. Food Microbiol.* 64, 295, 2001.
38. Anonymous, Microbiology of food and animal feeding stuffs—Horizonal method for the detection and enumeration of *Listeria monocytogenes*—Part 1: Detection method, International Standard ISO 11290-1 ed. International Organization for Standardization, Geneva, 1996.
39. Gorski, L., Flaherty, D., and Mandrell, R. E., Competitive fitness of *Listeria monocytogenes* serotype 1/2a and 4b strains in mixed cultures with and without food in the U.S. Food and Drug Administration enrichment protocol, *Appl. Environ. Microbiol.* 72, 776, 2006.
40. Curtis, G. D. W., Nichols, W. W., and Falla, T. J., Selective agents for *Listeria* can inhibit their growth, *Lett. Appl. Microbiol.* 8, 169, 1989.
41. Seeliger, H. P. R. and Höhne, K., Serotyping of *Listeria monocytogenes* and related species, *Methods Microbiol.* 13, 31, 1979.
42. Seeliger, H. P. R. and Langer, B., Serological analysis of the genus *Listeria.* Its values and limitations, *Int. J. Food Microbiol.* 8, 245, 1989.
43. Kozak, J. et al., Prevalence of *Listeria monocytogenes* in foods: Incidence in dairy products, *Food Cont.* 7, 215, 1996.
44. Jay, J. M., Prevalence of *Listeria* spp. in meat and poultry products, *Food Cont.* 7, 209, 1996.
45. Curiale, M. S. and Lewus, C., Detection of *Listeria monocytogenes* in samples containing *Listeria innocua*, *J. Food Prot.* 57, 1048, 1994.
46. Petran, R.-L. and Swanson, K. M. J., Simultaneous growth of *Listeria monocytogenes* and *Listeria innocua*, *J. Food Prot.* 56, 616, 1993.
47. MacDonald, F. and Swanson, K. M. J., Important differences between the generation times of *Listeria monocytogenes* and *Listeria innocua* in two *Listeria* enrichment broths, *J. Dairy Res.* 61, 433, 1994.
48. Camilli, A., Tilney, L. G., and Portnoy, D. A., Dual roles of *plcA* in *Listeria monocytogenes* pathogenesis, *Mol. Microbiol.* 8, 143, 1993.
49. Leimeister-Wachter, M., Domann, E., and Chakraborty, T., The expression of virulence genes in *Listeria monocytogenes* is thermoregulated, *J. Bacteriol.* 174, 947, 1992.
50. Smith, G. A. et al., The two distinct phospholipases C of *Listeria monocytogenes* have overlapping roles in escape from a vacuole and cell-to-cell spread, *Infect. Immun.* 63, 4231, 1995.
51. Notermans, S. H. W. et al., Phosphatidylinositol-specific phospholipase C activity as a marker to distinguish between pathogenic and nonpathogenic *Listeria* species, *Appl. Environ. Microbiol.* 57, 2666, 1991.
52. Vázquez-Boland, J. A. et al., Purification and characterization of two *Listeria ivanovii* cytolysins, a sphingomyelinase C and a thiol-activiated toxin (ivanolysin O), *Infect. Immun.* 57, 3928, 1989.
53. Anonymous, Microbiology of food and animal feeding stuffs—Horizontal method for the detection and enumeration of *Listeria monocytogenes*—Part 1: Detection method Amendment 1: Modification of the isolation media and the hemolysis test, and inclusion of precision data, International Standard ISO 11290-1 ed. International Organization for Standardization, Geneva, 2004.
54. Restaino, L. et al., Isolation and detection of *Listeria monocytogenes* using fluorogenic and chromogenic substrates for phosphatidylinositol-specific phospholipase C, *J. Food Prot.* 62, 244, 1999.
55. Jinneman, K. C. et al., Evaluation and interlaboratory validation of a selective agar for phosphatidylinositol-specific phospholipase C activity using a chromogenic substrate to detect *Listeria monocytogenes* from foods, *J. Food Prot.* 66, 441, 2003.
56. Vlaemynck, G., Lafarge, V., and Scotter, S., Improvement of the detection of *Listeria monocytogenes* by the application of ALOA, a diagnostic, chromogenic isolation medium, *J. Appl. Bacteriol.* 88, 430, 2000.
57. Becker, B. et al., Comparison of two chromogenic media for the detection of *Listeria monocytogenes* with the plating media recommended by EN/DIN 11290-1, *Int. J. Food Microbiol.* 109, 127, 2006.
58. Greenwood, M. et al., Evaluation of chromogenic media for the detection of *Listeria* species in food, *J. Appl. Microbiol.* 99, 1340, 2005.
59. Johansson, T., Enhanced detection and enumeration of *Listeria monocytogenes* from foodstuffs and food-processing environments, *Int. J. Food Microbiol.* 40, 77, 1998.
60. Kells, J. and Gilmour, A., Incidence of *Listeria monocytogenes* in two milk processing environments, and assessment of *Listeria monocytogenes* blood agar for isolation, *Int. J. Food Microbiol.* 91, 167, 2004.
61. Pinto, M. et al., Comparison of Oxford agar, PALCAM and *Listeria monocytogenes* blood agar for the recovery of *L. monocytogenes* from foods and environmental samples, *Food Cont.* 12, 511, 2001.
62. Gorski, L. (unpublished data), 2007.

63. Gonzalez-Zorn, B. et al., The *smcL* gene of *Listeria ivanovii* encodes a sphingomyelinase C that mediates bacterial escape from the phagocytic vacuole, *Molec. Microbiol.* 33, 510, 1999.
64. Volokov, D. et al., Discovery of natural atypical nonhemolytic *Listeria seeligeri* isolates, *Appl. Environ. Microbiol.* 72, 2439, 2006.
65. Johnson, J. et al., Natural atypical *Listeria innocua* strains with *Listeria monocytogenes* pathogenicity island 1 genes, *Appl. Environ. Microbiol.* 70, 4256, 2004.
66. McKellar, R. C., Use of the CAMP test for identification of *Listeria monocytogenes, Appl. Environ. Microbiol.* 60, 4219, 1994.
67. Groves, R. D. and Welshimer, H. J., Separation of pathogenic from apathogenic *Listeria monocytogenes* by three *in vitro* reactions, *J. Clin. Microbiol.* 5, 559, 1977.
68. Roberts, A. et al., Genetic and phenotypic characterization of *Listeria monocytogenes* lineage III, *Microbiology* 152, 685, 2006.
69. Liu, D. et al., *Listeria monocytogenes* subgroups IIIA, IIIB, and IIIC delineate genetically distinct populations with varied pathogenic potential, *J. Clin. Microbiol.* 44, 4229, 2006.
70. Sado, P. N. et al., Identification of *Listeria monocytogenes* from unpasteurized apple juice using rapid test kits, *J. Food Prot.* 61, 1199, 1998.
71. Anonymous, Isolation and identification of *Listeria monocytogenes* from red meat, poultry, egg, and environmental samples (http://www.fsis.usda.gov/PDF/MLG_8_05.pdf), 2006.
72. Barbour, A. H., Rampling, A., and Hormaeche, C. E., Variation in the infectivity of *Listeria monocytogenes* isolates following intragastric inoculation of mice, *Infect. Immun.* 69, 4657, 2001.
73. Kim, S. H. et al., Oral inoculation of A/J mice for detection of invasiveness differences between *Listeria monocytogenes* epidemic and environmental strains, *Infect. Immun.* 72, 4318, 2004.
74. Norton, D. M. et al., Characterization and pathogenic potential of *Listeria monocytogenes* isolates from the smoked fish industry, *Appl. Environ. Microbiol.* 67, 646, 2001.
75. Nightingale, K. K. et al., Select *Listeria monocytogenes* subtypes commonly found in foods carry distinct nonsense mutations in *inlA*, leading to expression of truncated and secreted internalin A, and are associated with a reduced invasion phenotype for human intestinal epithelial cells, *Appl. Environ. Microbiol.* 71, 8764, 2005.
76. Erdenlig, S., Ainsworth, A. J., and Austin, F. W., Pathogenicity and production of virulence factors by *Listeria monocytogenes* isolates from channel catfish, *J. Food Prot.* 63, 613, 2000.
77. Tappero, J. W. et al., Reduction in the incidence of human listeriosis in the United States, *JAMA* 273, 1118, 1995.
78. Schuchat, A., Swaminathan, B., and Broome, C. V., Epidemiology of human listeriosis, *Clin. Microbiol. Rev.* 4, 169, 1991.
79. Palumbo, J. D. et al., Serotyping of *Listeria monocytogenes* by enzyme-linked immunosorbent assay and identification of mixed-serotype cultures by colony immunoblotting, *J. Clin. Microbiol.* 41, 564, 2003.
80. Uchikawa, K., Sekikawa, I., and Azuma, I., Structural studies on teichoic acids in cell walls of several serotypes of *Listeria monocytogenes, J. Biochem.* 99, 315, 1986.
81. Clark, E. E. et al., Absence of serotype-specific surface antigen and altered teichoic acid glycosylation among epidemic-associated strains of *Listeria monocytogenes, J. Clin. Microbiol.* 38, 3856, 2000.
82. Schönberg, A. et al., Serotyping of 80 strains from the WHO multicenter international typing of *Listeria monocytogenes, Int. J. Food Microbiol.* 32, 279, 1996.
83. Doumith, M. et al., Differentiation of the major *Listeria monocytogenes* serovars by multiplex PCR, *J. Clin. Microbiol.* 42, 3819, 2004.
84. Borucki, M. K. and Call, D. R., *Listeria monocytogenes* serotype identification by PCR, *J. Clin. Microbiol.* 41, 5537, 2003.
85. Nadon, C. A. et al., Correlations between molecular subtyping and serotyping of *Listeria monocytogenes, J. Clin. Microbiol.* 39, 2704, 2001.
86. Besse, N. G. et al., Evolution of *Listeria* populations in food samples undergoing enrichment culturing, *Intl. J. Food Microbiol.*, 104, 123, 2005.
87. Yokoyama, E., Maruyama, S., and Katsube, Y., Production of bacteriocin-like-substance by *Listeria innocua* against *Listeria monocytogenes, Int. J. Food Microbiol.* 40, 133, 1998.
88. Cornu, M., Kalmokoff, M., and Flandrois, J.-P., Modeling the competitive growth of *Listeria monocytogenes* and *Listeria innocua* in enrichment broths, *Int. J. Food Microbiol.* 73, 261, 2002.
89. Kalmokoff, M. L. et al., Identification of a new plasmid-encoded *sec*-dependent bacteriocin produced by *Listeria innocua* 743, *Appl. Environ. Microbiol.* 67, 4041, 2001.

90. Winkowski, K., Crandall, A. D., and Montville, T. J., Inhibition of *Listeria monocytogenes* by *Lactobacillus bavaricus* MN in beef systems at refrigeration temperatures, *Appl. Environ. Microbiol.* 59, 2552, 1993.
91. Bennick, M. H. J. et al., Biopreservation in modified atmosphere stored mungbean sprouts: The use of vegetable-associated bacteriocinigenic lactic acid bacteria to control the growth of *Listeria monocytogenes*, *Lett. Appl. Microbiol.* 28, 226, 1999.
92. Bruhn, J. B., Vogel, B. F., and Gram, L., Bias in the *Listeria monocytogenes* enrichment procedure: Lineage 2 strains outcompete lineage 1 strains in University of Vermont selective enrichments, *Appl. Env. Microbiol.* 71, 961, 2005.
93. Beuchat, L. R. et al., Growth and thermal inactivation of *Listeria monocytogenes* in cabbage and cabbage juice, *Can. J. Microbiol.* 32, 791, 1986.
94. Golden, D. A., Beuchat, L. R., and Brackett, R. E., Inactivation and injury of *Listeria monocytogenes* as affected by heating and freezing, *Food Microbiol.* 5, 17, 1988.
95. Golden, D. A., Beuchat, L. R., and Brackett, R. E., Evaluation of selective direct plating media for their suitability to recover uninjured, heat-injured, and freeze-injured *Listeria monocytogenes* from foods, *Appl. Environ. Microbiol.* 54, 1451, 1988.
96. Ngutter, C. and Donnelly, C., Nitrite-induced injury of *Listeria monocytogenes* and the effect of selective versus nonselective recovery procedures on its isolation from frankfurters, *J. Food Prot.* 66, 2252, 2003.
97. Busch, S. V. and Donnelly, C. W., Development of a repair-enrichment broth for resuscitation of heat-injured *Listeria monocytogenes* and *Listeria innocua*, *Appl. Environ. Microbiol.* 58, 14, 1992.
98. Osborne, C. M. and Bremer, P. J., Development of a technique to quantify the effectiveness of enrichment regimes in recovering “stressed” *Listeria* cells, *J. Food Prot.* 65, 1122, 2002.
99. Silk, T. M., Roth, T. M. T., and Donnelly, C. W., Comparison of growth kinetics for healthy and heat-injured *Listeria monocytogenes* in eight enrichment broths, *J. Food Prot.* 65, 1333, 2002.
100. Buchanan, R. L. et al., *Listeria* methods development research at the Eastern Regional Research Center, U.S. Department of Agriculture, *J. Assoc. Anal. Chem.* 71, 651, 1988.
101. Varabioff, Y., Incidence and recovery of *Listeria* from chicken with a pre-enrichment technique, *J. Food Prot.* 53, 555, 1990.
102. Ferron, P. and Michard, J., Distribution of *Listeria* spp. in confectioners’ pastries from western France: Comparison of enrichment methods, *Int. J. Food Microbiol.* 18, 289, 1993.
103. McClain, D. and Lee, W. H., Development of USDA-FSIS method for isolation of *Listeria monocytogenes* from raw meat and poultry, *J. Assoc. Anal. Chem.* 71, 660, 1988.
104. Vlaemynck, G. M. and Moermans, R., Comparison of EB and Fraser enrichment broths for the detection of *Listeria* spp. and *Listeria monocytogenes* in raw-milk dairy products and environmental samples, *J. Food Prot.* 59, 1172, 1996.
105. Pritchard, T. J. and Donnelly, C. W., Combined secondary enrichment of primary enrichment broths increases *Listeria* detection, *J. Food Prot.* 62, 532, 1999.
106. Duffy, G. et al., Comparison of selective and nonelective enrichment media in the detection of *Listeria monocytogenes* from meat containing *Listeria innocua*, *J. Appl. Microbiol.* 90, 994, 2001.
107. Kathariou, S., *Listeria monocytogenes* virulence and pathogenicity, a food safety perspective, *J. Food Prot.* 65, 1811, 2002.
108. Loncarevic, S., Tham, W., and Danielsson-Tham, M.-L., The clones of *Listeria monocytogenes* detected in food depend on the method used, *Lett. Appl. Microbiol.* 22, 381, 1996.
109. Buncic, S. et al., Can food-related environmental factors induce different behavior in two key serovars, 4b and 1/2a, of *Listeria monocytogenes*? *Int. J. Food Microbiol.* 65, 201, 2001.
110. Lianaou, A. et al., Growth and stress resistance variation in culture broth among *Listeria monocytogenes* strains of various serotypes and origins, *J. Food Prot.* 69, 2640, 2006.
111. Nelson, K. E. et al., Whole genome comparisons of serotype 4b and 1/2a strains of the food-borne pathogen *Listeria monocytogenes* reveal new insights into the core genome components of this species, *Nuc. Acids Res.* 32, 2386, 2004.
112. Capita, R. et al., Occurrence of *Listeria* species in retail poultry meat and comparison of a cultural/immunoassay for their detection, *Int. J. Food Microbiol.* 65, 75, 2001.
113. Waak, E., Tham, W., and Danielsson-Tham, M.-L., Comparison of the ISO and IDF methods for detection of *Listeria monocytogenes* in blue veined cheese, *Int. Dairy J.* 9, 149, 1999.
114. Gunasinghe, C. P. G. L., Henderson, C., and Rutter, M. A., Comparative study of two plating media (PALCAM and Oxford) for detection of *Listeria* species in a range of meat products following a variety of enrichment procedures, *Lett. Appl. Microbiol.* 18, 156, 1994.

115. Yan, Z., Gurtler, J. B., and Kornacki, J. L., A solid agar overlay method for recovery of heat-injured *Listeria monocytogenes*, *J. Food Prot.* 69, 428, 2006.
116. Williams, R. C. and Golden, D. A., Influence of modified atmospheric storage, lactic acid, and NaCl on survival of sublethally heat-injured *Listeria monocytogenes*, *Int. J. Food Microbiol.* 64, 379, 2001.
117. Kornacki, J. L. et al., Evaluation of several modifications of an ecometric technique for assessment of media performance, *J. Food Prot.* 66, 1727, 2003.
118. Gorski, L., Palumbo, J. D., and Mandrell, R. E., Attachment of *Listeria monocytogenes* to radish tissue is dependent upon temperature and flagellar motility, *Appl. Environ. Microbiol.* 69, 258, 2003.
119. Sheridan, J. J. et al., The use of selective and nonselective enrichment broths for the isolation of *Listeria* species from meat, *Food Microbiol.* 11, 439, 1994.
120. Leclercq, A., Atypical colonial morphology and low recoveries of *Listeria monocytogenes* strains on Oxford, PALCAM, Rapid'L. mono and ALOA solid media, *J. Microbiol. Meth.* 57, 251, 2004.
121. Gracieux, P. et al., Hypovirulent *Listeria monocytogenes* strains are less frequently recovered than virulent strains on PALCAM and Rapid'*L. mono* media, *Int. J. Food Microbiol.* 83, 133, 2003.
122. Ryser, E. T. et al., Recovery of different *Listeria* ribotypes from naturally contaminated, raw refrigerated meat and poultry products with two primary enrichment media, *Appl. Environ. Microbiol.* 62, 1781, 1996.
123. Duarte, G. et al., Efficiency of four secondary enrichment protocols in differentiation and isolation of *Listeria* spp. and *Listeria monocytogenes* from smoked fish processing chains, *Int. J. Food Microbiol.* 52, 163, 1999.
124. Flanders, K. J., Pritchard, T. J., and Donnelly, C. W., Enhanced recovery of *Listeria* from dairy-plant processing environments through combined use of repair enrichment and selective enrichment/detection procedures, *J. Food Prot.* 58, 404, 1995.
125. Bauwens, L., Vercammen, F., and Hertsens, A., Detection of pathogenic *Listeria* spp. in zoo animal feces: Use of immunomagnetic separation and a chromogenic isolation medium, *Vet. Microbiol.* 91, 115, 2003.
126. Nannapeneni, R. et al., Reactivities of genus-specific monoclonal antibody EM-6E11 against *Listeria* species and serotypes of *Listeria monocytogenes* grown in nonselective and selective enrichment broth media, *J. Food Prot.* 61, 1195, 1998.
127. Geng, T., Hahm, B.-K., and Bhunia, A. K., Selective enrichment media affect the antibody-based detection of stress-exposed *Listeria monocytogenes* due to differential expression of antibody-reactive antigens identified by protein sequencing, *J. Food Prot.* 69, 1879, 2006.
128. Kämpfer, P., Differentiation of *Corynebacterium* spp., *Listeria* spp., and related organisms using fluorogenic substrates, *J. Clin. Microbiol.* 30, 1067, 1992.
129. Clark, A. G. and McLauchlin, J., Simple color tests based on an alanyl peptidase reaction which differentiate *Listeria monocytogenes* from other *Listeria* species, *J. Clin. Microbiol.* 35, 2155, 1997.
130. Carroll, S. A. et al., A colony lift immunoassay for the specific identification and quantification of *Listeria monocytogenes*, *J. Microbiol. Methods* 41, 145, 2000.
131. Carroll, S. A. et al., Development and evaluation of a 24-hour method for the detection and quantification of *Listeria monocytogenes* in meat products, *J. Food Prot.* 63, 347, 2000.
132. Entis, P. and Lerner, I., Twenty-four-hour direct presumptive enumeration of *Listeria monocytogenes* in food and environmental samples using the ISO-GRID method with LM-137 agar, *J. Food Prot.* 63, 354, 2000.
133. Fugett, E. et al., International Life Sciences Institute of North America *Listeria monocytogenes* strain collection: Development of standard *Listeria monocytogenes* strain sets for research and validation studies, *J. Food Prot.* 69, 2929, 2006.

6 Genotypic Identification

Dongyou Liu, Mark L. Lawrence, A. Jerald Ainsworth, and Frank W. Austin

CONTENTS

6.1 INTRODUCTION

The genus *Listeria* comprises six closely related, Gram-positive bacterial species (i.e., *L. monocytogenes, L. ivanovii, L. seeligeri, L. innocua, L. welshimeri,* and *L. grayi*), which share many similar morphological, biochemical, and molecular characteristics.[1–4] *Listeria* species are ubiquitously distributed in the environment and frequently found in various manufactured food products; *L. monocytogenes* strains are remarkably tolerant to a broad spectrum of external stresses (including wide pH, osmolarity, and temperature ranges)[5] and are capable of causing especially severe illness to vulnerable human population groups (e.g., infants, pregnant women, the elderly, and immunocompromised individuals),[4] and listeriosis often produces nonspecific initial clinical manifestations (e.g., flu-like symptoms and gastroenteritis) with potentially fatal consequences (e.g., meningitis, encephalitis, septicemia, and occasional death).[3] Thus, it is imperative that specific, sensitive, and speedy laboratory procedures are available for prompt detection and accurate identification of *L. monocytogenes* from nonpathogenic *Listeria* and other contaminating bacteria. This is essential to the timely implementation of antibiotic therapy, thus preventing an otherwise easily treated malaise from developing into a life-threatening disease.

The early generation laboratory diagnostic tests for *L. monocytogenes* are largely phenotype-based and usually assess *L. monocytogenes* gene products by biological, biochemical, and serological means (see chapter 5). As the phenotypic properties of *Listeria* bacteria may vary with the constantly changing external conditions, growth phase, and spontaneous genetic alterations, use of the phenotype-based diagnostic procedures may lead to equivocal results at times. In addition, since many phenotypic tests are dependent on lengthy *in vitro* culture procedures, they are notoriously time consuming, thus delaying result availability. To circumvent the drawbacks of the phenotype-based procedures, new generation genotype-based methods targeting the nucleic acids (DNA or RNA) of *Listeria* bacteria have been developed over the preceding decade. As nucleic acids (particularly DNA) are intrinsically more stable than proteins and less prone to influences by outside factors, the genotypic (or molecular) diagnostic procedures are much more precise and less variable than the phenotype-based methods. Furthermore, as many of the genotypic techniques involve template and/or signal amplification, they are also much faster and more sensitive. Application of these state-of-the-art genotypic identification procedures has the potential to overcome the problems associated with the phenotype-based tests, providing unprecedented levels of sensitivity, specificity, and speed for laboratory detection and identification of *Listeria* bacteria.

Not surprisingly, a plethora of genotype-based tests that direct at a variety of genetic elements (including nonspecific genomic targets, shared, or specific gene targets) have been designed and described.[6,7] The testing formats have also evolved from the initial relatively unsophisticated non-amplified DNA hybridization procedures to the recent all-encompassing nucleic acid amplification technologies (e.g., polymerase chain reaction [PCR]) coupled with a real-time detection capacity. The impact of these advancements can be clearly felt by the widespread acceptance and adoption of numerous nucleic acid amplification and detection methods in both research and clinical laboratories throughout the world for enhanced identification and differentiation of *Listeria* bacteria. Nevertheless, due to the fact that many molecular assays rely on enzymatic reaction (e.g., use of DNA polymerase) for template amplification (which is prone to interference from inhibitory substances present in clinical, food, and environmental specimens), sensitive, reliable, and consistent genetic detection and identification of *Listeria* bacteria have been achieved only with purified nucleic acids or cultured isolates as starting materials—notwithstanding the enormous efforts that have been directed so far toward the development of innovative and efficient specimen handling procedures to reduce and eliminate inhibitory elements in clinical, food, and environmental specimens.

By first examining the genetic targets on which various genotypic methods for *Listeria* diagnosis are based, this chapter reviews the types of genotype-based methods that have been developed and refined for identification of *L. monocytogenes* from non*monocytogenes Listeria* species. This is followed by a description of the practical challenges and potential solutions in applying genotypic tests for detection and identification of *Listeria* bacteria from various starting materials (purified nucleic acids, cultured isolates, and clinical and food specimens) that are encountered in routine situations. The chapter ends with a discussion on the future research requirements in our endeavors to further enhance the genotypic detection and identification of *L. monocytogenes*, thus facilitating the improved control and prevention of human listeriosis.

6.2 DETECTION TARGETS

6.2.1 NONSPECIFIC GENOMIC TARGETS

6.2.1.1 Genomic Composition and Sequence

Like many other biological organisms, *Listeria* genomes are composed of double-stranded DNA sequences, which are in turn built up with four basic types of nucleotides: guanine (G), cytosine (C), adenine (A), and thymine (T). The proportion of guanine and cytosine (or G+C content, usually expressed as a percentage) ranges from 20 to 80% in bacterial genomes and tends to remain constant among closely related species and not significantly affected by chromosomal dearrangements, gene loss, or gene duplication. For instance, the average G+C content in *Listeria* genomes at about 37% is clearly distinct from that in *Escherichia coli* at about 50%. Thus, the G+C content (%) can be exploited for determination of the taxonomical status (or species identity) of *Listeria* isolates and for differentiation of *Listeria* species.

With guanine–cytosine pairing in the double-stranded DNA being linked by three hydrogen bonds and adenine–thymine pairing connected with two such bonds, the G-C pairs are notably stronger and more resistant to denaturation by high temperature than the A-T pairs. While double-stranded DNA sequences will dissociate (denature) to produce two complementary single-stranded DNA sequences under certain physical conditions, such as high temperature or high pH, those having a higher GC content tend to be more resistant to heat denaturation than those having a lower GC content. The two complementary single-stranded DNA sequences will reassociate (renature) under a lower temperature and a high salt concentration to form DNA/DNA heteroduplexes if they share a sequence homology of at least 80%. The remaining nucleotide divergence (up to 20%) may be spread out between 0 and 100% upon reassociation.

Thus, a bacterial species can be defined as an entity in which members have a DNA–DNA homology value of 70% or more at optimal conditions (e.g., 60°C) and 55% or more at less than optimal conditions (e.g., 75°C) and whose DNA contains 6% or less divergence in related nucleotide sequences, in addition to a certain degree of phenotypic consistency. The natural tendency of DNA to denature and reassociate under defined conditions forms the basis of DNA–DNA hybridization analysis, with which the overall ability of two genomic sequences to form heteroduplexes is measured. By assessing the degree of genetic relatedness between genomic DNA sequences, the genetic distance between two prokaryotic species can be determined, and species with similarity values can be arranged in a phylogenetic tree, thus verifying the species identity of the bacteria concerned.[8,9]

6.2.1.2 Random Primer Sites

Bacterial genomes contain multiple short nucleotide repeats that may bind the 3′ ends of random primers, enabling subsequent PCR amplification at a relatively low temperature (e.g., 36°C). Given that a putative three-nucleotide sequence is generally present once in a 64-nucleotide sequence (4^3 permutations) in bacterial genomes, it is highly likely that two such sequences may occur within a certain distance of each other and in opposing orientation. A random primer that shows no complete homology to a genome may have a perfect match of two to three nucleotides from the 3′ end of the primer to the template strand to allow annealing and priming complementary strand synthesis by DNA polymerase under a reduced temperature.[10–12] The existence of these random trinucleotide sites and their recognition by random primers at less stringent annealing temperature are the likely mechanisms behind random amplified polymorphism of DNA (RAPD) or arbitrarily primed PCR (AR-PCR) that has been proven useful for species-specific identification of *Listeria* bacteria and subtyping of *L. monocytogenes* strains.[13–19]

6.2.1.3 Repetitive Elements

*R*epetitive *e*xtragenic *p*alindromes (REP) are randomly dispersed, repetitive sequence elements of 35–40 bp with an inverted repeat; *e*nterobacterial *r*epetitive *i*ntergenic *c*onsensus sequences (ERIC) are intergenic repeat units of 124–147 bp with a highly conserved central inverted repeat. Both REP and ERIC sequences exist in a wide range of bacterial genomes including *Listeria*, which represent potential primer binding sites for PCR amplification to achieve species and strain discrimination.[20,21] For instance, use of ERIC primers (ERIC1R and ERIC2) in ERIC-PCR resulted in the formation of specific band profiles that facilitated the identification of *Listeria* species. Combining REP-PCR with ERIC-PCR, it was feasible to distinguish *L. monocytogenes* serotypes 1/2a, 1/2b, 1/2c, 3b, and 4b. Moreover, within the serotype 1/2a, REP-PCR appeared more discriminative than ERIC-PCR, and had a discriminative potential that is comparable to RAPD using three to four random primers.[20] Additional support on the usefulness of REP-PCR for *L. monocytogenes* subtyping was garnered by several more recent studies.[22,23]

*V*ariable-*n*umber *t*andem *r*epeats (VNTRs), also known as *s*imple *s*equence *r*epeats (SSRs) or microsatellites, are a class of short DNA sequence motifs that are tandemly repeated at certain specific locations. These sequences are abundant throughout most bacterial genomes, and play vital roles in both transcriptional and translational control of gene expression. A subgroup of VNTRs is termed the mononucleotide repeats (MNRs) and are useful for bacterial typing through multiple locus VNTR analysis (or MLVA).[24] *L. monocytogenes* EGD-e genome possesses tens of thousands of perfect SSR tracts, with 94% being MNRs. Most of the MNR tracts in *L. monocytogenes* EGD-e genome are short and do not exceed 9 bp. The variation in MNR tracts enables PCR amplification of distinct products that are useful for separation between *L. monocytogenes* virulent and nonvirulent strain groups and further discrimination among *L. monocytogenes* isolates.[25]

6.2.1.4 Restriction Enzyme Sites

Restriction endonucleases are enzymes that recognize and cleave precise sequences of nucleotides (usually of four to eight nucleotides in length) in double-stranded DNA. A single nucleotide alteration in a recognition site often renders it unrecognizable by the restriction enzyme at that point, leading to changed DNA band patterns after separation in agarose gel electrophoresis and/or detection in Southern blot. This feature has been exploited in several genotype-based identification and subtyping techniques for *L. monocytogenes* strains, including restriction fragment length polymorphism (RFLP) analysis and ribotyping.[26–30] The main advantages of using restriction enzyme analysis for *L. monocytogenes* identification and subtyping are attributable to its relatively high reproducibility and ease of performance. However, a perceived drawback of the restriction enzyme analysis is that it tends to display very complex band patterns with genomic DNA. Fortunately, this deficiency can be resolved by probing the Southern-transferred DNA fragments with labeled bacterial ribosomal operons encoding for 16S and/or 23S ribosomal RNA (rRNA) (i.e., ribotyping).

6.2.2 Shared Gene Targets

6.2.2.1 Ribosomal RNA Genes

Ribosome is an essential cellular organelle that is responsible for protein synthesis in all living organisms. As the key components of the ribosome, rRNA molecules possess two complex folded structures of differing sizes, whose main functions are to provide a mechanism for decoding messenger RNA (mRNA) into amino acids (at the center of a small ribosomal subunit) and to interact

FIGURE 6.1 Organization of prokaryotic ribosomal RNA genes. The shaded areas within 16S and 23S rRNA genes indicate regions of conservation, and the clear areas indicate regions of relative variability. IGS = intergenic sequence.

with the transfer RNA (tRNA) during translation by providing petidyltransferase activity (large subunit). Being a fragment with a sedimentation coefficient of 16S, the prokaryotic small-subunit rRNA molecule is commonly termed as 16S rRNA, whereas the large-subunit rRNA is made up of two fragments (i.e., 23S rRNA and 5S rRNA) with sedimentation coefficiencies of 23S and 5S. Although one or more copies of the rRNA operon may be dispersed in the prokaryotic genomes, it appears that *Listeria* species harbor six such copies.[2,3] Because of its critical role in cellular function and maintenance, rRNA is not only the most conserved (i.e., the least variable) gene in all cells, but also the most abundant (with each living cell containing 10^4–10^5 copies of the 5S, 16S, and 23S rRNA molecules). While the precise rate of change in the rRNA gene sequences is unknown, it does help demarcate evolutionary distance and relatedness of biological organisms. For this reason, the rRNAs (and their genomic coding sequences' rDNAs) are often subjected to sequencing analysis to identify and confirm an organism's taxonomic status and species identity, and to estimate rates of species divergence.

Among the three prokaryotic rRNA molecules, 16S rRNA (or 16S rDNA) has been most frequently applied for bacterial characterization. With an approximate length of 1550 nucleotides, the 16S rRNA molecule contains alternating regions of sequence conservation and heterogeneity (Figure 6.1). The conserved regions are useful as targets for phylogenetic determination of higher taxonomic orders (e.g., phylum, family, and genus) of most bacteria, while the regions of sequence diversity are valuable for characterizing isolates to the genus or species level (with detection of sequence identity over 97% in the 16S rRNA gene being regarded as the identical species).[28,31–40] Due to its relative straightforwardness, comparison of the bacterial 16S rRNA gene sequence has emerged as a preferred genetic technique for taxonomic investigations. 16S rRNA gene sequence analysis is especially invaluable for identification of poorly described or phenotypically aberrant strains,[41] and this is highlighted by a recent report on *L. monocytogenes* identification using amplified rDNA restriction analysis (ARDRA).[28]

On the other hand, having nearly double the size of 16S rRNA, the 23S rRNA molecule (ca. 2900 nucleotides) includes a larger number of phytogenetically meaningful sites, and thus offers a supplemental target for achieving a higher degree of differentiation (Figure 6.1).[30,35,42,43] Further, as the rRNA intergenic spacer (IGS) regions are not as essential as the actual rRNA genes, they tend to experience more genetic drift and consequently are not as highly conserved. This makes the rRNA IGS an additional tool for bacterial determination. Indeed, sequence analysis of the IGS regions of 16S–23S rRNA operons has been extremely rewarding in respect to the genotypic identification of bacteria species including *Listeria*.[44–46] Interestingly, *Listeria* 16S–23S IGS regions are of two different sizes: a smaller one of about 340 bp and a larger one of 550–590 bp. The small rRNA IGS regions *of L. innocua, L. ivanovii, L. seeligeri, L. welshimeri*, and *L. grayi* show 83–99% homology to that of *L. monocytogenes*; the large rRNA IGS region of *L. monocytogenes* demonstrates a 81–96% similarity to those of non*monocytogenes Listeria* species.[46]

6.2.2.2 Other Shared Genes

Apart from rRNA molecules, other genetic elements such as housekeeping genes (*groESL, rpoB, recA, gyrB*, and *prs*) can be utilized to characterize isolates to the genus or species level (Table 6.1).[47–52] As nucleotide sequence conservation among housekeeping genes is invariably lower than that detected in rRNA, such heterogeneity often makes them ideal for determining taxonomic relationships among

TABLE 6.1
***Listeria* Shared Gene Targets**

Specificity	Target Gene	Protein	Ref.
Listeria genus and *species*	16S rRNA gene	Unknown	28, 32, 33, 35–37, 39
Listeria genus and *species*	23S rRNA gene	Petidyltransferase	30, 35, 42, 43
Listeria genus and *species*	16S/23S rRNA intergenic regions	Unknown	45, 46, 59
Listeria genus	*groESL*	Heat shocking protein	47
Listeria genus	*rpoB*	Housekeeping gene	52
Listeria genus	*recA*	Housekeeping gene	48, 51
Listeria genus	*gyrB*	Housekeeping gene	50, 51
Listeria genus	*prs*	Putative phosphoribosyl pyrophosphate synthase	49, 50
Listeria genus and *species*	*iap (or p60)*	Invasion associated protein	53, 54, 56, 57, 60, 61
Listeria genus and *species*	*flaA*	Flagellin A	55
L. monocytogenes and *L. innocua*	*fbp*	Fibronectin-binding protein	58

closely related bacteria. Further, genes encoding for invasion-associated protein (*iap*), flagellin A (*flaA*), and fibronectin-binding protein (*fbp*) are also useful for genotypic identification of *Listeria* species (Table 6.1).[53–58] For example, rapid distinction of *L. monocytogenes* from other *Listeria* species was attainable with a multiplex PCR that detected nucleotide differences at the *iap* gene.[54]

6.2.3 Specific Gene Targets

6.2.3.1 Species-Specific Genes

Besides the aforementioned nonspecific genomic sequences and shared genes, *Listeria* genomes also harbor many unique genes that have enormous diagnostic appeal. In contrast to assays targeting a shared gene, where extreme care is required to ascertain the exact sizes of the amplified products with minute differences,[53,54,58] detection of *Listeria* genes unique to individual species provides an independent and highly definitive means of confirming the species identities.[62–66] In cases where the co-presence of several *Listeria* species complicates the identification of *L. monocytogenes*, the availability of PCR assays for uniquely present, species-specific genes is desirable to help clarify the issue. Toward this end, a large number of species-specific genes have been identified following the recent completion of whole genome sequences of several *L. monocytogenes* and one *L. innocua* strains.[2,3] As a consequence, many molecular assays targeting these species-specific genes (e.g., *inlA*, *inlB*, delayed-type hypersensitivity gene, aminopeptide gene, and putative transcriptional regulator gene *lmo0733*) have been devised for improved differentiation and independent confirmation of *L. monocytogenes* from other *Listeria* species and common bacteria (Table 6.2).

6.2.3.2 Group-Specific Genes

The publication of *L. monocytogenes* genome sequences[2,3] has also permitted the selection of various group- and virulence-specific genes for rapid identification and virulence determination of *L. monocytogenes*[49,86–91] (Table 6.3). The benefit of utilizing the group- or virulence-specific primers and probes (e.g., *lmo0733, lmo2821, lmo1134,* and *lmo2672*) in the molecular discrimination of lineage III from other lineages has been convincingly shown, as *L. monocytogenes* serotypes 4a and 4c strains belonging to lineage III reacted differently with these primers and probes in PCR and Southern blot in comparison with other serotypes belonging to lineages I and II.[92,93] Furthermore, the group- and virulence-specific primers and probes have been instrumental in helping validate

TABLE 6.2
***Listeria* Species-Specific Gene Targets**

Specificity	Target Gene	Protein	Ref.
L. monocytogenes	*hly*	Listeriolysin (LLO)	61, 67–73
	plcA	Phosphatidylinositol-phospholipase C (PI-PLC)	29, 74
	plcB	Phosphatidylcholine-phospholipase C (PC-PLC)	29, 74
	actA	ActA	29, 75
	mpl	Metalloprotease	76, 77
	prfA	Transcriptional regulator PrfA	78
	inlA	Internalin A	79
	inlB	Internalin B	80
	clpE	Clp ATPase	74
	lmaA/dth18	LmaA antigen/delayed-type hypersensitivity protein	81
	lmo0733	Putative transcriptional regulator	82
	lmo2234	Unknown protein	83
	pepC	Aminopeptidase C	83
L. ivanovii	*liv22-228*	Putative N-acetylmuramidase	64
L. seeligeri	*lse24-315*	Putative internalin	65
L. innocua	*lin0464*	Putative transcriptional regulator	62
	lin2483	Putative transporter	84
L. welshimeri	*lwe7-571*	Putative phosphotransferase system enzyme IIBC	63
L. grayi	*lgr20-246*	Putative oxidoreductase	66

TABLE 6.3
***L. monocytogenes* Group-Specific Gene Targets**

Specificity	Target Gene	Protein	Ref.
All *L. monocytogenes* serotypes but 4a	*lmo2821*	Internalin J	86
All *L. monocytogenes* serotypes but 4a and some 4c	*lmo2470*	Putative internalin	86
All *L. monocytogenes* serotypes but 4a and some 4c	*inlC*	Internalin C	86a
All *L. monocytogenes* serotypes but 4a and some 4c	*lmo2672*	Putative transcriptional regulator	86
All *L. monocytogenes* serotypes but 4a and 4c	*lmo1134*	Putative transcriptional regulator	86
L. monocytogenes serotypes 1/2b, 3b, 4b, 4d, 4e, and 7	ORF2819	Putative transcriptional regulator	49
L. monocytogenes serotypes 4b, 4d, and 4e	ORF2110	Putative secreted protein	49
L. monocytogenes serotypes 4b, 4d, and 4e	ORF2372	Similar to techoic acid protein precursor	91
L. monocytogenes serotypes 1/2a, 1/2c, 3a, and 3c	*lmo0737*	Unknown	49
L. monocytogenes serotypes 1/2a, 1/2c, 3a, and 3c	*lmo0171*	Cell surface protein	91
L. monocytogenes serotypes 1/2c and 3c	*lmo1118*	Unknown	49
L. monocytogenes serotypes 1/2b and 3b	Gene region flanking gltA-gltB	Unknown	87
L. monocytogenes serotypes 1/2a and 3a	*flaA*	Flagellin A	87

L. monocytogenes subgroups IIIA, IIIB, and IIIC as genetically distinct populations. Consequently, it becomes apparent that subgroup IIIA is made up of typical rhamnose-positive avirulent serotype 4a and virulent serotype 4c strains, subgroup IIIC of atypical rhamnose-negative virulent serotype 4c strains, and subgroup IIIB of atypical rhamnose-negative virulent non-4a and non-4c strains—some of which may be related to serotype 7. There is a distinct possibility that subgroup IIIB (including serotype 7) may represent a novel subspecies within *L. monocytogenes*.[93]

6.3 DETECTION METHODS

6.3.1 Nonamplified Methods

6.3.1.1 Determination of G+C Content

A conventional method to determine the G+C content of bacterial genomes is through measurement of thermal denaturation (melting) temperature (T_m) of the DNA double helix with a spectrophotometer. The absorbance of the DNA at a wavelength of 260 nm increases abruptly when the double-stranded DNA converts into single strands upon heat treatment. Alternatively, the G+C content can be ascertained rapidly and effectively by monitoring fluorescent SYBR green intensity during DNA denaturation in a capillary tube. As SYBR green preferentially binds to double-stranded DNA, heat denaturation of DNA bound to this dye will lead to a noticeable decrease in fluorescent signal, which can be conveniently documented with a light cycler.[94] In spite of the tedious nature associated with acquiring a reliable reading, the mol% G+C content is considered as a highly desirable component in the description and establishment of a new taxon.[9]

6.3.1.2 DNA–DNA Hybridization

In the original DNA–DNA hybridization protocol, genomic DNA from a reference strain is sheared by sonication and then mixed with radioactively labeled, sheared genomic DNA from a test strain. The mixture is heated at 100°C for 3–4 min and maintained at 60 or 75°C for 16 h. After eliminating the unbound, single-stranded DNA via a hydroxyapatite column, the bound double-stranded DNA is eluted, from which reassociation value is determined by counting the remaining radioactivity.[95] In the subsequent protocols, enzyme, biotin, and SYBR green dye have been used in place of radioactive isotope. Indeed, taking advantage of SYBR green's affinity for double-stranded DNA, an easy and rapid fluorimetric method has been developed for estimation of DNA–DNA relatedness between microbial strains via a quantitative, real-time PCR.[96] With the DNA–DNA relatedness at species level being set at 70%, DNA–DNA hybridization has proven its value in clarifying the taxonomic status of *Listeria* species.[31,97] However, due to the fact that the DNA–DNA hybridization technique remains technically demanding and that DNA reassociation values do not necessarily represent actual sequence identity or gene content differences, this method is not suited for routine diagnostic applications.

6.3.1.3 Gene Probe Hybridization

Gene probe is a short stretch of specific single-stranded nucleic acid that is labeled with enzyme, fluorescence, biotin, or radioactive isotope and applied for detection of a complementary nucleic acid target sequence that is bound to a solid surface. To conduct the gene probe hybridization, *Listeria* DNA is first isolated from individual strains and spotted on a supporting matrix (e.g., nitrocellulose filter, nylon membrane, and microtiter plate). It is then hybridized with a labeled *Listeria*-specific gene probe (which can be a genomic fragment or synthesized from 16S and 23S rRNA or other genetic elements) and subsequently detected with an appropriate substrate or autoradiography. As this procedure assesses the differences of *Listeria* species at the genetic level, it is unquestionably more specific than the phenotype-based methods.[60,71,98]

Making full use of this feature, several commercial gene probe assays for *L. monocytogenes* have been marketed, including Accuprobe (Gen-Probe Inc., San Diego, California), which focuses on virulence gene mRNA for detection of viable cells; GeneTrak DNA hybridization kit (Neogen Corporation, Lansing, Michigan), which utilizes horseradish peroxidase-labeled probe for colorimetric detection; and the vermicon identification technology (VIT®, Munich, Germany), which detects intracellular RNA target with fluorescently labeled oligonucleotide probe.[99] However, because gene probe hybridization does not involve nucleic acid amplification, it often displays limited sensitivity (requiring at least 10^4 copies of target gene per microliter for reliable detection without signal amplification, and improving to as low as 500 copies of target gene per microliter with signal amplification), which is generally inadequate for direct detection of *L. monocytogenes* from most clinical/food samples. The advent of nucleic acid amplification technology has thus drastically reduced the attraction of gene probe hybridization systems in the routine diagnosis of listeriosis.

6.3.1.4 Southern Blot

As a variant of gene probe hybridization, Southern blot identifies and locates agarose gel-separated DNA sequences that are complementary to a gene probe. In this procedure, genomic DNA is first digested with restriction enzymes into fragments of various sizes and then separated via agarose gel electrophoresis. The agarose gel containing separated DNA fragments is treated with alkali (e.g., NaOH) to convert the double-stranded DNA into single strands, facilitating DNA binding to the membrane and subsequent hybridization with gene probe. The denatured DNA is transferred (or blotted) from the gel onto the membrane by capillary action, and the membrane containing DNA fragments is then baked (in the case of nitrocellulose) or exposed to ultraviolet radiation (nylon) to fix (or cross-link) the DNA to the membrane. The membrane is then hybridized with a DNA or RNA probe, which is tagged with a fluorescent or chromogenic dye, or a radioactive isotope. After washing away excess probe, the hybridization pattern is visualized by colorimetric development on the membrane for a chromogenic probe, or by autoradiography on x-ray film for a radioactive or fluorescent probe. Southern blot has been used in a variety of studies for differentiation of *Listeria* species, for clarification of *L. monocytogenes* serotype 4b strains belonging to lineages I and III, and also for confirmation of *L. monocytogenes* subgroups IIIA, IIIB, and IIIC in recent studies.[92,93]

6.3.1.5 Fluorescence In Situ Hybridization (FISH)

FISH is another variant of gene probe hybridization that uses a fluorescent molecule to label a nucleic acid probe for identification of a specific complementary sequence in a chromosome spread so that the site becomes visible under a microscope. FISH is useful for highlighting the locations of genes, subchromosome regions, entire chromosomes, or specific DNA sequences, as well as for elucidating the phylogenetic status of bacterial isolates. To undertake FISH analysis, *L. monocytogenes* cells are fixed onto a slide, membrane filter, or microtiter well with ethanol. The fixed cells are then permeabilized with proteinase K so that fluorescent probes (labeled with Cy3, Cy5, or FLUOS) can go across the cell membrane. After incubation with the labeled probes that bind to complementary sequences, the labeled cells are visualized under a fluorescence microscope. Using labeled primers specific for the 16S rRNA genes, it is possible to assign *Listeria* bacteria reliably to the genus and species levels.[36,100]

6.3.2 Amplified Methods

6.3.2.1 PCR Amplification

Polymerase chain reaction (PCR). In vitro nucleic acid amplification is a more recent innovation in the genotype-based detection technology that has become a workhorse for laboratory diagnosis of microbial diseases and genetic mutations. Among several elegant and distinct approaches, PCR is the first and utmost widely applied technique in both research and clinical laboratories due to its robustness and versatility. In its simplest (or conventional) form, PCR amplification relies on two single-stranded oligonucleotide primers (usually of 20–30 bases in length) that flank the front and rear ends of a specific DNA target, a thermostable DNA polymerase that is capable of synthesizing the specific DNA, and double-stranded DNA to function as a template for DNA polymerase. The PCR amplification process usually begins at a high temperature (e.g., 94°C) to convert the double-stranded template DNA into single strands, followed by a relatively low temperature (e.g., 55°C) to enable annealing between the single-stranded primer and the single-stranded template, and then at 72°C to enable DNA polymerase extending along the template (Figure 6.2). The whole denaturing, annealing, and extending process is repeated 25–30 times so that one copy of DNA template is turned into billions of copies within 3–4 h.

As PCR has the ability to selectively amplify specific targets present in low concentrations (theoretically down to a single copy of DNA template), it demonstrates not only exquisite specificity and unsurpassed sensitivity, but also a rapid turnover time and amenableness to automation for high-throughput testing, in addition to its capacity for identifying both cultured and noncultivatable organisms. The resulting PCR products can be separated by gel electrophoresis and detected with a DNA stain or, alternatively, by labeled probes, DNA sequencing, microarray, and other related techniques.[33,37,74] Since its introduction, PCR has made an immense impact on all areas of biological

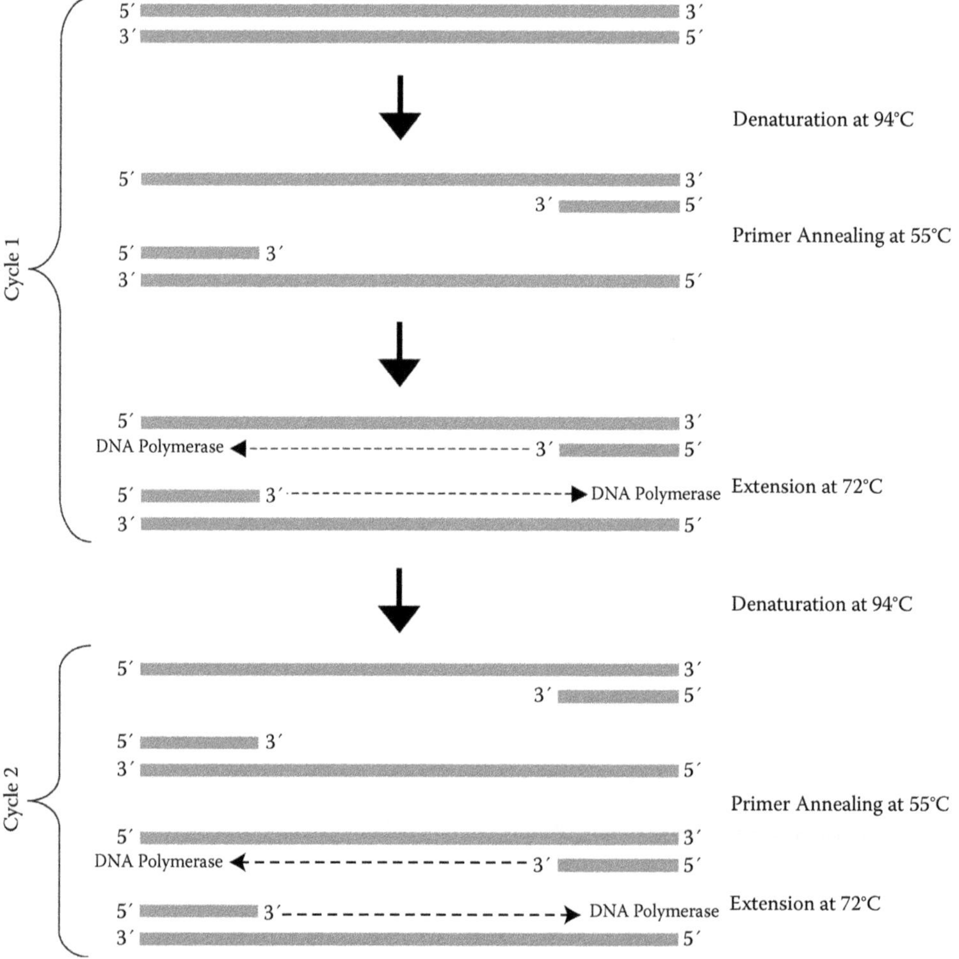

FIGURE 6.2 Schematic overview of polymerase chain reaction (PCR).

research and applications. Because of its ability to generate a billionfold increase in DNA copies from a single molecule, PCR demonstrates a superior sensitivity (needing only about 10 pg DNA for reliable detection based on agarose gel staining) in comparison with gene probe hybridization, which requires a much larger quantity of target DNA or RNA. PCR is now considered as a highly sensitive, versatile, and reproducible technique for identification of *Listeria* species from other bacteria,[101,102] with several commercial PCR tests available for routine applications (e.g., the BAX® PCR system [Qualicon, Wilmington, Delaware], the Probelia® assay [Sanofi Diagnostic Pasteur, Marne la Coquette, France], LightCycler foodproof [Roche Diagnostics, Indianapolis, Indiana], and Warnex L. mono Assay [Warnex Inc., Laval, Quebec, Canada]).[103–107]

However, as dependable and powerful as a diagnostic procedure can be, PCR is by no means without its drawbacks. One obvious risk is that PCR may be liable to carry over contamination from the previously amplified products, leading to potential false positive results. Apart from ensuring that separate rooms/spaces are allocated for PCR setup and product detection, a common strategy to avoid such contamination is replacing dTTP with dUTP during PCR amplification, producing uracil-containing DNA (U-DNA). The PCR amplicons that are synthesized with dUTP can be eliminated by an enzyme called uracil-DNA-glycosylase, and thus only unamplified test DNA, which is resistant to uracil-DNA-glycosylase digestion, remains in the tube for subsequent PCR experiment. Another notable shortcoming of PCR-based assay is its inefficiency in amplifying the target of interest directly from clinical, food, and environmental samples, as DNA polymerase used in PCR is often inhibited by various interfering substances present in these samples.

Nested PCR. Being a variant of conventional PCR, nested PCR employs two sets of oligonucleotide primers in separate reactions, with the aim to further enhance assay sensitivity and specificity, which is especially beneficial in dealing with specimens containing a low number of target organisms. After a first-round amplification with one set of primers involving 15–30 cycles, the resultant products are transferred to a new tube for a second-round amplification with another set of primers that are internal to the first-round products. Owing to this dual amplification process and to the fact that the second set of primers verifies the first-round products, nested PCR is intrinsically more sensitive and specific than conventional one-round PCR.[108–110] For example, using primers from the 16S rRNA conserved regions in a first step, and those from the species-specific regions in a second, nested PCR accomplishes a detection level of 4 fg DNA per reaction, corresponding to about one *L. monocytogenes* genome per reaction tube.[110] Similarly, employing two sets of primers from the invasion-associated protein gene (*iap*), as few as 10 cells can be detected with nested PCR. However, apart from taking extra time for a second-round amplification, nested PCR demands special attention in order to eliminate the potential risk of contamination during transfer of the first-round products to the second-round reaction tube.

Multiplex PCR. As another adaptation of conventional PCR, multiplex PCR introduces two or more sets of primer pairs specific for different targets in the same tube, resulting in the amplification of multiple gene products at the same time. This modification makes it feasible to include internal controls or to detect multiple pathogens from a single specimen. In its original format, the amplification products from multiplex PCR are separated by agarose gel electrophoresis and visualized with a stain. While it is an effective approach, this testing platform is not conducive to high-throughput screening. However, with the incorporation of different fluorescent dyes into individual gene primer sets, it is now possible to detect the multiplex PCR products in a convenient, real-time manner. For example, using primers complementary to the highly conserved *Listeria* 23S rRNA gene and *L. monocytogenes*-specific *hly* gene, both *Listeria* spp. and *L. monocytogenes* bacteria could be distinguished simultaneously via a real-time PCR.[84]

In a more elaborate version of multiplex PCR, universal primers derived from the prokaryotic 23S rRNA were labeled with biotin. The amplified PCR products were denatured and subsequently mixed with fluorescently labeled microspheres, each displaying pathogen-specific oligonucleotides and each having a different spectral pattern. After addition of streptavidin-R-phycoerythrin, which interacts with the biotin label on the amplified DNA and whose phycoerythrin fluoresces at a different wavelength than the microspheres, the microspheres were sorted according to their spectral signatures, and in the process the various pathogen DNA fragments were differentiated.[42] Application of a multiplex PCR assay to selectively amplify a shared *iap* gene also facilitated discrimination of all six *Listeria* species in a single test.[54] However, given that several primer pairs are included in the same tube, it is necessary to carefully design and optimize empirically the primers in order to ensure that these primers share similar annealing temperatures and do not interfere with each other in multiplex PCR format.[49,87,88,111–117]

Reverse transcription-PCR (RT-PCR). Instead of targeting DNA as is the case with conventional PCR, RT-PCR takes advantage of a reverse transcriptase to create complementary DNA (cDNA) from a messenger RNA (mRNA) template corresponding to a transcribed gene in the presence of a complementary primer. The cDNA is then amplified in another reaction using oligonucleotide primers and DNA polymerase under normal PCR conditions. This two-step amplification process can be streamlined into a one-tube reaction through the use of *T*th DNA polymerase, which possesses both reverse transcriptase and DNA polymerase activities. RT-PCR provides a highly sensitive technique with which a very low copy number of mRNA molecules can be amplified and detected. The presence of specific mRNA sequences is a tangible indicator of cell viability, as mRNA disintegrates quickly (within 1–20 min in a majority of cases) after cell death; DNA is comparatively stable and can last for years, depending on storage conditions.[118–120] Indeed, RT-PCR amplification of mRNA from heat-injured *L. monocytogenes* Scott A was preferable to direct PCR in an attempt to avoid false positives from dead cells.[119] Furthermore, using lysates obtained with a commercial lysis buffer, RT-PCR enabled detection of low numbers of *L. monocytogenes* cells from artificially contaminated chicken meat samples.[120]

Real-time PCR. In this procedure, a fluorescent dye (e.g., a double-stranded DNA intercalating dye SYBR green) is used to monitor the PCR amplification process in real time. SYBR green has an emission spectrum that is 50- to 100-fold brighter when it binds to double-stranded DNA. As the double-stranded DNA is synthesized during PCR, an increase in fluorescence correlates to an increasing concentration of PCR products, which can be determined instantly and in real time, with reference to a standard sample. In addition, when using SYBR green for real-time multiplex PCR, discrimination of amplicons from different genes is also possible if these gene products have sufficiently different T_m values. A melting curve analysis can be performed after PCR, using the SYBR green as a fluorescent marker. As the melting point is reached, the DNA denatures and the fluorescence decreases sharply. The plotting of fluorescence versus temperature in a graph assists calculation of the melting temperature for each product. Pathogen-specific amplicons having different melting points can be compared to standards to ascertain which pathogens are present in the sample. Since the fluorescence is measured in a LightCycler, real-time PCR negates the need for agarose gel electrophoresis, making the results available shortly after completion of PCR.[121–130]

Quantitative PCR (Q-PCR). Being an indispensable tool for pathogen confirmation, conventional PCR does not usually reveal the precise quantity of pathogen in the testing samples. Quantitative PCR has thus been designed specifically to fulfill this function, via the use of an internal or external standard, either competitively or noncompetitively. Currently, three common detection formats for Q-PCR exist: analysis via agarose gel electrophoresis, inclusion of SYBR green dye, and use of fluorescent reporter probe. Although detection of Q-PCR products by agarose gel electrophoresis is economical, use of SYBR green and fluorescent reporter probe in a real-time Q-PCR has gained increasing popularity these days, due to their enhanced sensitivity and convenience.

In a competitively based Q-PCR that employs an external standard, a competitor fragment of DNA that matches the gene to be amplified is introduced into the sample. Since the competitor fragment is a deletion mutant that can be amplified by the same primers being used to amplify the target DNA, the competitor fragment will be distinguished from the pathogen gene fragment by its smaller size. To determine the level of pathogen contamination by Q-PCR, DNA purified from a clinical sample is serially diluted and added to a constant amount of competitor DNA. Q-PCR is performed and the intensity of the pathogen gene's signal is compared to that of the competitor DNA via a stained agarose gel. The number of cells in the original sample can be estimated by the intensity of the full-length PCR product (from the pathogen) as compared to the intensity of the smaller, competitive PCR product using a standard curve.[131] Another version of competitive PCR involves use of a restriction enzyme site in the *L. monocytogenes* target gene. The competitor fragment is mutated to remove this particular restriction endonuclease site so that after PCR and digestion with the enzyme, the competitor amplicon presents as undigested, and thus a slightly larger molecule than that from the target of interest.[131]

More recent versions of Q-PCR (also known as 5′ nuclease PCR or fluorescent resonance energy transfer [FRET]-based Q-PCR) rely on the endogenous 5′ → 3′ nuclease activity of thermostable DNA polymerase (e.g., Taq, Tth, and Tfl DNA polymerases) to generate a quantifiable signal by hydrolysis of a dual-fluorophore-labeled oligonucleotide probe during amplification. The oligonucleotide probe has a covalently attached fluorescent reporter dye and quencher dye and is added directly in the PCR master mix. The fluorogenic reporter probe is released by the DNA polymerase only when it is hybridized to the amplicon, thus providing a signal-dependent quantitative measure of template concentration.[72] The fluorogenic 5′ nuclease PCR has been used successfully to detect *L. monocytogenes* with both DNA and RNA as target sequences.[61,72,85,132–136] A modification of the 5′ nuclease assay to detect mRNA facilitates monitoring of *L. monocytogenes* cell viability.[133] Overall, the Q-PCR system is highly sensitive and specific for *L. monocytogenes* with detection limits of 10^4 colony forming units (CFUs) per millimeter after 35 cycles and 10^2 CFUs per millimeter after 45 cycles.

Random amplified polymorphic DNA (RAPD). RAPD, also referred to as arbitrary primed PCR (AR-PCR), employs short random sequence primers (usually 10 bases long for AR-PCR and 10–15 bases long for RAPD) to hybridize with *Listeria* genomic DNA sequences at relatively low temperature (e.g., 36°C), generating amplified products of distinct sizes, which can then be separated by agarose gel electrophoresis. The feasibility of RAPD is built on the premise that if two random sites are located within a certain distance (say, a few hundred bases to several kilobases) of each other in the opposite orientation and recognized by random primers, a PCR product with a molecular length corresponding to the distance between the two primer binding sites can be generated.[10–12] As the number and location of the random primer sites vary for different *Listeria* species and strains, the resultant amplification products will form characteristic band patterns upon agarose gel electrophoresis, on which *Listeria* species and strains can be effectively distinguished. Apart from requiring no prior knowledge of the target DNA sequences, RAPD is more economical and faster than other identification and typing methods, and thus it is suitable for examining multiple strains (<50). However, the discriminatory ability of RAPD is sometimes inconsistent,[13,16,137,138] in addition to its requirement of cultured isolates or purified DNA to prevent formation of bands from other contaminating organisms present in the starting material.

6.3.2.2 Detection of PCR Products

Agarose gel electrophoresis. Agarose is a polysaccharide essentially made up of very long chains of cross-linked galactopyranose residues, which provides a convenient, inexpensive medium for electrophoretic separation and semiquantitation of biological molecules such as DNA and RNA. Agarose (0.5–3%) is usually prepared and utilized in one of the two buffers (i.e., Tris-acetate with EDTA [TAE] or Tris-borate with EDTA [TBE]) for effective separation of DNA molecules in the range of 100–20,000 bp. During electrophoresis, the negatively charged agarose remains stationary, while the positive ions in agarose migrate toward the cathode. Water moves along with the positive ions, in opposite direction to the negatively charged DNA molecules that migrate toward the anode. With the migration rate of DNA molecules being dependent on their size, the smaller DNA molecules move faster toward the anode than larger DNA molecules do. As a result, the smaller DNA molecules are located on the bottom of the gel, and the larger DNA molecules are situated on the top of the gel. The DNA molecules can then be visualized with a DNA stain (e.g., ethidium bromide, gelstar, or SYBR green) and their sizes estimated with reference to a known DNA molecular weight marker.

Polyacrylamide gel electrophoresis (PAGE). Polyacrylamide is a cross-linked polymer of acrylamide, the concentration (typically between 3.5 and 20%) of which dictates the length of the polymer chains and subsequent resolving power. Because oxygen inhibits the polymerization process, polyacrylamide gel must be prepared between glass plates (or cylinders). Polyacrylamide gels have a rather small range of separation, but very high resolving power. In the case of DNA, polyacrylamide is useful for separating fragments of fewer than about 500 bp. However, under appropriate conditions, fragments of DNA differing in length by a single base pair are easily resolved. Therefore, PAGE provides a versatile, gentle, high-resolution method for fractionation and physical–chemical characterization of molecules on the basis of size, conformation, and net charge.

Among the various PAGE-based procedures for detection of PCR products, single-strand conformational polymorphism analysis (SSCP), denaturing gradient gel electrophoresis (DGGE), and temperature gradient gel electrophoresis (TGGE) are most commonly used. SSCP is capable of detecting single nucleotide variations in PCR amplified products. PCR is performed with primers from a region of suspected polymorphism. Variations in the physical conformation of PCR products are reflected in differential gel migration. Early SSCP employs nondenaturing polyacrylamide gels for analysis, but more recently capillary electrophoresis (CE) is used in conjunction. In SSCP-CE, separation of analytes is carried out in silica capillaries filled with a sieving matrix. DNA injected into the sieving matrix is separated in accordance to its size under high voltage. DNA fragments are detected under UV light when they pass a detector at the end of the capillary.[139] The process is fully automated and lends itself to high-throughput screening. *L. monocytogenes* strains and serotypes can be effectively determined on the basis of SSCP analysis of the 16S rDNA and other genes (e.g., *hly* and *iap*).[139,140]

Similarly, DGGE works by applying a small sample of DNA (or RNA) to a polyacrylamide gel that contains a denaturing agent such as urea or formamide. A change in the nucleotide sequence of a PCR product will result in variation in the migration rate of the denatured DNA in the gel. Rather than partially melting in a continuous zipper-like manner, most fragments melt in a stepwise process. Discrete portions or domains of the fragment suddenly become single stranded within a very narrow range of denaturing conditions. Because of this distinctive quality of DNA when placed in denaturing gel, it is possible to discern differences in DNA sequences or mutations of various genes by comparing the melting behavior of the polymorphic DNA fragments side by side on denaturing gradient gels. Sequence differences in otherwise identical fragments often cause them to partially melt at different positions in the gradient and therefore "stop" at different positions in the gel. Using PCR targeting the *iap* gene from the five *Listeria* species followed by DGGE, *L. monocytogenes, L. innocua, L. welshimeri, L. seeligeri,* and *L. ivanovii* can be effectively identified.[57]

Furthermore, TGGE applies a temperature gradient across the gel that partially denatures double-stranded DNA. Differences in the nucleotide sequence will lead to different melting behavior of the DNA and, hence, DNA fragments will migrate at a different rate through the gel.[37] A microtemperature gradient gel electrophoresis (mu-TGGE) has been used for the rapid subtyping of *L. monocytogenes* strains.[141]

Restriction fragment length polymorphism (RFLP). RFLP analysis exploits the differences of restriction enzyme sites in DNA to discriminate *L. monocytogenes* species and subtypes. After PCR amplification of a particular target gene (e.g., *L. monocytogenes* housekeeping or virulence-associated genes *hly*, *actA,* and *inlA*), the products are digested with select restriction enzymes. The restriction fragments are separated by gel electrophoresis, and the genetic relatedness among *L. monocytogenes* and its subtypes is established by comparison of the number and size of fragments.[27] Based on RFLP analysis of PCR-amplified fragments of the 23S rRNA gene, *Listeria* bacteria have been defined to the species level.[30] This is achieved by amplification of two 23S rRNA fragments (a 460-bp S1 and an 890-bp S2) from boiled DNA followed by excision of an S2 fragment with restriction enzymes *Xmn*I or *Cfo*I and, if needed, digestion of an S1 fragment with either *Alu*I or *Cla*I. The PCR-RFLP method provides a rapid, easy-to-use, inexpensive, and reliable identification of the six *Listeria* species.[30]

Enzyme-linked immunosorbent assay (ELISA). ELISA is a common immunological technique that has been applied for confirming the presence of an antibody or an antigen in a sample. With high sensitivity and being amenable to automation, ELISA has been adapted for detection of PCR products.[77,142–144] In a commonly used version of PCR-ELISA, streptavidin-coated microtiter plate is incubated with a biotinylated capture probe (or oligonucleotide) with specificity for the *L. monocytogenes mpl* gene. Aliquots of a PCR products generated with digoxigenin-labeled primers (also derived from the *mpl* gene) are denatured in NaOH and subsequently hybridized to the capture probe. Specific hybridization products are then visualized by a colorimetric detection system based on an antidigoxigenin horseradish peroxidase conjugate in the presence of a chromogenic substrate solution. After stopping the enzyme reaction with H_2SO_4, the absorbance is measured at 450 nm in an ELISA reader. This specific assay for *L. monocytogenes* takes about 5–6 h to complete.[77,143]

For rapid and inexpensive detection of PCR amplicons, a novel microsphere agglutination assay has been developed. PCR is carried out using biotinylated forward and reverse primers, and the amplified DNA fragments are applied without purification to agglutinate streptavidin-coated microspheres (5.7 μm in diameter). Agglutination can be detected visually in 2 min without any additional equipment or reagents. Using the *hly* gene-specific biotinylated primers, *L. monocytogenes* cells can also be easily detected and identified in the background of several other common bacterial species. This protocol considerably reduces the time and cost of diagnostic PCR experiments for *L. monocytogenes* and is potentially useful for clinical and field applications.[145]

DNA sequencing. DNA sequencing is a process of determining the nucleotide order of a given DNA fragment (or DNA sequence), which generally begins with PCR amplification of a sample DNA directed at genetic regions of interest, followed by sequencing reactions. In the classical chain termination sequencing method (or Sanger method), a DNA sample is divided into four separate samples. Each of the four samples has a primer and four normal deoxynucleotides (dATP, dGTP, dCTP, and dTTP), DNA polymerase, and only one of the four dideoxynucleotides (ddATP, ddGTP, ddCTP, and ddTTP) lacking a crucial 3′-OH group as terminating base added to it, with primer or the dideoxynucleotides containing either a radioactive label or a fluorescent tag. A series of DNA fragments are produced with random lengths containing one of the four tags. The DNA is then denatured and the resulting fragments are separated (with a resolution of just one nucleotide) by gel electrophoresis from longest to shortest. Depending on whether the primers or dideoxynucleotides are labeled with radioactive isotope or fluorescent tag, the DNA bands can be detected by exposure to x-rays or UV-light and the DNA sequence can be directly read off the gel. The 16S, 23S, and 5S rRNA genes and their intergenic spacer regions are the most extensively used in the sequencing experiments for identification and typing purposes. Sequencing analysis of multiple genes (e.g., housekeeping genes and genes coding for virulence factors) forms the scientific basis for multilocus sequence typing (MLST). Combining PCR with DNA sequencing offers a more accurate and faster means of identifying *L. monocytogenes* bacteria than culture-based identification. Using PCR and pyrosequencing together enables rapid and specific differentiation between highly diverse *L. monocytogenes* subgroups.[126]

DNA microarray. Evolved from Southern blot, DNA microarray (also known as DNA chip, gene, or genome chip, or gene array) is a collection of microscopic DNA spots affixed to a solid surface (e.g., glass, plastic, or silicon chip), usually 100–200 μ*M* in size situated 200–500 μ*M* apart, forming an array for the purpose of simultaneous pathogen identification. The attached DNA segments are known as probes, thousands of which can be loaded on a single DNA microarray. To perform DNA microarray, probes (either in the form of oligonucleotide or PCR amplified DNA fragment) specific for unique portions of the 16S rRNA gene are coated on glass, and universal primers (one of which contains a fluorescent label) are used in PCR to amplify all the 16S rRNA genes present in a sample. The amplified DNA fragments bind only to the probes for which they have a complementary sequence. Pathogens are identified by the pattern of fluorescing spots in the array.[146–148] Alternatively, multiple primer sets can be employed simultaneously to amplify a number of pathogen-specific genes (e.g., *iap, hlyA, inlB, plcA, plcB,* and *clpE*) in a multiplex PCR.[74,149] Although planar DNA microarray based on solid surface has decent resolution and is genetically informative, its output is still relatively low together with multiple technically challenging steps. For this reason, a suspension microarray technology is developed to overcome the technical limitations associated with the planar microarrays and achieve high-throughput subtyping of *L. monocytogenes* strains.[150]

6.3.2.3 Other Amplification Procedures

Nucleic acid sequence-based amplification (NASBA). As an alternative technique to RT-PCR for RNA analysis, NASBA involves three enzymes (i.e., reverse transcriptase, RNaseH, and T7 RNA polymerase) in one tube (containing both dNTPs and NTPs) for amplification from single-stranded RNA templates. The first primer includes a recognition sequence for T7 RNA polymerase to facilitate its binding with the RNA template, allowing the reverse transcriptase to synthesize a cDNA strand. After digesting the RNA by the RNase, the second primer binds to the cDNA, permitting the reverse transcriptase to make a double-stranded cDNA copy of the original sequence. The new double-stranded cDNA then functions as a template for transcription by the T7 RNA polymerase to produce thousands of RNA transcripts, which can act as new templates for the subsequent reaction. With the reaction being performed at a relatively low, single temperature (normally 41°C), the genomic DNA from the target microorganism remains double stranded and does not involve as a template for amplification. NASBA obviates the necessity for DNase treatment, which is frequently required when using RT-PCR for RNA amplification, and thus it offers a convenient way for specific detection of viable cells. As the product of an NASBA reaction is mainly single-stranded RNA, it may be detected by gel electrophoresis followed by ethidium bromide staining. To ensure product specificity, a confirmatory step, generally involving probe hybridization, is usually employed.[151,152]

The NASBA is useful for detection of *L. monocytogenes* 16S rRNA and various mRNAs, allowing specific detection of viable cells.[152] NASBA targeting the 16S rRNA sequences and *hly* mRNA has been successfully applied to detect viable *L. monocytogenes* in meat or seafood.[132,152–154] Using 16S rRNA sequences as the target in NASBA, 10^6 *L. monocytogenes* CFUs in 1 ml of cell suspension are detectable.[153] Similarly, using the inducible *hly* mRNA as target for amplification followed by solid-phase probe, a detection limit of *L. monocytogenes* 500 CFUs per reaction is consistently achieved by this method. With 48-h enrichment in a selective broth, *L. monocytogenes* at levels down to 0.2 CFU/g is detected by the method.[132]

Ligase chain reaction (LCR). LCR is an alternative method to PCR for DNA amplification in which DNA ligase instead of DNA polymerase is employed to amplify the DNA fragment of interest. In this procedure, two oligonucleotide primers (or probes) attach in a head-to-tail fashion (i.e., the 3′ end of one probe abutting the 5′ end of the second) to each strand, and the 3′ end of one primer (probe) is then joined with the 5′ end of the other primer (probe) by DNA ligase to form a duplicate of one strand of the target. After converting these newly formed double strands into two single strands by heat in a thermal cycler, a second primer pair, complementary to the first pair, then anneals to this duplicated strand (and the original target strand) as a template for further ligation reaction. Repeating the process leads to exponential multiplication of ligation products, which can be detected by means of a tag or label attached to the oligonucleotide primers. LCR has been successfully applied for identification of *L. monocytogenes*; when used after a target amplification method (e.g., PCR), LCR is also highly sensitive for the detection of point mutations in *L. monocytogenes* because of the greater discriminatory power of ligation over primer extension.[155,156]

It is clear from the preceding discussion that, while the nonamplified molecular methods (e.g., DNA–DNA hybridization and gene probes) have been useful for specific identification and confirmation of *L. monocytogenes*, they generally lack sensitivity, speed, and convenience—all of which are important for clinical diagnostic applications. Because these tests do not undergo nucleic acid amplification, they invariably require large quantities of purified DNA as starting material for result consistency. With the development of new molecular methods where both template and signal multiplications often take place, it becomes feasible to detect and identify *L. monocytogenes* in a rapid, precise, and sensitive manner. Among the various nucleic acid amplification procedures, PCR appears to be more robust, versatile, and straightforward than other amplification methods (e.g., NASBA and LCR), achieving a detection limit of 10 ng of nucleic acids per reaction via stained agarose gel, and an even higher sensitivity via real-time detection format, or further signal amplification by ELISA and other techniques. Therefore, it is no surprise that PCR (and its variants) has played an increasingly dominant role in the laboratory diagnosis of *L. monocytogenes* in many parts of the world since its inception.

When designing experiments for detection and monitoring of *L. monocytogenes* mRNA, consideration should be given to selection of appropriate targets and specific growth conditions, as *L. monocytogenes* is noted for its ability to tolerate external stresses and adjust its level of expression accordingly. For example, using total RNA isolated from *L. monocytogenes* enriched for 1 h in RT-PCR followed by Southern blotting, a much lower number (10–15) of *L. monocytogenes* cells/ml in pure culture was detected with the *iap* gene than with the *hly* and *prfA* genes as targets, suggesting differential expression of these genes under the condition.[118]

6.4 TEMPLATE REQUIREMENTS AND PREPARATIONS

To achieve optimal test performance and result consistency, nucleic acid–based detection procedures often demand distinct types of templates as starting materials. While FISH and gene probes may work satisfactorily even with minimally processed or crude specimens, other techniques such as DNA–DNA hybridization are reliant on DNA preparations of relatively high purity for consistent outcomes. On the other hand, PCR-based assays can cope with specimens ranging from purified nucleic acids to relatively impure samples, as long as inhibitors to DNA polymerase are removed prior to test commencement.

6.4.1 Nucleic Acids

6.4.1.1 DNA

As the early generation molecular identification methods (e.g., determination of G+C content, DNA–DNA hybridization, and Southern blot) for *L. monocytogenes* do not involve nucleic acid amplification, they are dependent on the availability of relatively large quantities of DNA extracted from cultured isolates as templates. The presence of nucleic acids from other microbial organisms often compromises the performance of these tests. The subsequent development of nucleic acid amplification technology such as PCR has facilitated the identification and detection of *L. monocytogenes* from minute amounts of material (theoretically down to one genome) with unsurpassed specificity. While these new generation molecular methods can work well with DNA preparations of lesser purity, purified DNA is usually desired during the initial stage of assay development to ascertain test sensitivity and specificity.

Conventional protocols for isolation of DNA from *L. monocytogenes* cultured isolates have often been lengthy and tedious and have required use of biohazard chemicals. The cultured bacteria are first lysed with lysozyme, contaminating proteins digested with proteinase K in the presence of detergents (e.g., Triton X-100 and sodium dodecyl sulfate [SDS]). Indeed, Triton X-100 is preferable to SDS for more efficient lysis of *Listeria* bacteria. The degraded proteins and other cell debris are then extracted with phenol/chloroform extraction, and finally the DNA is precipitated and washed with ethanol.[82] To simplify and speed up the DNA purification process, many innovative techniques based on filter column and glass beads have been designed that obviate the necessity to use hazardous reagents.[115,127,128,135,157,158] These techniques have formed the basis of a number of commercial kits for rapid and reliable purification of DNA from *L. monocytogenes* cultured isolates (e.g., DNeasy kit). However, given a relatively low level of target microorganisms that are commonly found in clinical, food, and environmental specimens, the utility of the nucleic acid purification kits in the direct detection and identification of *L. monocytogenes* from these samples is somewhat restricted without first going through other separation and concentration steps outlined next.

6.4.1.2 RNA

One notable disadvantage of the DNA-based methods is their inability to distinguish between living and dead organisms, as DNA is relatively stable and tends to be intact after the cells become nonviable. This disadvantage of the DNA-based methods can be overcome by employing assays that target mRNA, such as RT-PCR and NABSA, as RNAs (especially mRNA) are short-lived, and often disintegrate within minutes of cell death due to the existence of endogenous nucleases that digest them very rapidly. Thus, the presence of *L. monocytogenes* mRNA in a given specimen provides a useful indicator for its viability and potential infectivity, which is of importance for the food processing industries and health authorities.

As the existing detection methods (RT-PCR and NABSA) for RNA (or mRNA) are unable to differentiate RNA from DNA, it is important that purified RNA (or mRNA) free of DNA contamination is used as a template to ensure that a positive outcome does not result from contaminating DNA.[120,152] For effective isolation of RNA from *L. monocytogenes*, the cultured bacteria need to be first lysed with lysozyme, as *L. monocytogenes* cells are somewhat resistant to guanidine thiocyanate treatment. This is followed by phenol/chloroform extraction and DNAse treatment to remove residual DNA. A commercial RNA isolation kit (i.e., TRI reagents) has been marketed on the basis of this procedure, which streamlines the RNA purification process considerably. In addition, similar to the DNA purification scene, many commercial kits based on filter columns and glass beads have been also developed for quick and efficient preparation of RNA from *L. monocytogenes* cultured isolates (e.g., RNeasy kit).

Use of a commercial lysis buffer also permitted the rapid extraction of bacterial DNA and RNA. The buffer provided an RT-PCR-compatible lysis; RT-PCR was carried out directly after DNase I treatment of crude bacterial lysates using *T*th polymerase for RT-PCR in a single tube. Untreated lysate was used for standard PCR. The procedure allowed the amplification of either mRNA or DNA of *L. monocytogenes* at a level similar to that obtained with purified nucleic acids. Using lysates obtained with this buffer, nested PCR and RT-PCR assays all detected low numbers of *L. monocytogenes* cells from artificially contaminated chicken meat samples. The use of a single buffer decreased the time needed for analysis, which is amenable to automation and real-time assays.[120]

There is an interesting recent development that uses a DNA stain (i.e., ethidium monoazide bromide [EMA]) to sequester DNA from dead cells, so that only DNA from viable cells is available for amplification. This offers a way to distinguish viable cells from dead ones without RNA isolation. Apparently, EMA selectively enters cells with damaged membranes, and binds to DNA. The DNA bound to EMA cannot be amplified by PCR, and thus only the DNA of intact live cells is amplified by PCR and detected. Using this stain, 10^2 *L. monocytogenes* CFUs/g in Gouda-like cheeses were detected by PCR. This technique also offers an effective separation of viable but not culturable cells from dead cells.[127] Additionally, using EMA prior to DNA extraction and real-time PCR prevented the amplification of DNA from bacteria with damaged cell walls, and thus allowed only the DNA from bacteria with intact membranes to be detected. EMA treatment resulted in a significant reduction ($P < .001$) in the number of coliforms detected compared to real-time PCR without EMA treatment. The assay has the potential to be applied on a routine basis to slaughterhouse lines for the detection of indicator organisms or specific pathogens.[129]

RT-PCR amplification of mRNA from heat-injured *L. monocytogenes* Scott A was clearly preferable to direct PCR in an attempt to avoid false positives from dead cells. The RT-PCR had a detection limit of 3×10^6 CFUs/g, compared to 3 CFUs/g for untreated controls, but it might not be suitable for the identification of all viable cells. The heat-injured survivors could be readily distinguished from total viable cells using selective media. As a result, combinations of molecular and visual methods including selective media improved detectability of heat-injured, viable *L. monocytogenes* Scott A.[119]

6.4.2 Cultured Isolates

Cultured isolates are useful as template for molecular tests that involve nucleic acid amplification, but they may not be suitable for many nonamplified molecular assays. Confirmation of *L. monocytogenes* from cultured isolates using nucleic acid amplification procedures is relatively straightforward, as inhibitory substances present in the original clinical, food, and environmental specimens have been mostly eliminated during the culturing processes. Like many other bacteria, *L. monocytogenes* isolates can be detected in PCR without pretreatment.[115] However, prior processing helps further decrease unknown elements that may interfere with DNA polymerase, since use of untreated bacteria directly in PCR may require additional $MgCl_2$ for enhanced PCR sensitivity and reproducibility. Not surprisingly, many rapid, low-cost, and easy-to-perform procedures for processing *L. monocytogenes* cultured isolates have been devised for use before the commencement of PCR amplification. These procedures often utilize high temperature (heat), detergent (Triton X-100), and alkali as well as alkaline polyethylene glycol (PEG) to break up the *L. monocytogenes* cell wall and release its nucleic acid contents.

Although boiling is an effective way to treat *L. monocytogenes* cultured isolates before PCR,[137] it can produce variable outcome as reported previously.[159] While treating *L. monocytogenes* isolates with lysozyme and proteinase K followed by boiling does not always generate consistent results, heat treatment in the presence of Triton X-100 works well.[159] In fact, by using Triton X-100 and heat treatment, about 10 *L. monocytogenes* CFUs can be readily detected by a PCR targeting listeriolysin O gene.[159] Alkaline treatment is also an easy and effective way to get PCR-ready DNA out of *L. monocytogenes* cultured isolates. By lysing *L. monocytogenes* cells with NaOH and neutralizing with Tris-HCl pH 7·0, detection limits of approximately 50 or 10 fg of DNA (corresponding to 10 or 2 genome equivalents) are achievable by PCR with ethidium bromide–stained agarose gels or Southern blots using nonradioactively labeled probes, respectively.[77] In addition, boiling *L. monocytogenes* cell pellets obtained from enrichment broths in 50 m*M* NaOH and 0.125% of sodium dodecyl sulphate (SDS) is also useful prior to PCR amplification and detection.[160]

Another simple method for preparing *L. monocytogenes* isolates before the start of PCR is through the use of polyethylene glycol (PEG), which is an organic solvent with a relatively low viscosity. PCR-ready DNA is obtained by heating cultured bacteria in alkaline PEG (i.e., 60% PEG 200 and 20 m*M* KOH) without further neutralization.[161] Recently, the relative performance of several rapid sample handling procedures for *L. monocytogenes* isolates has been comparatively evaluated (Figure 6.3) (D. Liu, unpublished data). It is apparent that boiling in distilled water or in 1% Triton X-100 followed by a brief spin is effective in procuring PCR-ready DNA from *L. monocytogenes* cultured isolates and that neutralization is a vital step for alkali and heat, or alkaline PEG and heat procedures to achieve consistent PCR test results (Figure 6.3). The latter results contradict the previously published reports that no neutralization is required in the alkali and heat, and alkaline PEG and heat protocols for preparation of PCR-ready DNA from bacterial isolates.[160,161]

Besides cultured isolates, the rapid processing procedures described above are also applicable to *L. monocytogenes*-enriched broth culture.[160] Although the enrichment broth culture may contain other microbial organisms, it will not present a significant problem as non-*Listeria* organisms are unlikely to be recognized by the molecular diagnostic tests targeting *L. monocytogenes*–specific genes. Therefore, coupled with short enrichment steps, these rapid sample handling procedures may provide a practical solution to the current difficulty of dealing with the direct detection and identification of *L. monocytogenes* from clinical, food, and environmental specimens using molecular diagnostic tests.

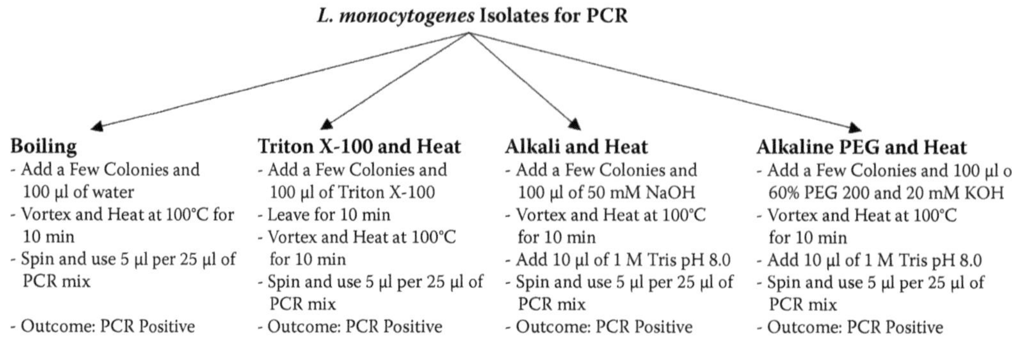

FIGURE 6.3 Comparison of rapid processing methods for *L. monocytogenes* isolates prior to PCR.

6.4.3 Nonenriched Specimens

Besides having microorganisms of interest, clinical, food and environmental specimens often contain many other substances, some of which may interfere with nucleic acid amplification processes (possibly through inhibition of thermostable DNA polymerase, or degradation and sequestration of available nucleic acids). For instance, several natural components (e.g., heme, leukocyte DNA, and immunoglobulin G)[162,163] and added anticoagulants (e.g., EDTA and heparin)[164,165] in blood have been shown to inhibit PCR amplification. In urine, urea is also found to be inhibitory to DNA polymerase.[166] In milk, proteinases and calcium ions have been identified as major sources of PCR inhibitors.[167,168] In soil and sediments, humic substances are responsible for interfering in PCR amplification.[169] Therefore, it is important that these known and uncharacterized interfering substances are removed prior to PCR amplification in order to ensure the authenticity of molecular testing results. Obviously, specimens originated from diverse sources will demand individualized sample preparation procedures with a differing degree of complexity for optimal efficiency. It is of interest to note that the inhibitory effect of calcium ions in milk can be modulated in part by addition of an extra quantity of magnesium ions (in the form of $MgCl_2$) in PCR mix,[168] whereas addition of 0.6% (wt/vol) bovine serum albumin to PCR mixtures appears to decrease the inhibitory action of blood, feces, or meat on DNA polymerase.[163]

Furthermore, clinical, food, and environmental specimens tend to be in large volume, with relatively low levels of *L. monocytogenes* organisms. Traditionally, *L. monocytogenes* clinical, food, and environmental samples usually go through a number of *in vitro* enrichment and isolation processes, which may delay the diagnostic result availability by 5–7 days or longer. Following the development of nucleic acid amplification technology (especially PCR), rapid confirmation of *L. monocytogenes* from cultured isolates has become possible. Nevertheless, given the tendency of the various known and uncharacterized substances to interfere with PCR amplification, it remains a challenge to detect and identify *L. monocytogenes* directly from clinical, food, and environmental samples even with molecular testing methods. Therefore, innovative specimen handling procedures are needed to eliminate the nonspecific inhibitory materials in the clinical, food, and environmental samples, and to concentrate the target organisms at the same time.

Preparation of clinical and food specimens without lengthy enrichment comes under the following categories: (1) physical separation and concentration involving filtration and centrifugation,[170–173] (2) capture by paramagnetic and immunobead,[158,174,175] and (3) attachment to other solid phases. Besides being easy to perform and free of biohazardous chemicals, many of these DNA purification procedures demonstrate a sensitivity that is comparable to the traditional method involving phenolchloroform extraction and ethanol precipitation.

6.4.3.1 Physical Separation and Concentration

Physical separation of bacterial organisms of interest from other contaminating substances in clinical, food, and environmental specimens, and subsequent concentration are often achieved through a combination of filtration and centrifugation. These procedures can be utilized as a stand-alone technique, or as an initial part of the comprehensive sample preparation scheme. Using an FTA filter (Fitzco, Inc.) made of a fibrous matrix impregnated with chelators and denaturants that effectively traps and lyses microorganisms on contact, microbial DNA is released and preserved intact within the membrane. After removal of cell debris and other nonbinding contaminants with wash solutions, the filters can be applied directly in PCR mixtures or used as a solid medium to store samples for later use. This filter-based procedure provides an efficient, sensitive, and uniform method for PCR template preparation, with a detection rate of 200 CFUs for *L. monocytogenes* bacteria.[171] Similarly, using a two-step method that incorporates one centrifugation step (119 × *g* for 15 min at 5°C) to remove large food particulates in potato salad filtrate and a second centrifugation step (11,950 × *g* for 10 min at 5°C) to concentrate the bacterial cells in the supernatant, it is possible to induce fivefold sample volume reduction and 1000-fold improvement in test sensitivity (from 10^6 CFUs/g [no sample processing] to 10^3 CFUs/g) upon subsequent DNA extraction, PCR amplification, and Southern blotting. This sample preparation method permits molecular detection of *L. monocytogenes* directly from food specimens.[173]

Next, a multistep sample preparation protocol is devised for recovery of *L. monocytogenes* in cheese. This protocol consists of a centrifugation step to precipitate large particles in cheese homogenate, passage of the supernatant over a sieve to further eliminate particles and fat, another centrifugation step to pellet the bacteria, and a final pronase digestion to degrade the remaining small food particles. It is clear that the protocol allows recovery of low numbers of *L. monocytogenes* (0.5–1.5 CFUs per 1 g of cheese) without prior enrichment as confirmed by Oxoid agar plating.[178] Further, through rapid concentration of bacteria on sponge swabs by vacuum filtration, a real-time PCR facilitates detection and quantification of *L. monocytogenes* in carcass and environmental swabs without enrichment, with results becoming available in fewer than 4 h.[128] In another comparative study involving eight pre-PCR sample preparation strategies, it was found that the optimal pre-PCR procedure involved filtration and DNA purification with the use of a commercial kit. This sample handling scheme allows detection of 10 CFUs of *L. monocytogenes* per gram of smoked salmon.[123] Moreover, using high-speed centrifugation (9700 *g*) followed by DNA extraction, PCR amplification, and amplicon confirmation by hybridization, 10–10^3 *L. monocytogenes* CFUs per 11 g yogurt and cheese are detected without prior cultural enrichment.[172]

Other separation and concentration techniques have also been reported for preparation of clinical, food, and environmental samples prior to molecular detection of *L. monocytogenes*. By using an aqueous two-phase system containing polyethylene glycol (PEG) 4000 and dextran 40, PCR inhibitors are restricted to the top polyethylene glycol phase and most *L. monocytogenes* cells (or DNA) are concentrated in the bottom dextran phase, which can then be collected for molecular detection.[176] Preparation of *L. monocytogenes* DNA by alcohol precipitation in the presence of NaI also facilitates the direct detection of *L. monocytogenes* from food, with a sensitivity of 10^3 CFUs per 0.5 g of soft cheese and minced meat.[179] On the whole, physical separation and concentration are useful for removing large particles in the clinical, food, and environmental specimens and reducing total sample volume to a workable quantity. Therefore, they form an integral part of sample handling procedures for recovery of *L. monocytogenes* organisms from a variety of specimens.

6.4.3.2 Capture by Paramagnetic and Immunobeads

Under chaotropic conditions, nucleic acids have an affinity for silica-coated beads (or glass beads, usually 3–10 μm in diameter), to which nucleic acids will bind and then be easily eluted with a low ionic strength buffer (or distilled water) after washing away any unbound materials. This unique feature has been exploited in many in-house and commercial nucleic acid purification systems to generate high-purity DNA and RNA for molecular applications. Extending this principle, paramagnetic beads are made of silica impregnated with iron, which demonstrates magnetic quality only upon exposure to a magnetic force. Paramagnetic beads can then be linked with specific oligonucleotides for direct binding with target nucleic acids and subsequent elution. Alternatively, these beads can be covalently attached to specific antibodies and other ligands for interaction with target proteins (which are located on the surface of microorganisms) followed by elution. These processes constitute the basis of nucleic acid capture by paramagnetic and immunobeads, resulting in highly purified nucleic acids or concentrated microorganisms for downstream experiments.[110]

A useful approach for capturing nucleic acids from clinical, food, and environmental specimens involves covalent immobilization of *L. monocytogenes*-specific oligonucleotides as capture probes. For example, a 21-mer oligonucleotide selected from a highly conserved region of the listeriolysin O gene (*hly*) (nt. 248–268) was coupled to silica-coated paramagnetic nanoparticles, and used to purify *L. monocytogenes* DNA from lysed bacteria. This magnetic capture procedure facilitated reliable detection of 10 CFUs per milliliter of *L. monocytogenes*-spiked milk without pre-enrichment, which compared favorably with a Dynabead M-280 Streptavidin-*bListeria* (Dynal, Invitrogen), detecting 10^2 CFUs/ml only.[175] Similarly, paramagnetic beads attached with a specific oligonucleotide from the listeriolysin O gene enabled isolation of listerial DNA from food homogenates and blood for examination in a semi-nested PCR targeting the *hly* gene, achieving a detection limit of between 1 and 10 CFUs of *L. monocytogenes* in 50 ml of a buffer solution.[174] In addition, utilizing the paramagnetic nanoparticles-based commercial *L. monocytogenes* DNA Isolation Kit: Milk (Diatheva, Italy), PCR-ready DNA was obtained after treatment of *L. monocytogenes* from milk with lysozyme, RNase A, and proteinase K followed by mixing with the paramagnetic particles in the presence of the specific DNA-binding buffer.[177] Moreover, employing commercial Dynabeads DNA DIRECT kit (Dynal), a detection level of 10^5 *L. monocytogenes* CFUs per milliliter of human cerebrospinal fluid (CSF) was achieved in PCR.[110] A combination of nonspecific binding of bacteria to paramagnetic beads with subsequent DNA purification based on the same beads also gave satisfactory results in a 5′-nuclease PCR for quantitative detection of *L. monocytogenes*.[72]

Another version of the capture technique utilizes paramagnetic beads coated with specific monoclonal antibodies (MAbs) to isolate *Listeria* bacteria from clinical, food, and environmental samples, and to eliminate PCR-inhibitory factors present. Upon lysis of the isolated bacteria, the supernatant containing the bacterial DNA is tested in PCR, with a detection rate of 40 CFUs per 25 g of spiked food without enrichment, and 1 CFU of *L. monocytogenes* per 1 g of cheese after enrichment for 24 h in Fraser broth.[180] On the other hand, using Dynabeads anti-*Listeria* (Dynal) leads to a recovery of 40–100 CFUs/ml of *L. monocytogenes* broth culture; after a 24-h enrichment, its sensitivity increases to 10 *L. monocytogenes* CFUs per 1 g of cheese.[178] In another study, protein A agarose beads (immunobeads) coated with a anti-*L. monocytogenes* MAb C11E9 were used to capture *Listeria* cells from a variety of samples followed by testing for cytopathogenic action on a murine hybridoma B-lymphocyte cell line (Ped-2E9) by Trypan blue staining (or by an alkaline phosphatase-based cytotoxicity assay). This approach facilitated isolation, detection, and confirmation of cytopathogenicity of viable *L. monocytogenes* within 28 h.[181] After immunomagnetic separation (IMS) of *Listeria* bacteria directly from ham followed by extraction of DNA and detection by a multiplex PCR targeting listeriolysin O gene and *Listeria* 23S rRNA gene, a detection limit as low as 1.1 *L. monocytogenes* cells per gram in a 25-g ham sample can be obtained.[73] An immunoseparation system utilizing polystyrene microbeads (3.8 μm in diameter) coated with covalently bound anti-*Listeria* genus-specific antibody has also proven valuable for recovery of *L. monocytogenes* from cheese.[182]

6.4.3.3 Attachment to Other Solid Surfaces

Apart from coating paramagnetic beads with specific oligonucleotides and antibodies, other ligands have also been experimented upon for separation of *L. monocytogenes* from contaminating materials in the clinical, food, and environmental samples. Indeed, it has been shown that immobilization of *Triticum vulgaris* lectin on paramagnetic microspheres enables recovery of 87–100% input of *L. monocytogenes* cells.[183] Overall, the paramagnetic and immunobeads are invaluable laboratory tools for capturing *L. monocytogenes* DNA and concentrating target microorganisms in spite of the fact that there is a lack of consensus on the selection of *L. monocytogenes*-specific oligonucleotides and antibodies for coating the paramagnetic beads. While some nonspecific bindings to the paramagnetic beads by other microorganisms are unavoidable,[115] in general these will not be a concern to the molecular assays that demonstrate specific recognition for *L. monocytogenes.*

6.4.4 Enriched Specimens

Since many clinical, food, and environmental samples may contain very few *L. monocytogenes* bacteria, using the aforementioned concentration and capture procedures will not always lead to satisfactory results in molecular tests. Therefore, in many cases, it is often helpful to employ enrichment media containing selective and elective agents (e.g., acriflavine, nalidixic acid, and cycloheximide) so that low numbers of *L. monocytogenes* in these samples can grow to a detectable level by the molecular assays.[57,59,105,107,144,160,184–193] Acriflavine inhibits RNA synthesis and mitochondriogenesis. At low pH values (<5.8) more acriflavine is bound, but growth-promoting effects are limited, since growth of this pathogen is restricted at low pH. Thus, enrichment protocols employing low acriflavine concentrations with an adequate buffer will favor the isolation of *L. monocytogenes.* Acriflavine is commonly used in combination with various other selective agents such as potassium thiocyanate, polymyxin B-sulfate, nalidixic acid, and cycloheximide. Nalidixic acid inhibits the DNA synthesis of cells and is often combined with other inhibitory agents, such as cycloheximide (actidione), which inhibits protein synthesis in eukaryotic cells by binding to 80S ribosomal RNA and prevents growth of most yeasts and molds. Both nalidixic acid and cycloheximide have no relevant effect on the growth of listeriae in enrichment media.[190]

Currently, three selective enrichment media (i.e., ISO 11290, FDA BAM, and USDA) incorporating acriflavine, nalidixic acid, and cycloheximide are in common use (see chapter 5), of which the USDA method appears suited for meat products and the FDA method is applicable for other food products (e.g., dairy products, seafood, and vegetables). These methods usually take 24–48 h to complete, and represent essential tools for enriching *L. monocytogenes* to levels that are detectable by routine diagnostic methods.[59,188] Experiments with other selective enrichment media (e.g., Fraser broth and buffered peptone water) have also been successful. There is ample evidence that use of selective enrichment media enables recovery of *L. monocytogenes* bacteria from clinical, food, and environmental specimens, facilitating more sensitive detection and identification by molecular diagnostic assays.

For instance, a short enrichment period before PCR amplification allowed detection of 10–10^2 *L. monocytogenes* CFUs per 1 g of chicken skin and soft cheese.[185] Employing a short culture enrichment step followed by isolation of bacterial cells, a multiplex PCR detected *L. monocytogenes* from foods within 48 h of sample receipt in a highly sensitive and cost-effective manner.[101] Similarly, PCR assay directing against the listeriolysin O gene (*hly*) showed a detection limit of 10^4 CFUs per gram of meat cultures enriched in buffered peptone water for 10 h at 30°C.[194] Next, after enrichment of *L. monocytogenes* from raw fish in *Listeria* enrichment broth (LEB) and from environmental samples in LMDS enrichment medium, the BAX for screening *L. monocytogenes* demonstrated a sensitivity of 84.8% and specificity of 100% with raw fish, and a 94.7% sensitivity and 97.4% specificity with environmental samples. While the BAX systems provided screening results in about 3 days, the use of LMPM allowed for *L. monocytogenes* isolation in 4–5 days.[104] In addition, using a one-step, recovery-enrichment broth, low levels of injured and uninjured *L. monocytogenes* cells could be consistently detected from ready-to-eat foods containing various background microflora.[195] Also, selective enrichment followed by DNA extraction produced a PCR assay sensitivity of 1–5 *L. monocytogenes* CFUs in 25 g of foods in 24 h.[39]

In another report, using PCR primers directed to the *hly* gene in 60 naturally contaminated foods led to a highest number of positives being recorded by PCR following a 24-h pre-enrichment step at 30°C and a 24-h enrichment step at 37°C.[157] With a PCR targeting *L. monocytogenes* internalin AB (*inl*AB) gene, the limit of detection of the pathogen in pure cultures was 10^5 cells per milliliter. When frankfurters were spiked with the pathogen, the limit of detection was improved to 10 cells in a 25-g sample, with the sample being first enriched for at least 16 h.[80] PCR assay detected 10^3 CFUs/0.5 ml milk using the *hlyA* gene as the target without enrichment, and it improved to 1 CFU if culture enrichment for 15 h was conducted first.[131] Using a combined enrichment/real-time PCR method, 7.5 *L. monocytogenes* CFUs per 25 ml of artificially contaminated raw milk, and 9, 1, and 1 CFUs per 15 g of artificially contaminated salmon, pate, and green-veined cheese, respectively, were detected.[130]

The enrichment step is often used along with other concentration procedures to enhance molecular detection and identification of *L. monocytogenes*. Following the enrichment step with universal pre-enrichment (UP) broth, if an immunomagnetic separation method using anti-*Listeria* immunomagnetic beads is performed prior to PCR, *L. monocytogenes* can be detected unambiguously from food samples such as milk, dairy, and meat products.[112] Through centrifugation and partial heat treatment after enrichment, *L. monocytogenes* is detected at 1 initial CFU without genomic DNA extraction in the culture and with artificially inoculated food samples including milk, chicken, ham, and pork.[196]

While full-strength enrichment media are unquestionably useful for recovery of *L. monocytogenes* bacteria from various specimens, recent studies indicate that nonselective or half-strength enrichment media may be valuable for helping recover sublethally stressed or injured *L. monocytogenes*.[184,197] For example, after pre-enrichment of samples in the modified USDA enrichment broth that omitted acriflavine and nalidixic acid and subsequent transfer of a larger volume of the initial culture broth to the secondary enrichment media, the recovery of low numbers of sublethally stressed *L. monocytogenes* was significantly increased.[184] Similarly, 24 h of enrichment in half Fraser broth followed by 16 h of enrichment in a full Fraser broth provided a basis for improving PCR-based detection of *L. monocytogenes* from dairy products and other foodstuffs.[198]

6.5 CONCLUSIONS AND PERSPECTIVES

L. monocytogenes is a facultative intracellular bacterium that has the capacity to cause serious illness in susceptible human population groups such as infants, pregnant women, the elderly, and immunocompromised persons. In order to limit the potential damage of listeriosis, it is critical to have fast and reliable diagnostic methods available for differentiation of *L. monocytogenes* from nonpathogenic *Listeria* and other contaminating bacteria, thus facilitating the implementation of prompt antibiotic treatment. Despite their being relatively slow and variable at times, the early generation phenotype-based tests for *Listeria* (chapter 5) have contributed to the control and prevention of listeriosis. However, rapid, precise, and sensitive identification of *L. monocytogenes* has only become a reality following the advent of new generation molecular methods. Indeed, the development of *in vitro* template amplification techniques such as PCR has revolutionized pathogen detection in both research and clinical laboratories. It is now feasible to confirm the identity of all six *Listeria* species and discrimination of *L. monocytogenes* from non*monocytogenes Listeria* species rapidly and precisely without resorting to the time-consuming measurement of phenotypic characteristics.

Although the nucleic acid amplification-based molecular techniques provide a highly sensitive, specific, and rapid means of detecting and identifying *L. monocytogenes* cultured isolates, the full potential of these methods has yet to be realized. This is mainly attributable to the fact that many molecular diagnostic assays are ineffective in dealing with clinical, food, and environmental specimens. One major reason accounting for this difficulty is that most of the molecular tests in use depend on enzymatic reaction for template amplification, which is prone to inhibition by various interfering substances that are present in unprocessed specimens. Therefore, the availability of rapid, efficient, and reliable sample preparation procedures is vital for further strengthening the performance of the molecular tests for direct detection and identification of *L. monocytogenes* from clinical, food, and environmental specimens. This will not only help shorten the time from specimen submission to result finalization, but also benefit the control and prevention program against listeriosis.

While purified *L. monocytogenes* nucleic acids and cultured isolates are ideal starting materials for molecular confirmation of listeriosis, over-reliance on these template sources may be unwise as it will lead to undesirable delay in result availability. Since several rapid sample handling procedures are applicable to both *L. monocytogenes* isolates and enriched broth culture, they may play a key part in helping solve the problem in regard to the inability of the molecular tests to function properly with unprocessed specimens. On the other hand, clinical, food, and environmental specimens often harbor very low levels of *L. monocytogenes* organisms, which are unlikely to be detected by current molecular tests even after undergoing physical separation and concentration as well as capture by paramagnetic beads. By putting these samples through an overnight enrichment step followed by rapid specimen preparation procedure and specific molecular detection, it is likely that a definitive result will be available within 24 h instead of 5–7 days, based on conventional enrichment and isolation procedures. Clearly, there is a need to develop and optimize physical separation and concentration protocols for various types of specimens (e.g., meat, milk, and plants) as well as enrichment procedures (e.g., the types and strengths of the media, and the length of enrichment). As to the possible existence of other microorganisms in the enrichment broth, it will not be an issue with *L. monocytogenes* specific molecular assays being used.

The capture technology based on paramagnetic and immunobeads represents a promising development in the extraction of PCR-ready nucleic acids and concentration of target microorganisms from clinical, food, and environmental specimens without enrichment. However, as it stands, there are some obvious deficiencies in the current procedures. The reported procedures for capturing nucleic acids often utilize oligonucleotides from a single gene sequence that are attached to paramagnetic beads as capture probe. The potential of using more than one type of capture probe (or multiple paramagnetic beads each linked with one type of oligonucleotide) has not been examined. In addition, the cell lysis protocols that have been described in the literature are unduly cumbersome, with most requiring lysozyme and proteinase K treatments. Perhaps, some rapid cell lysis methods (e.g., heat in the presence of Triton X-100, or alkali and heat) may provide a more practical alternative to the lengthy procedures that are in common use. Similarly, only paramagnetic beads attached with a single antibody type have been used to concentrate the target organisms. There is a scope to explore multiple antibody types on paramagnetic beads (or multiple paramagnetic beads, each linked with distinct antibodies). Furthermore, after concentration of target organisms, some simpler cell lysis protocols may be evaluated for subsequent molecular detection and identification.

As discussed earlier, enrichment of clinical, food, and environmental specimens will continue to be an indispensable tool for improving laboratory detection of *L. monocytogenes*. Although many enrichment media have been successfully applied for selective growth of *L. monocytogenes* bacteria from various types of samples, there have not been enough comparative studies undertaken to determine the optimal types and strengths of enrichment media or minimal length of enrichment for individual specimen types. In addition, there is a need to further investigate the most appropriate combination of physical separation and concentration procedures for various types of specimens prior to the enrichment step. The availability of these data will help streamline the sample preparation procedures for clinical, food, and environmental specimens and enhance molecular detection and identification of *L. monocytogenes,* ultimately contributing to the more effective control and prevention of listeriosis.

ACKNOWLEDGMENTS

Listeria research in our laboratory was supported by the U.S. Department of Agriculture Agricultural Research Service (Agreement No. 58-6202-5-083).

REFERENCES

1. Gouin, E., Mengaud, J., and Cossart, P., The virulence gene cluster of *Listeria monocytogenes* is also present in *Listeria ivanovii*, an animal pathogen, and *Listeria seeligeri*, a nonpathogenic species, *Infect. Immun.,* 62, 3550, 1994.
2. Glaser, P. et al., Comparative genomics of *Listeria* species, *Science,* 294, 849, 2001.
3. Nelson, K.E. et al., Whole genome comparisons of serotype 4b and 1/2a strains of the food-borne pathogen *Listeria monocytogenes* reveal new insights into the core genome components of this species, *Nucleic Acids Res.,* 32, 2386, 2004.
4. Vazquez-Boland, J.A. et al., *Listeria* pathogenesis and molecular virulence determinants, *Clin. Microbiol. Rev.,* 14, 584, 2001.
5. Liu, D. et al., Comparative assessment of acid, alkali and salt tolerance in *Listeria monocytogenes* virulent and avirulent strains, *FEMS Microbiol. Lett.,* 243, 373, 2005.
6. Liu, D., *Listeria monocytogenes*: Comparative interpretation of mouse virulence assay, *FEMS Microbiol. Lett.,* 233, 159, 2004.
7. Liu, D., Identification, subtyping and virulence determination of *Listeria monocytogenes* in important food-borne pathogen, *J. Med. Microbiol.,* 55, 645, 2006.
8. Stackebrandt, E., and Goebel, B.M., A place for DNA–DNA reassociation and 16S ribosomal-RNA sequence-analysis in the present species definition in bacteriology, *Int. J. Syst. Bacteriol.,* 44, 846, 1994.
9. Stackebrandt, E. et al., Report of the ad hoc committee for the reevaluation of the species definition in bacteriology, *Int. J. Syst. Evol. Microbiol.,* 52, 1043, 2002.
10. Welsh, J., and McClelland, M., Fingerprinting genomes using PCR with arbitrary primers, *Nucleic Acids Res.,* 18, 7213, 1990.
11. Williams, J.G. et al., DNA polymorphisms amplified by arbitrary primers are useful genetic markers, *Nucleic Acids Res.,* 18, 6531, 1990.
12. Caetano-Anolles, G., Basam, B.J., and Greshoff, P.M., DNA amplification fingerprinting using very short arbitrary oligonucleotide primers, *Bio/Technology,* 9, 553, 1991.
13. Farber, J., and Addison, C., RAPD typing for distinguishing species and strains in the genus *Listeria, J. Appl. Bacteriol.,* 77, 242, 1994.
14. Black, S.F. et al., Rapid RAPD analysis for distinguishing *Listeria* species and *Listeria monocytogenes* serotypes using a capillary air thermal cycler, *Lett. Appl. Microbiol.,* 20, 188, 1995.
15. Boerlin, P. et al., Typing *Listeria monocytogenes*: A comparison of random amplification of polymorphic DNA with five other methods, *Res. Microbiol.,* 146, 35, 1995.
16. O'Donoghue, K. et al., Typing of *Listeria monocytogenes* by random amplified polymorphic DNA (RAPD) analysis, *Int. J. Food Microbiol.,* 27, 245, 1995.
17. Wagner, M., Maderner, A., and Brandl, E., Development of a multiple primer RAPD assay as a tool for phylogenetic analysis in *Listeria* spp. strains isolated from milk product associated epidemics, sporadic cases of listeriosis and dairy environments, *Int. J. Food Microbiol.,* 52, 29, 1999.
18. Mereghetti, L. et al., Combined ribotyping and random multiprimer DNA analysis to probe the population structure of *Listeria monocytogenes, Appl. Environ. Microbiol.,* 68, 2849, 2002.
19. Vogel, B.F. et al., High-resolution genotyping of *Listeria monocytogenes* by fluorescent amplified fragment length polymorphism analysis compared to pulsed-field gel electrophoresis, random amplified polymorphic DNA analysis, ribotyping, and PCR-restriction fragment length polymorphism analysis, *J. Food Prot.,* 67, 1656, 2004.
20. Jersek, B. et al., Repetitive element sequence-based PCR for species and strain discrimination in the genus *Listeria, Lett. Appl. Microbiol.,* 23, 55, 1996.

21. Jersek, B. et al., Typing of *Listeria monocytogenes* strains by repetitive element sequence-based PCR, *J. Clin. Microbiol.,* 37, 103, 1999.
22. Van Kessel, J.S. et al., Subtyping *Listeria monocytogenes* from bulk tank milk using automated repetitive element-based PCR, *J. Food Prot.,* 68, 2707, 2005.
23. Chou C.H., and Wang, C., Genetic relatedness between *Listeria monocytogenes* isolates from seafood and humans using PFGE and REP-PCR, *Int. J. Food Microbiol.,* 110, 135, 2006.
24. Denoeud, F., and Vergnaud, G., Identification of polymorphic tandem repeats by direct comparison of genome sequence from different bacterial strains: A Web-based resource, *BMC Bioinformatics,* 5, 4, 2004.
25. Danin-Poleg, Y. et al., Towards the definition of pathogenic microbe, *Int. J. Food Microbiol.,* 112, 236, 2006.
26. Comi, G. et al., A RE-PCR method to distinguish *Listeria monocytogenes* serovars, *FEMS Immunol. Med. Microbiol.,* 18, 99, 1997.
27. Wiedmann, M. et al., Ribotypes and virulence gene polymorphisms suggest three distinct *Listeria monocytogenes* lineages with differences in pathogenic potential, *Infect. Immun.,* 65, 2707, 1997.
28. Vaneechoutte, M. et al., Comparison of PCR-based DNA fingerprinting techniques for the identification of *Listeria* species and their use for atypical *Listeria* isolates, *Int. J. Syst. Bacteriol.*, 48, 127, 1998.
29. Jaradat, Z.W., Schutze, G.E., and Bhunia, A.K., Genetic homogeneity among *Listeria monocytogenes* strains from infected patients and meat products from two geographic locations determined by phenotyping, ribotyping and PCR analysis of virulence genes, *Int. J. Food Microbiol.,* 76, 1, 2002.
30. Paillard, D. et al., Rapid identification of *Listeria* species by using restriction fragment length polymorphism of PCR-amplified 23S rRNA gene fragments, *Appl. Environ. Microbiol.,* 69, 6386, 2003.
31. Rocourt, J. et al., Assignment of *Listeria grayi* and *Listeria murrayi* to a single species, *Listeria grayi,* with a revised description of *Listeria grayi, Int. J. Syst. Bacteriol.*, 42, 171, 1992.
32. Wang, R.F, Cao, W.W., and Johnson, M.G., 16S rRNA-based probes and polymerase chain reaction method to detect *Listeria monocytogenes* cells added to foods, *Appl. Environ. Microbiol.*, 58, 2827, 1992.
33. Wang, R.F. et al., A 16S rRNA-based DNA probe and PCR method specific for *Listeria ivanovii, FEMS Microbiol. Lett.,* 106, 85, 1993.
34. Czajka, J. et al., Differentiation of *Listeria monocytogenes* and *Listeria innocua* by 16S rRNA genes and intraspecies discrimination of *Listeria monocytogenes* strains by random amplified polymorphic DNA polymorphisms, *Appl. Environ. Microbiol.,* 59, 304, 1993.
35. Sallen, B. et al., Comparative analysis of 16S and 23S rRNA sequences of *Listeria* species, *Int. J. Syst. Evol. Bacteriol.,* 46, 669, 1996.
36. Wagner, M. et al., In situ detection of a virulence factor mRNA and 16S rRNA in *Listeria monocytogenes, FEMS Microbiol. Lett.,* 160, 159, 1998.
37. Manzano, M. et al., Temperature gradient gel electrophoresis of the amplified product of a small 16S rRNA gene fragment for the identification of *Listeria* species isolated from food, *J. Food Prot.,* 63, 659, 2000.
38. Garrec, N. et al., Heteroduplex mobility assay for the identification of *Listeria* spp. and *Listeria monocytogenes* strains: Application to characterization of strains from sludge and food samples, *FEMS Immunol. Med. Microbiol.,* 38, 57, 2003.
39. Somer, L., and Kashi, Y., A PCR method based on 16S rRNA sequence for simultaneous detection of the genus *Listeria* and the species *Listeria monocytogenes* in food products, *J. Food Prot.,* 66, 1658, 2003.
40. Brehm-Stecher, B.F., Hyldig-Nielsen, J.J., and Johnson, E.A., Design and evaluation of 16S rRNA-targeted peptide nucleic acid probes for whole-cell detection of members of the genus *Listeria*, *Appl. Environ. Microbiol.,* 71, 5451, 2005.
41. Clarridge, J.E., III, Impact of 16S rRNA gene sequence analysis for identification of bacteria on clinical microbiology and infectious diseases, *Clin. Microbiol. Rev.*, 17, 840, 2004.
42. Dunbar, S.A. et al., Quantitative, multiplexed detection of bacterial pathogens: DNA and protein applications of the Luminex LabMAP™ system, *J. Microbiol. Methods,* 53, 245, 2003.
43. Rodríguez-Lázaro, D., Hernández, M., and Pla, M., Simultaneous quantitative detection of *Listeria* spp. and *Listeria monocytogenes* using a duplex real-time PCR-based assay, *FEMS Microbiol. Lett.,* 233, 257, 2004.
44. Thompson, D.E. et al., Studies on the ribosomal RNA operons of *Listeria monocytogenes*, *FEMS Microbiol. Lett.,* 75, 219, 1992.
45. Graham, T. et al., Genus- and species-specific detection of *Listeria monocytogenes* using polymerase chain reaction assays targeting the 16S/23S intergenic spacer region of the rRNA operon, *Can. J. Microbiol.,* 42, 1155, 1996.

46. Graham, T.A. et al., Inter- and intraspecies comparison of the 16S-23S rRNA operon intergenic spacer regions of six *Listeria* spp., *Int. J. Syst. Bacteriol.,* 47, 863, 1997.
47. Gahan, C.G.M., O'Mahony, J., and Hill, C., Characterization of the *groESL* operon in *Listeria monocytogenes*: Utilization of two reporter systems (*gfp* and *hly*) for evaluating *in vivo* expression, *Infect. Immun.,* 69, 3924, 2001.
48. Cai, S. et al., Rational design of DNA sequence-based strategies for subtyping *Listeria monocytogenes, J. Clin. Microbiol.,* 40, 3319, 2002.
49. Doumith, M. et al., Differentiation of the major *Listeria monocytogenes* serovars by multiplex PCR, *J. Clin. Microbiol.,* 42, 3819, 2004.
50. Meinersmann, R.J. et al., Multilocus sequence typing of *Listeria monocytogenes* by use of hypervariable genes reveals clonal and recombination histories of three lineages, *Appl. Environ. Microbiol.,* 70, 2193, 2004.
51. Revazishvili, T. et al., Comparative analysis of multilocus sequence typing and pulsed-field gel electrophoresis for characterizing *Listeria monocytogenes* strains isolated from environmental and clinical sources, *J. Clin. Microbiol.*, 42, 276, 2004.
52. Sue, D. et al., SigmaB-dependent gene induction and expression in *Listeria monocytogenes* during osmotic and acid stress conditions simulating the intestinal environment, *Microbiology,* 150, 3843, 2004.
53. Bubert, A., Kohler, S., and Goebel, W., The homologous and heterologous regions within the *iap* gene allow genus- and species-specific identification of *Listeria* spp. by polymerase chain reaction, *Appl. Environ. Microbiol.,* 58, 2625, 1992.
54. Bubert, A. et al., Detection and differentiation of *Listeria* spp. by a single reaction based multiplex PCR, *Appl. Environ. Microbiol.,* 65, 4688, 1999.
55. Gray, D.I., and Kroll, R.G., Polymerase chain reaction amplification of the *flaA* gene for the rapid identification of *Listeria* spp., *Lett. Appl. Microbiol.,* 20, 65, 1995.
56. Hein, I. et al., Detection and quantification of the *iap* gene of *Listeria monocytogenes* and *Listeria innocua* by a new real-time quantitative PCR assay, *Res. Microbiol.,* 152, 37, 2001.
57. Cocolin, L. et al., Direct identification in food samples of *Listeria* spp. and *Listeria monocytogenes* by molecular methods, *Appl. Environ. Microbiol.,* 68, 6273, 2002.
58. Gilot, P., and Content, J., Specific identification of *Listeria welshimeri* and *Listeria monocytogenes* by PCR assays targeting a gene encoding a fibronectin-binding protein, *J. Clin. Microbiol.,* 40, 698, 2002.
59. O'Connor, L. et al., Rapid polymerase chain reaction/DNA probe membrane-based assay for the detection of *Listeria* and *Listeria monocytogenes* in food, *J. Food Prot.,* 63, 337, 2000.
60. Kohler, S. et al., The gene coding for protein p60 of *Listeria monocytogenes* and its use as a specific probe for *Listeria monocytogenes. Infect. Immun.,* 58, 1943, 1990.
61. Koo, K., and Jaykus, L.A., Detection of *Listeria monocytogenes* from a model food by fluorescence resonance energy transfer-based PCR with an asymmetric fluorogenic probe set, *Appl. Environ. Microbiol.,* 69, 1082, 2003.
62. Liu, D. et al., Identification of *Listeria innocua* by PCR targeting a putative transcriptional regulator gene, *FEMS Microbiol. Lett.,* 203, 205, 2003.
63. Liu, D. et al., Identification of a gene encoding a putative phosphotransferase system enzyme IIBC in *Listeria welshimeri* and its application for diagnostic PCR, *Lett. Appl. Microbiol.,* 38, 151, 2004.
64. Liu, D. et al., PCR detection of a putative N-acetylmuramidase gene from *Listeria ivanovii* facilitates its rapid identification, *Vet. Microbiol.,* 101, 83, 2004.
65. Liu, D. et al., Species-specific PCR determination of *Listeria seeligeri, Res. Microbiol.*, 155, 741, 2004.
66. Liu, D. et al., Isolation and PCR amplification of a species-specific, oxidoreductase-coding gene region in *Listeria grayi, Can. J. Microbiol.,* 51, 95, 2005.
67. Furrer, B. et al., Detection and identification of *Listeria monocytogenes* in cooked sausage products and in milk by *in vitro* amplification of hemolysin gene fragments, *J. Appl. Bacteriol.,* 70, 372, 1991.
68. Johnson, W. et al., Detection of genes coding for listeriolysin and *Listeria monocytogenes* antigen A (lmA) in *Listeria* spp. by the polymerase chain reaction, *Microbial. Pathog.,* 12, 79, 1992.
69. Lehner, A. et al., A rapid differentiation of *Listeria monocytogenes* by use of PCR–SSCP in the listeriolysin O (*hlyA*) locus, *J. Microbiol. Methods,* 34, 165, 1999.
70. Bsat, N., and. Batt, C.A., A combined modified reverse dot-blot and nested PCR assay for the specific nonradioactive detection of *Listeria monocytogenes, Mol. Cell Probes,* 7, 199, 1993.
71. Blais, B.W., and Phillippe, L.M., A simple RNA probe system for analysis of *Listeria monocytogenes* polymerase chain reaction products, *Appl. Environ. Microbiol.,* 59, 2795, 1993.

72. Nogva, H.K. et al., Application of 5′-nuclease PCR for quantitative detection of *Listeria monocytogenes* in pure cultures, water, skim milk, and unpasteurized whole milk, *Appl. Environ. Microbiol.*, 66, 4266, 2000.
73. Hudson, J.A. et al., Rapid detection of *Listeria monocytogenes* in ham samples using immunomagnetic separation followed by polymerase chain reaction, *J. Appl. Microbiol.*, 90, 614, 2001.
74. Volokhov, D. et al., Identification of *Listeria* species by microarray-based assay, *J. Clin. Microbiol.*, 40, 4720, 2002.
75. Longhi, C. et al., Detection of *Listeria monocytogenes* in Italian-style soft cheeses, *J. Appl. Microbiol.*, 94, 879, 2003.
76. Nishibori, T. et al., Correlation between the presence of virulence-associated genes as determined by PCR and actual virulence to mice in various strains of *Listeria* spp., *Microbiol. Immunol.*, 39, 343, 1995.
77. Scheu, P., Gasch, A., and Berghof, K., Rapid detection of *Listeria monocytogenes* by PCR-ELISA, *Lett. Appl. Microbiol.*, 29, 416, 1999.
78. Wernars, K. et al., Suitability of the *prfA* gene, which encodes a regulator of virulence genes in *Listeria monocytogenes*, in the identification of pathogenic *Listeria* spp., *Appl. Environ. Microbiol.*, 58, 765, 1992.
79. Poyart, C., Trieu-Cuot, P., and Berche, P., The *inlA* gene required for cell invasion is conserved and specific to *Listeria monocytogenes, Microbiology,* 142, 173, 1996.
80. Pangallo, D. et al., Detection of *Listeria monocytogenes* by polymerase chain reaction oriented to *inlB* gene, *New Microbiol.*, 24, 333, 2001.
81. Schaferkordt, S., and Chakraborty, T., Identification, cloning and characterization of the *lma* operon, whose products are unique to *Listeria monocytogenes, J. Bacteriol.*, 179, 2707, 1997.
82. Liu, D. et al., Use of PCR primers derived from a putative transcriptional regulator gene for species-specific identification of *Listeria monocytogenes, Int. J. Food Microbiol.*, 91, 297, 2004.
83. Zhang, W. et al., The BAX PCR for screening *Listeria monocytogenes* targets a partial putative gene *lmo2234, J. Food Prot.*, 67, 1507, 2004.
84. Rodríguez-Lázaro, D. et al., Quantitative detection of *Listeria monocytogenes* and *Listeria innocua* by real-time PCR: Assessment of *hly*, *iap*, and *lin02483* targets and AmpliFluor technology, *Appl. Environ. Microbiol.*, 70, 1366, 2004.
85. Winters, D.K., Maloney, T.P., and Johnson, M.G., Rapid detection of *Listeria monocytogenes* by a PCR assay specific for an aminopeptidase, *Mol. Cell Probes,* 13, 127, 1999.
86. Liu, D. et al., Characterization of virulent and avirulent *Listeria monocytogenes* strains by PCR amplification of putative transcriptional regulator and internalin genes, *J. Med. Microbiol.*, 52, 1066, 2003.
86a. Liu, D. et al., A multiplex PCR for species- and virulence-specific determination of *Listeria monocytogenes, J. Microbiol. Methods*, 71, 133, 2007.
87. Borucki, M.K., and Call, D.R., *Listeria monocytogenes* serotype identification by PCR, *J. Clin. Microbiol.,* 41, 5537, 2003.
88. Doumith, M. et al., Multicenter validation of a multiplex PCR assay for differentiating the major *Listeria monocytogenes* serovars 1/2a, 1/2b, 1/2c, and 4b: Toward an international standard, *J. Food Prot.*, 68, 2648, 2005.
89. Sabet, C. et al., LPXTG protein InlJ, a newly identified internalin involved in *Listeria monocytogenes* virulence, *Infect. Immun.*, 73, 6912, 2005.
90. Tsai, Y.H. et al., *Listeria monocytogenes* internalins are highly diverse and evolved by recombination and positive selection, *Infect. Genet. Evol.*, 6, 378, 2006.
91. Zhang, W., and Knabel, S.J., Multiplex PCR assay simplifies serotyping and sequence typing of *Listeria monocytogenes* associated with human outbreaks, *J. Food Prot.*, 68, 1907, 2005.
92. Liu, D. et al., *Listeria monocytogenes* serotype 4b strains belonging to lineages I and III possess distinct molecular features, *J. Clin. Microbiol.*, 44, 214, 2006.
93. Liu, D. et al., *Listeria monocytogenes* subgroups IIIA, IIIB and IIIC delineate genetically distinct populations with varied pathogenic potential, *J. Clin. Microbiol.*, 44, 2229, 2006.
94. Xu, H.X. et al., A rapid method for determining the G+C content of bacterial chromosomes by monitoring fluorescence intensity during DNA denaturation in a capillary tube, *Int. J. Syst. Evol. Microbiol.*, 50, 1463, 2000.
95. Brenner, D.J. et al., *Escherichia vulneris*: A new species of Enterobacteriaceae associated with human wounds, *J. Clin. Microbiol.*, 15, 1133, 1982.
96. Gonzalez, J.M., and Sai-Jimenez, C., A simple fluorimetric method for the estimation of DNA–DNA relatedness between closely related microorganisms by thermal denaturation temperatures, *Extremophiles,* 9, 75, 2005.

97. Hartford, H., and Sneath, P.H., Optical DNA–DNA homology in the genus *Listeria*, *Int. J. Syst. Bacteriol.,* 43, 26, 1993.
98. Klinger, J.D. et al., Comparative studies of nucleic acid hybridization assay for *Listeria* in foods, *J. AOAC Int.,* 71, 669, 1988.
99. Stephan, R., Schumacher, S., and Zychowska, M.A., The VIT technology for rapid detection of *Listeria monocytogenes* and other *Listeria* spp., *Int. J. Food Microbiol.*, 89, 287, 2003.
100. Schmid, M. et al., Nucleic acid-based, cultivation-independent detection of *Listeria* spp. and genotypes of *Listeria* monocytogenes, *FEMS Immunol. Med. Microbiol.,* 35, 215, 2003.
101. Bansal, N.S. et al., Multiplex PCR assay for the routine detection of *Listeria* in food, *Int. J. Food Microbiol.,* 33, 293, 1996.
102. Aznar R., and Alarcón, B., On the specificity of PCR detection of *Listeria monocytogenes* in food: A comparison of published primers, *Syst. Appl. Microbiol.,* 25, 109, 2002.
103. Norton, D.M. et al., Application of the BAX for screening/genus *Listeria* polymerase chain reaction system for monitoring *Listeria* species in cold-smoked fish and in the smoked fish processing environment, *J. Food Prot.,* 63, 343, 2000.
104. Hoffman, A.D., and Weidmann, M., Comparative evaluation of culture and BAX polymerase chain reaction-based detection methods for *Listeria* spp. and *Listeria monocytogenes* in environmental and raw fish samples, *J. Food Prot.,* 64, 1521, 2001.
105. Wan, J. et al., Detection of *Listeria monocytogenes* in salmon using Probelia polymerase chain reaction system, *J. Food Prot.,* 66, 436, 2003.
106. Silbernagel, K. et al., Evaluation of the BAX System for detection of *Listeria monocytogenes*, *J. AOAC Int.,* 87, 395, 2004.
107. Junge, B., and Berghof-Jager, K., Roche/BIOTECON diagnostic LightCycler foodproof *L. monocytogenes* detection kit in combination with ShortPrep foodproof II kit. Performance-tested method 070401, *J. AOAC Int.,* 89, 374, 2006.
108. Herman, L.M., De Block, J.H., and Moermans, R.J., Direct detection of *Listeria monocytogenes* in 25 milliliters of raw milk by a two-step PCR with nested primers, *Appl. Environ. Microbiol.,* 61, 817, 1995.
109. Olcen, P. et al., Rapid diagnosis of bacterial meningitis by a semi-nested PCR strategy, *Scand. J. Infect. Dis.,* 27, 537, 1995.
110. Backman, A. et al., Evaluation of an extended diagnostic PCR assay for detection and verification of the common causes of bacterial meningitis in CSF and other biological samples, *Mol. Cell Probes,* 13, 49, 1999.
111. Lawrence, L.M., and Gilmour, A., Incidence of *Listeria* spp., *Listeria monocytogenes* in a poultry processing environment and in poultry products and their rapid confirmation by multiplex PCR, *Appl. Environ. Microbiol.,* 60, 4600, 1994.
112. Hsih, H.Y., and Tsen, H.Y., Combination of immunomagnetic separation and polymerase chain reaction for the simultaneous detection of *Listeria monocytogenes* and *Salmonella* spp. in food samples, *J. Food Protect.,* 64, 1744, 2001.
113. Kanuganti, S.R. et al., Detection of *Listeria monocytogenes* in pigs and pork, *J. Food Prot.,* 65, 1470, 2002.
114. Alarcon, B. et al., Simultaneous and sensitive detection of three food-borne pathogens by multiplex PCR, capillary gel electrophoresis, and laser-induced fluorescence, *J. Agric. Food Chem.,* 52, 7180, 2004.
115. Jofre, A. et al., Simultaneous detection of *Listeria monocytogenes* and *Salmonella* by multiplex PCR in cooked ham, *Food Microbiol.,* 22, 109, 2005.
116. Kim, J. et al., Simultaneous detection by PCR of *Escherichia coli, Listeria monocytogenes* and *Salmonella typhimurium* in artificially inoculated wheat grain, *Int. J. Food Microbiol.*, 111, 21, 2006.
117. Park, Y.S., Lee, S.R., and Kim, Y.G., Detection of *Escherichia coli* O157:H7, *Salmonella* spp., *Staphylococcus aureus* and *Listeria monocytogenes* in kimchi by multiplex polymerase chain reaction (mPCR), *J. Microbiol.,* 44, 92, 2006.
118. Klein, P.G., and Juneja, V.K., Sensitive detection of viable *Listeria monocytogenes* by reverse transcription-PCR, *Appl. Environ. Microbiol.,* 63, 4441, 1997.
119. Novak, J.S., and Juneja, V.K., Detection of heat injury in *Listeria monocytogenes* Scott A, *J. Food Protect.,* 64, 1739, 2001.
120. Navas, J., Ortiz, S., and Martinez-Suarez, J.V., Simultaneous detection of *Listeria monocytogenes* in chicken meat enrichments by PCR and reverse-transcription PCR without DNA/RNA isolation, *J. Food Prot.,* 68, 407, 2005.
121. Jothikumar, N., Wang, X., and Griffiths, N.W., Real-time multiplex SYBR green I-based PCR assay for simultaneous detection of *Salmonella* serovars and *Listeria monocytogenes*, *J. Food Prot.,* 66, 2141, 2003.

122. Rodríguez-Lázaro, D. et al., Rapid quantitative detection of *Listeria monocytogenes* in meat products by real-time PCR, *Appl. Environ. Microbiol.,* 70, 6299, 2004.
123. Rodríguez-Lázaro, D. et al., Rapid quantitative detection of *Listeria monocytogenes* in salmon products: Evaluation of pre-real-time PCR strategies, *J. Food Prot.,* 68, 1467, 2005.
124. Rodríguez-Lázaro, D. et al., A novel real-time PCR for *Listeria monocytogenes* that monitors analytical performance via internal amplification control, *Appl. Environ. Microbiol.*, 71, 9008, 2005.
125. Wang, X., Jothikumar, N., and Griffiths, M.W., Enrichment and DNA extraction protocols for the simultaneous detection of *Salmonella* and *Listeria monocytogenes* in raw sausage meat with multiplex real-time PCR, *J. Food Prot.,* 67, 189, 2004.
126. Jordan, J.A., and Durso, M.B., Real-time polymerase chain reaction for detecting bacterial DNA directly from blood of neonates being evaluated for sepsis, *J. Mol. Diagn.,* 7, 575, 2005.
127. Rudi, K. et al., Detection of viable and dead *Listeria monocytogenes* on Gouda-like cheeses by real-time PCR, *Lett. Appl. Microbiol.,* 40, 301, 2005.
128. Berrada, H. et al., Quantification of *Listeria monocytogenes* in salads by real-time quantitative PCR, *Int. J. Food Microbiol.,* 107, 202, 2006.
129. Guy, R.A. et al., A rapid molecular-based assay for direct quantification of viable bacteria in slaughterhouses, *J. Food Prot.,* 69, 1265, 2006.
130. Rossmanith, P. et al., Detection of *Listeria monocytogenes* in food using a combined enrichment/real-time PCR method targeting the *prfA* gene, *Res. Microbiol.,* 157, 763, 2006.
131. Choi, W.S., and Hong, C.-H., Rapid enumeration of *Listeria monocytogenes* in milk using competitive PCR, *Int. J. Food Microbiol.,* 84, 79, 2003.
132. Blais, B.W. et al., A nucleic acid sequence-based amplification system for detection of *Listeria monocytogenes hlyA* sequences, *Appl. Environ. Microbiol.,* 63, 310, 1997.
133. Norton, D.M., and Batt, C.A., Detection of viable *Listeria monocytogenes* with a 5′ nuclease PCR assay, *Appl. Environ. Microbiol.,* 65, 2122, 1999.
134. Lunge, V.R. et al., Factors affecting the performance of 5′ nuclease PCR assays for *Listeria monocytogenes* detection, *J. Micriobiol. Methods*, 51, 361, 2002.
135. Guilbaud, M. et al., Quantitative detection of *Listeria monocytogenes* in biofilms by real-time PCR, *Appl. Environ. Microbiol.,* 71, 2190, 2005.
136. Oravcova, K. et al., Detection and quantification of *Listeria monocytogenes* by 5′-nuclease polymerase chain reaction targeting the *actA* gene, *Lett. Appl. Microbiol.*, 42, 15, 2006.
137. Lawrence, L.M., Harvey, J., and Gilmour, A., Development of a random amplification of polymorphic DNA typing method for *Listeria monocytogenes*, *Appl. Environ. Microbiol.,* 59, 3117, 1993.
138. Franciosa, G. et al., Characterization of *Listeria monocytogenes* strains involved in invasive and non-invasive listeriosis outbreaks by PCR-based fingerprinting techniques, *Appl. Environ. Microbiol.,* 67, 1793, 2001.
139. Widjojoatmodjo, M.N., Fluit, A.C., and Verhoef, J., Rapid identification of bacteria by PCR-single-strand conformation polymorphism, *J. Clin. Microbiol.,* 32, 3002, 1994.
140. Manzano, M. et al., Single-strand conformation polymorphism (SSCP) analysis of *Listeria monocytogenes iap* gene as tool to detect different serogroups, *Mol. Cell Probes,* 11, 459, 1997.
141. Tominaga, T., Rapid discrimination of *Listeria monocytogenes* strains by microtemperature gradient gel electrophoresis, *J. Clin. Microbiol.,* 44, 2199, 2006.
142. Kerdahi, K.F., and Istafanos, P.F., Rapid determination of *Listeria monocytogenes* by automated enzyme-linked immunoassay and nonradioactive DNA probe, *J. AOAC Int.,* 83, 86, 2000.
143. Garrec, N. et al., Comparison of a cultural method with ListerScreen plus Rapid'*L. mono* or PCR-ELISA methods for the enumeration of *L. monocytogenes* in naturally contaminated sewage sludge, *J. Microbiol. Methods,* 55, 763, 2003b.
144. Oyarzabal, O.A., Behnke, N.M., and Mozola, M.A., Validation of a microwell DNA probe assay for detection of *Listeria* spp. in foods. Performance-tested method 010403, *J. AOAC Int.,* 89, 651, 2006.
145. Wu, S.J., Chan, A., and Kado, C.I., Detection of PCR amplicons from bacterial pathogens using microsphere agglutination. *J. Microbiol. Methods,* 56, 395, 2004.
146. Call, D.R., Borucki, M.K., and Loge, F.J., Detection of bacterial pathogens in environmental samples using DNA microarrays, *J. Microbiol. Methods,* 53, 235, 2003.
147. Borucki, M.K. et al., Selective discrimination of *Listeria monocytogenes* epidemic strains by a mixed-genome DNA microarray compared to discrimination by pulse-filed gel electrophoresis, *J. Clin. Microbiol.,* 42, 5270, 2004.

148. Jin, L.Q. et al., Detection and identification of intestinal pathogenic bacteria by hybridization to oligonucleotide microarrays, *World J. Gastroenterol.,* 11, 7615, 2005.
149. Sergeev, N. et al., Multipathogen oligonucleotide microarray for environmental and biodefense applications, *Biosens. Bioelectron.,* 20, 684, 2004.
150. Borucki, M.K. et al., Suspension microarray with dendrimer signal amplification allows direct and high-throughput subtyping of *Listeria monocytogenes* from genomic DNA, *J. Clin. Microbiol.,* 43, 3255, 2005.
151. Kievits, T. et al., NASBA isothermal enzymatic *in vitro* nucleic acid amplification optimized for the diagnosis of HIV-1 infection, *J. Virol. Methods,* 35, 273, 1991.
152. Cook, N., The use of NASBA for the detection of microbial pathogens in food and environmental samples, *J. Microbiol. Methods,* 53, 165, 2003.
153. Uyttendaele, M. et al., Development of NASBA, a nucleic acid amplification system, for identification of *Listeria monocytogenes* and comparison to ELISA and a modified FDA method, *Int. J. Food Microbiol.,* 27, 77, 1995.
154. Hough, A.J. et al., Rapid enumeration of *Listeria monocytogenes* in artificially contaminated cabbage using real-time polymerase chain reaction, *J. Food Prot.,* 65, 1329, 2002.
155. Wiedmann, M. et al., Discrimination *of Listeria monocytogenes* from other *Listeria* species by ligase chain reaction, *Appl. Environ. Microbiol.,* 58, 3443, 1992.
156. Wiedmann, M., Barany, F., and Batt, C.A., Detection of *Listeria monocytogenes* with a nonisotopic polymerase chain reaction-coupled ligase chain reaction assay, *Appl. Environ. Microbiol.,* 59, 2743, 1993.
157. Aznar R., and Alarcón, B., PCR detection of *Listeria monocytogenes*: A study of multiple factors affecting sensitivity, *J. Appl. Microbiol.,* 95, 958, 2003.
158. Amagliani, G. et al., Direct detection of *Listeria monocytogenes* from milk by magnetic-based DNA isolation and PCR, *Food Microbiol.,* 21, 597, 2004.
159. Agersborg, A., Dahl, R., and Martinez, I., Sample preparation and DNA extraction procedures for polymerase chain reaction identification of *L. monocytogenes* in seafoods, *Int. J. Food Microbiol.,* 35, 275, 1997.
160. Rijpens, N., and Herman, L., Comparison of selective and nonselective primary enrichments for the detection of *Listeria monocytogenes* in cheese, *Int. J. Food Microbiol.,* 94, 15, 2004.
161. Chomczynski, P., and Rymaszewski, M., Alkaline polyethylene glycol-based method for direct PCR from bacteria, eukaryotic tissue samples, and whole blood, *BioTechniques,* 40, 454, 2006.
162. Akane, A. et al., Identification of the heme compound copurified with deoxyribonucleic acid (DNA) from blood strains, a major inhibitor of polymerase chain reaction (PCR) amplification, *J. Forensic Sci.,* 39, 362, 1994.
163. Abu Al-Soud, W., and Radstrom, P., Effects of amplification facilitators on diagnostic PCR in the presence of blood, feces, and meat, *J. Clin. Microbiol.,* 38, 4463, 2000.
164. Wang, J.T. et al., Effects of anticoagulants and storage of blood samples on efficacy of the polymerase chain reaction assay for hepatitis C virus, *J. Clin. Microbiol.*, 30, 750, 1992.
165. Satsangi, J. et al., Effect of heparin on polymerase chain reaction, *Lancet,* 343, 1509, 1994.
166. Khan, G. et al., Inhibitory effects of urine on the polymerase chain reaction for cytomegalovirus DNA, *J. Clin. Pathol.*, 44, 360, 1991.
167. Powell, H.A. et al., Proteinase inhibition of the detection of *Listeria monocytogenes* in milk using the polymerase chain reaction, *Lett. Appl. Microbiol.,* 18, 59, 1994.
168. Bickley, J. et al., Polymerase chain reaction (PCR) detection of *Listeria monocytogenes* in diluted milk and reversal of PCR inhibition caused by calcium ions, *Lett. Appl. Microbiol.,* 22, 153, 1996.
169. Tsai, Y-L., and Olson, B.H., Detection of low numbers of bacterial cells in soils and sediments by polymerase chain reaction, *Appl. Environ. Microbiol.,* 58, 754, 1992.
170. Simon, M.C., Gray, D.I., and Cook, N., DNA extraction and PCR methods for the detection of *Listeria monocytogenes* in cold-smoked salmon, *Appl. Environ. Microbiol.,* 62, 822, 1996.
171. Lampel, K.A., Orlandi, P.A., and Kornegay, L., Improved template preparation for PCR-based assays for detection of food-borne bacterial pathogens, *Appl. Environ. Microbiol.,* 66, 4539, 2000.
172. Stevens, K.A., and Jaykus, L.A., Direct detection of bacterial pathogens in representative dairy products using a combined bacterial concentration-PCR approach, *J. Appl. Microbiol.,* 97, 1115, 2004.
173. Isonhood, J., Drake, M., and Jaykus, L.A., Upstream sample processing facilitates PCR detection of *Listeria monocytogenes* in mayonnaise-based ready-to-eat (RTE) salads, *Food Microbiol.,* 23, 584, 2006.
174. Niederhauser, C., Luthy, J., and Candrian, U., Direct detection of *Listeria monocytogenes* using paramagnetic bead DNA extraction and enzymatic DNA amplification, *Mol. Cell Probes,* 8, 223, 1994.

175. Amagliani, G. et al., Development of a magnetic capture hybridization-PCR assay for *Listeria monocytogenes* direct detection in milk samples, *J. Appl. Microbiol.,* 100, 375, 2006.
176. Lantz, P.G. et al., Enhanced sensitivity in PCR detection of *Listeria monocytogenes* in soft cheese through use of an aqueous two-phase system as a sample preparation method, *Appl. Environ. Microbiol.,* 60, 3416, 1994.
177. Amagliani, G. et al., A combination of diagnostic tools for rapid screening of ovine listeriosis, *Res. Vet. Sci.,* 81, 185, 2006.
178. Uyttendaele, M., Van Hoorde, I., and Debevere, J., The use of immunomagnetic separation (IMS) as a tool in a sample preparation method for direct detection of *L. monocytogenes* in cheese, *Int. J. Food Microbiol.,* 54, 205, 2000.
179. Makino, S., Okada, Y., and Maruyama, T., A new method for direct detection of *Listeria monocytogenes* from foods by PCR, *Appl. Environ. Microbiol.* 61, 3745, 1995.
180. Fluit, A.C. et al., Detection of *Listeria monocytogenes* in cheese with the magnetic immuno-polymerase chain reaction assay, *Appl. Environ. Microbiol.,* 59, 1289, 1993.
181. Gray, K.M., and Bhunia, A.K., Specific detection of cytopathogenic *Listeria monocytogenes* using a two-step method of immunoseparation and cytotoxicity analysis, *J. Microbiol. Methods,* 60, 259, 2005.
182. Kaclikova, E. et al., Separation of *Listeria* from cheese and enrichment media using antibody-coated microbeads and centrifugation, *J. Microbiol. Methods,* 46, 63, 2001.
183. Payne, M.J. et al., The use of immobilized lectins in the separation of *Staphylococcus aureus, Escherichia coli, Listeria* and *Salmonella* spp. from pure cultures and foods, *J. Appl. Bacteriol.,* 73, 41, 1992.
184. Lammerding, A.M., and Doyle, M.P., Evaluation of enrichment procedures for recovering *Listeria monocytogenes* from dairy products, *Int. J. Food Microbiol.,* 9, 249, 1989.
185. Fitter, S., Heuzenroeder, M., and Thomas, C.J., A combined PCR and selective enrichment method for rapid detection of *Listeria monocytogenes, J. Appl. Bacteriol.,* 73, 53, 1992.
186. Partis, L. et al., Inhibitory effects of enrichment media on the Accuprobe test for *Listeria monocytogenes*, *Appl. Environ. Microbiol.,* 60, 1693, 1994.
187. Loncarevic, S., Tham, W., and Danielson-Tham, M.L., The clones of *Listeria monocytogenes* detected in food depend on the method used, *Lett. Appl. Microbiol.,* 22, 381, 1996.
188. Ingianni, A. et al., Rapid detection of *Listeria monocytogenes* in foods, by a combination of PCR and DNA probe, *Mol. Cell. Probes,* 15, 275, 2001.
189. Bauwens, L., Vercammen, F., and Hertsens, A., Detection of pathogenic *Listeria* spp. in zoo animal feces: Use of immunomagnetic separation and a chromogenic isolation medium, *Vet. Microbiol.,* 91, 115, 2003.
190. Beumer, R.R., and Hazeleger, W.C., *Listeria monocytogenes*: Diagnostic problems, *FEMS Immunol. Med. Microbiol.,* 35, 191, 2003.
191. Wallace, F.M., Call, J.E., and Luchansky, J.B., Validation of the USDA/ARS package rinse method for recovery of *Listeria monocytogenes* from naturally contaminated, commercially prepared frankfurters, *J. Food Prot.,* 66, 1920, 2003.
192. Duvall, R.E. et al., Improved DNA probe detection of *Listeria monocytogenes* in enrichment culture after physical-chemical fractionation. *J. AOAC Int.,* 89, 172, 2006.
193. Ueda, S., Maruyama, T., and Kuwabara, Y., Detection of *Listeria monocytogenes* from food samples by PCR after IMS-plating, *Biocontrol. Sci.,* 11, 129, 2006.
194. Duffy, G. et al., The development of a combined surface adhesion and polymerase chain reaction technique in the rapid detection of *Listeria monocytogenes* in meat and poultry, *Int. J. Food Microbiol.,* 49, 151, 1999.
195. Knabel, S.J., Optimized, one-step, recovery-enrichment broth for enhanced detection of *Listeria monocytogenes* in pasteurized milk and hot dogs, *J. AOAC Int.,* 85, 501, 2002.
196. Moon, G.S., Kim, W.J., and Shin, W.S., Optimization of rapid detection of *Escherichia coli* O157:H7 and *Listeria monocytogenes* by PCR and application to field test, *J. Food Prot.,* 67, 1634, 2004.
197. Silk, T.M., Roth, T.M., and Donnelly, C.W., Comparison of growth kinetics for healthy and heat-injured *Listeria monocytogenes* in eight enrichment broths, *J. Food Prot.,* 65, 1333, 2002.
198. D'Agostino, M. et al., A validated PCR-based method to detect *Listeria monocytogenes* using raw milk as food model—Towards an international standard, *J. Food Prot.,* 67, 1646, 2004.

7 Strain Typing

Yi Chen and Stephen J. Knabel

CONTENTS

7.1 INTRODUCTION

7.1.1 DEFINITIONS AND PRINCIPLES

The purpose of this chapter is to describe the principles and advancements in strain typing methods that have been applied to subtyping *Listeria monocytogenes*. Strain typing is the process by which a bacterial species can be further separated into different subgroups or strains. A direct function of strain typing is to discriminate between different strains that belong to the same genus and/or species. Typing techniques have undergone extensive improvements and many new methods have led to significantly enhanced performance according to various criteria. Our understanding of epidemiology, evolution, phylogenetics, and the population genetics of *L. monocytogenes* has greatly evolved with the development of these advanced strain typing techniques.

Applications of strain typing techniques. Three major applications of strain typing are in taxonomy, epidemiology, and phylogeny (evolutionary genetics). Taxonomy, also known as (bio)systematics, is the practice and science of classification of organisms based on their common characteristics.[1]

A species is considered a group of isolates with a common origin or ancestry as demonstrated by a discrete typing unit. However, it is still controversial as to whether a clear and strict species definition should be used, as various pieces of evidence for overlapping populations have been discovered.[1] Good strain typing data always reveal variations at the subspecies level; therefore, analysis of typing data using phylogenetic approaches is useful for taxonomy. Ribosomal RNA genes have traditionally been used as taxonomic markers because these genes are present in all cells, show little or no evidence of horizontal gene transfer, and possess both conserved and variable domains.[2] However, many new strain typing methods have provided further insights into the already assigned species, and many species have been reassigned based on new strain typing data. Therefore, selection of proper targets, quality of the typing data, and analysis of these data are critical for correct classification of microbial species.

New strain typing strategies have evolved that improve discrimination at the subspecies level. The process of strain typing at the subspecies level is often referred to as subtyping. Evolutionary genetics involves the analysis of subtyping data with the goal of inferring evolutionary history (phylogeny) that can explain these subspecies variations.[3] Population genetics involves studying the relationships of various subgroups within a population (genus and/or species). Many phylogenetic computational algorithms have been developed to infer evolutionary history based on subtyping data. Different algorithms do not always produce concordant results, especially when the targets of the subtyping methods undergo extensive intergenomic DNA recombination via conjugation, transformation, transduction, and/or intragenomic recombination.[1] However, recently developed strain typing techniques have greatly enhanced our understanding of the evolution and population genetics of *L. monocytogenes*.[4–6]

Although strain typing methods have been used to study the taxonomy and phylogeny of *L. monocytogenes*, they have mostly been applied to studying the epidemiology of listeriosis and most literature has focused on utilizing strain typing techniques for this purpose. Understanding the basic principles of epidemiology is important for scientists who are developing subtyping strategies. Unlike evolutionary genetics, which focuses on long-term evolution, molecular epidemiology focuses on short-term evolution of microorganisms in order to track the transmission of certain clonal groups during weeks, months, or a few years. Also, a big difference between molecular epidemiology and evolutionary genetics is that molecular epidemiology is somewhat empirical as will be discussed later. Riley[7] gave a detailed review of the principles of molecular epidemiology of infectious diseases. Some basic principles of molecular epidemiology discussed in that book have been expanded and incorporated into this chapter.

Important definitions and basic principles of molecular epidemiology. Epidemiology is defined as the study of the distribution and determinants of infectious diseases.[7] Some goals of epidemiology include the identification of physical sources, routes of transmission of infectious agents, distribution and genetic relationships of different subgroups, etc. Some basic terms and/or concepts in epidemiology are defined in Table 7.1 and should be reviewed before considering epidemiological subtyping. Some of these definitions were proposed by the European Study Group of Epidemiological Markers (ESGEM) and the Molecular Typing Working Group of the Society for Healthcare Epidemiology of America,[8] some were proposed by Riley,[7] and some are proposed in this chapter.

The terms "epidemic" and "outbreak" are often used interchangeably; however, to avoid confusion, the two concepts can be distinguished to differentiate between long-term and short-term spread of an epidemic clone, respectively. The United States Centers for Disease Control and Prevention (CDC) has defined all isolates belonging to the same outbreak as the outbreak strain.[9] However, evidence has shown that isolates involved in the same outbreak can exhibit slightly different subtypes and thus may be assigned to different strains.[9] Given the existence of these slightly different genetic subtypes within epidemic clones and the assumption that all isolates associated with an outbreak are clonally related, we propose "outbreak clone" (Table 7.1) as a more appropriate term than "outbreak strain" when referring to these slightly different strains within an outbreak.

TABLE 7.1
Important Definitions in Epidemiology

Term	Definition	Ref.
Isolate	A collection of cells derived from a primary colony growing on solid media on which the source of the isolates was inoculated	8
Strain	Isolates that have distinct phenotypic and/or genotypic characteristics from other isolates from the same species	8
Source strain	The original strain that spreads and causes an outbreak or epidemic	This chapter
Clone	A strain or group of strains descended asexually from a single ancestral cell (source strain) that has identical or similar phenotypes or genotypes as identified by a specific strain typing method	8
Outbreak	An acute appearance of a cluster of an illness caused by an outbreak clone that occurs in numbers in excess of what is expected for that time and place	7
Outbreak clone	A strain or group of strains descended asexually from a single ancestral cell (source strain) that is involved in one outbreak	This chapter
Epidemic	One or more outbreaks caused by an epidemic clone that survives and spreads over a long period of time	7
Epidemic clone	A strain or group of strains descended asexually from a single ancestral cell (source strain) that is involved in one epidemic and can include several outbreak clones	7
Pandemic	An epidemic that spreads globally that may last for years	7

Before subtyping techniques were developed and applied to the field of epidemiology, epidemiologists relied solely on conventional epidemiological data to track outbreaks. Conventional epidemiological surveillance systems monitor the unusual temporal increases in the number of cases. However, these surveillance systems have some limitations that can be overcome by strain typing methods. A good example of the use of strain typing techniques in an outbreak investigation was described by Bender et al.[10] The authors incorporated pulsed-field gel electrophoresis (PFGE) into the surveillance system for *Salmonella enterica* serotype typhimurium in the state of Minnesota from 1994 to 1998 and concluded that PFGE patterns aided them in assigning priorities for treating patients, focusing investigations, and avoiding unnecessary investigation of concurrent increases in unrelated patterns. In contrast, conventional surveillance caused unnecessary investigation of epidemiologically unrelated cases that were temporally related. Routine PFGE typing allowed the detection of four out of six community-based outbreaks that would not have been detected by traditional surveillance methods, because PFGE efficiently separated outbreak-related isolates from sporadic-case isolates. On the other hand, conventional surveillance systems did not have enough sensitivity to detect outbreaks involving common serotypes, especially in those outbreaks with few cases spanning a period of several weeks. This was confirmed by the fact that conventional surveillance systems failed to detect any outbreaks of these serotypes from 1990 to 1993.[10] PFGE typing allowed investigators to intervene to stop transmission from a variety of sources of infection. In one case PFGE data were instrumental in the recall of a contaminated commercial product.[10] The most commonly used conventional epidemiological approach is called the case-control study, which uses a questionnaire designed to determine the likely source of an outbreak. One problem with the case-control study is that patients often cannot remember what they ate and/or when they ate it. This is especially a problem with listeriosis outbreaks, because the incubation period for humane listeriosis can vary from 7 days up to 60 days. Also, due to the long incubation period of listeriosis and the fact that only specific portions of the population are susceptible, listeriosis outbreaks often appear to occur over a wide geographic and temporal range (i.e., multistate outbreaks). Thus, timely detection of the outbreak is often problematic using conventional epidemiological approaches.[11] Molecular subtyping of *L. monocytogenes* isolates from patients can allow rapid detection of widespread clusters of listeriosis cases and facilitate outbreak recognition and control. Therefore, strain typing techniques have dramatically enhanced our ability to investigate listeriosis outbreaks.[11]

A basic assumption in the field of molecular epidemiology is that isolates that are part of the same chain of transmission are the descendents of the source strain and thus can be referred to as a clone.[12] An empirical guideline is that epidemic isolates within a clonal group always have identical or very similar subtypes and the threshold similarity used to define a clonal group varies between different species and the strain typing techniques utilized. In reality, bacteria are not always totally clonal because recombination constantly alters their clonal population structure.[13] During the spread of the source strain or even during the isolation, passage, and storage of isolates in laboratories, various changes due to recombination can lead to minor genetic variations in the genomes of isolates within the same epidemic clone. Therefore, a good epidemiological subtyping scheme must be able to demonstrate the close relatedness of these clonal subtypes even though they possess minor genetic differences.

It is thus important that epidemiological typing methods target markers that are epidemiologically relevant and the speed at which the molecular markers undergo genetic changes should match the scope and purpose of the relevant study. Long-term epidemiological studies and short-term outbreak investigations may require the targeting of different markers with different variability. To properly select stable and epidemiologically relevant markers for epidemiological typing, an understanding of the physiology, evolution, and genomic structure of the target bacteria is essential. It is notable that the evolutionary relationship between different bacterial isolates might not necessarily be concordant with the epidemiological relatedness in the case of outbreaks with two or more independent sources. However, in the case of the more common single-source food-borne disease outbreaks like listeriosis, the evolutionary relationship is usually concordant with the epidemiological relatedness of bacterial strains.

There are two other applications of molecular subtyping methods related to epidemiology. First, some subtyping methods can differentiate virulent and nonvirulent bacterial strains and identify subtypes and clonal groups that have the ability to cause human disease. Identification, characterization, and tracking of these virulent epidemic strains/clones facilitate the prevention and control of potential outbreaks. Second, subtyping data can also be used to detect the routes by which food-borne pathogens are transmitted throughout the food system. Strains in the same chain of transmission in food processing plants are presumed to be clonally related and have identical or close subtypes. For example, various strain typing techniques have been developed to detect routes of transmission of *L. monocytogenes* in shrimp,[14] meat,[15] salmon,[16] and vegetable processing plants,[17] and fish smoking and slaughter houses.[18] Identifying the routes by which *L. monocytogenes* is transmitted to finished products in processing plants will facilitate establishment of effective intervention strategies to prevent contamination. These types of studies can also improve our understanding of the ecology of *L. monocytogenes* in food processing plants, which can also help lead to more effective intervention strategies. However, routine testing of *L. monocytogenes* in food processing environments is not mandated by USDA/FDA[19]; therefore, a limited number of *L. monocytogenes* isolates from food processing environments are available for this type of study.

7.1.2 Performance Criteria of Strain Typing Techniques

This chapter focuses on molecular epidemiology, which can be defined as the use of molecular biology to study the distribution and determinants of infectious diseases.[7] The following performance criteria have been proposed for molecular epidemiological typing methods.

7.1.2.1 Typeability

Strains in the targeted population can be typed by a specific subtyping method, which means the markers targeted by the subtyping methods need to be present in as many strains as possible. There are some examples of low typeability: (1) Some strains do not have H antigens and thus cannot be assigned a serotype, and (2) strains that do not have plasmids cannot be subtyped by plasmid profiles.

7.1.2.2 Reproducibility

A subtyping method should have both interlaboratory and intralaboratory reproducibility. This criterion is related to the stability of the genetic markers, which is affected by the type of genetic changes involved. For example, single base mutations tend to be relatively rare and thus are more stable, while recombination events due to mobile genetic elements such as plasmids, bacteriophages, and transposons occur more frequently and thus are relatively less stable and reproducible. These latter recombination events are known to occur during regular laboratory handling of bacterial cultures and thus tend to affect both inter- and intralaboratory reproducibility.

7.1.2.3 Discriminatory Power (D)

Discriminatory power describes the ability of a subtyping system to generate distinct and discrete units of information from unrelated isolates.[8] In order to track the routes of transmission of outbreaks, high discriminatory power is needed to separate outbreak-related isolates from nonrelated isolates. Therefore, discriminatory power is one of the most important criteria for molecular subtyping methods. In addition, in many cases not all strains in one species of a bacterial pathogen are virulent. Subtyping schemes are needed to differentiate pathogenic from nonpathogenic strains, track virulent clones, and identify subpopulations that are specific to certain hosts or environments.[7] Early subtyping methods generally lacked enough discriminatory power to separate epidemiologically unrelated isolates and therefore discriminatory power has been the main criterion of various subtyping studies in the past three decades. Hunter et al.[20] proposed a numerical index (Simpson's index; see later discussion) for discriminatory power that is based on the probability that two unrelated strains sampled from the test population will be placed into different typing groups. This index has subsequently been used in various studies to compare the relative discriminatory power of different strain-typing techniques:

$$\text{Simpson's Index} \quad D = 1 - \frac{1}{N(N-1)} \sum_{j=1}^{S} n_j \left(n_j - 1\right),$$

where N = total number of test strains; S = total number of subtypes; and n_j = the number of strains belonging to the jth subtype.

7.1.2.4 Epidemiological Concordance (E)

Epidemiological concordance, sometimes referred to as epidemiological relevance, describes the ability of a typing system to correctly classify into the same clone all epidemiologically related isolates from a well-described outbreak.[8] Therefore, an epidemiologically concordant subtyping method should be able to: (1) cluster isolates that are epidemiologically related with a particular epidemic/outbreak, and (2) separate these isolates from those that are not related to the same epidemic/outbreak.[21] Epidemiological concordance drew less attention than discriminatory power in early studies of molecular subtyping. Many epidemiological typing studies only evaluated discriminatory power without validating the epidemiological concordance of their methods. However, high epidemiological concordance is critical for early recognition of outbreaks. Epidemiologists rely on high epidemiological concordance to detect a cluster of isolates with closely related subtypes for early recognition of outbreaks. A subtyping method with too high discriminatory power may fail to detect such clusters.[22] Therefore, too much or too little discriminatory power is not desirable.

It should always be remembered that a valid typing technique is ultimately judged by its ability to generate results that consistently explain observations made in the real world.[23]

7.1.2.5 Portability

With the development of international and national food manufacturing and distribution systems, many food-borne outbreaks appear as multistate outbreaks. Some epidemics are even multinational. Therefore, the combined effort of health agencies from all different geographic areas is needed to control and investigate these outbreaks/epidemics. This requires that subtyping data be easily exchanged electronically.

7.1.2.6 Practical Concerns (Ease of Use, Cost, and High Throughput)

Practical concerns are also important issues that need to be addressed for all subtyping methods. A subtyping strategy that utilizes complicated and expensive technologies may have excellent discriminatory power and epidemiological relevance, but the accessibility of this subtyping scheme may be limited, especially for small food processing plants, local health agencies, and community microbiology laboratories.[24] In this case, many local outbreaks may fail to be identified rapidly and multistate outbreaks may originate from these local outbreaks. Therefore, simple and inexpensive molecular subtyping methods have important roles in epidemiological investigations. High throughput is another important criterion for molecular subtyping methods. High-throughput methods can analyze a large number of samples at the same time and the cost per sample is usually much lower than with low-throughput methods. High-throughput methods can be used by state and federal health agencies to analyze large sample sets. Another type of high-throughput method, like whole genome microarray analysis, measures variations in large portions of the whole genome, but the number of isolates that can be tested is limited due to the high costs associated with microarrays.[24]

To evaluate the performance criteria of subtyping schemes, the first important step is to select the proper strains for validation. The test strains need to include isolates from a few well-characterized outbreaks and also isolates that are not related to these outbreaks. These latter isolates should be geographically and temporally distinct from those that are associated with outbreaks. It is also preferable that these isolates come from various sources and lineages from the whole population and are not epidemiologically related. A common mistake in developing subtyping methods is to attempt to achieve maximum discriminatory power using all isolates in a collection even though some isolates may be related. This is especially important when developing subtyping methods for epidemiological investigations, because it is critical that these methods not discriminate between those isolates that are epidemiologically related. However, it is often not clear whether or not isolates are truly related.[7] Therefore, isolates can only be presumed to be related or unrelated based on their origin and source information. When selecting well-characterized outbreaks for evaluation of new molecular subtyping methods, it is recommended that several outbreaks caused by a single epidemic clone be selected. Evidence has shown that some epidemic clones of certain microorganisms (i.e., *L. monocytogenes*) caused multiple outbreaks that were separated by several years.[25] Successful identification of these epidemic clones will greatly contribute to studies on the distribution, risk factors, and molecular determinants of infectious diseases.

A sufficient number of diverse isolates representing the entire population is needed in order to be confident in the conclusions reached in these studies. In the case of *L. monocytogenes*, a validation study would ideally select isolates from all three genetic lineages and all serotypes except those serotypes that are very rare. More serotype 1/2a, 1/2b, and 4b isolates should be selected, because they are associated with most clinical cases. The selection of isolates should include those from clinical, food, environmental, and animal sources. The International Life Sciences Institute (ILSI) has a standard *L. monocytogenes* strain collection that includes 25 genetically diverse isolates representing all three genetic lineages and 21 isolates from nine major listeriosis outbreaks.[26] These isolates have been well characterized by serotyping, ribotyping, and PFGE. The 25 genetically diverse isolates are classified into 23 ribotypes and 25 pulsotypes and all isolates within the same outbreak have identical ribotypes and pulsotypes. To allow comparison between laboratories it is strongly recommended that subtyping studies utilize the ILSI standard strain collection. There is no universal rule regarding the total number of isolates used in a given study; however, many recent subtyping studies analyzed more than 100 isolates.[6,27,28]

7.1.3 The Evolution of Molecular Markers for *L. monocytogenes*

Different strain typing techniques often have different targets selected by scientists based on certain criteria. These targets are often molecular markers, which are polymorphic in the whole population and thus each variant is specific to a certain subtype. The targets determine the performance of strain typing techniques and good strain typing techniques always target molecular markers that match the purpose of the study.

Early strain typing methods targeted phenotypes such as growth and morphological and biochemical characteristics (i.e., sugar fermentation) to differentiate bacterial strains. However, phenotypes can be affected by many confounding variables, such as growth conditions, and therefore are often not stable and reproducible markers. With the development of molecular biology techniques, scientists were able to target genetic markers of microorganisms that are not easily affected by laboratory handling and culture conditions. These genotypic methods enhanced discriminatory power because DNA polymorphisms that do not change the phenotype could be detected.

All DNA polymorphisms between different isolates are determined by differences in their genomic sequences. However, when genotypic methods were first developed, direct measurement of DNA sequences was not practical due to the high costs involved. Therefore, scientists utilized technologies such as arbitrary primers, restriction enzymes, hybridization, and polymerase chain reaction (PCR) to convert DNA sequence polymorphisms into electrophoretic gel patterns or hybridization patterns. With the development of DNA sequencing technologies, subtyping methods emerged that directly targeted partial genomic DNA sequences. Now subtyping methods have evolved to target a substantial portion of the whole genome and their performance has been greatly enhanced by various advanced technologies.

7.2 STRAIN TYPING TECHNIQUES

Strain typing methods have been developed and improved over the last three decades, with a general shift from phenotype-based to genotype-based strategies. Every method has its advantages and disadvantages and, thus, it seems that no one method is perfect for all species and purposes. In this chapter, methods that have proved useful for *L. monocytogenes* typing will be discussed.

7.2.1 Phenotypic Methods

Phenotypic methods distinguish isolates by their phenotypes, which are a result of gene expression. These methods are generally only useful for species identification because their discriminatory power within the same species is limited.[7] Subtyping based on biochemical characteristics can be costly and time consuming, depending on the biochemical tests selected (i.e., serotyping) and typeability is still limited for many methods. Also, the metabolic activities of microorganisms are greatly affected by growth conditions. Many confounding variables associated with biochemical characteristics may provide false discrimination.[29] For example, antimicrobial susceptibility typing has not been widely applied to *L. monocytogenes* because it does not always provide consistent results.[30] Overall, phenotypic methods are often not reproducible, because phenotypes are due to gene expressions, which are influenced by growth conditions and can suffer from phenotype switching of bacteria.[31] Common phenotypes used for subtyping *L. monocytogenes* included, but were not limited to, surface antigen structures, virulence, plasmid profiles, phage profiles, and sugar metabolism profiles. Major phenotypic methods used for typing *L. monocytogenes* at the subspecies level will be discussed next.

7.2.1.1 Serotyping

Serotyping is one of the most commonly used phenotypic methods for subtyping Gram-negative bacterial pathogens such as *Samonella* and pathogenic *Escherichia coli*. Serotyping is also an important tool for epidemiological studies of *L. monocytogenes*. Serotyping targets antigenic variations on cell surfaces, including somatic (O), capsular (K), and flagellar (H) antigens. Different tertiary structures of these antigens react with specific polyclonal and monoclonal antibodies from the bloodstream of an animal host. Different strains can then be assigned serotypes based on their reaction patterns with a panel of antibodies. Major limitations of serotyping are that it is expensive and time consuming due to the need to maintain and handle polyclonal or monoclonal antisera, and therefore it is mostly accessible to large reference laboratories that perform routine serotyping. Another problem associated with serotyping is that antigens from different strains may cross-react with different antibodies and some strains may not express typing antigens on their surfaces.[7]

However, serotyping has performed well for subtyping *L. monocytogenes*. Serotyping of *L. monocytogenes* was started as early as 1969.[32] In 1998, McLauchlin and Jones[33] described a standard serotyping protocol for *L. monocytogenes* that identified 13 serotypes (1/2a, 1/2b, 1/2c, 3a, 3b, 3c, 4a, 4b, 4ab, 4c, 4d, 4e, 7) based on the combination of O antigen and H antigen structures (see chapter 5). Most isolates reported in the literature are typeable using this serotyping protocol and cross-reactions between serotypes of *L. monocytogenes* seem not to be a major problem. Nonetheless, serotyping of *L. monocytogenes* yields relatively low discriminatory power. For example, most listeriosis outbreaks are caused by strains from serotypes 4b, 1/2a, and 1/2b, which makes serotyping of limited use for epidemiological investigation purposes.[25] In 1996, the World Health Organization (WHO) performed a multicenter international subtyping study for *L. monocytogenes* using serotyping.[34] The interlaboratory reproducibility was only 61.3% and the intralaboratory reproducibility ranged from 33 to 100% depending on different serotypes, with a medium value of 91%. Serotyping has been used to evaluate the epidemiological concordance of many other molecular subtyping methods. Molecular subtyping methods that provide data concordant with serotyping are generally believed to be epidemiologically and phylogenetically meaningful.[5,27,35]

7.2.1.2 Phage Typing

Before genotypic subtyping methods were widely utilized, phage typing was an important epidemiological typing tool for *L. monocytogenes*. Phage typing separates bacterial pathogens into subspecies based on their susceptibility to lysis by a panel of bacteriophages.[36] Previous studies have identified more than 40 *L. monocytogenes* phages with different host ranges.[37] Isolates are differentiated based on their patterns of susceptibility to different phages. In 1983, Ralovich et al.[38] demonstrated the usefulness of phage typing for epidemiological investigation of *L. monocytogenes* and showed that phage typing and serotyping were concordant. Although phage typing was still found to have limited discriminatory power, it improved subtyping by providing further discrimination between strains with the same serotype, especially serotype 4b, which includes most listeriosis outbreak-associated isolates. In 1984, Audurier et al.[39] utilized 27 phages to subtype listeriosis outbreak isolates and concluded that phage typing had acceptable reproducibility and discriminatory power. Rocourt and her colleagues established the International Center for *Listeria* Phage Typing to study listeriosis outbreaks.[40] Phage typing was the first subtyping method to show that listeriosis outbreaks were mostly food borne.[39]

In 1996, WHO performed a multicenter international subtyping study on phage typing using an international phage set and concluded that phage typing had many disadvantages.[41] For example, not all *L. monocytogenes* strains were typeable by phage typing and inter- and intralaboratory reproducibilities were relatively low because of some unstable phages. Phage typing was also labor intensive and required specialized skills. The study finally concluded that it was difficult to compare phage typing results between different laboratories and therefore standardization was required. However, phage typing has the advantage of allowing analysis of a large set of isolates. For example, Rocourt et al.[42] analyzed more than 16,000 isolates by phage typing in 1 year.

7.2.1.3 Plasmid Profiling

Plasmid profile typing was evaluated for subtyping *L. monocytogenes* in some studies. It was of limited use because not all *L. monocytogenes* isolates contain plasmids and thus are not stable targets. Lebrun et al.[43] found that around 28% of *L. monocytogenes* test strains contained plasmids and among them only 13% of outbreak strains contained plasmids. Isolates can gain or lose plasmids during regular laboratory operations. For example, in the 1998 multistate listeriosis hot dog outbreak, some outbreak isolates contained plasmids while others did not.[44] However, McLauchlin et al.[45] reported that plasmid subtyping can be combined with antibiotic resistance typing to provide acceptable discriminatory power for subtyping *L. monocytogenes*.

The phenotypic methods just described—serotyping, phage typing, and plasmid profiling—have provided some degree of discrimination of *L. monocytogenes* at the subspecies level and have enhanced our understanding of the epidemiology of listeriosis outbreaks. However, the typeability, reproducibility, and discriminatory power of these methods are still not satisfactory in many cases.

7.2.1.4 Multilocus Enzyme Electrophoresis (MEE)

Phenotypic methods underwent a major breakthrough with the invention of multilocus enzyme electrophoresis (MEE) typing. MEE differentiates different isolates by the electrophoretic mobility of major metabolic enzymes.[46] Briefly, water-soluble enzymes are obtained from cell culture, separated by electrophoresis, and stained. The amino acid sequences determine the electrostatic charges of the proteins and therefore variations in amino acid sequences may be reflected by differences in the mobility of the enzymes. The different mobilities of different enzymes produce a protein banding pattern (subtype) unique to each isolate. MEE has been applied to a variety of bacterial pathogens such as *Escherichia coli*, *Salmonella* spp., *Haemophilus influenzae*, *Neisseria meningitides*, and *Streptococcus* spp. and provides very good discriminatory power. In one of the earliest applications of MEE for *L. monocytogenes*, Piffaretti et al.[47] identified 45 subtypes from 175 isolates of *L. monocytogenes*. MEE was then used to help identify routes of transmission of human listeriosis outbreaks and sporadic cases. Boerlin et al.[48] utilized MEE to analyze 181 isolates of *L. monocytogenes* from a variety of sources and concluded that food processing environments were important sources of contamination for meat products.

Overall, MEE has enhanced typeability, reproducibility, and discriminatory power, compared to earlier phenotypic typing methods. In 1989, MEE was used to clarify the population structure of *L. monocytogenes* and indicated that *L. monocytogenes* can be classified into at least two genetic divisions: Division I contains serotype 1/2a and division II contains serotypes 1/2b, 3b, and 4b.[49] This finding was later confirmed and expanded using many other subtyping schemes.[6,27] To assess the reproducibility and discriminatory power of MEE for subtyping *L. monocytogenes*, Caugant et al.[50] performed an international multicenter study and revealed that reproducibility was still a problem for all laboratories. Of the 11 pairs of duplicate strains, only 3–8 were identified as identical in seven laboratories. Some presumably epidemiologically unrelated strains were found to possess identical MEE types. The discriminatory power of MEE (from 0.827 to 0.925 in different laboratories) was relatively low for *L. monocytogenes* compared to other bacterial pathogens, probably because of the highly clonal nature of *L. monocytogenes*.

Many other phenotypic methods have been applied to *L. monocytogenes*, such as biotyping,[51] antibiotic-resistance typing,[52] and monocine typing.[52,53] However, they have not been widely applied and therefore are not discussed here. Readers are referred to the original articles in these references for details.

7.2.2 Genotypic Methods

In general, molecular subtyping techniques can be divided into five categories based on the genetic markers and technologies utilized based on: (1) PCR, (2) restriction pattern, (3) repetitive element, (4) hybridization, and (5) sequence.

7.2.2.1 PCR-Based Typing

Random amplified polymorphic DNA (RAPD). RAPD is a PCR technique widely used for subtyping various bacterial pathogens. Unlike conventional PCR, the target PCR products of RAPD are unknown and arbitrary PCR primers are used. The arbitrary primers are usually 10 mer long and are designed by the researcher or randomly generated by computer. The arbitrary primer can simultaneously anneal to multiple sites in the whole genome and generate multiple PCR amplicons. These products can then be separated by gel electrophoresis and the banding patterns of different isolates compared. Since different strains have different whole genome DNA sequences, RAPD-PCR is expected to generate different banding patterns between different strains.

Williams et al.[54] and Welsh et al.[55] first described this technique and found that polymorphisms generated by RAPD can be useful markers for subtyping microorganisms. In 1992, Mazurier et al.[56] evaluated the epidemiological relevance of RAPD using well-characterized outbreak isolates of *L. monocytogenes* and found that RAPD correctly classified 92 out of 102 isolates into corresponding epidemic groups. However, the discriminatory power of RAPD was not very high in some cases. For example, 31 RAPD profiles were generated from 91 *L. monocytogenes* isolates in another study. Identical RAPD profiles were observed in isolates from different serotypes, especially isolates belonging to serogroup 4, and RAPD could not efficiently distinguish some serotype 4b isolates. Czajka et al.[57] compared serotyping, MEE, and RAPD using multiple arbitrary primers for subtyping isolates from listeriosis cases and found that RAPD with multiple arbitrary primers had enhanced discriminatory power and could separate isolates with identical serotypes or MEE types. In 1996, WHO performed a multicenter international study on RAPD typing of *L. monocytogenes.*[58] Like some other subtyping methods such as phage typing and MEE, reproducibility remained a problem for RAPD. The overall concordance between results from different laboratories varied from 32 to 85%. The low reproducibility was mainly due to the inconsistent quality and concentration of DNA template, the variable quality of the reagents and thermocyclers, and variability in the skills of the operators.[58] Therefore, although RAPD is rapid and simple, the difficulty with standardization limits it application for molecular subtyping of *L. monocytogenes.*

Multiplex PCR. The early prototypes of multiplex PCR did not provide sufficient discriminatory power for subtyping purposes, but Nightingale et al.[59] combined multiplex-PCR with *sigB* sequencing for subtyping *L. monocytogenes* and improved the discriminatory power to 0.91. Chen et al.[44] developed a multiplex scheme for simultaneous species identification of *L. monocytogenes* and genotyping of three epidemic clones of *L. monocytogenes* that caused major listeriosis outbreaks. While lacking discriminatory power, multiplex PCR excels in its low cost and simplicity, which make it potentially useful as a first-step subtyping method for screening large numbers of isolates during outbreak investigations.

7.2.2.2 Restriction Pattern-Based Typing

Amplified fragment length polymorphism (AFLP). AFLP is a highly discriminatory subtyping method first described by Zabeau and Vos in 1993.[60] With AFLP, genomic DNA is purified and treated with two restriction enzymes and then two different restriction-specific adaptors are ligated to ends of the restriction fragments. PCR primers, which are complementary to the adaptors, are designed to selectively amplify the ligated restriction fragments. The PCR amplicons are then analyzed by gel electrophoresis and gel patterns (polymorphisms between and within restriction sites) are used to assign subtypes. Ripabelli et al.[61] and Guerra et al.[62] both developed AFLP schemes for subtyping *L. monocytogenes* and found that, although not discriminatory enough, AFLP results were congruent with serotyping, phage typing, and other subtyping methods, thus confirming the three genetic lineages of *L. monocytogenes*.

Timonen et al.[63] subsequently improved AFLP by careful selection of restriction enzymes. The discriminatory power of their AFLP scheme was greater than 0.999 and the results were congruent with PFGE. They concluded that AFLP is a highly discriminatory and reproducible subtyping scheme that is faster and less labor intensive than PFGE. Fonnesbech et al.[64] systematically compared AFLP with PFGE, ribotyping, RAPD, and RFLP using the same set of *L. monocytogenes* isolates. The authors showed that AFLP provided discriminatory power comparable to PFGE, RAPD, and RFLP; however, none of these methods provided a completely correct picture of the clonal relationship of *L. monocytogenes* isolates. They concluded that these molecular subtyping methods needed to be combined with each other to identify the clonal relationship between different *L. monocytogenes* isolates. A major disadvantage of AFLP is that it requires DNA purification, restriction, and ligation of adaptors, which are technically demanding. Internal variabilities due to incomplete digestion and/or ligation are also known to affect the final banding pattern.[7]

Restriction fragment length polymorphism (RFLP). RFLP is another subtyping technique that targets the polymorphisms within and between restriction sites. Briefly, genomic DNA is purified from cell cultures and cut into fragments using restriction enzymes, and the fragments are separated using gel electrophoresis. Different strains can differ in the distances between restriction sites or in the sequences in the restriction sites, and thus yield different gel patterns. The number of restriction sites in the whole genome varies from 10 to 1000 depending on the type of restriction enzyme used. Some "frequent cutter" restriction enzymes can produce more than 1000 fragments with different sizes. Some "rare cutters" can produce around 10 fragments with sizes ranging from 500 to 800,000 bp.

Because conventional gel electrophoresis is of limited resolution, there are generally three different strategies to perform RFLP analysis. First, PCR can be used to amplify a specific section of the whole genome and the section is then analyzed by RFLP using frequent-cutting restriction enzymes. Second, the whole genome can be analyzed using frequent-cutting restriction enzymes followed by gel electrophoresis and Southern blotting using probes specific to certain genes. When the probes target rDNA genes the method is known as ribotyping.[65] Third, the whole genome can be analyzed using rare-cutting restriction enzymes, which yield fragments with sizes up to 800 kbp (macrorestriction). Conventional gel electrophoresis is not able to resolve these large fragments; therefore (as discussed later), a special technique, called pulsed-field gel electrophoresis (PFGE), is required to accurately separate these large fragments.

PCR-RFLP. PCR-RFLP was used for molecular subtyping of *L. monocytogenes* in some early studies.[66,67] Different genes are selected for this analysis, among which virulence genes and important surface protein genes are the most popular when subtyping *L. monocytogenes*.[67,68] Multiple restriction enzymes are used to enhance the discriminatory power of PCR-RFLP. PCR-RFLP can generate data containing phylogenetic signals when virulence genes are targeted. For example, in 1992, PCR-RFLP analysis of four virulence genes classified *L. monocytogenes* into two subdivisions: Division I contained serotypes 1/2a, 1/2c, and 3c and division II containing serotypes 1/2b, 3b, and 4b.[68] This classification expanded previous findings concerning the population structure of *L. monocytogenes* by MEE[49] and was subsequently confirmed by many other molecular subtyping methods.

Ribotyping. In ribotyping, chromosomal DNA is digested by a frequent-cutting restriction enzyme and small DNA fragments are produced. DNA is then analyzed by Southern blotting with rRNA probes to generate unique banding patterns. An automated ribotyping system, the DuPont Qualicon RiboPrinter®, is used to generate and analyze these banding patterns. Automated ribotyping has been applied to subtype a variety of food-borne pathogens, including *L. monocytogenes*.[35,67,69,70] A Web-based database (www.pathogentracker.net) has been developed to allow exchange of ribotyping data and other subtyping data from various food-borne pathogens, including *L. monocytogenes*. In 1996, WHO performed a multicenter evaluation of ribotyping using well-characterized epidemiologically related *L. monocytogenes* isolates.[71] The discriminatory power of ribotyping for *L. monocytogenes*, as measured by Simpson's index, ranged from 0.83 to 0.88 for six laboratories, indicating relatively poor discriminatory power (<0.95) according to the consensus subtyping guideline proposed by ESGEM.[8] Ribotyping was found to provide unacceptable discriminatory power for serotype 4b isolates of *L. monocytogenes*, which are involved in many listeriosis outbreaks.[70]

The relatively low discriminatory power of ribotyping for subtyping *L. monocytogenes* was subsequently confirmed by Sauders et al.[22] and Nightingale et al.,[5] who found that identical ribotype could be found in different clonal groups of *L. monocytogenes* and, thus, ribotyping tended to overestimate the clusters of listeriosis isolates. However, ribotyping had good intra- and interlaboratory reproducibility.[70] In 1997, Wiedmann et al.[67] utilized ribotyping and PCR-RFLP based on virulence genes to further separate isolates that were previously identified as belonging to division II of *L. monocytogenes* into two distinct genetic lineages containing mainly epidemic and animal isolates, respectively. This finding led to our current understanding of the population structure of *L. monocytogenes*: Lineage I contains serotypes 1/2b, 4b, and 3b; lineage II contains serotypes 1/2a, 1/2c, and 3c; and lineage III contains serotypes 4a, 4b, and 4c. Ribotyping has also been used to successfully determine the distribution of *L. monocytogenes* strains in food processing plants.[69,72]

Pulse field gel electrophoresis (PFGE). PCR-RFLP and ribotyping have enhanced discriminatory power, epidemiological concordance, reproducibility, and typeability compared to phenotypic methods. These methods have aided the epidemiological investigation of listeriosis outbreaks for many years and have also clarified the population structure of *L. monocytogenes*. However, their discriminatory power can be insufficient for epidemiological investigation in some situations. This is probably because they only target restriction enzyme polymorphism in several genes in the whole genome. PFGE targets restriction enzyme polymorphisms spanning the whole genome and thus improved the overall performance of fragment-based subtyping methods, especially in terms of discriminatory power. As a result, macrorestriction PFGE is currently CDC's gold standard subtyping method for investigating the molecular epidemiology of food-borne pathogens, including *L. monocytogenes*.

Schwartz and Cantor[73] first described the use of PFGE as a method to separate large DNA fragments that could not be resolved by conventional gel electrophoresis. Under conventional gel electrophoresis conditions, DNA fragments above 30–50 kb migrate with similar mobility and cannot be accurately separated. During PFGE the electric field is periodically reoriented, which forces DNA fragments to change direction and thus large fragments can be separated from each other according to their molecular weight (Figure 7.1). To prevent mechanical shearing, chromosomal DNA used for PFGE analysis must be prepared using special procedures. To accomplish this, a bacterial suspension is mixed with melted agarose and immobilized after the agarose solidifies, creating an agarose plug. Bacterial cells within the agarose plug are then subject to a lysis solution and then bacterial genomic DNA is digested by a rare-cutting restriction enzyme (macrorestriction). The plugs containing the purified and digested DNA are then placed into wells in agarose gel and separated by PFGE. The preceding steps are relatively labor intensive, which is why PFGE is often time consuming compared to other subtyping methods.

Since it was invented in the late 1980s, PFGE has proven to be very useful for investigating food-borne disease outbreaks, including those due to *L. monocytogenes*. In 1991, Brosch et al.[75] described for the first time the application of PFGE using ApaI, SmaI, and NotI for subtyping *L. monocytogenes* serotype 4b isolates and showed that PFGE provided very high discriminatory power. Buchrieser et al.[76] subsequently applied PFGE to *L. monocytogenes* serogroups 1/2 and 3 and obtained satisfactory discriminatory power. In 1993, Buchrieser et al.[77] applied PFGE to listeriosis outbreak investigations and demonstrated its utility in grouping epidemiologically related isolates and tracking outbreak sources. In 1996, WHO performed a multicenter *L. monocytogenes* PFGE subtyping study using ApaI, SmaI, and AscI.[78] The interlaboratory reproducibility ranged from 79 to 90% and most epidemiologically related isolates were correctly clustered by all four laboratories.

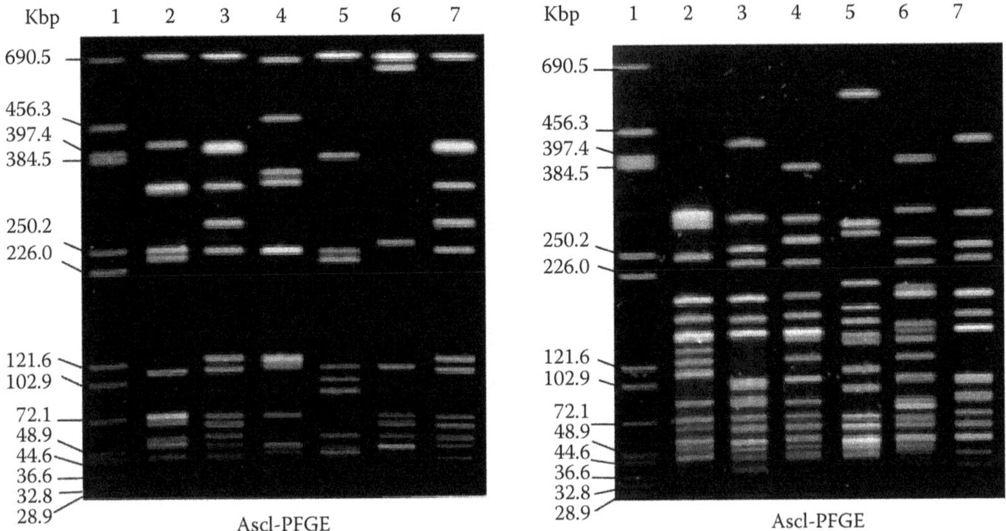

FIGURE 7.1 Typical PFGE gel image of *L. monocytogenes*. (Adapted from Graves, L.M. and Swaminathan, B., *Int. J. Food Microbiol.*, 65, 55, 2001.)

This study confirmed the results of previous studies that PFGE is a highly discriminatory and reproducible method for subtyping *L. monocytogenes* and that PFGE represents a major improvement over earlier fragment-based subtyping methods.

In 1996 the U.S. CDC established PulseNet, a national subtyping and surveillance system for the detection and tracking of food-borne outbreaks in the United States involving public health and food regulatory agency laboratories coordinated by CDC. In this network, participating laboratories utilize standardized PFGE protocols to subtype food-borne pathogens such as *Escherichia coli* O157:H7, *Salmonella*, *Shigella*, *Listeria*, and *Campylobacter*. PFGE patterns are submitted electronically to a shared database at CDC, which allows rapid exchange and comparison of nationwide subtyping data. After PulseNet was introduced, many more outbreaks were detected and sources successfully identified. All 50 U.S. states participated in the PulseNet system in 2001 and the number of PFGE patterns submitted to the database reached 270,000 in 2005.[79] The PulseNet system has now expanded to include Canada, Europe, Asia Pacific, and Latin America (http://www.cdc.gov/pulsenet/).

Graves et al.[74] described CDC's standardized PFGE protocol for subtyping *L. monocytogenes* using two restriction enzymes. The primary enzyme, AscI (Figure 7.2), is used for screening the relatedness of different isolates. AscI PFGE banding patterns have fewer bands with larger sizes and thus are more easily analyzed than those generated by the secondary enzyme, ApaI (Figure 7.3), which is used to confirm the results of the primary enzyme and achieve further discrimination. Insertion and deletion of mobile genetic elements in *L. monocytogenes* is known to affect PFGE patterns.[80] Since small fragments generated by ApaI-PFGE are comparable to the size of many deletions and insertions, ApaI-PFGE patterns can be affected by these changes and therefore are not as stable as those generated using AscI.[9]

Although PFGE provides excellent discriminatory power for subtyping *L. monocytogenes*, PFGE banding patterns are ambiguous and sometimes difficult to interpret accurately. In 1995, Tenover et al.[81] proposed the following criteria for interpretation of PFGE banding patterns:

Isolates that are *indistinguishable* from the source strain are *part* of the outbreak.
Isolates that differ from the outbreak strain by *two to three band differences* in PFGE banding patterns are *probably part* of the outbreak.

Cleavage Position	Length of Sequence (bp)	Length of Sequence (bp) (Sorted)
461298	461301	729598
508175	46877	461301
547300	39125	354158
548754	1454	330926
618675	69921	236971
855646	236971	235616
902725	47079	119797
1632323	729598	116627
1748183	115860	115860
2102341	354158	69921
2337957	235616	47079
2457754	119797	46877
2574381	116627	39125
2905307	330926	1454

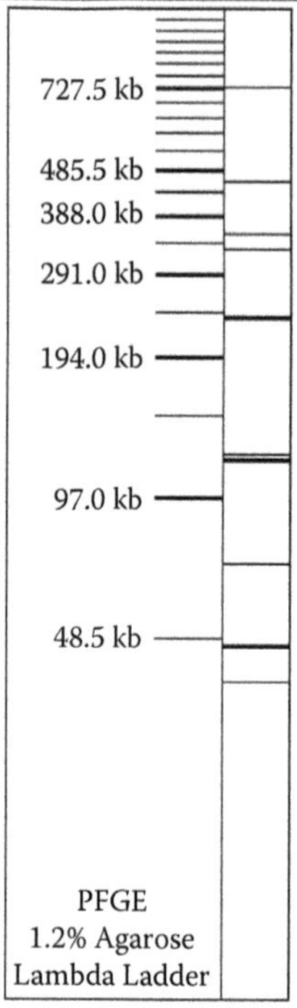

PFGE

727.5 kb
485.5 kb
388.0 kb
291.0 kb
194.0 kb
97.0 kb
48.5 kb

PFGE
1.2% Agarose
Lambda Ladder

FIGURE 7.2 In silico restriction map and PFGE of AscI based on the whole genome sequence of *L. monocytogenes* serotype 4b strain F2365. (Generated from http://insilico.ehu.es.)

Isolates that differ from the outbreak strain by *four to six band differences* in PFGE banding patterns are *possibly part* of the outbreak.

Isolates that differ from the outbreak strain by *more than seven band differences* in PFGE banding patterns are *not part* of the outbreak.

The "Tenover criteria" were subsequently used in many studies in the 10 years following their publication. However, various studies have since shown that the Tenover criteria do not apply in all situations. For example, Barrett et al.[82] found that different PFGE patterns were present in a single chain of *Escherichia coli* O157 transmission and suggested that the source of this outbreak would likely have been unidentified if epidemiologists had relied solely on PFGE data. In this case, PFGE was too discriminatory and separated isolates in the same outbreak clone. Sauders et al.[22] found that when subtyping outbreak isolates of *L. monocytogenes*, PFGE was too discriminatory and thus may miss clusters caused by clonally related isolates that may not have an identical pulsotype.

On the other hand, an identical pulsotype is sometimes detected in two samples that are not epidemiologically linked.[83] In the case of some other food-borne pathogens, PFGE did not provide enough discriminatory power.[84,85] PFGE utilizes gel banding patterns to infer the genetic relatedness of different bacterial isolates; however, Davis et al.[86] found that similarity coefficients between two gel patterns were not good measures of genetic relatedness, because matching bands do not always represent homologous genetic material and there are limitations to the power of PFGE to resolve bands of nearly identical size. The authors suggested that combination of analyses using six or more restriction enzymes might provide a reliable estimate of genetic relatedness without reference to epidemiological data. Singer et al.[87] explored the possible reasons for the discrepancies

Cleavage Position	Length of Sequence (bp)	Length of Sequence (bp) (Sorted)
58116	58122	368385
58814	698	282259
60601	1787	248438
64135	3534	223595
64833	698	187451
66620	1787	162376
138830	72210	148409
171093	32263	146058
270990	99897	144158
388108	117118	138020
532266	144158	117118
814525	282259	102192
1182910	368385	99897
1183568	658	81826
1242029	58461	72210
1247109	5080	58461
1385129	138020	58122
1392746	7617	49259
1541155	148409	48918
1542942	1787	47688
1543640	698	37469
1645832	102192	32263
1647619	1787	15351
1648317	698	7617
1651403	3086	5080
1838854	187451	3534
2062449	223595	3086
2208507	146058	2038
2210545	2038	2038
2211243	698	1787
2260161	48918	1787
2309420	49259	1787
2391246	81826	1787
2438934	47688	698
2440972	2038	698
2441670	698	698
2479139	37469	698
2641515	162376	698
2656866	15351	698
2905304	248438	658

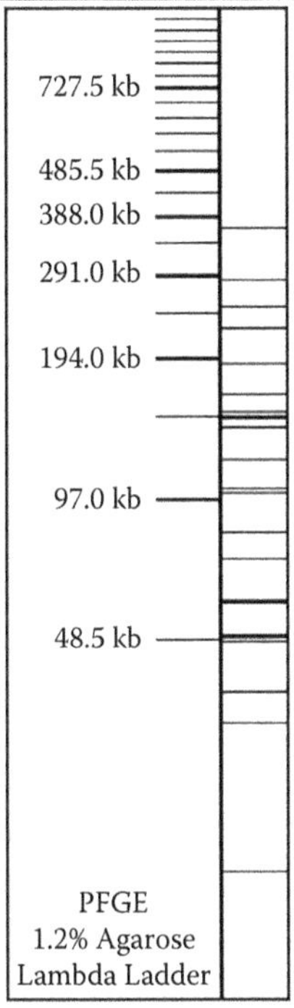

FIGURE 7.3 In silico restriction map and PFGE of ApaI based on the whole genome sequence of *L. monocytogenes* serotype 4b strain F2365. (Generated from http://insilico.ehu.es.)

between PFGE banding patterns and genetic relationships using computer-simulated populations of *Escherichia coli* isolates with a known genetic relationship and found that the performance of PFGE depended on the restriction enzyme used, but the use of multiple enzymes significantly improved the correlation between PFGE banding patterns and the true phylogenetic relationship. Because of these limitations, scientists have recommended that the Tenover criteria be applied only to short-term outbreak investigations (<1 year) when there is already an implied epidemiological linkage.[82]

Fragment-based subtyping methods that utilize PCR technology sometimes suffer from poor reproducibility due to the internal variability of PCR and restriction analyses. The same PCR test on the same culture in different operations may generate slightly different patterns due to the variability of primers, enzymes, buffer, thermocycler, and DNA template. Gel electrophoresis also has internal variability such as uneven lane-to-lane migration of DNA fragments and variation in intensity of bands in separate runs.[7] The presence of multiple bands with similar sizes also confounds the gel pattern analysis. Therefore, many commercial software packages have been developed to aid gel banding pattern recognition and analysis. However, the analysis of gel banding patterns using commercial software is not always reliable and the usefulness and reliability of currently available software is still being debated.

Cardinali et al.[88] compared three analytical systems for DNA banding patterns of *Cryptococcus neoformans* and found that different algorithms provided slightly different topologies using the same set of isolates. However, Gerner-Smidt et al.[89] concluded that the computer software was robust and performed well after evaluating two commercial software packages. In 2002, Rementeria et al.[90] conducted a thorough comparison and evaluation of three commercial software packages for analysis of gel banding patterns for RAPD and PFGE. The authors found general agreement between different software and visual observation, but slight discrepancies still existed. The computerized analysis of different gel images must go through a normalization process that needs to be supervised by operators and all programs require operators to make decisions at some steps; thus, the final results are subjective. Singer et al.[87] found that subjectivity can possibly influence the divergence between gel banding patterns and the true genetic relationship of isolates. A commonly used algorithm for analysis of banding pattern data is UPGMA (unweighted pair group method with arithmetic mean) analysis, which is based on the number of different bands and the number of common bands. However, UPGMA itself is not a good algorithm for inferring the genetic relationship between different bacterial strains.[91] In 2004, Duck et al.[92] showed that parameters of computer software need to be optimized for each species to compensate for the various intra- and intergel variations in PFGE libraries, and that the algorithms used for gel analysis still need to be improved. Therefore, van Belkum et al.[1] suggested that a binary output (numbers or characters) is preferred over gel banding patterns for molecular subtyping strategies.

Another limitation of fragment-based methods such as AFLP, RAPD, and PFGE is that the mechanisms of the variations detected by these methods are poorly understood, which makes it difficult to infer genomic changes between different isolates from gel banding patterns. This limitation was recently overcome by the application of sequence-based typing approaches, because DNA sequences can be used for direct analysis of evolutionary relationship between different bacterial isolates. As a result of these limitations, the CDC started to look for second-generation subtyping methods with both high discriminatory power and high epidemiological relevance for the PulseNet system.[79,93]

Despite the disadvantages discussed here, fragment-based subtyping methods, especially PFGE, have proven very useful in many circumstances and have greatly facilitated the epidemiological investigation of listeriosis outbreaks. Another major advantage of PFGE is that it has been used to subtype tens of thousands of isolates worldwide in the past few decades with the implementation of PulseNet. Therefore, the very large database of clinical, food, and environmental isolates will greatly aid any ongoing or future epidemiological investigations and also help us understand the long-term transmission of important infectious agents.

7.2.2.3 Repetitive Element-Based Typing

Repetitive-PCR (Rep-PCR). Rep-PCR typing has been applied to the epidemiological typing of *L. monocytogenes* in numerous studies. Jersek et al.[94] developed a Rep-PCR scheme targeting short, repetitive extragenic palindromic (REP) elements and enterobacterial repetitive intergenic consensus (ERIC) sequences in *L. monocytogenes* and found that its discriminatory power was very high (0.98). Chou et al.[95] compared PFGE and Rep-PCR using a set of 128 *L. monocytogenes* isolates from human and various seafood sources and showed that Rep-PCR had similar discriminatory power to PFGE. Van Kessel et al.[96] further improved Rep-PCR by automation.

Multiple-locus variable-number tandem repeat analysis (MLVA). The development of MLVA, a revolutionary breakthrough in subtyping methods, is based on repetitive elements. MLVA is one of CDC's candidates for second-generation molecular subtyping. MLVA targets tandem repeat (TR) polymorphisms in the genomes of different bacterial pathogens. PCR primers are designed to amplify all possible TRs in the chromosome based on whole genome sequences. The size and number of repeats at each loci are then analyzed by computer and combinations of these repeats define MLVA types (MTs). Tandem repeats are well recognized as containing phylogenetic signals because the repeats sometimes are targets of evolutionary events such as mutation and recombination and these evolutionary events may change the size and number of the repeats. The number of such repeats at a specific locus is similar among isolates that are closely related and varies among unrelated isolates. TRs also correlate with many genomic changes essential for bacterial survival under stress conditions. Such changes include deletions, insertions, and mutations that affect gene regulation, antigenic shifts, and inactivation of mismatch repair systems.[93] TRs actually play an important role in the adaptation of bacteria, especially those with small genomes.[93] Therefore, MLVA is expected to provide relatively accurate information about the genetic relatedness between different bacterial strains.

Unlike PFGE, the targets of MLVA are specific TRs that can be PCR amplified using primers designed based on whole genome sequences. Thus, MLVA is easier to interpret than PFGE, because the fragments generated by MLVA are of known size and sequence. In addition, the essential steps in MLVA are multiplex PCR and capillary electrophoresis, which are very easy to perform, standardize, and automate, making MLVA a potentially high-throughput subtyping strategy. The final results in MLVA are sizes of each TR loci (Figure 7.4) and thus it is easier to compare than gel banding patterns generated by other fragment-based methods. Like other subtyping schemes, stability of targets is important for development of reproducible and epidemiologically relevant MLVA schemes. Some TRs can be very unstable and thus can potentially separate isolates within the same outbreak clone, which would confound the study of long-term epidemiology. Some extremely unstable TRs may even change during regular laboratory culturing and thus affect the reproducibility of MLVA.

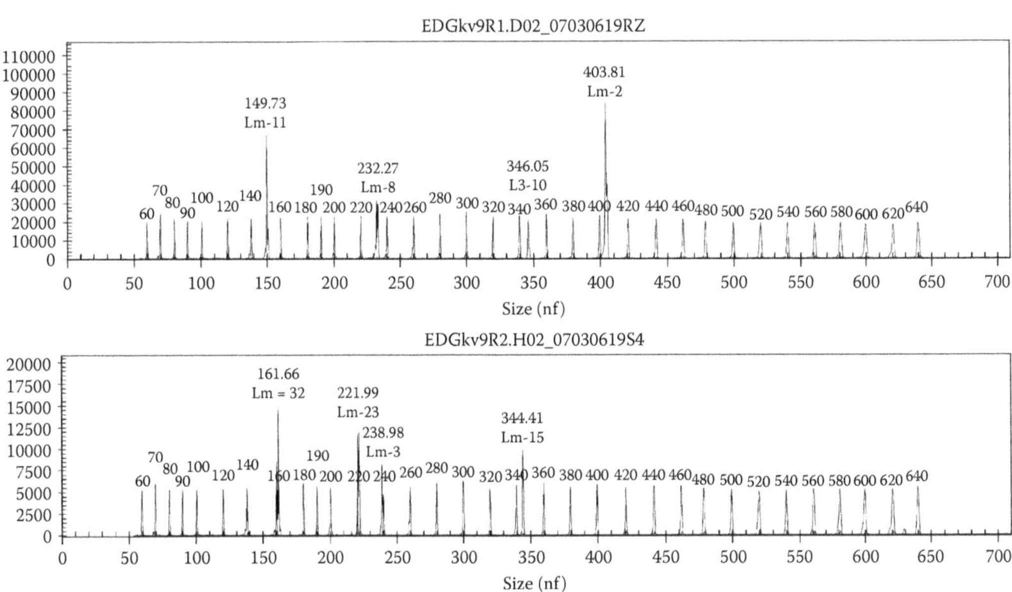

FIGURE 7.4 *Listeria monocytogenes* MLVA. Two multiplex reactions of four loci each performed on the CEQ 8000 (Beckman Coulter). The numbers above the peaks and on the X axis represent the size in nucleotides (nt) of each fragment. The fragments with sizes of multiples of 10 nts are size standards. Other fragments represent amplicons from multiplex PCR designed to amplify each repeat. LM-2, LM-11, LM-32, LM-23, LM-3, and LM-15 are six tandem repeat loci analyzed by multiplex PCR. The sizes of PCR amplicons are used to calculate the number of repeats in each locus. The Y axis represents intensity of fluorescent signals in fluoresecent units. (Image provided by Kate Volpe Sperry from North Carolina State Laboratory of Public Health, Raleigh, North Carolina.)

Another potential drawback of MLVA is that since the primers for amplifying the TRs are designed based on currently available whole genome sequences, it may not be possible to successfully amplify all TRs from all strains in the same species; thus, typeability might be an issue. For example, an insertion within a TR would confound the analysis of the size of the TR. Therefore, selection of TRs for MLVA typing and design of PCR primers are critical to an epidemiologically relevant MLVA scheme. Intensive evaluation and validation are also needed for each MLVA scheme. MLVA has been applied to many food-borne pathogens, such as *Enterococcus faecium, Escherichia coli, Salmonella* spp., *Bacillus anthracis,* and *L. monocytogenes,* and has been proven to yield very high discriminatory power. Hyytia-Trees et al.[93] evaluated the epidemiological relevance of an MLVA scheme for *Escherichia coli* O157:H7 and claimed MLVA possessed promising epidemiological relevance by correctly clustering isolates belonging to eight well-characterized outbreaks.

In 2006, the first MLVA scheme for subtyping *L. monocytogenes* was described by Murphy et al.[98] This MLVA scheme targeted six TRs and correctly separated unrelated food isolates and clustered all serotype 4b isolates. MLVA was also shown to be able to discriminate isolates of the same serotype and correlated with PFGE data from the same set of isolates. However, the study used only 45 isolates of *L. monocytogenes.* More well-characterized isolates are needed to further evaluate the utility of MLVA typing for epidemiological investigation of *L. monocytogenes.* Although MLVA is a fragment-based method, the utilization of meaningful molecular markers, PCR, and capillary electrophoresis generates a more phylogenetically meaningful and nonambiguous output, which provides a major evolution over other fragment-based subtyping methods.

7.2.2.4 Hybridization-Based Typing

Another type of molecular subtyping method is based on hybridization using a panel of subtype-specific DNA probes. Different strains have their own specific genomic regions and may react with the panel of probes to generate different patterns. Liu et al.[99] developed a Southern-hybridization scheme using several serotype-, lineage-, and virulence-specific probes and correctly classified lineage III isolates of *L. monocytogenes* into three subgroups. The performance of hybridization-based typing methods is determined by the number and selection of appropriate specific probes. When the number of probes is high, the technology is referred to as DNA array technology. Rudi et al.[100] developed a DNA array scheme for subtyping *L. monocytogenes* using 29 sequence-specific oligonucleotide probes targeting virulence genes such as *hlyA*, *iap*, *flaA*, *inlA,* and *actA*. The method provided comparable discriminatory power to AFLP. Doumith et al.[80] developed an array typing scheme and demonstrated its usefulness in epidemiological investigation of listeriosis outbreaks. The authors concluded that a big advantage of the array typing scheme over PFGE is that the genetic basis of strain variations can be inferred from the hybridization patterns.

DNA microarrays. DNA microarray is an extension of the hybridization-based array technology. It is a powerful tool for molecular subtyping microorganisms. Thousands of oligonucleotide probes are designed based on whole genome sequences and spotted onto small solid surfaces, such as glass, plastic, and silicon chips. The genomic DNA of "unknown" isolates is extracted, labeled, and hybridized to the oligonucleotide array. DNA microarrays can detect the presence or absence of genomic regions that are complementary to the oligonucleotide probes, which allows identification of sequences unique to each strain. The length of the oligonucleotide probes can vary, depending upon the different markers targeted. For example, if the lengths of the oligonucleotides are small enough, microarrays can detect single-nucleotide mutations. The biggest advantage of microarrays over other methods is that they can simultaneously detect variations (i.e., chromosomal rearrangements and insertion/deletions) throughout the entire genomes of many isolates. Microarrays can be used as a high-throughput subtyping tool with a low cost per region. However, the overall cost of microarrays is still very high.

Data analysis is a major challenge of microarray technology. When genomic DNA hybridizes to the array chip, not all regions generate absolute positive and negative signals. Random partial hybridization creates noise, which can pose a big problem regarding the reproducibility of microarray data. Therefore, how to filter out this noise and target only the true signals is the key to successful microarray analysis. Different materials and the labeling effects of different dyes can generate artifacts, and therefore microarray procedures must be standardized and data need to be normalized. A program—Minimum Information about a Microarray Experiment (MIAME)—was proposed to facilitate reproducible and unambiguous interpretation of microarray data.[101] The proposed principles include experimental design, samples used, extract preparation and labeling, hybridization procedures and parameters, measurement data, specifications, and array design.[101]

Microarrays have been widely used for molecular classification and genomic mutation studies. Borucki et al.[102] developed a mixed-genome microarray containing 629 probes for subtyping *L. monocytogenes*. The mixed-genome array provided high epidemiological relevance by correctly grouping isolates from well-identified outbreaks and was more discriminatory than ribotyping and MLST using six housekeeping genes. The microarray generated data that were congruent with PFGE, ribotyping, and serotyping and also had acceptable reproducibility. The authors concluded that microarrays seem to be a promising method for subtyping *L. monocytogenes*. Kathariou et al.[103] used a microarray based on chromosomal sequences to confirm that the 1998 and 2002 listeriosis outbreaks were caused by a single epidemic clone, ECII.

Another big advantage of microarray technology is that it does not require prior knowledge of genome sequences of test strains and it will reveal a large number of previously unidentified genomic regions that are specific to various species and subspecies and therefore may be functionally important. Doumith et al.[4] developed a DNA microarray based on whole genome sequences of two *L. monocytogenes* strains and microarray analysis identified many lineage and serotype-specific genomic regions in more than 100 different test strains of *L. monocytogenes*.

Major phenotypic and genotypic methods have been discussed regarding their applications, advantages, and disadvantages. In some studies, phenotypic and genotypic methods were combined to achieve better discrimination. For example, Liu et al.[99] combined PCR, Southern blotting, and sugar metabolism to subtype *L. monocytogenes* lineage III isolates. Rudi et al.[104] described an interesting approach of combining genotypes and gene expression patterns of *hly* genes for subtyping *L. monocytogenes*. The significance of this study was that the subtyping scheme directly targeted virulence properties. However, the phylogenetic trees produced by genotype and gene expression patterns were not consistent.

7.2.2.5 DNA Sequence-Based Typing

From the mid-1990s, DNA sequence-based subtyping methods have gained in popularity due to the increased availability of whole genome sequences, high and unambiguous information content of sequence data, increased cost efficiency, speed of DNA sequencing, and the ability to analyze and share sequence data via the Internet.[105] In general, sequence-based methods are more reproducible and accurate than fragment-based methods, because PCR and DNA sequencing are inherently specific and highly informative, while fragment-based methods tend to be more ambiguous and less informative. However, there are thousands of genes in a bacterial genome and it is difficult to find a gene that can represent the whole organism. Genes that undergo intensive recombination are especially unsuitable for inferring genetic relatedness of different strains. Sequence typing strategies that target multiple genes that are not subject to recombination were subsequently developed to prevent this problem.[106]

One of the earliest applications of sequence-based typing of *L. monocytogenes* was described by Rasmussen et al.[107] in 1995. Three virulence genes (*fla*, *hly*, and *iap*) and the 23S rRNA gene were sequenced in a set of *L. monocytogenes* isolates from various serotypes and origins and the results were congruent with PFGE, serotyping, and MEE. More importantly, the evolutionary analysis based on DNA sequences revealed for the first time that *L. monocytogenes* could be divided into three evolutionary lineages.

Multilocus sequence typing (MLST). MLST, which evolved from MEE, was developed by Maiden et al.[108] in 1998. Unlike MEE, which targets the electrophoretic mobility of multiple enzymes, MLST directly targets the sequences of housekeeping genes, which are essential for cell survival and reproduction. In brief, 6–14 genes are selected from the whole genome of the bacterial species, internal fragments (400–600 bp) of each gene are sequenced, and the concatenated sequences are used for discrimination between different strains. To avoid problems with horizontal gene transfer, the 6–14 genes are generally selected so that they are evenly distributed throughout the genome (Table 7.2). Fragments with a size of 400–600 bp are selected because they can be easily sequenced by rapid automated sequencing (Figure 7.5).

One of the major advantages of sequence-based methods over fragment-based methods is that DNA sequences can be used for direct and reliable phylogenetic analysis and therefore are expected to provide more accurate information on the relatedness of different strains. Sequence data are also more portable and unambiguous than fragment-based data and thus are easier to compare and interpret. MLST is not only suitable for epidemiological typing, but also suitable for studying the population structure and genetics of bacterial species, which in turn can help scientists understand the distribution and long-term spread of epidemic clones. MLST has been applied to various food-borne pathogens, including *Staphylococcus aureus, Streptococcus pyogenes, E. coli* O157:H7, *Salmonella* spp., and *L. monocytogenes,* and has greatly facilitated our understanding of the epidemiology and the population genetics of these species.

Salcedo et al.[109] developed an MLST scheme using nine housekeeping genes for *L. monocytogenes* and identified 29 sequence types from 62 isolates. They found that MLST results were congruent with PFGE results. Meinersmann et al.[110] developed an MLST scheme for subtyping *L. monocytogenes* using hypervariable housekeeping genes, which improved discriminatory power. Selection of proper genes is the key to an MLST scheme. Housekeeping genes generally evolve relatively slowly and thus are highly conserved. Therefore, while they are good markers for studying the population structure of bacterial pathogens, they lack the ability to discriminate between closely related strains. To enhance the discriminatory power of MLST, Cai et al.[111] incorporated virulence genes into an MLST scheme for subtyping *L. monocytogenes* and separated 15 isolates into 14 sequence types.

Zhang et al.[112] subsequently developed an MLST scheme termed multivirulence-locus sequence typing (MVLST), which was based solely on virulence gene sequences. MVLST was able to separate 28 diverse *L. monocytogenes* isolates into 28 sequence types and was more discriminatory than ApaI-PFGE. Virulence genes are believed to be more susceptible to recombination than housekeeping genes and that is probably why virulence genes were not at first incorporated into MLST schemes to study the molecular epidemiology of *L. monocytogenes.*[113] Chen et al.[27] evaluated the epidemiological relevance of MVLST using 106 well-identified outbreak and nonoutbreak isolates of *L. monocytogenes* and showed that the discriminatory power of MVLST was 0.99 according to Simpson's index and the epidemiological concordance was 1.00 according to the formula provided by ESGEM.[8] In that study, MVLST correctly identified three known epidemic clones (ECI, ECII, ECIII), redefined another epidemic clone (ECIV), and correctly clustered isolates into the three known evolutionary lineages (I, II, III) of *L. monocytogenes.* While MVLST results were congruent with most fragment-based methods, MVLST was the first sequence typing scheme that identified the three known epidemic clones of *L. monocytogenes.* Unlike other fragment-based methods, MVLST does not need to be combined with other subtyping methods to provide accurate information on epidemic clones, serotypes, and lineages of *L. monocytogenes.*[27]

Sequence-based typing is a newly developed subtyping method that targets polymorphisms in DNA sequences, of which the single nucleotide polymorphism (SNP) is an important type. While whole genome sequencing is still relatively expensive and impractical for molecular epidemiology, direct detection of SNPs has evolved as a next-generation subtyping strategy. SNP typing takes advantage of rapid advancements in genomic sequencing technologies. Unlike direct sequencing of a fragment in the genome, SNP typing directly targets SNPs in the whole genome and thus has

TABLE 7.2
Genes and Intergenic Regions of *L. monocytogenes* Targeted in Various MLST Schemes

Category	Gene	Encoded Protein	Size of Gene Fragment Analyzed (bp)	Percentage of Polymorphic Sites (%)	Ref.
Housekeeping genes	*prs*	Phosphoribosyl synthetase	957	4.9	111
	recA	Recombinase A	7047	10.2	111
	abcZ	ABC transporter	552	6.1	109
	bglA	Beta-glucosidase	417	5.0	109
	cat	Catalase	501	7.8	109
	dapE	Succinyl diaminopimelate dessucinylase	480	8.3	109
	dat	D-amino acid aminotransferase	484	12.4	109
	ldh	L-lactate dehydrogenase	354	4.3	109
	lhkA	Histidine kinase	488	3.5	109
	pgm	Phosphoglucomutase	364	4.4	109
	sod	Superoxide dismutase	420	2.9	109
	gyrB	DNA gyrase subunit B	627	6.3	110
	hisJ	Histidinol phosphate phosphatase	530	13.9	110
	cbiE	Precorrin methylase	537	19.3	110
	rnhB	Ribonuclease H rnh	275	14.6	110
	truB	tRNA pseudouridine 55 synthase	880	20.4	110
	ribC	Riboflavin kinase and FAD synthase	885	17.5	110
	comEA	Integral membrane protein ComEA	482	19.3	110
	purM	Phosphoribosylaminoimidazole synthetase	720	20.8	110
	aroE	5-Enolpyruvylshikimate-3-phosphate synthase	951	20.4	110
	hisC	Histidinol-phosphate aminotransferase	809	17.4	110
	addB	ATP-dependent deoxyribonuclease (subunit B)	1048	17.7	110
	gap	Glyceraldehyde 3-phosphate dehydrogenase	569	1.6	5
	Prs	Phosphoribosyl pyrophosphate synthetase	663	6.9	5
	purM	Phosphoribosylaminoimidazole synthetase	714	20.4	5
	ribC	Riboflavin kinase and FAD synthase	639	15.9	5
	betL	Glycine betaine transporter BetL	650	7.8	115
	gyrB[a]	DNA gyrase subunit B	708	4.9	115
	pgm	Phosphoglycerate mutase	768	4.4	115
	recA	Recombinase A	655	7.3	115
Virulence-associated genes	*actA*	Protein involved in actin tail formation	1929	14.3	111
	inlA	Internalin A	2235	8.3	111
	sigB	Stress responsive alternative sigma factor B	780	10.1	111
	inlB	Internalin B	433	12.24	112
	inlC	Internalin C	418	8.31	112
	prfA	listeriolysin positive regulatory protein	469	8.10	112
	lisR	Two-component response regulator	448	14.1	112
	clpP	ATP-dependent Clp protease proteolytic subunit	419	9.6	112
	dal	Alanine racemase	441	19.1	112
	actA	Protein involved in actin tail formation	561	16.5	5
	inlA	Internalin A	771	9.3	5
	sigB	Stress responsive alternative sigma factor B	666	5.7	5
	actA	Protein involved in actin tail formation	963	10.1	115
	hlyA	Listeriolysin O precursor	830	3.8	115

(continued on next page)

TABLE 7.2 (continued)
Genes and Intergenic Regions of *L. monocytogenes* Targeted in Various MLST Schemes

Category	Gene	Encoded Protein	Size of Gene Fragment Analyzed (bp)	Percentage of Polymorphic Sites (%)	Ref.
Intergenic regions	*Mpl-hly*	Intergenic	335	8.4	111
	Hly-plc	Intergenic	242	10.7	111

[a] The same genes that are analyzed in different studies were listed in different rows.

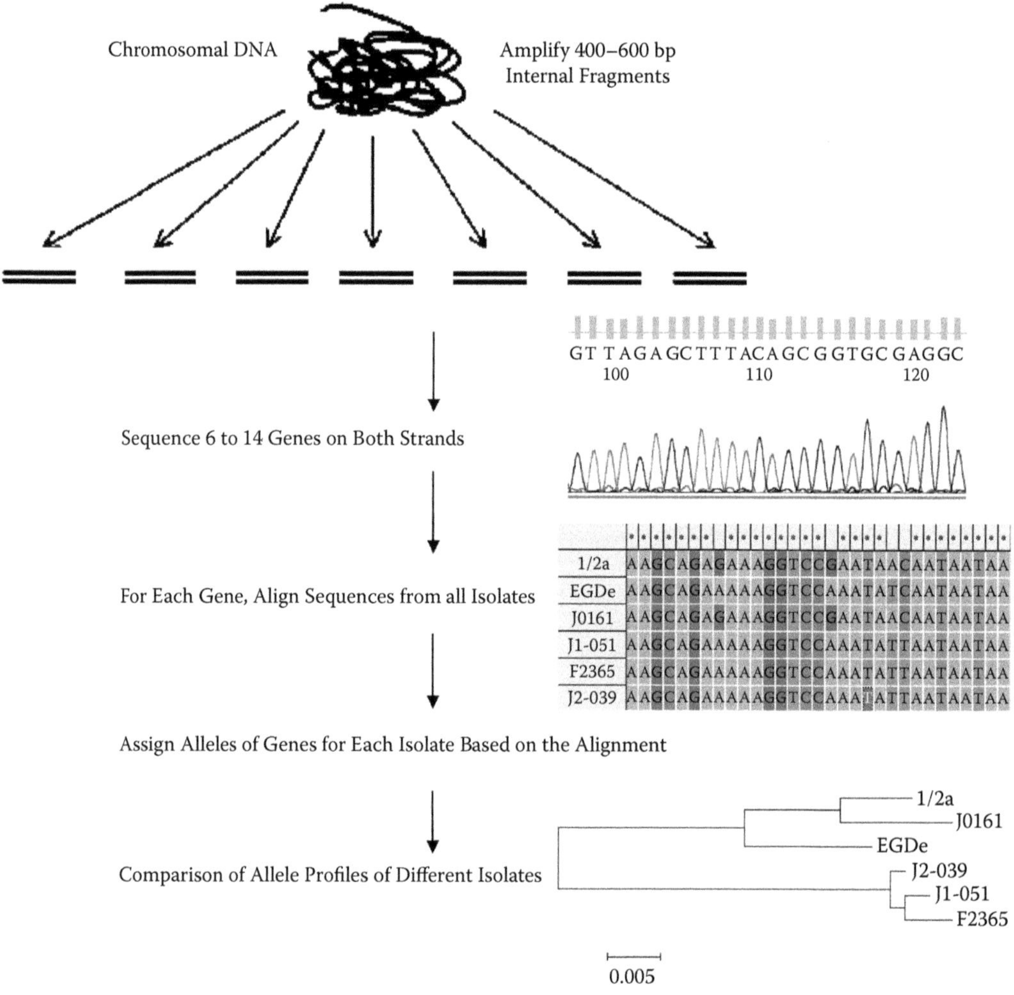

FIGURE 7.5 Flow diagram of MLST. (Adapted from Feil, E.J. and Spratt, B.G., *Annu. Rev. Microbiol.*, 55, 561, 2001.)

the potential to be more rapid and cost efficient than MLST-based schemes. SNP typing lends itself to high-throughput formats in contrast to traditional subtyping methods, such as PFGE and MLST. Major SNP typing techniques are discussed next. Readers can refer to Sobrino et al.[116] and Dearlove

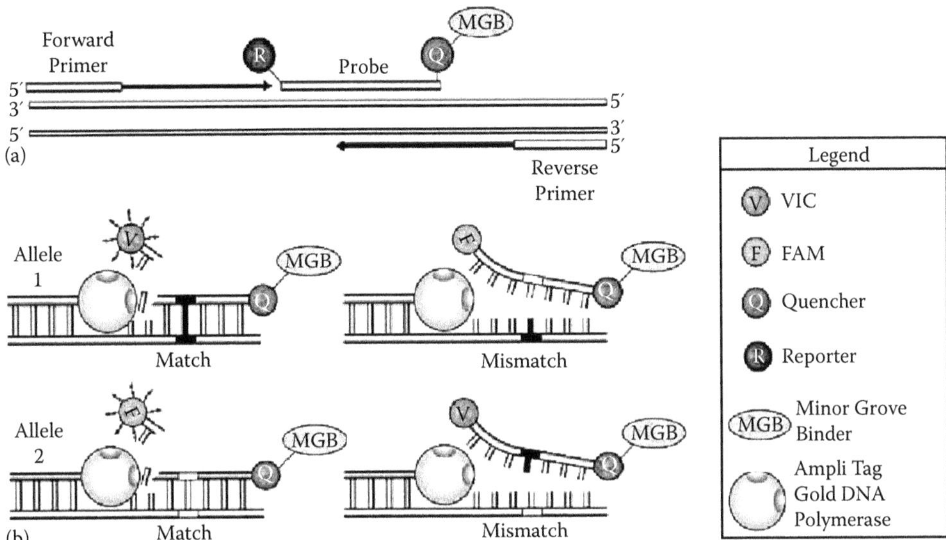

FIGURE 7.6 Flow diagram of Taqman SNP typing. (a) Primer pair amplifies the target and the fluorescent labeled probe bind to the middle of the PCR amplicon. The probe has a fluorescent molecule (reporter) attached facing the forward primer and a quencher is attached on the other side of the probe. The fluorescent molecule does not show fluorescence because of the nearby quencher. (b) If the probe sequence matches the PCR amplicon, probe binds to the amplicon and the extension of forward primer cuts the fluorescent molecule from the probe. Fluorescent molecule shows fluorescence after released and stay away from the quencher. If the probe sequence has a mismatch to the PCR amplicon, probe does not bind to the amplicon and the extension of forward primer does not cut the fluorescent molecule from the probe. The match probe and mismatch probe are labeled by different fluorescent molecules (VIC and FAM, respectively). (De la Vega, F.M. et al., *Mutat. Res.*, 573, 111, 2005.)

et al.[117] for a detailed review of current well-developed SNP typing techniques. There are two major types of SNP typing techniques: allele-specific hybridization and primer extension.

Hybridization-based SNP typing. Allele-specific hybridization is based on hybridization to genomic regions that differ at the SNP site using probes specific to each allele. Taqman® SNP assay (Applied Biosystems, California) is a representative of this type of SNP technique.[118] The Taqman assay targets binary SNP sites and is based on the 5′ nuclease activity of Taq polymerase. Two probes that differ at the SNP site are designed; one matches the test allele of the targeted SNP and the other matches the control allele. The two probes are labeled by different fluorescent dyes attached to their 5′ end and a quencher attached to its 3′ end. The quencher inhibits the fluorescence by the fluorescence resonance energy transfer (FRET) probe. A primer binds upstream of the targeted SNP and Taq polymerase initiates DNA extension. If the probe matches the test strain, then DNA synthesis extends to the SNP site and Taq polymerase cleaves and releases the 5′ fluorescent end of the FRET probe, which results in fluorescence (Figure 7.6). The Taqman assay is fast and accurate and permits high throughput. It can detect up to 100 SNPs within hours. The advantages of this assay compared to other SNP assays are that it only requires a one-step enzymatic reaction and all reactions use universal PCR conditions. The workflow is very simple and can be easily automated. However, the Taqman assay cannot be multiplexed and therefore the cost is high compared to some other SNP detection methods. Some other SNP typing methods based on allele-specific hybridization include but are not limited to LightCycler®,[119] molecular beacons,[120] and dynamic allele-specific hybridization (DASH).[121]

An extension of allele-specific hybridization is microarray-based hybridization, which is also called microarray resequencing.[122] In this strategy, hundreds to thousands of allele-specific oligonucleotides are designed and attached to a solid support and are then hybridized with fluorescent genomic fragments containing SNP sites. This method can detect many SNPs from a large portion of the whole genome. Zhang et al.[123] developed an SNP typing scheme using microarray resequencing that targeted 906 SNPs in *E. coli* O157. This typing scheme allowed discrimination of 11 outbreak-associated isolates and provided insights on the genomic diversity and evolution of *E. coli* O157. One disadvantage of this method is that it is technically demanding with an overall high cost. The GeneChip® (Affymetrix) system improves the performance of this technology by utilizing a tiling strategy[124] and has been applied to genotyping various organisms.

There are two variants of hybridization-based SNP typing methods. One is allele-specific oligonucleotide ligation.[125] In brief, a probe common to both alleles is designed and binds to the immediately downstream regions of the SNP, and two allele-specific probes with the 3′ end nucleotide complementary to each allele of the SNP compete to anneal to the DNA regions immediately upstream of the common probe. The allele-specific probe matching the SNP will ligate to the common probe in the presence of ligase. The ligation products are then amplified by PCR. SNPlex™ genotyping system (Applied Biosystems, California) is a representative of this technology. Allele-specific ligation products are detected by PCR using ZipChute probes, which are fluorescently labeled and hybridized to the complementary sequences that are part of allele-specific amplicons. These probes are eluted and detected by electrophoresis using Applied Biosystem's 3730 or 3730*xl* DNA Analyzers. The major advantage of the SNplex system is that it can detect up to 48 SNPs per reaction and therefore it is suitable for large-scale SNP analysis. However, the design and optimization of the multiplex system is time consuming. De la Vega et al.[126] reported a comparison between Taqman and SNPlex typing. The concordance between these two methods was 99%. The overall cost of SNPlex typing is higher than pyrosequencing and Taqman typing, but the cost per SNP is lower. Other allele-specific ligation-based SNP strategies include but are not limited to Illumina genotyping (Illumina Inc., www.illumina.com), genotyping using Padlock probes,[127] and sequence-coded separation.[128]

The other variant of hybridization-based SNP typing methods is based on invasive cleavage. The Invader® assay (Third Wave™ Technologies, www.twt.com) is a representative of this SNP typing strategy currently used in many clinical diagnostics. An invading probe complementary to the upstream sequence of the SNP site binds to the PCR-amplified DNA template. Another allele-specific probe is designed so that the 3′ region is complementary to the downstream sequence of the SNP site (including the SNP site) and the 5′ region (arm) is not complementary to the DNA template. When the allele-specific probe binds to the DNA template, it will overlap with the invading probe and the structure can be recognized and cleaved by the Flap endonuclease, releasing the 5′ arm of the allele-specific probe. The arm sequence then binds to a complementary FRET probe and forms an invasive cleavage structure. A cleavage enzyme recognizes the structure and cuts the FRET probe releasing the fluorescent dye. The advantage of invader assays is that the PCR amplification of template DNA and subsequent SNP detection occur in the same tube and thus is labor efficient; however, invader assays cannot be multiplexed.

Primer extension-based SNP typing. Primer extension is based on the principle of PCR technology. A primer binds to the DNA region upstream of the SNP site and is extended by a single ddNTP specific to the SNP site using DNA polymerase. This technology is also called minisequencing. SNaPshot™ multiplex SNP typing (Applied Biosystems) is one of the most common commercial technologies based on minisequencing. In this typing system, several fragments of genomic DNA are targeted and amplified by multiplex PCR. The SNaPshot primers are designed to anneal immediately upstream of the SNP site and are extended by one fluorescently labeled dideoxy nucleotide (ddNTP). Different ddNTPs are labeled by different dyes and emit light with different wavelengths. The reaction is performed using a DNA sequencer and subtypes are determined by the color and size of fluorescent fragments. Several primers can be multiplexed into one SNaPshot reaction for simultaneous detection of several SNPs. Hommais et al.[129] developed a SNaPshot typing system targeting 13 SNPs for subtyping *Escherichia coli* and showed that the results of SNaPshot typing

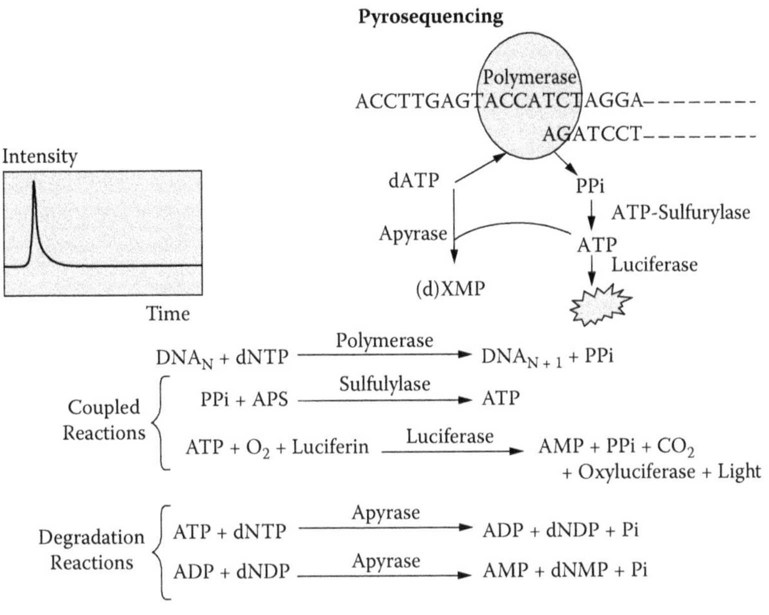

FIGURE 7.7 Flow diagram of pyrosequencing. A primer anneals to one strand of a PCR amplified DNA template, and they are incubated with DNA polymerase, ATP sulfurylase, luciferase and apyrase, adenosine 5′ phosphosulfate (APS), and luciferin. Deoxynubleotide triphosphates (dNTP) is then added to the reaction and incorporated into the DNA template by DNA polymerase. The incorporation releases pyrophosphate (PPi). ATP sulfurylase then converts PPi to ATP in the presence of adenosine 5′ phosphosulfate. Luciferase then converts this ATP into luciferin that generates visible light which can be detected by a charge coupled device (CCD) camera. Unincorporated dNTPs and excess ATP are degraded by apyrase. Addition of dNTPs is performed one at a time. As the process continues, the complementary DNA strand is entended and the nucleotide sequence is determined from the visible light signals. (Adapted from www.pyrosequencing.com.)

were consistent with results obtained by MEE and ribotyping. The SNaPshot multiplex system has a relatively low cost, but it requires prior PCR amplifications and purification of genomic regions containing SNP sites. Other minisequencing-based SNP detection methods include but are not limited to matrix assisted laser desorption/ionization time of flight mass spectrometry (MALDI-TOF MS),[130] Sequenom MassARRAY,[131] arrayed primer extension (APEX),[132] and Luminex xMAP fluorescent polystyrene microspheres (Luminex Corporation).[28] Ducey et al.[133] developed an SNP typing method utilizing Luminex xMAP technology for subtyping *L. monocytogenes* isolates and the strategy provided good discriminatory power when analyzing lineage I isolates of *L. monocytogenes*.

Pyrosequencing is another popular minisequencing SNP detection method based on primer extension. The technology utilizes an enzyme cascade system and generates a light signal whenever a dNTP binds to the DNA template (Figure 7.7). In brief, sequencing primer is mixed with DNA template, DNA polymerase, ATP sulfurylase, luciferase and apyrase, adenosine 5′ phosphosulfate (APS), and luciferin. Deoxyribonucleotide triphosphates (dNTP) are then added to the reaction, one base solution (A, T, G, or C) after the other with washing in between. If a specific nucleotide base can bind to the DNA template, equimolar pyrophosphate (PPi) will be released and subsequently converted to ATP by ATP sulfurylase in the presence of APS. ATP then helps convert luciferin to oxyluciferin, which generates visible light. The intensity of the light (represented by the height of the peak in light analysis software) is proportional to the number of dNTPs incorporated. Pyrosequencing allows inexpensive, fast, and accurate sequencing of genomic regions up to 50 bp long and is a good alternative for DNA sequencing. It is especially suitable for SNPs that are close together. Actually, an ultrahigh-throughput automated pyrosequencing system, Genome Sequencer 20™ System, was developed by 454 Life Sciences (Branford, Connecticut) and has been used for whole genome sequencing. Whole genome sequences of *L. monocytogenes* have recently been completed using the Genome Sequencer 20 System in a short time and at low cost.[134]

Selection of the appropriate SNP typing strategy. Various SNP typing technologies have been developed in the past few years. Different SNP typing techniques have their own merits and disadvantages. Practical concerns are critical factors when choosing an SNP typing assay because the cost of SNP typing is still relatively high compared to conventional typing techniques such as PFGE and MLST. Common considerations include cost (overall cost and cost per SNP), labor and time efficiency, multiplexing, level of throughput, and instrument requirements. Allele-specific hybridization-based techniques like the Taqman assay are easy to perform because PCR and hybridization occur in the same reaction and samples do not need prior and post treatment. However, their ability to be multiplexed is limited and thus the cost can be high when a large number of SNPs need to be analyzed. Pyrosequencing and invader assays are both easy to perform, but cannot be easily multiplexed. However, their cost is lower than allele-specific hybridization-based technologies. Minisequencing assays like SNaPshot have a relatively low cost and can be multiplexed to further reduce cost, thus making it useful for screening large numbers of SNPs.

However, the optimization of multiplex systems can be time consuming and multiplexing will reduce the flexibility of combining different SNPs. Also, DNA templates need to be preamplified and purified before minisequencing reactions are carried out. Pati et al.[135] performed a direct comparison between invader assay, SNapshot typing, and pyrosequencing and found that invader typing was the most accurate with the lowest cost. SNaPshot typing and pyrosequencing had similar performance, but the cost of SNaPshot typing was higher than that for the other two methods. SNaPshot uses standard DNA sequencers; Taqman uses standard real-time thermocyclers and invader assays use a standard fluorescent spectrophotometer. These instruments are readily available in many research and public health laboratories. In contrast, microarray-based analyses require special instruments that are not widely available. Other major disadvantages of microarray-based SNP assays are problems associated with reproducibility and data analysis as discussed in the microarray section. It is expected that with the development of genomic technologies, the throughput of SNP typing assays will increase rapidly. The cost of setting up equipment and reagents such as probes and chips is also expected to decrease dramatically in the near future.[117]

Careful selection of highly informative SNPs and design of SNP typing primers are the key to accurate and cost-efficient SNP typing schemes. Although whole genome sequence data and MLST data are increasingly available for SNP identification, SNPs that are identified based on only a few whole genome sequences may not be representative of the entire population. The genomic locations of the SNPs are also important. SNPs harbored in repeat regions or highly recombinogenic regions would probably not be useful for epidemiology or evolutionary genetics. The MLST database is another useful source for SNP identification. The DNA alignments in the MLST database generally contain sequences from more than 50 isolates and therefore SNPs identified based on these alignments would be expected to provide high discriminatory power. However, an internal problem with any SNP typing strategy is that, with the evolution of genomes, new SNPs may emerge and, therefore, SNP typing schemes based on currently available SNPs may not be able to detect those new SNPs. SNP typing still cannot replace DNA sequencing for epidemiological typing.

The preceding new high-throughput typing technologies provide the advantage of targeting many more markers than conventional molecular subtyping methods and therefore are expected to be more discriminatory. They can also overcome the loci bias that MLST may have. However, their overall high cost may compromise their wide application in nationwide surveillance systems such as PulseNet. For epidemiological typing purposes, if a confined region of a whole genome can provide epidemiologically relevant data, then targeting whole genome variations may be unnecessary and/or confounding. For example, some genomic variations may be random and may not correlate with the epidemic properties of the bacterial species; incorporation of these variations may increase "noise" and confound data interpretation.[12]

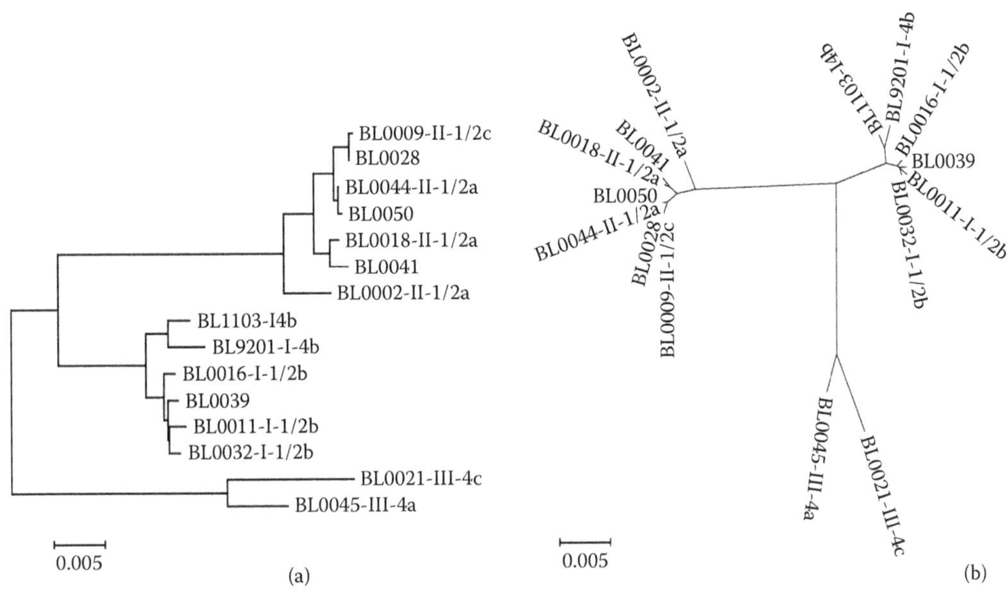

FIGURE 7.8 Conventional (a) and radial (b) dendrograms of neighbor-joining trees of a group of *L. monocytogenes* isolates. (Chen, Y. et al., *J. Clin. Microbiol.*, 2007.) *Note*: The two unrooted trees are equivalent.

7.3 IDENTIFICATION OF GENETIC RELATEDNESS OF STRAINS

Determination of genetic relatedness and cluster analysis between different bacterial strains is important for epidemiological investigation. This genetic relatedness is often determined using phylogenetic analysis. Cluster analysis of epidemic isolates is needed for scientists to recognize an outbreak from seemingly unrelated sporadic cases. Cluster analysis is also needed to identify the routes of transmission of the source strain from a large number of background strains. This is especially important because genetic relationships identified by molecular subtyping methods may reveal hidden routes of transmission that cannot be identified by conventional epidemiology. Furthermore, phylogenetic analysis is needed if we want to estimate the congruence of different subtyping methods. Phylogenetic analysis of the genetic relationship of bacterial isolates is important for outbreak detection and source tracking. For single-source outbreaks, the clonal relationship between isolates is determined using phylogenetic analysis. Phylogenetic analysis is also needed to identify the outbreak/epidemic clone from various background isolates and/or to study the emergence and long-term transmission of certain infectious agents. Furthermore, a phylogenetic analysis is needed if we want to estimate the congruence of different subtyping methods. Therefore, it is necessary for molecular epidemiologists to have a basic understanding of the principles of evolutionary analysis (i.e., phylogenetic tree construction).

However, it should be remembered that minor genetic variations in outbreak clones occur in a relatively short time span (days to months), whereas major changes typically span thousands to millions of years during the long-term evolution of species. Phylogenetic trees of isolates are often presented as either radial or conventional dendrograms (Figure 7.8). There are many algorithms proposed for tree construction and basic principles and commonly used methods will be discussed here. Various computer software packages have been developed for cluster analysis (Table 7.3). Readers can refer to Karlin and Cardon,[136] Nei and Kumar,[91] and Whelan et al.[137] for details. Major software programs used for evolutionary analysis were discussed by Excoffier et al.[138] Generally, phylogenetic analysis algorithms are based on a set of assumptions regarding the dynamics of the evolutionary process and the major task is to convert the molecular data sets (containing A, T, G, C,

TABLE 7.3
Commonly Used Software Packages for Phylogenetic Analysis

Software	Website	Major Functions
MEGA	http://www.megasoftware.net	NJ,[a] ME,[b] MP,[c] UPGMA[d]
Clustal X	http://bips.u-strasbg.fr/fr/Documentation/ClustalX/	NJ[a]
PHYLIP	http://evolution.gs.washington.edu/phylip.html	ME,[b] MP,[c] ML[e]
PAUP*	http://paup.csit.fsu.edu/	ME,[b] LS,[f] MP,[c] ML,[e] Bayesian
MacClade	http://macclade.org/	Macintosh friendly
PAL	http://www.cebl.auckland.ac.nz/pal-project/	NJ,[a] LS,[f] ML[e]
Bionumerics	http://www.applied-maths.com/bionumerics/bionumerics.htm	UPGMA,[d] NJ,[a] MP,[c] ML[e]
DAMBE	http://dambe.bio.uottawa.ca/software.asp	NJ,[a] MP,[c] ML[e]

[a] NJ, Neighbour Joining
[b] ME, Minimum Evolution
[c] MP, Maximum Parsimony
[d] UPGMA, Unweighted Pair Group Method with Arithmetic Mean
[e] ML, Maximum Likelihood
[f] LS, Lease Square

or 20 amino acids), gel electrophoresis banding patterns, or hybridization patterns into evolutionary distances represented by branch lengths in a tree.

There are two major approaches for analyzing strain relatedness: phenetic and cladistic approaches.[7] Phenetic approaches do not make assumptions on the overall evolutionary process and isolates are clustered based on overall similarities, which are given numerical values using mathematical formulas. Two major phenetic approaches are the unweighted pair-group method using arithmetic averages (UPGMA) and the neighbor-joining (NJ) method. Before discussing these two tree building methods, it is necessary to discuss the algorithms used to transfer molecular data (DNA sequence or gel banding pattern) to numerical similarities.

Phenetic approaches first convert DNA sequence data or gel banding patterns to numerical values of genetic distance. Various mathematical formulas for calculating evolutionary distance have been explored. Although there are still controversies regarding which methods are the most accurate and reliable, some methods are generally believed to be useful for calculating genetic distances from DNA sequence data. The simplest method of calculating divergence between two microorganisms is using the number of different nucleotides between homologous sequences. A ratio of the number of different nucleotides between two sequences to the total number of nucleotides is calculated and called p distance. Another model, the Jukes and Cantor's model,[139] modifies the equation of p distance and is expected to be more accurate.

However, the preceding methods assume that (1) substitutions are random and the nucleotide frequency is 0.25, (2) each substitution is independent of the preceding nucleotide, and (3) the substitution rates are equivalent. This ignores the difference between transition and transversion mutations. Some of these assumptions are easily violated in the real evolutionary process, especially when evolutionary distances between highly divergent taxa are calculated. A simple example is that the rate of transition is often higher than the rate of transversion. Kimura's Two-Parameter model[140] was developed to take the difference between transition and transversion into account. However, even transitional mutations (or transversional mutations) may occur at different rates. In summary, the substitution rates of different nucleotide pairs (i.e., A to T, G to C, A to G, T to C) can all be different and can also be influenced by the nucleotide frequencies of A, T, G, and C in specific sequences.

More recent models such as Nei and Tajima's model,[141] Tamura's model,[142] and Tamura and Nei's model[143] combined parameters for differences in substitution rates and differences in nucleotide frequencies to estimate substitution rates. These models generally performed well; however, they all ignored the fact that the rates of substitution also vary with the genomic locations of the nucleotide sites. For this reason, gamma distances for each of these models were developed for estimating rates of substitution.[91] In recent years, more complicated models, such as logDet[144] and paralinear models[145] have been proposed. Interestingly, these complicated models do not necessarily perform better than much simpler models even though they significantly increase the computational load.[91]

A variety of formulas have been developed to calculate indices that measure genetic distance or similarity between different gel banding patterns or hybridization patterns. The calculation is based on the presence and absence of positive signals (bands or hybridization) between two strains (*i* and *j*). Several index calculation methods have been proposed[7,146]:

Dice index or coefficient: $$S_D = \frac{n_i + n_j}{2n_{ij} + n_i + n_j}$$

Jaccard index: $$S_J = \frac{n_i}{n_{ij} + n_i + n_j}$$

Ochiai index: $$S_O = \frac{n_{ij}}{\sqrt{(n_{ij} + n_i)(n_{ij} + n_j)}}$$

Sneath and Sokal index: $$S_{SS} = \frac{2(n_{ij} + n_0)}{2(n_{ij} + n_0) + n_i + n_j}$$

Simple matching index: $$S = \frac{n_{ij} + n_0}{n_{ij} + n_i + n_j + n_0}$$

where

n_{ij} = number of common positive signals present in both patterns; the positive signals (bands or hybridization) are listed according to their positions; presence of a positive signal in the same position of two strains is called a common positive signal

n_i = number of positive signals present only in strain *i*

n_j = number of positive signals present only in strain *j*

n_0 = number of common negative signals in both strains (knowledge of all possible positions is needed to calculate this value)

It is generally believed that when analyzing gel electrophoretic banding patterns the number of bands absent from both strains (n_0) does not provide meaningful results and therefore the Sneath and Sokal index and the Simple matching index are used less frequently.[7,146] In contrast, the number of common negative signals is used when calculating indices based on hybridization patterns. As discussed before, the similarity indices calculated by computer software need to be visually cross-examined by investigators, and the strain relatedness determined by similarity indices is only a reference for inferring the epidemiological relationship of different bacterial isolates.[7] Again, scientists need to combine conventional epidemiological data with similarity indices to determine the epidemiological relationship between bacterial strains. A common question in cluster analysis using gel banding patterns is what threshold value of similarity should be used to define an outbreak clone or epidemic clone, and it certainly depends on the discriminatory power of the subtyping methods. For example, when ribotyping is used for *L. monocytogenes*, the threshold value should probably be 100% for identifying clones, but PFGE based on different restriction enzymes may have different threshold values.

UPGMA is one of the earliest and simplest algorithms for cluster analysis of different strains based on the numerical values of genetic distance.[91] UPGMA is a sequential clustering algorithm, relationships between taxonomic units are identified in order of similarity, and the phylogenetic tree is built stepwise. UPGMA calculates the average genetic distance between different clusters, and different taxa in the same cluster are considered to have identical genetic distances to a taxon outside the cluster (which is called ultrametric). This may not hold true in real-world situations. As stated earlier, UPGMA is commonly used to analyze gel banding patterns obtained by ribotyping, RAPD, and PFGE; however, it does not always construct reliable trees, especially when the taxa are not closely related and/or when there is recombination between two sequences. However, if a validated phenogram containing a collection of well-characterized isolates is available, UPGMA is useful for assigning any new taxa into the phenogram.

Neighbor joining (NJ) is one of the most commonly used methods for tree construction.[91] It is a bottom-up clustering method used for the stepwise construction of phylogenetic trees. Neighbor-joining has been shown to construct trees that are statistically consistent under many models of evolution, and therefore it is believed to construct the true tree with a high probability given the existence of sufficient data. The NJ algorithm is very efficient, and very large data sets can be analyzed in a relatively short period of time, which is computationally prohibitive for other phylogenetic analysis algorithms such as maximum parsimony and maximum likelihood. Other phenetic algorithms include but are not limited to minimum evolution (ME) and least square (LE). Readers are referred to Nei and Kumar [91] for detailed descriptions of these algorithms.

Another approach for constructing phylogenetic trees is the cladistic approach by which the similarity of different taxa is determined by a certain evolutionary model or theory instead of mathematical values. Two typical approaches include maximum parsimony (MP) and maximum likelihood (ML).

Maximum parsimony is a widely used tree building method.[147] This method constructs a tree that requires the smallest number of character state transitions (i.e., nucleotide substitutions, or changes in antibiotic resistance) to explain the whole evolutionary process. While MP is useful in many circumstances (e.g., morphological data), it is known to tend to construct incorrect topologies if the number of sequences is large and the number of nucleotides is small. Also, MP cannot always construct statistically consistent trees in the case of short-branch or long-branch attractions. MP is also not as computationally efficient as NJ.

Maximum likelihood is another popular method used to make inferences about parameters of the underlying probability distribution from a given data set.[147] This method calculates the probability that the proposed topology and the evolutionary process generate the observed data set. The topology and the evolutionary process with a higher probability of producing the observed data set are preferred and the method searches for the tree with the highest probability or likelihood. ML considers the rate of nucleotide substitution. The biggest advantage of ML over other tree building methods is that it is statistically meaningful and is very robust to many violations of the assumptions in the evolutionary model. However, it is computationally intensive and impractical for large data sets. ML is generally used to infer phylogenetic relationships using nucleotide sequences and is not suitable for analyzing gel banding patterns or hybridization patterns. However, a variant of the ML method, Bayesian analysis, has recently been developed for relatedness analysis based on gel patterns or hybridization patterns.[148]

Because no true phylogenetic topologies are available, it is impossible to determine which tree building method produces the most accurate tree and, in many cases, different tree building algorithms will generate dendrograms with different structures based on the same set of molecular data. If possible, multiple tree construction methods should be used on the same data set, and if they all produce a similar topology, then this "consensus" tree is expected to be reliable.

When recombination is frequent among the whole population, the genetic relationship between closely related isolates can still be reliably estimated because they are descendents of a recent ancestor and are expected to have less recombination. However, the genetic relationship of those distantly related isolates can be obscured by recombination over the long term. Fortunately, epidemiologists do not always need information on the relatedness between major lineages of the whole population. What is most important is the ability to cluster identical or closely related isolates in order to understand the spread of a few epidemic clones.

7.4 CONCLUSIONS AND PERSPECTIVES

This chapter attempted to highlight the basic principles and applications of strain typing techniques in the fields of taxonomy, epidemiology, and population genetics with a focus on molecular epidemiology. Strain typing techniques have evolved by targeting various genetic markers with more advanced molecular biology technologies. The discriminatory power, epidemiological concordance, reproducibility, and typeability of strain typing methods have recently been significantly enhanced due to numerous advances in genomics, molecular technologies, instrumentation, and computer and software capabilities. Genetic markers have evolved from a few protein structures/gene sequences to whole genome SNPs or hybridizations.

Understanding the strengths and weaknesses of each subtyping method is not only critical to the selection of the appropriate subtyping method, but also enhances study design and data interpretation. A summary of common subtyping methods with their relative performances in various proposed criteria is presented in Table 7.4. With the rapid development of novel molecular biology technologies and our enhanced understanding of the genomic structures and functions of bacterial pathogens, more advanced subtyping strategies that target highly relevant molecular makers are expected to be developed. Therefore, understanding the basic principles of proper marker selection and correct interpretation of molecular subtyping data is of critical importance. This chapter has provided some insights into what performance characteristics a subtyping method should possess (e.g., level of discrimination) when applied to specific subtyping applications.

Combining multiple subtyping techniques is often beneficial since none of them is perfect for all applications. MLST based on housekeeping genes can be used to identify the population structure of *L. monocytogenes*. Therefore, if one wants to classify an unknown *L. monocytogenes* isolate into certain genetic lineage, MLST using housekeeping genes can be performed. If one wants to further characterize the isolate to determine if it belongs to one of the major epidemic clones of *L. monocytogenes*, MVLST would be a good choice. If the subtyping purpose is to determine if the isolate is associated with any specific listeriosis outbreaks, PFGE or MLVA can be used. The choice of specific subtyping techniques also depends on the availability of laboratory facilities. Some subtyping methods have special requirements of instruments. For example, PFGE requires PFGE units and MLST requires DNA sequencing facility. In laboratories with no access to these facilities, combination of several simple fragment-based subtyping methods such as multiplex PCR, ribotyping, and PCR-RFLP may be a valuable alternative for fairly accurate estimation of the population structure and identification of epidemic clones of *L. monocytogenes*. Combining subtyping schemes typically increases discriminatory power and provides much more accurate clustering of isolates. In large federal or state laboratories where routine testing of large numbers of samples is performed, SNP typing would be more cost and labor efficient than many other subtyping methods.

It is notable that the introduction of molecular subtyping methods has greatly changed the scope of epidemiology and facilitated the understanding of population structure and the long-term and short-term evolution and transmission of *L. monocytogenes* and various other infectious agents. The choice of specific strain typing method depends on the scope and purpose of the investigation. The ultimate goal of strain typing is often to identify the epidemiological relationship between bacterial strains. Different computational algorithms and computer software packages have been developed to identify strain relatedness based on output data generated by various subtyping methods. However, the strain relatedness revealed by specific strain typing methods is not necessarily the actual epidemiological relationship of different bacterial strains. In many cases, strain typing data are used as a basis to support conventional epidemiological findings during outbreak investigations.[7]

TABLE 7.4
Comparison of Major Strain Typing Methods in Terms of Performance on Various Criteria

Subtyping Methods	Targets	Type of Data Output	Discriminatory Power	Epidemiological Concordance	Reproducibility	Typeability	Cost
Serotyping	Surface antigen structures	Character	Low	Low	Medium	Medium	Medium to high
Phage typing	Susceptibility to phages	Binary	Low	Low to medium	Low to medium	Medium	
Plasmid profile	Number and size of plasmids	Banding pattern	Low	Low to medium	Medium	Low	Low
MEE	Size of major metabolic enzymes	Protein gel banding pattern	Moderate	Medium	High	High	Medium
RAPD	Polymorphism within and between arbitrary priming regions	DNA gel banding pattern	Moderate	Medium	Medium	High	Medium
Rep-PCR	Repetitive elements	DNA gel banding pattern	Moderate	Medium	Medium to high	High	Low
MLVA	Variable number of tandem repeats	Number	High	High	Medium to high	High	Medium
PFGE	Polymorphism within/between restriction sites in whole genome	DNA gel banding pattern	High	Medium to high	Medium to high	High	Medium
AFLP	Polymorphism within/between restriction sites	DNA gel banding pattern	Moderate to high	Medium	Medium to high	High	Medium
PCR-RFLP	Polymorphism within/between restriction sites in specific genes	DNA gel banding pattern	Moderate	Medium	High	High	Medium
Automated ribotyping	Polymorphism within/between restriction sites in RNA genes	DNA gel banding pattern	Moderate	Medium	High	High	Medium
Microarray	Specific genomic regions	Hybridization pattern	High	High	High	High	High
MLST	Housekeeping gene sequences	DNA sequences	Moderate to high	Medium to high	High	High	Medium to high
MVLST	Virulence gene sequences	DNA sequences	High	High	High	High	Medium to high
SNP typing	Single nucleotide substitutions	Single nucleotide substitution	High	Medium to high	High	High	High

However, strain typing data can reveal epidemiological relationships that cannot be or were not detected by conventional epidemiology. Therefore, it should always be kept in mind that molecular epidemiology and conventional epidemiology must complement and support one another in order to ensure accurate epidemiology.

REFERENCES

1. van Belkum, A. et al., Role of genomic typing in taxonomy, evolutionary genetics, and microbial epidemiology, *Clin. Microbiol. Rev.*, 14, 547, 2001.
2. Lane, D.J. et al., Rapid determination of 16S ribosomal RNA sequences for phylogenetic analyses, *Proc. Natl. Acad. Sci. USA*, 82, 6955, 1985.
3. Gurtler, V. and Mayall, B.C., Genomic approaches to typing, taxonomy and evolution of bacterial isolates, *Int. J. Syst. Evol. Microbiol.*, 51, 3, 2001.
4. Doumith, M. et al., New aspects regarding evolution and virulence of *Listeria monocytogenes* revealed by comparative genomics and DNA arrays, *Infect. Immun.*, 72, 1072, 2004.
5. Nightingale, K.K., Windham, K., and Wiedmann, M., Evolution and molecular phylogeny of *Listeria monocytogenes* isolated from human and animal listeriosis cases and foods, *J. Bacteriol.*, 187, 5537, 2005.
6. Ward, T.J. et al., Intraspecific phylogeny and lineage group identification based on the *prfA* virulence gene cluster of *Listeria monocytogenes*, *J. Bacteriol.*, 186, 4994, 2004.
7. Riley, L.W., *Molecular epidemiology of infectious diseases: Principles and practices*, ASM Press, Washington, D.C., 2004.
8. Struelens, M.J., Consensus guidelines for appropriate use and evaluation of microbial epidemiologic typing systems, *Clin. Microbiol. Infect.*, 2, 2, 1996.
9. Graves, L.M. et al., Microbiological aspects of the investigation that traced the 1998 outbreak of listeriosis in the United States to contaminated hot dogs and establishment of molecular subtyping-based surveillance for *Listeria monocytogenes* in the PulseNet network, *J. Clin. Microbiol.*, 43, 2350, 2005.
10. Bender, J.B. et al., Use of molecular subtyping in surveillance for *Salmonella enterica* serotype *typhimurium*, *N. Engl. J. Med.*, 344, 189, 2001.
11. Wiedmann, M., Molecular subtyping methods for *Listeria monocytogenes*, *J. AOAC,* 85, 524, 2002.
12. Wassenaar, T.M., Molecular typing of pathogens, *Berl. Munch. Tierarztl. Wochenschr.*, 116, 447, 2003.
13. Feil, E.J. et al., Recombination within natural populations of pathogenic bacteria: Short-term empirical estimates and long-term phylogenetic consequences, *Proc. Natl. Acad. Sci. USA*, 98, 182, 2001.
14. Destro, M.T., Leitao, M., and Farber, J.M., Use of molecular typing methods to trace the dissemination of *Listeria monocytogenes* in a shrimp processing plant, *Appl. Environ. Microbiol.*, 62, 1852, 1996.
15. Senczek, D., Stephan, R., and Untermann, F., Pulsed-field gel electrophoresis (PFGE) typing of *Listeria* strains isolated from a meat processing plant over a 2-year period, *Int. J. Food Microbiol.*, 62, 155, 2000.
16. Dauphin, G., Ragimbeau, C., and Malle, P., Use of PFGE typing for tracing contamination with *Listeria monocytogenes* in three cold-smoked salmon processing plants, *Int. J. Food. Microbiol.*, 64, 51, 2001.
17. Aguado, V., Vitas, A.I., and Garcia-Jalon, I., Characterization of *Listeria monocytogenes* and *Listeria innocua* from a vegetable processing plant by RAPD and REA, *Int. J. Food Microbiol.*, 90, 341, 2004.
18. Wulff, G. et al., One group of genetically similar *Listeria monocytogenes* strains frequently dominates and persists in several fish slaughter- and smokehouses, *Appl. Environ. Microbiol.*, 72, 4313, 2006.
19. Tompkin, R.B., Control of *Listeria monocytogenes* in the food-processing environment, *J. Food Prot.*, 65, 709, 2002.
20. Hunter, P.R. and Gaston, M.A., Numerical index of the discriminatory ability of typing systems: An application of Simpson's index of diversity, *J. Clin. Microbiol.*, 26, 2465, 1988.
21. Chen, Y., Zhang, W., and Knabel, S.J., Multivirulence-locus sequence typing clarifies epidemiology of recent listeriosis outbreaks in the United States, *J. Clin. Microbiol.*, 43, 5291, 2005.
22. Sauders, B.D. et al., Molecular subtyping to detect human listeriosis clusters, *Emerg. Infect. Dis.*, 9, 672, 2003.
23. Foxman, B. et al., Choosing an appropriate bacterial typing technique for epidemiologic studies, *Epidemiol. Perspect. Innov.*, 2, 10, 2005.
24. Van Belkum, A., High-throughput epidemiologic typing in clinical microbiology, *Clin. Microbiol. Infect.*, 9, 86, 2003.
25. Kathariou, S., *Listeria monocytogenes* virulence and pathogenicity, a food safety perspective, *J. Food Prot.*, 65, 1811, 2002.

26. Fugett, E., Forte E., Nnoka, C., and Wiedmann, M., International Life Sciences Institute of North America *Listeria monocytogenes* strain collection: Development of standard *Listeria monocytogenes* strain sets for research and validation studies, *J. Food Prot.*, 69, 2929, 2006.
27. Chen, Y., Zhang, W., and Knabel, S.J., Multivirulence-locus sequence typing identifies single nucleotide polymorphisms which differentiate epidemic clones and outbreak strains of *Listeria monocytogenes*, *J. Clin. Microbiol.*, 2007.
28. Ducey, T.F. et al., A single-nucleotide-polymorphism-based multilocus genotyping assay for subtyping lineage I isolates of *Listeria monocytogenes*, *Appl. Environ. Microbiol.*, 73, 133, 2007.
29 Tenover, F.C. et al., Comparison of traditional and molecular methods of typing isolates of *Staphylococcus aureus*, *J. Clin. Microbiol.*, 32, 407, 1994.
30. Romanova, N., Favrin, S., and Griffiths, M.W., Sensitivity of *Listeria monocytogenes* to sanitizers used in the meat processing industry, *Appl. Environ. Microbiol.*, 68, 6405, 2002.
31 Tardif, G. et al., Spontaneous switching of the sucrose-promoted colony phenotype in *Streptococcus sanguis*, *Infect. Immun.*, 57, 3945, 1989.
32. Seeliger, H.P., Finger, H., and Klutsch, J., Studies on the importance of the antigen-fixation test for the serodiagnosis of *Listeria monocytogenes* infections, *Zentralbl. Bakteriol.*, 211, 215, 1969.
33. McLauchlin, J. and Jones, D., *Erysipelothrix* and *Listeria*. In *Topley and Wilson's microbiology and microbial infections Volume 2, Systematic bacteriology*, 9th ed. (p. 683). Arnold, London, 1998.
34. Schonberg, A. et al., Serotyping of 80 strains from the WHO multicenter international typing study of *Listeria monocytogenes*, *Int. J. Food Microbiol.*, 32, 279, 1996.
35. Nadon, C.A. et al., Correlations between molecular subtyping and serotyping of *Listeria monocytogenes*, *J. Clin. Microbiol.*, 39, 2704, 2001.
36. Loessner, M.J. and Busse, M., Bacteriophage typing of *Listeria* species, *Appl. Environ. Microbiol.*, 56, 1912, 1990.
37. Capita, R. et al., Evaluation of the international phage typing set and some experimental phages for typing of *Listeria monocytogenes* from poultry in Spain, *J. Appl. Microbiol.*, 92, 90, 2002.
38. Ralovich, B., Ewan, E.P., and Emody, L., Alteration of phage- and biotypes of *Listeria* strains, *Acta. Microbiol. Hung.*, 33, 19, 1986.
39. Audurier, A. et al., A phage typing system for *Listeria monocytogenes* and its use in epidemiological studies, *Clin. Invest. Med.*, 7, 229, 1984.
40. Rocourt, J. and Catimel, B., International phage typing center for *Listeria*: Report for 1987, *Acta. Microbiol. Hung.*, 36, 225, 1989.
41. Bille, J. and Rocourt, J., WHO International multicenter *Listeria monocytogenes* subtyping study—Rationale and setup of the study, *Int. J. Food Microbiol.*, 32, 251, 1996.
42. Rocourt, J. et al., A multicenter study on the phage typing of *Listeria monocytogenes*, *Zentralbl. Bakteriol. Mikrobiol. Hyg.*, 259, 489, 1985.
43. Lebrun, M. et al., Plasmids in *Listeria monocytogenes* in relation to cadmium resistance, *Appl. Environ. Microbiol.*, 58, 3183, 1992.
44. Chen, Y., Unpublished data, 2007.
45. McLauchlin, J. et al., Subtyping of *Listeria monocytogenes* on the basis of plasmid profiles and arsenic and cadmium susceptibility, *J. Appl. Microbiol.*, 83, 381, 1997.
46. Bibb, W.F. et al., Analysis of clinical and food-borne isolates of *Listeria monocytogenes* in the United States by multilocus enzyme electrophoresis and application of the method to epidemiologic investigations, *Appl. Environ. Microbiol.*, 56, 2133, 1990.
47. Piffaretti, J.C. et al., Genetic characterization of clones of the bacterium *Listeria monocytogenes* causing epidemic disease, *Proc. Natl. Acad. Sci. USA*, 86, 3818, 1989.
48. Boerlin, P. and Piffaretti, J.C., Typing of human, animal, food, and environmental isolates of *Listeria monocytogenes* by multilocus enzyme electrophoresis, *Appl. Environ. Microbiol.*, 57, 1624, 1991.
49. Bibb, W.F. et al., Analysis of *Listeria monocytogenes* by multilocus enzyme electrophoresis and application of the method to epidemiologic investigations, *Int. J. Food Microbiol.*, 8, 233, 1989.
50. Caugant, D.A. et al., Multilocus enzyme electrophoresis for characterization of *Listeria monocytogenes* isolates: Results of an international comparative study, *Int. J. Food Microbiol.*, 32, 301, 1996.
51. Notermans, S. et al., Specific gene probe for detection of biotyped and serotyped *Listeria* strains, *Appl. Environ. Microbiol.*, 55, 902, 1989.
52. Harvey, J. and Gilmour, A., Characterization of recurrent and sporadic *Listeria monocytogenes* isolates from raw milk and nondairy foods by pulsed-field gel electrophoresis, monocin typing, plasmid profiling, and cadmium and antibiotic resistance determination, *Appl. Environ. Microbiol.*, 67, 840, 2001.

53. Wilhelms, D. and Sandow, D., Preliminary studies on monocine typing of *Listeria monocytogenes* strains, *Acta. Microbiol. Hung.*, 36, 235, 1989.
54. Williams, J.G. et al., DNA polymorphisms amplified by arbitrary primers are useful as genetic markers, *Nucleic Acids Res.*, 18, 6531, 1990.
55. Welsh, J. and McClelland, M., Fingerprinting genomes using PCR with arbitrary primers, *Nucleic Acids Res.*, 18, 7213, 1990.
56. Mazurier, S.I. and Wernars, K., Typing of *Listeria* strains by random amplification of polymorphic DNA, *Res. Microbiol.*, 143, 499, 1992.
57. Czajka, J. and Batt, C.A., Verification of causal relationships between *Listeria monocytogenes* isolates implicated in food-borne outbreaks of listeriosis by randomly amplified polymorphic DNA patterns, *J. Clin. Microbiol.*, 32, 1280, 1994.
58. Wernars, K. et al., The WHO multicenter study on *Listeria monocytogenes* subtyping: Random amplification of polymorphic DNA (RAPD), *Int. J. Food Microbiol.*, 32, 325, 1996.
59. Nightingale, K. et al., Combined *sigB* allelic typing and multiplex PCR provide improved discriminatory power and reliability for *Listeria monocytogenes* molecular serotyping, *J. Microbiol. Methods*, 68, 52, 2007.
60. Zabeau, M. and Vos, P., European Patent Application, EP 0534858, 1993.
61. Ripabelli, G., McLauchin, J., and Threlfall, E.J., Amplified fragment length polymorphism (AFLP) analysis of *Listeria monocytogenes*, *Syst. Appl. Microbiol.*, 23, 132, 2000.
62. Guerra, M.M., Bernardo, F., and McLauchlin, J., Amplified fragment length polymorphism (AFLP) analysis of *Listeria monocytogenes*, *Syst. Appl. Microbiol.*, 25, 456, 2002.
63. Keto-Timonen, R.O., Autio, T.J., and Korkeala, H.J., An improved amplified fragment length polymorphism (AFLP) protocol for discrimination of *Listeria* isolates, *Syst. Appl. Microbiol.*, 26, 236, 2003.
64. Fonnesbech Vogel, B. et al., High-resolution genotyping of *Listeria monocytogenes* by fluorescent amplified fragment length polymorphism analysis compared to pulsed-field gel electrophoresis, random amplified polymorphic DNA analysis, ribotyping, and PCR-restriction fragment length polymorphism analysis, *J. Food Prot.*, 67, 1656, 2004.
65. Jacquet, C. et al., Investigations related to the epidemic strain involved in the French listeriosis outbreak in 1992, *Appl. Environ. Microbiol.*, 61, 2242, 1995.
66. Paillard, D. et al., Rapid identification of *Listeria* species by using restriction fragment length polymorphism of PCR-amplified 23S rRNA gene fragments, *Appl. Environ. Microbiol.*, 69, 6386, 2003.
67. Wiedmann, M. et al., Ribotypes and virulence gene polymorphisms suggest three distinct *Listeria monocytogenes* lineages with differences in pathogenic potential, *Infect. Immun.*, 65, 2707, 1997.
68. Vines, A. et al., Restriction fragment length polymorphism in four virulence-associated genes of *Listeria monocytogenes*, *Res. Microbiol.*, 143, 281, 1992.
69. Destro, M.T., Leitao, M.F., and Farber, J.M., Use of molecular typing methods to trace the dissemination of *Listeria monocytogenes* in a shrimp processing plant, *Appl. Environ. Microbiol.*, 62, 705, 1996.
70. De Cesare, A. et al., Automated ribotyping using different enzymes to improve discrimination of *Listeria monocytogenes* isolates, with a particular focus on serotype 4b strains, *J. Clin. Microbiol.*, 39, 3002, 2001.
71. Swaminathan, B. et al., WHO-sponsored international collaborative study to evaluate methods for subtyping *Listeria monocytogenes*: Restriction fragment length polymorphism (RFLP) analysis using ribotyping and Southern hybridization with two probes derived from *Listeria monocytogenes* chromosome, *Int. J. Food Microbiol.*, 32, 263, 1996.
72. Kabuki, D.Y. et al., Molecular subtyping and tracking of *Listeria monocytogenes* in Latin-style fresh-cheese processing plants, *J. Dairy Sci.*, 87, 2803, 2004.
73. Schwartz, D.C. and Cantor, C.R., Separation of yeast chromosome-sized DNAs by pulsed field gradient gel electrophoresis, *Cell*, 37, 67, 1984.
74. Graves, L.M. and Swaminathan, B., PulseNet standardized protocol for subtyping *Listeria monocytogenes* by macrorestriction and pulsed-field gel electrophoresis, *Int. J. Food Microbiol.*, 65, 55, 2001.
75. Brosch, R., Buchrieser, C., and Rocourt, J., Subtyping of *Listeria monocytogenes* serovar 4b by use of low-frequency-cleavage restriction endonucleases and pulsed-field gel electrophoresis, *Res. Microbiol.*, 142, 667, 1991.
76. Buchrieser, C., Brosch, R., and Rocourt, J., Use of pulsed field gel electrophoresis to compare large DNA-restriction fragments of *Listeria monocytogenes* strains belonging to serogroups 1/2 and 3, *Int. J. Food Microbiol.*, 14, 297, 1991.
77. Buchrieser, C. et al., Pulsed-field gel electrophoresis applied for comparing *Listeria monocytogenes* strains involved in outbreaks, *Can. J. Microbiol.*, 39, 395, 1993.

78. Brosch, R. et al., Genomic fingerprinting of 80 strains from the WHO multicenter international typing study of *Listeria monocytogenes* via pulsed-field gel electrophoresis (PFGE), *Int. J. Food Microbiol.*, 32, 343, 1996.
79. Gerner-Smidt, P. et al., PulseNet USA: A five-year update, *Foodborne Pathog. Dis.*, 3, 9, 2006.
80. Doumith, M. et al., Use of DNA arrays for the analysis of outbreak-related strains of *Listeria monocytogenes, Int. J. Med. Microbiol.*, 296, 559, 2006.
81. Tenover, F.C. et al., Interpreting chromosomal DNA restriction patterns produced by pulsed-field gel electrophoresis: Criteria for bacterial strain typing, *J. Clin. Microbiol.*, 33, 2233, 1995.
82. Barrett, T.J., Gerner-Smidt, P., and Swaminathan, B., Interpretation of pulsed-field gel electrophoresis patterns in food-borne disease investigations and surveillance, *Foodborne Pathog. Dis.*, 3, 20, 2006.
83. Maslanka, S.E. et al., Molecular subtyping of *Clostridium perfringens* by pulsed-field gel electrophoresis to facilitate food-borne-disease outbreak investigations, *J. Clin. Microbiol.*, 37, 2209, 1999.
84. Liebana, E. et al., Multiple genetic typing of *Salmonella enterica* serotype typhimurium isolates of different phage types (DT104, U302, DT204b, and DT49) from animals and humans in England, Wales, and Northern Ireland, *J. Clin. Microbiol.*, 40, 4450, 2002.
85. Clark, C.G. et al., Subtyping of *Salmonella enterica* serotype enteritidis strains by manual and automated PstI-SphI ribotyping, *J. Clin. Microbiol.*, 41, 27, 2003.
86. Davis, M.A. et al., Evaluation of pulsed-field gel electrophoresis as a tool for determining the degree of genetic relatedness between strains of *Escherichia coli* O157:H7, *J. Clin. Microbiol.*, 41, 1843, 2003.
87. Singer, R.S., Sischo, W.M., and Carpenter, T.E., Exploration of biases that affect the interpretation of restriction fragment patterns produced by pulsed-field gel electrophoresis, *J. Clin. Microbiol.*, 42, 5502, 2004.
88. Cardinali, G. et al., Multicenter comparison of three different analytical systems for evaluation of DNA banding patterns from *Cryptococcus neoformans*, *J. Clin. Microbiol.*, 40, 2095, 2002.
89. Gerner-Smidt, P. et al., Computerized analysis of restriction fragment length polymorphism patterns: Comparative evaluation of two commercial software packages, *J. Clin. Microbiol.*, 36, 1318, 1998.
90. Rementeria, A. et al., Comparative evaluation of three commercial software packages for analysis of DNA polymorphism patterns, *Clin. Microbiol. Infect.*, 7, 331, 2001.
91. Nei, M. and Kumar, S., *Molecular evolution and phylogenetics*, Oxford University Press, Oxford, 2000.
92. Duck, W.M. et al., Optimization of computer software settings improves accuracy of pulsed-field gel electrophoresis macrorestriction fragment pattern analysis, *J. Clin. Microbiol.*, 41, 3035, 2003.
93. Hyytia-Trees, E. et al., Second generation subtyping: A proposed PulseNet protocol for multiple-locus variable-number tandem repeat analysis of Shiga toxin-producing *Escherichia coli* O157 (STEC O157), *Foodborne Pathog. Dis.*, 3, 118, 2006.
94. Jersek, B. et al., Typing of *Listeria monocytogenes* strains by repetitive element sequence-based PCR, *J. Clin. Microbiol.*, 37, 103, 1999.
95. Chou, C.H. and Wang, C., Genetic relatedness between *Listeria monocytogenes* isolates from seafood and humans using PFGE and REP-PCR, *Int. J. Food Microbiol.*, 110, 135, 2006.
96. Van Kessel, J.S. et al., Subtyping *Listeria monocytogenes* from bulk tank milk using automated repetitive element-based PCR, *J. Food Prot.*, 68, 2707, 2005.
97. Volpe, K.E. et al., Multiple-locus variable number tandem repeat analysis as a subtyping tool for *Listeria monocytogenes*. Poster 2007 ISOPOL, 115, 187, 2007.
98. Murphy, M. et al., Development and application of multiple-locus variable number of tandem repeat analysis (MLVA) to subtype a collection of *Listeria monocytogenes*, *Int. J. Food Microbiol.*, 2006.
99. Liu, D. et al., *Listeria monocytogenes* subgroups IIIA, IIIB, and IIIC delineate genetically distinct populations with varied pathogenic potential, *J. Clin. Microbiol.*, 44, 4229, 2006.
100. Rudi, K., Katla, T., and Naterstad, K., Multilocus fingerprinting of *Listeria monocytogenes* by sequence-specific labeling of DNA probes combined with array hybridization, *FEMS. Microbiol. Lett.*, 220, 9, 2003.
101. Brazma, A. et al., Minimum information about a microarray experiment (MIAME)—Toward standards for microarray data, *Nat. Genet.*, 29, 365, 2001.
102. Borucki, M.K. et al., Selective discrimination of *Listeria monocytogenes* epidemic strains by a mixed-genome DNA microarray compared to discrimination by pulsed-field gel electrophoresis, ribotyping, and multilocus sequence typing, *J. Clin. Microbiol.*, 42, 5270, 2004.
103. Kathariou, S. et al., Involvement of closely related strains of a new clonal group of *Listeria monocytogenes* in the 1998–99 and 2002 multistate outbreaks of food-borne listeriosis in the United States, *Foodborne Pathog. Dis.*, 3, 292, 2006.

104. Rudi, K. et al., Subtyping *Listeria monocytogenes* through the combined analyses of genotype and expression of the *hlyA* virulence determinant, *J. Appl. Microbiol.*, 94, 720, 2003.
105. Chan, M.S., Maiden, M.C., and Spratt, B.G., Database-driven multi locus sequence typing (MLST) of bacterial pathogens, *Bioinformatics*, 17, 1077, 2001.
106. Lemee, L. et al., Multilocus sequence analysis and comparative evolution of virulence-associated genes and housekeeping genes of *Clostridium difficile*, *Microbiology*, 151, 3171, 2005.
107. Rasmussen, O.F. et al., *Listeria monocytogenes* exists in at least three evolutionary lines: Evidence from flagellin, invasive associated protein and listeriolysin O genes, *Microbiology*, 141, 2053, 1995.
108. Maiden, M.C. et al., Multilocus sequence typing: A portable approach to the identification of clones within populations of pathogenic microorganisms, *Proc. Natl. Acad. Sci. USA*, 95, 3140, 1998.
109. Salcedo, C. et al., Development of a multilocus sequence typing method for analysis of *Listeria monocytogenes* clones, *J. Clin. Microbiol.*, 41, 757, 2003.
110. Meinersmann, R.J. et al., Multilocus sequence typing of *Listeria monocytogenes* by use of hypervariable genes reveals clonal and recombination histories of three lineages, *Appl. Environ. Microbiol.*, 70, 2193, 2004.
111. Cai, S. et al., Rational design of DNA sequence-based strategies for subtyping *Listeria monocytogenes*, *J. Clin. Microbiol.*, 40, 3319, 2002.
112. Zhang, W., Jayarao, B.M., and Knabel, S.J., Multivirulence-locus sequence typing of *Listeria monocytogenes*, *Appl. Environ. Microbiol.*, 70, 913, 2004.
113. Cooper, J.E. and Feil, E.J., Multilocus sequence typing—What is resolved? *Trends Microbiol.*, 12, 373, 2004.
114. Feil, E.J. and Spratt, B.G., Recombination and the population structures of bacterial pathogens, *Annu. Rev. Microbiol.*, 55, 561, 2001.
115. Revazishvili, T. et al., Comparative analysis of multilocus sequence typing and pulsed-field gel electrophoresis for characterizing *Listeria monocytogenes* strains isolated from environmental and clinical sources, *J. Clin. Microbiol.*, 42, 276, 2004.
116. Sobrino, B., Brion, M., and Carracedo, A., SNPs in forensic genetics: A review on SNP typing methodologies, *Forensic Sci. Int.*, 154, 181, 2005.
117. Dearlove, A.M., High-throughput genotyping technologies, *Brief Funct. Genomic Proteomic*, 1, 139, 2002.
118. Livak, K.J. et al., Oligonucleotides with fluorescent dyes at opposite ends provide a quenched probe system useful for detecting PCR product and nucleic acid hybridization, *PCR Methods Appl.*, 4, 357, 1995.
119. Lareu, M. et al., The use of the LightCycler for the detection of Y chromosome SNPs, *Forensic Sci. Int.*, 118, 163, 2001.
120. Tyagi, S., Marras, S.A., and Kramer, F.R., Wavelength-shifting molecular beacons, *Nat. Biotechnol.*, 18, 1191, 2000.
121. Howell, W.M. et al., Dynamic allele-specific hybridization. A new method for scoring single nucleotide polymorphisms, *Nat. Biotechnol.*, 17, 87, 1999.
122. Sobrino, B. and Carracedo, A., SNP typing in forensic genetics: A review, *Methods Mol. Biol.*, 297, 107, 2005.
123. Zhang, W. et al., Probing genomic diversity and evolution of *Escherichia coli* O157 by single nucleotide polymorphisms, *Genome Res.*, 16, 757, 2006.
124. Wang, D.G. et al., Large-scale identification, mapping, and genotyping of single-nucleotide polymorphisms in the human genome, *Science*, 280, 1077, 1998.
125. Landegren, U. et al., A ligase-mediated gene detection technique, *Science*, 241, 1077, 1988.
126. De la Vega, F.M. et al., Assessment of two flexible and compatible SNP genotyping platforms: TaqMan SNP genotyping assays and the SNPlex genotyping system, *Mutat. Res.*, 573, 111, 2005.
127. Nilsson, M. et al., Padlock probes: Circularizing oligonucleotides for localized DNA detection, *Science*, 265, 2085, 1994.
128. Grossman, P.D. et al., High-density multiplex detection of nucleic acid sequences: Oligonucleotide ligation assay and sequence-coded separation, *Nucleic Acids Res.*, 22, 4527, 1994.
129. Hommais, F. et al., Single-nucleotide polymorphism phylotyping of *Escherichia coli*, *Appl. Environ. Microbiol.*, 71, 4784, 2005.
130. Haff, L.A. and Smirnov, I.P., Single-nucleotide polymorphism identification assays using a thermostable DNA polymerase and delayed extraction MALDI-TOF mass spectrometry, *Genome Res.*, 7, 378, 1997.
131. Sauer, S. et al., A novel procedure for efficient genotyping of single nucleotide polymorphisms, *Nucleic Acids Res.*, 28, 13, 2000.

132. Shumaker, J.M., Metspalu, A., and Caskey, C.T., Mutation detection by solid phase primer extension, *Hum. Mutat.*, 7, 346, 1996.
133. O'Driscoll, B., Gahan, C.G., and Hill, C., Adaptive acid tolerance response in *Listeria monocytogenes*: Isolation of an acid-tolerant mutant which demonstrates increased virulence, *Appl. Environ. Microbiol.*, 62, 1693, 1996.
134. Nyren, P., The history of pyrosequencing(r), *Methods Mol. Biol.*, 373, 1, 2006.
135. Pati, N. et al., A comparison between SNaPshot, pyrosequencing, and biplex invader SNP genotyping methods: Accuracy, cost, and throughput, *J. Biochem. Biophys. Methods*, 60, 1, 2004.
136. Karlin, S. and Cardon, L.R., Computational DNA sequence analysis, *Annu. Rev. Microbiol.*, 48, 619, 1994.
137. Whelan, S., Lio, P., and Goldman, N., Molecular phylogenetics: State-of-the-art methods for looking into the past, *Trends Genet.*, 17, 262, 2001.
138. Excoffier, L. and Heckel, G., Computer programs for population genetics data analysis: A survival guide, *Nat. Rev. Genet.*, 7, 745, 2006.
139. Jukes, T.H. and Cantor, C., Mammalian protein metabolism. In *Evolution of protein molecules*, Academic Press, New York, 1969.
140. Kimura, M., A simple method for estimating evolutionary rates of base substitutions through comparative studies of nucleotide sequences, *J. Mol. Evol.*, 16, 111, 1980.
141. Nei, M. and Tajima, F., Evolutionary change of restriction cleavage sites and phylogenetic inference for man and apes, *Mol. Biol. Evol.*, 2, 189, 1985.
142. Tamura, K., Estimation of the number of nucleotide substitutions when there are strong transition–transversion and G+C-content biases, *Mol. Biol. Evol.*, 9, 678, 1992.
143. Tamura, K. and Nei, M., Estimation of the number of nucleotide substitutions in the control region of mitochondrial DNA in humans and chimpanzees, *Mol. Biol. Evol.*, 10, 512, 1993.
144. Zharkikh, A., Estimation of evolutionary distances between nucleotide sequences, *J. Mol. Evol.*, 39, 315, 1994.
145. Lake, J.A., Reconstructing evolutionary trees from DNA and protein sequences: Paralinear distances, *Proc. Natl. Acad. Sci. USA*, 91, 1455, 1994.
146. Carrico, J.A. et al., Assessment of band-based similarity coefficients for automatic type and subtype classification of microbial isolates analyzed by pulsed-field gel electrophoresis, *J. Clin. Microbiol.*, 43, 5483, 2005.
147. Persing, D.H. et al., *Molecular microbiology: Diagnostic principles and practice,* ASM Press, Washington, D.C., 2003.
148. Yang, Z. and Rannala, B., Bayesian phylogenetic inference using DNA sequences: A Markov chain Monte Carlo method, *Mol. Biol. Evol.*, 14, 717, 1997.

8 Virulence Determination

Sylvie M. Roche, Philippe Velge, and Dongyou Liu

CONTENTS

8.1 INTRODUCTION

Listeria monocytogenes is a facultative intracellular bacterium that has the capacity to enter, survive, and multiply not only in phagocytic but also in nonphagocytic cells, and to cross a host's intestinal, blood–brain and fetoplacental barriers.[1,2] Being tolerant of external stresses (including extreme pH, temperature, and osmolarity), this bacterium endures many food processing procedures. In line with the current trend toward an increased consumption of convenient, ready-to-eat or heat-and-eat food products nowadays, *L. monocytogenes* has thus become an important source of human food-borne infections. While all human population groups are susceptible to *L. monocytogenes,* infants, pregnant women, and elderly and immunocompromised individuals are especially vulnerable to listeriosis due to their generally weakened immune status. The initial clinical symptoms of human listeriosis are often mild and nonspecific (e.g., chills, fatigue, headache, muscular and joint pain, and gastroenteritis); however, failure to undertake prompt antibiotic treatment of the infection may have severe consequences, with septicemia, meningitis, abortions, and, occasionally, death being the usual outcomes.[2,3]

For a considerable period, *L. monocytogenes* has been regarded as pathogenic at the species level, with a generally accepted belief that all *L. monocytogenes* isolates are potentially virulent and capable of causing diseases. However, from the experimental data collected over the recent years, it becomes clear that *L. monocytogenes* demonstrates enormous serotype/strain variation in virulence and pathogenicity. Whereas many epidemic strains are unquestionably highly infective and sometimes deadly, others (especially those from food and environmental specimens) show limited capability to establish infection and are relatively avirulent.[4–7] Of the 12 *L. monocytogenes* serotypes (i.e., 1/2a, 1/2b, 1/2c, 3a, 3b, 3c, 4a, 4b, 4c, 4d, 4e, and 7), only 3 (i.e., 1/2a, 1/2b, and 4b) are implicated in human listeriosis since these serotypes frequently account for over 96% of *L. monocytogenes* isolations from human clinical cases. Other serotypes (especially 4a and 4c)

TABLE 8.1
Distribution of *L. monocytogenes* Serovars in Human Clinical Cases and Foods

Serovar	Percentages of Strains from Human Clinical Cases[a]	Percentages of Strains from Foods[b]
1/2a	27	64
1/2b	20	9
1/2c	4	12
4b	49	8
Others	<1	7

[a] Eurosurveillance 2001–2003. Mandatory notification of cases. In France, the Centre National de Référence des *Listeria* (Institut Pasteur, Paris) centralizes and characterizes *L. monocytogenes* strains from all sources and notifies the Institut de Veille Sanitaire of all *L. monocytogenes* strains of human origin.

[b] AFSSA 2003–2004. Voluntary submissions of *L. monocytogenes* strains isolated from foods to the laboratory Agence Française de Sécurité Sanitaire in Maisons-Alfort, France.

commonly isolated from animal, food, or environmental specimens are seldom associated with human *L. monocytogenes* infections. This is supported by the surveillance data collected in France during the years of 2001–2004 (Table 8.1).

These findings suggest that many *L. monocytogenes* isolates from animals, foods, and the environment may have reduced virulence, and some may be totally nonpathogenic. Therefore, the availability of laboratory techniques to rapidly and precisely differentiate pathogenic from nonpathogenic *L. monocytogenes* strains not only is vital to the control campaign against listeriosis, but may also play a role in decreasing the number of potential food product recalls and allaying consumer concerns.

The early laboratory procedures for determining the pathogenic potential of *L. monocytogenes* were predominantly *in vivo* bioassays that utilized various animal models such as mice. Despite their relatively high cost and the need to sacrifice animals, these techniques have provided an invaluable means of defining *L. monocytogenes* pathogenicity, in addition to generating new insights on the bacterial pathogenesis, host–pathogen interaction, and immune response of this bacterium, which have ultimately contributed to the more effective prevention and control of listeriosis. Given the notable shortcomings of *in vivo* bioassays, attempts have been made to employ *in vitro* cell assays as a low-cost and easy-to-perform alternative. Following the identification and characterization of a number of key virulence-associated genes and proteins in recent years, novel techniques have been developed for improved evaluation of *L. monocytogenes* pathogenicity. While these latter techniques offer a more definitive measurement on the several important aspects of *L. monocytogenes* virulence, they have yet to surpass the *in vivo* bioassays in the overall performance for virulence determination of this bacterium.

This chapter begins with a concise review of the virulence-associated genes and proteins that are critical to *L. monocytogenes* infection and pathogenicity, and that have underscored the development of new virulence testing procedures. It then proceeds to an update on the key features of various laboratory procedures that have been designed and applied for enhanced determination of *L. monocytogenes* virulence. In the final section, perspectives on future research needs are discussed with the aim of achieving a more precise laboratory definition of *L. monocytogenes* virulence.

8.2 VIRULENCE-ASSOCIATED GENES AND PROTEINS

L. monocytogenes is an extraordinary bacterium that has accumulated a large battery of molecular machineries to facilitate its survival under arduous external conditions and to contribute to its success as a food-borne intracellular pathogen. Upon ingestion by a host via contaminated foods, *L. monocytogenes* first withstands exposure to the host's proteolytic enzymes, a highly acidic environment, and bile salts in addition to nonspecific inflammatory attacks in the stomach. The ability of *L. monocytogenes* to pass through this initial phase is reliant primarily on the normal functioning of a protein subunit of RNA polymerase (RNAP)—alternative sigma factor σ^B (encoded by *sigB*)—that is involved in the regulation of stress-response genes (*opuCA, lmo1421,* and *bsh*) and related proteins, since mutation in *sigB* gene invariably leads to a reduction in *L. monocytogenes* survival under acidic conditions.

Shortly afterwards, *L. monocytogenes* attaches to and enters host cells both passively through phagocytosis and actively through actions of a family of surface proteins known as internalins, a distinct feature of which is their shared leucine-rich repeats (LRR). Of the 25 internalins or internalin-like proteins in *L. monocytogenes* that have been identified and predicted by sequence homology and other analysis, InlA and InlB are the most important. Being covalently bound to the surface of the bacterium through its C-terminal LPXTG motif, InlA (encoded by *inlA*) reacts with host adhesion protein E-cadherin, leading to local cytoskeletal rearrangements, and thus enabling *L. monocytogenes'* entry into epithelial cells. Being noncovalently attached to the surface of the bacterium by an association of its GW domain with bacterial cell-wall lipoteichoic acids, InlB (encoded by *inlB*) recognizes its cellular receptor Met as well as its co-receptor C1q-R (which also functions as a co-receptor for complement component C1q), and thus facilitates *L. monocytogenes'* internalization into hepatocytes, fibroblasts, and epithelioid cells.

Besides InlA and InlB, other internalins (notably InlC and InlJ) also contribute to *L. monocytogenes* invasion during the later (i.e., postintestinal) stages of infection, since alteration in *inlC* or *inlJ* (*lmo2821*) genes negatively impacts *L. monocytogenes* virulence expression beyond the intestinal phase. Furthermore, to gain efficacious entry into eukaryotic cells, *L. monocytogenes* may also employ other factors such as clathrin-mediated endocytosis. Undoubtedly, being inside host cells helps *L. monocytogenes* avoid detection and elimination by a host's humoral immune surveillance network.[3]

Following a brief period in single-layer membrane vacuoles, *L. monocytogenes* migrates to the cytoplasm, a process that is aided by two virulence-associated molecules: listeriolysin O (LLO) and phosphatidylinositol-phospholipase C (PI-PLC). LLO (encoded by *hly*) is a pore-forming, thiol-activated toxin that is an essential component for the successful completion of the *L. monocytogenes* life cycle. In particular, amino acids 91–99 in the LLO protein contribute greatly to LLO stability and thus *L. monocytogenes* virulence, since strains containing mutated *hly* gene are absolutely avirulent. In collaboration with phosphatidylcholine-phospholipase C (PC-PLC, a protein encoded by *plcB*), PI-PLC (encoded by *plcA*) assists LLO in the lysis of the primary vacuoles. Once in the cytoplasm, *L. monocytogenes* replicates rapidly. With the production of another surface protein, ActA (encoded by *actA*), that is cotranscribed with PC-PLC, *L. monocytogenes* begins a cell-to-cell spread process through the formation of the actin structures (called actin comet tails) that propel the bacterium from the cytoplasm toward the cytoplasmic membrane. *L. monocytogenes* is then enveloped in filopodium-like structures that are engulfed by adjacent cells, leading to the formation of secondary double-layer membrane vacuoles. Lysis of the secondary double-layer membrane vacuoles initiates a new infection cycle that is dependent on LLO and PC-PLC, which is in turn activated by Mpl (a metalloprotease encoded by *mpl*)[3] (Table 8.2).

Interestingly, the genes encoding the virulence-associated proteins PI-PLC, LLO, Mpl, ActA, and PC-PLC are located in a 9.6-kb virulence gene cluster that comprises a pleiotropic virulence regulator PrfA (encoded by *prfA*), which is responsible for initiating transcription of many *L. monocytogenes*

TABLE 8.2
Confirmed and Putative Virulence-Associated Proteins in *L. monocytogenes*

Proteins	Function and Involvement	Ref.
	Regulation	
PrfA	Central virulence regulator	19
SigmaB	General stress transcription factor	20
CesR/CesK	Two-component signal transduction responding to the presence of cell wall–acting antibiotics and affecting beta-lactam resistance	21
VirS/VirR	Two-component system involved in virulence in mice as well as cell invasion in cell-culture experiments; VirR appears to control virulence by a global regulation of surface component modifications	22
CheY/CheA	The cheYA operon abolishes response to oxygen gradients and reduces the number of flagella	23, 24
AgrA/AgrC	The deletion of agrA leads to an attenuation of the virulence in the mouse but does not affect the ability of the pathogen to invade and multiply in cells *in vitro*	15
Stp	Regulating EF-Tu and controlling bacterial survival in the infected host	25
DegU	Pleiotropic regulator involved in expression of both motility at low temperature and *in vivo* virulence in mice	10, 11
	Attachment and Invasion	
FbpA	Fibronectin-binding protein required for efficient liver and/or intestinal colonization of mice; chaperone or escort protein for InlB and LLO	26
Ami	Autolytic amidase required for the cell-binding activity	27
InlA	Entry into cells expressing its receptor, the E-cadherin	28, 29
SrtA	SrtA abolishes anchoring of InlA to the bacterial surface, and its deletion impairs colonization of the liver and spleen after oral inoculation in mice	30, 31
InlB	Entry into cells expressing one of the receptors gC1qR, HGF-SF, or Met and the glycosaminoglycanes (GAGs)	32, 33
InlC or Irp	An InlC deletion mutant shows reduced virulence when tested in an i.v. mouse model, but its intracellular replication in Caco-2 and J774 cells appears to be comparable with that of the wild-type strain	34, 35
InlC2, D, E, F	InlC2, InlD, InlE, and InlF null mutants are not affected for entry into any of the cell lines tested, raising the possibility that these genes are needed for an aspect of pathogenicity other than invasion	36
InlG, H, E	Mutants show decreased *in vivo* virulence but do not affect the entry into epithelial cells or intracellular multiplication	37
InlI and InlJ	An InlJ deletion mutant, in contrast to InlI, is significantly attenuated in virulence after i.v. infection of mice or oral inoculation of mice expressing human-E cadherin	17
P60 (Iap)	Required for the invasion into nonprofessional phagocytic 3T6 mouse fibroblast cells, and also implicated in bacteria division	38
LpeA	Lipoprotein favors the entry into nonprofessional phagocytes but not macrophages	39
Auto	Autolysin required for entry into cultured nonphagocytic eukaryotic cell; a mutant shows reduced virulence after i.v. inoculation of mice and oral infection of guinea pig; independent of PrfA	40
Vip	Anchored to the *Listeria* cell wall by sortase A and required for entry into some mammalian cells; exploiting Gp96 as a receptor for cell invasion and/or signaling events that may interfere with the host immune response during the infection	41
IspC	Autolysin and target of the humoral immune response	42
	Lysis of Vacuoles	
LLO	Required for escape from vacuole by lysis of the phagosome membrane	43, 44
SvpA	Both secreted in culture supernatants and surface exposed, promoting bacterial escape from phagosomes of macrophages; a mutant is strongly attenuated in the mouse; independent of PrfA, controlled by a MecA-dependent regulatory network	45

TABLE 8.2 (continued)
Confirmed and Putative Virulence-Associated Proteins in *L. monocytogenes*

Proteins	Function and Involvement	Ref.
PC-PLC	Activated by proteolytic cleavage involving Mpl or by cellular proteases; required for the lysis of the double-membrane vacuole; a mutant is less virulent after i.v. inoculation in mice and strongly attenuated in the intracranial model of infection	46–48
Mpl	Required for the maturation of PC-PLC; although mpl mutant does not affect its uptake and the intracellular growth *in vitro*; its virulence is strongly impaired in mice	49, 50
	Intracellular Multiplication	
Hpt	Impaired listerial intracytosolic proliferation and attenuated virulence in mice	51, 52
LplA1	Required for growth within the host cytosol; a mutant is less virulent in mice	53
	Cell-to-Cell Spread	
ActA	Surface protein necessary for actin assembly, which is involved in cell to cell spread	54
	Other Functions	
HtrA	Involved in stress responses (low pH and penicillin G); required for full virulence in mice; dependent upon the LisRK two-component sensor-kinase	14
Bsh	Involved in the intestinal and hepatic phases of listeriosis: mutants show decreased resistance to bile *in vitro*, reduced bacterial fecal carriage after oral infection of the guinea pigs, and reduced virulence and liver colonization after i.v. inoculation of mice	55
Hfq	Required for *L. monocytogenes* tolerance of osmotic and ethanol stress and important for long-term survival under amino acid–limiting conditions; contributing to pathogenesis in mice, yet playing no role in the infection of cultured cell lines; dependent on σB	56
ClpC	Stress protein promoting early bacterial escape from the phagosome of macrophages, and thus virulence	57
ClpE	Involved in cell division and virulence; required for prolonged survival at 42°C	58
ClpP	Involved in proteolysis and required for growth under stress conditions	59
Frm	Ferritin-like protein	60
L2537	Responsible for teichoic acid biogenesis	12
MogR	Tightly represses expression of flagellin	13
SOD	Superoxide dismutases	61

virulence-associated genes. The *prfA* gene is situated immediately downstream of and sometimes cotranscribed with the *plcA* gene, and it contains two separate transcription binding sites in its promoter region, with *prfAp1* being a 14-bp palindromic sequence (i.e., TTAACATTTGTTAA, also called PrfA box) and *prfAp2* (i.e., TTGTTACT-N_{14}-GGGTAT) strongly resembling the consensus sequence from the –35 and –10 regions of the alternative sigma factor σ^B-dependent promoters. The PrfA box (or *prfAp1*) is also found upstream of the transcriptional start site of each other PrfA regulated operon. The *prfA* gene appears to be also partially regulated by σ^B via *prfAp2,* in addition to its self-regulation. It is of note that some of the virulence-associated genes are differentially regulated at changing temperatures (e.g., at 20 or 37°C) that accompany the *L. monocytogenes* switch from environment to host or vice versa. In addition, the partially σ^B-dependent *prfAp2* promoter region accounts for the majority of prfA transcripts in both intra- and extracellular bacteria.

The genes encoding InlA and InlB are positioned elsewhere in the genome. As *inlA* and *inlB* genes possess a shared transcription binding site similar to that (i.e., PrfA box) recognized by PrfA, they may also be regulated in part by PrfA. It has also been shown that σ^B may play an accessory role in *L. monocytogenes* invasion of human host cells through controlling expression of InlA and InlB, which are fundamental to *L. monocytogenes* passage across the gastrointestinal phase in the guinea pig, but are not important for systemic spread of the organism. The significance of the PrfA-regulated virulence-associated genes, *sigB* and *inlA/inlB*, in *L. monocytogenes* virulence is evidenced by the experimental observations that alterations (e.g., nucleotide deletion, insertion, or substitution) in *prfA, plcA, plcB, hly, actA,* or *inlA* genes often lead to disruption of infection cycle and reduction in virulence and pathogenicity. The involvement of 5′-untranslated regions (5′-UTR) in the regulation of these virulence-associated genes also highlights the significance of post-transcriptional control for *L. monocytogenes* virulence.

It is noteworthy that the *prfA* virulence gene cluster and *inlA/inlB* genes have been identified prior to the whole genome sequences of *Listeria* strains becoming available. Because of their involvement in the distinct phases of the *L. monocytogenes* infectious cycle—attachment, invasion, lysis of the primary vacuole, intracellular motility, cell-to-cell spread, and lysis of the two-membrane vacuole—and of their close proximity in the genome location, these virulence-associated genes stand out in the genomic mutation screening and thus become the first to be characterized. The roles of these virulence-associated genes and proteins have been confirmed by deletion of the underlying genes and construction of mutants followed by assessment of the mutant strains through specific *in vitro* methods or global *in vivo* methods.

The publication of *L. monocytogenes* genome sequences[8,9] has further facilitated identification of many novel, but less prominent virulence-specific genes in this bacterium. Of these novel *L. monocytogenes* virulence-associated genes that have been verified by experimental gene deletions and subsequent examination in animal models, *lmo2515* (or *degU*) encodes a putative regulator for flagellin synthesis,[10,11] *lmo2537* is responsible for teichoic acid biogenesis,[12] *mogR* gene encodes a protein that tightly represses expression of flagellin,[13] and *htrA* gene encodes a serine protease HtrA[14] (Table 8.2). Other *L. monocytogenes* genes and proteins with functions in metabolism, stress response, and regulation (e.g., two-component systems and putative eukaryotic-like phosphatases) have also been characterized. Indeed, the number of confirmed and putative virulence-associated proteins are constantly on the increase (Table 8.2), some of which are specific to *in vivo* virulence (e.g., AgrA),[15] while others are uniquely present in known virulent strains (e.g., InlJ).[16,17]

Interestingly, the *inlJ* (or *lmo2821*) gene encoding a novel internalin (InlJ) exists only in strains capable of causing mouse mortalities, implying its potential role in *L. monocytogenes* virulence.[16] Subsequent study pinpoints InlJ's role in assisting *L. monocytogenes*' successful passage through the intestinal barrier as well as involvement in the later stages of infection.[17] The importance of this gene in *L. monocytogenes* virulence is also highlighted by the findings that *inlJ* (or *lmo2821*) processes distinct genetic structures among various serotypes, with genomic DNA from the more pathogenic serotypes (e.g., 1/2a, 1/2b, 1/2c, and 4b) showing a 5.0-kb *Hin*dIII fragment, DNA from the less pathogenic serotypes (e.g., 3a and 4c) having a 1.5- or 2.0-kb *Hin*dIII fragment, and DNA from the nonpathogenic serotypes (i.e., 4a) forming no band in Southern blotting using an *inlJ* (or *mo2821*) probe.[18] These results suggest that *L. monocytogenes* pathogenic serotypes may maintain a fully functional *inlJ* gene, ensuring their total expression of virulence, whereas less virulent serotypes possess a somewhat disrupted *inlJ* gene, hampering their pathogenicity. But for the majority of other putative virulence-associated proteins listed in Table 8.2, no correlation has yet to be established between the quantity of these proteins and the actual virulence levels of *L. monocytogenes* strains. Thus, the potential of these proteins and their corresponding genes for virulence determination of *L. monocytogenes* remains unclear.

8.3 LABORATORY DETERMINATION OF VIRULENCE

The fact that *L. monocytogenes* displays notable serotype/strain variation in virulence has necessitated the development of laboratory procedures for rapid discrimination of strains with pathogenic potential from those without. The earlier prototype assays for *L. monocytogenes* virulence are generally nonspecific, and dependent on the use of animal models, cell cultures, and chicken embryos. The later generation techniques, on the other hand, focus on the virulence-associated genes and proteins of *L. monocytogenes*, and may show a comparatively higher specificity.

8.3.1 Nonspecific Methods

8.3.1.1 *In Vivo* Bioassays

Based on various experimental animal models, *in vivo* bioassays provide a useful approach for evaluating *L. monocytogenes* pathogenic potential. A key feature of the *in vivo* bioassays is that they cover all potential virulence determinants—both known and uncharacterized—that underpin *L. monocytogenes* pathogenesis.[7,62,63] Of the animal models utilized, the murine model became the most widely adopted for several reasons: good adaptation to captivity, high fecundity, short gestation period, rapid infection progress, ready accessibility, ease of handling, and relatively low expense. Nonetheless, with a constantly escalating cost and ever stringent animal experimentation ethics, the *in vivo* bioassays (including mouse model) are rarely (if at all) applied in routine laboratory testing, apart from serving as a necessary reference in the validation of any newly developed *in vitro* procedures for *L. monocytogenes* virulence.

Mice: General considerations. The *in vivo* mouse bioassay has proven extremely valuable for virulence assessment and for investigation on the pathogenic effects and invasion mechanisms of *L. monocytogenes*. The mouse virulence assay is usually performed by inoculating groups of mice with various doses (usually of three to four different dilutions) of *L. monocytogenes* bacteria via different routes, including intravenous (i.v.),[64] intraperitoneal (i.p.),[65] intranasal, intragastric (oral),[66] subcutaneous (s.c.),[67] conjunctival,[68] intracerebral,[69] or intrarectal.[70] The corresponding dosages are estimated through enumeration of colony forming units (CFUs) on agar plates. The virulence of a given *L. monocytogenes* strain is then determined by the resultant mouse mortality in correlation with the CFU and often expressed as medium lethal dose (LD_{50}).[71,72] The LD_{50} can also be obtained on the basis of a probit dose–response model,[73] taking a log transformation of dose rates and the total number of dead mice.[74] Alternatively, the virulence of a tested *L. monocytogenes* strain can be ascertained by enumerating on agar plates the number of *L. monocytogenes* bacteria that reach the spleen and liver 2–3 days after an experimental infection begins.

Despite having a significant bearing on LD_{50} calculation, CFU estimation for individual *L. monocytogenes* strains through plate counts is a notably delicate task whose outcome is prone to vary with minute changes in handling procedures. Further, *L. monocytogenes* strains with differing levels of virulence also tend to display significant discrepancy in growth rates on nonselective media (e.g., BHI agar), with strains causing more mouse mortalities producing a higher number of CFUs than those causing fewer or no mouse mortalities. As an indirect and unexpected consequence, LD_{50} often exaggerates the pathogenicity of more virulent strains and diminishes the pathogenicity of less virulent strains. Therefore, it is not uncommon that CFUs for a given *L. monocytogenes* strain vary from run to run and from laboratory to laboratory, leading to differing LD_{50} values for an identical test strain.

To overcome this obvious drawback of the CFU estimation, which has a disproportionally large influence on LD_{50} calculation, relative virulence (%) has been put forward as a practical and more direct alternative for interpretation of mouse virulence assay data for *L. monocytogenes*.[75] The relative virulence is calculated by dividing the number of dead mice by the total number of mice tested for a given strain using a known virulent strain (e.g., EGD) as a reference. Since the relative virulence is independent of CFU, it not only avoids the pitfalls that are inherent in the CFU estimation, but also gives a straightforward and thus more accurate assessment of *L. monocytogenes* virulence. Thus, while the LD_{50} values are an imprecise and somewhat ambiguous indicator for *L. monocytogenes* virulence, the relative virulence has no such ambiguity and reflects more or less the true virulence potential of *L. monocytogenes* strains. In addition, with a requirement for fewer mouse dosage groups, the relative virulence is also more economical to conduct than the LD_{50} calculation, augmenting and upholding the increasingly accepted, ethical practice to reduce the number of animals used in biomedical research.

Effect of inoculation routes on infectivity. Inoculation of mice by different routes generally leads to a systemic infection, with invariable involvement of liver and spleen[76] and less frequent involvement of other organs such as the heart, lungs, brain, and, in pregnant mice, the uterus.[74,77,78] From the data available, it seems that the routes of inoculation may exert a noticeable influence on the infectivity of *L. monocytogenes* strains, with the intravenous route being the most efficient way to induce listeriosis, followed by intraperitoneal, intranasal, subcutaneous, and intragastral routes.

Among the common routes of inoculation, intraperitoneal injection of mice is probably one of the easiest to perform, and often gives a consistent outcome. Notwithstanding its technical complexity, intravenous inoculation of mice is also widely applied, which has been instrumental in the earlier elucidation of cell-mediated immunity.[62,63] Although intragastral (oral) inoculation represents the natural route of *L. monocytogenes* infection, it appears to be a very inefficient way to establish a systemic listeriosis, with few reproducible results.[67] This is due possibly to the fact that the translocation of *L. monocytogenes* across the intestinal barrier is low, which is comparable to that of the closely related nonpathogenic species *L. innocua*.[79] Further, wild-type mouse strains appear to lack a human-like E-cadherin receptor, which is critical for murine interaction with *L. monocytogenes* species-specific InlA to facilitate bacteria translocation of the intestinal barrier.[80] Besides its procedural simplicity, subcutaneous injection allows a better discrimination between low-virulence and virulent strains. This may be related to the fact that, in this model, bacteria must translocate from the lymphatic system to the blood system in order to colonize spleens.[6,7] Application of the subcutaneous mouse model has led to the experimental validation of *L. monocytogenes* low-virulence field strains that showed reduced LD_{50} values, poor recovery of virulence after *in vivo* passages, sporadic colonization in various organs (e.g., popliteal lymph nodes, spleen, liver, and lungs), and decreased plaque-forming ability in the plaque-forming assay in comparison with virulent control strains.[74]

Variation in susceptibility among mouse strains. Inbred mouse strains have been favored for the studies of *L. monocytogenes* virulence and pathogenicity because of their genetic uniformity. However, various inbred mouse strains often show differences in their resistance/susceptibility to *L. monocytogenes*, which may range from being totally resistant to being highly susceptible. By comparing several inbred mouse strains for their susceptibility to listeriosis through intravenous inoculation of an *L. monocytogenes* clinical isolate, it was found that CBA/H, BALB/cJ, DBA/iJ, A/J, and CBH/HeJ mice are much less resistant than SJL/WEHI and C57BL/6J mice[81] (Table 8.3). Apparently, the susceptibility of the A/J mouse strain to *L. monocytogenes* can be traced to a two-base-pair deletion in its *c5* gene that encodes for the C5 complement protein. The C5 protein helps recruit macrophages and polymorphonuclear cells to the site of infection. With the underlying defect in the *c5* gene, the A/J mouse strain is unable to produce C5 protein, and thus shows a much lowered inflammatory response to listeriosis. Moreover, the status of two genetic loci (i.e., *Listr1*

TABLE 8.3
Relative Susceptibility of Inbred Mouse Strains to *L. monocytogenes*

Mouse Strain	Intravenous Dosage (CFUs)	Mortality % (day 5 p.i.)	LD_{50}
CBA/H	1.0×10^5	100	5.0×10^2
BALB/cJ	1.0×10^5	100	3.9×10^3
DBA/iJ	1.0×10^5	100	1.0×10^4
A/J	1.0×10^5	100	1.0×10^4
C3H/HeJ	1.0×10^5	100	4.0×10^4
SJL/WEHI	1.0×10^5	0	2.5×10^5
C57BL/6J	1.0×10^5	0	9.0×10^5

Source: Cheers, C. et al., *Immunol. Cell Biol.*, 77, 324, 1999.

and *Listr2*) on murine chromosome 5 has also been implicated in the varied resistance/susceptibility to *L. monocytogenes* in several mouse strains.

The relative susceptibility of certain inbred mouse strains to *L. monocytogenes* infection can be usefully exploited for enhanced virulence determination and pathogenesis studies. For example, use of immunocompromised mouse strain provides a better discrimination between pathogenic and nonpathogenic strains of *L. monocytogenes*.[82] The relevance of this model to the natural infection is supported by the well-established predilection of *L. monocytogenes* for immunocompromised hosts. In addition, a carrageenan-treated murine model has been approved by the Food and Drug Administration (United States) as a laboratory procedure to determine the virulence of *Listeria* isolates. Recently, it has been shown that transgenic mice that harbor the human form of E-cadherin in their enterocytes are much more susceptible to oral infection, since it allows effective interaction between *L. monocytogenes* InlA and E-cadherin. This is the first animal model to accommodate a specific bacterial virulence factor.[83] Transgenic mice possessing human E-cadherin in all tissues are currently being validated (P. Cossart, personal communication). Even though InlB has also been shown to promote species-specific entry and ruffling through its Met interaction in human and mouse cells,[84] the existing mouse models may be inappropriate to the study of the precise role of InlB or its synergy with InlA in *L. monocytogenes* infection.

Variation in virulence among serotypes. Given that 3 (i.e., 1/2a, 1/2b, and 4b) of the 12 *L. monocytogenes* serotypes dominate the clinical isolations from human listeriosis cases (Table 8.1), it is conceivable that *L. monocytogenes* serotypes/strains have varied virulence potential. The varying ability of *L. monocytogenes* serotypes to infect mammalian hosts has been confirmed via mouse models. In fact, through intragastric injection, *L. monocytogenes* epidemic strains appeared to show higher invasiveness than environmental strains.[85] Via intraperitoneal inoculation, *L. monocytogenes* serotypes 1/2a and 4c strains tended to result in higher mortalities than other serotypes and serotype 4a strains were nonpathogenic.[16,75] Examination of a collection of representative *L. monocytogenes* serotypes/strains in A/J mice by the intraperitoneal route indicates that

serotype 4a strains (e.g., HCC23, HCC25, and ATCC 19114) are unable to produce mouse mortality (0%) and are thus truly nonpathogenic
serotype 4b, 4d, and 4e strains show moderate pathogenicity (40–70%)
serotypes 1/2a and 4c are exceedingly virulent (100%) (using relative virulence as a criterion)[16,75] (Table 8.4)

That is, possibly with the exception of serotype 4a, all other *L. monocytogenes* serotypes have the potential to cause mortality in the A/J mouse strain via intraperitoneal inoculation. The fact that serotype 4b strain (i.e., ATCC 19115) generated a somewhat lower mouse mortality than serotype 1/2a strain (i.e., EGD) appears to be in contradiction with the human listeriosis situation where serotype 4b is as potent as (if not more so than) serotype 1/2a in disease-inducing ability. This discrepancy could be related to the route of injection, the origin of the strains, or the status if *inlB* gene.

It is also of interest to note that *L. monocytogenes* avirulent serotype 4a strains with a relative virulence of 0% are negative by PCR, whereas *L. monocytogenes* virulent strains with a relative virulence of 30–100% are positive by PCR targeting *inlJ* (or *lmo2821*) gene. Being present in *L. monocytogenes* strains/serotypes that are capable of causing human listerial outbreaks and mouse mortality, but absent in avirulent, nonpathogenic strains/serotype, *lmo2821* (*inlJ*) therefore offers an ideal target for laboratory differentiation of virulent from avirulent *L. monocytogenes* strains[16,75] (see section 8.3.2.2).

Taken together, the preceding findings may have practical implications. Apart from serotypes 1/2a, 1/2b, and 4b, other *L. monocytogenes* serotypes are not commonly found in human listeriosis cases, due presumably to their inability to efficiently cross human intestinal barrier. Therefore, these latter serotypes would not normally cause problems to healthy individuals when incidentally ingested via contaminated foods. However, the latter serotypes may have the potential to produce

TABLE 8.4
Association of *L. monocytogenes* Serotypes and Relative Virulence

Strain	Source	Serovar	LD_{50}[a]	Relative Virulence (%)[b]
HCC8	Catfish brain	1	$<7.0 \times 10^8$	70
EGD	Guinea pig	1/2a	$<1.1 \times 10^7$	100
ATCC 19112	Human	2	1.6×10^9	30
ATCC 19114	Human	4a	1.9×10^{10}	0
HCC25	Catfish kidney	4a	3.5×10^{10}	0
ATCC 19115	Human	4b	6.0×10^8	70
ATCC 19116	Chicken	4c	2.6×10^8	100
874	Cow brain	4c	$<8.0 \times 10^7$	100
ATCC 19117	Sheep	4d	8.8×10^8	40
ATCC 19118	Chicken	4e	7.8×10^9	50
R2-142	Food	7	$<5 \times 10^7$	100

[a] LD_{50} (medium lethal dose) values were determined in A/J mouse strain.
[b] Relative virulence (%) is calculated by dividing the number of dead mice by the total number of mice tested using EGD as reference.
Source: Liu, D., *FEMS Microbiol. Lett.*, 233, 159, 2004; Liu, D. et al., *J. Clin. Microbiol.*, 44, 2229, 2006.

disease in humans whose immune functions are suppressed, or who are exposed to these serotypes via intravenous transfusions or open wounds.

Other animals. Although mice have been the first and most preferred model for *in vivo* study of *L. monocytogenes* due to their easy access and comparatively low cost, several other animal species (e.g., rat, gerbil, guinea pig, rabbit, and nonhuman primate) have also been utilized in listeriosis research. In fact, *in vivo* bioassays involving nonmurine animal models often provide pertinent details on certain specific aspects of *L. monocytogenes* pathogenicity that are not readily obtainable from the mouse model. For example, gerbils are an appropriate model for *L. monocytogenes* rhobencephalitis and are also valuable for study on the molecular mechanisms of bacterial crossing of the blood–brain barrier. Guinea pigs appear to be suitable for oral infection with *L. monocytogenes*, as they contain a human-like E-cadherin receptor that is vital for InlA-mediated internalization. The recent description of a transgenic mouse strain that expresses human E-cadherin on the surface of its enterocytes renders investigation on the invasive role of InlA in murine hosts possible.[80,83] Guinea pigs and rabbits are naturally sensitive to *L. monocytogenes* infection,[86–88] and in 1934 Anton's conjunctival test was based on these animal models, which developed purulent conjunctivitis 1–5 days after inoculation.[89]

To understand a human infection, it is necessary to have an animal model in which the infectious agent has the same cell and tissue tropism as in humans, and the observed effects should be the same as in humans. The animals that develop a disease closely resembling human listeriosis are not normal laboratory animals such as rodents but rather farm animals such as sheep, cattle, and goats as well as nonhuman primates. In fact, recent data suggest that nonhuman primates simulate humans more closely in their reaction to the *L. monocytogenes* pathogen than other animal models.

Another useful approach is to inoculate the chorio-allantoic membrane of embryonated eggs. This was first proposed in 1940 by Paterson, who showed that all chick embryos died within 72 h of systemic contamination with *L. monocytogenes* bacterium.[90] Many years later, Basher et al. described the inoculation of the allantoic sac of fertile hens' eggs.[91] The chick embryo models allow assessment of *L. monocytogenes* isolates[92] on the basis of 50% lethal dose or percentage mortality in infected embryos.[93,94] Using a pathogenicity test on 14-day-old chick embryos, Avery and Buncic distinguished strains responsible for human listeriosis from those isolated from meat.[95] Similarly, Olier et al. observed differences between human fecal carriage isolates and virulent strains.[96] Embryonated eggs have also been used to recover viable but not culturable (VBNC) *L. monocytogenes* strains, which exhibited the same level of virulence as before their entry in VBNC state (Cappelier, *J. Vet. Med. Res.*, in press).[96a]

8.3.1.2 *In Vitro* Cell Assays

General considerations. In vitro cell assays have been developed as an economical and technically simple alternative to the *in vivo* mouse assay for *L. monocytogenes* virulence. These assays are based on the premise that *L. monocytogenes* has the capacity to enter, survive, and multiply not only in phagocytic but also in nonphagocytic cells.[2] In fact, *L. monocytogenes* is extremely well equipped with various specialized molecules to complete these demanding tasks, such as internalins (notably InA and InlB) to aid cell invasion,[28,97] LLO to help escape from phagocytic vacuoles,[98,99] ActA to facilitate cell-to-cell spread,[54] and phospholipases (PC-PLC and PI-PLC) to act in synergy with LLO and ActA for vacuole escape and cell-to-cell spread.[99]

To date, a large number of the established mammalian cell lines have been employed in the *in vitro* cell assays for *L. monocytogenes* virulence (Table 8.5). Occasionally, bone marrow–derived macrophages are harvested from mice and used directly in such assays. These cell lines are chosen to mimic stage (or phase)-specific barriers to *L. monocytogenes* during the infection. For example, to resemble the intestinal barrier to *L. monocytogenes*, enterocyte-like Caco-2, adenocarcinoma HT-29, epithelial Henle 407, and L2 cell lines are often employed. Similarly, to simulate the barriers after *L. monocytogenes* passes through the intestinal phase, hepatocyte Hep-G2, macrophage-like J774, and fibroblast L929 cell lines are used. Further, hybridoma cell lines have also been explored satisfactorily to evaluate *L. monocytogenes* virulence.[100]

In vitro cell assays are usually conducted by incubation of a known amount of *L. monocytogenes* bacteria (in suspension) on confluent mammalian cell monolayers that are originated from a set number of seeding cells. Extracellular bacteria that have not adhered or entered the mammalian cells are removed by washing and/or killed by gentamycin that does not affect and get into

TABLE 8.5
Mammalian Cell Lines for *In Vitro* Assessment of *L. monocytogenes* Virulence

Cell Line	Lineage	Exemplary Study	Ref.
L2	Mouse fibroblast	Intracellular growth	101
HeLa	Human cervix carcinoma	Differentiation of *Listeria* spp.	102
Caco-2	Human colon adenocarcinoma	InlA	29, 103, 104
L2071	Embryonic fibroblast cells transfected with the human E-cahderin	InlA	105
Henle 407	Human embryonic intestine	PC-PLC	106
TIB 73	Mouse embryonic liver	LpeA	39
RPMI-4788	Human intestinal enterocyte	Adherence and invasion	107
HT-29	Human colon adenocarcinoma	Global estimation of the level of virulence	7
Hep-G2	Human hepatocyte carcinoma	Ami	108
3T3	Mouse embryo fibroblast	LLO	109
MDCK	Canine kidney	Cellular tight junctions—study on ActA	110, 111
Vero	African green monkey kidney (fibroblast)	InlB	112
HBMEC	Human brain microvascular endothelial cells	InlB	29
J774.A1	Mouse monocyte/macrophage	Intracellular growth	113
L929	Mouse connective tissue	Antibiotical studies	114

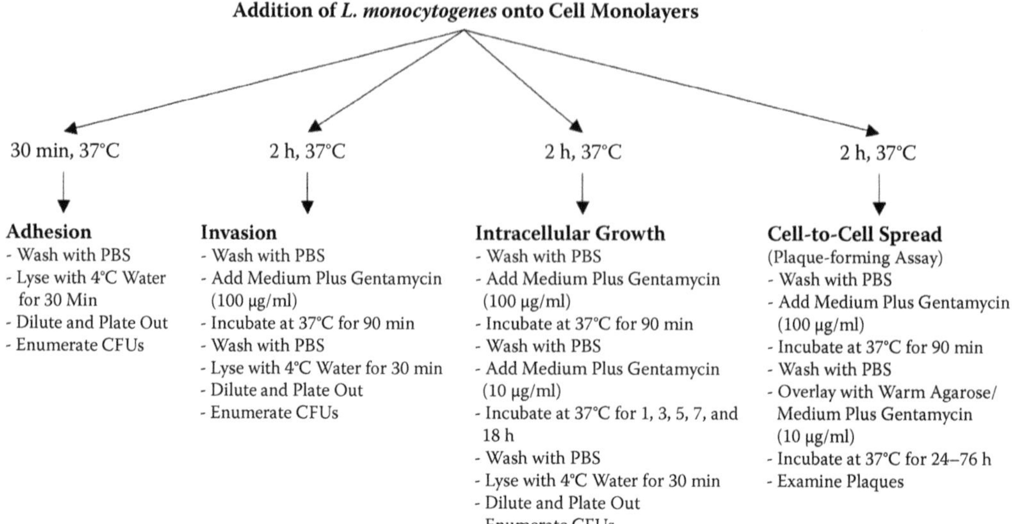

FIGURE 8.1 Common *in vitro* cell assay protocols for *L. monocytogenes* virulence. Variations in incubation time and gentamycin concentration permit assessment of distinct aspects of listerial pathogenicity.

mammalian cells at specified concentrations.[115–117] The remaining/surviving *L. monocytogenes* is then enumerated on solid agar medium. Depending on the specific testing purposes, four *in vitro* cell assay protocols (i.e., adhesion, invasion, intracellular growth, and cell-to-cell spread, which is also known as plaque-forming assay [PFA]) can be carried out.[118] By varying the lengths of additional incubation time and concentrations of gentamycin, it is possible to estimate *L. monocytogenes* capacity and efficiency to adhere and/or invade mammalian cells, replicate intracellularly, spread to neighboring cells, cause cytopathogenic damages, and form plaques—and thus to determine *L. monocytogenes* pathogenic potential (Figure 8.1).

For example, in the adhesion assay, *L. monocytogenes* is allowed to interact with mammalian cells for a short period (i.e., 30 min), which may be sufficient for the bacterium to attach but not long enough for it to enter the mammalian cells. The unattached (nonadhered) bacteria are removed by washing and what have remained may represent those that have managed to adhere to mammalian cells. The ratio of adhered and nonadhered *L. monocytogenes* bacteria can then be worked out with the original number of bacteria deposited as a reference. In the invasion assay, the interaction time between *L. monocytogenes* and mammalian cells is extended to 2 h, which may be adequate for the bacterium to enter the mammalian cells but not sufficient for it to settle in and begin replication intracellularly. The extracellular bacteria are washed away and those that have adhered to the mammalian cell surfaces are then killed with gentamycin (at 100 μg/ml for 90 min). The bacteria recovered will thus represent those that have successfully invaded the mammalian cells. In the intracellular growth assay, after interaction with mammalian cells for 2 h, the extracellular bacteria are washed away and those that have adhered to the mammalian cell surfaces are then killed with gentamycin (at 100 μg/ml for 90 min). The cell monolayers containing *L. monocytogenes* are incubated for 1, 3, 5, 7, and 18 h in the presence of gentamycin (at 10 μg/ml), which will eliminate residual extracellular bacteria but have limited impact on intracellular bacteria. As *L. monocytogenes* normally has a doubling time of about 40–60 min, it will show considerable growth during the incubation period, and produce increasing numbers of CFUs on agar plates.[6] Taking the invasion rate as base line, we could deduce the multiplication factor of the strain. In the cell-to-cell spread assay (or PFA), *L. monocytogenes* is transferred onto mammalian cell monolayers for 2 h. The extracellular bacteria are killed with gentamycin (at 100 μg/ml for 90 min), and the monolayers are then overlaid with warm, low-temperature-melting agarose containing culture medium and gentamycin (at 10 μg/ml) for 24–72 h.[6,7] This prolonged incubation will facilitate *L. monocytogenes'* spreading to neighboring cells and form plaques that can be detected optically, when the incubation period is after 48–72 h, or under microscope when the incubation period is only 24 h.[101] The diameter, number, and morphology of the resultant plaques often reflect *L. monocytogenes* ability to spread to neighboring cells and thus its virulence potential.

Assessment of virulence. The *L. monocytogenes* infection cycle consists of several distinct phases: adhesion, invasion, escape from phagocytic vacuoles, intracellular replication, and dissemination of the bacteria by direct cell-to-cell spread. The inability of *L. monocytogenes* strains to complete any of these phases will lead to the disruption of its life cycle, reducing or abrogating its pathogenicity and virulence. The four *in vitro* cell assay protocols for *L. monocytogenes* virulence are designed essentially to monitor these infection processes, and therefore any strains harboring the defective virulence-associated genes that result in truncated LLO, ActA, InlA, InlB, PI-PLC, and PC-PLC proteins may show aberrant results in the *in vitro* cell assays in comparison with the wild-type control strain, and be identified.

Naturally, *L. monocytogenes* wild-type virulent strains tend to show higher efficiency in adhering and entering Caco-2 and other cells, escaping from vacuoles, undergoing intracellular growth, and spreading to neighboring cells than mutated/avirulent strains. Also, *L. monocytogenes* virulent strains often produce more severe cytopathogenic damage in Caco-2 cells than mutated/avirulent strains. For example, originating from an infected rabbit many years ago, *L. monocytogenes* strain ATCC 15313 was previously hemolytic. After successive laboratory subculturing, this strain has become nonhemolytic due to a mutation in its *hly* gene that led to a truncated LLO. This strain is completely avirulent in mouse model and fails to replicate intracellularly (Figure 8.2), as its defective LLO is unable to assist its escape from the primary vacuole. Similarly, it has become known that some *L. monocytogenes* strains isolated from human carriers display lowered virulence in both *in vivo* and *in vitro* assays because of their production of a truncated InlA protein.[96]

Although numerous attempts have been made in the preceding decades to develop and apply *in vitro* cell assays for *L. monocytogenes* virulence, few have resulted in satisfactory outcomes. Yet we do not dispose of a global assay to assess the virulence of the strains; for example, the PFA could be performed to confirm the inactivity of key virulence-related genes and the embryonated eggs to detect the truncated internalins. Taking into account the nonhemolytic strains, the truncated InlA strains (about 30% of the *L. monocytogenes* strains) (K. Nightingale, personal communication, Isopol 2007), and the low-virulence strains according to the plaque-forming assay (8–21%), which contained the mutated PrfA strains (about 25% of the low-virulence strains),[118] the overall number of low-virulence strains may be higher than we have actually observed.

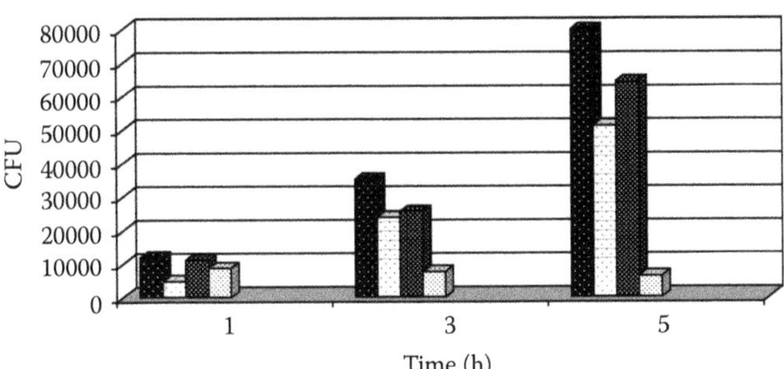

FIGURE 8.2 *L. monocytogenes* intracellular growth in macrophage-like J774 cells. The CFUs shown represent the numbers of bacteria (beginning with 5×10^6) that remained or grew within the J774 cells (in 12-well plates) after 1, 3, and 5 h in the presence of 10 μg/ml gentamycin. The order of strains at each time point from left is virulent strain EGD, avirulent strains HCC23 and ATCC 19114, and nonhemolytic strain ATCC 15313.

While three serotypes (1/2a, 1/2b, and 4b) are responsible for causing over 96% of human listeriosis cases, what is perplexing is why and how other serotypes often fail to produce diseases via contaminated foods despite sharing many molecular, biochemical, and morphological features with the known pathogenic serotypes 1/2a, 1/2b, and 4b (Table 8.1). The fact that virtually all serotypes but 4a are capable of causing mortality in A/J mice via intraperitoneal route suggests their inherent pathogenic potential (Table 8.4), but there may be additional unknown genetic differences that separate the less virulent from the more virulent serotypes. Since the current *in vitro* cell assays do not necessarily cover the genes that are expressed only *in vivo* or are involved in resistance to the immune response, it is no surprise that usually the *in vitro* cell assays can only give some qualitative but not quantitative measurement on *L. monocytogenes* virulence.

Even naturally avirulent serotype 4a strains (HCC23 and ATCC 19114), which are infrequently isolated, behave similarly to virulent EGD strain in the intracellular growth assay. As illustrated in Figure 8.2, starting with a similar amount of bacteria (approximately 5×10^6), the numbers of EGD, HCC23, and ATCC 15313 strains that remained in the macrophage-like J774 cells after gentamycin treatment (at 10 μg/ml) for 1 h were 12×10^3, 4.8×10^3, and 8.8×10^3, respectively—possibly reflecting the differential capacity of these strains to enter the J774 cells with EGD being the most efficient, ATCC 15313 the next efficient, and HCC23 the least efficient. After 3 h, the numbers of EGD and HCC23 tripled (i.e., 35.2×10^3 and 24×10^3), but the number of ATCC 15313 remained essentially the same (at 7.6×10^3). After 5 h, the numbers of EGD and HCC23 doubled to 80×10^3 and 51.2×10^3, respectively, while the number of ATCC 15313 was slightly reduced (at 6.4×10^3) (Figure 8.2).

These results indicate that both EGD and HCC23 underwent a successful intracellular expansion (taking about 60 min to double during the earlier stage and somewhat longer during the later stage, which is perhaps limited by the space and nutrients available within the cytoplasm), whereas ATCC 15313 was unable to multiply intracellularly, probably due to its inability to escape from vacuoles and enter the cytoplasm in the absence of a functional LLO protein. However, serotype 4a strains harbor a 105-bp deletion in their *actA* gene, which results in the production of a nonfunctional ActA protein that is 5 kDa smaller than that from the virulent strain EGD.

Thus, it is clear that, in dealing with *L. monocytogenes* wild-type strains/isolates that contain no defects in the PrfA-regulated virulence gene cluster, the *in vitro* adhesion, invasion, and intracellular growth assays may have limited value for virulence determination. It is worth mentioning that Roche et al. succeeded in identifying numerous *L. monocytogenes* strains with the plaque-forming assay that contained a functional PrfA-regulated virulence gene cluster but showed poor colonization in mice after s.c. and i.v. injections with an LD_{50} close to that of avirulent strains.[74]

Combined with a subcutaneous mouse inoculation, the PFA[7] provides a good estimation of the virulence level of the strains. This screening method gives results in 24 h, and when performed along with the mouse assay, it offers a useful approach for virulence assessment of *L. monocytogenes.* In a study involving 48 clinical listeriosis cases (47 *L. monocytogenes* and 1 *L. seeligeri*), it was found that the sensitivity of the subcutaneous test was 0.96 and specificity was 1. With one exception (a clinical case of *L. seeligeri*; sensitivity 0.98), the PFA successfully detected the virulent strains only (specificity 1). This testing procedure allowed the strains to be classified as low-virulence or virulent field strains. The low virulence of these strains was confirmed later by several other methods such as LD_{50}[74] and phenotypic and genotypic characterizations.[118,119] A recent publication provided additional support on the usefulness of PFA for *L. monocytogenes* virulence, in which 68 (99.8%) of 69 isolates from food and food-processing environments and all 13 clinical isolates were found to be virulent by the PFA and subcutaneous inoculation of mice.

Identification of virulence factors. In vitro cell assays have been also used as a screening method for *L. monocytogenes* mutants that were generated with transposon mutagenesis, signature-tagged mutagenesis and comparative genomics between pathogenic and nonpathogenic *Listeria* species.[43,44,120] Depending on the screening methods, the *in vitro* assays can cover large numbers of mutants more easily than the *in vivo* methods.

In short, although the *in vitro* cell assays (especially the cell-to-cell spread assay) are useful for examining the virulence of *L. monocytogenes* strains, they can be time consuming, and sometimes variable (particularly with isolates whose virulence falls between virulent and avirulent extremes), which necessitates further standardization. For this reason, apart from their utility as a supplemental tool in listeriosis research, the *in vitro* cell assays have not been widely adopted in clinical laboratories as a routine procedure for determining *L. monocytogenes* virulence and pathogenicity.

8.3.1.3 Other Techniques

Other model systems that have been used to identify the host factors essential to the intracellular survival of *L. monocytogenes* include *Drosophila melanogaster*[121] and *Caenorhabditis elegans*.[122] The *D. melanogaster* model has been applied previously in a large number of genetic and immunological studies. Recently, genome-wide RNA-interference screens in *D. melanogaster* S2 cells, which are macrophage-like cells, have revealed many new host factors that are important for cellular entry, escape from vacuole, and bacterial growth.[123,124]

8.3.2 Specific Methods

8.3.2.1 Typing Procedures

L. monocytogenes strains can be subgrouped on the basis of serological reactions between somatic (O)/flagellar (H) antigens and their corresponding antisera.[125] The initial purpose of the serotyping was to differentiate the strains, but it is now also used for tracking strains involved in disease outbreaks as most human illnesses result from infection with 3 of the 13 *L. monocytogenes* serotypes (1/2a, 1/2b, and 4b), with serotypes 4b being the most predominant in epidemics.[126] Although no relationship between serotype and virulence has been conclusively established, it appears that 70% of the strains recovered from contaminated foods belong to serogroup 1/2[127] and 50% of the human clinical isolates belong to serotype 4b. Subsequent development of multiplex PCR assays allowed either the *L. monocytogenes* strains to be separated into four groups: I (1/2b, 3b, 4b), II (1/2c), III (1/2a), and IV (3c),[128] or to differentiate strains 1/2a, 1/2b, 1/2c, and 4b[129] in a rapid and more precise manner.

However, the mechanisms by which subtypes tend to be associated with human cases are not linked to known virulence factors. A recent study using different *L. monocytogenes* strains revealed that, in contrast to strains from other serovars, serotype 4b epidemic strains appeared to be able to cause systemic infection in mice infected orally, suggesting that there might be serovar-specific virulence factors.[130] Ribotyping or PFGE (pulsed-field gel electrophoresis) has also been employed to group *L. monocytogenes* into clusters or lineages associated with human or animal disease,[131,132] but so far no direct relationship has been observed between these lineages and their virulence level in the *L. monocytogenes* strains concerned. Thus, even though these typing methods are mostly based on *L. monocytogenes* specific genes/proteins, their roles in virulence assessment can be only regarded as supplementary.

8.3.2.2 Detection of Virulence-Associated Proteins and Genes

Virulence-associated proteins. Several known virulence-associated proteins such as LLO, PC-PLC, PI-PLC, InlA, and InlB are important for *L. monocytogenes* pathogenicity, since alternations (e.g., nucleotide deletion or substitution) in the corresponding genes often lead to attenuation of the strains concerned. Not surprisingly, these proteins have been targeted by a number of laboratory procedures for assessment of *L. monocytogenes* virulence.[133–136] To undertake the testing, suitable substrates are incorporated into culture media. For example, sheep blood is added for assessment of the hemolytic activity of LLO, egg yolk for examination of the lecithinase activity of PC-PLC, and chromogenic medium for PI-PLC activity (see chapter 5). After incubation, the activity of these proteins can be detected by variation in zone clearing or color changes.

In general, these tests have proven useful for separation of pathogenic from nonpathogenic *Listeria* species, and also for discrimination of wild-type strains from mutant strains. For example, when an *L. monocytogenes* strain produces no PrfA or a truncated PrfA, it often becomes nonhemolytic and avirulent. This type of mutant strain can be easily identified by assays targeting LLO activity. However, attempts to use these virulence-associated proteins as targets for differentiation of wild-type virulent from avirulent *L. monocytogenes* strains have met minimal success, because these proteins are produced by both virulent and avirulent strains, albeit at varying quantities. Thus, while *in vitro* demonstration of LLO, PC-PLC, and PI-PLC activities often provide some tangible evidence as to the pathogenic potential of *L. monocytogenes* strains and an effective exclusion of non*monocytogenes Listeria* species, the dependability of these measurements as reliable virulence indicators for wild-type strains is far from conclusive. In fact, a direct correlation between the presence of these proteins and the level of *in vivo* virulence of *L. monocytogenes* strains has not been unequivocally established.

Nonetheless, analysis of *L. monocytogenes* virulence-associated proteins may be usefully exploited for other purposes. For example, when investigating strains responsible for human and animal clinical cases, Jacquet et al. found a link between the polymorphism of some virulence-associated factors (Ami, InlB, InlA, and LLO) and the animal or human origin of the strains.[134] They demonstrated that all serotype 4b strains of the study, responsible for most major food-borne outbreaks of invasive listeriosis and for numerous sporadic cases, expressed full-length InlA, suggesting that this serovar is associated with higher pathogenic potential.[134]

Virulence-associated genes. Similar to the virulence protein detection, targeting key virulence-associated genes such as *inlA, inlB, actA, hly, mpl, plcA, plcB,* and *iap*[137–145] by PCR and other molecular methods has not established a definitive and convincing relationship between the presence of these genes in *L. monocytogenes* and their virulence potential. These target genes are distributed throughout the species, being present in both wild-type virulent and avirulent strains. After examining the virulence characteristics of *L. monocytogenes* based on the presence of virulence-associated genes, the production of LLO, PI-PLC, and PC-PLC and attachment, entry, and replication within the Caco-2 cells, it was noted that the presence of virulence genes was not associated with the production of virulence-associated proteins *in vitro*, while virulence protein production in *L. monocytogenes* was unrelated, or marginally related, to the ability to invade the Caco-2 cells. Therefore, PCR detection of *hly, plcA,* and *plcB* was unable to consistently ascertain the differences in the virulence properties of the strains.

In another report, *Listeria* isolates from cattle were examined by PCR for virulence-associated genes (*prfA, plcA, hly, actA,* and *iap*) followed by pathogenicity testing with the PI-PLC assay and by mice and chick-embryo inoculation. While all *L. monocytogenes* isolates were hemolytic and positive for the *hly* gene, they showed varied reactions in the assays for *plcA* gene and PI-PLC activity. Thus, detection of multiple virulence-associated genes, in combination with *in vitro* pathogenicity tests, may be required for confirming the pathogenic potential of *L. monocytogenes* strains.

Despite the fact that some naturally virulence-attenuated *L. monocytogenes* strains (e.g., isolates from human carrier cases) often harbor mutations in their *prfA, hly, actA,* or *inlA* genes—leading to the expression of truncated or nonfunctional PrfA, LLO, ActA, or InlA proteins—detecting these gene mutations may not be an ideal way for assessing *L. monocytogenes* virulence. For example, Jacquet et al. showed that the clinical strains expressed full-length InlA far more frequently (96%) than did strains recovered from food products (65%). All strains from pregnancy-related cases (100%), strains from patients with central nervous system infections (98%), and strains from patients with bacteremia (93%) also expressed full-length InlA. Moreover, all strains belonging to serovar 4b, the most frequently implicated serovar in human listeriosis, expressed full-length InlA.[136] Thus, screening for genetic alternations in multiple *L. monocytogenes* genes does not take into account all the virulence attenuated strains. However, development of a multiplex PCR assay targeting several key mutations (e.g., truncated InlA, mutated PrfA, and mutated PI-PLC, which is a marker of Group III[118] low-virulence strains [S. Roche, unpublished data]), could be useful and increase test efficiency.

There is no doubt that detection of virulence-specific genes present only in virulent strains and absent in avirulent strains constitutes another more rational approach for *L. monocytogenes* virulence testing. With the whole genome sequences of several *L. monocytogenes* strains becoming available,[8,9] a panel of novel virulence-specific genes with potential for improved determination of *L. monocytogenes* virulence and pathogenicity have been identified through comparison of virulent and avirulent genomic DNA libraries.[16] These genes encode putative transcriptional regulators (i.e., *lmo0833, lmo1116, lmo1134,* and *lmo2672*), putative internalins (i.e., *lmo2821* and *lmo2470*), and unknown proteins (i.e., *lmo0834* and *lmo1188*). It is particularly noteworthy that the putative internalin gene *lmo2821* (later redesignated as *inlJ*) stands out as an excellent target for rapid differentiation of *L. monocytogenes* virulent strains capable of producing mouse mortalities from avirulent strains that produce no mouse mortalities.[16,75] As discussed earlier (see section 8.2), *inlJ* (or *lmo2821*) revealed a 5.0-kb *Hin*dIII fragment in Southern blot with genomic DNA from the more pathogenic serotypes (e.g., 1/2a, 1/2b, 1/2c, and 4b); a 1.5- or 2.0-kb *Hin*dIII fragment with DNA from the less pathogenic serotypes (e.g., 3a and 4c); and no band with DNA from the nonpathogenic serotypes (i.e., 4a),[18] indicating the need of having an intact *inlJ* (or *lmo2821*) for optimal expression of *L. monocytogenes* virulence.

The notion that the more pathogenic serotypes demonstrate genetic differences from the less pathogenic serotypes also gains credence from analysis on a newly identified species-specific gene *lmo0733*. Being present in all *L. monocytogenes* serotypes, *lmo0733* is detected as a *Hin*dIII band of 5.0 or 6.0 kb in the more pathogenic serotypes (e.g., 1/2a, 1/2c, and 4b), and a *Hin*dIII band of 1.0 or 1.5 kb in the less pathogenic or nonpathogenic serotypes (e.g., 4a and 4c).[18] Thus, apart from maintaining an uninterrupted *inlJ* gene, *L. monocytogenes* pathogenic serotypes may also require a full-length copy of *lmo0733* gene for successful establishment in the hosts.

While the presence or absence of internalin gene *inlJ* provides a useful indicator for the potential virulence or avirulence of *L. monocytogenes* strains, a more recent study showed that there are some unusual lineage IIIB strains (e.g., serotype 7 strain R2-142) that are negative for *inlJ* by PCR and Southern blot, yet have the capacity to cause mouse mortality via intraperitoneal inoculation.[145] This suggests that *inlJ* is not absolutely required for *L. monocytogenes* virulence through intraperitoneal inoculation, despite the fact that *inlJ* contributes to listerial breach of a host's intestinal phase and to later stages of infection, since an *inlJ* deletion mutant shows reduced virulence in both hEcad mice via oral route and wild-type mice via intravenous route. Considering that only three serotypes (1/2a, 1/2b, and 4b) are isolated from human listeriosis cases, the lineage IIIB strains without *inlJ* gene may have difficulty in crossing the host intestinal barrier during infection via a conventional oral route. However, it is possible that the lineage IIIB strains containing no *inlJ* gene may cause illness when they are transmitted via open wounds and fluid transfusions.

To provide an effective coverage for the lineage IIIB strains that lack *inlJ* gene, but have the capacity to cause mouse mortality via intraperitoneal route, a search for additional virulence markers for improved detection of potentially virulent strains of *L. monocytogenes* was undertaken recently. A previous report indicated that InlC may contribute to a late stage of *L. monocytogenes* infection instead of the uptake of the bacterium by nonprofessional phagocytic cells, as it was strongly transcribed in the cytoplasm of murine phagocytic J774 cells where *inlA* was poorly transcribed. A mutant strain containing *inlC* gene deletion displayed lowered virulence in an intravenous mouse model.[34] By using PCR primers designed from the *inlC* gene, it was noted that the *inlC* gene is present in all serotype 1/2a, 1/2b, 1/2c, 3a, 3b, 3c, 4b, 4d, and 4e strains and some serotype 4c strains, but absent in serotype 4a strains. The lineage IIIB strains without *inlJ* gene, although capable of causing mouse mortality, were also positive by PCR assay for the *inlC* gene (D. Liu, unpublished data) (Table 8.6 and Table 8.7).[145a]

By combining the *inlA* gene primers (for species-specific detection) with the *inlC* and *inlJ* gene primers (for virulence determination) in a multiplex PCR format, it became possible to achieve a

TABLE 8.6
Identities of *L. monocytogenes* Internalin Gene Primers

Gene	Coding Sequences	Primer Sequences (5′ → 3′)	Nucleotide Positions	Expected PCR Product (bp)
inlA	94534–96936	ACGAGTAACGGGACAAATGC	94612–94631	800
		CCCGACAGTGGTGCTAGATT	95411–95392	
inlC	107200–108090	AATTCCCACAGGACACAACC	107306–107325	517
		CGGGAATGCAATTTTTCACTA	107822–107802	
inlJ	188153–19070	TGTAACCCCGCTTACACAGTT	188989–189009	238
		AGCGGCTTGGCAGTCTAATA	189226–189207	

rapid and simultaneous confirmation of *L. monocytogenes* species identity and virulence potential (D. Liu, unpublished data)[145a] (Table 8.6, Table 8.7, and Figure 8.3). That is, the species identity of the *L. monocytogenes* strains can be validated through the formation of an 800-bp band with the *inlA* gene primers, and the virulence potential of these strains is ascertained by the production of 517- and/or 238-bp bands with the *inlC* and *inlJ* gene primers in the multiplex PCR (Table 8.6, Table 8.7, and Figure 8.3). Interestingly, *L. monocytogenes* strains with capacity to cause mortalities in A/J mice via intraperitoneal route (showing relative virulence of 40–100%) are invariably identified by the *inlC* and/or *inlJ* primers, whereas the nonpathogenic strains with a relative virulence of 0% (i.e., serotype 4a strains ATCC 19114 and HCC23 as well as IIIA strain X1-002, which may also be a typical serotype 4a strain) were negative with the *inlC* and *inlJ* primers (Table 8.7). Although *L. ivanovii* strains were also reactive with the *inlC* primers, they could be excluded as not *L. monocytogenes* by their negative reaction with the *inlA* primers in the multiplex PCR (Table 8.7 and Figure 8.3).

However, it is worth emphasizing that while the presence of *inlC* and/or *inlJ* genes in a given *L. monocytogenes* strain suggests its potential virulence inclination and its capacity to produce mouse mortality via intraperitoneal route, it does not indicate its likelihood to cause disease in humans through oral ingestion. Given that three serotypes (1/2a, 1/2b, and 4b) are responsible for most of the human listeriosis cases, many serotypes (e.g., 1/2c, 3a, 3b, 3c, 4c, 4d, and 4e) that are recognized as having virulence potential by the multiplex PCR may experience difficulty breaching a host's intestinal and other barriers during the course of infection. Obviously, for an *L. monocytogenes* strain to be fully infective to humans via oral ingestion, it requires input from many other known and uncharacterized virulence genes and proteins, which need further identification and examination. Nonetheless, an implication from these results is that even though *L. monocytogenes* serotypes other than 1/2a, 1/2b, and 4b are not commonly found in human listeriosis cases—due presumably to their inability to go across human intestinal and other barriers efficiently, they (possibly with the exception of serotype 4a) may have the potential to produce listeriosis in humans whose immune responses are compromised, or who are inadvertently exposed to these serotypes through intravenous transfusions or via open wounds.

The recent development of microarray technology offers potential for rapid and extensive analysis of *L. monocytogenes* gene activation in host–pathogen relationships, which is clearly advantageous over the current genomic and transcriptional procedures. For example, examination of the DNA content by the existing genomic techniques merely indicates whether a gene is present or not, without indicating a single nucleotide substitution that silences that gene. Transcription analysis depends largely on the test condition; if the bacteria are cultured in conditions that do not allow gene expression, no workable result will be obtained. The microarray technology incorporates the key features of both genomic and transcriptional procedures, and is poised to have a much bigger impact on the future investigation of the in/activation of various *L. monocytogenes* virulence-associated genes during the distinct phases of infection in mammalian hosts as well as under various strenuous external conditions.

TABLE 8.7
Multiplex PCR Analysis of *Listeria* Strains Using *inlA, inlC,* and *inlJ* Gene Primers

Strain	Source	Serotype (Subgroup)	Multiplex PCR *inlA* (800 bp)	*inlC* (517 bp)	*inlJ* (238 bp)	Relative Virulence (%)[a]
L. monocytogenes EGD	Guinea pig	1/2a	+	+	+	100
F6854	Turkey frank	1/2a	+	+	+	ND
RM3158	Human	1/2b	+	+	+	ND
RM3368	Environment	1/2b	+	+	+	ND
RM3017	Blood	1/2c	+	+	+	ND
RM3367	Environment	1/2c	+	+	+	ND
RM3026	Food	3a	+	+	+	ND
RM3162	Human	3a	+	+	+	ND
RM3836	Beef frank	3b	+	+	+	ND
RM3845	Hot dog	3b	+	+	+	ND
RM3027	Chicken	3c	+	+	+	ND
RM3159	Human	3c	+	+	+	ND
ATCC 19114	Ruminant brain	4a	+	–	–	0
HCC23	Catfish	4a	+	–	–	0
ATCC 19115	Human	4b	+	+	+	70
RM3177	Human	4b	+	+	+	ND
ATCC 19116	Chicken	4c	+	–	+	100
874	Cow brain	4c	+	+	+	100
ATCC 19117	Sheep	4d	+	+	+	40
RM3108	Chicken	4d	+	+	+	ND
ATCC 19118	Chicken	4e	+	+	+	50
RM2218	Oyster	4e	+	+	+	ND
F2-458	Human	(IIIA)	+	+	+	100
J2-074	Animal	(IIIA)	+	+	+	100
X1-002	Food	(IIIA)	+	–	–	0
F2-086	Human	(IIIB)	+	+	–	100
R2-142	Food	(IIIB)	+	+	–	100
F2-208	Human	(IIIC)	+	+	+	100
F2-270	Human	(IIIC)	+	+	+	100
L. ivanovii ATCC 19119	Sheep		–	+	–	ND
L. innocua						
ATCC 33090	Cow brain	6a	–	–	–	ND
L. seeligeri						
ATCC 35967	Soil		–	–	–	ND
L. welshimeri						
ATCC 43550	Soil	1/2b	–	–	–	ND
L. grayi						
ATCC 25400	Corn leaves/stalks		–	–	–	ND

Notes: The relative virulence (%) was determined via i.p. inoculation of A/J mice and calculated by dividing the number of dead mice by the number of tested and then multiplying by 100. ND = not done.

Source: Liu, D., *FEMS Microbiol. Lett.*, 233, 159, 2004; Liu, D. et al., *J. Microbiol. Methods*, 71, 133, 2007.

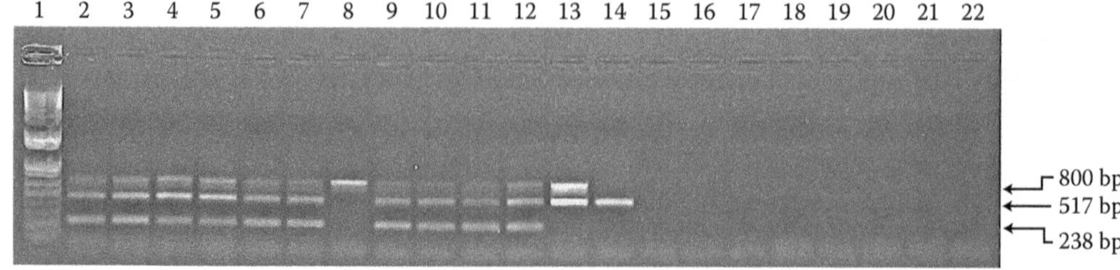

FIGURE 8.3 Agarose gel electrophoresis of PCR products generated with *inlA*, *inlC*, and *inlJ* gene primers. Lane 1, DNA molecular weight marker (1 kb plus, Life Technologies, Invitrogen, Carlsbad, California); lanes 2–13, PCR products amplified from DNA of *L. monocytogenes* serotype 1/2a (EGD), serotype 1/2b (RM3368), serotype 1/2c (RM3017), serotype 3a (RM3026), serotype 3b (RM3845), serotype 3c (RM3159), serotype 4a (ATCC 19114), serotype 4b (ATCC 19115), serotype 4c (874), serotype 4d (ATCC 19117), serotype 4e (ATCC 19118), and lineage IIIB (R2-142) strains; lane 14, *L. ivanovii* ATCC 19119; lane 15, *L. innocua* ATCC 33090; lane 16, *L. seeligeri* ATCC 35967; lane 17, *L. welshimeri* ATCC 43550; lane 18, *L. grayi* ATCC 25400; lane 19, *Enterococcus faecalis* ATCC 29212; lane 20, *Salmonella enterica* serovar typhimurium ATCC 14028; lane 21, *Staphylococcus aureus* ATCC 25923; and lane 22, negative control with no DNA template. The expected *inlA, inlC,* and *inlJ* gene products are of 800, 517, and 238 bp in size, respectively (as indicated on the right).

8.3.2.3 DNA Sequencing Analysis

Through comparative genomic analysis, it is apparent that *L. monocytogenes* virulent and avirulent strains not only process many unique genes or genetic regions of their own, but also have distinct nucleotide compositions among their shared genes. This is supported by the identification of virulence-specific putative internalin and transcriptional regulator genes from *L. monocytogenes* virulent strains, on the one hand, and the confirmation of nucleotide mutations in the *hly* gene in ATCC 15313, on the other (Genbank accession No. AY750700). Previously, nucleotide sequence analysis has not been widely applied for routine identification and virulence determination of *L. monocytogenes* bacteria due to its high cost. With the expense of DNA sequencing coming down dramatically in recent years, this technique has been increasingly used for in-depth examination of *L. monocytogenes* genes and genomes for identification and virulence determination purposes. Indeed, DNA sequencing analysis of the *prfA* virulence gene cluster from a naturally avirulent catfish strain HCC23 has revealed numerous nucleotide changes in this important gene region, which have led to amino acid substitutions and deletions in the translated proteins of PrfA, PI-PLC, LLO, Mpl, ActA, and PC-PLC (Genbank accession nos. DG118415 and AY878649).

One of the whole, relatively small percentages of changes has been noted in the translated proteins of PrfA (2/237 or 0.8%), PI-PLC (8/317 or 2.5%), LLO (3/527 or 0.5%), Mpl (18/510 or 3.5%), and PC-PLC (9/289 or 3.1%) between EGD-e and HCC23 (Table 8.8). These findings may help explain why the previous *in vitro* adhesion, invasion, and intracellular growth assays have been unsuccessful in discrimination of virulent *L. monocytogenes* from naturally avirulent serotype 4a strains. The same can be attributed to the inability of protein (or DNA) assays targeting PI-PLC, LLO, and PC-PLC to accurately determine *L. monocytogenes* virulence. However, a much higher percentage of amino acid variation is observed in the translated protein of ActA (81/639 or 12%) between EGD-e and HCC23 (Table 8.8). Whereas the implication of the amino acid changes in the PrfA, PI-PLC, LLO, Mpl, and PC-PLC proteins from avirulent strain HCC23 is by no means obvious at a first glance—apart from the fact that HCC23 has been shown to display a much lower level (about 50% or less) of hemolytic activity in comparison with EGD—the significance of a drastic amino acid change in the ActA protein from HCC23 is easy to fathom. In fact, the disappearance of the first six amino acids (i.e., MGLNRF) in the leader sequence of HCC23 ActA may possibly compromise its function, and the deletion of a stretch of 35 amino acids (i.e., PTDEELRLALPETP MLLGFNAPATSEPSSFEFPPP at positions of 304–338) effectively removes two of the four copies

TABLE 8.8
Amino Acid Variations in the PrfA-Regulated Proteins between *L. monocytogenes* Virulent EGD-e and Avirulent HCC23 Strains

		Amino Acid				Amino Acid	
Protein	**Position**	**EGD-e**	**HCC23**	**Protein**	**Position**	**EGD-e**	**HCC23**
PrfA (237 aa)	165	A	T	ActA (639 aa)	304–338	PTDEELRLALP-ETPMLLGFNAP-ATSEPSSFEFPPP	Deleted
	197	N	K		347	I	M
PI-PLC (317 aa)	17	V	I		352	S	P
	57	S	N		360	R	S
	61	I	L		368	N	S
	178	N	T		375	Q	E
	190	I	V		459	A	T
	211	H	R		476	T	A
	220	S	P		479	K	E
	235	K	T		481	S	P
LLO (527 aa)	14	V	I		497	V	A
	438	V	I		498	T	S
	523	K	S		501	P	I
Mpl (510 aa)	31	K	R		505	K	N
	36	T	I		507	A	S
	42	P	H		512	E	A
	115	R	K		515	A	I
	122	I	M		518	P	A
	146	V	I		522	V	A
	157	L	I		524	R	G
	182	V	I		528	T	A
	186	V	A		537	T	P
	203	R	P		539	K	N
	251	K	N		542	N	D
	253	N	K		554	A	V
	256	T	A		557	S	R
	325	Q	R		558	D	N
	423	E	A		572	E	G
	427	L	I		576	S	P
	438	Y	F		580	A	V
	475	S	A		584	N	K
ActA (639 aa)	1–6	MGLNRF	Deleted		608	P	S
	73	K	E		609	G	A
	82	R	K		621	I	M
	95	E	A		639	N	S
	165	P	L	PC-PLC (289 aa)	13	I	T
	171	V	A		26	N	S
	179	E	A		34	Q	K
	182	A	T		35	T	P
	184	A	T		81	N	D
	200	S	T		93	K	N
	244	S	G		126	R	K
	257	S	N		150	T	A
	267	P	A		215	A	V

Source: Adapted from Genbank accession Nos. AL591978, DQ118415, and AY878649.

TABLE 8.9
Detection of Amino Acid Substitutions in *L. monocytogenes* PrfA, InlA, PC-PLC, and PI-PLC Proteins

	Substitution			
Phenotypic Group (strain)	**PrfA**	**PC-PLC**	**PI-PLC**	**InlA**
Group Ia				
(CHU860776, CNL895793, CNL895803, CNL895804, CNL89580t6, CNL895809, SO49, and AF10)	K220T			
Group Ib				
(BO18, BO38, and AF95)	Δ174-237			
Group II				
(454)		D61E, L183F, Q216K, A223V		
Group III				
(CNL895807, CNL895795, 416, 417, and BO43)			I17V, F119Y, T262A	No InlA

Source: Roche, S.M. et al., *Appl. Environ. Microbiol.*, 71, 6039, 2005.

of proline-rich repeats (DFPPPPTDEEL) that control *L. monocytogenes* movement between cells (with each repeat contributing about 2 μm/min).

Moreover, the amino acid changes at position 267 (from P > A) may also impact negatively a third copy of proline-rich repeat. Interestingly, the amino acid variations in the ActA protein are not restricted to HCC23, as other *L. monocytogenes* serotype 4a strains also appear to harbor similarly altered *actA* genes. Therefore, it is no surprise that *L. monocytogenes* serotype 4a strain L99 is shown to produce an ActA protein that is 5 kDa smaller than that from virulent strain EGD, and it encounters difficulty in spreading to neighboring cells. In the cell-to-cell spread assay, serotype 4a strains are unable to spread to neighboring cells.[146] It is likely that the changes in the ActA protein may be one of the key factors that adversely impact the ability of *L. monocytogenes* serotype 4a strains to undergo cell-to-cell spread and to cause mouse mortality via *in vivo* bioassays. In addition, a lack of the *inlJ* (*lmo2821*) gene in HCC23 also contribute to its inefficiency on crossing the host intestinal barrier.[16,75] These findings offer tangible support for the use of the cell-to-cell (plaque-forming) assay (or its derivatives) in the differentiation of virulent *L. monocytogenes* from naturally avirulent serotype 4a strains.

In the same way, many low-virulence strains were shown to exhibit nucleotide changes in *prfA*, *plcA*, *plcB*, or *inlA*, which have resulted in amino acid substitutions and deletions in the translated proteins. For instance, the phenotypic Groups I, II, and III low-virulence strains (Table 8.9)[118] were first detected by a plaque-forming assay on HT-29 human cells, and then their low virulence was confirmed after subcutaneous injection into mice. Of the 11 Group I strains that did not enter cells, showed no phospholipase activity, and exhibited a mutated PrfA, 8 strains harbored a single amino acid substitution K220T in PrfA and the other three had a truncated PrfA, PrfA Δ174-237. These genetic modifications could explain the low virulence of Group I strains, because mutated PrfA proteins were nonfunctional and inactive. This K220T mutation was further characterized by analysis of its capacity to form a complex PrfA K220T–RNA polymerase–DNA complex.[119] Groups II and III strains entered cells but did not form plaques. Group II strains had low PC-PLC activity, whereas Group III strains had low PI-PLC activity. Several substitutions were observed for five out of the six group III strains in the *plcA* gene and for one strain out of the three Group II strains in the *plcB* gene. There is no doubt that further DNA sequencing analysis of other genes will yield additional insights on the molecular mechanisms of *L. monocytogenes* virulence.

8.4 CONCLUSIONS AND PERSPECTIVES

L. monocytogenes is an intracellular bacterial pathogen that has an uncanny ability to survive in a variety of environments and to prosper within mammalian hosts. There are few other intracellular pathogens that employ such a diverse collection of specialized molecules to assist in adhesion and entry into host cells, escape from vacuoles, intracellular replication, and cell-to-cell spread without being totally annihilated by the host during the process. Extensive investigations over the previous two decades have led to the identification of many virulence-associated molecules from this bacterium, the most notable of which is the PrfA-regulated virulence cluster. Other important molecules such as internalins and σ^B-regulated proteins have also been characterized shortly thereafter. With the recent publication of the complete genome sequences of several *L. monocytogenes* strains, an increasing number of virulence-related proteins are being reported (Table 8.2). The detailed examination of these virulence-associated molecules has not only yielded valuable insights on the molecular mechanisms of *L. monocytogenes* pathogenicity, but also contributed to the development of new generation techniques for improved virulence determination.

In vivo bioassays based on the use of laboratory animal models (in particular mice) represent a first-generation technique for assessing *L. monocytogenes* virulence. Because the mouse virulence assay offers comprehensive coverage of all potential virulence determinants, its performance and usefulness for virulence assessment of *L. monocytogenes* have yet to be surpassed in spite of many worthwhile attempts involving *in vitro* cell assays and detection of virulence-associated genes and proteins. While calculation of LD_{50} values for the mouse virulence assay has relied on CFU enumeration, which suffers from the drawback of being sometimes variable, use of the recently described relative virulence (%) provides a much more consistent alternative for its resultant interpretation, in addition to reducing the number of mouse groups that are required for conventional LD_{50} calculation.

Despite the fact that wild-type mice lack appropriate E-cadherin receptor for *L. monocytogenes* InlA, thus compromising its value as a model for listeriosis via oral route, use of the mouse virulence assay has nonetheless improved our understanding on *L. monocytogenes* pathogenesis and its virulence variations among distinct serotypes/strains. Undoubtedly, in the absence of comparative *in vitro* methodologies that provide a viable alternative, the *in vivo* bioassays (especially the mouse virulence assay) will continue to play a vital role in the studies on the mechanisms of *L. monocytogenes* infection and also in the validation of any new laboratory tests developed for assessing *L. monocytogenes* pathogenic potential. It will be of interest to see if new *in vivo* assays can be developed in future that allow the specific screening of new genes only involved in the *in vivo* virulence (e.g., *agrA*[15] and *inlJ*).[17,75]

With the elucidation of the *L. monocytogenes* infection cycle, which encompasses adhesion/invasion, escape from vacuole, intracellular multiplication, and cell-to-cell spread, it becomes possible to design appropriate *in vitro* cell assays that monitor these distinct stages of infection. Any *L. monocytogenes* strains encountering difficulty in one or more of these phases will be unable to complete the infection cycle, and show low or no pathogenicity since they will be subordinated and/or eliminated by the host. The four *in vitro* cell assay protocols (i.e., adhesion, invasion, intracellular growth, and cell-to-cell spread assays) have proven valuable for separation of pathogenic from nonpathogenic *Listeria* species, as the latter either completely lack or only produce defective PrfA regulated virulence-associated molecules. These assays have also been useful for detection of virulence attenuated *L. monocytogenes* strains that harbor mutations in the key virulence-associated genes, resulting in non- or subfunctional PrfA, LLO, InlA, InlB, or other virulence-associated proteins.

However, only the *in vitro* cell-to-cell spread assay (or plaque-forming assay) has shown promise for global evaluation of the level of virulence of *L. monocytogenes* wild-type strains, as naturally virulent and avirulent strains often behave similarly in the *in vitro* adhesion, invasion, and intracellular growth assays. At present, the cell-to-cell spread (plaque-forming) assay is notoriously slow (requiring a few days), in addition to its occasional lack of reproducibility. To be useful in the routine testing situation, further improvement is clearly needed in the visualization of plaques formed.

Moreover, future development of alternative methods for detecting specific molecules involved in cell-to-cell spread (e.g., ActA and its corresponding gene *actA*) may one day negate the necessity to conduct *in vitro* cell assay for *L. monocytogenes* virulence.

The recent development of a multiplex PCR incorporating *inlA* primers for species-specific confirmation and *inlC* and *inlJ* primers for virulence determination has enhanced our capability for laboratory determination of *L. monocytogenes* pathogenicity (Figure 8.3). The species identity of *L. monocytogenes* is validated by the formation of an 800-bp band using the *inlA* primers, whereas its virulence potential is ascertained by the production of 517- and/or 238-bp bands using the *inlC* and *inlJ* gene primers. The inclusion of both the *inlC* and *inlJ* gene primers in the multiplex PCR provides a double verification of *L. monocytogenes* virulence for most strains. The usefulness of this multiplex PCR for virulence determination is supported by the observations that *L. monocytogenes* virulent strains with the potential to cause mouse mortalities via intraperitoneal inoculation of mice were positive with the *inlC* and/or *inlJ* primers, while naturally avirulent strains unable to produce mouse mortality were negative with these primers (Table 8.7). Even though *L. ivanovii* strains also harbor a gene with homology to *L. monocytogenes inlC* and cross-react with *L. monocytogenes inlC* primers, they can be readily excluded through their negative reaction with *L. monocytogenes inlA* primers incorporated in the multiplex PCR.

Thus, it appears that an optimal strategy for virulence assessment of *L. monocytogenes* strains/isolates is to firstly apply a multiplex PCR targeting *inlA, inlC,* and *inlJ* genes, which will help verify the species identity of these strains, and separate potentially virulent from naturally avirulent strains. This is followed by an improved cell-to-cell spread (or plaque-forming) assay, which will discriminate *inlC*- and/or the *inlJ*-positive strains with capability to spread to neighboring cells from those without. Obviously, *L. monocytogenes* strains harboring mutations in PrfA-regulated gene clusters and *inlA/inB* genes will be flagged out by the cell-to-cell spread assay, as they may encounter difficulty in adhering to/entering the mammalian cells, escaping from vacuoles, multiplying in cytoplasm, and thus spreading to next cells without fully functional PrfA, LLO, PC-PLC, PI-PLC, ActA, and InA/InlB proteins. This is exemplified by an *hly* mutant strain ATCC 15313, which is capable of entering the host cells but unable to escape from vacuoles and thus undergoing intracellular replication (Figure 8.2). To further pinpoint the cause of *L. monocytogenes* low virulence or avirulence, a real-time, multiplex PCR targeting specific mutations in PrfA (e.g., K220T and Δ174–237), PC-PLC (e.g., D61E, L183F, Q216K, and A223V), PI-PLC (e.g., I17V, F119Y, and T262), and ActA (e.g., Δ304–338) may then applied if necessary. Finally, a mouse model will provide an ultimate means of defining the true virulence of *L. monocytogenes* strains concerned.

Based on the *in vitro* and *in vivo* mouse assay results, it is clear that the number of low-virulence strains (i.e., attenuated-virulence strains) may have been underestimated in the past. By including strains that are nonhemolytic, have truncated InlA (about 30% of the *L. monocytogenes* strains) (K. Nightingale, personal communication, Isopol 2007), contain mutated PrfA (about 25% of the low-virulence strains), and show lowered plaque-forming ability (8–21%),[118] the low-virulence strains may easily account for 50% of the *L. monocytogenes* strains in the field. One of the possible reasons for the underestimation of the number of low-virulence *L. monocytogenes* strains is the types of culture/growth media that are currently employed for listerial isolation. There is some tangible evidence that these media tend to favor the growth of high-virulence strains over that of the low-virulence strains.[75,147] Further research is definitely warranted in this area so that more efficient culture/growth media can be optimized and used for unbiased isolation of *L. monocytogenes* low-virulence strains.

In the past 2–3 years, different communities (e.g., The European Union, Canada, and the United States) have formulated directives regarding the quantitative assessment of relative food safety risks. To prevent food-borne *L. monocytogenes* infection, the objective is to limit the contamination to fewer than 100 *Listeria* per gram of food products at the time of consumption, without any distinction between *Listeria* species. Considering that non*monocytogenes Listeria* species do not usually produce diseases in humans, lumping them together with *L. monocytogenes* will unquestionably lead to a larger number of potential food product recalls and increase economical burdens on food processors. While this policy could be justified in the past when detection and differentiation of *L. monocytogenes* and non*monocytogenes Listeria* species relied on slow and laborious phenotypic techniques, it is becoming increasingly difficult to sustain with rapid and precise molecular testing procedures being readily available and widely applied.

In addition, until relatively recently, *L. monocytogenes* has been regarded as pathogenic at the species level, which implies that all *L. monocytogenes* strains are potentially pathogenic. This is understandable as it has been based on our previously limited knowledge on *L. monocytogenes* and its virulence mechanisms, and our inability to rapidly and reliably distinguish virulent from avirulent strains. Accordingly, any manufactured food products that have been found to contain *L. monocytogenes* (irrespective of its virulence potential) are withdrawn from the market. Perhaps, with an increasing realization that some *L. monocytogenes* strains (e.g., those belonging to serotype 4a) are intrinsically nonpathogenic, together with the availability of rapid, sensitive, and precise tests (e.g., PCR targeting *inlC* and *inlJ* genes) for differentiation of virulent from avirulent strains, a stage is set to reconsider the strict tolerance guideline for *L. monocytogenes* in food products so that manufactured food products containing only naturally avirulent strains of *L. monocytogenes* need not be taken off the shelves. This will not only eliminate unnecessary food product recalls and reduce economic losses, but also provide an assurance to consumers on the safety of food products being marketed. Future development of assays that enable more precise quantification of the virulence potential of *L. monocytogenes* strains will also be a worthwhile contribution to the enhanced risk assessment and management of listeriosis.

REFERENCES

1. Farber, J.M. and Peterkin, P.I., *Listeria monocytogenes*, a food-borne pathogen, *Microbiol. Rev.,* 55, 476, 1991.
2. Cossart, P. and Lecuit, M., Interactions of *Listeria monocytogenes* with mammalian cells during entry and actin-based movement: Bacterial factors, cellular ligands and signaling, *EMBO J.,* 17, 3797, 1998.
3. Vazquez-Boland, J.A. et al., *Listeria* pathogenesis and molecular virulence determinants, *Clin. Microbiol. Rev.,* 14, 584, 2001.
4. Conner, D.E. et al., Pathogenicity of food-borne, environmental and clinical isolates of *Listeria monocytogenes* in mice, *J. Food Sci.,* 54, 1553, 1989.
5. Tabouret, M. et al., Pathogenicity of *Listeria monocytogenes* isolates in immunocompromised mice in relation to listeriolysin production, *J. Med. Microbiol.,* 34, 13, 1991.
6. Van Langendonck, N. et al., Tissue culture assays using Caco-2 cell line differentiate virulent from nonvirulent *Listeria monocytogenes* strains, *J. Appl. Microbiol.,* 85, 337, 1998.
7. Roche, S.M. et al., Assessment of the virulence of *Listeria monocytogenes*: Agreement between a plaque-forming assay with HT-29 cells and infection of immunocompetent mice, *Int. J. Food Microbiol.,* 68, 33, 2001.
8. Glaser, P. et al., Comparative genomics of *Listeria* species, *Science,* 294, 849, 2001.
9. Nelson, K.E. et al., Whole genome comparisons of serotype 4b and 1/2a strains of the food-borne pathogen *Listeria monocytogenes* reveal new insights into the core genome components of this species, *Nucleic Acids Res.,* 32, 2386, 2004.
10. Knudsen, G.M., Olsen, J.E., and Dons, L., Characterization of DegU, a response regulator in *Listeria monocytogenes*, involved in regulation of motility and contributes to virulence, *FEMS Microbiol. Lett.,* 240, 171, 2004.
11. Williams, T. et al., Response regulator DegU of *Listeria monocytogenes* regulates the expression of flagella-specific genes, *FEMS Microbiol. Lett.,* 252, 287, 2005.
12. Dubail, I. et al., Identification of an essential gene of *Listeria monocytogenes* involved in teichoic acid biogenesis, *J. Bacteriol.,* 188, 6580, 2006.
13. Shen, A. and Higgins, D.E., The MogR transcriptional repressor regulates nonhierarchical expression of flagellar motility genes and virulence in *Listeria monocytogenes*, *PLoS Pathog.,* 2, e30, 2006.
14. Stack, H.M. et al., Role for HtrA in stress induction and virulence potential in *Listeria monocytogenes*, *Appl. Environ. Microbiol.,* 71, 4241, 2005.
15. Autret, N. et al., Identification of the agr locus of *Listeria monocytogenes*: Role in bacterial virulence, *Infect. Immun.,* 71, 4463, 2003.

16. Liu, D. et al., Characterization of virulent and avirulent *Listeria monocytogenes* strains by PCR amplification of putative transcriptional regulator and internalin genes, *J. Med. Microbiol.,* 52, 1065, 2003.
17. Sabet, C. et al., LPXTG protein InlJ, a newly identified internalin involved in *Listeria monocytogenes* virulence, *Infect. Immun.,* 73, 6912, 2005.
18. Liu, D. et al., *Listeria monocytogenes* serotype 4b strains belonging to lineages I and III possess distinct molecular features, *J. Clin. Microbiol.,* 44, 214, 2006.
19. Kreft, J. and Vazquez-Boland, J.A., Regulation of virulence genes in *Listeria*, *Int. J. Med. Microbiol.,* 291, 145, 2001.
20. Wiedmann, M. et al., General stress transcription factor sigma(B) and its role in acid tolerance and virulence of *Listeria monocytogenes*, *J. Bacteriol.,* 180, 3650, 1998.
21. Kallipolitis, B.H. et al., CesRK, a two-component signal transduction system in *Listeria monocytogenes*, responds to the presence of cell wall-acting antibiotics and affects beta-lactam resistance, *Antimicrob. Agents Chemother.,* 47, 3421, 2003.
22. Mandin, P. et al., VirR, a response regulator critical for *Listeria monocytogenes* virulence, *Mol. Microbiol.,* 57, 1367, 2005.
23. Flanary, P.L. et al., Insertional inactivation of the *Listeria monocytogenes* cheYA operon abolishes response to oxygen gradients and reduces the number of flagella, *Can. J. Microbiol.,* 45, 646, 1999.
24. Dons, L. et al., Role of flagellin and the two-component CheA/CheY system of *Listeria monocytogenes* in host cell invasion and virulence, *Infect. Immunol.,* 72, 3237, 2004.
25. Archambaud, C. et al., Translation elongation factor EF-Tu is a target for Stp, a serine-threonine phosphatase involved in virulence of *Listeria monocytogenes*, *Mol. Microbiol.,* 56, 383, 2005.
26. Dramsi, S. et al., FbpA, a novel multifunctional *Listeria monocytogenes* virulence factor, *Mol. Microbiol.,* 53, 639, 2004.
27. Milohanic, E. et al., The autolysin Ami contributes to the adhesion of *Listeria monocytogenes* to eukaryotic cells via its cell wall anchor, *Mol. Microbiol.,* 39, 1212, 2001.
28. Gaillard, J.L. et al., Entry of *L. monocytogenes* into cells is mediated by internalin, a repeat protein reminiscent of surface antigens from Gram-positive cocci, *Cell,* 65, 1127, 1991.
29. Bergmann, B. et al., InlA- but not InlB-mediated internalization of *Listeria monocytogenes* by nonphagocytic mammalian cells needs the support of other internalins, *Mol. Microbiol.,* 43, 557, 2002.
30. Garandeau, C. et al., The sortase SrtA of *Listeria monocytogenes* is involved in processing of internalin and in virulence, *Infect. Immun.,* 70, 1382, 2002.
31. Bierne, H. et al., Inactivation of the srtA gene in *Listeria monocytogenes* inhibits anchoring of surface proteins and affects virulence, *Mol. Microbiol.,* 43, 869, 2002.
32. Dramsi, S. et al., Entry of *Listeria monocytogenes* into hepatocytes requires expression of inlB, a surface protein of the internalin multigene family, *Mol. Microbiol.,* 16, 251, 1995.
33. Bierne, H. and Cossart, P., InlB, a surface protein of *Listeria monocytogenes* that behaves as an invasin and a growth factor, *J. Cell Sci.,* 115, 3357, 2002.
34. Engelbrecht, F. et al., A new prfa-regulated gene of *Listeria monocytogenes* encoding a small, secreted protein which belongs to the family of internalins, *Mol. Microbiol.,* 21, 823, 1996.
35. Lingnau, A. et al., Identification and purification of novel internalin-related proteins in *Listeria monocytogenes* and *Listeria ivanovii*, *Infect. Immun.,* 64, 1002, 1996.
36. Dramsi, S. et al., Identification of four new members of the internalin multigene family of *Listeria monocytogenes* EGD, *Infect. Immun.,* 65, 1615, 1997.
37. Raffelsbauer, D. et al., The gene cluster *inlC*2DE of *Listeria monocytogenes* contains additional new internalin genes and is important for virulence in mice, *Mol. Gen. Genet.,* 260, 144, 1998.
38. Kuhn, M. and Goebel, W., Identification of an extracellular protein of *Listeria monocytogenes* possibly involved in intracellular uptake by mammalian cells, *Infect. Immun.,* 57, 55, 1989.
39. Reglier-Poupet, H. et al., Identification of LpeA, a PsaA-like membrane protein that promotes cell entry by *Listeria monocytogenes*, *Infect. Immun.,* 71, 474, 2003.
40. Cabanes, D. et al., Auto, a surface associated autolysin of *Listeria monocytogenes* required for entry into eukaryotic cells and virulence, *Mol. Microbiol.,* 51, 1601, 2004.
41. Cabanes, D. et al., Gp96 is a receptor for a novel *Listeria monocytogenes* virulence factor, Vip, a surface protein, *EMBO J.,* 2005.
42. Wang, L. and Lin, M., Identification of IspC, an 86-kilodalton protein target of humoral immune response to infection with *Listeria monocytogenes* serotype 4b, as a novel surface autolysin, *J. Bacteriol.,* 189, 2046, 2007.

43. Gaillard, J.L., Berche, P., and Sansonetti, P., Transposon mutagenesis as a tool to study the role of hemolysin in the virulence of *Listeria monocytogenes*, *Infect. Immun.*, 52, 50, 1986.
44. Cossart, P. et al., Listeriolysin O is essential for virulence of *Listeria monocytogenes*: Direct evidence obtained by gene complementation, *Infect. Immun.*, 57, 3629, 1989.
45. Borezee, E. et al., SvpA, a novel surface virulence-associated protein required for intracellular survival of *Listeria monocytogenes*, *Microbiology*, 147, 2913, 2001.
46. Geoffroy, C. et al., Purification and characterization of an extracellular 29-kilodalton phospholipase C from *Listeria monocytogenes*, *Infect. Immun.*, 59, 2382, 1991.
47. Marquis, H., Doshi, V., and Portnoy, D.A., The broad-range phospholipase c and a metalloprotease mediate listeriolysin o-independent escape of *Listeria monocytogenes* from a primary vacuole in human epithelial cells, *Infect. Immun.*, 63, 4531, 1995.
48. Schluter, D. et al., Phosphatidylcholine-specific phospholipase c from *Listeria monocytogenes* is an important virulence factor in murine cerebral listeriosis, *Infect. Immun.*, 66, 5930, 1998.
49. Raveneau, J. et al., Reduced virulence of a *Listeria monocytogenes* phospholipase-deficient mutant obtained by transposon insertion into the zinc metalloprotease gene, *Infect. Immun.*, 60, 916, 1992.
50. Poyart, C. et al., The zinc metalloprotease of *Listeria monocytogenes* is required for maturation of phosphatidylcholine phospholipase-c—Direct evidence obtained by gene complementation, *Infect. Immun.*, 61, 1576, 1993.
51. Goetz, M. et al., Microinjection and growth of bacteria in the cytosol of mammalian host cells, *Proc. Natl. Acad. Sci. USA*, 98, 12221, 2001.
52. Chico-Calero, I. et al., Hpt, a bacterial homolog of the microsomal glucose-6-phosphate translocase, mediates rapid intracellular proliferation in *Listeria*, *Proc. Natl. Acad. Sci. USA*, 99, 431, 2002.
53. O'Riordan, M., Moors, M.A., and Portnoy, D.A., *Listeria* intracellular growth and virulence require host-derived lipoic acid, *Science*, 302, 462, 2003.
54. Kocks, C. et al., *L. monocytogenes*-induced actin assembly requires the *actA* gene product, a surface protein, *Cell*, 68, 521, 1992.
55. Dussurget, O. et al., *Listeria monocytogenes* bile salt hydrolase is a PrfA-regulated virulence factor involved in the intestinal and hepatic phases of listeriosis, *Mol. Microbiol.*, 45, 1095, 2002.
56. Christiansen, J.K. et al., The RNA-binding protein Hfq of *Listeria monocytogenes*: Role in stress tolerance and virulence, *J. Bacteriol.*, 186, 3355, 2004.
57. Rouquette, C. et al., The ClpC ATPase of *Listeria monocytogenes* is a general stress protein required for virulence and promoting early bacterial escape from the phagosome of macrophages, *Mol. Microbiol.*, 27, 1235, 1998.
58. Nair, S. et al., ClpE, a novel member of the HSP100 family, is involved in cell division and virulence of *Listeria monocytogenes*, *Mol. Microbiol.*, 31, 185, 1999.
59. Gaillot, O. et al., The ClpP serine protease is essential for the intracellular parasitism and virulence of *Listeria monocytogenes*, *Mol. Microbiol.*, 35, 1286, 2000.
60. Mohamed, W. et al., The ferritin-like protein Frm is a target for the humoral immune response to *Listeria monocytogenes* and is required for efficient bacterial survival, *Mol. Genet. Genomics*, 275, 344, 2006.
61. Archambaud, C. et al., Control of *Listeria* superoxide dismutase by phosphorylation, *J. Biol. Chem.*, 2006.
62. Mackaness, G.B., Cellular resistance to infection, *J. Exp. Med.*, 116, 381, 1962.
63. Mackaness, G.B. and Hill, W.C., The effect of antilymphocyte globulin on cell-mediated resistance to infection, *J. Exp. Med.*, 129, 993, 1969.
64. Brosch, R. et al., Virulence heterogeneity of *Listeria monocytogenes* strains from various sources (food, human, animal) in immunocompetent mice and its association with typing characteristics, *J. Food Prot.*, 56, 296, 1993.
65. Pine, L., Malcolm, G.B., and Plikaytis, B.D., *Listeria monocytogenes* intragastric and intraperitoneal approximate 50% lethal doses for mice are comparable, but death occurs earlier by intragastric feeding, *Infect. Immun.*, 58, 2940, 1990.
66. Munder, A. et al., Pulmonary microbial infection in mice: Comparison of different application methods and correlation of bacterial numbers and histopathology, *Exp. Toxicol. Pathol.*, 54, 127, 2002.
67. Audurier, A. et al., Experimental infection of mice with *Listeria monocytogenes* and *L. innocua*, *Ann. Inst. Pasteur Microbiol.*, 131 B, 47, 1980.
68. Stams, A., Studies in cases of experimental eye infections with *Listeria monocytogenes*, *Albrecht Von Graefes Arch. Klin. Exp. Ophthalmol.*, 173, 1, 1967.
69. Schluter, D. et al., Intracerebral targets and immunomodulation of murine *Listeria monocytogenes* meningoencephalitis, *J. Neuropathol. Exp. Neurol.*, 55, 14, 1996.

70. Nishikawa, S. et al., Systemic dissemination by intrarectal infection with *Listeria monocytogenes* in mice, *Microbiol. Immunol.*, 42, 325, 1998.
71. Reed, L.J. and Muench, H., A simple method of estimating fifty per cent endpoints, *Am. J. Hyg.*, 27, 493, 1938.
72. Welkos, S. and O'Brien, A., Determination of median lethal and infectious doses in animal model systems, *Methods Enzymol.*, 235, 29, 1994.
73. Finney, D.J., *Probit analyses*, 3rd ed., Cambridge University Press, London, 1971.
74. Roche, S.M. et al., Experimental validation of low virulence in field strains of *Listeria monocytogenes*, *Infect. Immun.*, 71, 3429, 2003.
75. Liu, D., *Listeria monocytogenes*: Comparative interpretation of mouse virulence assay, *FEMS Microbiol. Lett.*, 233, 159, 2004.
76. Mandel, T.E. and Cheers, C., Resistance and susceptibility of mice to bacterial infection: Histopathology of listeriosis in resistant and susceptible strains, *Infect. Immun.*, 30, 851, 1980.
77. Gray, M.L. and Killinger, A.H., *Listeria monocytogenes* and listeric infections, *Bacteriol. Rev.*, 30, 309, 1966.
78. Pohjanvirta, R. and Huttunen, T., Some aspects of murine experimental listeriosis, *Acta Vet. Scand.*, 26, 563, 1985.
79. Lecuit, M. and Cossart, P., Genetically modified animal models for human infections: The *Listeria* paradigm, *Trends Mol. Med.*, 8, 537, 2002.
80. Lecuit, M. et al., A single amino acid in E-cadherin responsible for host specificity towards the human pathogen *Listeria monocytogenes*, *EMBO J.*, 18, 3956, 1999.
81. Cheers, C. et al., Use of recombinant viruses to deliver cytokines influencing the course of experimental bacterial infection, *Immunol. Cell Biol.*, 77, 324, 1999.
82. Stelma, G.N. et al., Pathogenicity test for *Listeria monocytogenes* using immunocompromised mice, *J. Clin. Microbiol.*, 25, 2085, 1987.
83. Lecuit, M. et al., A transgenic model for listeriosis: Role of internalin in crossing the intestinal barrier, *Science*, 292, 1722, 2001.
84. Khelef, N. et al., Species specificity of the *Listeria monocytogenes* InlB protein, *Cell Microbiol.*, 8, 457, 2006.
85. Barbour, A.H., Rampling, A., and Hormaeche, C.E., Variation in the infectivity of *Listeria monocytogenes* isolates following intragastric inoculation of mice, *Infect. Immun.*, 69, 4657, 2001.
86. Murray, E.G.D., Webb, R.A., and Swann, M.B.R., A disease of rabbits characterized by a large mononuclear leucocytosis, caused by a hitherto undescribed bacillus *Bacterium monocytogenes*, *J. Pathol. Bacteriol.*, 29, 407, 1926.
87. Waseem, M., Vahidy, R., and Khan, M.A., Correlation between production of listeriolysin O by variants of *Listeria monocytogenes* and their virulence for rabbits, *Zentralbl. Bakteriol.*, 282, 384, 1995.
88. Pron, B. et al., Comprehensive study of the intestinal stage of listeriosis in a rat ligated ileal loop system, *Infect. Immun.*, 66, 747, 1998.
89. Anton, W., Kritish-experimenteller Beitrag zur Biologie des Bakterium monocytogenes. Mit besonderer Berücksichtigung seiner Beziehung zur infektiösen Mononukleose des Menschen, *Zentralbl. Bakteriol.*, 131, 89, 1934.
90. Paterson, J.S., Experimental infection of the chick embryo with organisms of the genus *Listerella*, *J. Pathol. Bacteriol.*, 51, 437, 1940.
91. Basher, H.A. et al., Pathogenesis and growth of *Listeria monocytogenes* in fertile hens' eggs, *Zentralbl. Bakteriol. Mikrobiol. Hyg. [A]*, 256, 477, 1984.
92. Notermans, S. et al., The chick embryo test agrees with the mouse bio-assay for assessment of the pathogenicity of *Listeria* species, *Lett. Appl. Microbiol.*, 13, 161, 1991.
93. Norrung, B. and Andersen, J.K., Variations in virulence between different electrophoretic types of *Listeria monocytogenes*, *Lett. Appl. Microbiol.*, 30, 228, 2000.
94. Norton, D.M. et al., Molecular studies on the ecology of *Listeria monocytogenes* in the smoked fish processing industry, *Appl. Environ. Microbiol.*, 67, 198, 2001.
95. Avery, S.M. and Buncic, S., Differences in pathogenicity for chick embryos and growth kinetics at 37°C between clinical and meat isolates of *Listeria monocytogenes* previously stored at 4°C, *Int. J. Food Microbiol.*, 34, 319, 1997.
96. Olier, M. et al., Assessment of the pathogenic potential of two *Listeria monocytogenes* human fecal carriage isolates, *Microbiology*, 148, 1855, 2002.

96a. Cappelier, J.M. et al., Avirulent viable but nonculturable cells of *Listeria monocytogenes* need the presence of an embryo to be recovered in egg yolk and regain virulence after recovery, *Vet. Res.*, 38, 573, 2007.
97. Lebrun, M. et al., Internalin must be on the bacterial surface to mediate entry of *Listeria monocytogenes* into epithelial cells, *Mol. Microbiol.,* 21, 579, 1996.
98. Portnoy, D.A., Jacks, P.S., and Hinrichs, D.J., Role of hemolysin for the intracellular growth of *Listeria monocytogenes, J. Exp. Med.,* 167, 1459, 1988.
99. Smith, G.A. et al., The two distinct phospholipases c of *Listeria monocytogenes* have overlapping roles in escape from a vacuole and cell-to-cell spread, *Infect. Immun.,* 63, 4231, 1995.
100. Bhunia, A.K. et al., A six-hour *in vitro* virulence assay for *Listeria monocytogenes* using myeloma and hybridoma cells from murine and human sources, *Microb. Patholog.,* 16, 99, 1994.
101. Sun, A.N., Camilli, A., and Portnoy, D.A., Isolation of *Listeria monocytogenes* small plaque mutants defective for intracellular growth and cell-to-cell spread, *Infect. Immun.,* 58, 3770, 1990.
102. Farber, J.M. and Speirs, J.I., Potential use of continuous cell lines to distinguish between pathogenic and nonpathogenic *Listeria* spp., *J. Clin. Microbiol.,* 25, 1463, 1987.
103. Gaillard, J.L. et al., *In vitro* model of penetration and intracellular growth of *Listeria monocytogenes* in the human enterocyte-like cell line Caco-2, *Infect. Immun.,* 55, 2822, 1987.
104. Pine, L. et al., Cytopathogenic effects in enterocytelike Caco-2 cells differentiate virulent from avirulent *Listeria* strains, *J. Clin. Microbiol.,* 29, 990, 1991.
105. Lecuit, M. et al., A role for alpha-and beta-catenins in bacterial uptake, *Proc. Natl. Acad. Sci. USA,* 97, 10008, 2000.
106. Grundling, A., Gonzalez, M.D., and Higgins, D.E., Requirement of the *Listeria monocytogenes* broad-range phospholipase PC-PLC during infection of human epithelial cells, *J. Bacteriol.,* 185, 6295, 2003.
107. Meyer, D.H. et al., Differences in invasion and adherence of *Listeria monocytogenes* with mammalian gut cells, *Food Microbiol.,* 9, 115, 1992.
108. Milohanic, E. et al., Sequence and binding activity of the autolysin-adhesin Ami from epidemic *Listeria monocytogenes* 4b, *Infect. Immun.,* 72, 4401, 2004.
109. Singh, R. et al., Fusion to listeriolysin O and delivery by *Listeria monocytogenes* enhances the immunogenicity of HER-2/neu and reveals subdominant epitopes in the FVB/N mouse, *J. Immun.,* 175, 3663, 2005.
110. Suarez, M. et al., A role for ActA in epithelial cell invasion by *Listeria monocytogenes, Cell Microbiol.,* 3, 853, 2001.
111. Pentecost, M. et al., *Listeria monocytogenes* invades the epithelial junctions at sites of cell extrusion, *PLoS Pathol.,* 2, e3, 2006.
112. Braun, L., Ohayon, H., and Cossart, P., The InlB protein of *Listeria monocytogenes* is sufficient to promote entry into mammalian cells, *Mol. Microbiol.,* 27, 1077, 1998.
113. Barry, R.A. et al., Pathogenicity and immunogenicity of *Listeria monocytogenes* small-plaque mutants defective for intracellular growth and cell-to-cell spread, *Infect. Immun.,* 60, 1625, 1992.
114. Nichterlein, T. and Hof, H., Effect of various antibiotics on *Listeria monocytogenes* multiplying in L-929 cells, *Infection,* 19, S234, 1991.
115. Devenish, J.A. and Schiemann, D.A., HeLa cell infection by *Yersinia enterocolitica*: Evidence for lack of intracellular multiplication and development of a new procedure for quantitative expression of infectivity, *Infect. Immun.,* 32, 48, 1981.
116. Velge, P. et al., Cell immortalization enhances *Listeria monocytogenes* invasion, *Med. Microbiol. Immunol.,* 183, 145, 1994.
117. Drevets, D.A. et al., Gentamicin kills intracellular *Listeria monocytogenes, Infect. Immun.,* 62, 1994. 2222.
118. Roche, S.M. et al., Investigation of specific substitutions in virulence genes characterizing phenotypic groups of low-virulence field strains of *Listeria monocytogenes, Appl. Environ. Microbiol.,* 71, 6039, 2005.
119. Velge, P. et al., A naturally occurring mutation K220T in the pleiotropic activator PrfA of *Listeria monocytogenes* results in a loss of virulence due to decreasing DNA-binding affinity, *Microbiology,* 153, 995, 2007.
120. Autret, N. et al., Identification of new genes involved in the virulence of *Listeria monocytogenes* by signature-tagged transposon mutagenesis, *Infect. Immun.,* 69, 2054, 2001.
121. Mansfield, B.E. et al., Exploration of host–pathogen interactions using *Listeria monocytogenes* and *Drosophila melanogaster, Cell Microbiol.,* 5, 901, 2003.
122. Thomsen, L.E. et al., *Caenorhabditis elegans* is a model host for *Listeria monocytogenes, Appl. Environ. Microbiol.,* 72, 1700, 2006.

123. Cheng, L.W. et al., Use of RNA interference in *Drosophila* S2 cells to identify host pathways controlling compartmentalization of an intracellular pathogen, *Proc. Natl. Acad. Sci. USA,* 102, 13646, 2005.
124. Agaisse, H. et al., Genome-wide RNAi screen for host factors required for intracellular bacterial infection, *Science,* 309, 1248, 2005.
125. Seeliger, H.P.R. and Höhne, K., Serotyping of *Listeria monocytogenes* and related species. In *Methods in microbiology,* Bergan, T. and Norris, J.R., eds., Academic Press, Inc., New York, 1979, 31.
126. Rocourt, J., Jacquet, C., and Reilly, A., Epidemiology of human listeriosis and seafoods, *Int. J. Food Microbiol.,* 62, 197, 2000.
127. Rocourt, J., Taxonomy of the *Listeria* genus and typing of *L. monocytogenes, Pathol. Biol.,* 44, 749, 1996.
128. Ortiz-Rivera, R.R., Swaminathan, B., and Fields, P.I., Identification of flagellar antigens of *Listeria monocytogenes* by sequence-specific amplification of *flaA* alleles, in IV International Meeting on Bacterial Epidemiological Markers, Helsinor, Denmark, 10–13 September, 1997, S102.
129. Doumith, M. et al., Differentiation of the major *Listeria monocytogenes* serovars by multiplex PCR, *J. Clin. Microbiol.,* 42, 3819, 2004.
130. Czuprynski, C.J., Faith, N.G., and Steinberg, H., Ability of the *Listeria monocytogenes* strain Scott A to cause systemic infection in mice infected by the intragastric route, *Appl. Environ. Microbiol.,* 68, 2893, 2002.
131. Jeffers, G.T. et al., Comparative genetic characterization of *Listeria monocytogenes* isolates from human and animal listeriosis cases, *Microbiology,* 147, 1095, 2001.
132. Swaminathan, B. et al., Pulsenet: The molecular subtyping network for food-borne bacterial disease surveillance, United States, *Emerg. Infect. Dis.,* 7, 382, 2001.
133. Kathariou, S. et al., Levels of *Listeria monocytogenes* hemolysin are not directly proportional to virulence in experimental infections of mice, *Infect. Immun.,* 56, 534, 1988.
134. Jacquet, C. et al., Expression of ActA, Ami, InlB, and listeriolysin O in *Listeria monocytogenes* of human and food origin, *Appl. Environ. Microbiol.,* 68, 616, 2002.
135. Kathariou, S., *Listeria monocytogenes* virulence and pathogenicity, a food safety perspective, *J. Food Prot.,* 65, 1811, 2002.
136. Jacquet, C. et al., A molecular marker for evaluating the pathogenic potential of food-borne *Listeria monocytogenes, J. Infect. Dis.,* 189, 2094, 2004.
137. Cheers, C. et al., Resistance and susceptibility of mice to bacterial infection: Course of listeriosis in resistant or susceptible mice, *Infect. Immun.,* 19, 763, 1978.
138. Mengaud, J. et al., Pleiotropic control of *Listeria monocytogenes* virulence factors by a gene that is autoregulated, *Mol. Microbiol.,* 5, 2273, 1991.
139. Wernars, K. et al., Suitability of the *prfA* gene, which encodes a regulator of virulence genes in *Listeria monocytogenes,* in the identification of pathogenic *Listeria* spp., *Appl. Environ. Microbiol.,* 58, 765, 1992.
140. Bhunia, A.K. et al., Frozen stored murine hybridoma cells can be used to determine the virulence of *Listeria monocytogenes, J. Clin. Microbiol.,* 33, 3349, 1995.
141. Nishibori, T. et al., Correlation between the presence of virulence-associated genes as determined by PCR and actual virulence to mice in various strains of *Listeria* spp., *Microbiol. Immunol.,* 39, 343, 1995.
142. Wiedmann, M. et al., Ribotypes and virulence gene polymorphisms suggest three distinct *Listeria monocytogenes* lineages with differences in pathogenic potential, *Infect. Immun.,* 65, 2707, 1997.
143. Jaradat, Z.W., Schutze, G.E., and Bhunia, A.K., Genetic homogeneity among *Listeria monocytogenes* strains from infected patients and meat products from two geographic locations determined by phenotyping, ribotyping and PCR analysis of virulence genes, *Int. J. Food Microbiol.,* 76, 1, 2002.
144. Borucki, M.K. et al., Selective discrimination of *Listeria monocytogenes* epidemic strains by a mixed-genome DNA microarray compared to discrimination by pulsed-field gel electrophoresis, ribotyping, and multilocus sequence typing, *J. Clin. Microbiol.,* 42, 5270, 2004.
145. Liu, D., Identification, subtyping and virulence determination of *Listeria monocytogenes,* an important food-borne pathogen, *J. Med. Microbiol.,* 55, 645, 2006.
145a. Liu, D. et al., A multiplex PCR for species- and virulence-specific determination of *Listeria monocytogenes, J. Microbiol. Methods,* 71, 133, 2007.
146. Chakraborty, T. et al., Naturally occurring virulence-attenuated isolates of *Listeria monocytogenes* capable of inducing long-term protection against infection by virulent strains of homologous and heterologous serotypes, *FEMS Immunol. Med. Microbiol.,* 10, 1, 1994.
147. Gracieux, P. et al., Hypovirulent *Listeria monocytogenes* strains are less frequently recovered than virulent strains on PALCAM and Rapid' *L. mono* media, *Int. J. Food Microbiol.,* 83, 133, 2003.

Section III

Genomics and Proteomics

9 Genetic Manipulations

Armelle Bigot and Alain Charbit

CONTENTS

9.1 INTRODUCTION

9.1.1 When Did It All Begin?

Listeria monocytogenes was officially discovered in 1924, when E. G. D. Murray isolated Gram-positive rods from the blood of infected rabbits.[1] This new agent was initially named *Bacterium monocytogenes* and the genus was renamed *Listeria* later by Pirie.[2] The first cases of human listeriosis were reported in 1929 in Denmark. However, from 1926 to 1950, the importance of *L. monocytogenes* as a life-threatening food-borne human pathogen was not realized. In a very important epidemiologic study, Seeliger[3] gathered a collection of about 6000 *Listeria* isolates. By using specific antibodies directed against O-antigens (teichoic acids) and H-antigens (flagella), more than 60% of the strains could be classified into distinct serovars (see chapter 5 for details). Notably, ca. 90% of the isolates deriving from human patients belong to either serovar 1 or 4. Later, *L. monocytogenes* could be distinguished from other *Listeria* species by biochemical methods.[4] Nearly all strains of *L. monocytogenes* were shown to be pathogenic.

In parallel, immunologists were interested in *L. monocytogenes* because an infection mimicking human listeriosis was easily reproducible in laboratory rodents. The pioneer work of Mackaness in the early 1960s demonstrated that *L. monocytogenes* was able to survive and multiply in macrophages:

> The mouse was found to be susceptible to *L. monocytogenes*, due to the capacity of the organism to survive and multiply in host macrophages. During the first 3 days of a primary infection the bacterial populations of spleen and liver were found to increase at a constant rate. On the 4th day of infection the host became hypersensitive to *Listeria* antigens and at the same time bacterial growth ceased. A rapid inactivation of the organism ensued. Convalescent mice were resistant to challenge, but no protective factor could be found in their serum.[5]

Since then, *L. monocytogenes* has been used in immunological research as a prototype intracellular parasite (see chapters 13–16). The 1980s marked the start of investigations into the molecular mechanisms responsible for *L. monocytogenes* intracellular parasitism and virulence. Genetic studies, mainly based on transposon mutagenesis (see later discussion), led between 1986 and 1989 to the discovery of the hemolysin gene, *hly*, encoding listeriolysin O (LLO) and to elucidation of the key role that LLO plays in phagosomal escape and intracytosolic bacterial proliferation. Over the past 15 years, *L. monocytogenes* has become a model organism both for immunological investigation and for the analysis of the molecular mechanisms of intracellular parasitism (see references 6–10 for reviews).

Obviously, the sequencing of the genomes of *L. monocytogenes* strain EGD-e and *L. innocua* by a European Consortium in 2001[11] opened a new era of research on the molecular pathogenesis of *Listeria* spp. The postgenomic period saw the development of genome-scale genetic studies as well as genomic, transcriptomic, and proteomic studies, generating constantly fresh insight on *monocytogenes* pathogenesis.

9.1.2 What Is Genetic Manipulation?

The classical genetic approach, also defined as "forward genetics," starts from a given phenotype to the identification of the corresponding genotype. In turn, the genotype gives access to the protein and allows subsequent functional analyses. With the availability of whole genome sequences of organisms, scientists can identify (or choose) candidate genes within an organism on the basis of sequence similarity with orthologs of known function in other organisms. Thus, it became possible to do "reverse genetics"—that is, to go from a gene (or protein sequence) back to the corresponding phenotype. The major conceptual difference between the two approaches is that, while the first one (forward) does not require any prior information on the gene to be identified, the second (reverse) is based on an educated guess of a possible function of the chosen gene. In most cases, both approaches must be used for a complete analysis of a biological function (such as adhesion and invasion, nutrient acquisition, and gene regulation).

The first step of a genetic analysis is generally the generation of banks of mutants to inactivate all the mutable genes of the bacterial genome. For this, tools for random mutagenesis must be available and appropriate screens or selections (i.e., the appropriate model and experimental conditions) must be defined. In a second step, other molecular genetic tools are used to perform more thorough analyses. In particular, allelic replacement allows the construction of clean chromosomal deletions as well as the introduction of allelic variants.

Both random and targeted strategies have been used to identify and characterize genes involved in the virulence of *L. monocytogenes.* While transposon mutagenesis has been initially used (see later discussion), the availability of *Listeria* genomic sequences saw the development of many targeted approaches aiming at systematically inactivating all the genes belonging to the same family (e.g., adhesins, regulators, and nutrient acquisition systems).

The direct proof that a gene inactivation is responsible for a mutant phenotype is generally brought by complementation studies, where the wild-type allele is reintroduced in the mutated strain (in *cis*, in the chromosome, or in *trans,* carried by a plasmid) and restores the wild-type phenotype. Alternatively, heterologous expression can be used to probe the function of the gene in another related bacterial species, including in nonpathogenic species from the same genus. Vectors for complementation and heterologous expression are available for *L. monocytogenes.* We will provide a nonexhaustive series of examples in this chapter.

Detailed analyses of proteins require the use of site-directed mutagenesis (to generate defined amino acid substitutions, deletions, insertions, domain swapping, etc.) within the polypeptide chain and must be combined with biochemical, biophysical, cellular, and immunological approaches. Several proteins of *L. monocytogenes* have been studied in greater detail, in particular, the proteins LLO and ActA. They will be discussed later.

9.1.3 Why Is Genetic Analysis Needed?

Specific attributes influence the outcome of an infection with a pathogenic *Listeria* strain. These can be grouped into three main categories: (1) virulence genes per se, (2) virulence-related genes, and (3) genes that are essential for pathogenesis (see references 12–15 for reviews).

One can define a virulence gene by two criteria: (1) as a gene that is present in pathogenic strains and absent (or inactive) in nonpathogenic strains, or (2) disruption of a virulence gene that should reduce or abolish virulence. In contrast, virulence-related genes might be common to both pathogenic and nonpathogenic strains and expressed outside the infected host, but still assist in the infectious process. Genes that are essential for pathogenesis can be either specifically required during infection (although still present in nonpathogenic species) or simply essential for the survival of the bacteria (in any condition; see section 9.4.4).

Several major technical advances led to the "explosion" of genetic studies on *L. monocytogenes* in the years 1986–1990. First, tissue culture models of infection were developed in a broad variety of cell lines. Second, the use of the transposable elements Tn*1545*, Tn*916*, and derivatives of Tn*917* allowed the easy construction of banks of mutants. Third, a series of vectors allowing complementation, allelic exchange, and site-specific plasmid integration have been devised.

We will discuss the approaches and tools available for gene inactivation in *L. monocytogenes* (random and targeted approaches). The plasmid vectors that are available for mutagenesis, complementation, and heterologous expression will be provided. Then, the genetic approaches and tools available to study gene expression and regulation will be presented. However, the list of genetic tools (plasmids and transposons) presented in this chapter is not exhaustive for two main reasons: (1) different laboratories constructed similar tools, and (2) many vectors used in other Gram-positive bacteria can be used with *Listeria.* The genetic aspects and tools concerning the host (cells and animal models) will only be evoked at the end of this chapter.

9.2 APPROACHES AND TOOLS AVAILABLE FOR GENE INACTIVATION AND GENE TRANSFER

This section deals with how to proceed with genetic analyses in *Listeria* species. A wide range of genetic techniques is now available to manipulate *L. monocytogenes*. Transposons (Tn*916*, Tn*917*, and Tn*1545*) have been shown to insert at random into the *L. monocytogenes* chromosome. Artificial transformation is possible, and a variety of plasmid vectors, developed for use in other Gram-positive bacteria, have proven to be stably maintained in *L. monocytogenes*. General transduction is also possible to transfer genes from one *Listeria* strain to another. First, we will recall briefly how the different sets of genetic tools available today have been developed for (or adapted to) *Listeria* and further improved.

9.2.1 Major Modes of Horizontal Gene Transfer in Bacteria

There are three major modes of transfer of genetic material between bacteria: conjugation, transduction, and transformation. When available, these modes of transfer provide very useful tools to perform genetic manipulations. Early studies have shown that conjugation and general transduction mechanisms existed in *Listeria*. In contrast, member of the genus *Listeria* are not naturally transformable, but artificial transformation allows the introduction of plasmid DNA.

9.2.1.1 Conjugation

This mode of transfer of DNA can be mediated by either conjugative plasmids or conjugative transposons. The first clinical strain of *L. monocytogenes* multiresistant to antibiotics was reported in 1988[16] and was associated with the presence of conjugative plasmids from enterococcal and streptococcal origins. Several studies have further described the transfer by conjugation of enterococcal and streptococcal plasmids and transposons carrying antibiotic resistance genes from *Enterococcus* or *Streptococcus* to *Listeria* as well as between species of *Listeria*.[16–23]

The conjugative plasmid pIP501,[24] initially detected in *Streptococcus agalactiae*, was the first plasmid reported to be transferable by conjugation to *L. monocytogenes* at a frequency of approximately 10^{-6}. Plasmid pIP501 was shown to be able to replicate in *Listeria* and to promote its own transfer between strains of *Listeria* and, from *Listeria*, back to *Streptococcus*.

Later, the broad-host-range plasmid pAMβ1 of *Enterococcus faecalis*, conferring resistance to erythromycin, was transferred successfully by conjugation from *E. faecalis* to *L. monocytogenes* at frequencies ranging from 10^{-4} to 10^{-8}.[25] pAMβ1 could replicate in *L. monocytogenes* and, in turn, be transferred by conjugation between strains of *L. monocytogenes* and from *L. monocytogenes* back to *E. faecalis*. A series of derivatives of pAMβ1 has been constructed allowing gene complementation studies, promoter dosage, etc.

The conjugative transposon Tn*916*, initially detected in *E. faecalis*, is a broad-host-range conjugative transposon carrying the *tetM* tetracycline resistance gene.[26] In a study by Vicente et al.,[27] transfer by conjugation of Tn*916* from *E. faecalis* to *L. innocua* was obtained at a frequency of 10^{-6}. Tn*916* was also shown to mediate its own transfer from *L. innocua* to other *Listeria* species and, from there, back to *E. faecalis* or to other species such as *S. agalactiae*, and *S. pneumoniae*.[28] Conjugative transfer of the Tn*916*-related transposon Tn*1545*, initially observed in *S. pneumoniae*, was also obtained from *E. faecalis to L. monocytogenes in vitro* and *in vivo*.[20] The Tn3-like transposon Tn*917* was also used to perform random mutagenesis into *L. monocytogenes*. The plasmid, carried on vector pTV1, was demonstrated to exhibit a high degree of insertional randomness and to generate very stable insertional mutations.[29]

A series of derivatives of Tn*916*, Tn*917*, and Tn*1545*, has been engineered (see section 9.2.2.1), allowing the construction of banks of mutants that led to the discovery of many genes involved in the pathogenesis of *L. monocytogenes*. They will be presented in greater detail in section 9.2.2.

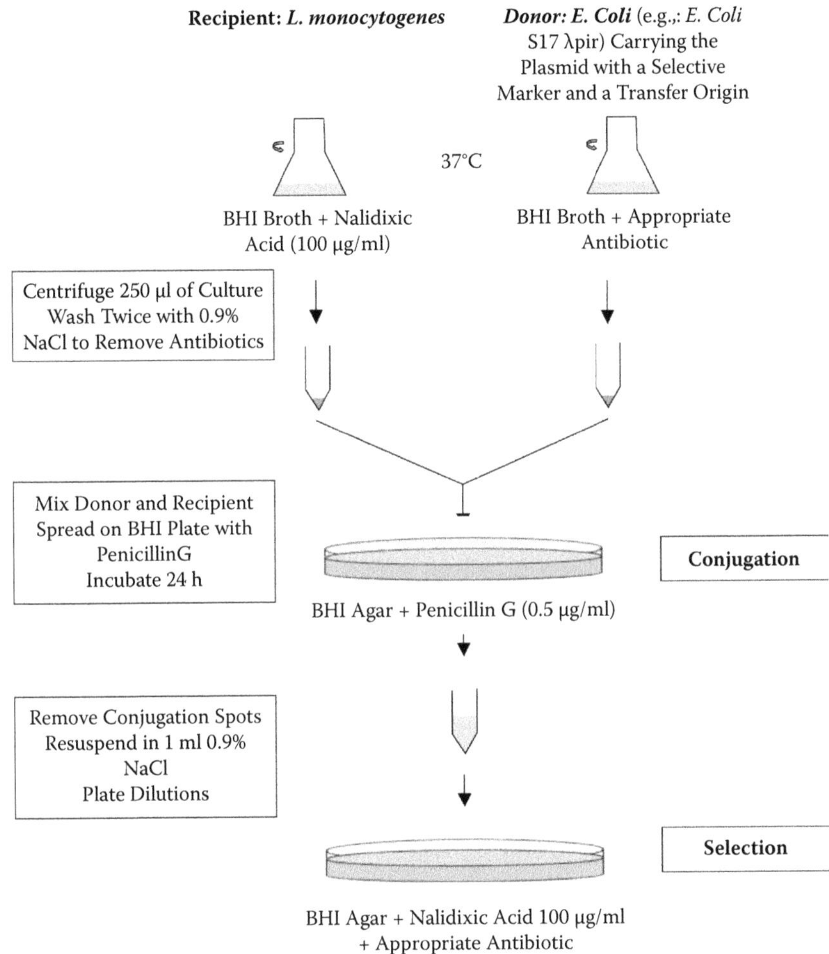

FIGURE 9.1 Protocol for conjugative transfer of mobilizable plasmid to *L. monocytogenes*.

Many plasmid vectors have been constructed with an origin of transfer (see Tables 9.1 and 9.2), allowing them to be transferred by conjugation from an *E. coli* donor to the *Listeria* recipient strain (see illustrated protocol for conjugation in Figure 9.1).

9.2.1.2 Transduction

The first report of the isolation of a bacteriophage capable of infecting *L. monocytogenes* was in 1945. Since then, many bacteriophages have been isolated for bacteriophage subtyping of *Listeria* spp. strains.[30,31] In 2000, Hodgson[32] reported the isolation and preliminary characterization of a series of bacteriophages isolated from environmental samples, from *L. monocytogenes* lysogens or from listeriophage collections. Of 59 bacteriophages tested, 34 proved to be capable of transduction. Hodgson demonstrated interstrain transduction and used transduction to test linkage between transposon insertions and mutant phenotypes in a variety of strains (see section 9.3.3).

9.2.1.3 Transformation

L. monocytogenes is not naturally transformable, although most of the competence genes identified in *B. subtilis* are present. However, artificial transformation (protoplast transformation and electroporation) proved to be feasible in *Listeria* species.[33] Both processes were rather inefficient as compared to yields obtained in bacteria such as *E. coli* or *Salmonella*. Later, an optimized procedure was developed, allowing frequencies of ca. 4 × 10^6 *Listeria* transformants per microgram of plasmid DNA.[34,35] Thus, transformation by electroporation and conjugation are the two methods employed to introduce recombinant plasmids into *Listeria* species.

9.2.2 Random Approaches

9.2.2.1 Transposon-Based Mutagenesis

Transposon (Tn) mutagenesis is a classical genetic approach to generate banks of mutants that has been used in numerous bacterial species. Once banks of mutants are generated (using a unique transposon or different transposons), two different strategies can be employed to identify genes of interest.

The first approach is based on a systematic individual analysis of all the clones generated without any prior screening or selection. This strategy was first exemplified by the work of Fields et al.[36] in 1986 that carried out a systematic analysis of a bank of Tn*10* insertion mutants of *Salmonella typhimurium* to identify genes that are required for intracellular survival. In that huge study, nearly 10,000 transposon insertion mutants were individually screened for the ability to survive in primary macrophage cultures. Different classes of mutants were identified, such as mutants with auxotrophic requirements or mutants with altered lipopolysaccharide structure or hypersensitivity to host bactericidal mechanisms.

The second and most commonly used approach is based on the screening of mutant clones within a bank that display a defect in a given experimental condition (such as impaired growth on a given medium or in a cell system and altered morphology). Variations of this strategy have also been developed to select rather than screen the mutants in the bank (i.e., by defining conditions where the mutants of interest have a survival advantage over the nonmutated clones).

Transposon-mediated insertional mutagenesis has been the first tool developed in *L. monocytogenes* to generate random banks of mutants. This approach was extensively used during the first 10 years of genetic analyses (1987–1997), by the introduction of the conjugative transposons Tn*1545*, Tn*916*, Tn*917*, and derivatives. Each of these transposons has been repeatedly used for random mutagenesis by different laboratories. Improvements for their delivery have been developed. All sorts of screens and various strains of *L. monocytogenes* were used. Altogether, these transposon-based mutageneses have led to the identification of most of the major virulence factors known today.

We will provide here a nonexhaustive series of examples, ranked by historical order, which led to the identification of many important virulence factors of *L. monocytogenes*. The technical improvements brought over the years of the transposon-based strategies will also be discussed. The hemolysin gene, *hly*, was the first virulence determinant to be identified by transposon mutagenesis and sequenced in *Listeria* spp.[37] Subsequently, the characterization of the *hly* locus led to discovery of the chromosomal virulence gene cluster encoding major determinants required for the intracellular life cycle of pathogenic *Listeria* spp. The *hly* gene has been identified by different teams and using three different transposons. Its discovery and the identification of the other genes of the virulence locus by transposon mutagenesis will be described in greater detail.

Identification of the hly gene, 1986–1989. In 1986, Gaillard et al. used transposon Tn*1545* to mutagenize *L. monocytogenes*. The transposon was delivered into *L. monocytogenes* at low frequency (10^{-8}) by conjugation from BM4140 (harboring this transposon) to EGD-SmR (a spontaneous mutant of EGD resistant to streptomycin). BM4140 is a derivative of *L. monocytogenes* strain LO17 carrying the 26-kilobase (kb) conjugative transposon Tn*1545* from *Streptococcus pneumoniae*. The exconjugants were selected on broth supplemented with streptomycin (Sm) and tetracyclin (Tc). The authors addressed the role of hemolysin secretion in bacterial virulence by screening a bank of 2500 mutants for the expression of hemolysin (characterized by the presence of a halo of β-hemolysis surrounding the colonies on blood agar plates). This procedure allowed them to identify one nonhemolytic mutant that was further characterized. The mutant strain appeared to be totally avirulent in the mouse model of infection; the lethal dose 50% (LD_{50}) of the strain was of $10^{9.6}$, as compared to $10^{6.2}$ for the parental EGD-SmR strain (by the intravenous route of infection). The transposon insertion was not determined in that first study.

One year later, Kathariou et al. reported a similar study, using transposon Tn*916* this time to mutagenize *L. monocytogenes* strain EGD.[38] Tn*916* was introduced into SmR spontaneous mutants of *Listeria* by conjugation with Tn*916*-containing strains of *Enterococcus faecalis* at higher frequency (10^{-6}). The *hly*-negative mutants were screened on blood agar plates and preliminarily characterized. Notably, Kathariou and coworkers also used banks of Tn*916* insertion mutants to screen phage- or monoclonal antibody-resistant mutants in several strains of *Listeria* as well as to screen mutants affected in motility or growth at low temperature.[39–43]

Mengaud et al. determined the exact nature of the *hly*-negative clones isolated in the first study of Gaillard et al., in 1987.[44] By using an antiserum raised against purified LLO, the authors demonstrated that the transposon insertion had led to the production of a truncated cytolysin and to loss of hemolytic activity. Furthermore, they determined by DNA sequencing that the site of Tn*1545* insertion lay within the *hly* open reading frame.

The Tn*3*-like transposon Tn*917*, which generates stable insertional mutations in *B. subtilis*,[45] has also been used in *L. monocytogenes* strain L028 to generate an *hly*-negative mutant.[46] The transposon Tn*917*, carried on the thermosensitive vector pTV1 (Figure 9.2), was transferred to *L. monocytogenes* by chemical transformation. The Tn*917* insertion mutants were obtained after passage at nonpermissive temperature (42°C) and selecting for Em^R (transposon determined). Loss of the plasmid was confirmed by loss of the plasmid-borne chloramphenicol resistance (i.e., by screening the Em^R Cm^S clones). Complementation studies unambiguously established that *hly* inactivation was specifically responsible for loss of hemolytic activity and loss of virulence.

The early 1990s. From 1990, a variety of tissue-culture assays has been developed to screen mutants with impaired intracellular growth. For example, the ability of *L. monocytogenes* to form plaques in the monolayer of tissue-culture cells corresponds to the ability to escape from the phagosome, to replicate in the cytoplasm, and to spread to adjacent cells. These plaque-forming defects were generally shown to correlate with attenuated virulence in the mouse model.

An elegant alternative approach for isolating intracellular growth-defective mutants was developed by Camilli et al. in 1989.[47] This approach is based on a penicillin selection that kills only replicating bacteria. Infection of tissue-culture monolayers in the presence of the antibiotic methicillin (able to diffuse through the membrane of the infected cell into its cytoplasm) enriches for mutants with intracellular growth defects.

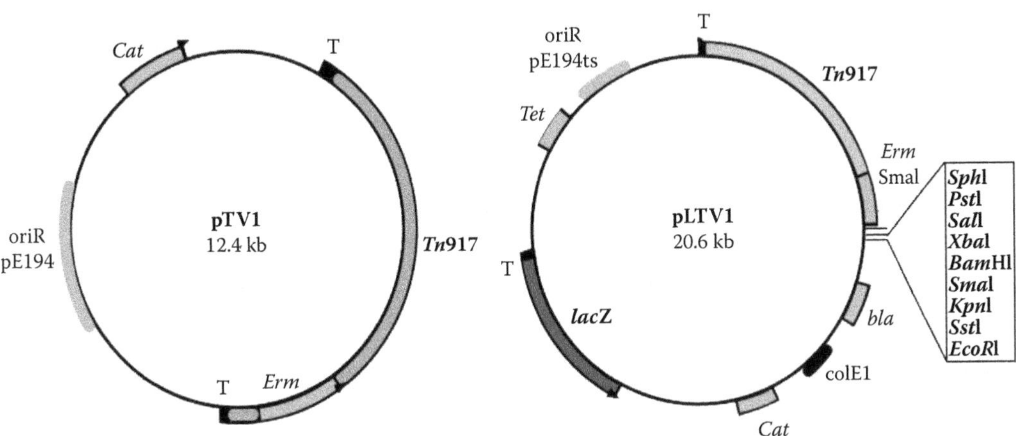

FIGURE 9.2 Vectors for transposon delivery.

Camilli et al. improved the Tn*917* transposon mutagenesis tool by constructing two Tn917 derivatives carried on a plasmid, designated pLTV1 (Figure 9.2) and pLTV3, respectively.[48] The recombinant plasmids carried (in addition to the Tn*917* derivative): (1) a Tet resistance (Tc^R) gene, to counterselect the loss of the plasmid; (2) a *lacZ* reporter gene, to monitor transcription of the inactivated gene (when insertion occurred in the correct orientation); (3) a thermosensitive pE194ts origin of replication, to replicate at permissive temperature in *L. monocytogenes*; and (4) a ColE1 origin of replication and appropriate antibiotic markers for expression in *E. coli*, to facilitate recovery in *E. coli* and analysis of the transposon insertion sites.

Using these vectors, transposon insertions were easily obtained by growing first *L. monocytogenes* harboring plasmid pLTV3 in broth containing erythromycin (Em), lincomycin (Lm), and Tc, at permissive temperature (30°C) to stationary phase. Then, the culture was diluted into fresh medium containing Em and Lm (but without Tc), and grown at nonpermissive temperature (41°C) until stationary phase. This treatment was sufficient to yield 90% of Em^R, Lm^R, and Tc^S clones, indicating loss of pLTV3 and transposon insertion into the chromosome. The randomness of Tn*917* insertions was demonstrated on several independent insertion libraries and screening for various kinds of insertional mutations. Auxotrophic mutations were isolated in at least eight distinct loci. Several insertions were also obtained within the *hly* gene.

Identification of prfA, mpl, plcA, plcB, ActA, inlA, and inlB genes, 1990–1992. Libraries of *L. monocytogenes* were generated as described[48] by using either Tn*917*-LTV3 (a derivative of Tn*917*-*lac*) or Tn*916*. A series of mutants forming small plaques on monolayers of L3 cells was screened (i.e., defective in cell-to-cell spreading). This strategy ultimately led to the identification of *prfA*, *plcA*, *mpl*, *actA*,and *plcB* genes of the virulence locus[49,50] (see also Portnoy et al.[12] for a review).

Screening a bank of 5000 Tn*1545* transposon-induced mutants on 2.5% egg yolk brain heart infusion agar isolated two types of lecithinase-deficient, attenuated-virulence mutants. The mutants corresponded to transposon insertion into either *mpl* gene, which encodes a zinc metalloprotease[51,52], or *actA* gene.[53] The same year Domann et al. cloned the *actA* gene and sequenced the region surrounding actA, identifying the *mpl* and *plcB* genes.[54] The construction of an isogenic Δ*actA* mutant strain (by plasmid insertion) established that the gene encoded the ActA polypeptide.

plcA-deficient mutants were initially isolated during a search for molecular determinants involved in cell-to-cell spread.[49] These *plcA* insertion mutants were defective in intracellular proliferation in macrophages and were 1000-fold less virulent in mice. However, this severe attenuation of virulence was due to a polar effect on the downstream gene *prfA* that is cotranscribed with *plcA*. Later studies with an in-frame *plcA* deletion mutant revealed only a slight reduction in virulence and in the ability to escape from the primary phagosomes of murine bone marrow macrophages. The *plcA* gene, which encodes a PI-specific PLC, was identified in *L. monocytogenes* by DNA sequencing of the region upstream from *hly* in 1991.[55,56]

During the same period, Gaillard et al. also identified by transposon mutagenesis the *inlAB* region.[57] Sequence analysis and complementation studies in *L. innocua* demonstrated the role of InlA as an invasin. InlA was the first member of the LPXTG family identified. The role of InlB in bacterial entry was demonstrated a few years later[58] by the construction and analysis of a chromosomal Δ*inlB* mutant. This study revealed that *inlB* encodes a surface protein, required for entry of *L. monocytogenes* into hepatocytes. The other members of the internalin multigenic family were identified later by using molecular genetic techniques[59] and DNA macroarray analyses.[60]

Notably, the isolation and analysis of spontaneous mutants may also lead to important findings. For example, the isolation and analysis of a spontaneous nonhemolytic mutant of *L. monocytogenes*, in 1990, allowed the identification of the *prfA* gene.[61] The 450-bp deletion mutant within *prfA* abolished *hly* expression as well as that of *mpl* and *plcB*, demonstrating for the first time that *prfA* determines the activator of these genes. Further analyses with transposon and site-specific integration mutations confirmed that PrfA controlled the expression of all the genes of the locus.[51]

9.2.2.2 Other Applications of Tn Mutagenesis

In 1994, Klarsfeld et al. used the Tn*917-lac* transposon in an *in vitro* study to identify genes that would be up-regulated during macrophagic infection.[62] A library of Tn*917-lac* insertion mutants was screened for transcriptional fusions to *lacZ* with higher expression inside a macrophage-like cell line than in a rich broth medium. Five genes with up to 100-fold induction inside cells were identified, including three genes involved in nucleotide biosynthesis and *plcA*, encoding PI-PLC.

In 1997, He and Luchansky reported the construction of two temperature-sensitive cloning and insertion vectors (pLUCH80 and pLUCH88) containing Tn*917* derivatives that carry *Not*I and *Sma*I recognition sites within transposon sequences.[63] These vectors, derived from plasmid pTV51Ts (carrying Tn*917* and the temperature-sensitive origin of replication pE194ts), were used to generate a physical map of the chromosome of the Scott A strain of *L. monocytogenes* by pulse-field gel electrophoresis.

9.2.2.3 Other Screens, Other Mutants

We have chosen several recent examples of transposon mutagenesis in *L. monocytogenes* to illustrate the broad applications of this simple and efficient mutagenesis approach. A vector, designated pTV1-OK, was constructed for Tn*917*-mediated mutagenesis and subsequent recovery of interrupted genes in *E. coli*.[64] The thermosensitive plasmid pTV1-OK is a derivative of the broad-host-range plasmid pWVO1 harboring Tn*917* and carrying a *repA*(Ts) origin of replication. It was used to mutagenize *L. monocytogenes* strain LO28 and the screening of pH sensitive mutant on the basis of reduced growth on agar adjusted to pH 5.5. This procedure led to the identification of a gene, designated *btlA* (bile tolerance locus, i.e., *lmo1417* in EGD-e) involved in resistance to bile salts and several other stresses.[65]

Wonderling et al. used pLTV3 for mutagenesis of *L. monocytogenes* strain 10403S and screened mutants with increased sensitivity to NaCl.[66] They identified a gene designated *htrA*, encoding a protein homologous to the serine protease of *E. coli* involved in stress response. More recently, Liu et al. also used pLTV3 for mutagenesis of *L. monocytogenes* strain 10403S and screened mutants with increased cold sensitivity.[67] They identified a gene (corresponding to *lmo1466* in EGD-e) encoding a putative transmembrane protein involved in the control of cellular (p)ppGpp levels. Xue et al. constructed transposon Tn*917* knockout libraries in *Listeria innocua* using pTLV3 and screened for mutants that are resistant to pediocin AcH.[68] This procedure led them to identify genes that are important for class IIa bacteriocin resistance in *Listeria* species.

9.2.2.4 Signature-Tagged Mutagenesis

All the transposon mutagenesis approaches described thus far rely on screening assays to identify or select transposon insertions that result in defects in growth or in intracellular survival or from banks of mutants. Signature tagged mutagenesis (STM) is an improvement of traditional transposon mutagenesis that allows the large-scale analysis of transposon-insertion mutants for the identification of virulence genes in pathogenic bacteria. STM is based on a negative selection of the mutants—that is, mutants that have lost the capacity to survive in a given host (see later discussion), allowing the discovery of virulence genes without prior indication of their nature or function.

Technically, pools of mutants can be screened at the same time. STM does not necessarily require a transposon as a delivery vehicle. DNA tags can also be included during allelic replacement (signature-tagged allele replacement is also called bar-coding).

STM is often used for *in vivo* screening, but can also be performed on *in vitro* systems like cell cultures (see Autret and Charbit[69] for a review). Very briefly, pools of mutants (designated pools of inoculation or PINO) are constituted and administered to the animal. Bacteria are recovered from the target organs at selected intervals after infection (designated recovered pools or PREC). The tags from both pools (the PINO and the PREC) can be specifically amplified by PCR. Then, by

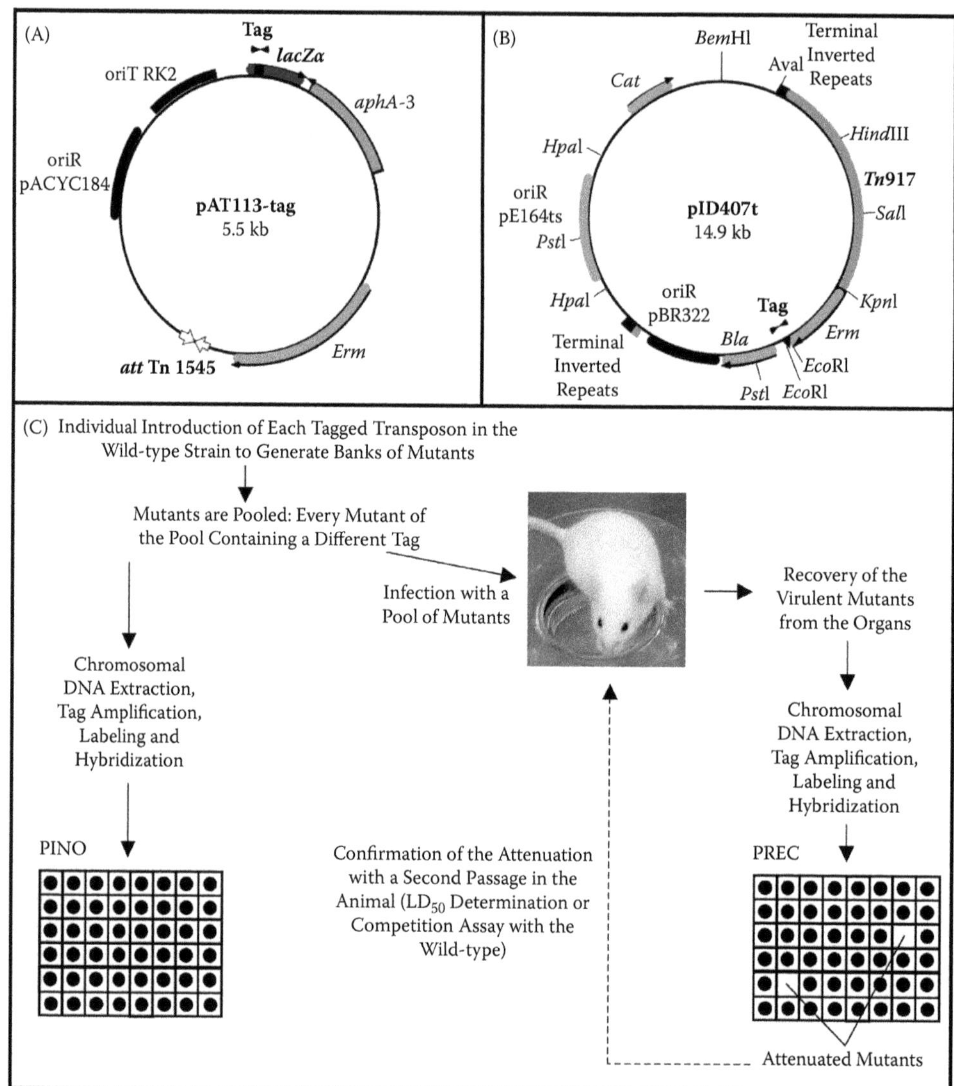

FIGURE 9.3 STM in *L. monocytogenes*. (A) pAT113-tag vector for tagged transposon delivery used by Autret et al.[70] (B) pID407t vector used by Dramsi et al.[73] (C) Screening method in STM.

dot-blot hybridization with PCR-amplified tags, it is possible to identify the mutants that are still present in the PREC and thus to infer which mutants were unable to persist *in vivo* (Figure 9.3).

STM was first applied to *L. monocytogenes* in our laboratory in 2001[70] to identify genes possibly involved in passage across the blood–brain barrier. For this, we used STM to identify mutants affected in their multiplication in the brains of infected animals. As a transposon delivery vehicle we have used pAT113 (Figure 9.3), a system that takes advantage of the transposition properties of Tn*1545* carried on a plasmid containing the origin of replication of pACYC184 (enabling it to replicate in *E. coli*), the attachment site of Tn*1545*, and erythromycin and kanamycin resistance genes.[71] Chromosomal integration of the tagged transposons was achieved by the presence, in the recipient *L. monocytogenes* strain EGD-e, of the transposon-encoded integrase provided in *trans* by plasmid pAT145. Forty-eight transposon-tagged recombinant plasmids were constructed by cloning double-stranded DNA tags into the *Eco*RI restriction site of plasmid pAT113. Each plasmid, carrying a different tag, was introduced by electroporation into EGD-e carrying pAT145, thus generating

48 banks of *L. monocytogenes* mutants. Of 48 mutants of *L. monocytogenes,* 41 pools were generated (a bank of ca. 2000 mutants) and tested for their capacity to penetrate and/or to multiply in the brains of infected mice. Each pool was injected into two mice at a dose of 10^6 bacteria per mouse. At this dose, the average number of bacteria reaches up to 10^5 per brain in mice infected with the virulent strain (EGD-e) at day 4. Infected brains were collected, and overnight cultures inoculated with brain homogenates were prepared. Chromosomal DNA from the resulting bacterial cultures (from the PREC) was extracted. The chromosomally inserted tags were amplified by PCR (with primers corresponding to the conserved portions flanking each tag). These probes were then radio labeled and used for hybridization on nylon membranes coated with the 48 tags (all represented in the PINO).

This procedure led us to identify 18 attenuated mutants, corresponding to 10 distinct loci (for each mutant, the site of Tn insertion was determined by DNA sequencing, using either reverse PCR or linker-mediated PCR). Five loci encoded putative cell wall components, and five loci involved proteins participating in various cellular processes. The four mutants that showed the highest LD_{50} values (>10^6, i.e., 1.5 log higher than the parental strain) were studied in greater detail. The *in vivo* kinetics of bacterial multiplication of these four mutants (monitored in the spleen and the liver) revealed that, in the four mutants, attenuation did not result in a defect in the early stage of multiplication in the spleen or in the liver. Moreover, in no case was invasion of the brain completely abolished, suggesting a defect of persistence in the brain rather than a defect in crossing of the blood–brain barrier.

A new round of STM in strain EGD-e, using the liver as a target organ,[72] allowed us to identify a new locus encoding a protein homologous to AgrA, the response regulator of a two-component system previously shown to be involved in the virulence of *S. aureus*. The production of several secreted proteins was modified in the *agrA* mutant of *L. monocytogenes* grown in broth, indicating that the *agr* locus influenced protein secretion. Inactivation of *agrA* did not affect the ability of the pathogen to invade and multiply in cells *in vitro*. However, the virulence of the *agrA* mutant was attenuated in the mouse, demonstrating for the first time a role for the *agr* locus in the virulence of *L. monocytogenes*.

More recently, STM was applied on *L. monocytogenes* to identify two novel virulence factors: the adhesin FbpA[73] and the regulatory protein VirR.[74] Both studies used plasmid pID407t (Figure 9.3), which takes advantage of the transposition properties of Tn*917*, as described previously for *S. aureus*,[75] to create banks of mutants in strain EGD. In both studies, 96 tagged plasmids, each carrying a different tag, were used to generate banks of insertion mutants. Pools of 96 differently tagged mutants were used to infect 7- to 8-week-old Swiss mice intravenously. Mutants were recovered from infected spleens and livers (PREC) 72 h after infection and compared with the inoculated pool (PINO) by hybridization.

In the first study by Dramsi et al.,[73] the screening of 2000 mutants led to the identification of 12 attenuated mutants, including 1 in gene *lmo1829* that was further analyzed. The protein showed strong homologies to atypical fibronectin-binding proteins (hence its name FbpA). FbpA was shown to be required for intestinal and liver colonization after oral infection of transgenic mice expressing human E-cadherin. FbpA bound fibronectin and increased adherence of wild-type *L. monocytogenes* to HEp-2 cells in the presence of exogenous fibronectin. Despite the lack of conventional secretion/anchoring signals, FbpA was detected on the bacterial surface FbpA, possibly acting as a chaperone or an escort protein for exported virulence factors.

In the second study by Mandin et al.,[74] one of the mutants had the transposon inserted in *virR*, a gene encoding a putative response regulator of a two-component system. A deletion of *virR* was subsequently constructed and analyzed. The authors showed that the deletion severely decreased virulence in mice as well as invasion in cell culture experiments. Using a transcriptomic approach, they identified 12 genes regulated by VirR, including the *dlt*–operon, involved in lipoteichoic acid biogenesis, a locus previously reported to be important for *L. monocytogenes* virulence.

TABLE 9.1
Plasmid Vectors for Transposon-Based Mutagenesis

Plasmid	Size	Gram +	Origin of Transfer	Description[a]	Ref.
pTV1	12.4	Yes	No	Carrying Tn*917*, Cm^R, Em^R	162
pTV1-OK	12.4	Yes, ts[b]	No	Carrying Tn*917*,Em^R, Km^R	64
pLTV1	20.6	Yes, ts	No	Carrying Tn*917-lac*, Cm^R, Em^R, Tc^R	48
pLTV3	22.1	Yes, ts	No	Carrying Tn*917-lac*, Cm^R, Em^R	48
pAT113	5.4	No	Yes	Transposable, *att*Tn*1545*, Em^R, Km^R	71
pID407	14.8	Yes, ts	No	Carrying Tn*917*, Cm^R, Em^R	75

[a] Antibiotics resistance is given for Gram-positive bacteria only (another selection may be used in *E. coli*. Cm^R, chloramphenicol resistance; Em^R, erythomycin resistance; Km^R, kanamycin resistance; and Tc^R, tetracycline resistance).
b ts = thermosensitive.

9.2.3 Targeted Approaches

In 2001, the completion of the *L. monocytogenes* genome-sequencing project identified 2853 open reading frames (ORFs) in the genome of EGD-e and assigned putative functions to 65% of them. This precious information allowed then the systematic construction of targeted mutants chosen by in silico analysis. Two approaches predominated in targeted gene disruption: insertional plasmid mutagenesis and allelic replacement. They are schematized in Figure 9.4.

9.2.3.1 Insertional Plasmid Mutagenesis

Plasmid mutagenesis refers to the insertion of a plasmid in a gene by a single crossing-over. Like transposon insertion, this leads to the inactivation of the gene in most of the cases. Historically, one of the first plasmid mutageneses obtained in *L. monocytogenes* was the inactivation of the *prfA* gene by insertion of the plasmid pAUL-A by homologous recombination in an internal region of the *prfA* gene. In this work, Chakraborty et al.[56] constructed pAUL-A, a 9.6-kb shuttle vector conferring erythromycin resistance to both *E. coli* and *Listeria*. This vector contains a multiple cloning site in the *lacZα* sequence allowing white/blue selection, onto solid indicator medium, of the colonies after cloning of the insert. The Gram-positive origin of replication carried by pAUL-A derives from the thermosensitive vector pE194ts. After electroporation, *Listeria* containing the pAUL-A recombinant plasmid are grown at permissive temperature (30°C). Chromosomal integration of the recombinant plasmid is obtained by homologous recombination after growth at restrictive temperature (between 37 and 42°C for pAUL-A). This leads to two separated truncated copies of the disrupted gene (see Figure 9.4). Plasmid pAUL-A also contains transcription terminators around the *lacZα*-cloning region. Therefore, only the first truncated part of the gene can be transcribed from its natural promoter.

We can see here that one of the possible drawbacks of such a method is the high probability of generating polar effect on the expression of the downstream genes, when the insertion occurs in a

FIGURE 9.4 Principles of targeted approaches for gene inactivation. For insertional plasmic mutagenesis (A), a plasmid (e.g., suicide vector) containing a stretch of nucleotide sequence with homology to the target gene is induced to a bacterial strain by a single cross-over and selected in the presence of an antibiotic. This leads to the inactivation of the target gene, affecting the expression of a fully functional protein. For allelic replacement (B), a thermosensitive plasmid containing a homologous nucleotide sequence to the target gene is induced in a bacterial strain by a first cross-over and selected in the presence of an antibiotic at a permissive temperature.

(continued on facing page.)

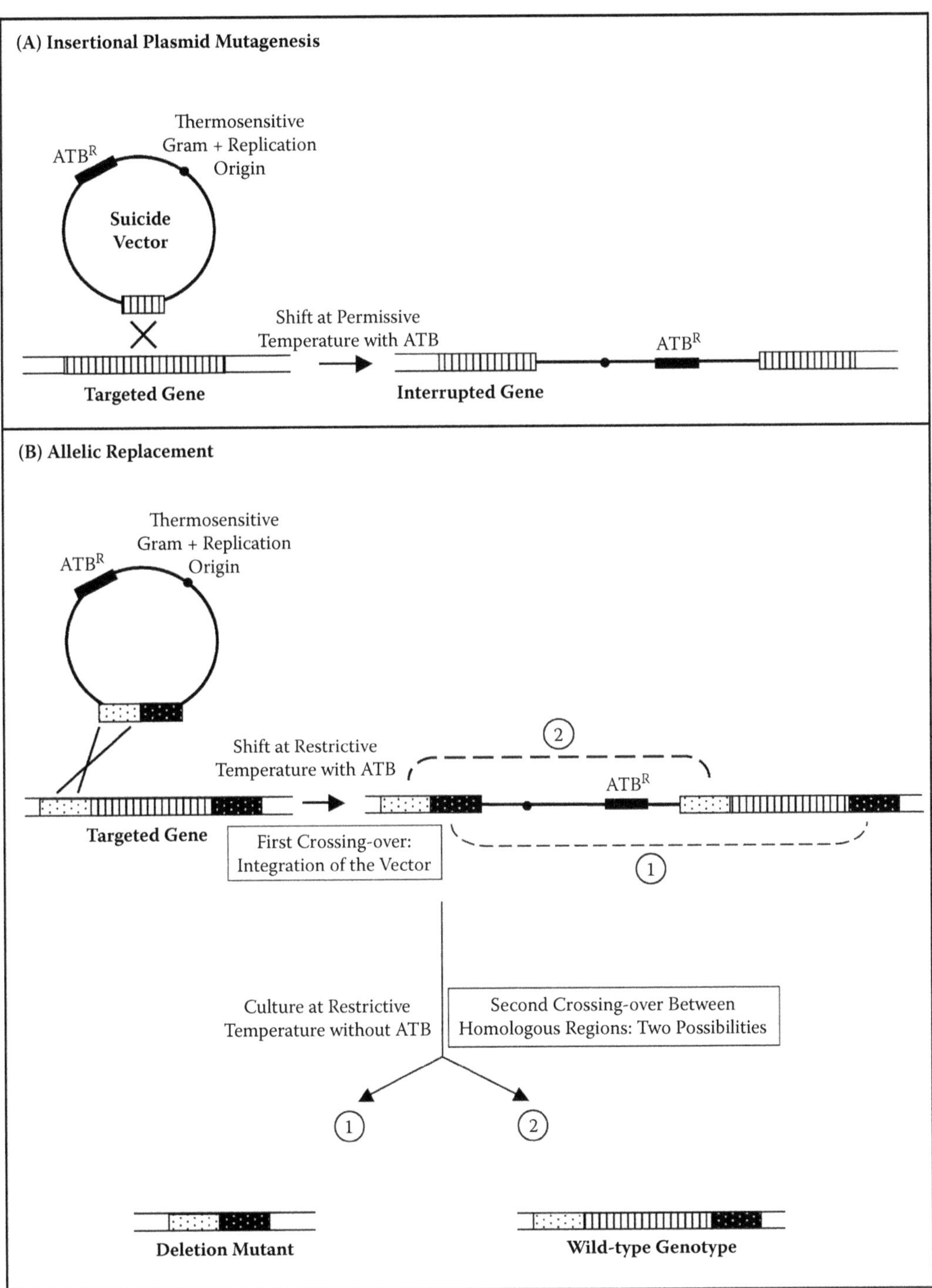

FIGURE 9.4 (continued). The bacterial strain that has successfully incorporated the plasmid is selected again in the absence of antibiotic at a restrictive temperature (i.e., a second cross-over) to remove any unwanted plasmid sequence. This leads to the so-called in-frame deletion, which avoids or reduces the potential polar effects of a deleted gene on downstream genes. ATB, antibiotic; ATB^R, antibiotic resistance gene.

gene not located at the end of an operon. However, even if nonpolar mutants have to be reconstructed afterwards, this is an easy and rapid procedure to inactivate target genes for screening purposes.

Another example of this strategy is given by the work of Rea et al.[76] with the disruption and analysis of six putative regulatory loci of *L. monocytogenes* EGD-e. The genes chosen correspond to a putative topoisomerase, a potential methyltransferase, a transcriptional regulator of the MarR family (*lmo0581, lmo2756, lmo0243,* and *lmo1618,* respectively), and two genes encoding the predicted ferric uptake regulators Fur (*lmo1956*) and PerR (*lmo1683*). Mutants were constructed by using the pORI19 integration strategy. Briefly, a central region of the target gene (amplified by PCR) is cloned into the multiple cloning site of pORI19 (RepA-). After electrotransfer, the recombinant plasmid replication is maintained at 30°C into *L. monocytogenes* containing plasmid pVE6007 (a temperature-sensitive vector providing a *repA*+ allele in *trans*). Subsequent growth at 42°C results in the curing of pVE6007 and integration of pORI19 by homologous recombination at the site of homology of the cloned *L. monocytogenes* DNA.

Sleator et al. have used the pORI19 vector system for the creation of a bank of random pORI19 insertion in the chromosome *L. monocytogenes* strain LO28.[77] Namely, chromosomal DNA from LO28 was partially digested with *Eco*RI and the resulting fragments were inserted into the *Eco*RI site of plasmid pORI19. The recombinant plasmids were first transformed into *E. coli* EC101 (RepA+). Em^R transformants were then pooled and plasmid DNA was extracted and used to transform *L. monocytogenes* LO28G (LO28 harboring the helper plasmid, pVE6007). Transformants were first incubated at permissive temperature (30°C for 180 min to induce expression of Emr-encoding genes). Then, to force loss of pVE6007 and chromosomal integration of pORI19 at the points of homology with the cloned insert, bacteria were incubated at nonpermissive temperature (42°C). The bank of 25,000 clones generated by this procedure corresponded to more than 10-fold the entire LO28 genome (taking an average insert size of 1.5 kb). In that study, the authors screened 2000 colonies to identify putative osmolyte-deficient transport mutants. For this, the pORI19 insertion mutant bank was tested by replica-plating a series of media, with or without NaCl, carnitine, and glycine betaine. This procedure led to the identification of the *Listeria* OpuC homologue involved in carnitine and glycine betaine uptake. This example shows that a given tool (here the pORI19) could be used for different approaches (random or targeted mutagenesis).

Notably, the systematic inactivation of all the open reading frames of *B. subtilis* was undertaken by Vagner et al. in 1998. This insertional plasmid mutagenesis project used the pMUTIN vector.[78] Although pMUTIN has not yet been used in *Listeria*, it contains several interesting features that would make it a suitable vector for *L. monocytogenes* systematic gene inactivation: (1) it is unable to replicate within Gram-positive bacteria; (2) the multiple cloning sites are followed by a reporter *lacZ* gene to monitor target gene expression; and (3) it contains an inducible SPAC promoter to control transcription of the downstream genes, preventing polar effects of the insertion.

9.2.3.2 Allelic Replacement

One way of avoiding or reducing polar effects on downstream genes is to obtain an allelic replacement of the targeted gene in the chromosome by an inactive version (in general, a deletion). Allelic replacement relies on two consecutive recombination steps schematized on Figure 9.4. The vector must carry a thermosensitive Gram-positive origin of replication. Most of the vectors designed used the thermosensitive mutation of the replication origin of plasmid pE194 (designed oriR pE194ts in the figures).[79] This approach allows the replacement of any chromosomal gene by any other gene. In most cases, it is used either to replace the target gene by an in-frame deletion of the entire gene or to substitute it by an antibiotic resistance gene.

Very briefly, the first step of the construction of an in-frame deletion in *L. monocytogenes* consists of cloning the deleted region in an appropriate thermosensitive shuttle plasmid. For this, the upstream and downstream regions immediately flanking the target gene are amplified by PCR. Typically, 600- to 900-bp fragments are amplified to ensure efficient homologous recombination. The

TABLE 9.2
Plasmid Vectors for Allelic Exchange, Complementation, and Gene Expression

Plasmid	Size (Kb)	Gram + Origin of Replication	Origin of Transfer	Description[a]	Ref.
pAUL-A	9.2	Yes, ts[b]	No	Allelic exchange or insertional mutagenesis, Em^R	56
pCON1	7.6	Yes, ts	Yes	Allelic exchange or insertional mutagenesis, Cm^R	161
pKSV7	6.9	Yes, ts	No	Allelic exchange or insertional mutagenesis, Cm^R	80
pMAD	9.7	Yes, ts	No	Allelic exchange or insertional mutagenesis, Em^R	82
pGF-EM	9.4	Yes, ts	Yes	Allelic exchange, GFP fusion, Cm^R	83
pAT18/19	6.6	Yes	Yes	Complementation, gene expression, Em^R	22
pAT28/29	6.7	Yes	Yes	Complementation, gene expression, Spc^R	84
pTCV-*lac*	12.0	Yes	Yes	Reporter system for *lacZ* fusions, Em^R, Km^R	85
pPL1/2	6.1	No	Yes	Integrative vector for complementation, Cm^R	89
pLIV1	13.4	Yes, ts	Yes	IPTG-inducible expresison, integrative, Em^R, Cm^R	90

[a] Antibiotics resistance is given for Gram-positive bacteria only (another selection may be used in *E. coli.* Cm^R, chloramphenicol resistance; Em^R, erythromycin resistance; Km^R, kanamycin resistance; and Spc^R, spectinomycin resistance).
b ts = thermosensitive.

amplified products, flanked by suitable restriction sites, are then ligated together and cloned into the shuttle vector. The recombinant plasmids are introduced into *Listeria* at a permissive temperature (e.g., 30°C) with antibiotic pressure. This leads to chromosomal integration of the recombinant plasmid by homologous recombination between the plasmid-borne region and the corresponding chromosomal region (this recombination can occur either in the upstream or downstream region of the target). Upon integration, a mutated and a wild-type copy of the gene coexist in the chromosome.

The next step, leading to allelic replacement, is obtained by shifting back the bacteria to a restrictive temperature (e.g., 42°C) in the absence of antibiotic pressure. This procedure allows, after a variable number of replication cycles, the spontaneous excision of the integrated plasmid via a second crossing-over. When the second recombination event occurs in the other flanking region over that of the first crossing-over, it leads to the excision of the wild-type gene (and thus to a mutant genotype, case 1 in Figure 9.4B). In contrast, when the second recombination event occurs in the same flanking region, it restores the wild-type genotype (case 2 in Figure 9.4B).

Many vectors harboring the thermosensitive pE194ts origin of replication can be used for allelic replacement strategies. The vectors more often used are: pKSV7,[80] pAUL-A[56] (described earlier), and pG+host5[81] or its derivatives. We will not describe them extensively here as their major characteristics are summarized in Table 9.2.

With plasmids like pAUL-A or pKSV7, the first step of allelic replacement (leading to chromosomal integration of the recombinant plasmid) is generally easy. Indeed, the thermosensitivity of these plasmids is stringent and after one or two passages at restrictive temperature (with antibiotic pressure), the plasmid is integrated in nearly all growing bacteria. In contrast, plasmid excision is often less frequent and can be time consuming because this event cannot be directly selected. Usually, the following screening procedure is applied. After five to ten passages at permissive temperature in broth (without antibiotic pressure), the recombinant bacteria are plated onto solid medium without antibiotic and then replica plated onto the same medium with antibiotic. The antibiotic-sensitive clones (corresponding to plasmid loss) must be then checked by PCR to identify to the mutated clones, since statistically 50% of the second crossing-over leads to a wild-type genotype.

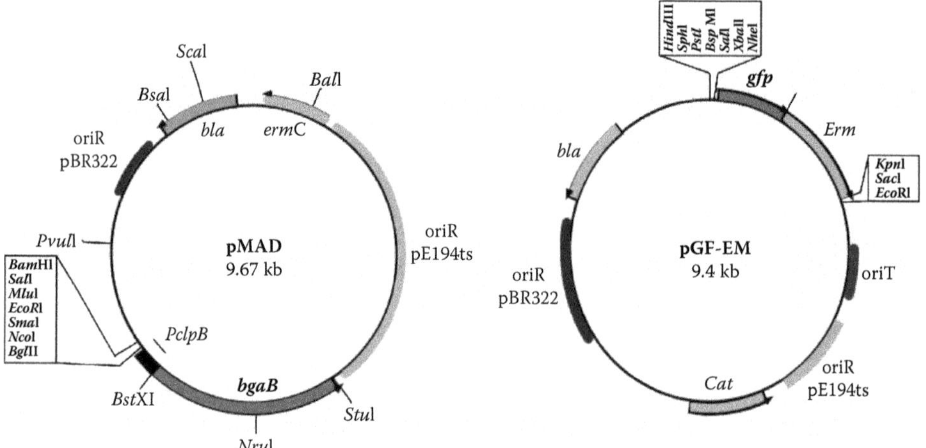

FIGURE 9.5 Vectors for allelic replacement in *L. monocytogenes*.

New vectors have been constructed to improve the efficiency and rapidity of these screening steps and to allow an easier detection of the recombinants.

pMAD.[82] Arnaud et al. constructed a shuttle vector, pMAD, for gene inactivation mutants in several Gram-positive bacteria (Figure 9.5). This vector, which encodes the β-galactosidase (*bga*B) gene under the control of a Gram-positive promoter (see map of the plasmid on Figure 9.5), allows a quick colorimetric blue/white discrimination of bacteria that have lost the plasmid, on X-Gal (5-bromo-4-chloro-3-indolyl-beta-D-galactopyranoside) indicator plates, facilitating clone identification during mutagenesis. Transformants and integrants are blue and Em^R. Clones that have lost the plasmid after passages at 30°C are Em^S and white. The plasmid was used in *S. aureus*, *L. monocytogenes*, and *Bacillus cereus* to efficiently construct allelic replacement mutants. It is worth mentioning that, like pAUL-A or pORI19, pMAD can be also used for simple gene disruption, by cloning an integral portion of the gene to be inactivated and integration of the recombinant plasmid by homologous recombination via a single crossing-over.

pGF-EM.[83] Li and Kathariou. developed a vector with an easier selection of the recombination process. With this plasmid (see map of the pGF-EM on Figure 9.5), chromosomal integration leads to the replacement of a targeted gene by the *gfp* gene and an erythromycin cassette. Two polylinkers, located respectively upstream of the promoterless *gfp* gene and downstream of the *erm* cassette, can be used to clone the upstream and downstream sequence of the targeted gene. The plasmid also harbors a chloramphenicol-acetyl-transferase gene, which allows, first, selection on chloramphenicol of the transformants at permissive temperature (30°C) and, second, of the bacteria with integrated plasmid after growth at restrictive temperature. After passages at 30°C, bacteria where the second crossing-over event has occurred can be selected on erythromycin (which eliminates the clones where the second crossing-over led to the restoration of a wild-type genotype). Under conditions where the promoter of the targeted gene is active, resulting mutant clones are Em^R, Cm^S, and GFP+. This vector also allows the construction of transcriptional GFP fusion and thus can serve as a reporter system for monitoring promoter activity.

9.3 OTHER GENE MANIPULATION METHODS

9.3.1 Intraspecies and Interspecies Complementation Studies

9.3.1.1 The Goals and the Vectors

Classically, complementation is used in *Listeria* species to demonstrate the direct relationship between a mutation generated in the bacterial chromosome by allelic replacement and the phenotype of the resulting mutant strain. For this, a plasmid-borne copy of the wild-type gene is introduced in the mutant strain (by conjugation or electroporation), either under its natural promoter control or under the control of a specific promoter (constitutive or inducible). The properties of the complemented strain are expected to be similar (if not identical) to those of the wild-type strain.

A number of complementation vectors have been described. They can be divided into two main categories: (1) the first category of vectors is composed of the replicative plasmids, which maintain themselves in the cytosol, allowing complementation in *trans* with several copies of the gene; (2) the second category of vectors allows the insertion of the complementing allele into the chromosome of the mutated strain, at a location that does not impair any biological activity of the bacterium (i.e., complementation in *cis*).

9.3.1.2 Complementation in *trans*

A series of high copy number plasmids, designated pAT18, 19, 28, and 29 (Figure 9.6A), have been constructed.[22] The mobilizable shuttle cloning vectors, pAT18 and pAT19, contain the replication origins of pUC and of pAMβ1. They carry an erythromycin-resistance-encoding gene expressed in Gram-negative and Gram-positive bacteria, the transfer origin of plasmid RK2, and the multiple cloning site and the *lacZα* reporter gene of pUC18 (pAT18) and pUC19 (pAT19). The efficient cloning of DNA inserts into the multiple cloning sites can be screened by alpha-complementation in *E. coli* carrying the *lacZ* delta M15 deletion. These 6.6-kb plasmids can be used for complementation studies. Plasmids pAT18, pAT19, and recombinant derivatives have been successfully transferred by conjugation from *E. coli* to various Gram-positive bacteria (including *B. subtilis*, *L. monocytogenes*, and *S. aureus*) at frequencies ranging from 10^{-6} to 10^{-9}. Plasmids pAT28 and 29[84] are pAT derivatives carrying spectinomycin resistance genes (instead of erythromycin-resistance genes).

Plasmid pTCV-lac[85] is a 12 kb derivative of the mobilizable shuttle vector pAT187, which contains the origin of replication of pBR322 and that of pAMβ1. Plasmid pTCV-lac replicates at a low copy number (3–5) in Gram-positive hosts (*Bacillus*, *Clostridia*, *Listeria*, *Enterococcus*, *Staphylococcus*, and *Streptococcus*). pTCV-lac contains a promoterless *lacZ* gene preceded by a multiple cloning site (see Figure 9.9). This plasmid can be used either for complementation studies or for the quantification of promoter activities using the *lacZ* gene as a reporter (β-agalactosidase assays; see section 9.4.1). Since it also contains the transfer origin of the IncP plasmid RK2, pTCV-lac can be transferred from an *E. coli* mobilizing donor to Gram-positive bacteria.

Various pTCV, pAT and other derivatives were later constructed. For example, Fortineau et al. constructed a series of GFP vectors for the detection of *L. monocytogenes* in infected cells.[86] The *gfp-mut1* gene, which encodes a red-shifted GFP, was transcriptionally fused to a strong *L. monocytogenes* promoter p*dlt* (the promoter of the *dlt* operon involved in lipoteichoic acid biogenesis) and inserted into various *E. coli–Listeria* shuttle vectors: (1) the integrative monocopy plasmid pAT113, (2) the low copy number plasmid pTCV-Ex1, and (3) the high copy number plasmid pAT18. Only the high copy number plasmid pNF8 (derived from pAT18) led to the high levels of GFP expression and enabled the detection of *Listeria* both in cultured cells and tissue sections by fluorescence microscopy. This system was later used to follow bacteria in infected cells by flow cytometry.[87]

More recently, Andersen et al.[88] adapted the pNF8 vector to develop a multicolor tagging system based on the synthesis of the cyan fluorescent protein (CFP), the yellow fluorescent protein (YFP), the DsRedExpress or the HcRed. Figure 9.7 describes plasmid pJEBAN2 expressing *cfp* and the visualization and discrimination of fluorescent bacteria under confocal microscope and of colonies placed under UV excitation.

9.3.1.3 Complementation in *cis*

Two types of plasmids are currently available to study gene *cis* complementation, allowing the chromosomal integration of the rescuing allele: pPL and pLIV plasmids, respectively.

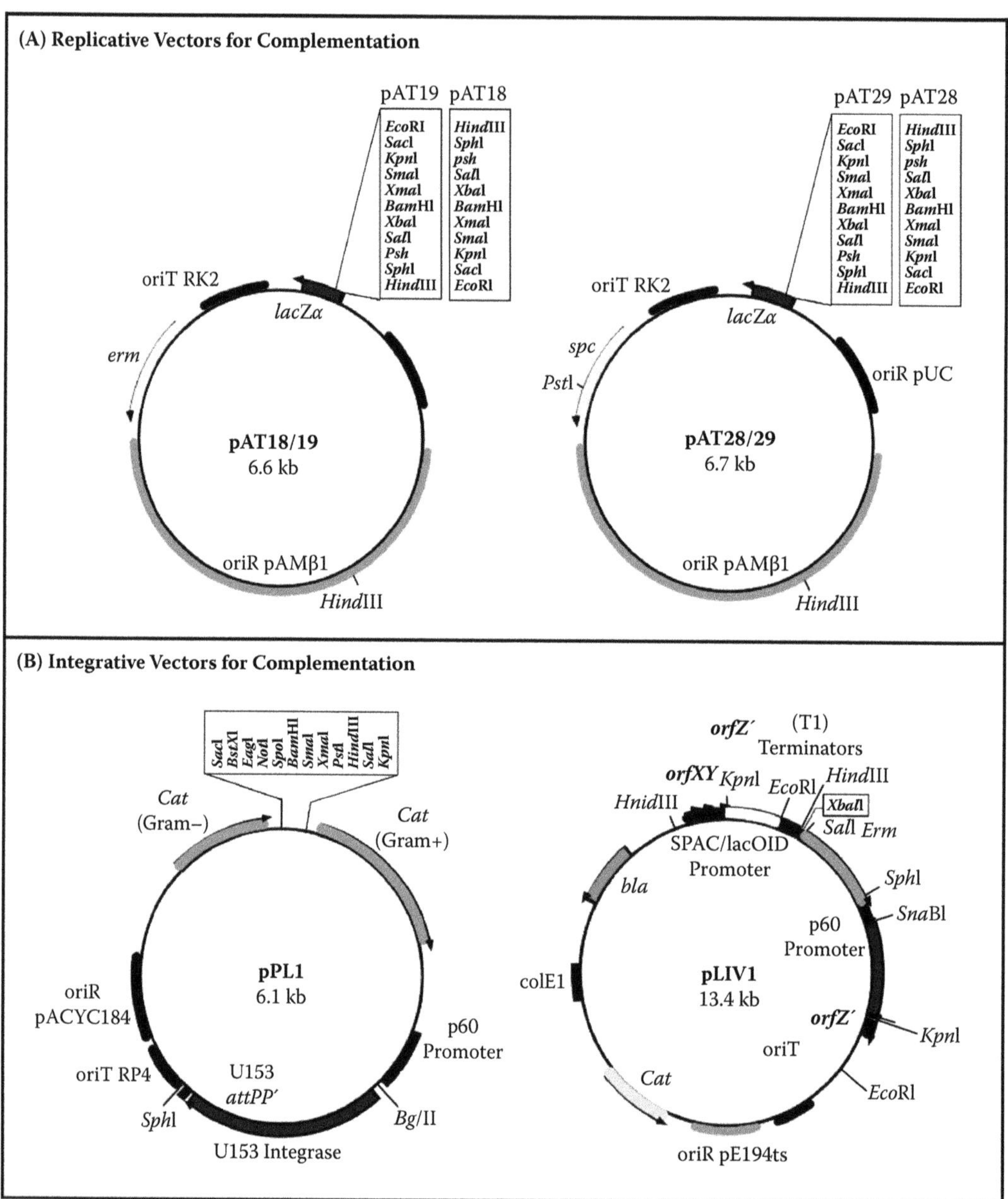

FIGURE 9.6 Vectors for complementation or gene expression in *L. monocytogenes*.

pPL1 and pPL2.[89] Two site-specific shuttle integration vectors were developed by Lauer et al., with two different chromosomal bacteriophage integration sites to facilitate strain construction in *L. monocytogenes*. The first vector, pPL1 (Figure 9.6B), utilizes the listeriophage U153 integrase and attachment site within the *comK* gene for chromosomal insertion. pPL1 can be directly conjugated from *E. coli* into *L. monocytogenes*, forming stable, single-copy integrants at a frequency of approximately 10^{-4} per donor cell in both 1/2 and 4b serogroups. The second vector, pPL2, utilizes the listeriophage PSA integration site in the 3′ end of an arginine tRNA gene. pPL2 shows the same frequency of integration as pPL1 in both 1/2 and 4b serogroups.

Chromosomal insertion downstream of the virulence locus (pLIV).[90] Plasmid pLIV1 was generated to allow chromosomal integration of IPTG-inducible genes in *L. monocytogenes*. The pLIV1

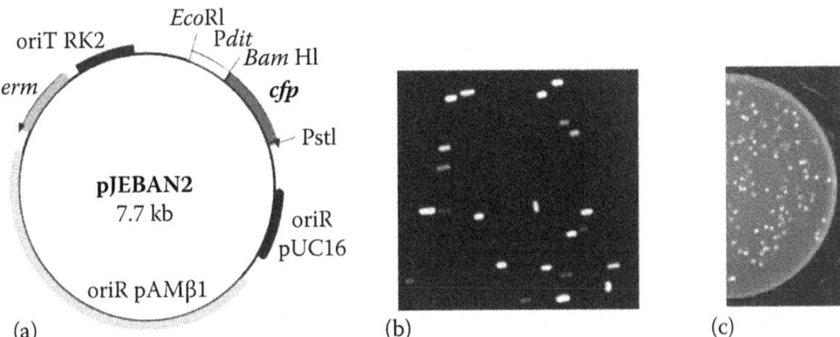

FIGURE 9.7 Fluorescence labeling system in *L. monocytogenes*. (a) Map of pJEBAN2 for CFP expression.[88] (b) Confocal visualization of fluorescent *L. monocytogenes* expressing CFP, YFP, or DsRedExpress. (c) Discrimination between CFP, YFP, and DSRedExpress expressing colonies under UV light (302 nm). (Pictures in (b) and (c) were taken from Andersen, J. B. et al., *BMC Microbiol.*, 6, 86, 2006.)

vector (Figure 9.6B) contains: (1) a ColE1 origin of replication and an ampicillin resistance gene, for cloning and selection in *E. coli*; (2) an origin of transfer, allowing conjugal transfer from *E. coli* to *L. monocytogenes*; (3) a chloramphenicol resistance gene, for plasmid selection in *L. monocytogenes*; and (4) a temperature-sensitive origin of replication, to allow insertion of the recombinant plasmids in the *L. monocytogenes* chromosome (at nonpermissive temperature).

A unique *Xba*I restriction site, immediately downstream of the IPTG-inducible promoter SPAC/lacOid (designated pSPAC), allows the regulated expression of the chosen gene with IPTG. The cloned inducible gene is followed by: (1) an erythromycin resistance gene, for the selection of inducible constructs on the chromosome; and (2) the *lacI* gene placed under the control of the constitutive *L. monocytogenes* p60 (*iap*) promoter, to ensure full repression of the inducible gene in the absence of IPTG.

In pLIV1, these three elements have been inserted, within the *orfZ* gene, into a 1578-bp region of *L. monocytogenes* designated *orfXYZ* locus (and corresponding to *lmo0206–lmo0208*, immediately downstream of *plcB*, the last gene of the virulence locus). This ingenious system thus allows the insertion of single copies of IPTG inducible genes in a "silent" (i.e., nonfunctional) region of the *L. monocytogenes* chromosome.

9.3.2 Selected Examples of Interspecies Complementation Studies

Several examples of intraspecies and interspecies complementation and site-directed mutagenesis studies will be described next, with a focus on the major virulence factors of *L. monocytogenes*, LLO, and ActA (see also chapter 4). Minimal background information on the rationale of the constructions and on the outcome of the experiments will be provided in each case.

9.3.2.1 Complementation among Related Cytolysins

LLO belongs to a family of cholesterol-dependent, pore-forming cytolysins (CDCs).[91,92] The best characterized members of the CDCs are perfringolysin O (PFO), pneumolysin (PLY), and streptolysin O (SLO), which are secreted by extracellular bacterial pathogens *Clostridium perfringens*, *Streptococcus pneumoniae* and *Streptococcus pyogenes*, respectively. In spite of many efforts, the three-dimensional structure of LLO is still unknown, whereas that of monomeric PFO was determined by x-ray crystallography a decade ago.[93] LLO is apparently unique among the CDCs to be able to disrupt a vacuolar membrane, without subsequently killing the host cell upon growth in the cytosol. Several salient characteristics of LLO have bee identified, including: (1) a pH-dependent activity,[94,95] (2) transcriptional and translational regulations of its cytosolic expression and/or activity,[96,97] and (3) a probably rapid degradation of LLO in the cytosol.[98,99] However, the molecular mechanism of LLO-dependent phagosomal escape of intracellular *L. monocytogenes* is not yet elucidated.

Expression in B. subtilis. Bielecki et al. showed in 1990 that the heterologous expression of the LLO-encoding *hly* gene from *L. monocytogenes* in *B. subtilis* conferred the capacity to this non-pathogenic bacterium to escape from the phagocytic vacuole of J774 macrophage-like cells and to proliferate within the cytoplasm of these cells.[100] Later, the structural genes encoding the cytolysins from *Streptococcus pyogenes* (*slo*, encoding SLO) and *Clostridium perfringens* (*pfo*, encoding PFO) were also expressed in *B. subtilis* and compared with LLO with regard to their capacity to promote intracellular growth of *B. subtilis*.[12] The *slo* and *pfo* genes were expressed under the IPTG-inducible pSPAC promoter control, as previously described for *hly*. Each plasmid was integrated into the *B. subtilis* chromosome by homologous recombination. Strikingly, PFO, but not SLO, promoted phagosomal escape and intramacrophagic multiplication of *B. subtilis*. Moreover, although *B. subtilis* expressing PFO was able to lyse the phagosomal membrane, there was evidence that PFO, but not LLO, might be exerting a deleterious effect on the host cell. These observations led to the conclusion that LLO is unique among cytolysins to allow efficient lysis of the host cell vacuole when produced in the phagocytic compartment, but without being cytotoxic when expressed by cytosolic bacteria.

Heterologous cytolysin expression in L. monocytogenes. One of the major differences between LLO and PFO is the relative lack of activity of LLO, compared with PFO, at neutral pH, whereas both cytolysins are active at an acidic pH. Portnoy[100a] and coworkers tested the properties of an *L. monocytogenes* strain expressing PFO instead of LLO. The strain, designated DP-L2221, generated by allelic replacement, had the *pfo* gene inserted in the *L. monocytogenes* chromosome immediately downstream from the promoter for *hly*. The recombinant strain was characterized for intracellular growth and cell-to-cell spread. It could escape from the phagocytic vacuoles, although with a reduced efficiency (64% that of the wild-type *L. monocytogenes*). It replicated intracellularly with a doubling time similar to that of the wild type for several hours. However, after 5 h, growth was aborted because of a cytotoxic effect on the host cell membrane permitting an influx of extracellular gentamicin. Moreover, the strain expressing PFO instead of LLO was completely avirulent in the mouse model.

The amino acid sequence of LLO is highly similar to that of ivanolysin O (ILO), the cytolysin secreted by the ruminant pathogen *L. ivanovii*. We have expressed the gene encoding ILO carried on a multicopy plasmid (i.e., in *trans*) under the control of the *hly* promoter, in a Δ*hly*-mutant of *L. monocytogenes* EGD-e.[101] ILO allowed efficient phagosomal escape of *L. monocytogenes* in both macrophages and hepatocytes. Whereas ILO expression was not cytotoxic, *in vivo*, the ILO-expressing strain was unable to persist in the spleen of infected animals, and thus highly attenuated.

The *B. anthracis* genome contains two genes encoding orthologues of LLO (and designated anthrolysin, ALO) and PI-PLC, respectively.[102] These two proteins were expressed in *L. monocytogenes* and their effects on intracellular growth and cell-to-cell spread were investigated by using the well-established *L. monocytogenes* pathogenesis system. For safety reasons, the authors used an *L. monocytogenes* Δ*dal* Δ*dat* double mutant strain, which requires D-alanine for growth as a host. In-frame allelic replacement of the *hly* and *plcA* genes by their *B. anthracis* counterparts was generated after PCR-mediated sequence overlap extension and by using derivatives of the shuttle vector pKSV7 for allelic exchange. This study indicated that ALO could functionally replace LLO in mediating escape of *L. monocytogenes* from the primary vacuole. However, as previously observed with PFO, it exerted a toxic effect on the host cell.

9.3.2.2 Site-Directed Mutagenesis

Many studies have used site-directed mutagenesis to elucidate the molecular mechanism of action of LLO and ActA during infection *by L. monocytogenes*. We will limit here, for each protein, to a few examples that provide major insight into their structure–functions relationship.

LLO. We will first describe the studies that focused on the role of a region rich in proline, glutamate, serine, and threonine residues (hence, designated PEST-like motif), located close to the N-terminus of the mature protein. Then, we will summarize the site-directed mutagenesis studies that led to the identification of the residues responsible for the pH-dependent LLO activity.

PEST motifs are thought to target eukaryotic proteins for phosphorylation and/or rapid degradation by the proteasome (see references 103–105 and references therein). Decatur and Portnoy analyzed the impact of a 26 amino acid deletion of the PEST-like motif on LLO cytosolic stability and bacterial virulence.[103] The mutated allele was constructed by using splice overlap extension PCR (SOE-PCR)[106] and introduced into the *L. monocytogenes* chromosome by allelic exchange. A series of LLO hybrid proteins, carrying mutation either within the PEST-like motif or in the flanking regions, were also constructed in our laboratory.[104,105] The recombinant LLO proteins were either plasmid-encoded (carried on pAT28-plasmids and expressed in a Δ*hly* EGD-e derivative) or inserted in the chromosome of the Δ*hly* recipient using the integrative vector pAT113 (see earlier discussion).

Altogether, these studies highlighted the critical role of the PEST-like region in bacterial virulence and suggested a role of this region in phagosomal escape. Very recently, Schnupf et al. showed that intracellular LLO was ubiquitinated and phosphorylated within the PEST-like sequence.[97] However, wild-type LLO and PEST region mutants had similarly short intracellular half-lives. The fact that PEST region mutants had higher intracellular LLO levels than wild-type bacteria suggests that, while not involved in proteasomal degradation, the PEST-like region might control LLO production in the cytosol.

In an elegant work, Glomski and coworkers[94,95] identified a single residue in LLO that increased its activity at a neutral pH and evaluated how this mutation affected *L. monocytogenes* pathogenesis. In these studies, all the mutations (constructed by SOE-PCR) were introduced into the chromosome of *L. monocytogenes* by allelic exchange, using pKSV7 plasmid (see earlier discussion). They first constructed a series of chimerical proteins by swapping the regions of dissimilarity within the fourth domain from PFO into LLO. These constructions indicated that specific determinants comprised within the D4 sequence of PFO were responsible for the absence of pH-dependent hemolytic activity of PFO. They could show that a single L461T amino acid substitution was sufficient to confer the pH activity profile of PFO onto LLO. Later, they compared the properties of single or double amino acid substitutions in LLO (LLO S44A, LLO L461T, and LLO S44A L461T), in terms of cytotoxicity and virulence. The observed opposite correlation between cytotoxicity and virulence led the authors to conclude that *L. monocytogenes* controls the cytolytic activity of LLO and restricts it to the phagosomal compartment to maintain its nutritionally rich intracytosolic niche and to avoid the extracellular defenses of the host.

ActA. Lasa et al. have constructed several *L. monocytogenes* strains expressing different domains of ActA and analyzed the ability of the different domains to trigger actin assembly and bacterial motility.[107,108] Theses studies showed that the amino-terminal part of ActA was critical for F-actin assembly and movement, whereas the internal proline-rich repeats and the carboxy-terminal domains were not essential (although *in vitro* motility assays revealed that mutants lacking the proline-rich repeats domain moved two times slower than the wild-type strain). Later, Lauer et al. performed a systematic mutational analysis of the amino-terminal domain of the ActA protein.[109] Twenty clustered charged-to-alanine mutations in the NH2-terminal domain of ActA were generated and introduced into the chromosomal *actA* gene. These specific mutations refined the two regions of ActA involved in Arp2/3 activation. They also identified mutations that caused motility rate defects and changed the ratio of intracellular bacteria associated with actin clouds and comet tails without affecting Arp2/3 activation, reflecting the complexity of ActA-dependent actin-based motility of *L. monocytogenes.*

More recently, Rafelski and Theriot have used a fusion of ActA to a monomeric red fluorescent protein (mRFP1) to examine how *L. monocytogenes* polarizes the ActA protein on its surface upon de novo induction *in vitro.*[110] They followed the RFP signal on the surface of live bacteria over several hours, using epifluorescence microscopy. The fusion protein consists of the ActA and monomeric RFP1 protein domains, linked by a 23-amino-acid-long flexible linker.

The *actA–rfp* construct was placed under the control of the endogenous *pactA* promoter, and integrated in the chromosome of *L. monocytogenes* at the *comK* locus, using pPL1, a site-specific phage integration vector for *L. monocytogenes* (see section 9.3.13). The recombinant plasmid was introduced by conjugation into two Δ*actA* derivatives of *L. monocytogenes* (in wild-type 10403S background or in SLCC-5764, a strain containing a constitutively activated mutant of PrfA). These studies showed that ActA is detected initially at distinct sites along the sides of the bacteria. The protein is then redistributed over the entire cell body through helical cell wall growth. Its accumulation at the bacterial poles occurs only after several generations.

9.3.3 Phages

Generalized transduction provides a convenient means to transfer genes (and mutations) from one genetic background to another. A series of transducing phages of *L. monocytogenes,* with various transduction capacities, has been isolated and characterized recently.[32] Transduction was successfully used in our laboratory for two different purposes.

In the first study,[111] we used the transducing properties of phage LMUP35 to monitor the impact of the deletion of the *mutSL* genes on the yield of homologous recombination. The *mutSL* genes belong to the methyl-directed mismatch repair system (MMR). It has been shown, in several bacterial species, that the production of intraspecies recombinants is inhibited to some extent by the MMR.[112,113] This prompted us to compare the frequencies of transduction of two regions of the chromosome of donor EGD strains, into recipient LO28 strains with a wild-type or a Δ*mutSL* background. The frequency of transductants, obtained with the lysates from the two donor strains, appeared to be 10- to 15-fold higher in the Δ*mutSL* background than in the wild-type background, indicating that *mutSL* genes reduce the yield of homologous recombination in *L. monocytogenes.* In another study,[72] we used phage LMUP35 to transfer a transposon insertion mutant generated from a given strain of *L. monocytogenes* EGD-e into another EGD-e strain, to avoid the possible presence of additional mutations in the initial mutated background. An indicative protocol for phage manipulation is given in Figure 9.8.

Other groups have also used transducing phages to transfer mutations or genetic markers from one genetic background to another. For example, Glomski and coworkers[94] used phage U153 to introduce an antibiotic resistance marker (Em^R) into a given strain of *L. monocytogenes* to allow its simple identification in a competition assay with a mutant of the same strain.

9.4 GENETIC APPROACHES AND TOOLS AVAILABLE FOR GENE EXPRESSION AND REGULATION

9.4.1 Reporter Systems

Various reporter systems have been developed in order to identify promoter regions or to follow listerial gene expression under various conditions. In most cases, the reporter systems are derived from the vectors that have been developed for gene complementation. For example, the pTCV-*lac* (Figure 9.9) described before (in which fusions to the *lacZ* gene can be achieved) has been used to identify the active promoter regions in the type-I signal peptidase operon.[114] In another study conducted by Christiansen et al., pTCV-lac was used to identify transcription terminators in small regulatory RNAs.[115]

Another category of reporter system uses the fluorescent properties of GFP or other fluorescent proteins. In fluorescent protein expressing plasmids (pNF8 or the pJEBAN series, for example), replacing the *pdlt* promoter by another promoter region can be carried out in order to follow its promoter activity by fluorescence measurement under various conditions. An example of such

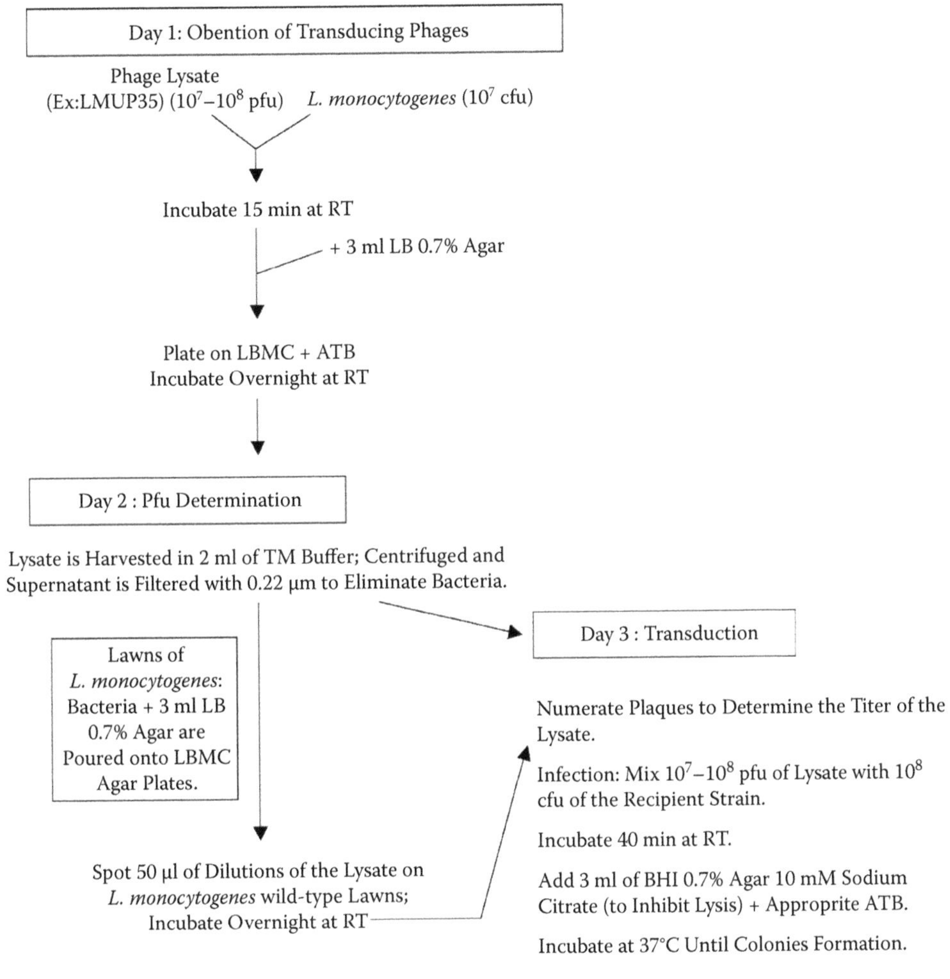

FIGURE 9.8 Protocol for phage transduction. Pfu: plaque-forming unit; RT: room temperature; ATB: appropriate antibiotic; LBMC: LB agar containing 10 m*M* $MgSO_4$ and 10 m*M* $CaCl_2$; TM buffer: 8 m*M* $MgSO_4$, 10 m*M* Tris HCl pH 8.0.

an analysis was carried out by Newton et al., who monitored the promoter activity of the region upstream the *svpA-srtB* locus of *L. monocytogenes*.[116] The *svpA-srtB* promoter region was cloned upstream of the GFP gene in pAT28. The new vector (pATgfp7) was electroporated into *L. monocytogenes* EGD-e and the fluorescence of the bacteria was measured in media with different iron concentrations.

Reporter systems based on β-galactosidase activity or on fluorescence measurements have drawbacks and cannot be applied easily to *in vivo* models. For example, in β-galactosidase measurements, bacteria must be recovered and lysed before the enzymatic assay. In fluorescent systems, naturally occurring fluorescence can lead to high background levels during measurements. In order to prevent those drawbacks, Bron et al. have adapted the bioluminescence reporter systems developed for other bacteria and constructed plasmid pPL2lux[117] (Figure 9.9). This plasmid is derived from the integrative plasmid pPL2 (described previously) and contains the *luxABCDE* operon. Promoter regions can be cloned using the *Sal*I-*Swa*I restriction sites allowing a translational fusion of listerial promoter with the *lux* operon. The *luxAB* genes encode the luciferase enzyme and the *luxCDE* genes encode a reductase involved in the production of the fatty aldehyde substrate of the luciferase. Therefore, in

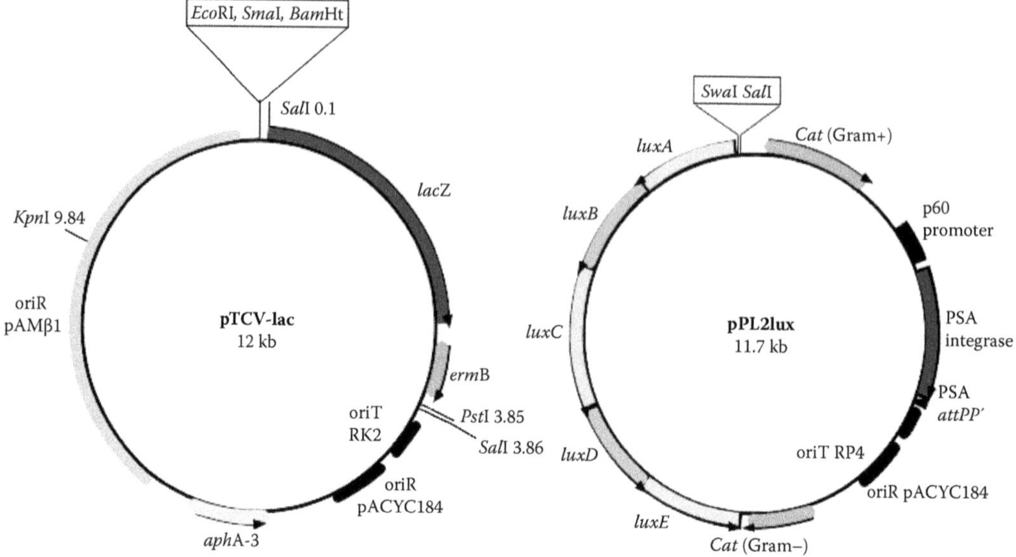

FIGURE 9.9 Reporter systems.

this system, there is no need of an exogenous substrate and the light emission can occur *in vitro* and *in vivo* and be detected in infected cell cultures or in intact murine organs.

9.4.2 *In Vivo* Expression Technologies (IVETs)

9.4.2.1 IVET

Mahan and collaborators developed, for the first time in 1993, a method allowing the *in vivo* selection of genes specifically induced in host tissues.[118] This positive selection procedure led to the identification of a number of new virulence factors in pathogenic Gram-positive and Gram-negative bacteria. Since then, several *in vivo* expression technology (IVET) methods have been adapted to identify promoter sequences specifically activated *in vivo*. The general principle consists of cloning a bank of chromosomal fragments upstream of a reporter gene. The expression of the putative promoter regions can be probed either in an animal model or in cell cultures, and thus lead to the identification of genes possibly required for *in vitro* or *in vivo* survival.

The original IVET system described by Mahan et al.[118] is based on the utilization of *purA* or *thyA* auxotrophy mutants of *Salmonella enterica* serovar typhimurium (i.e., mutants defective in the de novo biosynthesis of purine or pyrimidine nucleotides). The assay consisted in complementing a Δ*purA* (or Δ*thyA*) chromosomal mutation by introducing (in *cis*) the wild-type allele under the control of a heterologous inducible promoter control. Practically, in a first step, a bank of random chromosomal fragments was cloned upstream of the reporter gene in a shuttle plasmid. In a second step, the bank of recombinant plasmids was transferred into the Δ*purA* (or Δ*thiA*) mutant strain of *S. typhimurium* and integrated in the chromosome by homologous recombination between the cloned fragment and the corresponding region of the host chromosome. Integration of a single copy of the transcriptional fusion into the chromosome avoids titration effects that might be observed with multicopy plasmids. The recombination event generates a duplication of the cloned DNA, and thus retains a functional copy of the wild-type gene that may be essential for survival of the bacterium. The recombinant strains were injected in mice and recovered in the spleens 3 days after infection. Only the complemented mutants were able to survive *in vivo* and corresponded to transcriptional fusions that allowed the efficient *in vivo* expression of the *purA* (or *thyA*) reporter gene. In the original system, the *lacZ* and *lacY* genes (producing β-galactosidase and Lac permease, respectively) were introduced downstream of the reporter gene to distinguish the constitutive promoters that led to the production of β-galactosidase in broth (on X-gal indicator plates). In that screen, the *in vivo* inducible promoters corresponded to clones recovered from infected animals that did not produce a detectable β-galactosidase activity on plates.

This selection procedure was applied later in *Pseudmonas aeruginosa*.[119–121] Variations around this theme have also been developed using metabolic genes as reporter genes, and a variety of other reporter genes have been used in subsequent IVET screens (see Rediers et al.[122] for a review), including antibiotic resistance, recombinase (also RIVET), or virulence factors.

9.4.2.2 DFI

The method designated differential fluorescence induction (DFI)[123] is also a promoter trap approach that uses the green fluorescence protein (GFP) as a reporter to monitor promoter activity. DFI allows identification of promoters preferentially active under given growth conditions on the basis of their ability to drive expression of a promoterless GFP gene. Fluorescent cells can be isolated and analyzed by flow cytometry. It is also possible to apply DFI *in vivo*.[124–126] In contrast to most IVET approaches, the transcriptional fusions are not integrated in the host chromosome but carried on a multicopy plasmid (in *trans*). At any rate, DFI and IVET do not allow the identification of genes whose expression is regulated post-transcriptionally.[127]

9.4.2.3 IVET and DFI Strategies in *L. monocytogenes*

Two IVET studies and one DFI study have been published about *L. monocytogenes*. The two IVET studies have used *hly* as a reporter gene[128,129] to identify *in vivo* inducible (*ivi*) genes or loci, using the mouse model of infection. We will recall the technical characteristics and differences of the two studies that may account for the lack of overlap in the *ivi* promoters identified (see also Figure 9.10). The *hly* encodes LLO, an essential virulence factor of *L. monocytogenes* (*hly* mutants that do not produce LLO fail to escape from the phagocytic vacuoles and are hence totally avirulent).

In the study performed in our lab, the promoterless *hly* gene was cloned into plasmids pTCV-lac and pAT28, two Gram-negative/Gram-positive shuttle vectors[84] able to replicate stably in *L. monocytogenes* (multicopy). Banks of recombinant plasmids were constructed by cloning chromosomal fragments of *L. monocytogenes* strain EGD-e (*Sau*3A digest) upstream of *hly*. These banks were introduced by electroporation into a Δ*hly* derivative of EGD-e.[130] The recombinant clones were first screened onto blood agar plates to eliminate the clones constitutively expressing LLO (on this medium, the colonies secreting LLO are surrounded by a halo of β-hemolysis). This procedure led to the selection of 1000 nonhemolytic clones (LLO–) out of 4000 clones tested. These clones were then assembled into pools of LLO– clones (100 pools of 10 clones). Each pool was then administered by the intravenous (i.v.) route to mice at a dose of 2×10^8 bacteria per mouse (at this dose the Δ*hly* host strain is avirulent). Only 9 pools out of 100 tested killed the infected mice or made them visibly ill within 3–10 days after injection. These nine pools were then further analyzed; each clone was individually inoculated into mice. Nine *ivi* clones inducing death or severe illness 3–4 days after injection were identified by this procedure. The *ivi* loci of *L. monocytogenes* included genes encoding (1) the *in vivo*-inducible phosphatidylinositol phospholipase C and (2) a putative *N*-acetylglucosamine epimerase, possibly involved in teichoic acid biosynthesis.

In the report by Gahan and Hill, the promoterless *hly* gene was cloned into plasmid pCOR2. Banks of recombinant plasmids were constructed by cloning chromosomal fragments of *L. monocytogenes* strain LO28 (*Sau*3A or *Xba*I digests) upstream of *hly*. These banks were introduced by electroporation into a Δ*hly* derivative of LO28. Plasmid pCOR2 has a nonfunctional origin of replication (R*epA*-) and thus needs the presence of a functional *repA* allele (in *trans*) to be maintained in *Listeria*. Therefore, the recipient *Listeria* strain used also harbored the thermosensitive plasmid pVE6007, carrying the complementing wild-type *repA* gene (i.e., LO28Δ*hly* pVE6007).

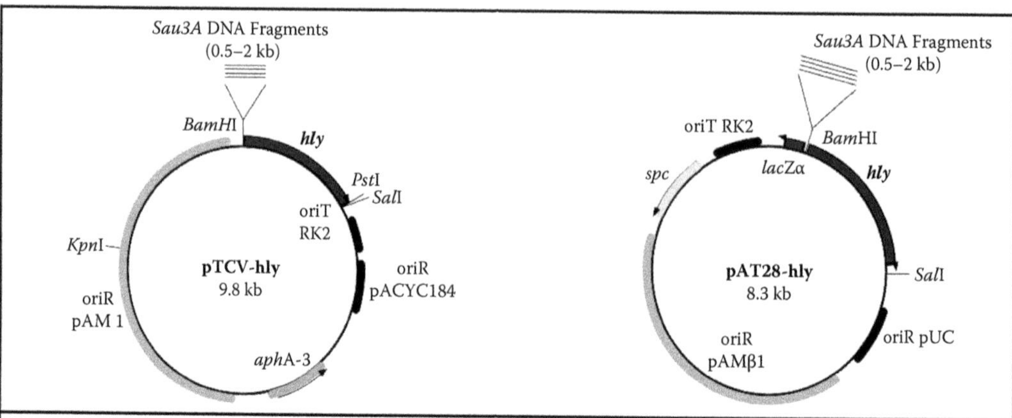

IVET-Trans-complementation (no Chromosomal Integration)
Bank: *Sau*3A Partial Digest of *L. monocytogenes* EGD-e Chromosomal DNA
Host Cell: *L. monocytogenes* EGDΔ*hly*
Adapted from Dubail et al. *Infect. Immun.* (2000)

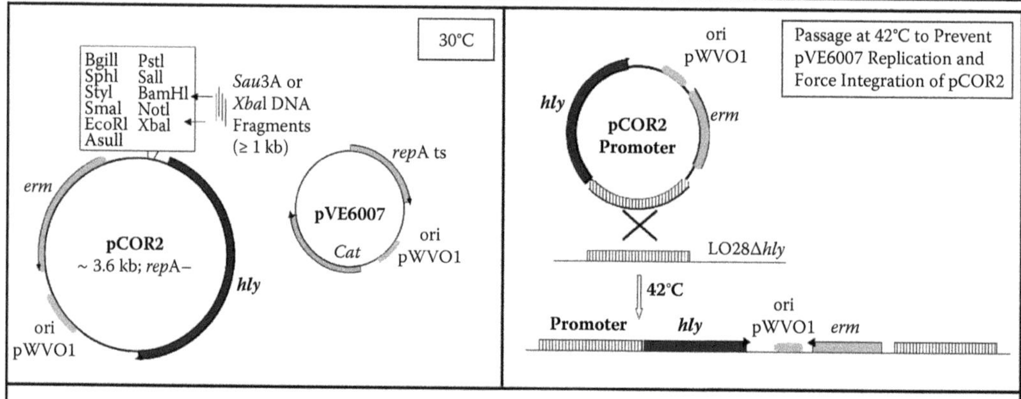

IVET-Cis-complementation (Chromosomal Integration)
Bank: *Sau*3A (Partial) or *Xba*I (Complete) Digest of *L. monocytogenes* LO28 Chromosomal DNA
Host Cell: *L. monocytogenes* LO28Δ*hly*
Adapted from Gahan and Hill. *Mol. Microbiol.* (2000)

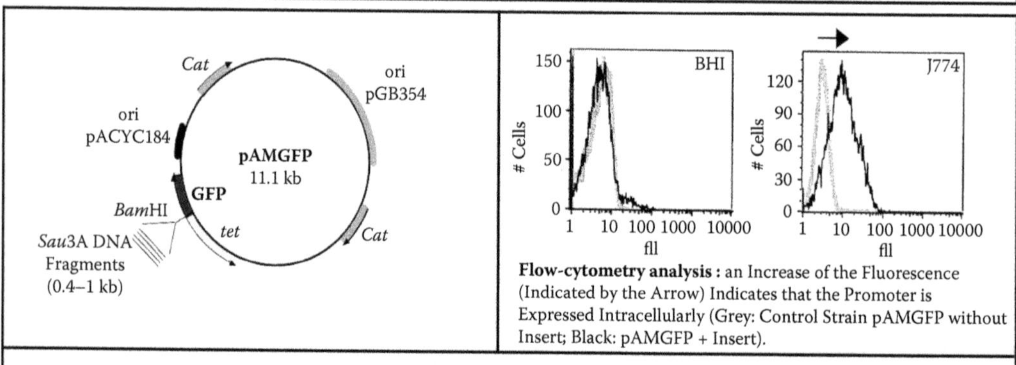

Flow-cytometry analysis : an Increase of the Fluorescence (Indicated by the Arrow) Indicates that the Promoter is Expressed Intracellularly (Grey: Control Strain pAMGFP without Insert; Black: pAMGFP + Insert).

DFI-Trans-complementation (no Chromosomal Integration)
Bank: *Sau*3A Digest (partial) of *L. monocytogenes* 10403s
Host Cell: *L. monocytogenes* 10403sΔ*act*A
Adapted from Wilson et al. *Infect. Immun.* (2001)

FIGURE 9.10 IVET and DFI in *L. monocytogenes*.

After transformation of the bank, the recombinant clones were first maintained at permissive temperature (30°C). Chromosomal integration of the recombinant plasmids was obtained after passage of the recombinant bacteria at nonpermissive temperature (preventing pVE6007 replication, thus allowing homologous recombination via the cloned sequence upstream of *hly*). Mice were infected by the intraperitoneal route at a dose of 5×10^7 bacteria per mouse and the surviving bacteria were recovered in infected spleens 2 days after infection. After five successive passages in mice of a bank of 800 clones, nine *ivi* loci were identified by this procedure. These loci included a DNA topoisomerase, a fumarate hydratase, a cellobiose transporter, and two proteins of unknown function (YpgQ, ipa-19d). Remarkably, the two *ivi* approaches that used *hly* as a reporter identified mainly genes encoding general biological functions.

Wilson et al. have used DFI, with a Δ*actA* strain of *L. monocytogenes* as a recipient.[131] A library of random chromosomal fragments (*Sau*3AI-digest) was cloned into plasmid pAMGFP, and the library was introduced into the *actA* mutant of strain DP-L1942. Approximately 5600 clones were pooled and the *ivi* genes were screened by FACS in the J774 mouse macrophage-like cell line (in two enrichment steps). Of 167 isolates tested, 42 clones were selected that showed low levels of GFP expression in BHI medium and higher levels of GFP expression within J774 cells. DNA sequence analysis of the region immediately upstream of *gfp* was determined. Notably, the promoter region of the *actA* gene itself was found repeatedly (17 times), confirming the validity of the cell-sorting technique itself but indicating a bias in the construction of the bank and/or the screen procedure. A broad spectrum of biological functions was identified, including putative epimerase, hemolysin, elongation factor, regulatory proteins, and a putative mannose PTS transporter.

9.4.3 *In Vivo* Imaging

In vivo imaging technologies include bioluminescence, positron emission tomography, micro x-ray tomography, and magnetic resonance imaging.[132–135] Using bioluminescence, bacteria have been modified to express luciferase and studied in animal models.[132] Bioluminescence allows quantitative detection of the bacteria, and has been now developed in many different models. However, this technique does not provide high resolution and spatial information and other imaging technologies have emerged. For example, very recently, Tournebize et al. applied MRI to successfully follow the development and regression of inflammatory lesions caused by infection by *Klebsiella pneumoniae* in mouse lungs.[135]

Hardy et al. were the first to use *in vivo* bioluminescence imaging (BLI) to study the spatiotemporal distribution of *L. monocytogenes* in infected mice.[136] They showed that *L. monocytogenes* could grow extracellularly in the lumen of the gall bladder of infected Balb/c mice. *L. monocytogenes* strains were made luminescent by chromosomal integration of a *lux-kan* transposon cassette. Briefly, the plasmid pAUL-A Tn*4001 luxABCDE* Km^R was used to transform *L. monocytogenes* strain 10403S (the hybrid *lux-kan* transposon is flanked by the inverted repeats derived from Tn*4001*). Transformants were selected on kanamycin plates and screened for brightness. The Km^R and bright clones correspond to the insertion of the cassette downstream of an active *L. monocytogenes* promoter. Chromosomal insertion sites were determined by inverse PCR. One bioluminescent clone (designated 2C), corresponding to an insertion into *flaA* (encoding flagellin), was chosen for further studies. The authors also constructed a bioluminescent *hly*-negative derivative of *L. monocytogenes* 10403S. For this, they transduced the 2C *lux* insertion into a Δ*hly* 10403S genetic background, using lysates of U153 or P35 transducing phage grown on the 2C strain. The bioluminescent LLO-negative strain was then used to infect mice. Strong signals were detected from the gall bladder, beginning 3–4 days after inoculation. Furthermore, the absence of internalins A and B did not affect the ability of *L. monocytogenes* to grow in the gall bladder, since a triple *hly*/*inlA*/*B* mutant retained this ability. These data led the authors to propose that *L. monocytogenes* grows transiently in the gall bladder, possibly escaping the immune system and spreading from the secretion of bile into the intestine.

9.4.4 Essential Genes

One "classical" genome-scale method that has not been applied yet to *L. monocytogenes* is the exhaustive identification of essential genes. The identification of essential genes in pathogenic bacteria is of considerable relevance because they are potential targets for novel antimicrobial agents.[137] Several experimental approaches have been developed recently for the large-scale identification of essential genes in a number of bacterial species. The systematic disruption of *orfs* has been undertaken by either transposon insertion tests or plasmid insertion mutagenesis. This second strategy works best with naturally competent organisms, such as *S. pneumoniae*[138] and *B. subtilis*.[139] A similar approach based on insertion–duplication mutagenesis was recently described to screen for essential genes of *S. typhimurium* by using a bank of genomic fragments cloned into a conditionally replicating vector.[140] A shotgun antisense RNA method has been also applied to identify conserved genes in *S. aureus*.[141] A list of all currently available essential genes has been compiled into a database that includes the essential genes identified in nine bacterial genomes (DEG, available at http://tubic.tju.edu.cn/deg/). To date, *lmo2537* represents the only experimentally demonstrated essential gene in *L. monocytogenes*.[142]

9.4.5 Other *In Vitro* Models

Many aspects of host–pathogen interactions have been studied using nonmammalian hosts such as *Dictyostelium discoideum*, *Caenorhabditis elegans*, or the fruit fly *Drosophila melanogaster*. Over the past 3 years, these model hosts have been used to dissect bacterial virulence mechanisms in a number of human pathogens, including *Vibrio cholerae*, *Yersinia pestis*, and *Staphylococcus aureus*. Host proteins required for the intracellular survival of *Mycobacterium* and *L. monocytogenes* have been unraveled using high-throughput dsRNA screens in a *Drosophila* cell culture system (see reviews in references 143–145).

In 2003, two studies first reported that *L. monocytogenes* was able to infect and replicate in the cytosol of *Drosophila* phagocytic cells. Cheng et al. demonstrated that infection of *Drosophila* S2 cell cultures and *in vitro* infection of mammalian macrophages required the same *L. monocytogenes* virulence factors.[146] Mansfield et al.[147] showed that both the Toll and IMD pathways were involved in fly resistance to *L. monocytogenes*. Although expression of virulence genes in *L. monocytogenes* had been previously shown to be induced only above 30°C, the environment in fly cells was apparently sufficient to stimulate virulence gene expression even at 25°C (the temperature used for fly cell infection).

In 2005, two studies using RNA interference in *Drosophila* cells were published.[148] Using a library of 7216 dsRNA (representing 50% of the predicted genes in the *Drosophila* genome), Cheng et al.[148] performed different screens that led to the identification of 116 genes, including several host pathways previously unrelated to listerial pathogenesis. Notably, approximately 42% of host genes identified were in host endocytic and vesicular protein trafficking pathways, suggesting that *L. monocytogenes* engages these pathways to promote uptake, phagosomal escape and intracytosolic growth. A screen for knockdowns able to bypass the LLO-dependence of *L. monocytogenes* phagosomal escape identified various components of vesicular trafficking complexes that control trafficking between multivesicular bodies and late endosomes. Altogether, the data obtained suggest that LLO would insert into the membrane of late endosomes, blocking their maturation and favoring the activity of phospholipases, ultimately leading to endosomal membrane destruction. On the other hand, intermediates of the sphingolipids pathway could act as signaling molecules and interfere with LLO activity.

In the second and much broader study, Agaisse et al.[149] used 21,300 dsRNA (covering > 95% of the *Drosophila* genome). Three classes of alterations of the infectious cell cycle were identified: (1) defects of entry and intracellular multiplication, (2) defects of phagosomal escape, or (3) increased rate of intracellular multiplication. A total of 358 candidate genes were identified, including 86 determining components of the ribosomes or of the proteosome. The other genes were mainly related to elements of vesicular trafficking, of signal transduction, or in the organization of the cytoskeleton.

Thus, these two studies identified genes involved in host endocytic and vesicular protein trafficking pathways. Altogether, the utilization of siRNA technology on S2 *Drosophila* cells proved a very useful model and provided important new insight on *Listeria*–eukaryotic host cell interactions. However, the intrinsic differences between the physiology of insect and mammalian cells may limit the interest of this model. Indeed, the growth rate of *L. monocytogenes* at 30°C is reduced in the S2 cytosol, as compared to that in the macrophage cytosol. Moreover, some mutants of *L. monocytogenes* unable to replicate in macrophage vacuoles were still capable of multiplying in S2 cell vacuoles.

9.5 CONCLUSIONS AND PERSPECTIVES

By the year 2000, many of the genetic tools described in this chapter had already been developed and/or adapted to *L. monocytogenes.*[150] In particular, all the vectors for allelic replacement, gene complementation, and gene inactivation (random and site directed) are available. In the past 6 years, most of the "genome-scale" studies, previously developed in various bacterial systems, have been adapted to *Listeria* pathogenesis—in particular, signature-tagged mutagenesis and *in vivo* expression technology. These adaptations have been greatly facilitated by the release of the genome sequences of *L. monocytogenes* EGD-e and *L. innocua* in 2001.[11] The availability of the genome sequence allows in turn a booming series of systematic disruption of predicted virulence-associated genes.

The postgenomic era also saw the development of other "omic" studies, such as proteomic and transcriptomic studies (see chapters 10–12). These new approaches provided novel global information on the regulation of protein expression in *L. monocytogenes* and its relation to bacterial virulence. They also generated increasing numbers of new possible candidate genes and proteins to study.

New approaches are now emerging in the field of bacterial pathogenesis, including *in vivo* transcriptome, as well as studies on the role of noncoding RNA. We will briefly evoke what has been already initiated with *Listeria* species, and what has been already developed in other pathogenic species that could/will be adapted soon to *Listeria.*

9.5.1 Toward *In Vivo* Transcriptomics

PrfA is the best-characterized pleiotropic transcriptional regulator of *L. monocytogenes* (see also chapter 4). In 2003, using high-density macroarrays, Milohanic et al. compared the transcriptome of *L. monocytogenes* EGD-e with that of two isogenic mutants (a *prfA* deletion mutant and a mutant expressing PrfA constitutively) grown in broth.[151] This study identified 63 new genes belonging to the *prfA* regulon. These genes could be classified into three distinct categories: (1) genes positively regulated by PrfA and preceded by a "PrfA box" (10 virulence genes previously known and 2 new genes); (2) genes positively regulated by PrfA but lacking a PrfA box (22 out of 53 being preceded by a putative sigmaB-dependent promoter); and (3) genes negatively regulated by PrfA (8). This work highlighted for the first time a role of repressor for PrfA and a possible interaction between the *prfA* and sigmaB regulons.

In 2006, two simultaneous transcriptomic studies determined the intracellular gene expression profile of *L. monocytogenes.*[152,153] In the first study, using mouse macrophages as a cellular model, Chatterjee et al. compared the expression profile of *L. monocytogenes* strain EGD-e in broth and in infected host cells.[152]. Using a double mutant Δ*hly*Δ*plcA* of *L. monocytogenes,* they could also compare gene expression profile of the bacteria inside the phagosomal and cytosolic compartments. They found that 17% of the total genome was involved in adaptation to the intracellular life (483 genes: 301 up-regulated, including 115 specifically in the cytosol, and 182 down-regulated, including 76 specifically in the cytosol). Notably, this study identified: (1) genes associated with virulence (including three new members of the *prfA* regulon); (2) genes involved in a variety of other general cellular function, such as general stress response, cell division, and cell wall structure, etc; and (3) many genes of unknown functions. The authors validated their transcriptomic data by performing quantitative real-time PCR analysis on a selected subset of genes. A series of chromosomal deletion mutants was also constructed (by allelic replacement using pAUL-A vector) and studied: three genes encoding putative regulators (*lmo1298, lmo2200, sigB*, and *ctsR*) and three PrfA-up-regulated genes identified by the transcriptomic study (*lmo0206, lmo0207*, and *prsA*). Intracellular growth of the *lmo0206, lmo0207, ctsR, lmo2200*, and *prsA*, mutant strains was impaired. Cell-to-cell spread was also affected in *lmo0206, prsA*, and *ctsR* mutants. In contrast, intracellular growth and cell-to-cell spread were unaffected in the *sigB* and *lmo1298* mutants. The exact role in *L. monocytogenes* pathogenesis of *lmo0206* and *lmo0207*, the two *orfs* lying immediately downstream of the virulence locus (and formerly designated *orfX, orfZ*), remains to be elucidated.

In the second study, Joseph et al. also used DNA microarray and real-time reverse transcriptase PCR analyses to investigate the transcriptional profile of intracellular *L. monocytogenes* following epithelial cell infection (Caco-2 cells). This study showed that approximately 19% of the genes were differentially expressed in cells relative to their level of transcription in broth. To validate their data, the authors also constructed a random insertion mutant library of *L. monocytogenes* and screened mutants for intracellular growth deficiency.

Briefly, a bank of *L. monocytogenes* EGD chromosomal fragments (200–400 bp) was constructed in the temperature-sensitive shuttle vector pLSV101 (a derivative of pLSV1). This vector replicates in *Listeria* at 30°C and is lost after several cell divisions at 42°C (due to mutations in the *repF* gene). Ligation samples were transformed into *E. coli.* Transformants were pooled and DNA isolated from different pools was transformed into EGD. The resulting *L. monocytogenes* EGD recombinant libraries were first selected at 30°C (nonintegrated recombinant pLSV101 plasmids). For each EGD recombinant, the shift to 42°C results in the chromosomal integration at the site corresponding to the inserted fragment. The insertion mutant library can be maintained by growth at 37°C (plasmid excision can be obtained by growing back the bacteria at 30°C). Approximately 760 such insertion mutants (ca. 13% of the genome) were tested in Caco-2 cells. This screening led to identification of 72 insertions that attenuated intracellular replication without affecting growth in broth at 37°C, among which 16 corresponded to genes that were intracellularly up-regulated.

Overall, these two studies, which combined transcriptomic, genetic, and cell biology approaches, provided novel interesting and complementary information on the strategies of intracellular survival of *L. monocytogenes.* They both revealed that virulence is dependent on a complex network of genes, many of whose functions are not yet understood.

At present, *in vivo* transcriptomic studies are not yet available for *L. monocytogenes*. However, this technology has been recently adapted to several model systems.[154,155] The critical step to successfully apply this technique relies upon the isolation of sufficient bacterial RNA from infected animals.[155] For this, the number of bacteria that can be isolated must be maximized. The choice of the target organ is thus important. In the case of *L. monocytogenes*, the spleen represents probably the most obvious candidate since loads up to 10^8 bacteria per organ can be reached. The procedure for extracting bacterial RNA from infected cells must also be adapted for each infection system to achieve maximal purity. The fact that bacteria are able to respond very rapidly to all sorts of environmental change implies that, ideally, the extraction of bacterial RNA should not involve any environmental change (such as cold or heat shock, and pressure variations, etc.). Finally, the short half-life of mRNAs also implies rapid extraction procedures.

9.5.2 Noncoding RNA

Noncoding RNAs (ncRNAs) have recently emerged as important regulators of a broad variety of biological processes, including bacterial pathogenesis. In prokaryotes and eukaryotes, strategies for finding new ncRNAs are still evolving—in particular, new bioinformatics programs are constantly being developed.[156–159] In the *E. coli* genome, 65 small ncRNAs have been identified so far. Many of these RNAs are involved in the control of global regulatory networks such as in stress responses. In general, regulatory RNAs act via one of two basic mechanisms: base-pairing with other nucleic acids (in particular, antisense RNA with mRNA) and binding to—and modifying the activity of—a protein or protein complexes.[160]

During the course of disease, bacterial pathogens have to respond and adapt to a series of different environmental conditions. This is particularly true for bacteria such as *L. monocytogenes* that can disseminate from their port of entry, break different physiological barriers, multiply in different organs and tissues, and provoke a systemic infection. For this purpose, bacteria have developed complex regulatory processes that allow temporal and spatial control of virulence factor expression. A growing number of small RNAs have been recently identified as key regulators of bacterial virulence, leading to the launching of several RNA research programs (http://www.projects.mfpl.ac.at/bacrnas/php/index.php?id=5). In particular, the program entitled "Noncoding RNAs in bacterial pathogenesis" (BACRNAs, http://www.projects.mfpl.ac.at/bacrnas/php/index.php?id=14) is focusing on the role of ncRNA in several pathogenicity bacteria, including *Listeria monocytogenes*.

In conclusion, integrative approaches combining genetic approaches with cellular biology, biochemistry, histopathology, and bioinformatics have led to the identification of key virulence factors of *L. monocytogenes*, of several specific cellular ligands, and of signaling events, thus providing a clear overview of the infectious process at the cellular level. New genes and regulatory networks that participate to the infection will be discovered in the near future. The development of new technologies such as *in vivo* imaging of the infectious process or *in vivo* transcriptomics is now required to expand *Listeria* research. In particular, the molecular mechanisms of crossing of the blood–brain and placental barriers remain to be unraveled.

ACKNOWLEDGMENTS

The authors are supported by CNRS, INSERM, and University Paris V.

REFERENCES

1. Murray, E., Webb, R., Sann, M., A disease of rabbit characterized by a large mononuclear leucocytis, caused by a hitherto undescribed bacillus *Bacterium monocytogenes*. *J. Pathol. Bacteriol.*, 29, 407, 1926.
2. Seeliger, H., Jones, D., *Genus Listeria Pirie*. In *Bergey's manual of systematic bacteriology*, P. Smeath, N. Mair, N. Sarpe and J. Holt (eds.). The Williams and Wilkins Co.: Baltimore, MD, 1235, 1940.
3. Seeliger, H., *Listeriosis*. Karger Verlag: Basel, 1961.
4. Rocourt, J., Schrettenbrunner, A., Seeliger, H. P., Biochemical differentiation of the *Listeria monocytogenes* (sensu lato) genomic groups. *Ann. Microbiol. (Paris)*, 134A, 65, 1983.
5. Mackaness, G. B., Cellular resistance to infection. *J. Exp. Med.*, 1962, 116, 381–406.
6. Vazquez-Boland, J. A. et al., Listeria pathogenesis and molecular virulence determinants. *Clin. Microbiol. Rev.*, 14, 584, 2001.
7. Hof, H., History and epidemiology of listeriosis. *FEMS Immunol. Med. Microbiol.*, 35, 199, 2003.
8. Bonazzi, M., Cossart, P., Bacterial entry into cells: A role for the endocytic machinery. *FEBS Lett.*, 580, 2962, 2006.
9. Hamon, M., Bierne, H., Cossart, P., *Listeria monocytogenes*: A multifaceted model. *Nat. Rev. Microbiol.*, 4, 423, 2006.
10. Pizarro-Cerda, J., Cossart, P., Subversion of cellular functions by *Listeria monocytogenes*. *J. Pathol.*, 208, 215, 2006.
11. Glaser, P. et al., Comparative genomics of *Listeria* species. *Science*, 294, 849, 2001.
12. Portnoy, D. A. et al., Molecular determinants of *Listeria monocytogenes* pathogenesis. *Infect. Immun.*, 60, 1263, 1992.

13. Dussurget, O., Pizarro-Cerda, J., Cossart, P., Molecular determinants of *Listeria monocytogenes* virulence. *Annu. Rev. Microbiol.*, 58, 587, 2004.
14. Finlay, B. B., Falkow, S., Common themes in microbial pathogenicity. *Microbiol. Rev.*, 53, 210, 1989.
15. Roberts, A. J., Wiedmann, M., Pathogen, host and environmental factors contributing to the pathogenesis of listeriosis. *Cell. Mol. Life Sci.*, 60, 904, 2003.
16. Poyart-Salmeron, C. et al., Transferable plasmid-mediated antibiotic resistance in *Listeria monocytogenes. Lancet*, 335, 1422, 1990.
17. Charpentier, E., Gerbaud, G., Courvalin, P., Presence of the *Listeria* tetracycline resistance gene tet(S) in *Enterococcus faecalis. Antimicrob. Agents Chemother.*, 38, 2330, 1994.
18. Charpentier, E., Gerbaud, G., Courvalin, P., Conjugative mobilization of the rolling-circle plasmid pIP823 from *Listeria monocytogenes* BM4293 among Gram-positive and Gram-negative bacteria. *J. Bacteriol.*, 181, 3368, 1999.
19. Courvalin, P., Carlier, C., Transposable multiple antibiotic resistance in *Streptococcus pneumoniae. Mol. Gen. Genet.*, 205, 291, 1986.
20. Doucet-Populaire, F. et al., Inducible transfer of conjugative transposon Tn*1545* from *Enterococcus faecalis* to *Listeria monocytogenes* in the digestive tracts of gnotobiotic mice. *Antimicrob. Agents Chemother.*, 35, 185, 1991.
21. Francois, B., Charles, M., Courvalin, P., Conjugative transfer of tet(S) between strains of *Enterococcus faecalis* is associated with the exchange of large fragments of chromosomal DNA. *Microbiology*, 143, 2145, 1997.
22. Trieu-Cuot, P. et al., Shuttle vectors containing a multiple cloning site and a *lacZ* alpha gene for conjugal transfer of DNA from *Escherichia coli* to Gram-positive bacteria. *Gene*, 102, 99, 1991.
23. Trieu-Cuot, P., Derlot, E., Courvalin, P., Enhanced conjugative transfer of plasmid DNA from *Escherichia coli* to *Staphylococcus aureus* and *Listeria monocytogenes. FEMS Microbiol. Lett.*, 109, 19, 1993.
24. Perez-Diaz, J. C., Vicente, M. F., Baquero, F., Plasmids in *Listeria. Plasmid*, 8, 112, 1982.
25. Flamm, R. K., Hinrichs, D. J., Thomashow, M. F., Introduction of pAM beta 1 into *Listeria monocytogenes* by conjugation and homology between native *Listeria monocytogenes* plasmids. *Infect. Immun.*, 44, 157, 1984.
26. Rice, L. B., Tn*916* family conjugative transposons and dissemination of antimicrobial resistance determinants. *Antimicrob. Agents Chemother.*, 42, 1871, 1998.
27. Vicente, M. F., Baquero, F., Perez-Diaz, J. C., Conjugative acquisition and expression of antibiotic resistance determinants in *Listeria* spp. *J. Antimicrob. Chemother.*, 21, 309, 1988.
28. Celli, J., Trieu-Cuot, P., Circularization of Tn*916* is required for expression of the transposon-encoded transfer functions: characterization of long tetracycline-inducible transcripts reading through the attachment site. *Mol. Microbiol.*, 28, 103, 1998.
29. Cossart, P. et al., Listeriolysin O is essential for virulence of *Listeria monocytogenes*: Direct evidence obtained by gene complementation. *Infect. Immun.*, 57, 3629, 1989.
30. Loessner, M. J., Busse, M., Bacteriophage typing of *Listeria* species. *Appl. Environ. Microbiol.*, 56, 1912, 1990.
31. Bille, J., Rocourt, J., WHO International Multicenter *Listeria monocytogenes* subtyping study—Rationale and setup of the study. *Int. J. Food Microbiol.*, 32, 251, 1996.
32. Hodgson, D. A., Generalized transduction of serotype 1/2 and serotype 4b strains of *Listeria monocytogenes. Mol. Microbiol.*, 35, 312, 2000.
33. Luchansky, J. B., Muriana, P. M., Klaenhammer, T. R., Application of electroporation for transfer of plasmid DNA to *Lactobacillus, Lactococcus, Leuconostoc, Listeria, Pediococcus, Bacillus, Staphylococcus, Enterococcus* and *Propionibacterium. Mol. Microbiol.*, 2, 637, 1988.
34. Alexander, J. E. et al., Development of an optimized system for electroporation of *Listeria* species. *Lett. Appl. Microbiol.*, 10(4), 179–181, 1990.
35. Park, S. F., Stewart, G. S., High efficiency transformation of *Listeria monocytogenes* by electroporation of penicillin-treated cells. *Gene*, 94, 129, 1990.
36. Fields, P. I. et al., Mutants of *Salmonella typhimurium* that cannot survive within the macrophage are avirulent. *Proc. Natl. Acad. Sci. USA*, 83, 5189, 1986.
37. Gaillard, J. L., Berche, P., Sansonetti, P., Transposon mutagenesis as a tool to study the role of hemolysin in the virulence of *Listeria monocytogenes. Infect. Immun.*, 52, 50, 1986.
38. Kathariou, S. et al., Tn*916*-induced mutations in the hemolysin determinant affecting virulence of *Listeria monocytogenes. J. Bacteriol.*, 169, 1291, 1987.

39. Promadej, N. et al., Cell wall teichoic acid glycosylation in *Listeria monocytogenes* serotype 4b requires gtcA, a novel, serogroup-specific gene. *J. Bacteriol.*, 181, 418, 1999.
40. Lan, Z., Fiedler, F., Kathariou, S., A sheep in wolf's clothing: *Listeria innocua* strains with teichoic acid-associated surface antigens and genes characteristic of *Listeria monocytogenes* serogroup 4. *J. Bacteriol.*, 182, 6161, 2000.
41. Lei, X. H. et al., A novel serotype-specific gene cassette (gltA-gltB) is required for expression of teichoic acid-associated surface antigens in *Listeria monocytogenes* of serotype 4b. *J. Bacteriol.*, 183, 1133, 2001.
42. Flanary, P. L. et al., Insertional inactivation of the *Listeria monocytogenes cheYA* operon abolishes response to oxygen gradients and reduces the number of flagella. *Can. J. Microbiol.*, 45, 646, 1999.
43. Zheng, W., Kathariou, S., Transposon-induced mutants of *Listeria monocytogenes* incapable of growth at low temperature (4°C). *FEMS Microbiol. Lett.*, 121, 287, 1994.
44. Mengaud, J. et al., Identification of the structural gene encoding the SH-activated hemolysin of *Listeria monocytogenes*: Listeriolysin O is homologous to streptolysin O and pneumolysin. *Infect. Immun.*, 55, 3225, 1987.
45. Youngman, P., Perkins, J. B., Losick, R., Construction of a cloning site near one end of Tn*917* into which foreign DNA may be inserted without affecting transposition in *Bacillus subtilis* or expression of the transposon-borne *erm* gene. *Plasmid*, 12, 1, 1984.
46. Cossart, P. et al., Listeriolysin O is essential for virulence of *Listeria monocytogenes*: Direct evidence obtained by gene complementation. *Infect. Immun.*, 57, 3629, 1989.
47. Camilli, A., Paynton, C. R., Portnoy, D. A., Intracellular methicillin selection of *Listeria monocytogenes* mutants unable to replicate in a macrophage cell line. *Proc. Natl. Acad. Sci. USA*, 86, 5522, 1989.
48. Camilli, A., Portnoy, A., Youngman, P., Insertional mutagenesis of *Listeria monocytogenes* with a novel Tn*917* derivative that allows direct cloning of DNA flanking transposon insertions. *J. Bacteriol.*, 172, 3738, 1990.
49. Sun, A. N., Camilli, A., Portnoy, D. A., Isolation of *Listeria monocytogenes* small-plaque mutants defective for intracellular growth and cell-to-cell spread. *Infect. Immun.*, 58, 3770, 1990.
50. Barry, R. A. et al., Pathogenicity and immunogenicity of *Listeria monocytogenes* small-plaque mutants defective for intracellular growth and cell-to-cell spread. *Infect. Immunl.*, 60, 1625, 1992.
51. Mengaud, J., Geoffroy, C., Cossart, P., Identification of a new operon involved in *Listeria monocytogenes* virulence: Its first gene encodes a protein homologous to bacterial metalloproteases. *Infect. Immun.*, 59, 1043, 1991.
52. Raveneau, J. et al., Reduced virulence of a *Listeria monocytogenes* phospholipase-deficient mutant obtained by transposon insertion into the zinc metalloprotease gene. *Infect. Immun.*, 60, 916, 1992.
53. Kocks, C. et al., *Listeria monocytogenes*-induced actin assembly requires the *actA* gene product, a surface protein. *Cell*, 68, 521, 1992.
54. Domann, E. et al., A novel bacterial virulence gene in *Listeria monocytogenes* required for host cell microfilament interaction with homology to the proline-rich region of vinculin. *EMBO J.*, 11, 1981, 1992.
55. Leimeister-Wachter, M., Domann, E., Chakraborty, T., Detection of a gene encoding a phosphatidylinositol-specific phospholipase C that is coordinately expressed with listeriolysin in *Listeria monocytogenes*. *Mol. Microbiol.*, 5, 361, 1991.
56. Chakraborty, T. et al., Coordinate regulation of virulence genes in *Listeria monocytogenes* requires the product of the *prfA* gene. *J. Bacteriol.*, 174, 568, 1992.
57. Gaillard, J. L. et al., Entry of *L. monocytogenes* into cells is mediated by internalin, a repeat protein reminiscent of surface antigens from Gram-positive cocci. *Cell*, 65, 1127, 1991.
58. Dramsi, S. et al., Entry of *Listeria monocytogenes* into hepatocytes requires expression of inlB, a surface protein of the internalin multigene family. *Mol. Microbiol.*, 16, 251, 1995.
59. Dramsi, S. et al., Identification of four new members of the internalin multigene family of *Listeria monocytogenes* EGD. *Infect. Immun.*, 65, 1615, 1997.
60. Sabet, C. et al., LPXTG protein InlJ, a newly identified internalin involved in *Listeria monocytogenes* virulence. *Infect. Immun.*, 73, 6912, 2005.
61. Leimeister-Wachter, M. et al., Identification of a gene that positively regulates expression of listeriolysin, the major virulence factor of *Listeria monocytogenes*. *Proc. Natl. Acad. Sci. USA*, 87, 8336, 1990.
62. Klarsfeld, A. D., Goossens, P. L., Cossart, P., Five *Listeria monocytogenes* genes preferentially expressed in infected mammalian cells: *plcA*, *purH*, *purD*, *pyrE* and an arginine ABC transporter gene, *arpJ*. *Mol. Microbiol.*, 13, 585, 1994.

63. He, W., Luchansky, J. B., Construction of the temperature-sensitive vectors pLUCH80 and pLUCH88 for delivery of Tn*917*: *Not*I/*Sma*I and use of these vectors to derive a circular map of *Listeria monocytogenes* Scott A, a serotype 4b isolate. *Appl. Environ. Microbiol.*, 63, 3480, 1997.
64. Gutierrez, J. A. et al., Insertional mutagenesis and recovery of interrupted genes of *Streptococcus mutans* by using transposon Tn*917*: Preliminary characterization of mutants displaying acid sensitivity and nutritional requirements. *J. Bacteriol.*, 178, 4166, 1996.
65. Begley, M., Hill, C., Gahan, C. G., Identification and disruption of *btlA*, a locus involved in bile tolerance and general stress resistance in *Listeria monocytogenes*. *FEMS Microbiol. Lett.*, 218, 31, 2003.
66. Wonderling, L. D., Wilkinson, B. J., Bayles, D. O., The *htrA* (degP) gene of *Listeria monocytogenes* 10403S is essential for optimal growth under stress conditions. *Appl. Environ. Microbiol.*, 70, 1935, 2004.
67. Liu, S. et al., A cold-sensitive *Listeria monocytogenes* mutant has a transposon insertion in a gene encoding a putative membrane protein and shows altered (p)ppGpp levels. *Appl. Environ. Microbiol.*, 72, 3955, 2006.
68. Xue, J. et al., Novel activator of mannose-specific phosphotransferase system permease expression in *Listeria innocua*, identified by screening for pediocin AcH resistance. *Appl. Environ. Microbiol.*, 71, 1283, 2005.
69. Autret, N., Charbit, A., Lessons from signature-tagged mutagenesis on the infectious mechanisms of pathogenic bacteria. *FEMS Microbiol. Rev.*, 29, 703, 2005.
70. Autret, N. et al., Identification of new genes involved in the virulence of *Listeria monocytogenes* by signature-tagged transposon mutagenesis. *Infect. Immun.*, 69, 2054, 2001.
71. Trieu-Cuot, P. et al., An integrative vector exploiting the transposition properties of Tn*1545* for insertional mutagenesis and cloning of genes from Gram-positive bacteria. *Gene*, 106, 21, 1991.
72. Autret, N. et al., Identification of the *agr* locus of *Listeria monocytogenes*: Role in bacterial virulence. *Infect. Immun.*, 71, 4463, 2003.
73. Dramsi, S. et al., FbpA, a novel multifunctional *Listeria monocytogenes* virulence factor. *Mol. Microbiol.*, 53, 639, 2004.
74. Mandin, P. et al., VirR, a response regulator critical for *Listeria monocytogenes* virulence. *Mol. Microbiol.*, 57, 1367, 2005.
75. Mei, J. M. et al., Identification of *Staphylococcus aureus* virulence genes in a murine model of bacteremia using signature-tagged mutagenesis. *Mol. Microbiol.*, 26, 399, 1997.
76. Rea, R. B., Gahan, C. G., Hill, C., Disruption of putative regulatory loci in *Listeria monocytogenes* demonstrates a significant role for Fur and PerR in virulence. *Infect. Immun.*, 72, 717, 2004.
77. Sleator, R. D. et al., Analysis of the role of OpuC, an osmolyte transport system, in salt tolerance and virulence potential of *Listeria monocytogenes*. *Appl. Environ. Microbiol.*, 67, 2692, 2001.
78. Vagner, V., Dervyn, E., Ehrlich, S. D., A vector for systematic gene inactivation in *Bacillus subtilis*. *Microbiology*, 144, 3097, 1998.
79. Villafane, R. et al., Replication control genes of plasmid pE194. *J. Bacteriol.*, 169, 4822, 1987.
80. Smith, K., Youngman, P., Use of a new integrational vector to investigate compartment-specific expression of the *Bacillus subtilis spoIIM* gene. *Biochimie*, 74, 705, 1992.
81. Biswas, I. et al., High-efficiency gene inactivation and replacement system for Gram-positive bacteria. *J. Bacteriol.*, 175, 3628, 1993.
82. Arnaud, M., Chastanet, A., Debarbouille, M., New vector for efficient allelic replacement in naturally nontransformable, low-GC-content, Gram-positive bacteria. *Appl. Environ. Microbiol.*, 70, 6887, 2004.
83. Li, G., Kathariou, S., An improved cloning vector for construction of gene replacements in *Listeria monocytogenes*. *Appl. Environ. Microbiol.*, 69, 3020, 2003.
84. Trieu-Cuot, P. et al., A pair of mobilizable shuttle vectors conferring resistance to spectinomycin for molecular cloning in *Escherichia coli* and in Gram-positive bacteria. *Nucleic Acid Res.*, 18, 4296, 1990.
85. Poyart, C., Trieu-Cuot, P., A broad host-range mobilizable shuttle vector for the construction of transcriptional fusions to β-galactosidase in Gram-positive bacteria. *FEMS Microbiol. Lett.*, 156, 193, 1997.
86. Fortineau, N. et al., Optimization of green fluorescent protein expression vectors for *in vitro* and *in vivo* detection of *Listeria monocytogenes*. *Res. Microbiol.*, 151, 353, 2000.
87. Join-Lambert, O. F. et al., *Listeria monocytogenes*-infected bone marrow myeloid cells promote bacterial invasion of the central nervous system. *Cell. Microbiol.*, 7, 167, 2005.
88. Andersen, J. B. et al., Construction of a multiple fluorescence labelling system for use in co-invasion studies of *Listeria monocytogenes*. *BMC Microbiol.*, 6, 86, 2006.
89. Lauer, P. et al., Construction, characterization, and use of two *Listeria monocytogenes* site-specific phage integration vectors. *J. Bacteriol.*, 184, 4177, 2002.

90. Dancz, C. E. et al., Inducible control of virulence gene expression in *Listeria monocytogenes*: Temporal requirement of listeriolysin O during intracellular infection. *J. Bacteriol.*, 184, 5935, 2002.
91. Alouf, J. E., Cholesterol-binding cytolytic protein toxins. *Int. J. Med. Microbiol.*, 290, 351, 2000.
92. Kayal, S., Charbit, A., Listeriolysin O: A key protein of *Listeria monocytogenes* with multiple functions. *FEMS Microbiol. Rev.*, 30, 514, 2006.
93. Rossjohn, J. et al., Structure of a cholesterol-binding, thiol-activated cytolysin and a model of its membrane form. *Cell*, 89, 685, 1997.
94. Glomski, I. J., Decatur, A. L., Portnoy, D. A., *Listeria monocytogenes* mutants that fail to compartmentalize listerolysin O activity are cytotoxic, avirulent, and unable to evade host extracellular defenses. *Infect. Immun.*, 71, 6754, 2003.
95. Glomski, I. J. et al., The *Listeria monocytogenes* hemolysin has an acidic pH optimum to compartmentalize activity and prevent damage to infected host cells. *J. Cell. Biol.*, 156, 1029, 2002.
96. Bubert, A. et al., Differential expression of *Listeria monocytogenes* virulence genes in mammalian host cells. *Mol. Gen. Genet.*, 261, 323, 1999.
97. Schnupf, P., Portnoy, D. A., Decatur, A. L., Phosphorylation, ubiquitination and degradation of listeriolysin O in mammalian cells: Role of the PEST-like sequence. *Cell. Microbiol.*, 8, 353, 2006.
98. Moors, M. A. et al., Expression of listeriolysin O and ActA by intracellular and extracellular *Listeria monocytogenes*. *Infect. Immun.*, 67, 131, 1999.
99. Villanueva, M. S., Beckers, C. J., Pamer, E. G., Infection with *Listeria monocytogenes* impairs sialic acid addition to host cell glycoproteins. *J. Exp. Med.*, 180, 2137, 1994.
100. Bielecki, J. et al., *Bacillus subtilis* expressing a hemolysin gene from *Listeria monocytogenes* grown in mammalian cells. *Nature*, 345, 175, 1990.
100a. Jones, S., Portnoy, D.A., Characterization of *Listeria monocytogenes* pathogenesis in a strain expressing perfringolysin O in place of listeriolysin O. *Infect. Immun.*, 62,5608, 1994.
101. Frehel, C. et al., Capacity of ivanolysin O to replace listeriolysin O in phagosomal escape and *in vivo* survival of *Listeria monocytogenes*. *Microbiology*, 149, 611, 2003.
102. Wei, Z., Zenewicz, L. A., Goldfine, H., *Listeria monocytogenes* phosphatidylinositol-specific phospholipase C has evolved for virulence by greatly reduced activity on GPI anchors. *Proc. Natl. Acad. Sci. USA*, 102, 12927, 2005.
103. Decatur, A. L., Portnoy, D. A., A PEST-like sequence in listeriolysin O essential for *Listeria monocytogenes* pathogenicity. *Science*, 290, 992, 2000.
104. Lety, M. A. et al., Critical role of the N-terminal residues of listeriolysin O in phagosomal escape and virulence of *Listeria monocytogenes*. *Mol. Microbiol.*, 46, 367, 2002.
105. Lety, M. A. et al., Identification of a PEST-like motif in listeriolysin O required for phagosomal escape and for virulence in *Listeria monocytogenes*. *Mol. Microbiol.*, 39, 1124, 2001.
106. Ho, S. N. et al., Site-directed mutagenesis by overlap extension using the polymerase chain reaction. *Gene*, 77, 51, 1989.
107. Lasa, I. et al., The amino-terminal part of ActA is critical for the actin-based motility of *Listeria monocytogenes*; the central proline-rich region acts as a stimulator. *Mol. Microbiol.*, 18, 425, 1995.
108. Lasa, I. et al., Identification of two regions in the N-terminal domain of ActA involved in the actin comet tail formation by *Listeria monocytogenes*. *EMBO J.*, 16, 1531, 1997.
109. Lauer, P. et al., Systematic mutational analysis of the amino-terminal domain of the *Listeria monocytogenes* ActA protein reveals novel functions in actin-based motility. *Mol. Microbiol.*, 42, 1163, 2001.
110. Rafelski, S. M., Theriot, J. A., Bacterial shape and ActA distribution affect initiation of *Listeria monocytogenes* actin-based motility. *Biophys. J.*, 89, 2146, 2005.
111. Merino, D. et al., A hypermutator phenotype attenuates the virulence of *Listeria monocytogenes* in a mouse model. *Mol. Microbiol.*, 44, 877, 2002.
112. Mejean, V., Claverys, J. P., DNA processing during entry in transformation of *Streptococcus pneumoniae*. *J. Biol. Chem.*, 268, 5594, 1993.
113. Humbert, O. et al., Homeologous recombination and mismatch repair during transformation in *Streptococcus pneumoniae*: Saturation of the Hex mismatch repair system. *Proc. Natl. Acad. Sci. USA*, 92, 9052, 1995.
114. Raynaud, C., Charbit, A., Regulation of expression of type I signal peptidases in *Listeria monocytogenes*. *Microbiology*, 151, 3769, 2005.
115. Christiansen, J. K. et al., Identification of small Hfq-binding RNAs in *Listeria monocytogenes*. *RNA*, 12, 1383, 2006.

116. Newton, S. M. et al., The *svpA-srtB* locus of *Listeria monocytogenes*: Fur-mediated iron regulation and effect on virulence. *Mol. Microbiol.*, 55, 927, 2005.
117. Bron, P. A. et al., Novel luciferase reporter system for *in vitro* and organ-specific monitoring of differential gene expression in *Listeria monocytogenes*. *Appl. Environ. Microbiol.*, 72, 2876, 2006.
118. Mahan, M. J., Slauch, J. M., Mekalanos, J. J., Selection of bacterial virulence genes that are specifically induced in host tissues. *Science*, 259, 686, 1993.
119. Handfield, M. et al., *In vivo*-induced genes in *Pseudomonas aeruginosa*. *Infect. Immun.*, 68, 2359, 2000.
120. Wang, J. et al., Isolation and characterization of *Pseudomonas aeruginosa* genes inducible by respiratory mucus derived from cystic fibrosis patients. *Mol. Microbiol.*, 22, 1005, 1996.
121. Wang, J. et al., Large-scale isolation of candidate virulence genes of *Pseudomonas aeruginosa* by *in vivo* selection. *Proc. Natl. Acad. Sci. USA*, 93, 10434, 1996.
122. Rediers, H. et al., Unraveling the secret lives of bacteria: Use of *in vivo* expression technology and differential fluorescence induction promoter traps as tools for exploring niche-specific gene expression. *Microbiol. Mol. Biol. Rev.*, 69, 217, 2005.
123. Valdivia, R. H., Falkow, S., Bacterial genetics by flow cytometry: Rapid isolation of *Salmonella typhimurium* acid-inducible promoters by differential fluorescence induction. *Mol. Microbiol.*, 22, 367, 1996.
124. Marra, A. et al., Differential fluorescence induction analysis of *Streptococcus pneumoniae* identifies genes involved in pathogenesis. *Infect. Immun.*, 70, 1422, 2002.
125. Marra, A. et al., *In vivo* characterization of the *psa* genes from *Streptococcus pneumoniae* in multiple models of infection. *Microbiology*, 148, 1483, 2002.
126. Schneider, W. P. et al., Virulence gene identification by differential fluorescence induction analysis of *Staphylococcus aureus* gene expression during infection-simulating culture. *Infect. Immun.*, 70, 1326, 2002.
127. Mahan, M. J. et al., Assessment of bacterial pathogenesis by analysis of gene expression in the host. *Annu. Rev. Genet.*, 34, 139, 2000.
128. Dubail, I., Berche, P., Charbit, A., Listeriolysin O as a reporter to identify constitutive and *in vivo*-inducible promoters in the pathogen *Listeria monocytogenes*. *Infect. Immun.*, 68, 3242, 2000.
129. Gahan, C. G., Hill, C., The use of listeriolysin to identify *in vivo* induced genes in the Gram-positive intracellular pathogen *Listeria monocytogenes*. *Mol. Microbiol.*, 36, 498, 2000.
130. Lingnau, A. et al., Expression of the *Listeria monocytogenes* EGD *inlA* and *inlB* genes, whose products mediate bacterial entry into tissue culture cell lines, by PrfA-dependent and -independent mechanisms. *Infect. Immun.*, 63, 3896, 1995.
131. Wilson, R. L. et al., Identification of *Listeria monocytogenes in vivo*-induced genes by fluorescence-activated cell sorting. *Infect. Immunol.*, 69, 5016, 2001.
132. Doyle, T. C., Burns, S. M., Contag, C. H., *In vivo* bioluminescence imaging for integrated studies of infection. *Cell. Microbiol.*, 6, 303, 2004.
133. Kesarwala, A. H. et al., Second-generation triple reporter for bioluminescence, micropositron emission tomography, and fluorescence imaging. *Mol. Imaging*, 5, 465, 2006.
134. Piwnica-Worms, D. et al., Single photon emission computed tomography and positron emission tomography imaging of multi-drug resistant P-glycoprotein-monitoring a transport activity important in cancer, blood–brain barrier function and Alzheimer's disease. *Neuroimaging Clin. N. Am.*, 16, 575, 2006.
135. Tournebize, R. et al., Magnetic resonance imaging of *Klebsiella pneumoniae*-induced pneumonia in mice. *Cell. Microbiol.*, 8, 33, 2006.
136. Hardy, J. et al., Extracellular replication of *Listeria monocytogenes* in the murine gall bladder. *Science*, 303, 851, 2004.
137. Zhang, R., Ou, H. Y., Zhang, C. T., DEG: A database of essential genes. *Nucleic Acids Res.*, 32, D271, 2004.
138. Zalacain, M. et al., A global approach to identify novel broad-spectrum antibacterial targets among proteins of unknown function. *J. Mol. Microbiol. Biotechnol.*, 6, 109, 2003.
139. Kobayashi, K. et al., Essential *Bacillus subtilis* genes. *Proc. Natl. Acad. Sci. USA*, 100, 4678, 2003.
140. Knuth, K. et al., Large-scale identification of essential *Salmonella* genes by trapping lethal insertions. *Mol. Microbiol.*, 51, 1729, 2004.
141. Forsyth, R. A. et al., A genome-wide strategy for the identification of essential genes in *Staphylococcus aureus*. *Mol. Microbiol.*, 43, 1387, 2002.
142. Dubail, I. et al., Identification of an essential gene of *Listeria monocytogenes* involved in teichoic acid biogenesis. *J. Bacteriol.*, 188, 6580, 2006.
143. Brennan, C. A., Anderson, K. V., *Drosophila*: The genetics of innate immune recognition and response. *Annu. Rev. Immunol.*, 22, 457, 2004.

144. Wang, L., Ligoxygakis, P., Pathogen recognition and signalling in the *Drosophila* innate immune response. *Immunobiology*, 211, 251, 2006.
145. Kurz, C. L., Ewbank, J. J., Infection in a dish: High-throughput analyses of bacterial pathogenesis. *Curr. Opin. Microbiol.*, 10, 10, 2007.
146. Cheng, L. W., Portnoy, D. A., *Drosophila* S2 cells: An alternative infection model for *Listeria monocytogenes. Cell. Microbiol.*, 5, 875, 2003.
147. Mansfield, B. E. et al., Exploration of host–pathogen interactions using *Listeria monocytogenes* and *Drosophila melanogaster. Cell. Microbiol.*, 5, 901, 2003.
148. Cheng, L. W. et al., Use of RNA interference in *Drosophila* S2 cells to identify host pathways controlling compartmentalization of an intracellular pathogen. *Proc. Natl. Acad. Sci. USA*, 102, 13646, 2005.
149. Agaisse, H. et al., Genome-wide RNAi screen for host factors required for intracellular bacterial infection. *Science*, 309, 1248, 2005.
150. Freitag, N. E., Genetic tools for use with *Listeria monocytogenes* in Gram-positive pathogens. In *Gram-positive pathogens*, V. A. Fischetti, R. P. N., J. J. Ferretti, D. A. Portnoy, and J. I. Rood (eds.). ASM Press: Washington, D.C., 488, 2000.
151. Milohanic, E. et al., Transcriptome analysis of *Listeria monocytogenes* identifies three groups of genes differently regulated by PrfA. *Mol. Microbiol.*, 47, 1613, 2003.
152. Chatterjee, S. S. et al., Intracellular gene expression profile of *Listeria monocytogenes. Infect. Immun.*, 74, 1323, 2006.
153. Joseph, B. et al., Identification of *Listeria monocytogenes* genes contributing to intracellular replication by expression profiling and mutant screening. *J. Bacteriol.*, 188, 556, 2006.
154. Wren, B. W., Microbial genome analysis: Insights into virulence, host adaptation and evolution. *Nat. Rev. Genet.*, 1, 30, 2000.
155. Hinton, J. C. et al., Benefits and pitfalls of using microarrays to monitor bacterial gene expression during infection. *Curr. Opin. Microbiol.*, 7, 277, 2004.
156. Wang, C. et al., PSoL: A positive sample only learning algorithm for finding noncoding RNA genes. *Bioinformatics*, 22, 2590, 2006.
157. Yachie, N. et al., Prediction of noncoding and antisense RNA genes in *Escherichia coli* with gapped Markov model. *Gene*, 372, 171, 2006.
158. Vogel, J., Sharma, C. M., How to find small noncoding RNAs in bacteria. *Biol. Chem.*, 386, 1219, 2005.
159. Majdalani, N., Vanderpool, C. K., Gottesman, S., Bacterial small RNA regulators. *Crit. Rev. Biochem. Mol. Biol.*, 40, 93, 2005.
160. Storz, G., Altuvia, S., Wassarman, K. M., An abundance of RNA regulators. *Annu. Rev. Biochem.*, 74, 199, 2005.
161. Behari, J., Youngman, P., Regulation of *hly* expression in *Listeria monocytogenes* by carbon sources and pH occurs through separate mechanisms mediated by PrfA. *Infect. Immun.*, 66, 3635, 1998.
162. Youngman, P. J., *Plasmid vector for recovering and exploiting Tn917 transpositions in B. subtilis and other Gram positive.* IRL Press: Oxford, 79, 1987.

10 Comparative Genomics and Evolution of Virulence

Sukhadeo Barbuddhe, Torsten Hain, and Trinad Chakraborty

CONTENTS

10.1 INTRODUCTION

Listeria spp. are rod shaped, Gram-positive, intracellular, facultative-anaerobic, nonsporulating rods and are motile at 10–25°C.[1–3] *Listeria* have been isolated from varied sources, including soil, water, plants, feces, decaying vegetables, meat, seafood, dairy products, and asymptomatic human and animal carriers.[4,5] The natural habitat of *Listeria* is decaying plant material, where they live as saprophytes.

The genus *Listeria* belongs to the phylum Firmicutes with low G+C-content, which also includes the genera *Bacillus*, *Clostridium*, *Enterococcus*, *Streptococcus*, and *Staphylococcus*. The genus *Listeria* consists of six species: *L. monocytogenes*, *L. innocua*, *L. welshimeri*, *L. ivanovii*, *L. seeligeri*, and *L. grayi*. Only two of the species, *L. monocytogenes* and *L. ivanovii*, are pathogenic. *L. monocytogenes* causes severe illnesses both in humans and in animals, whereas *L. ivanovii* is almost always associated with infections in animals. There are 13 described serotypes within *L. monocytogenes*, with serotypes 1/2a, 1/2b, and 4b accounting for 95% of human infections. *L. monocytogenes* survives under harsh conditions that are encountered in the food chain, such as high salt concentrations and extremes of pH and temperature.

Human listeriosis is a food-borne disease and it has been estimated that 99% of all human listeriosis cases are caused by consumption of contaminated food products.[6] The clinical features of listeriosis include meningitis, meningoencephalitis, septicemia, abortion, perinatal infections, and gastroenteritis.[7] The occurrence of listeriosis is quite low: 2–15 cases per million of the population per year. However, the high mortality rate of about 20–30% in those developing listeriosis (typically pregnant women, the elderly, and immunocompromised persons) makes *L. monocytogenes* a serious health threat.[6,8] Listeriosis in animals is predominantly a food-borne disease that is often transmitted by consumption of spoiled silage, causing abortions, stillbirths, and neonatal septicemia in sheep and cattle.[9–11]

The sequencing of *Listeria* genomes has opened new avenues toward understanding of the molecular mechanisms of *L. monocytogenes* virulence in humans and survival of this bacterium in food and in the environment. Apart from understanding the molecular basis of host–pathogen interactions in disease, the sequencing data obtained from *Listeria* species provide us with an excellent opportunity to understand the evolution of virulence within the genus, mechanisms by which signals are sensed while transiting between different environments, the genes, and their products involved in these transitions.

10.2 GENOMICS OF *LISTERIA*

Currently, the complete genome sequences of *L. monocytogenes* EGD-e (serotype 1/2a),[12] *L. monocytogenes* F2365 (serotype 4b),[13] *L. innocua* CLIP 11262 (serotype 6a),[12] and *L. welshimeri* SLCC 5334 (serotype 6b)[14] are published. Additionally, the incomplete genomes of *L. monocytogenes* F6854 (serotype 1/2a) and *L. monocytogenes* H7858 (serotype 4b)[13] are available. The genome sequences of *L. seeligeri* SLCC 3954 (serotype 1/2b), *L. ivanovii* PAM 55 (serotype 5), and *L. monocytogenes* serotype 4a strains have recently been completed and the genome sequencing of *L. grayi* is nearing completion[15] (http://www.genomesonline.org). Furthermore, a large number of partially sequenced genomes comprising different *L. monocytogenes* serotypes are also presently accessible (http://www.broad.mit.edu/seq/msc/).

10.2.1 Comparative Genomics

All of the *Listeria* genomes sequenced to date are circular chromosomes with sizes that vary between 2.7 and 3.0 Mb. Approximately 89% of the sequences of the different *Listeria* genomes are coding sequences, and 62.5% of these have an assigned function. Although strain- and serotype-specific genes were identified, all *Listeria* genomes revealed a high synteny in gene organization and content. The lack of inversions or shifting of large genome segments could be attributed to the low occurrence of transposons and insertion sequence (IS) elements present in the sequenced *Listeria* genomes.[16] The comparative features of sequenced *Listeria* genomes are summarized in Table 10.1 and represented in Figure 10.1.

L. monocytogenes EGD-e contains one circular chromosome of 2,944,528 base pairs (bp) with an average G+C content of 39%, and *L. innocua* CLIP 11262 chromosome has a similar size (3,011,209 bp) and a similar G+C content (37%). The completely sequenced genome of *L. monocytogenes* strain F2365 comprises a chromosome of 2,905,310 bp in length with an average G+C content of 38%.[13] In comparison with the genomes of *L. monocytogenes* and *L. innocua*, *L. welshimeri* (2,814,130 bp) has one of the smallest genomes within the genus *Listeria*[14] and has the lowest G+C content (36.4%) among currently sequenced *Listeria* spp.

There are a total of 2847 predicted coding regions in the genome of *L. monocytogenes* F2365[13] compared to 2853 protein-coding genes in *L. monocytogenes* EGDe chromosome, 2973 in *L. innocua* CLIP 11262 chromosome, and 2780 in *L. welshimeri*.[12,14] Comparative analysis using cluster analysis allows prediction of a common pool comprising a core set of 2254 conserved coding sequences (CDS) among *Listeria* spp. No function could be predicted for 35.3% of *L. monocytogenes* genes and 37% of *L. innocua* and *L. welshimeri* genes.

TABLE 10.1
General Features of *Listeria* Genomes

Feature	*L. monocytogenes* EGD-e	*L. monocytogenes* F2365	*L. innocua* CLIP 11262	*L. welshimeri* SLCC 5334
Serotype	1/2a	4b	6a	6b
Size of the chromosome (bp)	2,944,528	2,905,310	3,011,209	2,814,130
G+C content (%)	39	38	37.4	36.4
G+C content of protein-coding genes (%)	38.4	38.5	37.8	36.7
No. of protein-coding genes	2853	2821	2973	2,780
No. of rRNA operons (16S-23S-5S)	6	6	6	6
No. of tRNA genes	67	67	66	66
Percentage coding	90.3%	88.4%	90.3%	88.7%
Prophages	1	0	5	1

The majority of genomic differences between *Listeria* species comprise phage insertions, transposable elements, scattered unique genes, and islands encoding proteins of mostly unknown function, as well as single nucleotide polymorphisms (SNPs) in many genes associated with virulence functions.

Listeria genomes encode many putative surface proteins belonging to six families and expansion of these families seems to be partly due to gene duplications. Among these surface proteins, internalins belong to a family of proteins characterized by an NH2-terminal domain containing leucine rich repeats (LRRs). The *L. monocytogenes* EGDe genome sequences reveal the presence of 41 proteins containing an LPXTG motif, 19 of which belong to the LRR/internalin family. Eleven

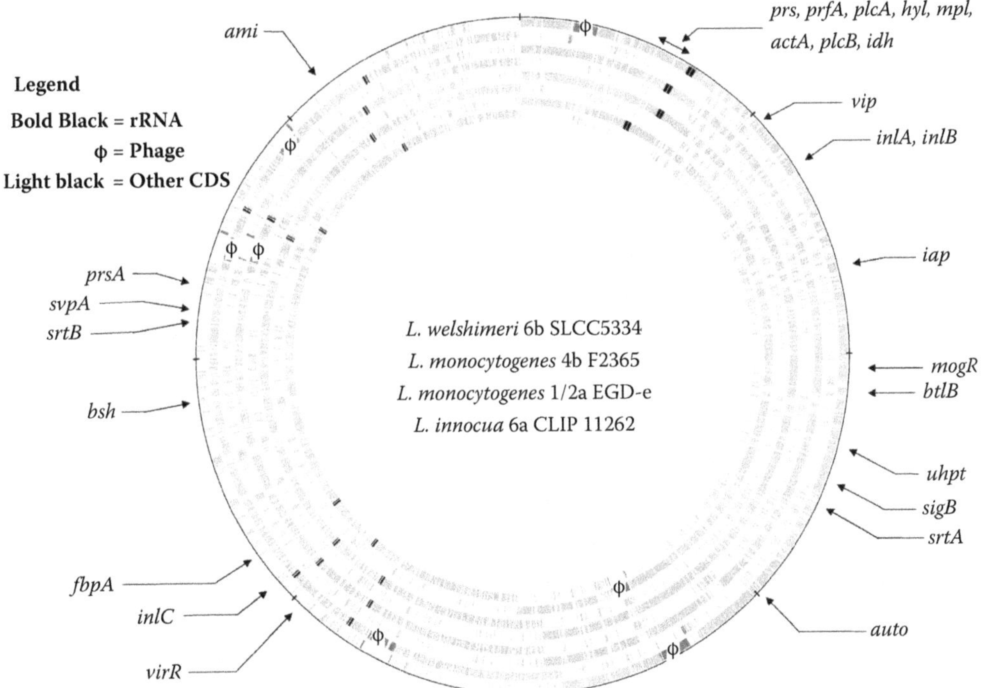

FIGURE 10.1 The comparative features of sequenced *Listeria* genomes. Outermost circle represents *L. innocua* CLIP 11262 followed by *L. monocytogenes* EGD-e, *L. monocytogenes* F2365, and *L. welshimeri* SLCC 5334. While *Listeria* genes with important functions are labeled, other coding sequences (CDS) are shown as vertical bars.

of those are absent from *L. innocua. L. monocytogenes* possesses more LPXTG proteins than any other Gram-positive bacteria whose genomes have been sequenced.[12]

An examination of the chromosomal loci harboring genes encoding LPXTG- and LRR-motif-containing proteins in *L. monocytogenes* revealed that many of these genes have been lost from the apathogenic species *L. innocua* and *L. welshimeri.* In contrast to the large number of internalins and LPXTG motif proteins present in *L. monocytogenes*, 17 and 20 out of the 25 LRR-motif-containing internalins of *L. monocytogenes* are absent in *L. innocua* and *L. welshimeri*, respectively.[14] For *L. welshimeri* and *L. innocua*, all of the currently studied internalin genes (*inlABCEFGHIJ*) were absent from their genome. In contrast to the genes encoding these surface protein families, genes coding for other cell wall–associated proteins, such as GW-motif-containing proteins, LysM-motif-containing proteins, and NLPC/P60-like proteins, were roughly constant in the genomes of all three species. The *L. welshimeri* genome contained the largest number of genes encoding lipoproteins, 69, versus 65 and 59 for *L. monocytogenes* and *L. innocua*. Unlike its counterpart in *L. monocytogenes*, the Ami protein in *L. welshimeri* is truncated and contains only two GW repeats, whereas the *L. monocytogenes* protein harbors four GW-motif-containing modules. Interestingly, the Ami protein in *L. innocua* lacks only one GW-containing repeat.[14]

Elements representing clustered, regularly interspaced palindromic repeats (CRISPRs) were identified at variable loci in the genomes of the *L. monocytogenes* serotype 1/2a strains and *L. innocua* CLIP 11262, but not in the genomes of the serotype 4b strains. The repeat sequence differs by only one nucleotide between *L. innocua* CLIP 11262 and *L. monocytogenes* strain F6854 at locus 1, but it is more variable at the two additional loci within strain F6854. The variable presence and absence of CRISPR elements in the *Listeria* lineage suggest that the presence of these elements may be the result of gene transfer events.[13]

The sequenced *L. monocytogenes* and *L. innocua* strains contain one and five prophages, respectively. *L. monocytogenes* strain F2365 does not have an intact prophage in the genome.[13] *L. welshimeri* harbors a prophage with strong homology to that in *L. innocua* located 2.6 Mb from its origin of replication, which is inserted within the region between the $tRNA^{Arg}$ and *ydeI* genes. In *L. ivanovii* the species-specific *Listeria* pathogenicity island 2 (LIPI-2) is also flanked by the genes for $tRNA^{Arg}$ and *ydeI,* suggesting that this region is an evolutionary "hot spot" of genome evolution for *Listeria* spp. A cluster of paralogous genes encoding for partial components of an F_0F_1-ATP synthase is translocated with respect to *L. monocytogenes, L. innocua,* and *L. welshimeri.*

The smaller size of the *L. welshimeri* genome is the result of deletions in genes involved in virulence and of "fitness" genes required for intracellular survival, transcription factors, LPXTG- and LRR-containing proteins, as well as 55 genes involved in carbohydrate transport and metabolism. Further analysis of the gene content of *L. monocytogenes* strains EGD-e (serotype 1/2a) and F2365 (serotype 4b), *L. innocua,* and *L. welshimeri* revealed 157 genes commonly absent in the sequenced *L. monocytogenes* F2365 serotype 4b, *L. innocua*, and *L. welshimeri* strains, but present in the *L. monocytogenes* EGD-e serotype 1/2a genome.[14] The localization of the deleted regions on the circular chromosomal map of *L. monocytogenes* indicates that the missing genes of *L. innocua* and *L. welshimeri* occur at the same loci within the chromosomes. Identification of strain-specific genes in each of the sequenced genomes suggests the gene expansion including horizontal gene transfer. Many of the genes present in *L. monocytogenes* and absent from *L. welshimeri* and *L. innocua* not only are the result of gene loss but also include genes acquired by horizontal gene transfer in *L. monocytogenes.* These genes include several insertions of unique genes that confer new adaptive properties for niche-specific environmental survival of *Listeria* spp. The examination of the *L. innocua* and *L. welshimeri* genomes also revealed that the majority of the genes acquired by horizontal transfer (also called alien genes) were located on the same region of the chromosome.

Gene clusters containing genes conferring bile and acid resistance present in *L. monocytogenes* are absent in *L. welshimeri* and *L. innocua*. However, only the *bilE* system is commonly present in all three species. The xylose operon in *L. welshimeri* consists of five genes, including a glycoside hydrolase family protein gene coding for a putative alpha-xylosidase, a xylose-proton symporter gene (*xylP*), a xylose isomerase gene (*xylA*), and a xylulose kinase gene (*xylB*) controlled by the transcriptional repressor *xylR* protein.[14] In *L. welshimeri* the presence of many uptake and utilization systems for energy sources found almost exclusively in plants and decaying vegetation suggests that the species has adapted to a saprophytic strategy to survive in its natural environment.[14]

Two specific secreted proteases encoding a trypsin-like serine/cysteine protease and an otherwise unassigned peptidase were detected for *L. welshimeri*.[14] Indeed, all of the *L. welshimeri* strains studied exhibit protease activity on skim milk agar plates, whereas *L. innocua* has no activity. The *L. monocytogenes* EGD-e strain was also protease positive. This novel property of *L. welshimeri* strains can be useful in distinguishing it from *L. innocua* in diagnostic tests.[14]

An unexpected genetic divergence between the sequences of an epidemic serovar 4b strain and nonepidemic *L. monocytogenes* EGD-e serovar 1/2a strain was observed, as about 8% of the sequences were serovar 4b specific.[17] These sequences comprise seven genes coding for surface proteins, two of which belong to the internalin family, and three genes coding for transcriptional regulators, all of which might be important in different steps of the infectious process.

Phylogenomic analysis based on whole genome sequences reveals the same phylogenetic relationship as determined by 16S-rRNA sequencing of *L. monocytogenes*, *L. innocua*, *L. welshimeri*, *L. seeligeri*, *L. ivanovii*, and *L. grayi*, which branch the genus in three main groups. The first group consists of the closely related species *L. monocytogenes*, *L. innocua*, and *L. welshimeri*; *L. welshimeri* exhibits the deepest branching of this group. *L. seeligeri* and *L. ivanovii* form the second group and *L. grayi* seems to be very distant from both groups.

Studies using analysis of 16S and 23S rRNA,[3] PCR-based DNA fingerprinting techniques,[7] or virulence locus and genome comparisons[7,12] indicated that a phylogenetically close relationship exists between *L. monocytogenes* and *L. innocua*, and that *L. innocua* possibly lost the virulence locus by deletion.[18] Sequence analysis of the different junction regions disclosed identical sequences among *L. monocytogenes* serovar 4a and the *L. innocua* strains, suggesting individual deletion events. The presence of these genes in the other *Listeria* species provides a clue that they were part of the genome of a common ancestor and that *L. innocua* evolved by successive gene loss from an ancestor of *L. monocytogenes* serogroup 4 strains.[17] The presence of the internalin gene in natural atypically hemolytic *L. innocua* strain suggests it is a direct descendent from *L. monocytogenes*.[19]

Excluding the prophage sequences, the genes specific to nonpathogenic strains (i.e., *L. innocua* CLIP 11262 and *L. welshimeri*) are absent from *L. monocytogenes* strain EGD-e as well as from the other three *L. monocytogenes* strains, suggesting that gene loss from a lineage ancestral to *L. monocytogenes* and *L. innocua* preceded the genomic diversification of *L. monocytogenes* into genomic divisions I and II.[14] The sequenced genomes exhibited high degree of similarity in the genomes in spite of isolation from different epidemiological backgrounds, having different genomic division, and serotype of the strains. These findings suggest that *L. monocytogenes* strains responsible for human and animal illness have high genomic stability and rely on a relatively small number of unique regions for antigenic diversity and epidemiologically relevant attributes.[13]

High-quality SNPs are found in the genomes of *L. monocytogenes* strains H7858 and F6854 when compared with *L. monocytogenes* strain F2365. SNPs result in a nonsynonymous (NS) change in amino acid sequence in *L. monocytogenes* strains H7858 and F6854. There are a higher number of NS-SNPs in cell envelope and cellular processes, as well as energy metabolism and transport.[13]

10.2.2 Plasmids

Frequently found in bacteria, plasmids are DNA molecules (whose sizes range from 1 to over 400 bp) that are separate from chromosome DNA. As plasmids possess at least an origin of replication (or *ori*), which is a starting point for DNA replication, they are capable of autonomous replication. That is, they can duplicate independently from the chromosomal DNA. In addition, plasmids have the ability to transfer to other bacteria via a process called conjugation (which provides a means for horizontal gene transfer). Plasmids often harbor genes or gene cassettes that confer a selective advantage to the parent bacteria. A plasmid that integrates into the chromosomal DNA is called an episome, which allows its duplication with every cell division of the host.

The *L. innocua* strain CLIP 11262 contains a plasmid pLI100 of 81,905 bp encoding genes required for resistance to heavy metals such as cadmium. The sequence of an *L. monocytogenes* plasmid, designated pLM80, from strain H7858 reflects its origin from *L. monocytogenes.* This 80-kb plasmid is populated by several different transposable elements that are not present in the chromosome, suggesting that the plasmid is a recent acquisition.[13] Plasmid pLM80 has a high level of sequence and gene organization similarity to the *L. innocua* CLIP 11262 plasmid pLI100.[12] The sequences of plasmids from *L. monocytogenes* J0161 serotype 1/2a, FSL J1-0194 serotype 1/2b, FSL R2-503 serotype 1/2b, FSL J2-071 serotype 4c, and FSL F2-515 serotype 1/2a are also available (http://www.broad.mit.edu/annotation/genome/listeria_group).

10.2.3 Comparative Phage Genomics

Bacteriophages (or phages) are virus-like agents that have the capacity to infect and destroy bacteria. Usually, they are made up of a nucleic acid core (e.g., ssRNA, dsRNA, ssDNA, or dsDNA measuring 5–500 kb in length) with an outer protein hull. Bacteriophages and plasmids may have played a role in gene acquisition in *Listeria*. Many bacteriophages specific for *Listeria* have been isolated (reviewed in Loessner and Calendar[20]), but despite the publication of the complete genome sequences of *L. monocytogenes* EGD-e and *L. innocua* CLIP11262,[12] the potential influence of temperate *Listeria* bacteriophages on the host cell phenotype is still not understood.

The DNA genomes of different *Listeria* bacteriophages infecting *L. monocytogenes* (A006, A500, A511, and P35), *L. innocua,* and *L. ivanovii* (B054, B025) were sequenced and analyzed. Phage A511, which can infect approximately 95% of all *L. monocytogenes* strains, belongs to serovars 1/2 and 4.[21,22]

The *Listeria* phages investigated feature genomes between 35.8 and 134.5 kb in size, with G+C contents between 35.5 and 40.8 mol%. Except for P35, the group of phages belonging to the *Siphoviridae* family (P35, A006, and A500) uncovers an overall similar genome organization, where open reading frames (ORFs) are organized into functional clusters in a life-cycle-specific manner, and are also reflected by the direction of transcription.[23,24] Phages A118 and A500 utilize programmed translational frame-shift to generate major capsid and tail proteins with different length C-termini, a decoding strategy first discovered in *Listeria* phage PSA.[20,24]

Phage B054 is a myovirus, and its larger genome features 80 predicted ORFs. On an overall basis, many of the B054 genes specify proteins with high similarity to phage-encoded proteins in the genome of *L. innocua* CLIP 11262,[12,13] pointing to a close relationship of B054 to the putative prophage ϕ11262.4 harbored by this strain.[13] The broad host range, strictly virulent phage P100, which can infect and kill a majority of *L. monocytogenes* strains, has a 131,384-bp genome and is predicted to encode 174 gene products and 18 tRNAs.[25] PSA, a temperate phage isolated from *L. monocytogenes* strain Scott A, consists of a linear 37,618-bp DNA featuring invariable, 3′-protruding single-stranded (cohesive) ends of 10 nucleotides.[24]

In contrast to the conserved genome arrangement observed for B054 and the *L. innocua* prophage ϕ11262.4, the genome of B025 apparently features a rather extensive mosaicism: the majority of the 65 predicted ORFs encode proteins with high similarity to *L. monocytogenes* phages A118, PSA, and to a (defective) prophage (prophage ϕ11262.6)[13] present on the bacterial chromosome of *L. innocua* CLIP 11262.[12] The P35 genome completely lacks any genes associated with establishing or controlling a lysogenic status.

The broad host range phage A511 belongs to the Myoviridae family.[26] Phage A511 features a 134.5-kb genome with a G+C content of 36 mol%, specifying 190 ORFs and 16 tRNAs. It has a relatively large genome highly similar to that of P100[25]; is unrelated to the smaller *Listeria* phages

A006, A500, P35, B025, and B054; and features intergenetic relationships to non-*Listeria* phages K and LP65.[27,28]

10.2.4 Population Structure and Lineages in *L. monocytogenes*

Although 12 serovars (i.e., 1/2a, 1/2b, 1/2c, 3a, 3b, 3c, 4a, 4b, 4c, 4d, 4e, and 7) are described for the species *L. monocytogenes*, about 98% of the strains isolated from patients are of serovars 1/2a, 1/2b, 1/2c, and 4b.[29] All major outbreaks of listeriosis and most of the sporadic cases are due to strains of serovar 4b, suggesting that strains of this serovar may possess unique virulence properties. These phenotypic differences among *L. monocytogenes* strains may be in part due to genetic differences among individual *L. monocytogenes* isolates.

Distinct patterns of presence or absence of genes among the *L. monocytogenes* and *Listeria* spp. strains tested were identified using a focused "*Listeria* biodiversity array" containing probes representative of specific sequences from two *L. monocytogenes* strains, EGD-e (serovar 1/2a) and CLIP80459 (4b), and one nonpathogenic *L. innocua* strain, CLIP11262.[17,29] These gene content patterns correlate with the previously defined three lineages (lineage I consisting of serovars 1/2a, 1/2c, 3a, 3c; lineage II of serovars 4b, 4d, 4e, 1/2b, 3b, 7; and lineage III of serovars 4a, 4c). This result together with the fact that many studies have found an association between *L. monocytogenes* serovars and various phenotypic characteristics, leading to the definition of three lineages among *L. monocytogenes* strains, suggests that these lineages mirror the evolution within the species due to an early divergence from ancestral *L. monocytogenes*.

Analysis of the distribution of the known virulence genes (*inlAB*, *prfA*, *plcA*, *hly*, *mpl*, *actA*, *plcB*, *uhpT*, and *bsh*) among the *L. monocytogenes* strains revealed that the known virulence factors are present in all the strains. However, the distribution of genes coding for putative surface proteins of the internalin/LPXTG/GW motif containing family belonging to the sequenced *Listeria* genomes (e.g., *L. monocytogenes* EGD-e of serovar 1/2a, and *L. monocytogenes* CLIP 80459) disclosed a pronounced heterogeneity, suggesting that some of them may play a role in virulence differences.[30]

10.2.5 Transcriptome-Based Profiling of Intracellular Gene Expression

L. monocytogenes is a facultative intracellular pathogen, which can breach the vacuolar compartment to gain access to the host cell cytosol. Identification of its genes that are expressed during invasive infection is important for understanding the infection processes. Examination of genome-wide adaptive changes during bacterial intracellular growth in eukaryotic cell lines uncovered survival properties of *L. monocytogenes* Δ*hly*Δ*plcA* and EGD-e strains inside cells and characteristics typical for adaptation and growth, respectively.[31] A total of 484 genes comprising ~17% of the total genome was differentially regulated to enable survival in the altered environment. Of these 484 genes, 301 were up-regulated and 182 were down-regulated during intracellular growth. Sixty-six genes were specifically up-regulated for adaptation in the vacuolar and 115 genes for growth in the cytosolic compartment of P388D1 cells.[31]

The expression profile of intracellular *L. monocytogenes* following invasion of epithelial cells was also investigated using whole genome DNA microarray.[32] Approximately 19% of the genes were found to be differentially expressed by at least 1.6-fold after 6 h of infection relative to their levels of transcription when grown in BHI medium. Among them were 279 up-regulated genes and 272 down-regulated genes, including species-specific genes, regulatory genes, and those of yet unknown functions. Strong transcriptional induction was observed for the PrfA-controlled virulence genes, confirming their indispensability for intracellular survival and replication. To validate the biological relevance of the intracellular gene expression profile, a random mutant library of *L. monocytogenes* was screened for intracellular growth deficiencies. Interfacing and extrapolation of the results of mutant screening and microarray data revealed that approximately 36% of all up-regulated genes were indeed required for listerial proliferation in the cytosol of epithelial cells.

Apart from the strong induction of all previously known PrfA regulated genes intracellularly, genes comprising the class I stress response (*dnaJ, dnaK, grpE, hrcA, groEL,* and *groES*); members of the CtsR-regulated class III stress responsive genes (*clpP, clpC, clpE, clpB*), which are ATP-dependent proteases; and 26 members of the class II stress response family were strongly induced during survival in both the vacuolar and cytosolic compartments of P388D1 cells. In addition, induction of listerial SOS responsive genes (*lexA, recA, radA*, lmo2676 [coding for a UV-damage repair protein], and *polIV* [lmo1975]) in the cytosol of the host cells was observed. Cumulatively these results suggest that *L. monocytogenes* may be experiencing stress during intracellular survival.

Approximately 30 up-regulated genes coding for ABC transporters, PTS systems, other transport systems, and genes known to be under carbon catabolite repression control (CCR) were identified, indicating that intracellularly replicating *Listeria* are mainly CCR repressed. Among the transcriptionally induced genes are those that are involved in the pentose phosphate cycle, providing evidence that this is a major catabolic pathway for the synthesis of necessary catabolic intermediates. Glycerol plays a role as carbon source for listerial growth in the cytosol of the host cells apart from glucose and phosphorylated glucose. The hypothesis is further strengthened by the strong up-regulation of *plcB* encoding a broad spectrum phospholipase, which could provide glycerol from host-derived phospholipids.[33] The up-regulation of the PrfA-dependent *hpt* gene encoding a hexose phosphate transporter and the down-regulation of the glycolysis genes together strengthened the assumption that the intracellular level of free glucose is low and that glucose is not the predominant carbohydrate source for *Listeria* growing inside host cells. A 55-kb gene cluster, ranging from lmo1142 (*pduS*) to lmo1208 (*cbiP*), encodes the factors for coenzyme B_{12} synthesis and for the B_{12}-dependent degradation of 1,2-propanediol and ethanolamine.[16] Ethanolamine, probably derived from phosphatidyl ethanolamine by the activity of PlcB, is used as an alternative nitrogen source for replication of *L. monocytogenes* as shown for *Salmonella typhimurium*.[34]

10.2.6 Regulons and Regulators

PrfA is a virulence regulator in *Listeria* and is the main switch of a regulon comprising virulence-associated loci scattered throughout the listerial chromosome, including members of the internalin multigene family.[35] A number of environmental and growth-phase dependent signals modulate expression of the virulence regulon via PrfA. The activating signals include high temperature (37°C),[36] stress conditions,[37] sequestration of extracellular growth medium components by activated charcoal,[38] contact with host cells,[39] and the eukaryotic cytoplasmic environment.[39,40]

The alternative sigma factor, sigmaB (σ^B), is a well-known regulator of the general stress response in Gram-positive bacteria such as *Listeria, Staphylococcus,* and *Bacillus.*[41,42] *L. monocytogenes* σ^B and PrfA are pleiotropic regulators of stress response and virulence gene expression. σ^B and PrfA contribute differentially to expression of the various internalins. Both σ^B and PrfA contribute to *inlA* and *inlB* transcription, while only PrfA contributes to *inlC* transcription and only σ^B contributes to *inlC2* and *inlD* transcription. The important role for σ^B in regulating expression of *L. monocytogenes* internalins suggests that exposure of this organism to environmental stress conditions, such as those encountered in the gastrointestinal tract, may activate internalin transcription. The σ^B operon consists of σ^B itself and seven regulators of σ^B genes (*rsb*). Interplay between σ^B and *prfA* also appears to be critical for regulating transcription of some virulence genes, including *inlA, inlB*, and *prfA*.[43]

In *L. monocytogenes*, a role for σ^B has been determined in response to several stresses; for example, it plays a role in acid resistance of stationary-phase cells, in oxidative and osmotic stress resistance, in the response to carbon starvation, and in growth at low temperatures.[41,44–45] *L. monocytogenes* displays an active acid tolerance response upon exposure to low, nonlethal pH and subsequent exposure to a lethal pH. Recently, the contribution of σ^B to growth-phase-dependent acid resistance and to the adaptive acid tolerance response in *L. monocytogenes* 10403S was analyzed by Ferreira et al.[46] Based on a specialized 208-gene microarray, 55 σ^B-regulated genes were identified in *L. monocytogenes*.[47]

A novel virulence regulator gene *virR* encoding a putative response regulator of a two-component system has recently been identified.[48] Deletion of *virR* severely decreased *L. monocytogenes* virulence in mice as well as invasion in cell-culture experiments. Using a transcriptomic approach, 12 genes regulated by VirR, including the *dlt*-operon, were identified. Another VirR-regulated gene homologous to *mprF*, which encodes a protein that modifies membrane phosphatidyl glycerol with l-lysine and with strong homology to the *mprF* of *Staphylococcus aureus*, has been identified.[49] *VirR* thus appears to control virulence by a global regulation of surface components' modifications and a second key virulence regulon in *L. monocytogenes*, after the *prfA* regulon.[48]

The alternative sigma factor σ^{54} is known to regulate several classes of genes involved in carbon and nitrogen metabolism, flagella biosynthesis, and virulence.[50] The role of the alternative σ^{54} factor, encoded by the *rpoN* gene, was investigated in *L. monocytogenes* by comparing the global gene expression of the wild-type EGD-e strain and an *rpoN* mutant.[51] Seventy-seven genes whose expression was modulated in the *rpoN* mutant as compared to the wild-type strain were identified. σ^{54} was also involved in the sensitivity to antibacterial peptides such as subclass IIa bacteriocins[52] and in osmotolerance in *L. monocytogenes*.[53]

L. monocytogenes has been found to encode a functional member of the *CodY* family of global regulatory proteins that is responsive to both GTP and branched chain amino acids,[54] indicating a role for *CodY* in *L. monocytogenes* in both carbon and nitrogen assimilation. The *degU* (lmo2515) gene encodes a putative response regulator,[55] which is a pleiotropic regulator involved in expression of both motility at low temperature and *in vivo* virulence in mice.[55,56]

Repression of flagellar motility genes in *L. monocytogenes* is mediated by a transcriptional repressor, MogR.[57] Repression by MogR is less stringent at low temperatures to allow for flagella production and motility. Deletion of *mogR* results in elevated transcription of *flaA*, encoding flagellin, relative to wild-type bacteria during broth culture at both room temperature and 37°C.[57] Intriguingly, despite producing high levels of *flaA* transcripts during growth at 37°C, MogR-negative bacteria produce reduced amounts of FlaA protein, indicating that *flaA* expression is subject to temperature-dependent post-transcriptional regulation.[57] MogR also represses *flaA* transcription irrespective of temperature during intracellular infection. This environment-specific repression of *flaA* by MogR may be necessary for full virulence, as MogR-negative bacteria exhibit a 250-fold increase in LD_{50} during *in vivo* infection of mice.[57] DNA binding and microarray analyses revealed that MogR represses transcription of all known flagellar motility genes by binding directly to a minimum of two TTTT-N5-AAAA recognition sites positioned within promoter regions.[58]

L. monocytogenes encodes only five sigma factors. The best characterized regulatory factor of *L. monocytogenes* is PrfA—a member of the Crp/Fnr family that is absent in *L. innocua*. PrfA activates most of the known virulence genes by binding to a palindromic PrfA recognition sequence (PrfA-box) located in the promoter region.[35] The Crp/Fnr family comprises 15 members in *L. monocytogenes* and 14 in *L. innocua*. The two largest families of regulatory proteins are the GntR-like regulators and the BglG-like antiterminators, many of which are associated with PTS. *Listeria* genomes encode histidine kinases and response regulators constituting two-component regulatory systems. The *L. monocytogenes and L. innocua* species express four different classes of stress proteins (of the HrcA- or σ^B-dependent Clp family)[59] and encode three paralogous cold shock proteins.

In *L. monocytogenes* and *L. innocua,* 209 and 203 transcriptional regulators (accounting for about 7% [209/2853 and 203/2821] of the total genomes) have been identified or predicted, respectively. Given that transcriptional regulators are specialized molecules responsible for controlling the expression of various proteins important for bacterial adaptation to niche environments and possibly for virulence, they may possess species-, group- or virulence specificity. Not surprisingly, comparative examination of *Listeria* DNA sequences led to the identification of species-specific transcriptional regulator genes in *L. monocytogenes* (e.g., *lmo0733*) and *L. innocua* (e.g., *lin0464*). In addition, several virulence-specific transcriptional regulator genes have also been isolated (e.g., *lmo0833, lmo1116, lmo1134,* and *lmo2672*). Using PCR primers derived from these genes resulted in the differentiation of *L. monocytogenes* virulent strains capable of causing mouse mortalities from avirulent serotype 4a strains that produce no mouse mortalities.

It was also of interest to note that even the newly identified, species-specific transcriptional regulator gene *lmo0733* demonstrates enormous structural variations between pathogenic and nonpathogenic *L. monocytogenes* serotypes. Based on Southern hybridization analysis, *lmo0733* is recognized as an *Hin*dIII band of 5.0 or 6.0 kb in the more pathogenic serotypes (e.g., 1/2a, 1/2c, and 4b) and as an *Hin*dIII band of 1.0 or 1.5 kb in the less pathogenic or nonpathogenic serotypes (e.g., 4a and 4c). These results suggest that apart from involvement by other virulence-associated genes and proteins, *L. monocytogenes* may require a complete copy of *lmo0733* gene for expression of full virulence, although the precise role of *lmo0733* in listeriosis still needs confirmation using mutant containing corresponding gene deletion. Thus, in addition to providing valuable targets for diagnostic application, subsequent investigation of species- and virulence-specific transcriptional regulator genes in *L. monocytogenes* may generate new details on the molecular mechanisms of listerial survival in their niche environments, as well as their pathogenicity, ultimately contributing to more effective control and prevention strategies against bacterial diseases.

10.2.7 Transporters

Profiling the metabolic capabilities encoded by the genome of *L. monocytogenes* strain F2365 uncovered a range of substrate utilization and transporter abilities—traits that are also present in other *L. monocytogenes* strains sequenced. The *L. monocytogenes* genomes have glycolysis and pentose phosphate pathways, and have genes for transport and utilization of a number of simple and complex sugars, including fructose, rhamnulose, rhamnose, glucose, mannose, chitin, sucrose, cellulose, pullulan, trehalose, and tagatose. These sugars are largely associated with the environments where *L. monocytogenes* is harbored, and conservation of genes for substrate utilization provides subtle clues regarding the survival and growth of the organism.[13] The ability of *Listeria* spp. to colonize and grow in a broad range of ecosystems correlates with the presence of 331 genes encoding different transport proteins (accounting for 11.6% of all predicted genes of *L. monocytogenes*). Interestingly, 88 (26%) of the 331 transporter genes are devoted to carbohydrate transport, mediated by phosphoenolpyruvate-dependent phosphotransferase systems (PTSs) and corresponding to 39 putative complete or incomplete enzyme II permeases.[12] Carbohydrates, in particular β-glucosides, have a remarkable impact on the virulence of *L. monocytogenes*. Eight enzyme II permeases, five of which are predicted to be specific for β-glucosides, are absent from *L. innocua*. The *L. monocytogenes* EGD-e-specific PTS could be implicated in virulence.[12]

10.2.8 Bioinformatics

Large-scale genomic analyses necessitate the development of bioinformatics tools for predictions, comparisons, and visualization. A bioinformatics tool, GenomeViz, is available to visualize the circular architecture of *Listeria* and other bacterial genomes.[60] This helps produce broad overviews of distinctive genomic features across several genomes. To visualize conservation of genes across related genomes, GECO, a browser-based program, was developed.[61] GECO provides visualization at the level of individual genes within pinned regions of multiple genomes, thus allowing for direct gene-for-gene comparisons within a selected region. The software package currently offers visualization for all available microbial genomes (~400).[61] A software pipeline, AUGUR, to automate analysis of all major surface protein types in Gram-positive bacteria and their comparison has recently been developed.[62] LEGER,[63] a program to support functional genome analyses by combining information obtained by applying bioinformatic methods and from public databases to improve the original annotations, was developed for listerial genomes.

10.2.9 Genomics-Based Diagnostics

Integration of the knowledge derived from genomics is predicted to have a tremendous impact on approaches to diagnostics for infectious diseases. Advances in genomics and proteomics increasingly contribute to the understanding of the pathogens, their virulence factors and species-specific niche. Epidemiological investigation requires identification of specific isolates, usually done by a combination of serotyping and subtyping using pulsed-field gel electrophoresis (PFGE). PulseNet USA, a molecular surveillance network for food-borne infections in the United States, has been instrumental in detection, investigation, and control of numerous outbreaks caused by food-borne pathogens. The method employed in the network is PFGE, which allows highly discriminatory molecular subtyping of *Listeria* spp.

Epidemiological studies and analysis of putative virulence genes have shown that *L. monocytogenes* has diverged into several phylogenetic divisions. Similar divergence has occurred for many genes that influence niche-specific fitness and virulence, and identifying these differences may offer new opportunities for the detection, treatment, and control of this important pathogen. Mixed-genome microarrays may act as a tool for deriving biologically useful information and for identifying and screening genetic markers for clinically important microbes[64] and, unlike PFGE, for identifying specific genes associated with the infecting pathogen.[65] Several microarray-based strategies were developed to differentiate among the six *Listeria* species[66] and for discrimination among *L. monocytogenes* serovars[65] and phylogenetic lineages.[67] Microarray-based assays were also applied to investigate the genome evolution within the genus *Listeria*[17] and for identification of natural atypical *L. innocua* strains harboring genes of the LIPI-1.[68] Thus, whole-genome sequence data have yielded new insights into epidemiological surveillance. Electronic database libraries of the different genomic profiles will enable continuous surveillance of infections and detection of possible infection clusters at an early stage.

10.3 EVOLUTION OF VIRULENCE

The relatively small number of species comprising the genus *Listeria* and the clear differences that separate pathogenic from nonpathogenic species make these bacteria attractive models for examining the evolution of pathogenicity in this genus. Also, the spectrum of ecological niches (i.e., from the abiotic environment to the intracellular compartments of the infected host cell occupied by these bacteria) raises questions as to how these features have evolved within these species.

Phylogenetic analysis based on the 16S and 23S-rRNA coding genes as well as the *prs*, *ldh*, *vclA*, *vclB,* and the *iap* gene indicated that five out of six *Listeria* species are divided in two lineages. *L. monocytogenes* and *L. innocua* form one group while the second group includes *L. welshimeri, L. ivanovii*, and *L. seeligeri*. In the latter group, *L. welshimeri* appears to be the most distant species while *L. grayi* forms the deepest branch within the genus.[69] The phylogenetic tree shows one topological conflict with a cluster analysis of the genus *Listeria* calculated from the results of MLEE.[70] The MLEE analysis placed *L. innocua* and *L. welshimeri* as sister species in the *L. monocytogenes* group while *L. seeligeri* and *L. ivanovii* formed the other group.

The genes responsible for the intracellular life cycle of the bacterium are located within a locus of the chromosome between *prs* and *ldh,* which is referred as LIPI-1 (*Listeria* pathogenicity island 1).[71] Following the lysis of vacuole mediated by two secreted proteins, listeriolysin (LLO) and phosphatidylinositol-specific phospholipase C (PlcA),[72,73] the bacterium replicates in the cytosol. The bacterial surface protein ActA leads to actin-based movement and finally to cell-to-cell spread where the bacterium upon contact with the plasma membrane induces pseudopod-like protrusions that invaginate into a neighboring cell and can be ingested.[74,75]

Remarkably, *L. monocytogenes* reveals host cell tropism due to the receptor specificity of InlA for human E-cadherin[76,77] and InlB for the globular head of the complement factor C1q (gC1q-R), together with the hepatocyte growth factor receptor (c-Met) and with glycosaminoglycans including heparan sulphate.[76] The lysis of the two plasma membranes is mediated by LLO together with the phophatidyl-choline-specific phospholipase C (PlcB), which enables the bacterium to infect the next cell.[71]

10.3.1 Virulence Gene Cluster

Pathogenicity of *L. monocytogenes* and *L. ivanovii* is enabled by an approximately 9-kb virulence gene cluster (vgc), which is alternatively referred as LIPI-1 (*Listeria* pathogenicity island 1).[71] The vgc is also present in a modified form in *L. seeligeri*. In these three species, the vgc is stably inserted at the same chromosomal position and there are no obvious traces of mobility genes, insertion sequences (ISs), direct repeats, and target sequences normally used for the integration of mobile genetic elements. The virulence gene cluster is, however, totally absent from the other three described *Listeria* species: *L. innocua*, *L. welshimeri*, and *L. grayi*, which are all nonpathogenic.[18,78] The vgc consists of three transcriptional units. The central position is occupied by the *hly* monocistron, encoding a pore-forming toxin LLO of the cholesterol-binding, "thiol-activated" family. Its product, LLO, is an essential virulence factor and its absence leads to total avirulence.[71] Upstream and divergent from *hly* lies the *plcA–prfA* operon. The *plcA* encodes a phosphatidylinositol-specific phospholipase C, which synergizes with LLO and PlcB in the destabilization of primary phagosomes.[79] PrfA is a virulence regulator in *Listeria* and is the main switch of a regulon comprising virulence-associated loci scattered throughout the listerial chromosome, including members of the internalin multigene family.[35]

Downstream from *hly* is a 5.7-kb operon comprising three genes: *mpl*, *actA*, and *plcB*.[71] The latter lecithinase requires a metalloprotease encoded by *mpl* for proper maturation. The *actA* gene, endowing *Listeria* with an actin assembly capability to drive intracellular movement within host cells, is part of the virulence gene cluster. Mpl is, like PlcB, a zinc-metalloprotease. PlcB cooperates with LLO in the disruption of the primary vacuoles formed after the phagocytosis of extracellular *Listeria*. The current organization of the virulence gene cluster loci of the six *Listeria* species raises questions about the evolutionary history of this chromosomal region. The vgc lies within the *ldh* and *prs* genes. *ldh* codes for lactate dehydrogenase (~310 amino acids). The region encoding the last 134 amino acids, representing the last two of the six conserved amino acid blocks, was present in all six *Listeria* species. *prs* encodes phosphoribosyl pyrophosphate synthetase. *vclB* is a conserved protein of unknown function found in all six listerial species and part of the vgc.

In addition to the vgc, other virulence genes have been identified that scatter elsewhere in the genome of *L. monocytogenes* and *L. ivanovii*. Multiple internalins have been identified in both *L. monocytogenes* and *L. ivanovii* and, by sequence analysis, also in *L. innocua*. In addition, the "invasion associated protein" (Iap), encoded by the *iap* gene, has been implicated to be important in maintaining the invasive phenotype in mouse fibroblasts, hepatocytes, and macrophages. However, Iap, also termed p60, acts as a murein hydrolase necessary for proper cell division.[80,81]

Another species-specific *Listeria* pathogenicity island 2 (LIPI-2) was described for *L. ivanovii*, which contains, apart from the membrane-damaging virulence factor SmcL, 10 genes encoding for internalin proteins located on a 22-kb chromosomal locus.[82,83] LIPI-2 is inserted into a tRNA locus and is unstable; half of it deletes at approximately 10^{-4} frequency with a portion of contiguous DNA. The region between the core genome loci *ysnB-tRNA*(arg) and *ydeI* flanking LIPI-2 contained distinct gene complements in the different *Listeria* spp. and even serovars of *L. monocytogenes*, including remnants of the PSA bacteriophage *int* gene in serovar 4b, indicating it is a hot spot for horizontal genome diversification. LIPI-2 is conserved in *L. ivanovii* subspecies *ivanovii* and *londoniensis*, suggesting an early acquisition during the species' evolution.[82] The expressional subordination of most LIPI-2 genes to the LIPI-1-encoded virulence regulator, PrfA, suggests that LIPI-2 is a more recent development than the LIPI-1. The LIPI-2 appears to play an important role in *L. ivanovii* as it is highly conserved and possibly had a significant impact on the organism's infectious ecology.[82]

The generally consistent tree topologies calculated from gene sequences of different chromosomal loci imply that the analyzed chromosomal regions including the genes flanking the virulence cluster have been shared by the common ancestor of the genus *Listeria*. Since the natural ecological niches of all Listeriae overlap—the six species are, for example, found in soil, rotting vegetation, sewage, contaminated waters of rivers and estuaries, and could also be detected in the intestinal tracts of healthy animals[84]—genetic exchange among the species is theoretically possible. Keeping in mind the phylogeny of the genus *Listeria*, one plausible scenario would be that one of the three respective *Listeria* species acquired the virulence gene cluster laterally[18] and subsequent changes in the progenitor strains led to the emergence of the other two species. In contrast to the classic definition of a "pathogenicity island" acquired en bloc from a foreign donor, the virulence gene cluster of the different listerial species does not show an atypical G+C content compared to housekeeping genes or the total genomic G+C values of the respective species.

The genetic structure and transcriptional organization of the virulence gene cluster are identical in *L. monocytogenes* and *L. ivanovii*.[78] The corresponding gene homologs, however, show a significant degree of divergence (73–78% identity at the DNA level[85,86]) compatible with the genetic distance separating the two pathogenic *Listeria* spp. This divergence is even more pronounced for the *actA* gene; its encoded protein is only 34% identical in the two species despite having the same function in actin-based motility.[87,88]

The absence of vgc in *L. innocua* is of particular relevance because this species is very closely related to *L. monocytogenes*. Thus, if the PrfA virulence gene cluster was indeed present in a common listerial ancestor, the only reasonable explanation for its absence in *L. innocua* is that it excised from the chromosome of this species at a later stage in the evolution of the genus *Listeria*. Recently, atypical isolates of *L. innocua* have been characterized that surprisingly contain the entire vcl locus embedded into an otherwise typical *L. innocua* genetic background.[68] It was postulated that these isolates are intermediates in the evolution to typical *L. innocua*, and that the cluster has originally been acquired from another microorganism by a transposition-like event. This was based on the finding of a 16-bp Tn1545 integration consensus sequence flanking the vcl locus.[68,89] In another study, natural atypical nonhemolytic *L. seeligeri* isolates have been discovered.[90] These isolates were identified as *L. welshimeri* by phenotypic tests, which exhibited discrepant genotypic properties in a well-validated *Listeria* species identification oligonucleotide microarray. However, the microarray identified these isolates as being atypical *hly*-negative *L. seeligeri* isolates. The *prs-prf*A cluster-*ldh* region of the *L. seeligeri* isolates is approximately threefold shorter due to the loss of *orf*D, *prf*A, *orf*E, *plc*A, *hly*, *orf*K, *mpl*, *act*A, *dplcB*, *plcB*, *orfH*, *orfX*, and *orfI*.[90] It was speculated that a free-living ancestor of *L. seeligeri* that did not contain virulence-associated genes gave rise to the currently prevailing typical hemolytic strains of *L. seeligeri* and that the nonhemolytic *L. seeligeri* strains described represent an atypical biotype. Alternatively, nonhemolytic biotype strains of *L. seeligeri* arose by secondary loss of the LIPI-1 virulence cluster genes by some type of genetic transfer. Theoretically, the secondary loss of the LIPI-1 region might have been possible through "genetic garbage" removal or adaptive gene loss.[91,92] As *L. seeligeri* is typically an avirulent free-living bacterium, the absence of such a large useless gene cluster might not have been vital for the survival of the atypical strains.

Although we cannot know what exactly constituted the "ancestral" virulence gene cluster, it must have included at least *prfA, plcA, hly, mpl, actA, plcB, vclX*, and *vclY*; the latter present completely in *L. seeligeri* but only as relict sequences in *L. monocytogenes* and *L. ivanovii*. The linkage of PrfA control with an internalin-like gene to the virulence gene cluster may represent an ancestral arrangement that gave rise to the PrfA-controlled, internalin genes found widely dispersed in the present day genomes of *L. monocytogenes* and *L. ivanovii*.[69] The origin of this virulence gene cluster remains a matter of speculation. From its genetic composition, it seems clear that its source was in the phylogenetic division of Gram-positive bacteria in which the genus *Listeria* is located. Deletion

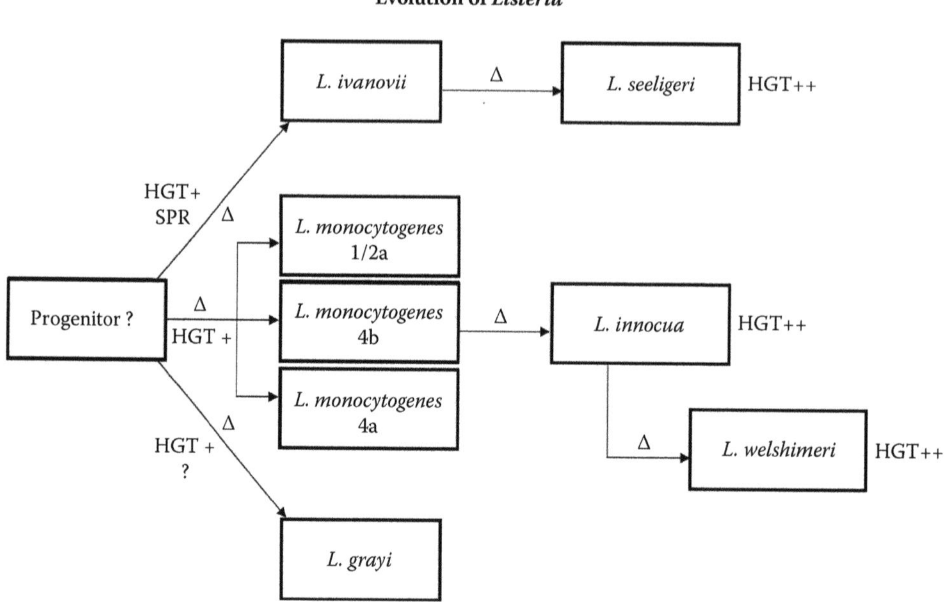

FIGURE 10.2 Schematic representation of evolution of the *Listeria* species. HGT: horizontal gene transfer; SPR: specific regions.

of the entire virulence gene cluster from *L. innocua* and *L. welshimeri* chromosomes is consistent with the gene cluster (or a substantial part of it) belonging to a horizontally transferable element.

Thus, a major evolutionary force leading to the generation of species within the genus *Listeria* probably involved genome reduction from a progenitor strain already harboring the virulence genes. The formation of distinct species and lineages was probably assisted by the gain of species- and strain-specific genes through horizontal gene transfer largely mediated by bacteriophage transduction. Figure 10.2 shows a schematic representation of the evolution of the *Listeria* species.

10.3.2 Novel Virulence Genes Specific to *L. monocytogenes*

The recent publication of the genome sequences of pathogenic and the nonpathogenic species of *Listeria* is a major step toward the identification and characterization of genetic loci involved in stress adaptation and virulence. However, functional genomics approaches are now required to identify stress-responsive systems from the large number (35.3%) of ORFs for which no function is currently known.

Major listerial virulence factors have been identified by classical genetics. The genetic basis of *L. monocytogenes* virulence can now be deciphered by exploitation of the genome sequences of serovars of *L. monocytogenes* and the closely related nonpathogenic species. It can be achieved by the analysis of genes that are present in *L. monocytogenes* and absent from nonpathogenic species, the study of genes that encode potentially interesting proteins, and the use of whole genome DNA arrays to screen a large population of strains from different *Listeria* species and to identify genes.

Efficient replication *in vivo* is essential for a microparasite to colonize its host and the understanding of the molecular mechanisms by which microbial pathogens grow within host tissues. *L. monocytogenes* exploits hexose phosphates (HP), mediated by Hpt, from the host cell as a source of carbon and energy to fuel fast intracellular growth.[93] Hpt is a bacterial homolog of the mammalian translocase that transports glucose-6-phosphate from the cytosol into the endoplasmic reticulum in the final step of gluconeogenesis and glycogenolysis, which upon entry into host cells induces a set of virulence factors required for listerial intracellular parasitism. Loss of Hpt resulted in impaired

L. monocytogenes intracytosolic proliferation and attenuated virulence in mice. Hpt is the first virulence factor specifically involved in the replication phase of a facultative intracellular pathogen.[93]

The ability to colonize the gall bladder has been shown to be an important feature of virulent *L. monocytogenes*. Comparison of the *L. monocytogenes* and *L. innocua* genomes has revealed the presence of an *L. monocytogenes*-specific putative gene encoding a bile salt hydrolase (BSH).[94] The *bsh* gene encodes a functional intracellular enzyme in all pathogenic *Listeria* species and is positively regulated by PrfA. Moreover, Bsh activity increases at low oxygen concentration. Deletion of *bsh* results in decreased resistance to bile *in vitro*, lowered bacterial fecal carriage after oral infection of guinea pigs, and reduced virulence and liver colonization after intravenous inoculation of mice.

A novel bile exclusion system that plays an essential role in intestinal colonization and virulence of *L. monocytogenes* has been identified.[95] In silico analysis of the *L. monocytogenes* EGDe genome disclosed a two-gene operon (formerly *opuB*), designated *bilE* (bile exclusion), exhibiting significant sequence similarity to members of the betaine-carnitine-choline transporter (BCCT) family. The operon is preceded by consensus σ^A- and σ^B-dependent promoter-binding sites and is transcriptionally up-regulated at elevated osmolarities and reduced temperatures. The *bilE* coordinately regulated by σ^B and PrfA, represents a new and important virulence factor in *L. monocytogenes*.[95] The *btlA* gene representing a membrane protein encodes a putative secondary transporter of the major facilitator superfamily. It is not involved in pathogenicity; however, it is essential for the maintenance of homeostasis under stress conditions, particularly bile.[96] Another bile-associated locus, *btlB*, also contributes to gastrointestinal persistence and bile tolerance of *L. monocytogenes*.[97] Analysis of deletion mutants revealed a role for all three genes in resisting the acute toxicity of bile and bile salts, particularly glycoconjugated bile salts at low pH.

Adherence of pathogenic microorganisms to the cell surface is a key event during infection. The autolysin Ami contributes to the adhesion of *L. monocytogenes* to eukaryotic cells via its cell wall anchor.[98] The *aut* gene, expressed independently of the virulence gene regulator PrfA, encoding a surface protein with an autolytic activity, is required for entry of *L. monocytogenes* into cultured nonphagocytic eukaryotic cells.[99] It contains an N-terminal autolysin domain and a C-terminal cell wall-anchoring domain made up of four GW modules.

FbpA, a fibronectin-binding protein, was shown to be required for intestinal and liver colonization after oral infection of transgenic mice expressing human E-cadherin. *fbpA* encodes a 570-amino-acid polypeptide that has strong homologies to atypical fibronectin-binding proteins. FbpA behaves as a chaperone or an escort protein for two important virulence factors, LLO and InlB, and appears as a novel multifunctional virulence factor of *L. monocytogenes*.[100]

Two genes encoding putative sortases, termed *srt*A and *srt*B, were identified in the genome of the intracellular pathogenic bacterium *L. monocytogenes*. Inactivation of *srtA* abolishes anchoring of the invasion protein InlA to the bacterial surface. It also prevents the proper sorting of several other peptidoglycan-associated LPXTG proteins.[101] The *srtB*, which seems to have only two substrates carrying an NXZTN motif instead of LPXTG, is not required for the infectious process in mice.[102]

L. monocytogenes encodes potential Fur-regulated iron transporters at 2.031 Mb (the *fur-fhu* region), 2.184 Mb (the *feo* region), 2.27 Mb (the *srt*B region), and 2.499 Mb (designated *hup*DGC region). Chromosomal deletions in the *fur-fhu* and *hup*DGC regions diminished iron uptake from ferric hydroxamates and hemin/hemoglobin, respectively. In the former locus, deletion of *fhu*D (lmo1959) or *fhu*C (lmo1960) strongly reduced [59Fe]-apoferrichrome uptake. Deletion of *hup*C (lmo2429) eliminated the uptake of hemin and hemoglobin, and decreased the virulence of *L. monocytogenes* 50-fold in mice.[103]

Flagellar structures have been shown to participate in virulence in a variety of intestinal pathogens. Two potential flagellar genes of *L. monocytogenes*—lmo0713, encoding a protein similar to the flagellar basal body component FliF, and lmo0716, encoding a protein similar to FliI—have been described.[104] Expression of *fli*F and *fli*I appears to be down-regulated at 37°C, like that of *fla*A, encoding flagellin. Deletion of either *fli*F or *fli*I abolishes bacterial motility and flagella production, impairs adhesion and entry into nonphagocytic epithelial cells, and also reduces uptake by bone marrow-derived macrophages.

10.3.3 Internalins

The first LPXTG protein identified in *L. monocytogenes* was internalin A (InlA).[105] InlA functions as an invasin, mediating bacterial internalization by epithelial cells. The host cell receptor for InlA is E-cadherin, and the InlA–E-cadherin interaction promotes specific binding and entry of *L. monocytogenes* into epithelial cells. The second invasion protein, internalin B (InlB), does not contain an LPXTG motif. InlA and InlB belong to a family of proteins characterized by an NH2-terminal domain containing LRRs. Seven other members of this family have already been identified.[71,106–108] These proteins are members of the so-called internalin multigene family, which contains proteins with an amino-terminal LRR domain followed by a conserved inter-repeat region (IR), several other repeats, and the LPXTG sorting signal. Interestingly, in one isolate the *inlH* gene is replaced by two internalin-encoding genes, *inlC2* and *inlD*, suggesting that *inlH* was generated by a recombination event.[106,109,110] The genome sequence of *L. monocytogenes* revealed, in addition to the five previously known LPXTG anchored internalin proteins, 14 ORFs that encode proteins also containing a signal peptide, an LRR domain, and a carboxy-terminal LPXTG motif. All in all, at least 25 internalins have been identified or predicted in *L. monocytogenes* serotype 1/2a strain EGD-e (Table 10.2).

Ami is an *L. monocytogenes* surface-associated protein implicated in the adhesion to eukaryotic cells and has eight GW modules. The presence of a higher number of GW modules in Ami compared with InlB is probably responsible for the more efficient anchoring of Ami to the bacterial cell surface.[111]

Except for InlB, which is loosely attached to the bacterial surface, and InlC, which is secreted, the five other LRR proteins have a Leu-Pro-X-Thr-Gly (LPXTG) motif that mediates their covalent linkage to peptidoglycan.[112] The *L. monocytogenes* genome sequence disclosed the presence of 41 proteins containing an LPXTG motif, 19 of which belong to the LRR/internalin family. *L. monocytogenes* contained more LPXTG proteins than any other Gram-positive bacteria whose genomes have been sequenced. InlB, absent from *L. innocua*, and the adhesion protein Ami are attached to lipoteichoic acid via GW modules.

L. monocytogenes expresses surface proteins covalently anchored to the peptidoglycan by sortase enzymes. Inactivation of *srtA* attenuates *Listeria* virulence in mice.[101] The postgenomic approach identified InlJ as a new virulence factor among the proteins belonging to the internalin family in *L. monocytogenes*[30] encoding an LRR protein that is structurally related to the listerial invasion factor internalin. The *inlJ* deletion mutant is significantly attenuated in virulence after intravenous infection of mice or oral inoculation of transgenic hEcad mice. The consensus sequence of the LRR defined a novel subfamily of cysteine-containing proteins belonging to the internalin family.[30]

By comparative genomics, *vip*, a gene that encodes an LPXTG surface protein absent from nonpathogenic *Listeria* species, has been identified.[113] It is positively regulated by PrfA. Vip is anchored to the *Listeria* cell wall by sortase A and is required for entry into some mammalian cells. The cellular receptor for Vip is the endoplasmic reticulum (ER) resident chaperone Gp96. The Vip-Gp96 interaction is critical for bacterial entry into some cells. Comparative infection studies using oral and intravenous inoculation of nontransgenic and transgenic mice expressing human E-cadherin demonstrated a role for Vip in *Listeria* virulence.[113]

The characterized internalins of *L. monocytogenes*, with the exception of InlC, are larger than those of *L. ivanovii* and are covalently bound or associated to the cell wall via their additional C-terminus.[112] The smaller, secreted internalins known so far are all under strict PrfA control,[107,114,115] whereas only two of the larger internalins of *L. monocytogenes* (InlAB) are partially controlled by PrfA.[108,109] Most of the known internalin genes of *L. monocytogenes* and *L. ivanovii* reside in numerous and diverse locations in their respective genomes. Many of these genes are

TABLE 10.2
Characteristics of *L. monocytogenes* EGD-e Internalin Family

Internalin	Subfamily[a]	Protein Size (a.a.)	Function	Ref.
InlG	LPXTG	490	Postintestinal colonization of host cells	113
InlE	LPXTG	499	Postintestinal colonization of host cells	105
Lmo1136	LPXTG	539	Unknown	
InlH	LPXTG	548	Postintestinal colonization of host cells	113
Lmo0610	LPXTG	589	Unknown	
Lmo1289	LPXTG	593	Unknown	
Lmo1290	LPXTG	598	Unknown	
Lmo0514	LPXTG	605	Unknown	
Lmo2026	LPXTG	626	Unknown	
Lmo0331	LPXTG	633	Unknown	
Lmo0732	LPXTG	638	Unknown	
Lmo0801	LPXTG	646	Unknown	
InlA	LPXTG	800	Entry into epithelial cells	100
InlF	LPXTG	821	Postintestinal colonization of host cells	113
Lmo0171	LPXTG	832	Unknown	
InlJ (Lmo2821)	LPXTG	851	Postintestinal colonization of host cells	30
Lmo2396	LPXTG	940	Unknown	
Lmo0327	LPXTG	1349	Unknown	
InlI (Lmo0333)	LPXTG	1778	Unknown	30
InlB	GW	630	Entry into hepatocytes, fibroblasts and epithelioid cells	71
InlC	Secreted	296	Postintestinal colonization of host cells	107
Lmo2445	Secreted	300	Unknown	
Lmo2027	Secreted	367	Unknown	
Lmo2470	Secreted	388	Unknown	
Lmo0549	Secreted	673	Unknown	

[a] In accordance with their confirmed and/or predicted spatial relationship with the bacterium, *L. monocytogenes* EGD-e internalins can be separated into three subfamilies. The LPXTG internalin subfamily anchors to the cell wall covalently; the GW internalin subfamily loosely interacts with the bacterial surface via GQ modules; and the secreted internalin subfamily possesses no surface-anchoring domains.

present in multiple, divergent, tandem copies: *inlAB,*[105] *inlGHE*[110] or *inlC2DE,*[109] *i-inlDC,*[116] and *i-inlFE.*[115] In contrast to InlA, the proteins InlE, InlF, InlG, and InlH are not involved in invasion, but are important for the colonization of host tissues *in vivo.*[110,117]

Recent studies[117,118] indicated that *L. monocytogenes* internalin genes are highly diverse with distinct evolutionary histories, their sequences cluster consistent with the phylogenetic lineages of *L. monocytogenes*, and both intragenic recombination and positive selection have contributed to the evolution of internalins. The diverse features of *L. monocytogenes* internalin genes can be usefully exploited for differentiation of virulent serotypes/strains from avirulent serotypes/strains. For example, *inlJ* has been found in all *L. monocytogenes* serotypes but 4a and possibly 7, and *inlC* in all *L. monocytogenes* serotypes but 4a and some 4c as well as in *L. ivanovii.* It appears that *L. monocytogenes* serotypes (all serotypes but 4a) harboring *inlJ* and/or *inlC* genes are capable of causing mouse mortality via intraperitonel route, while serotype 4a without these genes is unable to do so. Therefore, by incorporating the *inlA* gene primers for species recognition and the *inlC* and *inlJ* primers for virulence determination in a multiplex PCR format, it becomes feasible to confirm the species identity and virulence potential of *L. monocytogenes* rapidly and simultaneously (see chapter 8).

Nonetheless, it should be noted that although detection of *inlC* and/or *inlJ* genes in a given *L. monocytogenes* strain by PCR implies its potential virulence and its ability to cause mouse mortality through intraperitoneal inoculation, it does not indicate the certainty of this strain to cause disease in humans via the conventional oral ingestion. Considering that three serotypes (1/2a, 1/2b, and 4b) account for a majority of *L. monocytogenes* isolates from human cases of listeriosis, many serotypes (e.g., 1/2c, 3a, 3b, 3c, 4c, 4d, and 4e) that are recognized by the multiplex PCR targeting *inlC* and *inlJ* genes may encounter difficulty crossing the host's intestinal or other barriers during infection. An implication from the PCR results is that while *L. monocytogenes* serotypes other than 1/2a, 1/2b, and 4b are not commonly associated with human listeriosis, due possibly to their inability to efficiently cross human intestinal and other barriers, they (with the possible exclusion of serotype 4a) may have the potential to cause disease in humans whose immune functions are suppressed, or who are exposed to these serotypes via intravenous transfusions or open wounds.

10.3.4 Surface Proteins

Several listerial surface proteins play key roles in listerial interactions with mammalian host cells. *L. monocytogenes* EGD-e encodes a total of 133 surface proteins: 41 LPXTG proteins, nine GW proteins with a signal peptide, 11 hydrophobic tail proteins, four p60-like proteins, and 68 lipoproteins. Thus, at least 4.7% of the coding capacity of the genome is dedicated to surface proteins, with 1.4% being LPXTG proteins. Moreover, 22.6% (30/133) of cell-surface protein-encoding genes are absent from *L. innocua* CLIP 11262, whereas only 9.5% of the total *L. monocytogenes* genes are absent from *L. innocua*, indicating that the main differences between these two species lie in their surface protein composition.[119] In *L. monocytogenes* surface proteins included proteins that, like ActA,[120] have a signal sequence and a hydrophobic COOH-terminal region that may anchor them to the cell membrane and p60-like proteins.[12] *L. monocytogenes* and *L. innocua* encoded 68 putative lipoproteins, predicted on the basis of their characteristic signal sequences. A summary of predicted surface proteins in the sequenced *Listeria* genomes is given in Table 10.3.

Apart from strains with a higher potential to cause disease, there exist strains that seem to have lower virulence in the species *L. monocytogenes.* Most interestingly, the gene content comparison using the DNA array showed that in *L. monocytogenes* of serovar 4a and 4c, belonging to lineage III and mainly described as animal pathogens,[121,122] 13 of the 25 *L. monocytogenes* specific surface proteins, including all known internalins except *inlAB,* were missing.[17] Surface proteins are known to be important for host–pathogen interactions. The fact that many surface proteins are missing from strains of this lineage might indicate adaptation to another niche, the animal production environment, than lineage II and I isolates. Analyses of over 400 strains identified among *L. monocytogenes* serovar 1/2a strains isolated from food processing plants, showed one strain dominant in pork products that was not identified among the human isolates tested. DNA array characterization of these

TABLE 10.3
Surface Proteins Predicted in *Listeria* Genomes by AUGUR

Protein Function	*L. innocua CLIP 11262*	*L. monocytogenes EGD-e*	*L. monocytogenes F2365*	*L. welshimeri SLCC 5334*
GW modules	8	9	8	9
Lipoproteins	59	65	60	69
LPXTG sorting signal	35	42	44	28
LRR region	19	25	25	8
LysM domain	9	6	6	6
NLPC/P60 domain	5	4	3	5

Source: Billion, A. et al., *Bioinformatics,* 22, 2819, 2006.

strains identified a specific genetic profile and the absence of five genes predicted to encode internalins and other cell surface proteins, similar to the lineage III isolates.[123] Thus, surface protein diversity may contribute to the adaptation of *L. monocytogenes* to different environments. The elucidation of the functions of the different surface proteins and the putative strain-specific characteristics that they confer will be one of the challenging questions of the future.

The *srtB* gene is also part of a locus that contains genes encoding LPXTG proteins and a ferrichrome transporter, but without a gene encoding a protein with an NPQTN motif. The lipoprotein OppA mediates the transport of oligopeptides and is required for bacterial growth at low temperature. OppA has been shown to contribute to intracellular survival in cultured macrophages *in vitro* as well as *in vivo*.[124] *L. monocytogenes* contains several other genes coding for lipoproteins that are anticipated to be involved in host–pathogen interactions, including Lmo1847 and Lmo1800. Lmo1800 is a putative tyrosine phosphatase.

Cell wall hydrolases, such as the invasion-associated protein (Iap, p60, or Cwh), a cell wall amidase (Ami), and a surface-associated autolysin (Auto), can play a direct role in the pathogenicity of *L. monocytogenes*.[98,99,125] Cell wall hydrolases are involved in various biological processes including cell division, cell separation, and competence for genetic transformation; sporulation; and the lytic action of some antibiotics. Another autolysin involved in *L. monocytogenes* cell division is the recently described 66-kDa cell surface protein MurA (NamA).[126,127] Deletion of the *murA* gene results in chain formation in exponential-growth-phase cultures. This murein hydrolase is important for cell separation and for generalized autolysis in *L. monocytogenes*.[126] Both p60 and MurA proteins carry LysM domains, responsible for attachment of the respective proteins to the cell wall, and are secreted out of the bacterial cell in a SecA2-dependent manner. Simultaneous deficiency of both MurA and p60 proteins generates a rough phenotype in *L. monocytogenes*.[128] The *secA2* gene has been described as a second *Listeria secA* gene associated with protein secretion and is responsible for the transport of a variety of extracellular proteins in *L. monocytogenes*.[129] An *L. monocytogenes secA2* deletion mutant displays a rough phenotype[128,129] and is defective in the secretion of at least 15 additional cell wall-associated or -secreted products in addition to p60/Iap and MurA/NamA. Thus, proteins transported by the SecA2-dependent pathway are assumed to be responsible for generation of long cell chains and rough colony morphology in *L. monocytogenes*. It is hypothesized that SecA2-dependent protein secretion plays a role in the colonization of environmental and host surfaces.

The abundance of *L. monocytogenes* surface proteins and the variety of anchoring systems are probably related to the ability of this bacterium to survive in diverse environments and to interact with a large variety of cell types. Strikingly, 25% of the LPXTG protein genes are preceded by a putative PrfA box, whereas PrfA boxes are only detected upstream of 10% of all *L. monocytogenes* genes; this strongly suggests that LPXTG proteins are involved in virulence.

10.4 CONCLUSIONS AND PERSPECTIVES

The recent completion of the genome sequences of several *Listeria* species/strains has provided unprecedented opportunities for the studies of listerial genomics and virulence mechanisms. Through comparative examination of *L. monocytogenes, L. innocua,* and *L. welshimeri* genomes, new insights have been gained on the evolution of virulence, tropism, and survival strategies used by these bacterial species to persist in the different ecological niches they occupy. Comparative transcriptome and proteome analysis investigating regulatory networks and subproteomes suggests common and distinct species- and strain-specific adaptive responses and implies subtle strategies used by different strains and species for survival and transmission in living and nonliving environments.

The era of comparative genome sequencing of strains representing all species of the genus *Listeria* is well under way with the prospect that the genome sequences of an *L. monocytogenes* serotype 4a strain and other *Listeria* species will become available in the near future. The elucidation of the nucleotide composition of an *L. monocytogenes* serotype 4a genome will be especially useful since serotype 4a appears to be the only one that is truly avirulent and unable to produce mouse mortality even via intraperitoneal inoculation. Although previous description of the *L. innocua* CLIP 11262 genome has been of tremendous help to the investigation of the *Listeria* evolution of virulence, there is no doubt that more pertinent data can be attained by direct genome comparison between virulent and naturally avirulent *L. monocytogenes* strains than by inferring from the genome sequences of two separate species.

Genomics has opened new avenues in which the search for both virulence-associated and true virulence factors can be pursued using comparative genomic approaches. The functional characterization of these putative factors is more complicated and will involve analysis of cellular interactions, but also the development of tissue- and organ-based models, noninvasive *in vivo* imaging techniques, and whole-genome-based technology to examine the spatiotemporal role of these factors during infection. Determination of the genome sequences of different *Listeria* serovars should provide us with a more comprehensive view of the nature, specificity, and putative functions of surface proteins. Transcriptomics, proteomics, and functional analysis will nevertheless be necessary to understand fully the molecular basis of the interactions of *L. monocytogenes* with its hosts. As many of the low G+C bacterial genera comprise important pathogens, such as *Staphylococcus*, *Streptococcus*, *Clostridia*, and *Bacillus*, information gained from the study of *Listeria* can further improve our understanding of the evolution of virulence in this group of bacteria. Such findings serve to direct the focus of research efforts to a relatively small number of specific genomic regions, to elucidate their possible involvement in virulence and adaptive physiology attributes of epidemic-associated bacteria. These studies can not only lead to the development of novel ways to interrupt transmission and prevent food-borne disease but also provide alternative targets for prophylactic and therapeutic strategies.

ACKNOWLEDGMENTS

Support for this review was provided by funds from the ERA-NET Pathogenomics Network, supported by the Bundesministerium für Bildung und Forschung (BMBF) and from the Sonderforschungsbereich 535 supported by the Deutsche Forschungsgemeinschaft to T. C. T. H. S. B. is a scholar of the Department of Biotechnology, Government of India. We also thank M. Loessner for providing supportive data on listerial bacteriophages, and Rohit Ghai, Carsten Kuenne, and Andre Billion for bioinformatics analysis.

REFERENCES

1. Collins, M.D. et al., Phylogenetic analysis of the genus *Listeria* based on reverse transcriptase sequencing of 16S rRNA, *Int. J. Syst. Bacteriol.*, 41, 240, 1991.
2. Rocourt, J., The genus *Listeria* and *Listeria monocytogenes*: Phylogenetic position, taxonomy, and identification, in *Listeria, listeriosis, and food safety*, 2nd ed., Ryser E.T. and Marth, E., eds., Marcel Dekker Inc., New York, 1999, 1–20.
3. Sallen, B. et al., Comparative analysis of 16S and 23S rRNA sequences of *Listeria* species, *Int. J. Syst. Bacteriol.*, 46, 669, 1996.
4. Seeliger, H.P. and Jones, D., Genus *Listeria*, in *Bergey's manual of systematic bacteriology 2*, Sneath, P.H., Mair, N.S., Sharpe, M.E. and Hold, J.G., eds., Williams and Wilkins, Baltimore, MD, 1986.
5. Weis, J. and Seeliger, H.P., Incidence of *Listeria monocytogenes* in nature, *Appl. Microbiol.*, 30, 29, 1975.
6. Mead, P.S. et al., Food-related illness and death in the United States, *Emerg. Infect. Dis.*, 5, 607, 1999.
7. Vazquez-Boland, J.A. et al., *Listeria* pathogenesis and molecular virulence determinants, *Clin. Microbiol. Rev.*, 14, 584, 2001.
8. Farber, J.M. and Peterkin, P.I., *Listeria monocytogenes*, a food-borne pathogen, *Microbiol. Rev.*, 55, 476, 1991.
9. Alexander, A.V. et al., Bovine abortions attributable to *Listeria ivanovii*: Four cases (1988–1990), *J. Am. Vet. Med. Assoc.*, 200, 711, 1992.

10. Low, J.C. and Donachie, W., A review of *Listeria monocytogenes* and listeriosis, *Vet. J.*, 153, 9, 1997.
11. Wesley, I.V., Listeriosis in animals, in *Listeria, listeriosis and food safety*, 2nd ed., Ryser, E.T. and Marth, E., eds., Marcel Dekker, New York, 1999, 39–73.
12. Glaser, P. et al., Comparative genomics of *Listeria* species, *Science*, 294, 849, 2001.
13. Nelson, K.E. et al., Whole genome comparisons of serotype 4b and 1/2a strains of the food-borne pathogen *Listeria monocytogenes* reveal new insights into the core genome components of this species, *Nucleic Acids Res.*, 32, 2386, 2004.
14. Hain, T. et al., Whole-genome sequence of *Listeria welshimeri* reveals common steps in genome reduction with *Listeria innocua* as compared to *Listeria monocytogenes, J. Bacteriol.*, 188, 7405, 2006.
15. Hain, T., Steinweg, C., and Chakraborty, T., Comparative and functional genomics of *Listeria* spp., *J. Biotechnol.*, 126, 37, 2006.
16. Buchrieser, C. et al., Comparison of the genome sequences of *Listeria monocytogenes* and *Listeria innocua*: Clues for evolution and pathogenicity, *FEMS Immunol. Med. Microbiol.*, 35, 207, 2003.
17. Doumith, M. et al., New aspects regarding evolution and virulence of *Listeria monocytogenes* revealed by comparative genomics and DNA arrays, *Infect. Immun.*, 72, 1072, 2004.
18. Chakraborty, T., Hain, T., and Domann, E., Genome organization and the evolution of the virulence gene locus in *Listeria* species, *Int. J. Med. Microbiol.*, 290, 167, 2000.
19. Volokhov, D.V. et al., The presence of the internalin gene in natural atypically hemolytic *Listeria innocua* strains suggests descent from *L. monocytogenes, Appl. Environ. Microbiol.*, 73, 1928, 2007.
20. Loessner, M.J. and Calendar, R., The *Listeria* bacteriophages, in *The bacteriophages*, Oxford University Press, New York, 2005, 593–601.
21. Loessner, M.J. and Busse, M., Bacteriophage typing of *Listeria* species, *Appl. Environ. Microbiol.*, 56, 1912, 1990.
22. Wendlinger, G., Loessner, M.J., and Scherer, S., Bacteriophage receptors on *Listeria monocytogenes* cells are the N-acetylglucosamine and rhamnose substituents of teichoic acids or the peptidoglycan itself, *Microbiology*, 142, 985, 1996.
23. Loessner, M.J. et al., Complete nucleotide sequence, molecular analysis and genome structure of bacteriophage A118 of *Listeria monocytogenes*: Implications for phage evolution, *Mol. Microbiol.*, 35, 324, 2000.
24. Zimmer, M. et al., Genome and proteome of *Listeria monocytogenes* phage PSA: An unusual case for programmed + 1 translational frameshifting in structural protein synthesis, *Mol. Microbiol.*, 50, 303, 2003.
25. Carlton, R.M. et al., Bacteriophage P100 for control of *Listeria monocytogenes* in foods: Genome sequence, bioinformatic analyses, oral toxicity study, and application, *Regul. Toxicol. Pharmacol.*, 43, 301, 2005.
26. Zink, R. and Loessner, M.J., Classification of virulent and temperate bacteriophages of *Listeria* spp. on the basis of morphology and protein analysis, *Appl. Environ. Microbiol.*, 58, 296, 1992.
27. Chibani-Chennoufi, S. et al., *Lactobacillus plantarum* bacteriophage LP65: A new member of the SPO1-like genus of the family Myoviridae, *J. Bacteriol.*, 186, 7069, 2004.
28. O'Flaherty, S. et al., Genome of staphylococcal phage K: A new lineage of Myoviridae infecting Gram-positive bacteria with a low G+C content, *J. Bacteriol.*, 186, 2862, 2004.
29. Jacquet, C. et al., Expression of ActA, Ami, InlB, and listeriolysin O in *Listeria monocytogenes* of human and food origin, *Appl. Environ. Microbiol.*, 68, 616, 2002.
30. Sabet, C. et al., LPXTG protein InlJ, a newly identified internalin involved in *Listeria monocytogenes* virulence, *Infect. Immun.*, 73, 6912, 2005.
31. Chatterjee, S.S. et al., Intracellular gene expression profile of *Listeria monocytogenes, Infect. Immun.*, 74, 1323, 2006.
32. Joseph, B. et al., Identification of *Listeria monocytogenes* genes contributing to intracellular replication by expression profiling and mutant screening, *J. Bacteriol.*, 188, 556, 2006.
33. Goetz, M. et al., Microinjection and growth of bacteria in the cytosol of mammalian host cells, *Proc. Natl. Acad. Sci. USA*, 98, 12221, 2001.
34. Heithoff, D.M. et al., Coordinate intracellular expression of *Salmonella* genes induced during infection, *J. Bacteriol.*, 181, 799, 1999.
35. Goebel, W. et al., eds. *Gram-positive pathogens*, American Society for Microbiology, Washington, D.C., 2000, 499–506.
36. Leimeister-Wachter, M., Domann, E., and Chakraborty, T., The expression of virulence genes in *Listeria monocytogenes* is thermoregulated, *J. Bacteriol.*, 174, 947, 1992.
37. Sokolovic, Z., Fuchs, A., and Goebel, W., Synthesis of species-specific stress proteins by virulent strains of *Listeria monocytogenes, Infect. Immun.*, 58, 3582, 1990.

38. Ripio, M.T. et al., Transcriptional activation of virulence genes in wild-type strains of *Listeria monocytogenes* in response to a change in the extracellular medium composition, *Res. Microbiol.*, 147, 371, 1996.
39. Renzoni, A., Cossart, P., and Dramsi, S., PrfA, the transcriptional activator of virulence genes, is upregulated during interaction of *Listeria monocytogenes* with mammalian cells and in eukaryotic cell extracts, *Mol. Microbiol.*, 34, 552, 1999.
40. Moors, M.A. et al., Expression of listeriolysin O and ActA by intracellular and extracellular *Listeria monocytogenes, Infect. Immun.*, 67, 131, 1999.
41. Becker, L.A. et al., Identification of the gene encoding the alternative sigma factor σ^B from *Listeria monocytogenes* and its role in osmotolerance, *J. Bacteriol.*, 180, 4547, 1998.
42. Haldenwang, W.G., The sigma factors of *Bacillus subtilis, Microbiol. Rev.*, 59, 1, 1995.
43. McGann, P., Wiedmann, M., and Boor, K.J., The alternative sigma factor {sigma}B and the virulence gene regulator PrfA both regulate transcription of *Listeria monocytogenes* internalins, *Appl. Environ. Microbiol.*, 73, 2919, 2007.
44. Ferreira, A., O'Byrne, C.P., and Boor, K.J., Role of sigma(B) in heat, ethanol, acid, and oxidative stress resistance and during carbon starvation in *Listeria monocytogenes, Appl. Environ. Microbiol.*, 67, 4454, 2001.
45. Wiedmann, M. et al., General stress transcription factor σ^B and its role in acid tolerance and virulence of *Listeria monocytogenes, J. Bacteriol.*, 180, 3650, 1998.
46. Ferreira, A. et al., Role of *Listeria monocytogenes* sigma(B) in survival of lethal acidic conditions and in the acquired acid tolerance response, *Appl. Environ. Microbiol.*, 69, 2692, 2003.
47. Kazmierczak, M.J. et al., *Listeria monocytogenes* σ^B regulates stress response and virulence functions, *J. Bacteriol.*, 185, 5722, 2003.
48. Mandin, P. et al., VirR, a response regulator critical for *Listeria monocytogenes* virulence, *Mol. Microbiol.*, 57, 1367, 2005.
49. Thedieck, K. et al., The MprF protein is required for lysinylation of phospholipids in listerial membranes and confers resistance to cationic antimicrobial peptides (CAMPs) on *Listeria monocytogenes, Mol. Microbiol.*, 62, 1325, 2006.
50. Studholme, D.J. and Buck, M., The biology of enhancer-dependent transcriptional regulation in bacteria: insights from genome sequences, *FEMS Microbiol. Lett.*, 186, 1, 2000.
51. Arous, S. et al., Global analysis of gene expression in an *rpoN* mutant of *Listeria monocytogenes, Microbiology,* 150, 1581, 2004.
52. Robichon, D. et al., The *rpoN* (s^{54}) gene from *Listeria monocytogenes* is involved in resistance to mesentericin Y105, an antibacterial peptide from *Leuconostoc mesenteroides, J. Bacteriol.*, 179, 7591, 1997.
53. Okada, Y. et al., The sigma factor *RpoN* (sigma54) is involved in osmotolerance in *Listeria monocytogenes, FEMS Microbiol. Lett.*, 263, 54, 2006.
54. Bennett, H.J. et al., Characterization of *relA* and *codY* mutants of *Listeria monocytogenes*: Identification of the *CodY* regulon and its role in virulence, *Mol. Microbiol.*, 63, 1453, 2007.
55. Knudsen, G.M., Olsen, J.E., and Dons, L., Characterization of DegU, a response regulator in *Listeria monocytogenes*, involved in regulation of motility and contributes to virulence, *FEMS Microbiol. Lett.*, 240, 171, 2004.
56. Williams, T. et al., Response regulator DegU of *Listeria monocytogenes* regulates the expression of flagella-specific genes, *FEMS Microbiol. Lett.*, 252, 287, 2005.
57. Grundling, A. et al., *Listeria monocytogenes* regulates flagellar motility gene expression through MogR, a transcriptional repressor required for virulence, *Proc. Natl. Acad. Sci. USA*, 101, 12318, 2004.
58. Shen, A. and Higgins, D.E., The MogR transcriptional repressor regulates nonhierarchal expression of flagellar motility genes and virulence in *Listeria monocytogenes, PLoS Pathog.*, 2, e30, 2006.
59. Nair, S., Milohanic, E., and Berche, P., *ClpC* ATPase is required for cell adhesion and invasion of *Listeria monocytogenes, Infect. Immun.*, 68, 7061, 2000.
60. Ghai, R., Hain, T., and Chakraborty, T., GenomeViz: Visualizing microbial genomes, *BMC Bioinformatics,* 5, 198, 2004.
61. Kuenne, C.T. et al., GECO—Linear visualization for comparative genomics, *Bioinformatics,* 23, 125, 2007.
62. Billion, A. et al., Augur—A computational pipeline for whole genome microbial surface protein prediction and classification, *Bioinformatics,* 22, 2819, 2006.
63. Dieterich, G. et al., LEGER: Knowledge database and visualization tool for comparative genomics of pathogenic and nonpathogenic *Listeria* species, *Nucleic Acids Res.*, 34, D402, 2006.
64. Call, D.R., Borucki, M.K., and Besser, T.E., Mixed-genome microarrays reveal multiple serotype and lineage-specific differences among strains of *Listeria monocytogenes, J. Clin. Microbiol.*, 41, 632, 2003.

65. Borucki, M.K. et al., Discrimination among *Listeria monocytogenes* isolates using a mixed genome DNA microarray, *Vet. Microbiol.*, 92, 351, 2003.
66. Volokhov, D. et al., Identification of *Listeria* species by microarray-based assay, *J. Clin. Microbiol.*, 40, 4720, 2002.
67. Zhang, C. et al., Genome diversification in phylogenetic lineages I and II of *Listeria monocytogenes*: Identification of segments unique to lineage II populations, *J. Bacteriol.*, 185, 5573, 2003.
68. Johnson, J. et al., Natural atypical *Listeria innocua* strains with *Listeria monocytogenes* pathogenicity island 1 genes, *Appl. Environ. Microbiol.*, 70, 4256, 2004.
69. Schmid, M.W. et al., Evolutionary history of the genus *Listeria* and its virulence genes, *Syst. Appl. Microbiol.*, 28, 1, 2005.
70. Boerlin, P., Rocourt, J., and Piffaretti, J.C., Taxonomy of the genus *Listeria* by using multilocus enzyme electrophoresis, *Int. J. Syst. Bacteriol.*, 41, 59, 1991.
71. Vazquez-Boland, J.A. et al., Pathogenicity islands and virulence evolution in *Listeria, Microbes. Infect.*, 3, 571, 2001.
72. Dramsi, S. and Cossart, P., Listeriolysin O: A genuine cytolysin optimized for an intracellular parasite, *J. Cell Biol.*, 156, 943, 2002.
73. Marquis, H., Doshi, V., and Portnoy, D.A., The broad-range phospholipase C and a metalloprotease mediate listeriolysin O-independent escape of *Listeria monocytogenes* from a primary vacuole in human epithelial cells, *Infect. Immun.*, 63, 4531, 1995.
74. Mounier, J. et al., Intracellular and cell to cell spread of *Listeria monocytogenes* involves interaction with F-actin in the enterocytelike cell line Caco-2, *Infect. Immun.*, 58, 1048, 1990.
75. Tilney, L.G. and Portnoy, D.A., Actin filaments and the growth, movement, and spread of the intracellular bacterial parasite, *Listeria monocytogenes*, *J. Cell Biol.*, 109, 1597, 1989.
76. Pizarro-Cerda, J., Sousa, S., and Cossart, P., Exploitation of host cell cytoskeleton and signaling during *Listeria monocytogenes* entry into mammalian cells, *C. R. Biol.*, 327, 115, 2004.
77. Schubert, W.D. et al., Structure of internalin, a major invasion protein of *Listeria monocytogenes*, in complex with its human receptor E-cadherin, *Cell*, 111, 825, 2002.
78. Gouin, E., Mengaud, J., and Cossart, P., The virulence gene cluster of *Listeria monocytogenes* is also present in *Listeria ivanovii*, an animal pathogen, and *Listeria seeligeri*, a nonpathogenic species, *Infect. Immun.*, 62, 3550, 1994.
79. Smith, G.A. et al., The two distinct phospholipases C of *Listeria monocytogenes* have overlapping roles in escape from a vacuole and cell-to-cell spread, *Infect. Immun.*, 63, 4231, 1995.
80. Bubert, A. et al., Structural and functional properties of the p60 proteins from different *Listeria* species, *J. Bacteriol.*, 174, 8166, 1992.
81. Hess, J. et al., *Listeria monocytogenes* p60 supports host cell invasion by and *in vivo* survival of attenuated *Salmonella typhimurium*, *Infect. Immun.*, 63, 2047, 1995.
82. Dominguez-Bernal, G. et al., A spontaneous genomic deletion in *Listeria ivanovii* identifies LIPI-2, a species-specific pathogenicity island encoding sphingomyelinase and numerous internalins, *Mol. Microbiol.*, 59, 415, 2006.
83. Gonzalez-Zorn, B. et al., SmcL, a novel membrane-damaging virulence factor in *Listeria*, *Int. J. Med. Microbiol.*, 290, 369, 2000.
84. Ivanek, R., Grohn, Y.T., and Wiedmann, M., *Listeria monocytogenes* in multiple habitats and host populations: Review of available data for mathematical modeling, *Foodborne Pathog. Dis.*, 3, 319, 2006.
85. Haas, A., Dumbsky, M., and Kreft, J., Listeriolysin genes: Complete sequence of *ilo* from *Listeria ivanovii* and of *lso* from *Listeria seeligeri*, *Biochim. Biophys. Acta*, 1130, 81, 1992.
86. Lampidis, R. et al., The virulence regulator protein of *Listeria ivanovii* is highly homologous to PrfA from *Listeria monocytogenes* and both belong to the Crp-Fnr family of transcription regulators, *Mol. Microbiol.*, 13, 141, 1994.
87. Gouin, E. et al., *iactA* of *Listeria ivanovii*, although distantly related to *Listeria monocytogenes actA*, restores actin tail formation in an *L. monocytogenes actA* mutant, *Infect. Immun.*, 63, 2729, 1995.
88. Kreft, J., Dumbsky, M., and Theiss, S., The actin-polymerization protein from *Listeria ivanovii* is a large repeat protein which shows only limited amino acid sequence homology to ActA from *Listeria monocytogenes*, *FEMS Microbiol. Lett.*, 126, 113, 1995.
89. Cai, S. and Wiedmann, M., Characterization of the *prfA* virulence gene cluster insertion site in nonhemolytic *Listeria* spp.: Probing the evolution of the *Listeria* virulence gene island, *Curr. Microbiol.*, 43, 271, 2001.

90. Volokhov, D. et al., Discovery of natural atypical nonhemolytic *Listeria seeligeri* isolates, *Appl. Environ. Microbiol.*, 72, 2439, 2006.
91. Doolittle R.F., Biodiversity: Microbial genomes multiply, *Nature,* 416, 697, 2002.
92. Moran, N.A., Tracing the evolution of gene loss in obligate bacterial symbionts, *Curr. Opin. Microbiol.*, 6, 512, 2003.
93. Chico-Calero, I. et al., Hpt, a bacterial homolog of the microsomal glucose-6-phosphate translocase, mediates rapid intracellular proliferation in *Listeria, Proc. Natl. Acad. Sci. USA,* 99, 431, 2002.
94. Dussurget, O. et al., *Listeria monocytogenes* bile salt hydrolase is a PrfA-regulated virulence factor involved in the intestinal and hepatic phases of listeriosis, *Mol. Microbiol.*, 45, 1095, 2002.
95. Sleator, R.D. et al., A PrfA-regulated bile exclusion system (BilE) is a novel virulence factor in *Listeria monocytogenes, Mol. Microbiol.*, 55, 1183, 2005.
96. Begley, M., Hill, C., and Gahan, C.G., Identification and disruption of btlA, a locus involved in bile tolerance and general stress resistance in *Listeria monocytogenes, FEMS Microbiol Lett.*, 218, 31, 2003.
97. Begley, M. et al., Contribution of three bile-associated loci, *bsh*, *pva*, and *btlB*, to gastrointestinal persistence and bile tolerance of *Listeria monocytogenes, Infect. Immun.*, 73, 894, 2005.
98. Milohanic, E. et al., The autolysin Ami contributes to the adhesion of *Listeria monocytogenes* to eukaryotic cells via its cell wall anchor, *Mol. Microbiol.*, 39, 1212, 2001.
99. Cabanes, D. et al., *Auto*, a surface associated autolysin of *Listeria monocytogenes* required for entry into eukaryotic cells and virulence, *Mol. Microbiol.*, 51, 1601, 2004.
100. Dramsi, S. et al., FbpA, a novel multifunctional *Listeria monocytogenes* virulence factor, *Mol. Microbiol.*, 53, 639, 2004.
101. Bierne, H. et al., Inactivation of the srtA gene in *Listeria monocytogenes* inhibits anchoring of surface proteins and affects virulence, *Mol. Microbiol.*, 43, 869, 2002.
102. Bierne, H. et al., Sortase B, a new class of sortase in *Listeria monocytogenes, J. Bacteriol.*, 186, 1972, 2004.
103. Jin, B. et al., Iron acquisition systems for ferric hydroxamates, hemin and hemoglobin in *Listeria monocytogenes, Mol. Microbiol.*, 59, 1185, 2006.
104. Bigot, A. et al., Role of *fliF* and *fliI* of *Listeria monocytogenes* in flagellar assembly and pathogenicity, *Infect. Immun.*, 73, 5530, 2005.
105. Gaillard, J.L. et al., Entry of *L. monocytogenes* into cells is mediated by internalin, a repeat protein reminiscent of surface antigens from Gram-positive cocci, *Cell,* 65, 1127, 1991.
106. Cossart, P. and Lecuit, M., Interactions of *Listeria monocytogenes* with mammalian cells during entry and actin-based movement: Bacterial factors, cellular ligands and signaling, *EMBO J.*, 17, 3797, 1998.
107. Domann, E. et al., Identification and characterization of a novel PrfA-regulated gene in *Listeria monocytogenes* whose product, IrpA, is highly homologous to internalin proteins, which contain leucine-rich repeats, *Infect. Immun.*, 65, 101, 1997.
108. Lingnau, A. et al., Expression of the *Listeria monocytogenes* EGD *inlA* and *inlB* genes, whose products mediate bacterial entry into tissue culture cell lines, by PrfA-dependent and -independent mechanisms, *Infect. Immun.*, 63, 3896, 1995.
109. Dramsi, S. et al., Identification of four new members of the internalin multigene family of *Listeria monocytogenes* EGD, *Infect. Immun.*, 65, 1615, 1997.
110. Raffelsbauer, D. et al., The gene cluster *inlC2DE* of *Listeria monocytogenes* contains additional new internalin genes and is important for virulence in mice, *Mol. Gen. Genet.*, 260, 144, 1998.
111. Braun, L. et al., InlB: An invasion protein of *Listeria monocytogenes* with a novel type of surface association, *Mol. Microbiol.*, 25, 285, 1997.
112. Navarre, W.W. and Schneewind, O., Surface proteins of Gram-positive bacteria and mechanisms of their targeting to the cell wall envelope, *Microbiol. Mol. Biol. Rev.*, 63, 174, 1999.
113. Cabanes, D. et al., Gp96 is a receptor for a novel *Listeria monocytogenes* virulence factor, Vip, a surface protein, *EMBO J.*, 24, 2827, 2007.
114. Engelbrecht, F. et al., A new PrfA-regulated gene of *Listeria monocytogenes* encoding a small, secreted protein which belongs to the family of internalins, *Mol. Microbiol.*, 21, 823, 1996.
115. Engelbrecht, F. et al., A novel PrfA-regulated chromosomal locus, which is specific for *Listeria ivanovii*, encodes two small, secreted internalins and contributes to virulence in mice, *Mol. Microbiol.*, 30, 405, 1998.
116. Engelbrecht, F. et al., Sequence comparison of the chromosomal regions encompassing the internalin C genes (*inlC*) of *Listeria monocytogenes* and *L. ivanovii, Mol. Gen. Genet.*, 257, 186, 1998.
117. Schubert, W.D. et al., Internalins from the human pathogen *Listeria monocytogenes* combine three distinct folds into a contiguous internalin domain, *J. Mol. Biol.*, 312, 783, 2001.

118. Tsai, Y.H. et al., *Listeria monocytogenes* internalins are highly diverse and evolved by recombination and positive selection, *Infect. Gent. Evol.*, 6, 378, 2006.
119. Cabanes, D. et al., Surface proteins and the pathogenic potential of *Listeria monocytogenes, Trends Microbiol.*, 10, 238, 2002.
120. Domann, E. et al., A novel bacterial virulence gene in *Listeria monocytogenes* required for host cell microfilament interaction with homology to the proline-rich region of vinculin, *EMBO J.*, 11, 1981, 1992.
121. Jeffers, G.T. et al., Comparative genetic characterization of *Listeria monocytogenes* isolates from human and animal listeriosis cases, *Microbiology*, 147, 1095, 2001.
122. Wiedmann, M. et al., Ribotypes and virulence gene polymorphisms suggest three distinct *Listeria monocytogenes* lineages with differences in pathogenic potential, *Infect. Immun.*, 65, 2707, 1997.
123. Hong, E. et al., Genetic diversity of *Listeria monocytogenes* populations present in patients and in pork products at the store distribution level in France in 2000–2001, *Int. J. Food Microbiol.*, 114, 187, 2007.
124. Borezee, E., Pellegrini, E., and Berche, P., OppA of *Listeria monocytogenes*, an oligopeptide-binding protein required for bacterial growth at low temperature and involved in intracellular survival, *Infect. Immun.*, 68, 7069, 2000.
125. Pilgrim, S. et al., Deletion of the gene encoding p60 in *Listeria monocytogenes* leads to abnormal cell division and loss of actin-based motility, *Infect. Immun.*, 71, 3473, 2003.
126. Carroll, S.A. et al., Identification and characterization of a peptidoglycan hydrolase, MurA, of *Listeria monocytogenes*, a muramidase needed for cell separation, *J. Bacteriol.*, 185, 6801, 2003.
127. Lenz, L.L. et al., SecA2-dependent secretion of autolytic enzymes promotes *Listeria monocytogenes* pathogenesis, *Proc. Natl. Acad. Sci. USA*, 100, 12432, 2003.
128. Machata S. et al., Simultaneous deficiency of both MurA and p60 proteins generates a rough phenotype in *Listeria monocytogenes, J. Bacteriol.*, 187, 8385, 2005.
129. Lenz, L.L. and Portnoy, D.A., Identification of a second *Listeria secA* gene associated with protein secretion and the rough phenotype, *Mol. Microbiol.*, 45, 1043, 2002.

11 Genomic Divisions/Lineages, Epidemic Clones, and Population Structure

Ying Cheng, Robin M. Siletzky, and Sophia Kathariou

CONTENTS

11.1 INTRODUCTION

Ever since the development of the serotyping scheme for *Listeria monocytogenes*, several lines of evidence have indicated that the species is partitioned into subpopulations consisting of strains sharing well-defined characteristics. Interestingly, in the case of *L. monocytogenes* serotypic designations have been proven useful and meaningful in ways far exceeding the subtyping and clinical laboratory applications originally envisioned for serotype determinations. As will be discussed later in this chapter, species partitioning into serotype-associated clusters has been a key element for advances in our understanding of the epidemiology, ecology, population structure, and evolution of the organism. In the past 25 years, we have witnessed impressive advances in strain subtyping tools and applications, as well as the unprecedented impact of genome sequence information from several strains of *Listeria*. Such advances have further validated serotype-associated partitioning within *L. monocytogenes* and have further led to identification and characterization of additional clonal groups of significant epidemiological and food safety interest, such as the epidemic-associated clones implicated in outbreaks of food-borne listeriosis.

11.2 GENOMIC DIVISIONS/LINEAGES

The serotyping scheme of Seeliger and Höhne, still in use today, has grouped *L. monocytogenes* isolates into three serogroups (1/2, 3, and 4) and 13 known serotypes (serovars).[1] Serotype designations reflect somatic antigens (e.g., 1/2, 3, and 4), based primarily on the composition of the teichoic acid of the cell wall, and flagellin antigens (e.g., a, b, c). Of the 12 serotypes, only a few are frequently encountered in human illness. The majority (>98%) of human infections involve strains of serotypes 1/2a, 1/2b, 1/2c, and 4b, with 1/2c accounting for noticeably fewer cases of illness than the other three.[2] Isolates of serotypes 1/2a, 1/2b, and 4b are also key contributors to animal listeriosis. However, animal listeriosis may also be frequently caused by strains of serotype 4a and 4c strains, which are rare among human cases of illness.[3]

Various molecular typing methods, including multiple locus enzyme electrophoresis (MLEE), ribotyping and other restriction fragment length polymorphism (RFLP)-based schemes, pulsed field gel electrophoresis (PFGE), polymerase chain reaction (PCR)-based subtyping, DNA sequencing (including complete genome sequencing), DNA arrays, etc., have been used to explore the population genetics and evolution of *L. monocytogenes*. Data generated with such subtyping methods have shown that the genetic structure of *L. monocytogenes* is highly clonal and have consistently identified three major divisions: lineage I (also designated Genomic Division II), lineage II (also designated Genomic Division I), and lineage III (Genomic Division III). These divisions are correlated with serotype-based groups. Strains of serotype 1/2a, 3a, 1/2c, and 3c are included in Genomic Division I (lineage II), whereas those of serotypes 1/2b, 3b, 4b, 4d, and 4e are grouped into Genomic Division II (lineage I). Genomic Division III (lineage III) consists of strains of serotype 4a and 4c, as well as certain strains of serotype 4b (the majority of serotype 4b are in lineage I).[3–9]

The terms employed for two of the major intraspecific, serotype-associated groups have varied among different publications, leading to substantial confusion in the literature and within the *Listeria* community. For instance, Genomic Division I referred to a major population cluster including strains of serotypes 1/2a, 3a, 1/2c, and 3c in the first population genetic descriptions of *L. monocytogenes*, which utilized MLEE.[6,10] Subsequent studies employing ribotyping detected the same serotype-associated groups, but used an alternative term, "lineage II," for what was referred to as "Genomic Division I" in the earlier studies.[3] Other investigators, employing high-density DNA arrays derived from whole genome sequence data, designated this same cluster as "lineage I."[5] Similar variations in terminology have been employed in terms referring to the population cluster consisting of the strains with the "b" flagellar antigens (1/2b, 3b, 4b, 4d, and 4e). This group was first referred to as "Genomic Division II,"[6,10] but subsequently designated "Lineage I" or "lineage 1" in certain publications[3,9] and "Lineage II" in others,[5,11] leading to unnecessary confusion.

To minimize further confusion, in this review chapter we will adopt the terminology most frequently represented in the literature (even though this does not do justice to historical appearance in the scientific literature of the terms referring to the population structure of *L. monocytogenes*): lineage I will be employed to refer to strains of serotypes 1/2b, 3b, 4d, 4e, and most strains of serotype 4b; lineage II will be employed for strains of serotype 1/2a, 1/2c, 3a, and 3c; and lineage III will refer to strains of serotype 4a, 4c, and certain strains of serotype 4b.

11.2.1 Lineage I

11.2.1.1 Molecular Identification of Lineage I

This major strain cluster within *L. monocytogenes* was initially identified by MLEE investigations, and found to include strains of serotype 1/2b, 3b, and 4b.[6] Lineage I was subsequently also detected by several approaches employing ribotyping, PFGE, amplified fragment length polymorphism (AFLP), repetitive element PCR (rep PCR and variable number tandem repeat [VNTR]-based typing), and sequence analysis of allelic variation in virulence genes (*hly*, *actA*, and *inlA*).[3,12–18] More recently, phylogenetic data generated by multiple locus sequence typing (MLST)-based approaches including housekeeping and/or virulence genes,[19,20] DNA sequencing of the *prfA* virulence gene cluster (*p*VGC), and DNA arrays based on complete genome sequence data further verified the presence of major, serotype-associated evolutionary lineages, including lineage I, within *L. monocytogenes*.[4,8,9,21]

11.2.1.2 Further Partitioning within Lineage I

High-resolution subtyping investigations of 44 strains of serotype 1/2a, 1/2b, and 4b using DNA microarrays constructed based on the genome of the serotype 1/2a strain 10403S revealed that lineage I is further partitioned, with strains of serotype 1/2b clustering into one group, and strains of serotype 4b into another.[21] Several genes (47 genes, in 16 contiguous segments) were absent from the genome of 1/2b and 4b strains, but present in strains of serotype 1/2a, and 9 genes were absent only from the genome of serotype 4b strains. Genes specifically absent from serotype 1/2b or 4b strains included some mediating expression of putative surface antigens and others with putative functions in regulation of virulence, and in most cases gene absence reflected deletion from the chromosome of ancestral strains. A model was presented suggesting that strains of serogroup 1/2 (serotypes 1/2a and 1/2b) were ancestral to those of 4b.[21]

A different DNA study that also used arrays and included strains of these and additional serotypes, as well as *L. innocua*, yielded similar subclusters within lineage I: one consisting of strains of serotype 1/2b and 3b and the other of strains of serotype 4b, 4d, and 4e.[5] Genes encoding putative surface-associated proteins (including several similar to internalins) and putative transcriptional regulators again were represented among those with differential representation among the different lineages and lineage I subclusters. The distribution data of lineage-specific genes were in agreement with a model according to which serotype 1/2b strains diversified from those of serotype 1/2a, evolved into serotype 4b by acquisition of genes conferring serotype 4b specificity (such as those mediating synthesis and decoration of teichoic acid), and subsequently evolved, primarily by gene loss, to the nonpathogenic species *L. innocua*.[5]

Several additional studies have suggested that nonpathogenic *Listeria* spp. arose from a pathogenic ancestor, and that *L. innocua*, the nonpathogenic species most closely related to *L. monocytogenes*, arose from a pathogenic ancestor primarily through gene loss.[11,22–25] Evolution of *L. innocua* from a serogroup 4 ancestor is in agreement with the long known similarity in teichoic composition and flagellin antigens between these organisms: flagellin antigens and the backbone of the teichoic acid have the same composition in serogroup 4 and *L. innocua*, with significant differences from the corresponding structures present in serogroup 1/2 or 3 strains.[1,26] These findings are also in agreement with the fact that certain *L. innocua* strains were found to harbor serotype 4b–specific teichoic acid decorations and phage specificity, along with genes such as *gtcA* and *gltA-gltB*, mediating decoration of the polymer with galactose and glucose.[27] In addition, genome sequencing and other studies have revealed several genes harbored in common by *L. monocytogenes* serotype 4b and *L. innocua*, but not by strains of serogroup 1/2.[28]

11.2.1.3 The Serotype 4b Complex (Serotypes 4b, 4d, 4e)

The findings of common genetic markers among serotype 4b, 4d, and 4e strains[5] are worthy of special note. Serotype 4d and 4e strains are relatively rare among clinical or food isolates of *L. monocytogenes*, but antigenically and genetically appear to be highly similar to 4b. In fact, it has been to date impossible to identify molecular or well-defined immunological tools that consistently differentiate serotype 4b strains from strains of serotype 4d and 4e.[29–32] The designation "serotype 4b complex" has been employed to refer to *L. monocytogenes* in the lineage I subcluster that includes serotype 4b, 4d, and 4e strains.[33]

As will be discussed later, a small fraction of serotype 4b strains cluster outside lineage I into a distinct lineage (lineage III), along with strains of serotypes 4a and 4c.[9] Even though such strains were detected among human clinical isolates of serotype 4b, they appear to be rare among food and human clinical isolates; their exact prevalence among strains of serotype 4b is currently not known. DNA-based analysis suggests that such serotype 4b lineage III strains have undergone loss or pronounced divergence of two genes that were otherwise conserved among, and specific for, strains of the serotype 4b complex.[34] It is quite possible that these strains differ from other serotype 4b complex strains in regard to presence or content of additional genes, and preliminary data from our laboratory support such divergence (Y. Cheng and S. Kathariou, unpublished data).

11.2.1.4 Lineage I versus Serotype 4b in the Context of Epidemiology and Pathogenesis

The presence within lineage I of the majority of serotype 4b strains confers significant epidemiological importance to this lineage. This is because serotype 4b strains have been implicated in the majority of documented food-borne outbreaks of listeriosis (the serotype 4b epidemic clones will be discussed later in the chapter) and, in addition, serotype 4b strains contribute to a significant fraction of sporadic disease.[2] This has often led to statements suggesting general predominance of lineage I in human illness, especially in epidemics.[3,5] However, it must be kept in mind that lineage I includes not only the serotype 4b complex but also strains of serotypes 1/2b and 3b, which are encountered among human isolates (1/2b much more so than 3b) but are not predominant among outbreaks. Furthermore, a significant contribution to human illness (especially sporadic listeriosis) is made by serotype 1/2a strains, which are members of lineage II.

Thus, it is not lineage I as a whole, but instead only serotype 4b strains within this lineage that are a predominant contributor to human listeriosis, especially in regard to outbreaks of food-borne disease. Furthermore, inferences to unusually high pathogenic potential of lineage I[3] may in fact apply correctly not to the entire lineage but to serotype 4b strains within this lineage that may have higher pathogenic potential,[3,35,36] possibly accounting for their predominance among human epidemics and sporadic disease. Serotype 1/2b strains remain understudied in terms of virulence, and serotype 3b is relatively rare among human illness. Thus, current data do not permit extrapolation of pathogenesis or epidemiological findings based on serotype 4b to the entire lineage I.

11.2.1.5 Subclusters within Lineage I: Serotypes 1/2b and 3b versus the Serotype 4b Complex (4b, 4d, 4e)

Even though serotype 1/2b and 3b strains have been repeatedly detected as a well-defined subcluster within lineage I,[5,21] unique attributes that may differentiate them from serotype 4b remain largely uncharacterized. DNA array hybridizations identified five open reading frames (ORFs) that were harbored by all surveyed 1/2b and 3b ($n = 20$) and 4b complex ($n = 27$) strains; in addition, three ORFs were harbored by serotype 4b complex strains, but not by strains of serotype 1/2b or 3b.[5]

These findings were valuable in the design of a PCR assay that can discriminate among strains of serotype 1/2b (or 3b) and those of the serotype 4b complex.[5] However, our understanding of genomic content and evolution of serotype 1/2b and 3b strains has been hampered by absence of large-scale genome sequence data; to date genome sequence analysis has been performed and published for only four *L. monocytogenes* strains (two of serotype 1/2a; EGD-e and F6854, and two of serotype 4b; F2365 and H7858).[24,37]

From this perspective, the recent genome sequencing of serotype 1/2b strains (FSL J1-194 from a sporadic human listeriosis; FSL R2-503 from a gastrointestinal listeriosis outbreak; FSL J1-175 from water; FSL J2-064 from food isolates) by the Broad Institute (http://www.broad.mit.edu) is an important development, since it will allow the beginning of detailed genome content comparisons between different members of lineage I. Serotype 1/2b strains are involved in sporadic human listeriosis and have also been implicated in outbreaks of febrile gastroenteritis,[38,39] although invasive listeriosis outbreaks associated with strains of this serotype have not been described. Thus, comparative genomic analysis is expected to be of special interest, given the apparent genomic relatedness but markedly different contributions to invasive illness between 1/2b and 4b strains.

11.2.2 Lineage II

11.2.2.1 Molecular Identification and Further Partitioning within Lineage II

Studies that led to the identification of lineage I also revealed the presence of the other major lineage, lineage II, which includes strains of serotype 1/2a, 3a, 1/2c, and 3c.[3,5,6,8,10,13,14,17,18] As with lineage I, high-resolution subtyping revealed the presence of additional clusters within lineage II. Two subclusters have been identified, which themselves agree with serotypic groupings: One includes strains of serotypes 1/2a and 3a, and the other includes strains of serotype 1/2c and 3c. As was the case with the lineage I subclusters, the lineage II subclusters correlated with the presence of the flagellin antigen (the antigenic moiety responsible for the "a" or "c" in the serotypic designation).

Several lineage II–specific marker genes were identified following DNA array hybridization analysis. As with the lineage I–specific marker genes, several of these genes encoded putative surface-associated proteins or regulators.[5] The majority of these genes were shared between the two subclusters (strains of serotype 1/2a and 3a vs. 1/2c and 3c). Only five genes in one contiguous fragment, and two genes in another fragment, were more frequently found among 1/2c or 3c strains versus 1/2a or 3a. However, genes exclusively harbored by all strains of one of the subclusters were not identified in this study. These findings suggest pronounced potential for gene transfer among serotype 1/2a and 3a, and serotype 1/2c and 3c strains, which may homogenize the gene pool and reduce the likelihood of establishment of genes exclusively harbored by one or the other of the subclusters. Nonetheless, such studies were still able to identify DNA sequences that could be used in a PCR-based format for putative serotype determinations of serotype 1/2a or 3a versus 1/2c or 3c.[5]

To date, the genome sequences of two strains of serotype 1/2a (EGD-e and F6854) have been determined and published.[24,37] The genome sequences of several additional serotype 1/2a strains have been determined by the Broad Institute initiative. These include J0161 and J2818 (clinical and food isolates linked to the consumption of sliced turkey, respectively),[40] F6900 (isolated from the same processing plant where J0161 and J2818 came 11 years later), 10403S,[41] FSL J2-003 (associated from ruminant listeriosis),[42] and FSL-F2-515 (the most common subtype among food isolates).[35] This has been a good development, especially since neither of the previously sequenced strains is likely to serve as an adequate representative of serotype 1/2a or lineage II organisms: EGD-e was obtained from an outbreak of laboratory animal illness, and F6854 was from a food associated with human listeriosis.

Serotype 1/2a is the serotype frequently identified among *L. monocytogenes* from foods and the food processing plant environment,[43,44] and genome sequence data from additional strains, especially strains representing food and environmental sources, would greatly complement and enhance the information currently available from the EGD-e and F6854 genomes. In addition, several studies suggest that lineage II isolates, especially those of serotype 1/2a, are noticeably more divergent than isolates of serotype 4b.[44] Thus, genome sequence data and array-based applications that include lineage- or strain-specific marker genes will further reveal the extent of genomic diversity among strains of this lineage and within the subclusters. Genome sequence information from serotype 1/2c or 3c strains would also be valuable in further evaluations of the extent of genome content diversity between the lineage II subclusters.

11.2.2.2 Lineage II versus Serotypes 1/2a and 1/2c in the Context of Epidemiology and Pathogenesis

Serotype 1/2a represents one of the three leading serotypes encountered among isolates from human listeriosis (the other two are 1/2b and 4b). In addition, there has been at least one major, multistate outbreak involving serotype 1/2a[40] as well as several outbreaks of febrile gastroenteritis.[45–47] Such contributions of serotype 1/2a strains in human illness suggest caution in discussions of lineage-specific disease burden and pathogenic potential.[3,5] As was the case with serotype 4b in the context of lineage I, in the case of lineage II it must be emphasized that there is no evidence for lineage-wide differences in pathogenicity or disease burden in comparison to other lineages. Instead, certain serotypes or clonal groups within the lineage may have attributes that reduce their pathogenicity, as evidenced by their low prevalence in human illness.

Substantial evidence exists that certain strains within this lineage produce a truncated internalin (the gene product of *inlA*) and, possibly for this and other reasons, they have attenuated virulence in certain animal and cell culture models. The discovery of such truncated internalin-producing strains was first made with the extensively characterized strain LO28, of serotype 1/2c.[48] Truncated internalin would be unable to recognize the E-cadherin receptor and thus would fail to mediate adequate uptake of the bacteria by the host cells.[48,49]

In a subsequent analysis of clinical and food-derived strains of serotype 1/2a, 1/2b, 1/2c, and 4b, it was found that all serotype 1/2b and 4b strains expressed full-length internalin, whereas none of the tested serotype 1/2c strains (17 from foods and eight clinical isolates) produced full-length internalin. Among serotype 1/2a strains, full-length internalin was produced by 97% of the clinical isolates, but only by 63% of those from foods.[49] For clinical isolates, full-length internalin production was associated with involvement of the bacteria in perinatal listeriosis, suggesting a role of internalin in breach of the fetal–placental barrier.[49] Such findings suggest pronounced serotype- and lineage-specific differences in distribution of strains expressing this virulence determinant and may partially explain the observed low prevalence of serotype 1/2c strains in human illness. Furthermore, they suggest the presence of a subset within serotype 1/2a, primarily from foods, which would be expected to be attenuated in internalin-mediated virulence. Such putative virulence-attenuated isolates from foods described in other studies.[35,50]

If such isolates with truncated internalin are indeed attenuated for human virulence, it is intriguing that they are noticeably more prevalent among serotype 1/2a isolates (especially from foods) and among isolates of serotype 1/2c (lineage II subcluster) than among 1/2b and the serotype 4b complex (lineage I subclusters). In the case of serotype 1/2a, the pronounced diversity among strains and the frequent presence of these strains in foods may facilitate establishment of mutations that may be advantageous (or at least not deleterious) in the food environment, even though they may result in virulence attenuation and are therefore under-represented among clinical isolates. For reasons that are not clear, such putative virulence-attenuating mutations are apparently infrequent among serotype 1/2b or 4b isolates, even those from foods. In either case, however, it must be emphasized that putative virulence attenuation within lineage II is not a lineage-wide attribute, but is, instead, a characteristic of certain strains, from certain sources. In the case of serotype 1/2a, the available data would suggest that a significant proportion of the population from foods is virulent, and this is in fact supported by the epidemiological evidence of serotype 1/2a strains' frequent involvement in human listeriosis.

11.2.3 Lineage III

Rasmussen et al.[7] identified for the first time a third *L. monocytogenes* phylogenetic lineage, lineage III, based on allelic analysis of *flaA*, *iap*, and *hly*. The existence of lineage III was further confirmed by ribotyping and sequence analysis of virulence genes, as well as comparative genomics and DNA sequencing of the controlled virulence gene cluster (*p*VGC).[3,5,9] When lineage III was initially identified, it was reported to contain serotypes 4a and 4c. Lineage III isolates were reported to be more frequently implicated in animal listeriosis than in human disease.[3] As described earlier, certain serotype 4b strains are also included in this lineage.[9] Furthermore, certain non-4a and non-4c strains that may be related to serotype 7 (designated "subgroup IIIB") have been reported to be included in lineage III.[51] Phylogenetic analysis of *sigB* and *actA* sequences indicated that lineage III represents three distinct subgroups, termed IIIA, IIIB, and IIIC. Unlike typical *L. monocytogenes*, all subgroup IIIB and IIIC isolates are unable to ferment rhamnose. The majority of subgroup IIIA isolates lack the putative virulence gene *lmaA*, which encodes a listerial antigen capable of eliciting a delayed-type hypersensitivity reaction in *Listeria*-immune mice. In contrast, most IIIB and all IIIC strains carry this gene.[52]

Results from DNA array hybridizations suggested that serotype 4a and 4c strains lacked several of the *L. monocytogenes*-specific genes detected in other strains. With the exception of *inlAB*, serotype 4a and 4c strains lacked other putative internalin genes. Sequence analysis suggested that the absence of such genes corresponded to deletions and also revealed that on several occasions the genomic arrangement or gene deletion was identical between serotype 4a *L. monocytogenes* and *L. innocua*.[5] Such findings led these investigators to speculate that *L. innocua* likely arose from a serogroup 4 organism, and that serotype 4a strains constitute the most identifiable link between *L. monocytogenes* and the nonpathogenic species *L. innocua*.[5] This has been supported by subsequent data based on *p*VGC sequence analysis, which suggest a sister-group relationship between lineage I and III and indicate that lineage I and III share a common ancestor.[9] This hypothesis is also consistent with phylogenetic data based on two housekeeping genes (*gap*, *prs*) and one stress gene (*sigB*),[53] as well as MLST data of nine genes (*hisJ, cbiE, truB, ribC, comEA, purA, aroE, hisC*, and *addB*).[54]

Intraperitoneal infections of mice suggested variable virulence among lineage III strains.[51] Possibly impaired virulence of certain serotype 4a and 4c strains (presumably lineage III) in animal models has also been described by others.[21,55] However, infection by such strains still conferred protective immunity to subsequent infection in mice, suggesting the potential of lineage III bacteria that may be present in foods to serve in the generation of immunity toward other, more virulent *L. monocytogenes* strains.

Lineage III strains are relatively infrequent among human isolates and their representation among food-derived strains may also be low, leading to the suggestion that the latter, rather than presumed low pathogenicity of the bacteria, accounts for their limited prevalence among human listeriosis cases.[9] These isolates may be more prevalent among animal listeriosis cases than among those from human clinical cases or from foods and food processing environments.[3,9] The ecology of lineage III strains remains poorly understood. One noticeable aspect of lineage III is the pronounced genetic diversity among strains. Mechanisms driving such diversity remain unknown, but it is tempting to speculate that specific niche adaptations may be involved (e.g., to animal hosts), and that the population structure and genomic diversity in this lineage may serve as sources of clues for genomic diversification and evolution in *L. monocytogenes* as a whole.

The bacteriological and genomic peculiarities of lineage III strains (e.g., rhamnose-negative status and genotypic characteristics) readily separate them from other *L. monocytogenes*, leading on certain occasions to the proposal for new taxonomic designations (subspecies or potentially new species) for these strains.[3,51] Genome sequencing data from serotype 4a and 4c strains, currently being obtained by the Broad Institute project will undoubtedly contribute to our understanding of the potentially unique and unusual attributes of lineage III.

11.2.3.1 Implications of Clonal Population Structure in *L. monocytogenes*

In early MLEE studies, the clonal nature of the population structure was supported by three lines of evidence, including (1) the lack of sharing of serotypes between two major primary lineages, (2) the occurrence of linkage disequilibrium (nonrandom association of alleles) among electrophoretic types (ETs) for many pairs of the enzyme loci assayed, and (3) close genetic relatedness of geographically unlinked isolates.[6] Significant linkage disequilibrium between lineage I and lineage II alleles was detected in two sets of data based on mixed-genome microarray hybridizations and MLST of nine housekeeping genes, respectively.[4,8] However, no evidence for linkage disequilibrium was detected when the analysis was performed separately in lineage I and lineage II. These data suggested that recombination is rare between strains belonging to different lineages, but gene exchanges within the lineages may be common, thus reducing evidence for clonal structure within lineage I and lineage II.

Even though linkage disequilibrium is much more pronounced between lineages than within them, several studies suggest significant clonality within lineage I. The early MLEE studies detected prominent epidemic-associated clonal groups within serotype 4b,[6,10] a finding that has been confirmed and extended in numerous other studies, and as additional outbreak-related and sporadic strains have been analyzed.[3,8,56] Clonality appears to be much higher within lineage I, suggesting limited horizontal gene transfer among the clonal groups in this lineage, and genetic variation within this lineage is noticeably reduced, in comparison to that detected in lineages II or III.[9] Data from various analyses employing different tools for detection of genetic diversity (PFGE, MLST, DNA sequencing, combined ribotyping and randomly amplified polymorphic DNA (RAPD), single nucleotide polymorphism analysis, DNA array hybridizations) suggest the presence of a relatively small number of clonal groups within lineage I and noticeably greater diversity and weaker clonal structure in lineage II.[53,54,57]

Evolutionary mechanisms responsible for the observed differences in genetic variation between the major lineages remain unidentified. One can speculate that, in the case of lineage I, relatively recent population bottlenecks and subsequent clonal expansion of the survivors may have resulted in the prevalence of the small number of disseminated clonal groups that is currently observed (especially in regard to serotype 4b). Such population bottlenecks may have resulted in response to niche contraction/loss—for instance, in response to population reductions or extinctions of a preferred animal host.

11.3 EPIDEMIC-ASSOCIATED STRAINS

Early epidemiological and population genetic studies employing MLEE showed that not all strains of *L. monocytogenes* were equally likely to be isolated from human outbreaks of listeriosis, and indicated that strains implicated in several major outbreaks were closely related genetically.[6,10] These observations were further confirmed by ribotyping, allelic analysis of virulence genes, and PFGE.[56,58,59] To date, most documented human outbreaks of food-borne listeriosis have indeed involved a small number of closely related strains, primarily of serotype 4b.[2]

The "epidemic clone" concept[60] is useful in referring to groups of genetically related isolates implicated in different, geographically and temporally unrelated epidemics, and presumably of a common ancestor. Currently, at least four epidemic-associated clonal groups have been identified and are designated Epidemic Clone I (ECI), Epidemic Clone II (ECII), Epidemic Clone III (ECIII), and Epidemic Clone IV (ECIV, earlier also designated ECIa)[2,20,61] (Table 11.1). Of these clonal

TABLE 11.1
Major Epidemic Clones Implicated in Food-borne Outbreaks of Listeriosis

Epidemic Clone	Location, Implicated Food, Year	Serotype	Ribotype	Sequence Type (MVLST)
Epidemic clone I (ECI)	Nova Scotia, coleslaw, 1981 California, Mexican-style cheese, 1985 Switzerland, soft cheese, 1983–1987 Denmark, cheese, 1985–1987 France, jellied pork tongue, 1992	4b	DUP-1038B	20
Epidemic clone II (ECII)	U.S., multistate, hot dogs, 1998–1999 U.S., multistate, turkey deli meats, 2002	4b	DUP-1044A	19
Epidemic clone III (ECIII)	U.S., multistate, turkey deli meats, 2000	1/2a	DUP- DUP-1053A	1
Epidemic clone IV (ECIV; formerly ECIa)	Boston, vegetables, 1979; Boston, milk, 1983; U.K. pate, 1989	4b	DUP-1042B	21
Epidemic clone V (ECV)	North Carolina, Mexican-style cheese, 2000–2001	4b	DUP-1042B	30
Not designated	Illinois, chocolate milk, 1994	1/2b	DUP-1051B	32

Source: Modified from Kathariou, S., *Foodborne outbreaks of listeriosis and epidemic-associated lineages of* Listeria monocytogenes. Iowa State University Press, Ames, 2003; and Chen, Y., Zhang, W., and Knabel, S.J., *J. Clin. Microbiol.*, 45, 835, 2007.

groups, ECI has been involved in numerous outbreaks and has been most extensively studied.[28] In contrast, ECII, implicated in the 1998–1999 and 2002 multistate listeriosis outbreaks,[62–64] and ECIII, implicated in a 2000 multistate outbreak,[65] were identified relatively recently and are less extensively characterized. The genome sequences of representatives of three epidemic clones of *L. monocytogenes* (strains F2365, H7858, and F6854, representing ECI, ECII, and ECIII, respectively) were published in 2004. A better understanding of the physiology, ecology, and genetics of these and other epidemic-associated clones will greatly facilitate the detection of outbreaks and will contribute to the elimination of outbreak sources.

11.3.1 Epidemic Clone I (ECI)

A cluster of closely related strains, designated Epidemic Clone I (ECI) was implicated in numerous outbreaks in Europe and North America, including those in Nova Scotia, Canada (coleslaw, 1981), California (Mexican-style cheese, 1985), France (pork tongue in aspic, 1992), and others (Table 11.1). Several genetic and phenotypic characteristics specific to ECI strains have been identified. For instance, these strains share a unique RFLP in a genomic region essential for low temperature (4°C) growth of *L. monocytogenes*[66] and harbor an additional unique RFLP in a genomic region containing a putative mannitol permease locus.[67] Furthermore, their genomic DNA is resistant to digestion by the enzyme Sau3AI, suggesting methylation of cytosine at GATC sites.[68] Employing subtractive hybridization against the genome of *L. monocytogenes* EGD-e (serotype 1/2a) and subsequent hybridizations of the fragments against a panel of strains, Herd and Kocks identified several genomic fragments specific to ECI (as well as several serotype 4b-specific fragments, shared between ECI and other serotype 4b strains in the panel).[28]

The genome sequence of strain F2365, an ECI isolate from the California outbreak of 1985 (Table 11.1), was published in 2004.[37] Comparative genomic analysis between the genome of this strain and the genomes of other *L. monocytogenes* strains and of *L. innocua* strain CLIP 11262 greatly facilitated identification of strain- and serotype-specific genes.[37] Such analysis confirmed the presence of the ECI-specific DNA fragments identified by Herd and Kocks[28] and further identified several genes and gene clusters harbored by F2365 but not by other strains, the genomes of which were sequenced, including the serotype 4b strain H7858, representing a different epidemic-associated clonal group (ECII).[37]

The 51 F2365-specific genes were found in eight gene clusters (2–15 ORFs); only one ORF (LMOf2365_0687) was found to be alone, flanked by genes conserved among different *L. monocytogenes* strains. The G+C content of most of these genes was in the range of 24–33%. This range is noticeably lower than the average for *L. monocytogenes* (38%), suggesting the possibility that these genes were acquired by horizontal transfer from currently unidentified bacteria. Sequence analysis suggested that several of the genes may encode surface-associated proteins or be involved in carbohydrate transport and metabolism.

Using DNA hybridizations with a panel of strains from outbreaks and other sources, we have confirmed that the majority of the F2365-specific genes and gene clusters were harbored by all other tested ECI strains, but not other strains of serotype 4b, or other serotypes. Exceptions were a cluster of genes shared between ECI strains and other lineage I strains (1/2b, 3b) as well as a cluster that is shared between ECI and several other serotype 4b strains, but divergent in another major epidemic clone, ECII.[69] One can speculate that the genes present in ECI and strains of serotype 1/2b or 3b, but not other serotype 4b strains, may represent genomic remnants of sequences that were harbored by an ancestral lineage ECI strain, and were removed, by gene loss, from many serotype 4b strains, but were maintained in ECI. Alternatively, such sequences may have been independently acquired by various subclusters of lineage I (e.g., ECI and serotype 1/2b or 3b) through horizontal transfer events. The atypical G+C content of the genes supports acquisition of the genes by horizontal transfer, whether such acquisition took place in an ancestral strain or subsequently.

One F2365 gene cluster included genes encoding a putative Sau3AI-like restriction endonuclease, a DNA methylase with GATC site specificity, and a putative DNA binding/regulatory protein.[37] These putative functions suggested that this gene cluster likely represented the genes responsible for the extensively documented ability of ECI strains to methylate their genomic DNA at cytosines of GATC sites, resulting in DNA that is resistant to digestion by Sau3AI.[68] The presence of a functional methylase-encoding gene with GATC specificity in this cluster was confirmed in our laboratory by construction of a mutant harboring a deletion of the entire cluster, demonstrating that the mutant's DNA was fully susceptible to Sau3AI-digestion. Deletion mutants in several other ECI-specific genes and gene clusters have been constructed in our laboratory. Characterization of these mutants is being pursued, in order to obtain further information on contributions of these genes to possibly enhanced pathogenicity of ECI strains and on the ability of these strains to survive and grow in specific environmental niches. With the exception of the DNA restriction-modification cassette discussed previously, the functional roles of the other ECI-specific genes and gene cassettes have not yet been determined.

11.3.2 Epidemic Clone II (ECII)

In 1998 and 1999, a new genotype of *L. monocytogenes* serotype 4b was implicated in multistate outbreaks of listeriosis in the United States that involved contaminated hot dogs and were responsible for 101 human cases, including 21 deaths.[62,63] Strains from these outbreaks exhibited unique ribotype and PFGE patterns that had been only rarely encountered before in the national PulseNet database. The 1998–1999 hot dog *L. monocytogenes* outbreak isolates were also found to have diversification in a chromosomal region ("region 18") that was otherwise conserved among other *L. monocytogenes* of serotype 4b.[69] Fragments in this region were first identified as putative serotype 4b-specific fragments in the subtractive hybridization studies of Herd and Kocks.[28] The panel of serotype 4b strains used in that study[28] did not include strains from the 1998–1999 outbreak, and the diversification was therefore not noted until later. Since DNA sequence content in this region is markedly different between F2365 and H7858, PCR with F2365-derived primers amplified the expected fragments from F2365 and other ECI strains, as well as most other tested serotype 4b strains, but not from the 1998–1999 outbreak strains. Similarly, the labeled PCR products did not yield hybridization signals when used as probes in Southern blots of DNA from the 1998–1999 outbreak strains.[69] The overall genetic features of the 1998–1999 outbreak strains (based on PFGE, ribotype, and the region 18 data) suggested that these strains represented a novel clonal group distinct from ECI, designated Epidemic Clone II (ECII).

In 2002, another multistate outbreak of listeriosis also involved bacteria of serotype 4b and was attributed to contaminated turkey deli meats. PFGE and ribotyping data revealed that the isolates from the 1998–1999 and 2002 outbreaks were closely related.[70] This was further investigated by hybridizations with genomic markers identified on the basis of the genome sequencing of one of the 1998–1999 outbreak strains (H7858) as well as by DNA array hybridizations and by sequence-based analysis.[20,71] The findings suggested that the 1998–1999 and 2002 outbreaks were in fact caused by closely related strains that were members of the same epidemic clone, ECII. It is indeed intriguing that such strains, previously rare among clinical or other isolates (and, in fact, not identified as a clonal group prior to 1998) resulted in two major, consecutive outbreaks.

The publication of the genome sequence of the ECII strain H7858 greatly facilitated the identification of genomic fragments unique to ECII strains. Comparative genome sequence analysis identified 61 H7858-specific genes. As was also the case with the 51 genes specific to F2365, most genes specific to H7858 were found in a small number of gene clusters, and included genes encoding putative wall- associated proteins and regulatory proteins.[37] The presence of phage-related genes within some of the larger clusters suggests that these clusters may represent prophage remnants in the chromosome.

The genome sequence also revealed the presence of the serotype 4b–specific region (region 18) that had diversified in ECII strains. This region included a number of small ORFs and was positioned between the *inlAB* locus, found to have important functions in cell invasion and virulence, and a large (ca. 6.5 kb) ORF encoding a putative wall-associated protein (*wap*). Comparative analysis with other genomes revealed that the wall-associated protein (WAP) was harbored by serotype 4b strains (a finding confirmed by Southern blot hybridizations in our laboratory) as well as by *L. innocua*. The identification of genes shared by serotype 4b and *L. innocua* strains, but absent from the genome of *L. monocytogenes* of other serotypes, is evidence of the genetic relatedness between serotype 4b *L. monocytogenes* and *L. innocua*, and supports the earlier discussed hypothesis that nonpathogenic *L. innocua* evolved from serogroup 4 strains by progressive gene loss.[5] It is intriguing that ECII strains have diversified from other serotype 4b strains in this region, flanked by key virulence genes (*inlAB*) and by a wall-associated protein gene conserved within serotype 4b. The mechanisms driving this diversification, and the functions of the genes in the region of divergence, remain to be elucidated. Mutants harboring deletions of region 18 in ECII and in ECI strains have been constructed in our laboratory, and these may prove valuable in this regard.

The H7858-specific genes identified through genome sequence comparison can facilitate the detection and further characterization of strains belonging to this clonal group. This was exemplified by the use of DNA-based reagents (primers and PCR products used as probes in hybridizations) derived from some of these ECII-specific genetic markers, including those related to diversification in genomic region 18, to confirm the close genetic relatedness between the isolates from the 1998–1999 and the 2002 outbreaks, discussed earlier.[71] However, not all H7858-specific genes are suitable for detection of ECII as a group. For instance, H7858-specific sequences detected in region 18 are also harbored by a small number of strains that appear to be unrelated to ECII based on PFGE.[71] Our laboratory has employed the H7858 genome sequence data to identify alternative genetic markers, which appear to be indeed strictly specific for ECII (i.e., they are harbored by confirmed ECII outbreak strains and by strains with identical, or highly similar, PFGE fingerprints). Application of these markers and other subtyping tools has permitted us to identify ECII strains among those from clinical isolates and from foods or the food processing environment. A study of serotype 4b complex isolates from environmental samples derived from two turkey processing plants indeed suggested the presence of ECII-like strains among these isolates.[33] Application of such genetic markers would be extremely useful in estimating the prevalence of ECII strains among other serotype 4b isolates from foods.[21,35,72]

11.3.3 Epidemic Clone III (ECIII)

In 2000, a multistate outbreak of listeriosis in the United States involved contaminated turkey deli meat products and resulted in 29 cases across 10 states.[65] These outbreak strains had indistinguishable PFGE pattern and the same ribotype. Unlike most other outbreaks, the implicated strains were serotype 1/2a, belonging to lineage II. An interesting finding was that the outbreak strains had the same genotype as a strain that was identified in an earlier, sporadic human listeriosis case associated with consumption of contaminated turkey franks, in 1988.[73] The products implicated in the 2000 multistate outbreak were from the same processing plant as the earlier isolate, suggesting that this strain had persisted there over several years, without detectable genotypic changes.[40,44]

The previously published genome sequence of *L. monocytogenes* EGD-e (serotype 1/2a) and the genome sequence of *L. monocytogenes* F6854 (from the 1988 hot dog isolate) will greatly facilitate the use of genotyping schemes to identify serotype 1/2a-specific and ECIII-specific genetic markers.[37] The genome sequence of F6854 revealed 97 strain-specific genes, many of which had attributes similar to those identified among F2365- and H7858-specific genes in terms of putative functions (cell surface association, transport and metabolism of carbohydrates, regulatory genes) and atypically low G+C content.[37] Furthermore, the Broad Institute *Listeria* genome sequence project has undertaken the genomic sequencing of three additional ECIII isolates: the clinical isolate from the 1988 case of listeriosis, along with a food-patient isolate pair from the 2000 multistate outbreak. Comparative analysis of these genome sequences will reveal possible sequence differences between food and patient isolates, as well as possible differences in the genome between the isolates from 1988 and those from 2000. Such studies may serve as a paradigm for investigations of genomic diversification of *L. monocytogenes* in foods versus infection, and throughout a specific time period.

11.3.4 Epidemic Clone IV (ECIV)

A common ribotype (DUP-1042B) was detected among serotype 4b strains implicated in three early outbreaks (the 1979 Boston outbreak epidemiologically correlated to contaminated vegetables; the 1983 Boston outbreak associated with milk, epidemiologically correlated with pasteurized milk consumption; and the 1989 U.K. outbreak arising from contaminated paté). The shared ribotype and findings of similarity in MLEE and RFLP profiles of the strains[6,35] had suggested that these three outbreaks involved closely related strains that were also similar to those of ECI and constituted a clonal group designated ECIa.[2]

Recent development of sequence-based subtyping (multilocus genotyping or MLGT), employing identification of single nucleotide polymorphisms at numerous genomic sites, and multivirulence-locus sequence typing (MVLST), employing MLST based on selected virulence genes, have permitted further subtyping of these strains.[20,61] MLGT and MVLST data suggested that the U.K. outbreak strains and the strains implicated in the two Boston outbreaks are sufficiently distinct from ECI that different clonal designations would be warranted. On the basis of MVLST, the Boston vegetable outbreak strains and the U.K. strains were assigned to one clonal group, designated Epidemic Clone IV (ECIV). The 1983 Boston milk outbreak strains were found to harbor a different sequence type and were therefore presumed to represent a different cluster, currently lacking a specific designation (Table 11.1). Even though the 1983 Boston milk outbreak strains were excluded from ECIV based on MVLST,[20] their precise relationship to ECIV remains unclear. Additional evaluations of their genomic relationship to ECIV and to other serotype 4b outbreak strains (e.g., by array analysis) would be useful.

ECIV has remained poorly characterized. Genomic fragments unique to ECI are absent in this epidemic clone,[28] and ECIV strains lack RFLPs and cytosine methylation at GATC sites that are characteristic of ECI.[66–68] Genetic and phenotypic features that may be unique to this clone have not yet been identified. For reasons that remain unclear, ECIV strains have not been encountered among subsequent outbreaks of listeriosis.

11.3.5 Epidemic Clone V (ECV)

In 2000, an outbreak of listeriosis among Hispanics (primarily pregnant women) in North Carolina was traced to contaminated Mexican-style cheese.[74] The implicated bacteria were of serotype 4b and were isolated from patients as well as from contaminated milk. Even though the ribotype of the organisms (DUP-1042B) had been encountered before among serotype 4b outbreak strains (Table 11.1), the genomic fingerprints of the North Carolina outbreak strains were highly unusual, having rarely been detected among clinical isolates in the PulseNet database prior to the outbreak.[74]

Extensive investigations of these outbreak isolates in our laboratory have indeed confirmed the genomic peculiarities of the organisms. We have found that these strains are similar to ECII in harboring diversification in region 18, and that the genome content in this region is highly similar between them and ECII strains. On the other hand, other genetic markers specific to ECII are absent from the genome of the North Carolina outbreak isolates. Furthermore, the isolates harbor at least one of the genetic markers present in ECI strains, but not in other strains of serotype 4b. The combination of unique genetic markers of these strains suggests that they represent a defined clonal group, which has been designated Epidemic Clone V (ECV). Both MLGT and MVLST indicated that ECV harbors unique nucleotide sequence polymorphism signatures not encountered among other outbreaks of listeriosis.[20,61]

In the course of subtyping of environmental (turkey processing plant environment) and clinical isolates obtained in the years subsequent to the outbreak, we have discovered that putative ECV strains continue to be present, both among clinical (presumably sporadic) and environmental isolates. Further strain typing with genetic markers showing distribution patterns unique to ECV will be useful in monitoring the prevalence of this unusual clonal group and in assessing the food safety and public health burden it is associated with.

11.3.6 Strains Implicated in Outbreaks of Febrile Gastroenteritis

L. monocytogenes has been implicated in several food-borne outbreaks of febrile gastroenteritis, without invasive illness and deaths.[39,45–47,75] With the exception of one of these outbreaks, involving corn salad contaminated with a strain of serotype 4b,[75] other febrile gastroenteritis outbreaks have involved strains of serotype 1/2a or 1/2b. In all documented cases, the implicated foods were contaminated at high levels, and the affected individuals typically were healthy and not among the categories at risk for invasive listeriosis.

Little is known about the potentially unique attributes of the strains implicated in such outbreaks. The Broad Institute has undertaken genome sequencing of two febrile gastroenteritis outbreak strains: the serotype 4b strain mentioned before[75] and the serotype 1/2b strain implicated in an outbreak traced to heavily contaminated chocolate milk. The genome sequencing findings will greatly contribute to identification of possibly unique virulence attributes of such strains (for instance, previously unknown determinants associated with diarrhea and intestinal inflammatory response) and will likely lead to identification of genetic markers that will permit accurate monitoring of such strains in foods and diarrheal stool samples. It is currently not known to what extent such strains may contribute to human diarrheal illness, many cases of which are currently unreported and undiagnosed.

11.3.7 Epidemic Clones: Prevalence in Foods and Virulence

The repeated involvement of some clonal groups (lineages) in major outbreaks has prompted the following hypotheses: these clones may (1) be particularly well adapted for growth and/or survival in foods and food processing plants, and (2) have high levels of pathogenicity to humans.

Currently, data on the prevalence of epidemic-associated strains in foods and food processing plants remain limited. A recent survey of 3063 ready-to-eat food samples revealed that 2.97% samples were positive for *L. monocytogenes*, and lineage I strains were more prevalent than lineage II (57 and 34%, respectively). However, the majority (50/57) of the lineage I isolates were of serotype 1/2b or 3b, with only 7 isolates (ca. 8% of the total) being of the serotype 4b complex.[72] In a different survey of isolates from foods (ready-to-eat meats, raw chicken, fresh produce), 41% of the isolates were of serotype 1/2b or 3b, and 32% of the isolates were of the serotype 4b complex.[76] However, prevalence of strains with epidemic clone markers was not reported in these studies. Isolates of the serotype 4b complex from various foods and from environmental samples of two turkey processing plants in the United States were found to harbor genetic markers typical of ECI and ECII, suggesting potential presence of these epidemic clonal groups in food and food processing plants.[33,77] As mentioned previously, more extensive screening of food isolates with ECI- and ECII-specific genetic markers will be invaluable in estimating the prevalence of these potentially problematic strains in foods and food processing environments.

Numerous investigations have pursued the issue of whether epidemic-associated strains may be more virulent than other strains of the species. A number of early studies using cell culture and animal models have failed to identify differences between serotype 4b epidemic-associated strains and sporadic clinical isolates of the same serotype.[2] However, in murine intragastric infections, strains derived from epidemics were found to be, as a group, more virulent than strains from foods.[36] The complications and challenges inherent in such animal models were exemplified by the fact that differences in virulence were noted in mice obtained from one supplier (Harlan), but not in mice of the same strain from a different supplier. Such findings suggest the care that needs to be applied in comparing data from different studies.

Analysis of a large panel of food and clinical isolates with ribotyping and *in vitro* cytopathogenicity assays showed that isolates of three epidemic lineage I ribotypes (DUP-1038B, DUP-1044A, and DUP-1042B, encountered in ECI, ECII, and ECIa/ECIV, respectively) had increased cytopathogenicity in a tissue culture plaque assay, compared to those of other ribotypes. All three epidemic ribotypes were overrepresented among human isolates as compared with food isolates. On the contrary, isolates with ribotype DUP-1053A (encountered in ECIII, which has serotype 1/2a) seemed less prevalent among human cases and had lower cytopathogenicity than the epidemic-associated lineage I ribotypes.[35] It is difficult from these data to assess whether the observed differences may have reflected differences among serotypes (e.g., 4b vs. 1/2a) or differences between epidemic-associated strains and those implicated in sporadic illness. Furthermore, the correlation between cell culture–based cytopathogenicity and human virulence remains speculative. A study of plaque formation in Caco-2 cells suggested that there was no significant correlation between cell culture infection outcomes and epidemiological background of the strains (e.g., epidemic/clinical vs. sporadic or food).[78]

11.4 RESERVOIRS AND ECOLOGY OF EPIDEMIC CLONES

The repeated involvement of the same epidemic clone in different outbreaks suggests that the implicated strains have specific reservoirs during the interval between outbreaks.[2] Specific environmental niches (currently unidentified), and animal or human carriers may serve as reservoirs for epidemic clones. However, transmission pathways of *L. monocytogenes* involving such sources have remained unclear. Little is currently known about the ecology and adaptations of *L. monocytogenes* and other listeriae in natural ecosystems or within the gastrointestinal microbiota of humans and other animals.

ECI strains were detected prior to their involvement in outbreaks, along with other sporadic strains that constituted the background levels of human listeriosis isolates.[2] The widespread incidence of this clonal group in human illness (both epidemics and apparently sporadic cases)[71,79] suggests that it could be ubiquitous in the environment, resulting in frequent introduction into foods. Supporting evidence for this has been obtained from surveys of isolates from foods, food processing plants, and other habitats. Such studies have indicated an unexpectedly high prevalence of strains harboring ECI-specific genomic fingerprints and genetic markers among strains of the serotype 4b complex.[33,77,79] This may reflect relatively high fitness of putative ECI strains in foods, compared to other serotype 4b strains, and may partially account for the repeated involvement of this clone in outbreaks.

A survey of environmental samples from two turkey processing plants in the United States showed that environmental and raw product samples from one plant repeatedly yielded isolates with genetic markers typical of either ECI or ECII. The finding suggested that such putative ECI and ECII *L. monocytogenes* serotype 4b isolates were established and persisted in the processing plants (or, alternatively, became repeatedly introduced from a currently unidentified source).[33] As mentioned previously, ECIII strains appeared to have persisted in the processing plant for more than a decade.[40] Taken together, such findings suggest that the food manufacturing facility may be one of the key reservoirs for these epidemic clone strains. It is clear, however, from this discussion that substantial gaps remain in our knowledge of the ecology and transmission of the *L. monocytogenes* epidemic clones, and that additional epidemiological and fundamental ecology investigations are needed in this regard.

11.5 GENETIC DIVERSITY OF OUTBREAK STRAINS

Although the strains involved in outbreaks are highly clonal, the potential for genome diversification may be significant in *L. monocytogenes*. Depending on the phenotypes that are affected and the strain characterization methods that are employed, such variants may or may not be easily detected genotypically and phenotypically. Of the clinical strains from the Nova Scotia outbreak, 27% lacked galactose substituents in the teichoic acid of the cell wall, which serves as a major 4b-specific antigen determinant and receptor for 4b-specific phage. Such variants were negative with serotype 4b–specific monoclonal antibodies and resistant to serotype 4b–specific phages. Similar variants with the same phenotypic characteristics were identified among isolates from both the California and Massachusetts outbreaks.[80] The mechanisms responsible for the establishment of monoclonal antibody c74.22-negative phenotype in epidemic strain populations remain to be elucidated. One possibility is that these phenotypes became established during isolation, passage, or storage of the bacteria in the laboratory. An alternative hypothesis is that the observed variants were selected during infection of their human host or in response to other selective pressures.

Recently, we identified variants of strains representing the California listeriosis outbreak (ECI) as well as the 1998–1999 hot-dog-caused listeriosis outbreak (ECII) that lacked recognition with the monoclonal antibody c74.22, specific for the serotype 4b complex, that also had acquired resistance to certain phages. The underlying mutations appeared to have been acquired and to have become established in the course of routine maintenance and passage of the strains in the laboratory, possibly due to enhanced fitness under laboratory conditions. The emergence and establishment of surface antigen variants demonstrate the potential for genetic and antigenic diversity among these strains and suggest, among other things, caution in interpreting phenotypes of serotype 4b strains and genetic constructs thereof.

Additional evidence for genetic diversification among isolates of the same epidemic clone—in fact, of the same outbreak—was provided by PFGE analysis of isolates from the 1998–1999 outbreak, which revealed minor (one or two bands, with one of the two enzymes employed for PFGE) but consistent variations in the PFGE profiles among isolates from this outbreak.[81] Genome diversification among strains from different patients from this outbreak was also reported in another study. Distinct patterns of nucleotide sequence polymorphism were detected for each strain, even though the strains shared the same ribotype and had a common PFGE pattern, with only two strains differing by a single band.[21]

The genome sequencing of strain H7858 (1998–1999 hot dog listeriosis outbreak) revealed the presence of a plasmid of ca. 80 kb, designated pLM80.[37] This plasmid harbored several insertion elements and putative transposons absent from the chromosome of the strain, leading to the speculation that it may represent a relatively recent acquisition by the ECII bacteria.[37] We have found that some of the 1998–1999 outbreak isolates lacked this plasmid, and in PFGE analysis with ApaI they also lacked a fragment of 80 kb (the plasmid harbors a single ApaI site). The absence of this 80-kb ApaI fragment led to some of the diversity in PFGE patterns detected among the 1998–1999 isolates.[81]

Interestingly, ECII strains from the 2002 outbreak also vary in presence or absence of a plasmid, which differs in size from pLM80 (being ca. 55 kb and tentatively designated pLM55), but which shares at least certain DNA fragments with pLM80. We have found that changes in plasmid content also led to diversity in PFGE patterns of these strains, following digestion by ApaI (Figure 11.1). Conditions leading to loss of the plasmids from ECII strains remain to be identified. As was also the case with serotype 4b surface antigen variants described earlier for other outbreak populations, it is not clear whether pLM80 and pLM55 are lost in the original isolates (e.g., the organisms in the food, processing plant environment, or patient sample) or in the course of laboratory isolation of the organisms, and subsequent maintenance and passage of the strains in the laboratory.

Although the data from different outbreak populations support the potential for relatively rapid diversification of the genome among the isolates, there is little evidence for a common underlying

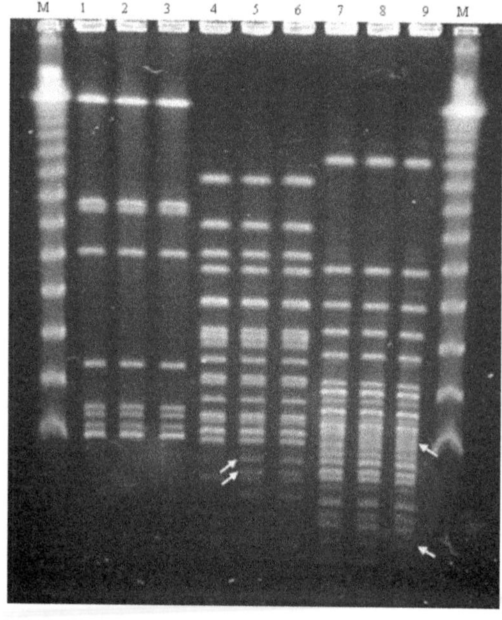

FIGURE 11.1 PFGE of ECII isolates from the 2002 listeriosis outbreak. Lanes 1, 4, 7: clinical isolate, ECII35, digested with AscI, ApaI, and SmaI, respectively; lanes 2, 5, 8: environmental isolate ECII07, digested with AscI, ApaI, and SmaI, respectively; lanes 3, 6, 9: environmental isolate ECII08, digested with AscI, ApaI, and SmaI, respectively; M: lambda concatemers, used as molecular size markers for PFGE. Arrows indicate fragments that hybridized with a plasmid-derived probe in Southern blots of the PFGE-separated fragments. Additional experiments have shown that plasmid pLM55 was harbored by both ECII07 and ECII08, but was absent from ECII35.

mechanism. Point mutations in genes involved in teichoic acid decoration with serotype-specific sugars such as galactose and glucose[82,83] may result in surface antigen, phage-resistant variants such as observed.[80] Plasmids frequently have intrinsic instability and may be lost following exposure of the bacteria to certain stress conditions (specific chemical agents, high temperature, etc.); it is conceivable that high temperatures associated with fever in listeriosis patients may result in plasmid-free derivatives of the original strain.

The observed diversity suggests caution in interpreting phenotypic and genetic findings derived from selected strains. Outbreaks are population events, and adequate analysis and characterization of the implicated organisms require inclusion of a representative population. This ideally should represent patient isolates as well as isolates from the food and processing plant implicated in the outbreak.

11.6 CONCLUSIONS AND PERSPECTIVES

11.6.1 Overall Trends from Genomic Comparisons

In order to address questions regarding the molecular basis of pathogenesis, phenotypic differences, and evolution of listeriae, the genomes of *L. monocytogenes* strain EGD-e (1/2a) and *L. innocua* CLIP 11262 (6a) were fully sequenced, followed by the genome sequencing of epidemic-associated *L. monocytogenes* strains F2365 (ECI), H7858 (ECII), and F6854 (ECIII). The genome sequence of *L. welshimeri* SLCC 5334 was recently published,[84] and the genome sequencing of other *Listeria* species (*L. seeligeri* SLCC 3954, *L. ivanovii* PAM 55, and *L. grayi*) have been undertaken.[11] Strain-specific genes have been identified in all sequenced genomes and, in the case of different strains of *L. monocytogenes,* sequencing of the genome of multiple strains of the same serotype has permitted the identification of serotype-specific genes as well.[37,84] Such efforts will be greatly enhanced by the additional genome sequencing projects that have been undertaken by the Broad Institute.

Although species-specific, serotype-specific, and strain-specific genes were identified through genome sequence analysis, the different genomes are remarkably similar in gene content and organization. The majority of genomic differences consist of phage insertions, transposable elements, scattered unique genes, and genomic islands, as well as single nucleotide polymorphisms (SNPs). Since *L. innocua* is the nonpathogenic species closest to *L. monocytogenes*, comparative analyses of the *L. monocytogenes* and *L. innocua* genomes have been extensively pursued to further elucidate virulence mechanisms and relevant adaptations, including the surface protein repertoire of pathogenic versus nonpathogenic *Listeria*.[11,24,85,86] Similarly, genes uniquely harbored by nonpathogenic listeriae may elucidate environmental adaptation mechanisms and attributes of the organism that may have characterized ancestral *Listeria* strains but have become diminished or lost in *L. monocytogenes*. Genes present in *L. innocua* CLIP 11262 that were absent from *L. monocytogenes* EGD-e were typically also absent from the genomes of the other three *L. monocytogenes* strains,[37] suggesting gene loss from a lineage ancestral to *L. monocytogenes* prior to the genomic diversification into genomic lineage I and lineage II.

11.6.2 Diversity in Surface Proteins

Listeria appears to dedicate a surprisingly large portion (at least ca. 5% in *L. monocytogenes* EGD-e) of the coding capacity of its genome to the biosynthesis of surface associated proteins.[35] Moreover, 22.6% of genes encoding putative cell-surface proteins in *L. monocytogenes* EGD-e are absent from *L. innocua*,[24] suggesting that differences between these two species, such as pathogenic potential, might be attributed to their surface protein composition.

Macroarray hybridizations and sequence analyses indicated that a group of genes encoding surface proteins that includes all known internalin genes (*inlA*, *inlB*, *inlG*, *inlH*, *inlE*, *inlC*, and *inlF*) is highly specific for the species *L. monocytogenes*. Each subgroup within each lineage of *L. monocytogenes* appeared to be characterized by a specific set of surface proteins. Data from a genomic subtraction study revealed that many epidemic-associated serotype 4b strain F4565–specific gene fragments had homologies to genes encoding surface proteins in *L. monocytogenes* or other Gram-positive bacteria,[28] and these data were confirmed and extended by subsequent genome sequencing projects.[37] Interestingly, in the rarely isolated *L. monocytogenes* serovar 4a strains, which are mostly animal pathogens, 13 of 25 *L. monocytogenes*-specific surface proteins, including all internalins except *inlAB*, were missing.[5] The lack of these surface proteins may decrease the disease potential of these strains for humans and may partially account for their lower prevalence in human illness (along with their apparent low prevalence among food isolates, described earlier).

The fact that different subgroups of each *L. monocytogenes* lineage contain different sets of surface proteins may also reflect possible differences in potential to cause illness or to adapt to different environmental niches. The functional analysis of different surface proteins and the putative strain–specific characteristics that they confer may provide additional insights into our understanding of possible tropism of *L. monocytogenes* toward different cell types and of the ubiquity of *L. monocytogenes* in the environment. Thus, comparative genome sequence analyses suggest that differences in the surface protein repertoire ("surfaceome") make important contributions to the observed differences in genome content among the *L. monocytogenes* serotypes and *Listeria* species. These findings suggest selection pressures for different combinations of cell surface characteristics in different species, lineages, serotypes, and epidemic-associated clonal groups, which perhaps impact the way the organisms interact at the cell surface with host cells and/or with the external environment.

ACKNOWLEDGMENTS

We sincerely apologize to colleagues whose work was not included in this review due to space constraints or inadvertent omission. Work from our laboratory that has been included in this review has been supported by the National Alliance for Food Safety and Security and by USDA-NRI. We are indebted to all members of our laboratory, especially Driss Elahanafi, Suleyman Yildirim, and Lili Yue, who provided some of the information summarized here. We thank E. Lanwermeyer for editorial assistance.

REFERENCES

1. Seeliger H.P.R. and Höhne, K., Serotypes of *Listeria monocytogenes* and related species. In *Methods in microbiology,* vol. 13, Bergan T., and Norris, J.R. (eds.), pp. 31–49. Academic Press, London, 1979.
2. Kathariou, S., *Food-borne outbreaks of listeriosis and epidemic-associated lineages of Listeria monocytogenes.* Iowa State University Press, Ames, 2003.
3. Wiedmann, M. et al., Ribotypes and virulence gene polymorphisms suggest three distinct *Listeria monocytogenes* lineages with differences in pathogenic potential, *Infect. Immun.*, 65, 2707, 1997.
4. Call, D.R., Borucki, M.K., and Besser, T.E., Mixed-genome microarrays reveal multiple serotype and lineage-specific differences among strains of *Listeria monocytogenes, J. Clin. Microbiol.*, 41, 632, 2003.
5. Doumith, M. et al., New aspects regarding evolution and virulence of *Listeria monocytogenes* revealed by comparative genomics and DNA arrays, *Infect. Immun.*, 72, 1072, 2004.
6. Piffaretti, J.C. et al., Genetic characterization of clones of the bacterium *Listeria monocytogenes* causing epidemic disease, *Proc. Natl. Acad. Sci. USA,* 86, 3818, 1989.
7. Rasmussen, O.F. et al., *Listeria monocytogenes* exists in at least three evolutionary lines: Evidence from flagellin, invasive associated protein and listeriolysin O genes, *Microbiology,* 141, 2053, 1995.
8. Salcedo, C. et al., Development of a multilocus sequence typing method for analysis of *Listeria monocytogenes* clones, *J. Clin. Microbiol.*, 41, 757, 2003.
9. Ward, T.J. et al., Intraspecific phylogeny and lineage group identification based on the *prfA* virulence gene cluster of *Listeria monocytogenes, J. Bacteriol.,* 186, 4994, 2004.
10. Bibb, W. F. et al., Analysis of clinical and food-borne isolates of *Listeria monocytogenes* in the United States by multilocus enzyme electrophoresis and application of the method to epidemiologic investigations, *Appl. Environ. Microbiol.*, 56, 2133, 1990.

11. Hain, T., Steinweg, C., and Chakraborty, T., Comparative and functional genomics of *Listeria* spp., *J. Biotechnol.*, 126, 37, 2006.
12. Brosch, R., Chen, J., and Luchansky, J.B., Pulsed-field fingerprinting of listeriae: Identification of genomic divisions for *Listeria monocytogenes* and their correlation with serovar, *Appl. Environ. Microbiol.*, 60, 2584, 1994.
13. Guerra, M.M., Bernardo, F., and McLauchlin, J., Amplified fragment length polymorphism (AFLP) analysis of *Listeria monocytogenes*, *Syst. Appl. Microbiol.*, 25, 456, 2002.
14. Jersek, B. et al., Typing of *Listeria monocytogenes* strains by repetitive element sequence-based PCR, *J. Clin. Microbiol.*, 37, 103, 1999.
15. Keto-Timonen, R.O., Autio, T.J., and Korkeala, H.J., An improved amplified fragment length polymorphism (AFLP) protocol for discrimination of *Listeria* isolates, *Syst. Appl. Microbiol.*, 26, 236, 2003.
16. Murphy, M. et al., Development and application of multiple-locus variable number of tandem repeat analysis (MLVA) to subtype a collection of *Listeria monocytogenes*, *Int. J. Food Microbiol.*, 115, 187, 2007.
17. Rasmussen, O.F. et al., *Listeria monocytogenes* isolates can be classified into two major types according to the sequence of the listeriolysin gene, *Infect. Immun.*, 59, 3945, 1991.
18. Ripabelli, G., McLauchin, J., and Threlfall, E.J., Amplified fragment length polymorphism (AFLP) analysis of *Listeria monocytogenes*, *Syst. Appl. Microbiol.*, 23, 132, 2000.
19. Chen, Y., Zhang, W., and Knabel, S.J., Multi-virulence-locus sequence typing clarifies epidemiology of recent listeriosis outbreaks in the United States, *J. Clin. Microbiol.*, 43, 5291, 2005.
20. Chen, Y., Zhang, W., and Knabel, S.J., Multi-virulence-locus sequence typing identifies single nucleotide polymorphisms which differentiate epidemic clones and outbreak strains of *Listeria monocytogenes*, *J. Clin. Microbiol.*, 45, 835, 2007.
21. Zhang, C. et al., Genome diversification in phylogenetic lineages I and II of *Listeria monocytogenes*: Identification of segments unique to lineage II populations, *J. Bacteriol.*, 185, 5573, 2003.
22. Cabanes, D. et al., Surface proteins and the pathogenic potential of *Listeria monocytogenes*, *Trends Microbiol.*, 10, 238, 2002.
23. Camilli, A., Tilney, L.G., and Portnoy, D.A., Dual roles of *plcA* in *Listeria monocytogenes* pathogenesis, *Mol. Microbiol.*, 8, 143, 1993.
24. Glaser, P. et al., Comparative genomics of *Listeria* species, *Science*, 294, 849, 2001.
25. Graves, L.M., Swaminathan, B., and Hunter, S.B., *Listeria, listeriosis and food safety,* 2nd ed. Marcel Dekker, Inc., New York, 1999.
26. Fiedler, F., Biochemistry of the cell surface of *Listeria* strains: A locating general view, *Infection,* 16 (Suppl. 2), S92, 1988.
27. Lan, Z., Fiedler, F., and Kathariou, S., A sheep in wolf's clothing: *Listeria innocua* strains with teichoic acid-associated surface antigens and genes characteristic of *Listeria monocytogenes* serogroup 4, *J. Bacteriol.*, 182, 6161, 2000.
28. Herd, M., and Kocks, C., Gene fragments distinguishing an epidemic-associated strain from a virulent prototype strain of *Listeria monocytogenes* belong to a distinct functional subset of genes and partially cross-hybridize with other *Listeria* species, *Infect. Immun.*, 69, 3972, 2001.
29. Doumith, M. et al., Differentiation of the major *Listeria monocytogenes* serovars by multiplex PCR, *J. Clin. Microbiol.*, 42, 3819, 2004.
30. Kathariou, S. et al., Monoclonal antibodies with a high degree of specificity for *Listeria monocytogenes* serotype 4b, *Appl. Environ. Microbiol.*, 60, 3548, 1994.
31. Lei, X.H., Promadej, N., and Kathariou, S., DNA fragments from regions involved in surface antigen expression specifically identify *Listeria monocytogenes* serovar 4 and a subset thereof: Cluster IIB (serotypes 4b, 4d, and 4e), *Appl. Environ. Microbiol.*, 63, 1077, 1997.
32. Palumbo, J.D. et al., Serotyping of *Listeria monocytogenes* by enzyme-linked immunosorbent assay and identification of mixed-serotype cultures by colony immunoblotting, *J. Clin. Microbiol.*, 41, 564, 2003.
33. Eifert, J.D. et al., Molecular characterization of *Listeria monocytogenes* of the serotype 4b complex (4b, 4d, 4e) from two turkey processing plants, *Foodborne Pathol. Dis.*, 2, 192, 2005.
34. Liu, D. et al., *Listeria monocytogenes* serotype 4b strains belonging to lineages I and III possess distinct molecular features, *J. Clin. Microbiol.*, 44, 214, 2006.
35. Gray, M.J. et al., *Listeria monocytogenes* isolates from foods and humans form distinct but overlapping populations, *Appl. Environ. Microbiol.*, 70, 5833, 2004.
36. Kim, S.H. et al., Oral inoculation of A/J mice for detection of invasiveness differences between *Listeria monocytogenes* epidemic and environmental strains, *Infect. Immun.*, 72, 4318, 2004.

37. Nelson, K.E. et al., Whole genome comparisons of serotype 4b and 1/2a strains of the food-borne pathogen *Listeria monocytogenes* reveal new insights into the core genome components of this species, *Nucleic Acids Res.*, 32, 2386, 2004.
38. Dalton, C.B. et al., An outbreak of gastroenteritis and fever due to *Listeria monocytogenes* in milk, *N. Engl. J. Med.*, 336, 100, 1997.
39. Salamina, G. et al., A food-borne outbreak of gastroenteritis involving *Listeria monocytogenes, Epidemiol. Infect.*, 117, 429, 1996.
40. Olsen, S.J. et al., Multistate outbreak of *Listeria monocytogenes* infection linked to delicatessen turkey meat, *Clin. Infect. Dis.*, 40, 962, 2005.
41. Edman, D.C., Pollock, M.B., and Hall, E.R., *Listeria monocytogenes* L forms. I. Induction maintenance, and biological characteristics, *J. Bacteriol.*, 96, 352, 1968.
42. Nightingale, K.K. et al., Ecology and transmission of *Listeria monocytogenes* infecting ruminants and in the farm environment, *Appl. Environ. Microbiol.*, 70, 4458, 2004.
43. Jacquet, C. et al., Expression of ActA, Ami, InlB, and listeriolysin O in *Listeria monocytogenes* of human and food origin, *Appl. Environ. Microbiol.*, 68, 616, 2002.
44. Kathariou, S., *Listeria monocytogenes* virulence and pathogenicity, a food safety perspective, *J. Food Prot.*, 65, 1811, 2002.
45. Carrique-Mas, J.J. et al., Febrile gastroenteritis after eating on-farm manufactured fresh cheese—An outbreak of listeriosis? *Epidemiol. Infect.*, 130, 79, 2003.
46. Frye, D.M. et al., An outbreak of febrile gastroenteritis associated with delicatessen meat contaminated with *Listeria monocytogenes, Clin. Infect. Dis.*, 35, 943, 2002.
47. Miettinen, M.K. et al., Molecular epidemiology of an outbreak of febrile gastroenteritis caused by *Listeria monocytogenes* in cold-smoked rainbow trout, *J. Clin. Microbiol.*, 37, 2358, 1999.
48. Jonquieres, R. et al., The *inlA* gene of *Listeria monocytogenes* LO28 harbors a nonsense mutation resulting in release of internalin, *Infect. Immun.*, 66, 3420, 1998.
49. Jacquet, C. et al., A molecular marker for evaluating the pathogenic potential of food-borne *Listeria monocytogenes, J. Infect. Dis.*, 189, 2094, 2004.
50. Roche, S.M. et al., Assessment of the virulence of *Listeria monocytogenes*: Agreement between a plaque-forming assay with HT-29 cells and infection of immunocompetent mice, *Int. J. Food Microbiol.*, 68, 33, 2001.
51. Liu, D. et al., *Listeria monocytogenes* subgroups IIIA, IIIB, and IIIC delineate genetically distinct populations with varied pathogenic potential, *J. Clin. Microbiol.*, 44, 4229, 2006.
52. Roberts, A. et al., Genetic and phenotypic characterization of *Listeria monocytogenes* lineage III, *Microbiology*, 152, 685, 2006.
53. Nightingale, K.K., Windham, K., and Wiedmann, M., Evolution and molecular phylogeny of *Listeria monocytogenes* isolated from human and animal listeriosis cases and foods, *J. Bacteriol.*, 187, 5537, 2005.
54. Meinersmann, R.J. et al., Multilocus sequence typing of *Listeria monocytogenes* by use of hypervariable genes reveals clonal and recombination histories of three lineages, *Appl. Environ. Microbiol.*, 70, 2193, 2004.
55. Chakraborty, T. et al., Naturally occurring virulence-attenuated isolates of *Listeria monocytogenes* capable of inducing long-term protection against infection by virulent strains of homologous and heterologous serotypes, *FEMS Immunol. Med. Microbiol.*, 10, 1, 1994.
56. Buchrieser, C. et al., Pulsed-field gel electrophoresis applied for comparing *Listeria monocytogenes* strains involved in outbreaks, *Can. J. Microbiol.*, 39, 395, 1993.
57. Mereghetti, L. et al., Combined ribotyping and random multiprimer DNA analysis to probe the population structure of *Listeria monocytogenes, Appl. Environ. Microbiol.*, 68, 2849, 2002.
58. Brosch, R. et al., Genomic fingerprinting of 80 strains from the WHO multicenter international typing study of *Listeria monocytogenes* via pulsed-field gel electrophoresis (PFGE), *Int. J. Food Microbiol.*, 32, 343, 1996.
59. Jacquet, C. et al., Investigations related to the epidemic strain involved in the French listeriosis outbreak in 1992, *Appl. Environ. Microbiol.*, 61, 2242, 1995.
60. Orskov, F., and Orskov, I., From the National Institutes of Health. Summary of a workshop on the clone concept in the epidemiology, taxonomy, and evolution of the Enterobacteriaceae and other bacteria, *J. Infect. Dis.*, 148, 346, 1983.
61. Ducey, T.F. et al., A single-nucleotide-polymorphism-based multilocus genotyping assay for subtyping lineage I isolates of *Listeria monocytogenes, Appl. Environ. Microbiol.*, 73, 133, 2007.

62. CDC, Multistate outbreak of listeriosis—United States, 1998–1999, *MMWR Morb. Mortal. Wkly. Rep.*, 47, 1085, 1999.
63. CDC, Multistate outbreak of listeriosis—United States, 1998, *MMWR Morb. Mortal. Wkly. Rep.*, 47, 1085, 1998.
64. CDC, Outbreak of listeriosis—northeastern United States, 2002, *MMWR Morb. Mortal. Wkly. Rep.*, 51, 950, 2002.
65. CDC, Multistate outbreak of listeriosis—United States, 2000, *MMWR Morb. Mortal. Wkly. Rep.*, 49, 1129, 2000.
66. Zheng, W., and Kathariou, S., Differentiation of epidemic-associated strains of *Listeria monocytogenes* by restriction fragment length polymorphism in a gene region essential for growth at low temperatures (4°C), *Appl. Environ. Microbiol.*, 61, 4310, 1995.
67. Tran, H. L., and Kathariou, S., Restriction fragment length polymorphisms detected with novel DNA probes differentiate among diverse lineages of serogroup 4 *Listeria monocytogenes* and identify four distinct lineages in serotype 4b, *Appl. Environ. Microbiol.*, 68, 59, 2002.
68. Zheng, W., and Kathariou, S., Host-mediated modification of Sau3AI restriction in *Listeria monocytogenes*: Prevalence in epidemic-associated strains, *Appl. Environ. Microbiol.*, 63, 3085, 1997.
69. Evans, M.R. et al., Genetic markers unique to *Listeria monocytogenes* serotype 4b differentiate epidemic clone II (hot dog outbreak strains) from other lineages, *Appl. Environ. Microbiol.*, 70, 2383, 2004.
70. CDC, Public health dispatch: Outbreak of listeriosis—northeastern United States, 2002, *MMWR Morb. Mortal. Wkly. Rep.*, 51, 950, 2002.
71. Kathariou, S. et al., Involvement of closely related strains of a new clonal group of *Listeria monocytogenes* in the 1998–1999 and 2002 multistate outbreaks of food-borne listeriosis in the United States, *Foodborne Pathog. Dis.*, 3, 292, 2006.
72. Shen, Y. et al., Isolation and characterization of *Listeria monocytogenes* isolates from ready-to-eat foods in Florida, *Appl. Environ. Microbiol.*, 72, 5073, 2006.
73. CDC, Listeriosis associated with consumption of turkey franks, *MMWR Morb. Mortal. Wkly. Rep.*, 38, 267, 1989.
74. MacDonald, P.D. et al., Outbreak of listeriosis among Mexican immigrants as a result of consumption of illicitly produced Mexican-style cheese, *Clin. Infect. Dis.*, 40, 677, 2005.
75. Aureli, P. et al., An outbreak of febrile gastroenteritis associated with corn contaminated by *Listeria monocytogenes, N. Engl. J. Med.*, 342, 1236, 2000.
76. Zhang, Y. et al., Characterization of *Listeria monocytogenes* isolated from retail foods, *Int. J. Food Microbiol.*, 113, 47, 2007.
77. Yildirim, S. et al., Epidemic clone I-specific genetic markers in strains of *Listeria monocytogenes* serotype 4b from foods, *Appl. Environ. Microbiol.*, 70, 4158, 2004.
78. Pine, L. et al., Cytopathogenic effects in enterocytelike Caco-2 cells differentiate virulent from avirulent *Listeria* strains, *J. Clin. Microbiol.*, 29, 990, 1991.
79. Fugett, E.B. et al., Pulsed-field gel electrophoresis (PFGE) analysis of temporally matched *Listeria monocytogenes* isolates from human clinical cases, foods, ruminant farms, and urban and natural environments reveals source-associated as well as widely distributed PFGE types, *J. Clin. Microbiol.*, 45, 865, 2007.
80. Clark, E.E. et al., Absence of serotype-specific surface antigen and altered teichoic acid glycosylation among epidemic-associated strains of *Listeria monocytogenes, J. Clin. Microbiol.*, 38, 3856, 2000.
81. Graves, L.M. et al., Microbiological aspects of the investigation that traced the 1998 outbreak of listeriosis in the United States to contaminated hot dogs and establishment of molecular subtyping-based surveillance for *Listeria monocytogenes* in the PulseNet network, *J. Clin. Microbiol.*, 43, 2350, 2005.
82. Lei, X.H. et al., A novel serotype-specific gene cassette (*gltA-gltB*) is required for expression of teichoic acid-associated surface antigens in *Listeria monocytogenes* of serotype 4b, *J. Bacteriol.*, 183, 1133, 2001.
83. Promadej, N. et al., Cell wall teichoic acid glycosylation in *Listeria monocytogenes* serotype 4b requires *gtcA*, a novel, serogroup-specific gene, *J. Bacteriol.*, 181, 418, 1999.
84. Hain, T. et al., Whole-genome sequence of *Listeria welshimeri* reveals common steps in genome reduction with *Listeria innocua* as compared to *Listeria monocytogenes, J. Bacteriol.*, 188, 7405, 2006.
85. Bierne, H., and Cossart, P., *Listeria monocytogenes* surface proteins: From genome predictions to function, *Microbiol. Mol. Biol. Rev.*, 71, 377, 2007.
86. Buchrieser, C. et al., Comparison of the genome sequences of *Listeria monocytogenes* and *Listeria innocua*: Clues for evolution and pathogenicity, *FEMS Immunol. Med. Microbiol.*, 35, 207, 2003.

12 Analysis of Cell Envelope Proteins

Mickaël Desvaux and Michel Hébraud

CONTENTS

12.1 INTRODUCTION

From a morphological point of view, the most fundamental dichotomy within prokaryotes (the term "prokaryotes" is used here in its primary etymological sense—that is, single-celled organisms without nuclei as opposed to eukaryotes, without any further phylogenetic considerations[1]) is between those bound by a single biological membrane (monoderm prokaryotes)—that is, the cytoplasmic membrane, and those bound by two concentric but topologically different membranes (diderm prokaryotes)—that is, the inner membrane (cytoplasmic membrane) and the asymmetric outer membrane.[2] In accordance with holistic and teleonomic concepts, organisms are far more than mere collections of genes,[3,4] and such difference in membrane organization, and thus cell compartmentation, is not trivial but has profound phylogenetic, structural, metabolic, and physiological implications. Based on the most recent advances in biological evolution and megaclassification of organisms,[5–7] monoderm prokaryotes are regrouped under the term Monodermata (also called Unibacteria), which essentially includes Archaea together with Posibacteria (formerly called Gram-positive bacteria).

It is worth stressing that the term "Gram-positive bacteria" is terminologically ambiguous, especially for researchers interested in aspects related to bacterial cell envelope (e.g., protein secretion or surface proteins).[8] From its origin, a positive or negative result given by the Gram staining method indicates whether or not bacteria retain the stain respectively. Later on, the difference in staining was related to profound divergence in structural organization of the cell envelope, briefly: (1) a cytoplasmic membrane surrounded by a thick cell wall in Gram-positive bacteria, and (2) a cytoplasmic membrane surrounded by a thin cell wall beneath the outer membrane in Gram-negative bacteria. Molecular analyses further revealed that, contrary to Gram-negative bacteria, Gram-positive bacteria correspond to a phylogenetically coherent grouping of prokaryotes within the domain Bacteria with phylum BXIII Firmicutes (low G+C mole percent) and phylum BXIV Actinobacteria (high G+C mole percent).[9,10] However, from Gram staining to cell envelope organization to taxonomic grouping, each step represents some approximations, which often result in misleading or incoherent statements in the literature. For example, some members of Firmicutes and Actinobacteria phyla do not retain Gram stain because of (1) the absence of a cell wall (e.g., bacteria from the genus *Mycoplasma*), (2) a too thin cell wall (e.g., some members of the genus *Clostridium*), or (3) the presence of a waxy outer sheath preventing penetration of the stain (e.g., species from the genus *Mycobacterium*).

Inversely, some bacteria not taxonomically related to Gram-positive bacteria retain the Gram stain (e.g., some members of the phylum BIV Deinococcus-Thermus). More confusingly, some bacteria clearly possessing a Gram-negative-like cell envelope architecture are in fact phylogenetically related to the taxonomic group of Gram-positive bacteria (e.g., *Thermotoga maritima* currently classified in phylum BII Thermotogae,[11] or *Fusobacterium nucleatum* belonging to phylum BXXI Fusobacteria).[12,13] Some other phyla regroup bacteria exhibit both cell envelope structures (Gram-negative-like or Gram-positive-like cell envelope)—for example, BVI Chloroflexi or BVII Thermomicrobia.[14] Even in some deep branches of the phylum Firmicutes, some bacteria clearly exhibit Gram-negative cell envelope ultrastructure (e.g., in genus *Desulfotomaculum*, *Selenomonas*, *Syntrophomonas,* or *Coprothermobacter*).[2] Therefore, it appears in numerous cases that the term "Gram-positive bacteria" cannot describe at once a particular Gram staining result, cell envelope organization, and taxonomic group; thus, when employing this term it is extremely important to specify what it refers to. Because of fewer terminological ambiguities, the terms "Monodermata" or "monoderm bacteria" will be preferred to describe prokaryotic cells surrounded by a single biological membrane but without any further phylogenetic considerations. For the purpose of the present review, the term "Gram-positive bacteria" will be used to describe bacteria with a cell envelope composed of (1) a cytoplasmic membrane, and (2) a cell wall composed at least of peptidoglycan.

Listeria species are monoderm bacteria possessing a thick cell wall retaining Gram stain and belonging to phylum Firmicutes, class Bacilli, order Bacillales, and family Listeriaceae,[9] and as such are Gram-positive bacteria in all meaning of the term. *L. monocytogenes* is undoubtedly the species that has attracted most attention, considering its frequent occurrence in food coupled with a high mortality rate.[15] Still, the genus *Listeria* comprises six species: (1) two pathogenic ones (*L. monocytogenes,* a human pathogen, and *L. ivanovii,* a ruminant pathogen), and (2) four nonpathogenic relatives (*L. innocua, L. seeligeri, L. welshimeri,* and *L. grayi.*)[16,17] Only two completed *L. monocytogenes* genome sequences are currently available—*L. monocytogenes* 1/2a EGD-e and 4b F2365[18,19]—but several other strains are unassembled[18] or sequenced (http://www.ncbi.nlm.nih.gov/genomes/lproks.cgi). Among other species, *L. innocua* CLIP11262[19] and *L. welshimeri* SLCC5334[20] are the only genomes available, but *L. ivanovii* PAM55, *L. seeligeri* SLCC3954, and *L. grayi* CLIP12515 are currently being sequenced.[17] Since the genomes of *L. monocytogenes* 1/2a F6854 and 4b H7858 are unfinished, some genes cannot be properly identified; also, final assembly of these genomic sequences may reveal homologues at a later date. Because no clear conclusion can be drawn from genomic analysis of unfinished genomes,[21] this review will only focus on completed genome sequences of *L. monocytogenes* strains.

12.2 PROTEIN SECRETION SYSTEMS

Within the cell envelope, *Listeria* species can exhibit a large variety of proteins; some of them can even interact with the cell surroundings and thus constitute the surfaceome (i.e., the subset of protein exposed on the bacterial cell surface). It is worth reminding that, on one hand, the cell wall is not an impermeable barrier and cell envelope proteins can interact with the environment without ever having a domain that leaves the confine of the cell wall[8] and that the extracellular milieu can penetrate the cell wall, so proteins do not necessarily need to poke out into the environment.[22] On the other hand, protein localization into the cell envelope is no guarantee that it is cell surface exposed *stricto sensu* as proteins can be masked by overlying components such as capsule polymer, for example.[8] Nevertheless, for the purpose of the present review, cell surface proteins will refer to gene products that are attached to the cell wall and/or cytoplasmic membrane and interacting with the external side, whereas cell envelope proteins will refer to all gene products present within the cell wall and/or the cytoplasmic membrane.

While cell surface proteins are systematically cell envelope proteins, the opposite is not necessarily true (e.g., proteins attached to the cytoplasmic membrane but interacting only with the cytoplasm). Still, all cell surface proteins (and most cell envelope proteins) must be first translocated to the cytoplasmic membrane via a protein secretion system before attaching to membrane or cell wall components and thus remaining in contact with the external side. Concerning the functions of cell envelope proteins, they are extremely diverse, ranging from transporters and enzymes involved in various metabolic pathways (such as carbohydrates, proteins, nucleotides, or lipids), signal transductions, and adhesion and colonization determinants, to virulence factors. It is worth stressing that among cell surface proteins, some so-called moonlighting proteins can be present.[23] Such proteins are multifunctional in the sense that they conduct enzymatic and/or nonenzymatic activities, sometimes taking part in widely divergent pathways, especially when present at different subcellular locations. For example, enolase, a cytoplasmic protein normally involved in glycolytic pathways, was found on the listerial cell surface, which can bind to human plasminogen.[24]

In Didermata (corresponding to Gram-negative bacteria, also called Negibacteria),[5,25] six major protein secretion systems (numbered from Type I to Type VI, i.e., T1SS to T6SS) are currently recognized and are restricted to these microorganisms.[26–29] In fact, protein secretion systems are categorized primarily by translocation mechanisms across the outermost lipid bilayer, which corresponds to the outer membrane in diderm bacteria but to the cytoplasmic membrane in monoderm prokaryotes. To date in monoderm bacteria, six systems are described as allowing protein secretion[30–33]—that is, protein transport from inside to outside cell cytoplasm—namely, (1) the Sec pathway (secretion, TC #3.A.5; TC#: transport classification number),[34] (2) the Tat pathway (twin-arginine translocation, TC #2.A.64), (3) the FEA (flagella export apparatus, TC #3.A.6.1), (4) the FPE (fimbrilin-protein exporter, TC #3.A.14), (5) the holins (hole-formers, TC#1.E.), and (6) the Wss (WXG100 secretion system, proteins with WXG motif of ~100 residues). To be complete, the MscL family (large conductance mechanosensitive ion channel, TC #1.A.22) and the putative Tad (tight adherence) apparatus could also be added to the list,[35,36] even though experimental evidence is not currently available in monoderm bacteria. Once translocated by one of these systems, a protein can remain associated to the cell envelope, be released into the extracellular milieu, or be translocated into a host cell.

As depicted in Figure 12.1, identification of protein secretion systems in *Listeria* involved screening of genome coding sequences (CDS) against various databases as well as bibliographic analyses. From there, Sec, Tat, FPE, FEA, holins, and Wss were identified in *L. monocytogenes*[37] (Figure 12.2). While some components of these secretion systems have been experimentally investigated, in *Listeria*, protein translocation per se has never been ascertained in any of them yet.

12.2.1 Sec

The presence, remarkable conservation, and essential nature of the Sec translocon in all living cells have given rise to the notion of a general secretory pathway (GSP) but also have led to confusing statements in the literature.[38] As illustrated in Figure 12.3, all components of the Sec translocon are encoded in *L. monocytogenes*. In addition to the SecYEG protein conducting channel, the signal

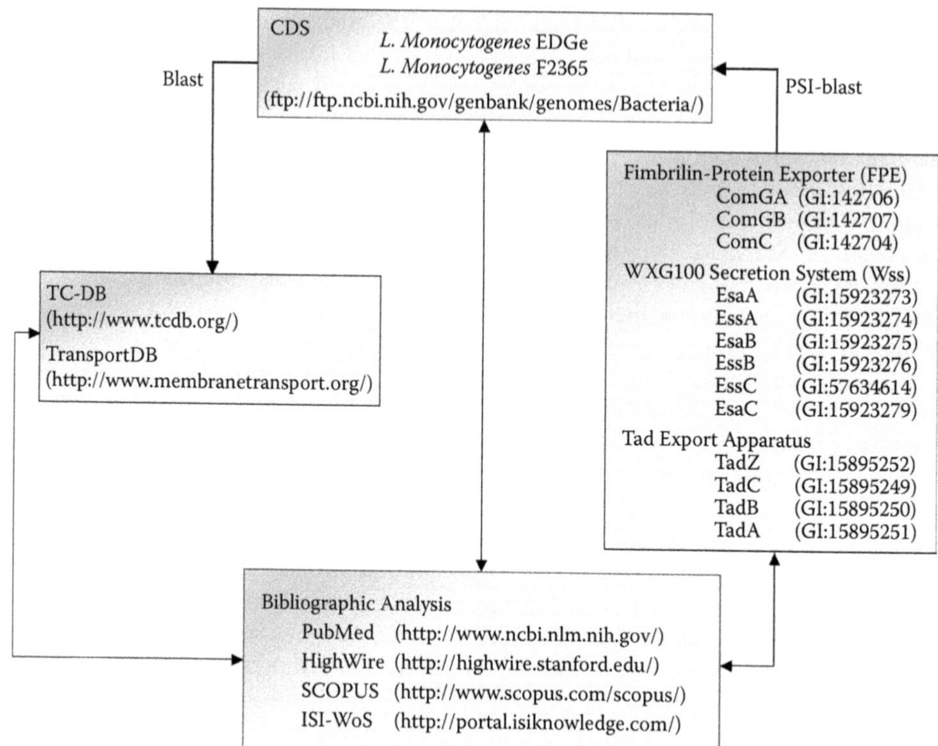

FIGURE 12.1 Genomic identification of protein secretion systems in *Listeria* species.[37] Prior to bioinformatic analysis, complete genome, coding sequences (CDS), and original annotation data sets were downloaded from GenBank. Each CDS was screened for the capacity to encode a component of a protein secretion system following BLAST against TCDB[146] and TransportDB.[248] These analyses revealed the presence of Sec components and partners as well as FEA subunits, Tat components and holins. MscL and ABC transporter truly implicated in protein secretion could not be identified. The identification of FPE was based on PSI-BLAST searches using GenBank amino acid sequences of ComGA, ComGB, and ComC from *B. subtilis* as queries. Similarly, Wss was identified using EsaA, EssA, EsaB, EssB, EssC, and EsaC from *S. aureus* as amino acid sequence queries. Using protein sequences of *Clostridium acetobutylicum* as queries,[33] Tad system components could not be identified. Overall, bibliographic analyses were also performed from various databases.

recognition particle (SRP) and the SRP receptor are ubiquitous and essential in all domains of life.[39] In *E. coli*, SRP interacts with nascent signal peptide for cotranslational translocation and specific integration of inner membrane proteins, whereas the targeting factor and chaperone SecB interacts with the mature part of the protein and allows post-translational translocation via Sec.[40] As in all Gram-positive bacteria,[41] SecB and CsaA (analogous to SecB in *B. subtilis*[30]) are absent from *L. monocytogenes*. In *E. coli*, three auxiliary proteins (SecD, SecF, and YajC) form a transmembrane complex loosely associated with SecYEG and increase the overall efficiency of protein translocation through the cytoplasmic membrane.[42]

Contrary to SecDF-YajC, the cytosolic ATPase SecA is essential to Sec-dependent translocation in bacteria as it provides the driving force for stepwise export of the protein.[43] A SecA paralogue (i.e., SecA2) has been identified in several Gram-positive bacteria including *L. monocytogenes*.[44] Contrary to *Streptococcus gordonii*, for example,[45] presence of SecA2 in *L. monocytogenes* is not accompanied by duplication of SecY. While SecA2 is not essential and its relationship with SRP/Sec is unknown, it clearly allows the secretion of a subset of proteins in *L. monocytogenes* (e.g., Iap,[44] NamA,[46] and FbpA[47]). Interestingly, the membrane protein FbpA lacks a putative N-terminal signal peptide. As in *B. subtilis*,[30] two paralogues of YidC could be identified in *L. monocytogenes*: SpoIIIJ and YqjG.[37] In *E coli*, the polytopic membrane protein YidC is necessary for cotranslational

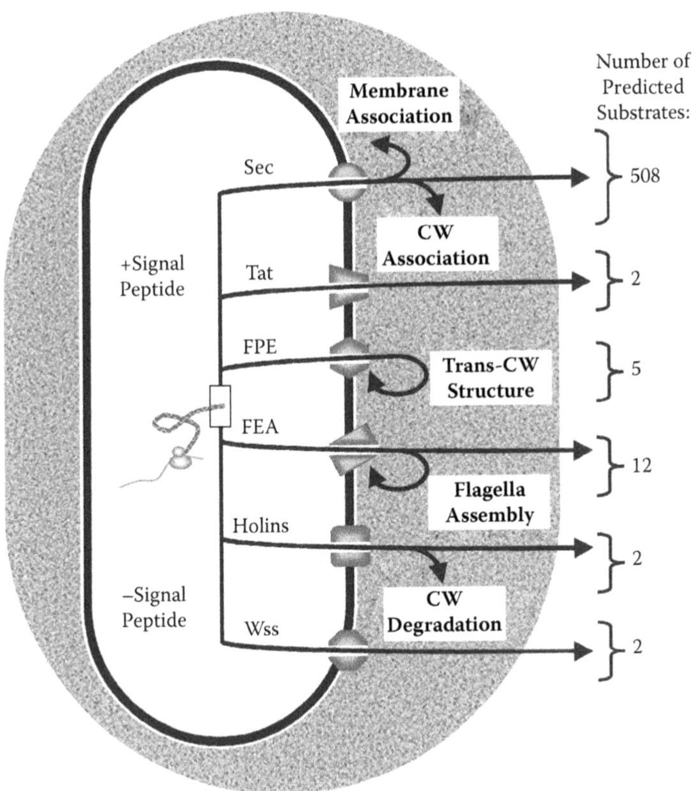

FIGURE 12.2 Schematic overview of protein secretion pathways in *L. monocytogenes* EGD-e.[37] Proteins to be translocated can exhibit (+) or not (–) an N-terminal signal peptide (with the exception of the Sec pathway, which can translocate proteins with or without signal peptide by alternative mechanisms). Proteins translocated via the Sec pathway remain membrane associated or cell wall associated, are released into the extracellular milieu, or may even be injected into an eukaryotic host cell. Proteins exported via Tat would most certainly be cell surfaced or released into the extracellular milieu. FPE would be involved in the formation of transcell-wall structures. FEA is involved in flagella assembly. Proteins exported by holins seem secreted into the extracellular milieu or involved in cell wall degradation. WXG100 proteins would be secreted into the extracellular milieu. The number of translocated proteins by each pathway is given from most recent estimations. CW, cell wall; Sec, secretion; FPE, fimbrilin-protein exporter; Tat, twin-arginine translocation; FEA, flagella export apparatus; Wss, WXG100 (proteins with WXG motif of ~100 amino acyl residues) secretion system.

insertion of all integral membrane proteins (IMPs).[48] YidC is a versatile pathway since it can be Sec-, SecA-, and/or SecB independent. In *B. subtilis*, studies have showed that SpoIIIJ and YqjG play a role in the folding of several secreted proteins and can work independently to insert integral membrane proteins.[49]

Signal peptide of translocated preprotein is cleaved by a membrane-bound signal peptidase (SPase). Different classes of N-terminal signal peptide are recognized and are cleaved by different types of SPases. Signal peptides of proteins targeted to Sec are of two classes: class 1 and class 2. Class 2 signal peptides are present in lipoproteins and are cleaved off by SPase II (for further details, see section 12.3.1.2). As depicted in Figure 12.3, precursor proteins exhibiting a class 1 signal peptide meet different fates; that is, they can (1) insert in the cytoplasmic membrane and thus become integral membrane proteins (for further details, see section 12.3.1.1), (2) remain attached covalently or noncovalently to cell wall components (for further details, see section 12.3.2), (3) be released into the extracellular milieu, or (4) be injected into a eukaryotic host cell via pores formed by Sec-secreted listeriolysin O in a process called cytolysin-mediated translocation (CMT).[50,51] It is worth noting that CMT has never been as yet reported in *Listeria*.

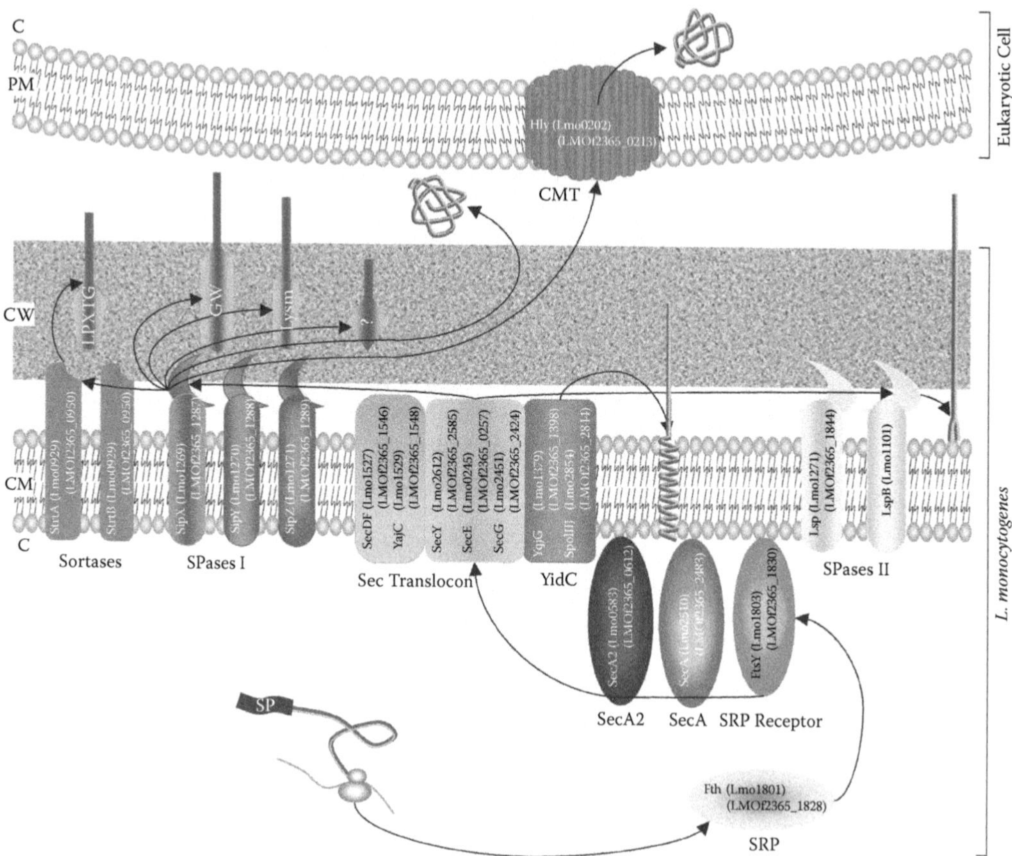

FIGURE 12.3 Schematic representation of the Sec pathway in *L. monocytogenes*.[37] N-terminal signal peptide is recognized by SRP before cotranslational translocation of the protein through the Sec translocon in a SecA-dependent manner. Some proteins with or without a signal peptide can also be translocated in a SecA2-dependent manner. Integral membrane proteins integrate into the CM via YidC homologues in Sec-dependent or -independent manner; such proteins bear stop-transfer sequence and can exhibit signal peptide or not, which can be cleaved or not. Lipoproteins, which bear signal peptide of class 2 cleavable by SPases II, are covalently attached to long-chain fatty acids of the CM. Proteins bearing class 1 signal peptide cleavable by SPases I are (1) secreted into the extracellular milieu or could even be injected into an eukaryotic host cell following CMT, thanks to pores formed by oligomerization of listeriolysin O; (2) bound to CW components via cell binding motifs (i.e., GW, LysM, or uncharacterized motifs); or (3) covalently attached to CW by sortases because of the presence of C-terminal LPXTG motif. C, cytosol; PM, plasmic membrane; CW, cell wall; EM, extracellular milieu; CM, cytoplasmic membrane; SP, signal peptide; SPase, signal peptidase; SRP, signal recognition particle; CMT, cytolysin mediated translocation.

In Gram-positive bacteria, some Sec-dependent signal peptides exhibit a YSIRK motif (PF04650) present at the beginning of the H-domain. This motif is required for efficient protein secretion and is systematically associated with an LPXTG motif, even though the opposite is not true. Class 1 signal peptides are not always cleaved as the H-domain can serve of transmembrane anchor domain as observed in SPases I. Three SPases I have been uncovered and characterized in *L. monocytogenes*: SipX, SipY, and SipZ.[52,53] Deletion of *sipY* genes had no detectable effect, whereas SipX and SipZ had overlapping substrate specificity.[52] *lsp* was demonstrated as encoding a genuine SPase II[54] and a second SPase II—LspB (Lmo1101)—was recently uncovered by genomic analysis but only in *L. monocytogenes* EGD-e.[37]

While some proteins cleaved by SPases I can remain noncovalently bound by various cell wall binding domains (for further details, see section 12.3.2.2), covalent attachment of proteins to cell wall requires sortases. Proteins emerging from the Sec apparatus and exhibiting an LPXTG-like motif C-terminally located (for further details, see section 12.3.2.1) are recognized by membrane-associated sortase.[55] Transpeptidase sortase attacks the TG bond of the LPXTG-like motif, capturing cleaved polypeptide as a thioester-linked acyl enzyme at its active site cystein residue.[56] Subsequently, this complex is resolved by the nucleophilic attack of the amino group of the cross-bridge within the lipid II precursor. Based on phylogenetic analyses, sortases are now classified into four classes, designated A, B, C, and D.[57] In *L. monocytogenes*, two sortases are present (SrtA and SrtB; Figure 12.3).

As observed in other Gram-positive bacteria, the sortase of class A (also called SrtA subfamily) in *L. monocytogenes* is encoded only once in the genome, resembles a Type II membrane protein, and is necessary for the anchoring of the majority of LPXTG-containing proteins.[58] The sortase of class B (SrtB subfamily) recognizes a particular type of sorting signal (i.e., an NXZTN motif), which suggests a lower stringency of the recognition motif of SrtB compared to SrtA.[59] Captivatingly, from investigations in *Streptococcus pyogenes* and *Staphylococcus aureus*, glycosylated LPXTGase, an enzyme that cleaves the C-terminal LPXTG motif, is the first enzyme found that is produced by nonribosomal peptide (NRP) synthesis.[60,61] It is known that NRP synthesis (and similarly related polyketide synthesis) occurs in Bacilli class, where NRPs are assembled in the cytoplasm by large megaproteins called NRP synthetases consisting of a series of active modules carrying out catalysis and modification of the tethered growing peptide chain.[62] However, investigations in *S. aureus* suggest that enzymes responsible for cell wall assembly may also be involved in the construction of LPXTGase.[61] Finally, it cannot be excluded that such a nonribosomally synthesized enzyme be also present and involved in LPXTG-like protein anchoring in *L. monocytogenes*.[37]

Substrates of the Sec system are generally considered as exhibiting an N-terminal signal peptide composed of three domains: (1) the N-domain contains positively charged amino terminus, (2) the H-domain is a hydrophobic core region, and (3) the C-domain contains the cleavage site.[63] It must be emphasized, however, that this is not the case for all proteins (e.g., some SecA2-dependent and/or YidC-dependent proteins). Still, the presence of an N-terminal signal peptide indicates a protein is targeted to membrane. Despite lack of amino acid sequence similarity, signal peptides can be detected with good accuracy by various documented and publicly available applications (Table 12.1). The first methods developed were SigCleave and SPScan, which were implementations of a simple weight matrix approach.[64] While SigCleave is part of the EMBOSS suite and also available by an interface on the World Wide Web, SPScan is only available as part of the GCG suite and thus requires ability to work under a Unix-like environment. Comparing the two programs, SPScan has clearly better predictive performance in terms of secretory protein and cleavage site recognition, especially for prokaryotic proteins.[65] Nearly a decade later, SignalP, a promising method based on a

TABLE 12.1
Bioinformatic Resources for Prediction of Bacterial N-Terminal Signal Peptides

Application	Method	Webserver	Ref.
SigCleave	Position weight matrix	http://bioweb.pasteur.fr/seqanal/interfaces/sigcleave.html	64
SPScan	Position weight matrix	none	64
SignalP	Neural network Hidden Markov model	http://www.cbs.dtu.dk/services/SignalP/	67
PrediSi	Position weight matrix	http://www.predisi.de/	69
SOSUIsignal	Global physicochemical analysis	http://bp.nuap.nagoya-u.ac.jp/sosui/sosuisignal/	70
Phobius	Hidden Markov model	http://phobius.binf.ku.dk/	71
PSORTb	Support vector machine Hidden Markov model	http://www.psort.org/psortb/	74
SPdb	BLAST	http://proline.bic.nus.edu.sg/spdb/	76

neural network, was released[66] and has undoubtedly become the most popular method for predicting N-terminal signal peptide. Since the first available version 1.1, SignalP has been substantially improved up to the latest version 3.0.[67]

While version 1.1 is definitively out of date, both versions 2.0 and 3.0 use either a neural network (NN) or HMM. When comparing SignalP v2.0-NN, -HMM, and SPScan, it appears that (1) SPScan predicts correctly more proteins as secreted than SignalP v2.0-NN or -HMM; (2) SignalP v2.0-NN and -HMM are superior in predicting the correct cleavage site; (3) SignalP v2.0-NN lags behind SPScan and SignalP v2.0-HMM in classifying correctly the proteins, the latter providing the best prediction; and (4) SignalP v2.0-NN is the best for predicting of the correct cleavage site.[65] In other words, these methods are complementary in predicting an N-terminal signal peptide. The main improvement in SignalP v3.0 is increased accuracy in prediction of signal peptidase cleavage sites.[67] In comparative analyses, SignalP v3.0 performs significantly better than other machine learning and HMM methods. Despite performance improvement in the latest SignalP v3.0, however, it appears that SignalP v2.0-NN remains the best signal prediction program.[68]

A position weight matrix approach was improved by a frequency correction, which takes into consideration the amino acid bias (i.e., PrediSi).[69] SOSUIsignal is a global structure analysis based on physicochemical features of the three signal peptide domains—N-, H-, and C-domains—and discriminates between cleavable and anchoring signal sequences.[70] Since a signal peptide contains a hydrophobic H-domain, there is a risk of erroneously identifying a transmembrane α-helix as a signal peptide or, conversely, classifying a protein with a signal peptide H-domain region as an IMP. In order to discriminate between the two, a combined TM topology and signal peptide predictor has been developed: Phobius.[71] Phobius significantly reduces false classifications of signal peptides compared to SignalP. Another machine learning approach used for prediction of signal peptides is the support vector machine (SVM), which can predict signal peptides with great accuracy.[72] Such an implementation of an SVM combined with an HMM is part of PSORTb,[73] now applicable to both Gram-positive and Gram-negative bacteria.[74] Finally, SPdb, a repository of experimentally determined and computationally predicted signal peptides, is also accessible via BLAST (basic local alignment search tool) search.[75,76]

It can be stressed again that these analyses only predict the presence of signal peptide, meaning that the protein is targeted to the cytoplasmic membrane. However, it does not necessarily mean the protein is translocated across the cytoplasmic membrane via Sec or released into the extracellular milieu. Indeed, proteins translocated via Tat or FPE also possess N-terminal signal peptides with additional features, which are not identified by the previous tools (Table 12.1). Thus, final prediction of a protein translocated via Sec requires additional inspections (see sections 12.2.2 and 12.2.3). Concerning proteins translocated by the Sec system and possessing a signal peptide, they can (1) be released into the extracellular medium or injected into a host cell, (2) remain associated to the cell wall by covalent or noncovalent interactions, or (3) remain associated to the cytoplasmic membrane by transmembrane domains (including the H-domain of uncleaved signal peptide) or be lipoproteins (see section 12.3). Thus, final localization prediction of Sec substrates requires a combination of tools for prediction of function, motifs, and TMDs. It is also recommended to combine these results with those from tools dedicated to prediction of protein subcellular localization in Gram-positive bacteria (Table 12.2).

NNPSL was the first tool developed for such prediction and is based on an NN.[77] SubLoc,[78] PSORTb,[73] CELLO[79] (recently extended to prediction in Gram-positive bacteria[80]), and LOCtree[81] are basically SVM. These tools have their own advantages and weaknesses,[82] and some of them, like PSORTb, combine a variety of individual predictors. Proteome Analyst is a novel type of machine-learning classifier that involves several steps in the prediction process, such as BLAST search against Swiss-Prot database and naïve Bayesian classifiers.[83] From the most recent studies on performance of prediction tools, PSORTb and Proteome Analyst achieve the highest overall precision.[84] Gpos-PLoc, another type of ensemble classifier, was recently developed where several basic

TABLE 12.2
Bioinformatic Resources for Prediction of Subcellular Localization of Proteins in Gram-Positive Bacteria

Application	Method	Webserver	Ref.
NNPSL	Neural network	http://www.doe-mbi.ucla.edu/~astrid/astrid.html	77
SubLoc	Support vector machine	http://www.bioinfo.tsinghua.edu.cn/SubLoc/	78
PSORTb	Support vector machine Ensemble classifier	http://www.psort.org/psortb/	73
CELLO	Support vector machine	http://cello.life.nctu.edu.tw/	80
LOCtree	Support vector machine	http://cubic.bioc.columbia.edu/services/loctree/	81
Proteome Analyst	Ensemble classifier	http://pa.cs.ualberta.ca:8080/pa/	83
Gpos-PLoc	Ensemble classifier	http://202.120.37.186/bioinf/Gpos/	85
DBSubLoc	BLAST	http://www.bioinfo.tsinghua.edu.cn/~guotao/intro.html	86
PSORTdb	BLAST	http://db.psort.org/	87
PA-GOSUB	BLAST	http://www.cs.ualberta.ca/~bioinfo/PA/GOSUB/	88
Augur	Ensemble classifier	http://bioinfo.mikrobio.med.uni-giessen.de/augur/	80

classifiers were fused and optimized for predicting subcellular localization of Gram-positive bacterial proteins.[85] Finally, several databases (derived from previously described prediction tools) are available following BLAST search (DBSubLoc,[86] PSORTdb,[87] and PA-GOSUB[88]). Augur is another database especially dedicated to protein localization on the cell surface of Gram-positive bacteria.[89] Once again, final prediction of secreted proteins (and localization) should combine results from these various bioinformatic tools.[84]

Using SignalP v2.0 to predict signal peptide region and TopPred v2.0 to exclude other transmembrane domains, 86 proteins were predicted as secreted into the extracellular medium from genomic analysis of *L. monocytogenes* EGD-e.[19] In *L. monocytogenes* F2365, 420 proteins were predicted with a putative N-terminal signal peptide, including 2 with a YSIRK motif.[18] Performing extensive genomic analyses, which combined results from SignalP v2.0, SigCleave, SOSUI, PSORT, and TMPinGS, the number of proteins bearing an N-terminal signal peptide was estimated at 525 in *L. monocytogenes* EGD-e, including 255 IMPs and 270 exported proteins where 121 would be released into the extracellular milieu.[90] All 14 virulence factors characterized so far in *L. monocytogenes* are most likely translocated via the Sec translocon.[37] Among the 121 proteins originally predicted as secreted via Sec and released into the extracellular milieu, a closer look revealed that four prepilins—that is, ComGC (Lmo1345), ComGD (Lmo1344), ComGE (Lmo1343), and ComGG (Lmo1341)—should be removed from the output since they would form trans-cell-wall structure following translocation via FPE.[37] Proteomic analysis of supernatant from liquid culture of *L. monocytogenes* EGD-e allowed the identification of 54 out of 117 proteins predicted as extracellular, including virulence factors Hly, PlcA, and PlcB.[90]

12.2.2 TAT

The term twin-arginine translocation (Tat) was coined from the systematic presence of the RR motif in signal peptide of proteins translocated via this secretion system.[91] The [ST]RRXFLK motif straddles the N-domain and H-domain of the N-terminal signal peptide.[92] Contrary to the Sec translocon, the main feature of this pathway is its ability to translocate proteins in a folded state. General knowledge on the precise succession and mechanistic events leading to protein secretion via this pathway remains rudimentary.[93] The generally accepted translocation model was first proposed by Mori and Cline,[94] where Tat translocation follows a cycle in which TatBC functions in the specific recognition of the substrate and TatA functions as the pore-forming unit. An alternative model

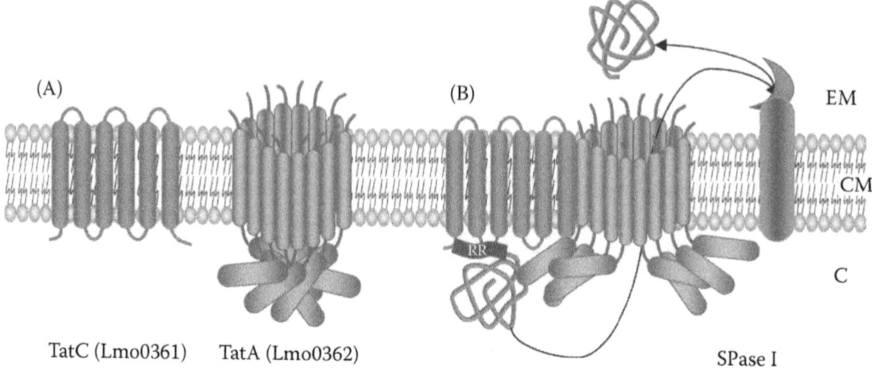

FIGURE 12.4 Tat translocon in *L. monocytogenes.*[37] During Tat secretion, the general model proposes a cyclical assembly of components. (A) In resting state, Tat machinery components are separately present in the cytoplasmic membrane (i.e., TatB and TatA). (B) Once Tat substrate protein precursor binds to the TatC in an energy-independent step, this complex associates with TatA in a step driven by the transmembrane proton electrochemical gradient. This association would persist until completion of protein transport across the membrane driven by proton motive force. Tat signal peptide is subsequently cleaved by SPase I and Tat machinery components disassemble, as depicted in (A). RR, twin-arginine motif in Tat signal peptide; EM, extracellular milieu; CM, cytoplasmic membrane; C, cytoplasm.

proposes that membrane integration could precede Tat-dependent translocation and the membrane targeting process may require ATP-dependent N-terminal unfolding-steps energy.[95]

Still, components of Tat translocon differ in number between Gram-negative and Gram-positive bacteria.[30] The most baffling difference is the absence of TatB from all Gram-positive bacteria sequenced so far, although it is an essential component of Tat translocon in *E. coli*, which is used as a paradigm.[96] As in most Gram-positive bacteria, Tat translocon in *L. monocytogenes* is encoded in one locus and is composed of only two proteins, TatA and TatC[37] (Figure 12.4). TatC is a large IMP generally considered as the primary site for signal-peptide recognition.[97] TatA is a membrane protein that oligomerizes to form a protein-conducting channel where the number of subunits would adjust in function of the Tat substrate size.[98] In TatA, a cytoplasmic lid region acts as a gate and would open following association of the TatC–substrate complex with TatA, then inducing conformation change and protein translocation driven by proton motive force. Translocated protein is finally released after cleavage by SPse I.[96] A Tat translocon does not seem to be systematically present in *L. monocytogenes*, as no component could be identified in *L. monocytogenes* F2365. The Tat system has never been experimentally investigated in *Listeria*; thus, its expression, functionality, involvement of one or three SPases I, and proteins secreted via this pathway remain unknown.[37]

Three tools are currently available to discern Tat substrates (Table 12.3) and TATFIND was the first program especially devoted to such identification.[99] In its original available version, TATFIND v1.2, prediction was based on two criteria: (1) presence of conserved Tat motif ZRRZZZ within the first 35 amino acid residues, where Z represents a defined set of permitted residues; and (2) presence

TABLE 12.3
Bioinformatic Resources for Prediction of Tat Signal Peptides

Application	Method	Webserver	Ref.
TATFIND	Physicochemical analysis and regular expression	http://signalfind.org/tatfind.html	100
TatP	Neural network and regular expression	http://www.cbs.dtu.dk/services/TatP/	101
TATPred	Naïve Bayesian network	http://www.jenner.ac.uk/logP/JennerlogPcalc.htm	102

of an uncharged stretch of at least 13 amino acids downstream of the twin arginine. In the latest version, TATFIND v1.4, search for a single charged residue in positions +2 and +5 relative to the RR was included.[100] TatP v1.0 incorporates signal peptide and cleavage site prediction based on a combination of two artificial neural networks followed by a postfiltering of the output based on regular expression RR[FGAVML][LITMVF].[101] Compared to TATFIND v1.2, TatP generates far fewer false positive but slightly more false negative predictions. TATPred is the latest algorithm based on naïve Bayesian network developed for prediction of Tat substrates.[102] Compared to TatP, TATPred appears as the most robust and reliable predictor with higher sensitivity of prediction.

According to TATFIND search, only two Tat substrates could be identified in *L. monocytogenes* EGD-e.[99] One of these putative Tat substrates, however, is also present in *L. monocytogenes* F2365, where the Tat system is not encoded.[37] These substrates have never been reported as present in the extracellular milieu of *L. monocytogenes.* While it has been long assumed that the RR motif was highly specific and conserved in Tat substrates, it must be stressed that substitutions of one arginine, or in some cases both arginines, by lysine[103] or that natural proteins harboring very distantly related RR motifs[104] could still permit targeting and translocation via the Tat pathway.[105] This indicates that Tat system specificity is more flexible than originally thought and, thus, presence of the Tat substrate cannot be systematically identified by bioinformatic analysis.

12.2.3 FPE

Components of fimbrilin-protein exporter (FPE) of Gram-positive bacteria are homologous to proteins required for secretion of substrate proteins in Gram-negative bacteria, namely, some ATPase and IMP components of the Type II protein secretion system (T2SS), Type 4 piliation sytem (Tfp), and Type IV protein secretion system (T4SS), as well as archaeal flagella.[106] As in all Gram-positive bacteria where it has been reported so far,[33,107,108] components of FPE in *L. monocytogenes* are encoded in a *comG* operon, except for *ComC* located elsewhere on the chromosome. Protein exporters of the FPE family consist of two constituents—ComGA and ComGB—that would function together in an ATP-hydrolysis-dependent export of proteins across the cytoplasmic membrane[109,110] (Figure 12.5). ComGA is an ATPase localized to the cytoplasmic side of the membrane that could participate in modeling of a pilus-like structure.[109] As a homologue to PilC of Tfp and PulF of T2SS,[109] ComGB is an IMP having three putative TMDs that could play the role of a protein-conducting channel.[111] ComC is a Type 4 prepilin peptidase involved in cleavage of N-terminal signal peptide of class 3;[112] this signal peptidase belongs to the aspartic acid protease family.[113]

While ComC is required for maturation, translocation, and assembly of prepilins, an initial translocation event across the cytoplasmic membrane has not been clearly elucidated. As prepilin signal peptide is cleaved at the cytoplasmic side between the N- and H-domains, prepilins are certainly not translocated by the Sec or Tat pathways and the hypothesis of ComGAB involvement is favored. However, YidC contribution cannot be excluded[30]; prepilins were originally thought to insert spontaneously in the membrane bilayer but with the current knowledge of membrane protein insertion this hypothesis should not be privileged (see section 12.3). Four Type 4 prepilins are encoded in *comG* locus by the *comGC*, *comGD*, *comGE,* and *comGG* genes[110]; ComGF is presumably an IMP. Once maturated and translocated, pilins form a trans-cell-wall macromolecular complex where monomers are covalently linked by disulphide bonds.[114] Since this structure is involved in bacterial competence and does not form a proper Type 4 pilus, it was named the competence pseudopilus.

In *B. subtilis*, Type 4 prepilins exhibit N-terminal signal peptides with a conserved motif [KR]G▼F[TSI][LTY][VLIP][EA] located between the N- and H-domains where ▼ indicates the predicted cleavage site.[110] In *Listeria*, the motif is slightly different—that is, [NPRS][GA]▼F[TS]L[VLP][EF]—and is found in five putative prepilins (i.e., ComGC, ComGD, ComGE, ComGF, and ComGG).[37] In *B. subtilis*, the highly conserved phenylalanine at position +1 is aminomethylated by ComC, which thus appears bifunctional as it is also involved in prepilin processing.[30] Using ScanProsite syntax,[115] a search for consensus motif [GA]F[TS]LX[EF] located between the N- and

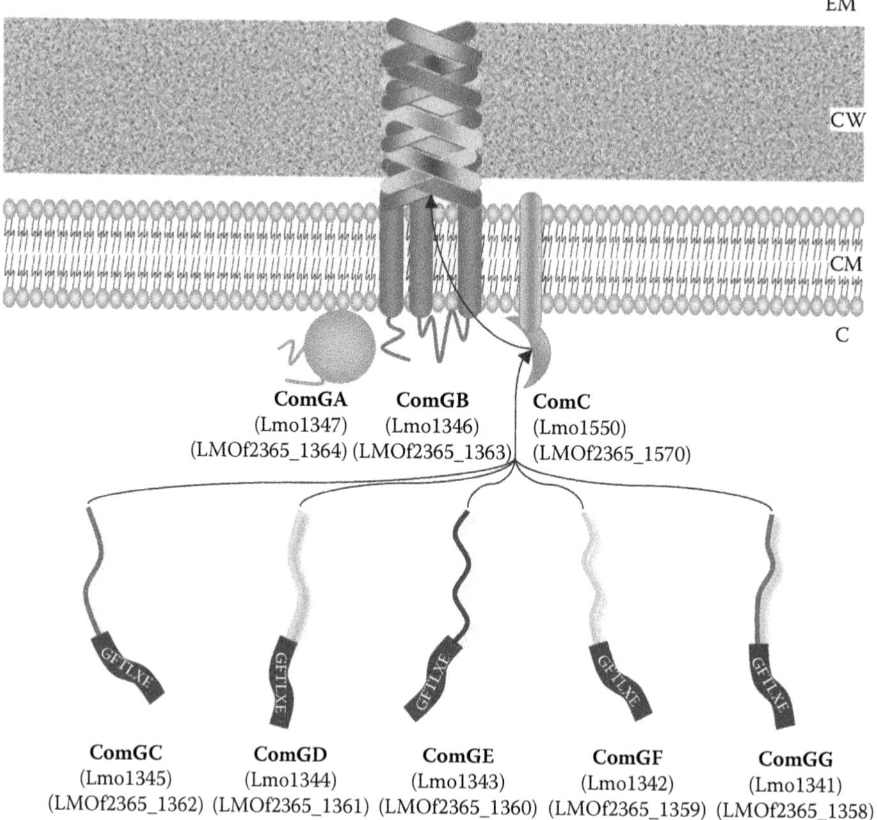

FIGURE 12.5 FPE in *L. monocytogenes*.[37] The prepilins initially float in the CM; initial insertion into the membrane is certainly Sec or Tat independent, but the involvement of ComGAB remains to be ascertained. After processing by ComC signal peptidase, ComGA and ComGB would be involved in assembly of pseudopilins to form a trans-cell-wall pilus-like structure. GFTLXE, conserved [GA]F[TS]LX[EF] motif in Type 4 prepilin signal peptide from *Listeria*; CW, cell wall; EM, extracellular milieu; CM, cytoplasmic membrane; C, cytoplasm.

H-domains of predicted signal peptide can thus be performed in order to identify putative FPE substrates in *Listeria*. The FPE system has never been experimentally investigated in *Listeria*; thus, its expression, functionality, and involvement in bacterial competence remain to be established.[108]

12.2.4 FEA

L. monocytogenes produces up to six peritricheous flagella, which are down-regulated at 37°C, although variation from one strain to another was reported.[116,117] Regulation of listerial flagella is not entirely understood and appears rather complex since at least five regulators involved in its expression have been identified so far: FlaR,[118] PrfA,[119] DegU,[120] MogR,[121] and GmaR (Lmo0688 also called WcaA).[122] Interestingly, the antirepressor GmaR is bifunctional since it also functions as a glycosyltransferase for flagellin FlaA[122] and glycosylation with β-O-linked N-acetylglucosamine was indeed established for FlaA.[123] This investigation constituted the first description of β-O-GlcNac post-translational modification on a prokaryotic protein, though flagella glycosylation is not essential for motility in *L. monocytogenes*.[124] As motility mediators, flagella are important in colonization of abiotic surfaces and host cell invasion but do not function as adhesins.[124,125]

Interestingly, FlaA was also demonstrated as exhibiting a peptidoglycan-hydrolyzing activity that might play a role during flagella assembly.[126,127] Indeed, some flagellar components are assembled on the bacterial cell surface where local digestion of cell wall sacculus might be required—namely, for (1) the rod proteins (i.e., FlgB, FlgC, FliE, and FlgG), (2) the hook/junction proteins (i.e., FliK, FlgD, FlgE, FlgK, and FlgL), and (3) the filament proteins FlaA and FliD. As in Gram-negative bacteria,[128–132] these proteins lack a cleavable N-terminal signal peptide and are presumably translocated by the flagella export apparatus (FEA) composed of FlhA, FlhB, FliH, FliL, FliP, FliO, and FliH. In *Listeria*, all flagella components are encoded in a single flagellar–motility–chemotaxis cluster of 41 genes,[37] where FEA and its potential substrates could be identified by homology search. In Didermata, T3SS refers to a secretion system where translocation apparatus is homologous to injectisomes (T3aSS) and flagella (T3bSS),[51,133] both of which are involved in secretion of extracellular proteins.[134,135] As already stressed, however, this terminology is restricted to Gram-negative bacteria. In monoderm bacteria, involvement of FEA in secretion of extracellular protein has only been suggested in *Bacillus thuringensis.*[136]

12.2.5 Holins

Holins (hole-formers) are small membrane proteins of phage origin that essentially control endolysin function in a process leading to bacterial apoptosis.[137–139] A current model for the holin-endolysin system proposes that holins accumulate in the cytoplasmic membrane, whereas endolysins accumulate in the cytoplasm[140,141] (Figure 12.6). At a programmed time, holins oligomerized to form pores in the cytoplasmic membrane, allowing release of endolysins into the extracytoplasmic space leading to cell lysis following cell wall degradation and membrane disruption. Homo-oligomeric pore complexes formed by holins would provide a passive but specific translocation system.[142] Generally, holin and its specific endolysin are genetically encoded in tandem. Some holin genes possess a dual start motif, which results in the expression of two distinct proteins with dramatically opposed function since one would promote autolysis (holin) and the other would inhibit it (antiholin).[143] Such regulation can also occur between proteins encoded at different loci (e.g., *lrgAB*/*cidAB* operons in *Staphylococcus aureus*).[144,145] Holins are an extremely diverse group of proteins with 23 distinct families recognized in TC-DB (transport classification database),[146] although they can be grouped into three classes based on membrane topologies.[147] Class 1 holins exhibit three helical TMDs, whereas class 2 holins have two TMDs. Besides classes 1 and 2, which cover most holins, a third class was identified on the basis of T protein of phage T4 where only a single TMD is present.[148]

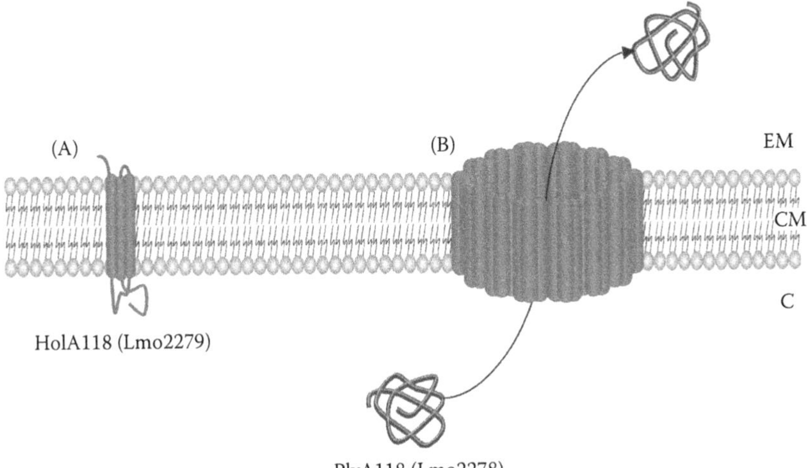

FIGURE 12.6 ϕA118 holin-endolysin system in *L. monocytogenes* EGD-e.[37] (A) HolA118 is a class 1 holin (i.e., with 3 TMDs). (B) HolA118 oligomerizes in the CM to form a pore allowing translocation and activation of endolysin Ply118. EM, extracellular milieu; CM, cytoplasmic membrane; C, cytoplasm.

The number of holins encoded in *Listeria* varies between strains and species[37]; only one holin could be identified in *L. monocytogenes* F2365, whereas five holins were found encoded in nonpathogenic strain *L. innocua* CLIP11262. However, only three distinct families of holins were identified in *L. monocytogenes* (i.e., as belonging to bacteriophage 118, TcdE, and bacteriophage 11 families). Holins of ϕA118 family (HolA118) were first identified and investigated in *L. monocytogenes* although a homologue is at least also encoded in *L. innocua*.[37,149] HolA118 is not encoded by all *L. monocytogenes* species as it is absent from *L. monocytogenes* F2365. Native holin HolA118 is a 93-amino-acid-long protein belonging to class 1, but its encoding gene is subjected to dual translational initiation, which leads to a second 83-amino-acid-long protein called HolA118(83) acting as an antiholin.[150] Gene encoding phage lysin of ϕA118 (PlyA118) systematically clusters with gene encoding HolA118.[18,151]The endolysin PlyA118 is an L-alanoyl-D-glutamate peptidase hydrolyzing the cross-linking bridges of cell wall peptidoglycan and thus responsible for bacterial lysis in a programmed cell death.

Holins belonging to TcdE family are encoded in all sequenced *Listeria* but as ϕ11 holins they have never been experimentally investigated. TcdE holin was investigated in *Clostridium difficile,* where toxigenic strains produce two large toxins, TcdA and TcdB, of major importance in bacterial virulence, which would be translocated across the cytoplasmic membrane by TcdE.[152,153] While in *C. difficile* all these genes are encoded within a pathogenicity locus, no genes coding for toxins or virulence factors could be identified in *Listeria.*[37] However, a putative autolysin lacking a signal peptide was systematically present (i.e., genes encoding Lmo0129 and LMOf2365_0147 in *L. monocytogenes* EGD-e and F2365, respectively). Although this particular holin family has never been investigated per se in *Listeria*, proteomic analysis in *L. monocytogenes* EGD-e disclosed the presence of Lmo0129 in supernatant of bacterial cultures, suggesting this secretion pathway is active in this species.[90] In *Listeria*, ϕ11 holins were only identified in unassembled genome of *L. monocytogenes* F6854 and nonpathogenic *L. innocua* CLIP11262, where they systematically clustered with genes encoding amidases presumably involved in cell wall degradation.[37]

12.2.6 Wss

Wss stands for proteins with WXG motif of ~100 residues (WXG100) in the secretion system.[154] WXG100 is a new superfamily of proteins around 100 amino acids long, possessing a coil–coil domain and bearing a conserved WXG motif located in the middle region. First identified members of this superfamily were paralogues ESAT-6 (early secreted antigen target of 6 kDa) and CFP-10 (culture filtrate protein 10) from *Mycobacterium tuberculosis.* While ESAT-6 and CFP-10 are specific and experimentally investigated proteins, WXG100 (PF06013) is an established and generic terminology more appropriate to describe protein members of this family, especially those that have not been experimentally investigated yet. No generic terminology for the different components of Wss apparatus has been established yet. Presence of a novel protein secretion system was clearly suggested by bioinformatic analysis.[154] In *B. subtilis*, genes encoding WXG100 proteins appeared to cluster systematically with *yukab*, which are predicted to encode membrane bound ATPases with FtsK/SpoIIIE domains. Similar genetic organization was observed in some *Corynebacterium, Mycobacterium, Streptomyces, Bacillus, Clostridium, Listeria,* and *Staphylococcus* species.[31,154] YukAB homologues appear encoded as single or two CDS. To date, Wss seems phylogenetically restricted to Gram-positive bacteria and has been experimentally investigated only in *M. tuberculosis, M. smegmatis,* and *S. aureus.*

In *Mycobacterium*, two WXG100 proteins are secreted: ESAT-6 and CFP-10.[155] Recently, a C-terminal signal sequence required for secretion via Wss was unraveled in CFP-10.[156] *Mycobacterium* Wss apparatus was named Snm (secretion in mycobacteria) and is composed at least of[155,157]: (1) Snm1, Snm2, and Snm6 containing NTP-binding motifs (where Snm1 and Snm2 are homologous to YukAB); (2) Snm4, which is an IMP; and (3) Snm5 and Snm7 with uncharacterized functions, and Snm8 (i.e., a membrane anchored serine protease). Snm permits translocation of ESAT-6 and CFP-10 as well as their heterodimerization.[158,159] In *S. aureus*, the Wss was named Ess (ESAT-6 secretion system) and is encoded in a locus composed of eight CDS including two WXG100 paralogues—EsxA (Ess extracellular protein A) and EsxB—as well asx[160]: (1) EssC (Ess protein C) homologous to YukAB; (2) EssA, EssB, and EsaA (ESAT-6 secretion accessory protein A), which are IMPs; and (3) EsaB and EsaC, which predict cytoplasmic chaperones. Compared to *B. subtilis*, where a putative Wss was primarily uncovered, EssB and EsaB appear homologous to YukC and YukD, respectively. In *S. aureus*, no homologue to Snm4, Snm5, Snm6, Snm7, or Snm8 was found, whereas in mycobacteria, no homologue to EssA, YukC, EsaA, and YukD could be identified. In both *M. tuberculosis* and *S. aureus*,[160,161] Wss is important and critical for bacterial pathogenicity, though the function of WXG100 proteins in virulence remains obscure.[162]

Synteny is highly conserved between Wss encoding loci of *S. aureus* and *L. monocytogenes*.[160] However, compared to *Mycobacterium* species or *S. aureus*, only a single copy of Wss locus is present in each sequenced *Listeria* genome.[37] Following homology with *S. aureus* and *Mycobacterium*, Wss in *L. monocytogenes* is represented in Figure 12.7. From one report,[163] it seems that WXG100 protein is not required for virulence of *L. monocytogenes*. Still, protein expression, system functionality, and involvement in bacterial virulence of Wss remain to be established.

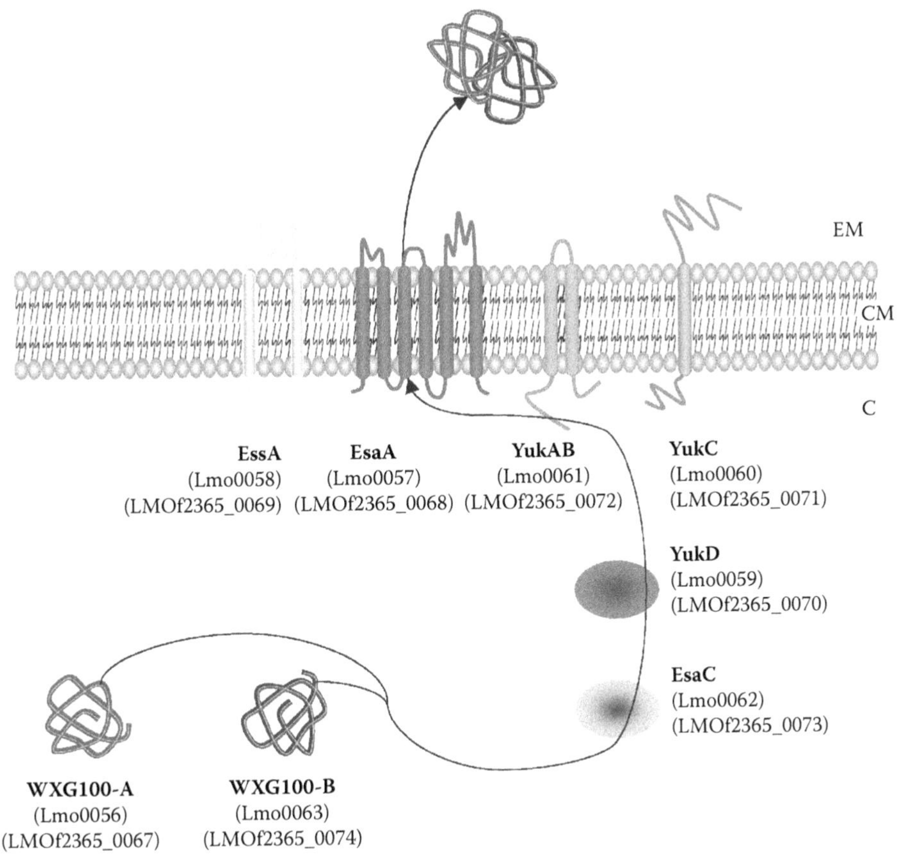

FIGURE 12.7 Wss in *L. monocytogenes*.[37] WXG100 proteins would interact with putative cytoplasmic chaperones YukD and EsaC before being translocated by the Wss apparatus constituted of EssA, EsaA, YukC, and YukAB; in the course of translocation the two WXG100 proteins would finally form a heterodimer. EM, extracellular milieu; CM, cytoplasmic membrane; C, cytoplasm.

12.3 CELL ENVELOPE-ASSOCIATED PROTEINS

The cell envelope of Gram-positive bacteria is primarily composed of a single biological membrane (i.e., the cytoplasmic membrane) and a cell wall made up of peptidoglycan (which, in turn, consists of linear polysaccharide chains cross-linked by short peptides).[164] Besides peptidoglycan, the rigid cell wall of Gram-positive bacteria contains large amounts of wall-associated polymers, also called "secondary" cell wall polymers (SCWPs), which can be classified into three distinct groups: (1) teichoic acids (i.e., polyol phosphate polymers, including lipoteichoic acids), (2) teichuronic acids, and (3) other neutral or acidic polysaccharides that cannot be assigned to the two former groups (e.g., lipoglycans).[55,165,166] The SCWPs, present in various proportions, are either covalently linked to the peptidoglycan backbone (i.e., teichoic acids) or tethered to a lipid anchor moiety. Except for teichoic and teichuronic acids, the structure and biosynthesis of other SCWPs are largely unknown. It must be stressed that in almost all phylogenetic branches of Archaea and Bacteria, the cell envelope is also constituted of a proteinaceous S-layer (regular crystalline surface layer), which forms the outermost cell-wall layer.[167] The S-layer entirely coats the bacterial cell surface and is composed of (glyco)proteins, which bind by noncovalent interactions to cell wall components and are arrayed in a two-dimensional lattice.[167] The S-layer, however, is not present in all Gram-positive bacteria as, it is absent from all members of *Listeria* genus. Within the cell envelope, proteins can associate with the cytoplasmic membrane or cell wall components.[8]

12.3.1 Membrane-Associated Proteins

Membrane-associated proteins include membrane integrated proteins as well as peripheral membrane proteins. Being different from membrane integrated proteins, peripheral membrane proteins do not possess membrane spanning domains. Membrane integrated proteins are anchored within the lipid bilayer and thus systematically exhibit hydrophobic transmembrane domains (TMDs), which are normally α-helices for proteins found in the cytoplasmic membrane. Peripheral membrane proteins include (1) lipoproteins, (2) subunits of membrane-associated complexes, and (3) proteins interacting with membrane components due to electrostatic and/or hydrophobic/steric properties.[168] Following recommendations of the Gene Ontology (GO) Consortium for describing location of cellular components (one of the three organizing principles of GO with biological process and molecular function),[169] two classes of membrane-related location are distinguished (Figure 12.8). First, intrinsic to plasma membrane (GO:0031226) refers to proteins with covalently attached moiety embedded in the cytoplasmic membrane, which splits into (1) integral to plasma membrane (GO:0005887) corresponding to membrane integrated proteins, where some part of the peptide sequence spans all or part of the cytoplasmic membrane; and (2) anchored to plasma membrane (GO:0046658) corresponding to proteins tethered to the cytoplasmic membrane by a nonpolypeptidic covalently attached anchor: lipoproteins. Second, extrinsic to plasma membrane (GO:0019897) refers to proteins neither anchored by covalent bonds to any moiety nor directly embedded in the cytoplasmic membrane; some of these proteins can be (1) primarily present in the cytoplasm (GO:0005737) but interact with membrane components, or (2) subcomponents localized within the protein complex (GO:0043234).

12.3.1.1 Integral Membrane Proteins

As already mentioned, all bacterial IMPs are presumably inserted into the cytoplasmic membrane via YidC homologues[48,170,171] (i.e., SpoIIIJ and YqjG in Gram-positive bacteria[30]; see section 12.2.1). The Sec-independent function of YidC homologues is conserved and essential for bacterial cell growth as it works like a membrane protein insertase.[172] YidC plays a major role in the folding step of transmembrane-spanning domains but the exact mechanism of functioning is not fully understood.[173] YidC would facilitate the insertion of membrane proteins by providing a special amphiphilic surface, which would overcome the repulsion of the hydrophobic protein segments by polar head groups. In addition, polar residues seem to be protected against the hydrocarbon core of

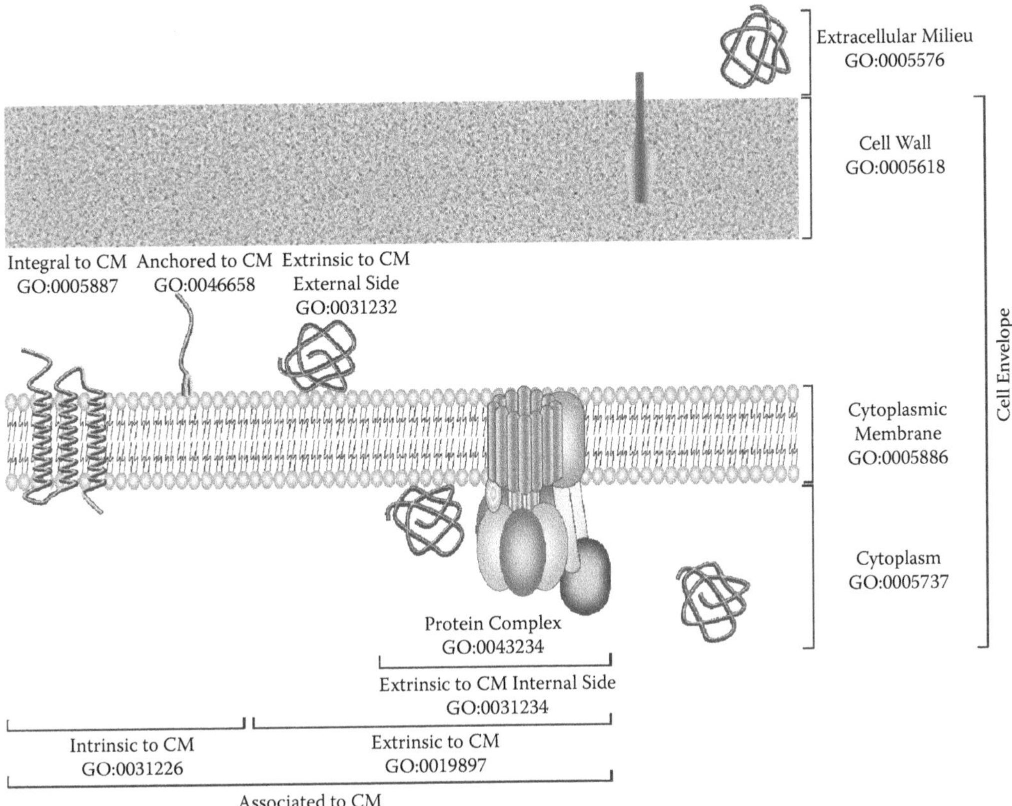

FIGURE 12.8 Description of protein localization following GO in Gram-positive bacteria.[179] In monoderm bacteria four subcellular compartments can be distinguished: (1) cytoplasm, (2) cytoplasmic membrane, (3) cell wall, and (4) extracellular milieu. A membrane-associated protein can be intrinsic or extrinsic. Proteins intrinsic to CM are either integral to membrane (i.e., integral membrane proteins) or anchored to CM, essentially lipoproteins with the restriction of lipoproteins having TMDs. Proteins extrinsic to CM can be located on the external or internal side of the CM (i.e., in exoplasmic or cytoplasmic compartment, respectively). They can interact more or less temporarily with membrane components or be part of a membrane protein complex (e.g., F_1F_0ATP synthetase δ subunit) as indicated on the schema. A protein can also have multiple final localization. Importantly, because cell wall of Gram-positive bacteria is permeable, extracellular milieu penetrates it. CM, cytoplasmic membrane; GO, gene ontology.

the membrane by YidC. In the case of Sec-dependent translocation, the protein would be stabilized and then folded by contact with YidC after leaving the Sec YEG channel. It is quite possible that the transmembrane segments could fold and interact with each other even within SecYEG–YidC machinery. It has been suggested that YidC functions as an assembly site for hydrophobic domains, so it may be necessary for its attaching to the individual subunits of the multisubunit membrane complex.[174] It is worth noting that, in *E. coli*, targeting, translocation, and insertion of IMPs are considered cotranslational and thus SRP dependent.[175]

Translocation of a polypeptide chain is promoted by signal peptides and interrupted by another type of topogenic element called stop-transfer sequence[176,177]; both types of topogenic sequences act as α-helical transmembrane domains. Multiple uncleaved signal peptides can be found all along the amino acid sequences; when located N-terminally, they can be cleaved or not. These types of topogenic elements have a C_{out}-N_{in} topology and when uncleaved are also called signal-anchors or Type II signals.[178] Similarly, one or more stop-transfer sequences with an N_{out}-C_{in} topology can be present in polypeptide chain and are also called Type II signals. Single-spanning membrane proteins are discriminated on the basis of Type I or Type II signal (Figure 12.9). Polytopic membrane

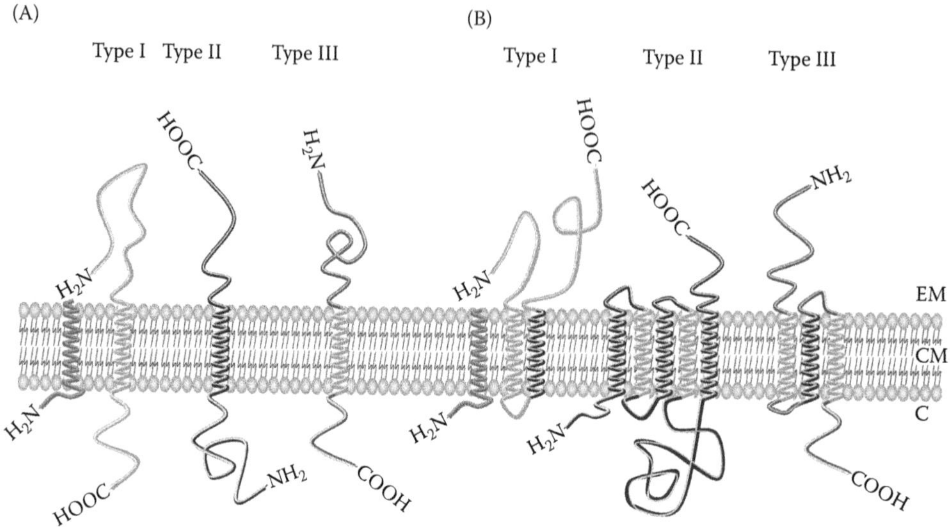

FIGURE 12.9 Classification and topology of IMP in cytoplasmic membrane.[179] (A) Three types of single-spanning membrane proteins can be discriminated: (1) Type I proteins possess a cleavable N-terminal signal peptide and thus have a Type I signal or stop-transfer sequence with N_{out}-C_{in} topology; (2) Type II proteins have a Type II signal or signal-anchor sequence with a C_{out}-N_{in} topology, which can correspond to an uncleavable N-terminal signal peptide; and (3) Type III proteins have reverse signal-anchor sequence (i.e., with a N_{out}-C_{in} topology) and are sometimes described as Type I proteins without a cleavable signal peptide since the reverse signal-anchor sequence is a Type I signal. (B) Three types of multispanning-membrane proteins (i.e., with a number of TMDs higher than 1) can be distinguished based on whether the most N-terminal TMD is either (1) cleaved by a SPase (i.e., Type I); (2) spans the membrane with an N_{out}-C_{in} orientation (i.e., Type II); or (3) have a C_{out}-N_{in} orientation. Various numbers of TMDs are present in multispanning-membrane proteins where alternates Type I and II signals alternate. EM, extracellular milieu; CM, cytoplasmic membrane; C, cytoplasm.

proteins are built up of a series of Types I and II modules that initiate and halt the translocation of the polypeptide chain. Such IMPs are classified on the basis of the orientation of the most N-terminal TMD spanning the lipid bilayer[179] (Figure 12.9).

Numerous tools are available to predict IMPs and their topology. Table 12.4 is an attempt to review all of them. These tools are based on various approaches, such as (1) statistical analyses (e.g., TMpred or TMSTAT; (2) hydrophobicity analyses (e.g., SOSUI or TopPred); (3) NNs (e.g., PREDTMR or PHDhtm); (4) SVMs (e.g., SVMtm); or (5) HMMs (e.g., HMMTOP or THUMBUP). Some of them—for example, ConPred or TUPS—combine results of several models. Readers are invited to study related publications listed in Table 12.4 in order to get further insight into the methods used. The total number of IMPs encoded in *L. monocytogenes* genomes is estimated to be 1204 and 733 in *L. monocytogenes* EGD-e and F2365, respectively.[18] Virulence factors ActA and SvpA are cell surface exposed IMPs of *L. monocytogenes* exhibiting a hydrophobic tail (i.e., a carboxyl terminal region containing a hydrophobic domain followed by positively charged residues).[180,181] Following genomic analysis, a total of 11 surface proteins with hydrophobic tails have been predicted in *L. monocytogenes* EGD-e.[19] Contrary to what is sometimes assumed, cell surface IMPs should not be restricted only to proteins with a hydrophobic tail[182]; indeed, depending on the number and organization of Type I and Type II signals in IMPs, final protein topology can result in the cell surface exposure of functional domains located not only N- or C-terminally but also in loops. With FbpA as an example,[47] it can be noticed that from experimental investigations, some proteins appear located within the cytoplasmic membrane despite the absence of predicted signal peptide and TMD.

TABLE 12.4
Bioinformatic Resources for Prediction of TMDs

Application	Method	Webserver	Ref.
TMpred	Statistical analysis	http://www.ch.embnet.org/software/TMPRED_form.html	249
TopPred	Hydrophobicity analysis	http://bioweb.pasteur.fr/seqanal/interfaces/toppred.html	250
PHDhtm	Neural network	http://npsa-pbil.ibcp.fr/cgi-bin/npsa_automat.pl?page=/NPSA/npsa_htm.html	251
DAS	Hydrophobicity analysis	http://www.sbc.su.se/~miklos/DAS/	252
TMAP	Statistical analysis	http://bioweb.pasteur.fr/seqanal/interfaces/tmap.html	253
TSEG	Hydrophobicity analysis	http://www.genome.jp/SIT/tsegdir/tseg_exe.html	254
TMHMM	Hidden Markov model	http://www.cbs.dtu.dk/services/TMHMM-2.0/	255
SOSUI	Hydrophobicity analysis	http://bp.nuap.nagoya-u.ac.jp/sosui/	256
PRED-TMR	Neural network	http://athina.biol.uoa.gr/PRED-TMR2/	257
kPROT	Statistical analysis	http://bioinfo.weizmann.ac.il/kPROT/	258
TMSTAT	Statistical analysis	http://bioinfo.mbb.yale.edu/tmstat/	259
HMMTOP	Hidden Markov model	http://www.enzim.hu/hmmtop/	260
TMFinder	Hydrophobicity analysis	http://www.bioinformatics-canada.org/TM/	261
DAS-TMfilter	Hydrophobicity analysis	http://www.enzim.hu/DAS/DAS.html	262
SPLIT	Statistical analysis	http://split.pmfst.hr/split/	263
THUMBUP	Hidden Markov model	http://sparks.informatics.iupui.edu/Softwares-Services_files/thumbup.htm	264
UMDHMM^{TMHP}	Hidden Markov model	http://sparks.informatics.iupui.edu/Softwares-Services_files/umdhmm.htm	264
TUPS	Combination	http://sparks.informatics.iupui.edu/Softwares-Services_files/tups.htm	264
BPROMPT	Bayesian belief network	http://www.jenner.ac.uk/BPROMPT/	265
SVMtm	Support vector machine	http://ccb.imb.uq.edu.au/svmtm/svmtm_predictor.shtml	266
ConPred	Combination	http://bioinfo.si.hirosaki-u.ac.jp/~ConPred2/	267
HMM-TM	Hidden Markov model	http://biophysics.biol.uoa.gr/HMM-TM/	268
MINNOU	Hydrophobicity analysis	http://minnou.cchmc.org/	269
MEMSAT	Statistical analysis	http://bioinf.cs.ucl.ac.uk/psipred/	270

12.3.1.2 Lipoproteins

In monoderm bacteria, lipoproteins are attached to the outer surface of the cytoplasmic membrane via a covalently bound lipid anchor.[32] Systematically, these proteins are first translocated in a Sec-dependent manner and thus possess N-terminal signal sequences. Such signal peptides, however, belong to class 2 as it exhibits a conserved lipobox motif in the C-domain.[30] It can be noticed, however, that in *E. coli* YidC plays an important role in targeting and translocation of some lipoproteins.[183] Lipobox includes invariably a cysteine residue located just after the cleavage site of signal peptide. After translocation of the prolipoprotein, a common post-translational modification involves a prolipoprotein diacylglyceryl transferase (Lgt), which adds an N-acyl diglyceride group from a glycerophospholipid to the SH-group of the lipobox cysteine.[184] This thioether linkage allows protein anchoring to the membrane, thanks to the insertion of the diacylglyceryl group into the lipid bilayer. Subsequently, SPase II (also called Lsp for lipoprotein signal peptidase) cleaves off the signal peptide and the cysteine becomes the N-terminal residue.[185] In contrast to *E. coli*, however, lipidation by Lgt (Lmo2482) in *Listeria* is neither essential for bacterial growth nor a prerequisite for activity of Lsp.[186] Once signal peptide is cleaved off, the amino-terminal cysteine residue is usually acylated at the free amino group by a phospholipid/apolipoprotein transacylase (Lnt), resulting

TABLE 12.5
Bioinformatic Resources for Prediction of Lipoproteins

Application	Method	Webserver	Ref.
ScanProsite	Profile (PS51257) search Pattern (G+LPP) search	http://www.expasy.org/tools/scanprosite/	189, 190
DOLOP	Pattern search	http://www.mrc-lmb.cam.ac.uk/genomes/dolop/	191
LipoP	Hidden Markov model	http://www.cbs.dtu.dk/services/LipoP/	192
SPEPlip	Neural network	http://gpcr.biocomp.unibo.it/predictors/	193
LipPred	Naïve Bayesian network	http://www.jenner.ac.uk/LipPred/	194

in protein anchoring to membrane long chain fatty acid.[187] This additional post-translational modification step, however, does not seem to be conserved in all bacteria,[184] as an *lnt* gene is apparently lacking from all sequenced members of Firmicutes phylum, and *Listeria* species is no exception.[188] It can be further noticed that some lipoproteins are IMPs integrated to the cytoplasmic membrane by TMDs in a YidC-dependent manner.[173]

As summarized in Table 12.5, several resources can be applied for genomic identification of lipoproteins. In PROSITE,[189] the lipobox motif was previously referred to as PS00013 and defined by the regular expression {DERK}(6)-[LIVMFWSTAG](2)-[LIVMFYSTAGCQ]-[AGS]-C, where {DERK}(6) means that none of the four amino acids are allowed in the first six positions relative to the cleavage site. The pattern had two additional rules: (1) the cysteine must be between positions 15 and 35, and (2) at least one positively charged residue (K or R) must be within one of the first seven N-terminal residues. This pattern (i.e., a qualitative motif description based on a regular expression-like syntax) is now replaced by a profile (i.e., a quantitative motif description based on the generalized profile syntax), referred to as PS51257 and defined as prokaryotic membrane lipoprotein lipid attachment site profile. Originally, the lipobox search using PS00013 was known to generate a significant proportion of false-positives, which prompted the need to improve the syntax of this regular expression. Thus, a refined pattern named G+LPP (for Gram-positive lipoprotein) and using PROSITE syntax—that is, [MV]-X(0,13)-[RK]-{DERKQ}(6,20)-[LIVMFESTAG]-[LVIAM][IVMSTAFG][AG]-C—was developed.[190] This pattern appears more specific for the identification of Gram-positive bacterial lipobox and allows greater discrimination against false positives compared with PS00013. Thus, lipobox can be predicted by scanning polypeptidic sequences for the presence of the PS51257 profile or G+LPP pattern using ScanProsite.[115]

However, correct sequence assignment as putative lipoprotein also requests that the lipobox is localized within an N-terminal Sec-dependent signal peptide where it covers H- and C-domains; a signal peptide can be predicted following analysis with previously described tools (Table 12.1). DOLOP compiles similar criteria by scanning query sequences for presence of (1) a lipobox within the first 40 residues from the N-terminus with the consensus as [LVI][ASTVI][ASG][C], (2) positively charged amino acid in the n-domain of the signal peptide, and (3) at least 7–22 residues between the predicted lipobox and the charged residue.[191] LipoP is based on an HMM and discriminates among lipoprotein signal peptides (cleaved by SPase II), other signal peptides (cleaved by SPase I), n-terminal membrane helices, and cytoplasmic protein following attribution of scores.[192] Despite having been primarily developed for Gram-negative bacteria, LipoP can also efficiently identify lipoproteins in Gram-positive bacteria. The only feeble point may be that when handling lipoproteins with transmembrane regions, the HMM misses, in some cases, the lipoprotein signal peptide. SPEPlip is an NN-based method for prediction of signal peptide and integrating a regular expression search based on PROSITE pattern.[193] LipPred is based on the test against a naïve Bayesian network allowing the identification of lipoprotein with a calculated index for prediction confidence.[194] When compared to other available methods, LipPred can be considered as the most accurate for detection of a lipoprotein signal sequence and SPase II cleavage site. Finally, lipoprotein-associated domains can be searched using HMMs from Pfam or Tigrfam: namely, in Firmicutes, PF00938, PF01347, PF01540, PF02030, PF03180, PF03202, PF03260, PF03304, PF03305, PF03330, PF03640, PF04200, PF05481, TIGR00363, TIGR00413, TIGR01742, TIGR01533, and TIGR02898. Proteins identified following this approach should, however, be considered with great care and ideally confirmed by methods listed previously for lipoprotein modification/processing motif.

Following a search for presence of PS00013 pattern and signal peptide using SignalP v2.0, 68 lipoproteins were originally identified in the genome of *L. monocytogenes* EGD-e.[19] Pfam and Tigrfam searches allowed the identification of 70 lipoproteins in *L. monocytogenes* F2365, whereas the number of lipoproteins was estimated at 63 in *L. monocytogenes* EGD-e.[18] Using the computational pipeline Augur,[89] where lipoproteins are identified on the basis of G+LPP pattern match, 65 lipoproteins were identified in *L. monocytogenes* EGD-e.[18] Generating a new HMM from 26 verified lipoproteins by proteomic analysis in *L. monocytogenes* EGD-e, the number of lipoproteins was reestimated down to 62 in *L. monocytogenes* EGD-e and 56 in *L. monocytogenes* F2365.[186] Despite a discrepancy in the predicted number, lipoproteins constitute the largest family of putative surface proteins in *Listeria* (68 out of 133 originally predicted in *L. monocytogenes* EGD-e).[19] These lipoproteins are putatively involved in various metabolic pathways (e.g., as substrate-binding components of ABC transport systems); remarkably, no biological function could be assigned for a large proportion. In spite of their predominance as surface proteins of Gram-positive bacteria, very few lipoproteins have been biochemically characterized.[195] In *Listeria*, only five have been more specifically investigated and thus confirmed, at least partially, in term of biological function: (1) TcsA (Lmo1388), a $CD4^+$ T cell–stimulating antigen presented by major histocompatibility complex class II molecules[196]; (2) GbuC (Lmo1016), a glycine betaine binding-protein part of an ABC transport system[197]; (3) the substrate–binding protein OpuCC (Lmo1426) part of an ABC L-carnitine transporter[198]; (4) OppA (Lmo2196), another ABC substrate binding protein mediating the transport of oligopeptides[199]; and (5) the virulence factor LpeA (Lmo1847) involved in bacterial entry into eukaryotic infected cells.[200]

12.3.1.3 Extrinsic Membrane Proteins

No bioinformatic tool is currently available to identify such proteins, which are most of the time primarily predicted as extracellular or cytoplasmic depending on their presence on the external or internal side of the cytoplasmic membrane (Figure 12.8). Thus, their identification requires a deep understanding of bacterial physiology and excellent general literature survey. Some of these proteins, which are not intrinsic to the cytoplasmic membrane, can be subunits of membrane protein complexes such as F_1F_0ATP synthetase (GO:0045260), fumarate reductase complex (GO:0045284), or ABC (ATP binding cassette) transporters (GO:0043190). Some other extrinsic membrane proteins can interact more or less temporarily with other membrane components, including other membrane-associated proteins. For example, in the SRP-dependent pathway, ribosomal proteins interact with Sec translocon in the course of cotranslational translocation,[201] or in two-component systems, response regulators interact with membrane-bound sensors.[202] It should also be noticed that some cytoplasmic proteins can associate with the lipid bilayer by weak interactions and by no means be functionally associated with membrane components. To date, the number of extrinsic membrane proteins has never been estimated in *L. monocytogenes*.

12.3.2 Cell Wall–Associated Proteins

Cell wall-associated proteins are either covalently linked to peptidoglycan when possessing a C-terminal LPXTG motif or noncovalently linked to cell wall components by a cell wall-binding domain (CBD).[8] In Gram-positive bacteria, six CWBDs are currently characterized: CWBD of Type 1 (CWBD1), CWBD of Type 2 (CWBD2), Lysin motif domain (LysM), GW modules, S-layer homology domain (SLHD), and WXL domain (WXL).[8,203] In *Listeria*, however, only proteins with LPXTG, LysM, GW, and WXL motifs have been identified so far.[182,203,204] In *L. monocytogenes*, proteins exhibiting such motifs systematically possess Sec-dependent N-terminal signal peptide. These motifs can be found using RPS-BLAST (reverse position-specific BLAST)[205] or HMM[206] from different databases—namely, InterPRO,[207] Pfam,[208] SMART,[209] TIGRfam,[210] and SuperFamily.[211]

12.3.2.1 LPXTG Motif

LPXTG motif (IPR001899, PF00746, TIGR01167) is found in proteins covalently attached to cell wall by sortases[55] (see section 12.2.1). This motif consists of a pattern varying around LPXTG, a hydrophobic domain, and a positively charged tail.[212] Cross-bridging of the protein to cell wall by sortase would occur in four steps.[213,214] Following translocation across the Sec apparatus, membrane-associated sortase recognizes the LPXTG motif and cleaves it before linking proteins to cell wall precursor lipid II.[215] The proteins thus linked to lipid II are further incorporated into the cell wall by transglycosylation and transpeptidation reactions that generate peptidoglycan. Forty-one proteins with LPXTG motif substrates of SrtA have been identified in *L. monocytogenes* EGD-e,[19] whereas two proteins with an NXZTN motif are recognized by SrtB.[181,216] In *L. monocytogenes* F2365, a total of 44 LPXTG-like proteins have been identified.[18] SrtA is required for bacterial virulence, as among its protein substrates several virulence factors have been identified, such as InlA.[59,217] Compared to SrtA, SrtB plays a minor role both in terms of number of proteins anchored to cell wall and involvement in bacterial virulence, although virulence factor SvpA is substrate to SrtB.[216]

12.3.2.2 Noncovalently Attached Cell Wall Proteins

Even though most cell-associated proteins contain an LPXTG motif in *Listeria*, several proteins bear domains involved in noncovalent attachment to the components of cell wall. Three motifs are clearly established as involved in noncovalent cell wall attachment in *Listeria*: LysM, GW, and WXL.[182,203,204] Besides these known attachment domains, other proteins found in the cell wall are retained by putative CWBDs, alternative or unknown mechanisms.

12.3.2.2.1 LysM

LysM (IPR002482, PF01476, SM00257, SSF54106) is a motif about 40 residues long and composed of three α-helices with a general peptidoglycan-binding function.[218] It is found in a variety of enzymes mostly involved in bacterial cell wall degradation. When present, this motif is often repeated several times in the protein sequence. Interestingly, proteins bearing this motif can attach to the surface of Gram-positive bacteria other than the ones that synthesized it.[218] In *L. monocytogenes*, several proteins bear LysM domains; among others, P60 (also called Iap or CwhA) can be cited as it is also considered as a virulence factor.[219] It is worth stressing here that contrary to previous assumption,[19,182] the NlpC/P60 domain should not be considered as the motif involved in cell wall binding *stricto sensu* (see section 12.3.2.2.4). In *L. monocytogenes* EGD-e and F2365 genomes, six LysM proteins, including P60 orthologue, were identified following bioinformatic analysis (Desvaux and Hébraud, unpublished data).

12.3.2.2.2 GW

The GW (SSF82057) module was originally identified in *L. monocytogenes* within internalin InlB.[220] This module comprises about 80 amino acids, contains a highly conserved dipeptide Gly-Trp, and interacts with lipoteichoic acids allowing cell surface attachment.[182] GW modules are found in multicopy, as in InlB, where three copies are present in the C-terminal region, or in Ami, where eight modules are present. It also appears that the higher the number of GW modules is, the stronger is the attachment to the bacterial cell wall[221,222]; proteins exhibiting only one GW module would not be surface attached at all. It is interesting to note that GW modules are related to the Src homology-3 (SH3) clan (CL0010) and, more specifically, prokaryotic SH3 of Type 3, or SH3b (IPR013247,

PF08239), but are unlikely to act as functional mimics of SH3 domains since proline-binding sites are blocked or destroyed in GW domains.[223] In *L. monocytogenes* EGD-e, nine GW proteins were identified, most of which (including Ami) exhibited an amidase domain.[19,204] This indicates that this class of protein would be mainly involved in cell wall degradation in *L. monocytogenes,* although Ami is also considered as a virulence factor involved in bacterial adhesion to infected cells.[182,224]

12.3.2.2.3 WXL

Following genomic analysis of *Enterococcus faecalis*, a novel C-terminal cell wall–binding motif was uncovered and named the WXL domain.[203] This conserved domain is characterized by a first WXL motif and an additional YXXX[LIV]TWXLXXXP motif found further downstream; the two WXL domains are separated by between 66 and 247 residues. The WXL domain was found in the CDS of several genomes of low G+C Gram-positive bacteria, including *L. monocytogenes*, where four proteins bearing such domain were identified.[203] In *E. faecalis*, it was demonstrated that the WXL domain is a determinant of bacterial subcellular protein. Indeed, its presence conferred the cell surface display of the protein, whereas specific deletions into the domains prevented its display. Moreover, neither proteins nor carbohydrates were necessary for cell wall attachment but the peptidoglycan was a binding ligand for WXL domain. As LysM, WXL can attach to the cell walls of a variety of Gram-positive bacteria. From genome-wide analysis of Gram-positive bacteria,[225] it appeared that genes encoding WXL proteins seem to cluster and that some N-terminal regions of these proteins are involved in utilization of plant complex polysaccharides. It was also suggested that some WXL proteins might mediate interactions between different bacteria species.[203] In *Listeria*, the physiological function of such proteins awaits to be established, and their presence on the bacterial cell surface remains to be demonstrated.

12.3.2.2.4 Other Noncovalently Attached Cell Wall Proteins

The ChW motif (IPR006637, SM00728, PF07538) stands for clostridial hydrophobic domain with a conserved W residue and was first uncovered from bioinformatic analysis of *Clostridium acetobutylicum* genome.[226] As GW, ChW contains a highly conserved Gly–Trp dipeptide motif and was suggested to be involved in cell surface attachment.[33] This repetitive domain can be found several times along the protein sequence (up to nine copies). ChW proteins were speculated to be part of a molecular complex on bacterial cell surface dedicated to degradation of polymer and surface adhesion.[226] One putative serine protease bearing three copies of ChW motif is encoded in the genome of *L. monocytogenes* F2365 (Desvaux and Hébraud, unpublished data); its expression, secretion, cell surface display, and function remain to be established. Similarly, the function of SH3b is as yet unknown, but *Staphylococcus simulans* lysostaphin contains such a domain in its C-terminal region.[227] Since this region mediates protein binding to the bacterial cell wall, SH3b may have this function.

It is important to note that despite the absence of cell wall–binding motifs in some enzymes involved in cell wall degradation, such proteins have affinity for cell wall components via their enzymatic active site. Thus, secreted proteins with cell wall degradation domains, such as NlpC/P60 (IPR000064, PF00877)[228] or N-acetylmuramoyl-L-alanine (IPR002508, PF01520), can be localized in the cell wall. However, such domains should not be considered as cell wall–binding motifs per se since primary function of these enzymes is to cleave cell wall components, then they can either find a new cleavage site or are released into the extracellular milieu. In *L. monocytogenes*, these enzymes are involved in numerous cellular processes.[126]

It is now well known that many metabolic enzymes can be surface localized in Gram-positive bacteria.[229] Such proteins lack the N-terminal signal peptide and are supposedly secreted by pathways alternative to the known ones. SecretomeP is a bioinformatic tool dedicated to the prediction of such proteins.[230] However, instead of secretion through the cytoplasmic membrane, these proteins could be released from the bacterial cell following autolysis and then attached to the cell surface of nonlysed cells. In *Streptococcus pneumoniae*, release of cytoplasmic proteins is triggered by competent cells and originates from lysis of noncompetent cells.[231] This tightly controlled phenomenon was named allolysis and involves several cell wall hydrolases. Once released and associated with bacterial cell surface, such proteins generally moonlight.[23] For example, in *S. pneumoniae*, the glycolytic enzyme glyceraldehyde 3-phosphate dehydrogenase exhibits plasmin(ogen)-binding activity on the bacterial cell surface and thus significantly enhances bacterial virulence.[232] In *L. monocytogenes*, several proteins primarily predicted as cytoplasmic were also identified in the cell wall fraction, including enolase, which was demonstrated to bind human plasminogen.[24] Functions of other cytoplasmic proteins found at this subcellular location remain to be elucidated as well as protein motifs involved in cell wall attachment.

12.4 CONCLUSIONS AND PERSPECTIVES

The proteomic technologies are certainly the most powerful and appropriate to provide global and accurate information on the expression, structure, and function of proteins. Since the first description of protein extraction and separation using two-dimensional gel electrophoresis (2-DE) in 1975,[233–235] many advances (use of new detergents, immobilized pH gradient, new apparatus for IsoElectroFocalization [first dimension], and SDS-PAGE [second dimension]) have been brought to improve protein solubilization and resolution as well as reproducibility and implementation of the techniques. Over the past two decades, 2-DE progressively became the classical method of choice to separate and compare complex mixtures of proteins and was mainly applied for soluble intracellular proteins. In the field of bacteriology, an increasing number of investigations using comparative proteomic approaches was devoted to the characterization of adaptive responses—namely, to various physicochemical stresses or to the effect of a gene mutation. At this time, however, protein identification was difficult and time consuming as it essentially involved Edman degradation and sequence alignment from short and on limited numbers of amino acid sequences. Consequently, these early studies generally remained quite descriptive and phenomenological.

However, two occurrences gave a considerable impetus to proteomic analyses: (1) the availability of ever growing amounts of genomic sequence data and (2) important advances in mass spectrometry technology for ionization and detection of large molecules such as peptides and proteins. Indeed, data obtained with mass spectrometry analysis—namely, peptide mass fingerprinting or fragmentation—could from then be matched against databases of all known gene products and thus greatly facilitated protein identification. The use of 2-DE and MS tools to separate and identify proteins is now widespread in all domains of life science. One of the consequences of this remarkable progress was the possibility to establish 2-DE databases containing several hundreds of identified protein spots available on proteome reference maps. Thus, the first bacterial 2-DE database was established on *E. coli* cytosoluble proteins separated into different pH gradients and regularly brought up to date.[236,237] Similar, but generally more limited, 2-DE databases are available for other bacterial species, including *L. monocytogenes* EGD-e[238] (http://www.clermont.inra.fr/proteome).

Another consequence was the possibility and need to investigate further the different cell compartments and, thus, following cell fractionation to explore thoroughly the different subcellular proteomes that, in Gram-positive bacteria, include (1) cytoplasmic proteins, (2) membrane–associated proteins, (3) cell wall associated proteins, and (4) proteins secreted in the extracellular milieu. While extracting and separating cytoplasmic and extracellular proteins could be achieved rather easily and efficiently, classical 2-DE procedures failed to give a good overview of proteins present in the cell envelope—that is, proteins associated with the cytoplasmic membrane or cell wall. Beyond the well-known limitations of 2-DE gel-based technology (the inability to separate or to reveal low-abundance proteins, high molecular mass and extreme p*I* [isoelectric point] proteins), other limitations appeared much more problematic for the cell surface subproteomes, due to the intrinsic properties of cell envelope–associated proteins. Indeed, multitransmembrane proteins are generally highly hydrophobic and are either almost impossible to solubilize during the extraction procedure or not recovered in the second dimension of 2-DE due to self-aggregation and irreversible precipitation in IEF. On the other hand, proteins noncovalently or *a fortiori* covalently attached to the cell wall peptidoglycan are very difficult to extract and require specific and laborious treatments not directly compatible with classical 2-DE separation. Altogether, this considerably hampers proteomic analysis in classical 2-DE gel-based technology.

Different strategies are now developed to tackle the difficulties of analyzing these cell envelope subproteomes. These strategies can associate different protocols of protein extraction with different techniques of separation and mass spectrometry.[239] Several studies have attempted to extract membrane–associated proteins of Gram-positive bacteria by combining protocols described for Gram-negative or eukaryote organisms.[240] Thus, the extraction procedures could include enzymatic treatment, fractionation of broken cell by centrifugation, use of chemical agents such as zwitterionic detergents for solubilization of hydrophobic proteins,[241] solvents for delipidation,[242] or protein extraction and separation.[243] For example, such a combinational approach has been used to efficiently characterize by 2-DE the cell-wall and membrane-associated subproteomes of the Gram-positive bacterium *Staphylococcus xylosus*.[244]

A different protocol originally developed for *Bacillus cereus*[245] has been applied for the global extraction of *L. monocytogenes* cell-surface proteins combining the protein solubilization by SDS with a classical SDS-PAGE separation.[246] The 1-DE and 2-DE separations were both used to characterize the cell wall subproteome of *L. monocytogenes*.[24] In this case, proteins were extracted by the sequential action of two salts at high concentration and their identification was performed by N-terminal sequencing and peptide mass fingerprint obtained with matrix assisted laser desorption ionization time of flight mass spectrometer (MALDI-TOF MS). It is interesting to note that among the 55 identified proteins, only 27 possessed a peptide signal, including 4 proteins with cell wall–binding motifs (2 GW proteins and 2 LysM proteins), 20 lipoproteins, and 3 proteins with no predictable surface association motif. The 28 remaining proteins without peptide signal were primarily predicted with cytoplasmic functions and nothing could explain how they managed to cross the cytoplasmic membrane or how they associated with bacterial cell wall. Such unusual localization of cytoplasmic proteins leads to the suggestion that they could moonlight on the bacterial cell surface, although no experimental evidence could back up such a hypothesis.

More recent strategies and technologies consist of analyzing peptide mixture obtained by tryptic digestion of cell envelope protein samples issued from stringent protocols extraction (e.g., combining cell disruption fractionation by centrifugation, and treatment with high concentration of SDS at 100°C, then at 80°C).[58,247] An alternative approach consists in "shaving" the bacterial surface with a specific protease (such as trypsin) to cleave surface-exposed proteins.[239] After lyophilization to remove SDS, the peptide hydrolysate is then separated by two-dimensional liquid chromatography coupled to tandem mass spectrometer (2-D LC MS/MS). This separation technique, termed "shotgun proteomic" or multidimensional protein identification technology (MudPIT), uses a two-dimensional liquid chromatography to separate a tryptic peptide mixture where a strong cation exchange is applied in the first dimension and a reverse phase is applied in the second dimension. The separated peptides are subjected online to analysis by fragmentation (MS/MS) in an electrospray ionization MS. A peptide fragmentation spectrum is further used to identify the original protein via query against databases.

Besides a significant gain of time, the use of 2-D LC MS/MS overcomes all limitations of gel-based 2-DE previously cited. Even hydrophobic proteins can be identified, thanks to amino acid cleavage sites accessible to tryptic digestion in exposed regions of the protein. These new approaches allowed extraction of proteins following treatments that are not always compatible with classical 2-DE. Consequently, the set of proteins identified with high number of transmembrane–spanning regions or LPXTG motif (i.e., covalently anchored to cell wall) has been significantly enlarged. The other development of 2-D LC MS/MS concerns the possibility to perform quantitative proteomics for comparative analysis of samples pretreated with amino acid tags or labels such as the ICAT™ (isotope-coded affinity tags), iTRAQ™, or SILAC technologies.[239]

The years to come will undoubtedly see the development and improvement of these new exploring methods of subproteomes. Everyone will have the possibility to map protein expression and to compare several biological samples with high throughput, sensibility, and resolution. In spite of this very attractive progress and considering at least its complementarity, the classical 2-DE technique remains irreplaceable. Indeed and contrary to LC MS/MS approaches, gel-based 2-DE allows one to separate simultaneously several hundreds of proteins at once and to visualize shifts due to post-translational modifications. The implementation of two or more complementary proteomic strategies would be one of the keys to generate valuable information on the role of cell envelope proteins in pathogenic processes, bacterial communication, sensing of and exchange with its environment, motility, and adhesion on and colonization of biotic or abiotic surfaces.

REFERENCES

1. Woese, C.R., A new biology for a new century, *Microbiol. Mol. Biol. Rev.*, 68, 173, 2004.
2. Shatalkin, A.I., Highest level of division in classification of organisms. 3. Monodermata and didermata, *Zh. Obshch. Biol.*, 65, 195, 2004.
3. Danchin, A., *The Delphic boat: What genomes tell us,* Harvard University Press, Cambridge, MA, 2003.
4. Monod, J., *Le hasard et la nécessité: Essai sur la philosophie naturelle de la biologie moderne,* Éditions du Seuil, Paris, 1970.
5. Cavalier-Smith, T., Rooting the tree of life by transition analyses, *Biol. Direct.*, 1, 19, 2006.
6. Cavalier-Smith, T., The neomuran origin of archaebacteria, the negibacterial root of the universal tree and bacterial megaclassification, *Int. J. Syst. Evol. Microbiol.*, 52, 7, 2002.
7. Gupta, R.S., The natural evolutionary relationships among prokaryotes, *Crit. Rev. Microbiol.*, 26, 111, 2000.
8. Desvaux, M. et al., Protein cell surface display in Gram-positive bacteria: From single protein to macromolecular protein structure, *FEMS Microbiol. Lett.*, 256, 1, 2006.
9. Garrity, G. M., *Bergey's manual of systematic bacteriology*, 2nd ed., Springer, Berlin, 2001.
10. Woese, C.R., Bacterial evolution, *Microbiol. Rev.*, 51, 221, 1987.
11. Snel, B., Huynen, M.A., and Dutilh, B.E., Genome trees and the nature of genome evolution, *Annu. Rev. Microbiol.*, 59, 191, 2005.
12. Gupta, R.S. and Griffiths, E., Critical issues in bacterial phylogeny, *Theor. Popul. Biol.*, 61, 423, 2002.
13. Mira, A. et al., Evolutionary relationships of *Fusobacterium nucleatum* based on phylogenetic analysis and comparative genomics, *BMC Evol. Biol.*, 4, 50, 2004.
14. Botero, L.M. et al., *Thermobaculum terrenum* gen. nov., sp. nov.: A nonphototrophic Gram-positive thermophile representing an environmental clone group related to the *Chloroflexi* (green nonsulfur bacteria) and *Thermomicrobia*, *Arch. Microbiol.*, 181, 269, 2004.
15. Roberts, A.J. and Wiedmann, M., Pathogen, host and environmental factors contributing to the pathogenesis of listeriosis, *Cell Mol. Life Sci.*, 60, 904, 2003.
16. Vaneechoutte, M. et al., Comparison of PCR-based DNA fingerprinting techniques for the identification of *Listeria* species and their use for atypical *Listeria* isolates, *Int. J. Syst. Bacteriol.*, 48, 127, 1998.
17. Hain, T., Steinweg, C., and Chakraborty, T., Comparative and functional genomics of *Listeria* spp., *J. Biotechnol.*, 126, 37, 2006.
18. Nelson, K.E. et al., Whole genome comparisons of serotype 4b and 1/2a strains of the food-borne pathogen *Listeria monocytogenes* reveal new insights into the core genome components of this species, *Nucl. Acids Res.*, 32, 2386, 2004.
19. Glaser, P. et al., Comparative genomics of *Listeria* species, *Science*, 294, 849, 2001.
20. Hain, T. et al., Whole-genome sequence of *Listeria welshimeri* reveals common steps in genome reduction with *Listeria innocua* as compared to *Listeria monocytogenes*, *J. Bacteriol.*, 188, 7405, 2006.
21. Fraser, C.M. et al., The value of complete microbial genome sequencing (you get what you pay for), *J. Bacteriol.*, 184, 6403, 2002.
22. Demchick, P. and Koch, A.L., The permeability of the wall fabric of *Escherichia coli* and *Bacillus subtilis*, *J. Bacteriol.*, 178, 768, 1996.
23. Jeffery, C.J., Moonlighting proteins, *Trends Biochem. Sci.*, 24, 8, 1999.
24. Schaumburg, J. et al., The cell wall subproteome of *Listeria monocytogenes*, *Proteomics*, 4, 2991, 2004.

25. Gupta, R.S., Protein phylogenies and signature sequences: A reappraisal of evolutionary relationships among archaebacteria, eubacteria, and eukaryotes, *Microbiol. Mol. Biol. Rev.*, 62, 1435, 1998.
26. Henderson, I.R. et al., Type V protein secretion pathway: The autotransporter story, *Microbiol. Mol. Biol. Rev.*, 68, 692, 2004.
27. Salmond, G.P. and Reeves, P.J., Membrane traffic wardens and protein secretion in Gram-negative bacteria, *Trends Biochem. Sci.*, 18, 7, 1993.
28. Henderson, I.R. et al., Renaming protein secretion in the Gram-negative bacteria, *Trends Microbiol.*, 8, 352, 2000.
29. Economou, A. et al., Secretion by numbers: Protein traffic in prokaryotes, *Mol. Microbiol.*, 62, 308, 2006.
30. Tjalsma, H. et al., Signal peptide-dependent protein transport in *Bacillus subtilis*: A genome-based survey of the secretome, *Microbiol. Mol. Biol. Rev.*, 64, 515, 2000.
31. Pallen, M.J., Chaudhuri, R.R., and Henderson, I.R., Genomic analysis of secretion systems, *Curr. Opin. Microbiol.*, 6, 519, 2003.
32. Tjalsma, H. et al., Proteomics of protein secretion by *Bacillus subtilis*: Separating the "secrets" of the secretome, *Microbiol. Mol. Biol. Rev.*, 68, 207, 2004.
33. Desvaux, M. et al., Genomic analysis of the protein secretion systems in *Clostridium acetobutylicum* ATCC824, *Biochim. Biophys. Acta-Mol. Cell Res.*, 1745, 223, 2005.
34. Busch, W. and Saier, M.H., Jr., The transporter classification (TC) system, *Crit. Rev. Biochem. Mol. Biol.*, 37, 287, 2002.
35. Ajouz, B. et al., Release of thioredoxin via the mechanosensitive channel MscL during osmotic downshock of *Escherichia coli* cells, *J. Biol. Chem.*, 273, 26670, 1998.
36. Kachlany, S.C. et al., Nonspecific adherence by *Actinobacillus actinomycetemcomitans* requires genes widespread in *Bacteria* and *Archaea*, *J. Bacteriol.*, 182, 6169, 2000.
37. Desvaux, M. and Hébraud, M., The protein secretion systems in *Listeria*: Inside-out bacterial virulence, *FEMS Microbiol. Rev.*, 30, 774, 2006.
38. Desvaux, M. et al., The general secretory pathway: A general misnomer? *Trends Microbiol.*, 12, 306, 2004.
39. Cao, T.B. and Saier, M.H., Jr., The general protein secretory pathway: Phylogenetic analyses leading to evolutionary conclusions, *Biochim. Biophys. Acta-Biomembr.*, 1609, 115, 2003.
40. Valent, Q.A. et al., The *Escherichia coli* SRP and SecB targeting pathways converge at the translocon, *EMBO J.*, 17, 2504, 1998.
41. Van Wely, K.H.M. et al., Translocation of proteins across the cell envelope of Gram-positive bacteria, *FEMS Microbiol. Rev.*, 25, 437, 2001.
42. de Keyzer, J., van der Does, C., and Driessen, A.J., The bacterial translocase: A dynamic protein channel complex, *Cell Mol. Life Sci.*, 60, 2034, 2003.
43. Economou, A. et al., SecA membrane cycling at SecYEG is driven by distinct ATP binding and hydrolysis events and is regulated by SecD and SecF, *Cell*, 83, 1171, 1995.
44. Lenz, L.L. and Portnoy, D.A., Identification of a second *Listeria secA* gene associated with protein secretion and the rough phenotype, *Mol. Microbiol.*, 45, 1043, 2002.
45. Bensing, B.A. and Sullam, P.M., An accessory *sec* locus of *Streptococcus gordonii* is required for export of the surface protein GspB and for normal levels of binding to human platelets, *Mol. Microbiol.*, 44, 1081, 2002.
46. Lenz, L.L. et al., SecA2-dependent secretion of autolytic enzymes promotes *Listeria monocytogenes* pathogenesis, *Proc. Natl. Acad. Sci. USA*, 100, 12432, 2003.
47. Dramsi, S. et al., FbpA, a novel multifunctional *Listeria monocytogenes* virulence factor, *Mol. Microbiol.*, 53, 639, 2004.
48. Froderberg, L. et al., Versatility of inner membrane protein biogenesis in *Escherichia coli*, *Mol. Microbiol.*, 47, 1015, 2003.
49. Tjalsma, H., Bron, S., and van Dijl, J.M., Complementary impact of paralogous Oxa1-like proteins of *Bacillus subtilis* on post-translocational stages in protein secretion, *J. Biol. Chem.*, 278, 15622, 2003.
50. Madden, J.C., Ruiz, N., and Caparon, M., Cytolysin-mediated translocation (CMT): A functional equivalent of Type III secretion in Gram-positive bacteria, *Cell*, 104, 143, 2001.
51. Desvaux, M. et al., Type III secretion: What's in a name? *Trends Microbiol.*, 14, 157, 2006.
52. Bonnemain, C. et al., Differential roles of multiple signal peptidases in the virulence of *Listeria monocytogenes*, *Mol. Microbiol.*, 51, 1251, 2004.

53. Raynaud, C. and Charbit, A., Regulation of expression of type I signal peptidases in *Listeria monocytogenes*, *Microbiology*, 151, 3769, 2005.
54. Reglier-Poupet, H. et al., Maturation of lipoproteins by type II signal peptidase is required for phagosomal escape of *Listeria monocytogenes*, *J. Biol. Chem.*, 278, 49469, 2003.
55. Navarre, W.W. and Schneewind, O., Surface proteins of Gram-positive bacteria and mechanisms of their targeting to the cell wall envelope, *Microbiol. Mol. Biol. Rev.*, 63, 174, 1999.
56. Mazmanian, S.K., Ton-That, H., and Schneewind, O., Sortase-catalyzed anchoring of surface proteins to the cell wall of *Staphylococcus aureus*, *Mol. Microbiol.*, 40, 1049, 2001.
57. Dramsi, S., Trieu-Cuot, P., and Bierne, H., Sorting sortases: A nomenclature proposal for the various sortases of Gram-positive bacteria, *Res. Microbiol.*, 156, 289, 2005.
58. Pucciarelli, M.G. et al., Identification of substrates of the *Listeria monocytogenes* sortases A and B by a nongel proteomic analysis, *Proteomics*, 5, 4808, 2005.
59. Bierne, H. et al., Inactivation of the *srtA* gene in *Listeria monocytogenes* inhibits anchoring of surface proteins and affects virulence, *Mol. Microbiol.*, 43, 869, 2002.
60. Lee, S.G., Pancholi, V., and Fischetti, V.A. , Characterization of unique glycosylated anchor endopeptidase that cleaves the LPXTG sequence motif of cell surface proteins of Gram-positive bacteria, *J. Biol. Chem.*, 277, 46912, 2002.
61. Lee, S.G. and Fischetti, V.A., Purification and characterization of LPXTGase from *Staphylococcus aureus*: The amino acid composition mirrors that found in the peptidoglycan, *J. Bacteriol.*, 188, 389, 2006.
62. Kleinkauf, H. and von Dohren, H., The nonribosomal peptide biosynthetic system—On the origins of structural diversity of peptides, cyclopeptides and related compounds, *Antonie Van Leeuwenhoek*, 67, 229, 1995.
63. Fekkes, P. and Driessen, A.J., Protein targeting to the bacterial cytoplasmic membrane, *Microbiol. Mol. Biol. Rev.*, 63, 161, 1999.
64. von Heijne, G., A new method for predicting signal sequence cleavage sites, *Nucleic Acids Res.*, 14, 4683, 1986.
65. Menne, K.M., Hermjakob, H., and Apweiler, R., A comparison of signal sequence prediction methods using a test set of signal peptides, *Bioinformatics*, 16, 741, 2000.
66. Nielsen, H. et al., Identification of prokaryotic and eukaryotic signal peptides and prediction of their cleavage sites, *Protein Eng.*, 10, 1, 1997.
67. Bendtsen, J.D. et al., Improved prediction of signal peptides: SignalP 3.0, *J. Mol. Biol.*, 340, 783, 2004.
68. Zhang, Z. and Henzel, W.J., Signal peptide prediction based on analysis of experimentally verified cleavage sites, *Protein Sci.*, 13, 2819, 2004.
69. Hiller, K. et al., PrediSi: Prediction of signal peptides and their cleavage positions, *Nucleic Acids Res.*, 32, W375, 2004.
70. Gomi, M., Sonoyama, M., and Mitaku, S., High performance system for signal peptide prediction: SOSUIsignal, *Chem-Bio Info. J.*, 4, 142, 2004.
71. Käll, L., Krogh, A., and Sonnhammer, E.L., A combined transmembrane topology and signal peptide prediction method, *J. Mol. Biol.*, 338, 1027, 2004.
72. Cai, Y.D., Lin, S.L., and Chou, K.C., Support vector machines for prediction of protein signal sequences and their cleavage sites, *Peptides*, 24, 159, 2003.
73. Gardy, J.L. et al., PSORT-B: Improving protein subcellular localization prediction for Gram-negative bacteria, *Nucleic Acids Res.*, 31, 3613, 2003.
74. Gardy, J.L. et al., PSORTb v.2.0: Expanded prediction of bacterial protein subcellular localization and insights gained from comparative proteome analysis, *Bioinformatics*, 21, 617, 2005.
75. Altschul, S.F. et al., Basic local alignment search tool, *J. Mol. Biol.*, 215, 403, 1990.
76. Choo, K.H., Tan, T.W., and Ranganathan, S., SPdb—A signal peptide database, *BMC Bioinformatics*, 6, 249, 2005.
77. Reinhardt, A. and Hubbard, T., Using neural networks for prediction of the subcellular location of proteins, *Nucleic Acids Res.*, 26, 2230, 1998.
78. Hua, S. and Sun, Z., Support vector machine approach for protein subcellular localization prediction, *Bioinformatics*, 17, 721, 2001.
79. Yu, C.S., Lin, C.J., and Hwang, J.K., Predicting subcellular localization of proteins for Gram-negative bacteria by support vector machines based on n-peptide compositions, *Protein Sci.*, 13, 1402, 2004.
80. Yu, C.S. et al., Prediction of protein subcellular localization, *Proteins*, 64, 643, 2006.

81. Nair, R. and Rost, B., Mimicking cellular sorting improves prediction of subcellular localization, *J. Mol. Biol.*, 348, 85, 2005.
82. Dönnes, P. and Höglund, A., Predicting protein subcellular localization: Past, present, and future, *Geno. Prot. Bioinfo.*, 2, 209, 2004.
83. Lu, Z. et al., Predicting subcellular localization of proteins using machine-learned classifiers, *Bioinformatics*, 20, 547, 2004.
84. Gardy, J.L. and Brinkman, F.S., Methods for predicting bacterial protein subcellular localization, *Nat. Rev. Microbiol.*, 4, 741, 2006.
85. Shen, H.B. and Chou, K.C., Gpos-PLoc: An ensemble classifier for predicting subcellular localization of Gram-positive bacterial proteins, *Protein Eng. Des. Sel.*, 20, 39, 2007.
86. Guo, T. et al., DBSubLoc: Database of protein subcellular localization, *Nucleic Acids Res.*, 32, D122, 2004.
87. Rey, S. et al., PSORTdb: A protein subcellular localization database for bacteria, *Nucleic Acids Res.*, 33, D164, 2005.
88. Lu, P. et al., PA-GOSUB: A searchable database of model organism protein sequences with their predicted gene ontology molecular function and subcellular localization, *Nucleic Acids Res.*, 33, D147, 2005.
89. Billion, A. et al., Augur—A computational pipeline for whole genome microbial surface protein prediction and classification, *Bioinformatics*, 22, 2819, 2006.
90. Trost, M. et al., Comparative proteome analysis of secretory proteins from pathogenic and nonpathogenic *Listeria* species, *Proteomics*, 5, 1544, 2005.
91. Berks, B.C., Palmer, T., and Sargent, F., Protein targeting by the bacterial twin-arginine translocation (Tat) pathway, *Curr. Opin. Microbiol.*, 8, 174, 2005.
92. Müller, M., Twin-arginine-specific protein export in *Escherichia coli*, *Res. Microbiol.*, 156, 131, 2005.
93. Lee, P.A., Tullman-Ercek, D., and Georgiou, G., The bacterial twin-arginine translocation pathway, *Annu. Rev. Microbiol.*, 60, 373, 2006.
94. Mori, H. and Cline, K., Post-translational protein translocation into thylakoids by the Sec and ΔpH-dependent pathways, *Biochim. Biophys. Acta-Mol. Cell Res.*, 1541, 80, 2001.
95. Bruser, T. and Sanders, C., An alternative model of the twin arginine translocation system, *Microbiol. Res.*, 158, 7, 2003.
96. Sargent, F., Berks, B.C., and Palmer, T., Pathfinders and trailblazers: A prokaryotic targeting system for transport of folded proteins, *FEMS Microbiol. Lett.*, 254, 198, 2006.
97. Alami, M. et al., Differential interactions between a twin-arginine signal peptide and its translocase in *Escherichia coli*, *Mol. Cell*, 12, 937, 2003.
98. Gohlke, U. et al., The TatA component of the twin-arginine protein transport system forms channel complexes of variable diameter, *Proc. Natl. Acad. Sci. USA*, 102, 10482, 2005.
99. Dilks, K. et al., Prokaryotic utilization of the twin-arginine translocation pathway: A genomic survey, *J. Bacteriol.*, 185, 1478, 2003.
100. Rose, R.W. et al., Adaptation of protein secretion to extremely high-salt conditions by extensive use of the twin-arginine translocation pathway, *Mol. Microbiol.*, 45, 943, 2002.
101. Bendtsen, J.D. et al., Prediction of twin-arginine signal peptides, *BMC Bioinformatics*, 6, 167, 2005.
102. Yen, M.R. et al., Sequence and phylogenetic analyses of the twin-arginine targeting (Tat) protein export system, *Arch. Microbiol.*, 177, 441, 2002.
103. Ize, B., Gerard, F., and Wu, L.F., *In vivo* assessment of the Tat signal peptide specificity in *Escherichia coli*, *Arch. Microbiol.*, 178, 548, 2002.
104. Ignatova, Z. et al., Unusual signal peptide directs penicillin amidase from *Escherichia coli* to the Tat translocation machinery, *Biochem. Biophys. Res. Commun.*, 291, 146, 2002.
105. Robinson, C. and Bolhuis, A., Tat-dependent protein targeting in prokaryotes and chloroplasts, *Biochim. Biophys. Acta-Mol. Cell Res.*, 1694, 135, 2004.
106. Peabody, C.R. et al., Type II protein secretion and its relationship to bacterial Type 4 pili and archaeal flagella, *Microbiology*, 149, 3051, 2003.
107. Chung, Y.S. and Dubnau, D., All seven *comG* open reading frames are required for DNA binding during transformation of competent *Bacillus subtilis*, *J. Bacteriol.*, 180, 41, 1998.
108. Claverys, J.P. and Martin, B., Bacterial "competence" genes: Signatures of active transformation, or only remnants? *Trends Microbiol.*, 11, 161, 2003.
109. Chen, I. and Dubnau, D., DNA uptake during bacterial transformation, *Nat. Rev. Microbiol.*, 2, 241, 2004.

110. Dubnau, D., Binding and transport of transforming DNA by *Bacillus subtilis*: The role of Type-4 pilin-like proteins—A review, *Gene*, 192, 191, 1997.
111. Filloux, A., The underlying mechanisms of Type II protein secretion, *Biochim. Biophys. Acta*, 1694, 163, 2004.
112. Chung, Y.S. and Dubnau, D., ComC is required for the processing and translocation of comGC, a pilin-like competence protein of *Bacillus subtilis*, *Mol. Microbiol.*, 15, 543, 1995.
113. LaPointe, C.F. and Taylor, R.K., The type 4 prepilin peptidases comprise a novel family of aspartic acid proteases, *J. Biol. Chem.*, 275, 1502, 2000.
114. Chen, I., Provvedi, R., and Dubnau, D., A macromolecular complex formed by a pilin-like protein in competent *Bacillus subtilis*, *J. Biol. Chem.*, 281, 21720, 2006.
115. de Castro, E. et al., ScanProsite: Detection of PROSITE signature matches and ProRule-associated functional and structural residues in proteins, *Nucleic Acids Res.*, 34, W362, 2006.
116. Peel, M., Donachie, W., and Shaw, A., Temperature-dependent expression of flagella of *Listeria monocytogenes* studied by electron microscopy, SDS-PAGE and western blotting, *J. Gen. Microbiol.*, 134, 2171, 1988.
117. Way, S.S. et al., Characterization of flagellin expression and its role in *Listeria monocytogenes* infection and immunity, *Cell Microbiol.*, 6, 235, 2004.
118. Sanchez-Campillo, M. et al., Modulation of DNA topology by *flaR*, a new gene from *Listeria monocytogenes*, *Mol. Microbiol.*, 18, 801, 1995.
119. Michel, E. et al., Characterization of a large motility gene cluster containing the *cheR*, *motAB* genes of *Listeria monocytogenes* and evidence that PrfA down-regulates motility genes, *FEMS Microbiol. Lett.*, 169, 341, 1998.
120. Knudsen, G.M., Olsen, J.E., and Dons, L., Characterization of DegU, a response regulator in *Listeria monocytogenes*, involved in regulation of motility and contributes to virulence, *FEMS Microbiol. Lett.*, 240, 171, 2004.
121. Grundling, A. et al., *Listeria monocytogenes* regulates flagellar motility gene expression through MogR, a transcriptional repressor required for virulence, *Proc. Natl. Acad. Sci. USA*, 101, 12318, 2004.
122. Shen, A. et al., A bifunctional O-GlcNAc transferase governs flagellar motility through antirepression, *Genes Dev.*, 20, 3283, 2006.
123. Schirm, M. et al., Flagellin from *Listeria monocytogenes* is glycosylated with β-O-linked N-acetyl-glucosamine, *J. Bacteriol.*, 186, 6721, 2004.
124. Lemon, K.P., Higgins, D.E., and Kolter, R., Flagellar motility is critical for *Listeria monocytogenes* biofilm formation, *J. Bacteriol.*, 189, 4418, 2007.
125. O'Neil, H.S. and Marquis, H., *Listeria monocytogenes* flagella are used for motility, not as adhesins, to increase host cell invasion, *Infect. Immun.*, 74, 6675, 2006.
126. Popowska, M., Analysis of the peptidoglycan hydrolases of *Listeria monocytogenes*: Multiple enzymes with multiple functions, *Pol. J. Microbiol.*, 53, 29, 2004.
127. Popowska, M. and Markiewicz, Z., Murein-hydrolyzing activity of flagellin FlaA of *Listeria monocytogenes*, *Pol. J. Microbiol.*, 53, 237, 2004.
128. Macnab, R.M., How bacteria assemble flagella, *Annu. Rev. Microbiol.*, 57, 77, 2003.
129. Aizawa, S.I., Bacterial flagella and Type III secretion systems, *FEMS Microbiol. Lett.*, 202, 157, 2001.
130. Journet, L., Hughes, K.T., and Cornelis, G.R., Type III secretion: A secretory pathway serving both motility and virulence, *Mol. Membr. Biol.* 22, 41, 2005.
131. Macnab, R.M., Type III flagellar protein export and flagellar assembly, *Biochim. Biophys. Acta-Mol. Cell Res.*, 1694, 207, 2004.
132. Minamino, T. and Namba, K., Self-assembly and Type III protein export of the bacterial flagellum, *J. Mol. Microbiol. Biotechnol.*, 7, 5, 2004.
133. Troisfontaines, P. and Cornelis, G.R., Type III secretion: More systems than you think, *Physiology (Bethesda)*, 20, 326, 2005.
134. Young, G.M., Schmiel, D.H., and Miller, V.L., A new pathway for the secretion of virulence factors by bacteria, the flagellar export apparatus functions as a protein-secretion system, *Proc. Natl. Acad. Sci. USA*, 96, 6456, 1999.
135. Cornelis, G.R., The Type III secretion injectisome, *Nat. Rev. Microbiol.*, 4, 811, 2006.
136. Ghelardi, E. et al., Requirement of *flhA* for swarming differentiation, flagellin export, and secretion of virulence-associated proteins in *Bacillus thuringensis*, *J. Bacteriol.*, 184, 6424, 2002.
137. Wang, I.N., Smith, D.L., and Young, R., Holins: The protein clocks of bacteriophage infections, *Annu. Rev. Microbiol.*, 54, 799, 2000.

138. Bayles, K.W., Are the molecular strategies that control apoptosis conserved in bacteria? *Trends Microbiol.*, 11, 306, 2003.
139. Young, R. and Blasi, U., Holins: Form and function in bacteriophage lysis, *FEMS Microbiol. Rev.*, 17, 191, 1995.
140. Bernhardt, T.G. et al., Breaking free: "Protein antibiotics" and phage lysis, *Res. Microbiol.*, 153, 493, 2002.
141. Loessner, M.J., Bacteriophage endolysins—Current state of research and applications, *Curr. Opin. Microbiol.*, 8, 480, 2005.
142. Ziedaite, G. et al., The holin protein of bacteriophage PRD1 forms a pore for small-molecule and endolysin translocation, *J. Bacteriol.*, 187, 5397, 2005.
143. Bläsi, U. and Young, R., Two beginnings for a single purpose: The dual-start holins in the regulation of phage lysis, *Mol. Microbiol.*, 21, 675, 1996.
144. Groicher, K.H. et al., The *Staphylococcus aureus* lrgAB operon modulates murein hydrolase activity and penicillin tolerance, *J. Bacteriol.*, 182, 1794, 2000.
145. Rice, K.C. et al., The *Staphylococcus aureus cidAB* operon: Evaluation of its role in regulation of murein hydrolase activity and penicillin tolerance, *J. Bacteriol.*, 185, 2635, 2003.
146. Saier, M.H., Jr., Tran, C.V., and Barabote, R.D., TC-DB: The transporter classification database for membrane transport protein analyses and information, *Nucleic Acids Res.*, 34, D181–D186, 2006.
147. Young, R., Bacteriophage holins: Deadly diversity, *J. Mol. Microbiol. Biotechnol.*, 4, 21, 2002.
148. Ramanculov, E. and Young, R., Functional analysis of the phage T4 holin in a lambda context, *Mol. Genet. Genomics*, 265, 345, 2001.
149. Loessner, M.J. et al., Complete nucleotide sequence, molecular analysis and genome structure of bacteriophage A118 of *Listeria monocytogenes*: Implications for phage evolution, *Mol. Microbiol.*, 35, 324, 2000.
150. Vukov, N. et al., Functional regulation of the *Listeria monocytogenes* bacteriophage A118 holin by an intragenic inhibitor lacking the first transmembrane domain, *Mol. Microbiol.*, 48, 173, 2003.
151. Loessner, M.J., Wendlinger, G., and Scherer, S., Heterogeneous endolysins in *Listeria monocytogenes* bacteriophages: A new class of enzymes and evidence for conserved holin genes within the siphoviral lysis cassettes, *Mol. Microbiol.*, 16, 1231, 1995.
152. Tan, K.S., Wee, B.Y., and Song, K.P., Evidence for holin function of *tcdE* gene in the pathogenicity of *Clostridium difficile*, *J. Med. Microbiol.*, 50, 613, 2001.
153. Mukherjee, K. et al., Proteins released during high toxin production in *Clostridium difficile*, *Microbiology*, 148, 2245, 2002.
154. Pallen, M.J., The ESAT-6/WXG100 superfamily—and a new Gram-positive secretion system? *Trends Microbiol.*, 10, 209, 2002.
155. Converse, S.E. and Cox, J.S., A protein secretion pathway critical for *Mycobacterium tuberculosis* virulence is conserved and functional in *Mycobacterium smegmatis*, *J. Bacteriol.*, 187, 1238, 2005.
156. Champion, P.A. et al., C-terminal signal sequence promotes virulence factor secretion in *Mycobacterium tuberculosis*, *Science*, 313, 1632, 2006.
157. Stanley, S.A. et al., Acute infection and macrophage subversion by *Mycobacterium tuberculosis* require a specialized secretion system, *Proc. Natl. Acad. Sci. USA*, 100, 13001, 2003.
158. Okkels, L.M. and Andersen, P., Protein–protein interactions of proteins from the ESAT-6 family of *Mycobacterium tuberculosis*, *J. Bacteriol.*, 186, 2487, 2004.
159. Renshaw, P.S. et al., Conclusive evidence that the major T-cell antigens of the *Mycobacterium tuberculosis* complex ESAT-6 and CFP-10 form a tight, 1:1 complex and characterization of the structural properties of ESAT-6, CFP-10, and the ESAT-6*CFP-10 complex. Implications for pathogenesis and virulence, *J. Biol. Chem.*, 277, 21598, 2002.
160. Burts, M.L. et al., EsxA and EsxB are secreted by an ESAT-6-like system that is required for the pathogenesis of *Staphylococcus aureus* infections, *Proc. Natl. Acad. Sci. USA*, 102, 1169, 2005.
161. Guinn, K.M. et al., Individual RD1-region genes are required for export of ESAT-6/CFP-10 and for virulence of *Mycobacterium tuberculosis*, *Mol. Microbiol.*, 51, 359, 2004.
162. Brodin, P. et al., ESAT-6 proteins: Protective antigens and virulence factors? *Trends Microbiol.*, 12, 500, 2004.
163. Way, S.S. and Wilson, C.B., The *Mycobacterium tuberculosis* ESAT-6 homologue in *Listeria monocytogenes* is dispensable for growth *in vitro* and *in vivo*, *Infect. Immun.*, 73, 6151, 2005.
164. Shockman, G.D. and Barrett, J.F., Structure, function, and assembly of cell walls of Gram-positive bacteria, *Annu. Rev. Microbiol.*, 37, 501, 1983.

165. Archibald, A.R., Structure and assembly of the cell wall in *Bacillus subtilis*, *Biochem. Soc. Trans.*, 13, 990, 1985.
166. Schäffer, C. and Messner, P., The structure of secondary cell wall polymers: How Gram-positive bacteria stick their cell walls together, *Microbiology*, 151, 643, 2005.
167. Sára, M. and Sleytr, U.B., S-Layer proteins, *J. Bacteriol.*, 182, 859, 2000.
168. Klein, C. et al., The membrane proteome of *Halobacterium salinarum*, *Proteomics*, 5, 180, 2005.
169. Harris, M.A. et al., The Gene Ontology (GO) database and informatics resource, *Nucleic Acids Res.*, 32, D258, 2004.
170. Samuelson, J.C. et al., YidC mediates membrane protein insertion in bacteria, *Nature*, 406, 637, 2000.
171. Luirink, J. et al., Biogenesis of inner membrane proteins in *Escherichia coli*, *Annu. Rev. Microbiol.*, 59, 329, 2005.
172. Dalbey, R.E. and Kuhn, A., YidC family members are involved in the membrane insertion, lateral integration, folding, and assembly of membrane proteins, *J. Cell Biol.*, 166, 769, 2004.
173. Van Bloois, E. et al., Distinct requirements for translocation of the N-tail and C-tail of the *Escherichia coli* inner membrane protein CyoA, *J. Biol. Chem.*, 281, 10002, 2006.
174. Van der Laan, M., Nouwen, N.P., and Driessen, A.J., YidC—An evolutionary conserved device for the assembly of energy-transducing membrane protein complexes, *Curr. Opin. Microbiol.*, 8, 182, 2005.
175. de Gier, J.W. and Luirink, J., The ribosome and YidC. New insights into the biogenesis of *Escherichia coli* inner membrane proteins, *EMBO Rep.*, 4, 939, 2003.
176. Dalbey, R.E. and Chen, M., Sec-translocase mediated membrane protein biogenesis, *Biochim. Biophys. Acta-Mol. Cell Res.*, 1694, 37, 2004.
177. von Heijne, G., Membrane-protein topology, *Nat. Rev. Mol. Cell Biol.*, 7, 909, 2006.
178. White, S.H. and von Heijne, G., The machinery of membrane protein assembly, *Curr. Opin. Struct. Biol.*, 14, 397, 2004.
179. Goder, V. and Spiess, M., Topogenesis of membrane proteins: determinants and dynamics, *FEBS Lett*, 504, 87, 2001.
180. Kocks, C. et al., *L. monocytogenes*-induced actin assembly requires the *actA* gene product, a surface protein, *Cell*, 68, 521, 1992.
181. Borezee, E. et al., SvpA, a novel surface virulence-associated protein required for intracellular survival of *Listeria monocytogenes*, *Microbiology*, 147, 2913, 2001.
182. Cabanes, D. et al., Surface proteins and the pathogenic potential of *Listeria monocytogenes*, *Trends Microbiol.*, 10, 238, 2002.
183. Froderberg, L. et al., Targeting and translocation of two lipoproteins in *Escherichia coli* via the SRP/Sec/YidC pathway, *J. Biol. Chem.*, 279, 31026, 2004.
184. Tjalsma, H. et al., The role of lipoprotein processing by signal peptidase II in the Gram-positive eubacterium *Bacillus subtilis.* Signal peptidase II is required for the efficient secretion of α-amylase, a nonlipoprotein, *J. Biol. Chem.*, 274, 1698, 1999.
185. Hayashi, S. and Wu, H.C., Lipoproteins in bacteria, *J. Bioenerg. Biomembr.*, 22, 451, 1990.
186. Baumgärtner, M. et al., Inactivation of Lgt allows systematic characterization of lipoproteins from *Listeria monocytogenes*, *J. Bacteriol.*, 189, 313, 2007.
187. Sankaran, K. and Wu, H.C., Lipid modification of bacterial prolipoprotein. Transfer of diacylglyceryl moiety from phosphatidylglycerol, *J. Biol. Chem.*, 269, 19701, 1994.
188. Garcia-Del Portillo, F. and Cossart, P., An important step in *Listeria* lipoprotein research, *J. Bacteriol.*, 189, 294, 2007.
189. Hulo, N. et al., The PROSITE database, *Nucleic Acids Res.*, 34, D227, 2006.
190. Sutcliffe, I. and Harrington, D.J., Pattern searches for the identification of putative lipoprotein genes in Gram-positive bacterial genomes, *Microbiology*, 148, 2065, 2002.
191. Babu, M.M. and Sankaran, K., DOLOP-database of bacterial lipoproteins, *Bioinformatics*, 18, 641, 2002.
192. Juncker, A.S. et al., Prediction of lipoprotein signal peptides in Gram-negative bacteria, *Protein Sci.*, 12, 1652, 2003.
193. Fariselli, P., Finocchiaro, G., and Casadio, R., SPEPlip: The detection of signal peptide and lipoprotein cleavage sites, *Bioinformatics*, 19, 2498, 2003.
194. Taylor, P.D. et al., LipPred: A Web server for accurate prediction of lipoprotein signal sequences and cleavage sites, *Bioinformation*, 1, 335, 2006.
195. Sutcliffe, I.C. and Russell, R.R., Lipoproteins of Gram-positive bacteria, *J. Bacteriol.*, 177, 1123, 1995.

196. Sanderson, S., Campbell, D.J., and Shastri, N., Identification of a CD4+ T cell-stimulating antigen of pathogenic bacteria by expression cloning, *J. Exp. Med.*, 182, 1751, 1995.
197. Ko, R. and Smith, L.T., Identification of an ATP-driven, osmoregulated glycine betaine transport system in *Listeria monocytogenes*, *Appl. Environ. Microbiol.*, 65, 4040, 1199.
198. Fraser, K.R. et al., Identification and characterization of an ATP binding cassette L-carnitine transporter in *Listeria monocytogenes*, *Appl. Environ. Microbiol.*, 66, 4696, 2000.
199. Borezee, E., Pellegrini, E., and Berche, P., OppA of *Listeria monocytogenes*, an oligopeptide-binding protein required for bacterial growth at low temperature and involved in intracellular survival, *Infect. Immun.*, 68, 7069, 2000.
200. Reglier-Poupet, H. et al., Identification of LpeA, a PsaA-like membrane protein that promotes cell entry by *Listeria monocytogenes*, *Infect. Immun.*, 71, 474, 2003.
201. Gal, L. et al., CelG from *Clostridium cellulolyticum*: A multidomain endoglucanase acting efficiently on crystalline cellulose, *J. Bacteriol.*, 179, 6595, 1997.
202. Gaudin, C. et al., CelE, a multidomain cellulase from *Clostridium cellulolyticum*: A key enzyme in the cellulosome? *J. Bacteriol.*, 182, 1910, 2000.
203. Brinster, S., Furlan, S., and Serror, P., C-terminal WXL domain mediates cell wall binding in *Enterococcus faecalis* and other Gram-positive bacteria, *J. Bacteriol.*, 189, 1244, 2007.
204. Popowska, M. and Markiewicz, Z., Classes and functions of *Listeria monocytogenes* surface proteins, *Pol. J. Microbiol.*, 53, 75, 2004.
205. Altschul, S.F. et al., Gapped BLAST and PSI-BLAST: A new generation of protein database search programs, *Nucleic Acids Res.*, 25, 3389, 1997.
206. Eddy, S.R., Hidden Markov models, *Curr. Opin. Struct. Biol.*, 6, 361, 1996.
207. Mulder, N.J. et al., New developments in the InterPro database, *Nucleic Acids Res.*, 35, D224, 2007.
208. Bateman, A. et al., The Pfam protein families database, *Nucleic Acids Res.*, 32, D138, 2004.
209. Schultz, J. et al., SMART, a simple modular architecture research tool: Identification of signaling domains, *Proc. Natl. Acad. Sci. USA*, 95, 5857, 1998.
210. Selengut, J.D. et al., TIGRFAMs and genome properties: Tools for the assignment of molecular function and biological process in prokaryotic genomes, *Nucleic Acids Res.*, 35, D260, 2007.
211. Wilson, D. et al., The SUPERFAMILY database in 2007: Families and functions, *Nucleic Acids Res.*, 35, D308, 2007.
212. Pallen, M.J. et al., An embarrassment of sortase—A richness of substrates? *Trends Microbiol.*, 9, 97, 2001.
213. Marraffini, L.A., Dedent, A.C., and Schneewind, O., Sortases and the art of anchoring proteins to the envelopes of Gram-positive bacteria, *Microbiol. Mol. Biol. Rev.*, 70, 192, 2006.
214. Paterson, G.K. and Mitchell, T.J., The biology of Gram-positive sortase enzymes, *Trends Microbiol.*, 12, 89, 2004.
215. Ton-That, H., Marraffini, L.A., and Schneewind, O., Protein sorting to the cell wall envelope of Gram-positive bacteria, *Biochim. Biophys. Acta-Mol. Cell Res.*, 1694, 269, 2004.
216. Bierne, H. et al., Sortase B, a new class of sortase in *Listeria monocytogenes*, *J. Bacteriol.*, 186, 1972, 2004.
217. Garandeau, C. et al., The sortase SrtA of *Listeria monocytogenes* is involved in processing of internalin and in virulence, *Infect. Immun.*, 70, 1382, 2002.
218. Steen, A. et al., Cell wall attachment of a widely distributed peptidoglycan binding domain is hindered by cell wall constituents, *J. Biol. Chem.*, 278, 23874, 2003.
219. Dussurget, O., Pizarro-Cerda, J., and Cossart, P., Molecular determinants of *Listeria monocytogenes* virulence, *Annu. Rev. Microbiol.*, 58, 587, 2004.
220. Gaillard, J.L. et al., Entry of *L. monocytogenes* into cells is mediated by internalin, a repeat protein reminiscent of surface antigens from Gram-positive cocci, *Cell*, 65, 1127, 1991.
221. Braun, L. et al., InlB: An invasion protein of *Listeria monocytogenes* with a novel type of surface association, *Mol. Microbiol.*, 25, 285, 1997.
222. Jonquieres, R. et al., Interaction between the protein InlB of *Listeria monocytogenes* and lipoteichoic acid: A novel mechanism of protein association at the surface of Gram-positive bacteria, *Mol. Microbiol.*, 34, 902, 1999.
223. Marino, M. et al., GW domains of the *Listeria monocytogenes* invasion protein InlB are SH3-like and mediate binding to host ligands, *EMBO J.*, 21, 5623, 2002.

224. Milohanic, E. et al., Transcriptome analysis of *Listeria monocytogenes* identifies three groups of genes differently regulated by PrfA, *Mol. Microbiol.*, 47, 1613, 2003.
225. Siezen, R. et al., *Lactobacillus plantarum* gene clusters encoding putative cell-surface protein complexes for carbohydrate utilization are conserved in specific Gram-positive bacteria, *BMC Genomics*, 7, 126, 2006.
226. Nölling, J. et al., Genome sequence and comparative analysis of the solvent-producing bacterium *Clostridium acetobutylicum*, *J. Bacteriol.*, 183, 4823, 2001.
227. Baba, T. and Schneewind, O., Target cell specificity of a bacteriocin molecule: A C-terminal signal directs lysostaphin to the cell wall of *Staphylococcus aureus*, *EMBO J.*, 15, 4789, 1996.
228. Anantharaman, V. and Aravind, L., Evolutionary history, structural features and biochemical diversity of the NlpC/P60 superfamily of enzymes, *Genome Biol.*, 4, R11, 2003.
229. Scott, J.R. and Barnett, T.C., Surface proteins of Gram-positive bacteria and how they get there, *Annu. Rev. Microbiol.*, 60, 397, 2006.
230. Bendtsen, J.D. et al., Nonclassical protein secretion in bacteria, *BMC Microbiol.*, 5, 58, 2005.
231. Guiral, S. et al., Competence-programmed predation of noncompetent cells in the human pathogen *Streptococcus pneumoniae*: Genetic requirements, *Proc. Natl. Acad. Sci. USA*, 102, 8710, 2005.
232. Bergmann, S., Rohde, M., and Hammerschmidt, S., Glyceraldehyde-3-phosphate dehydrogenase of *Streptococcus pneumoniae* is a surface-displayed plasminogen-binding protein, *Infect. Immun.*, 72, 2416, 2004.
233. Klose, J., Protein mapping by combined isoelectric focusing and electrophoresis of mouse tissues. A novel approach to testing for induced point mutations in mammals, *Humangenetik*, 26, 231, 1975.
234. O'Farrell, P.H., High resolution two-dimensional electrophoresis of proteins, *J. Biol. Chem.*, 250, 4007, 1975.
235. Scheele, G.A., Two-dimensional gel analysis of soluble proteins. Characterization of guinea pig exocrine pancreatic proteins, *J. Biol. Chem.*, 250, 5375, 1975.
236. VanBogelen, R.A. et al., The gene-protein database of *Escherichia coli*: Edition 5, *Electrophoresis*, 13, 1014, 1992.
237. Pasquali, C. et al., Two-dimensional gel electrophoresis of *Escherichia coli* homogenates: The *Escherichia coli* SWISS-2DPAGE database, *Electrophoresis*, 17, 547, 1996.
238. Folio, P. et al., Two-dimensional electrophoresis database of *Listeria monocytogenes* EGDe proteome and proteomic analysis of mid-log and stationary growth phase cells, *Proteomics*, 4, 3187, 2004.
239. Cordwell, S.J., Technologies for bacterial surface proteomics, *Curr. Opin. Microbiol.*, 9, 320, 2006.
240. Santoni, V., Molloy, M., and Rabilloud, T., Membrane proteins and proteomics: Un amour impossible? *Electrophoresis*, 21, 1054, 2000.
241. Santoni, V. et al., Membrane proteomics: use of additive main effects with multiplicative interaction model to classify plasma membrane proteins according to their solubility and electrophoretic properties, *Electrophoresis*, 21, 3329, 2000.
242. Mastro, R. and Hall, M., Protein delipidation and precipitation by tri-n-butylphosphate, acetone, and methanol treatment for isoelectric focusing and two-dimensional gel electrophoresis, *Anal. Biochem.*, 273, 313, 1999.
243. Deshusses, J.M. et al., Exploitation of specific properties of trifluoroethanol for extraction and separation of membrane proteins, *Proteomics*, 3, 1418, 2003.
244. Planchon, S. et al., Proteomic analysis of cell envelope from *Staphylococcus xylosus* C2a, a coagulase-negative *Staphylococcus*, *J. Proteome Res.*, 6, 3566, 2007.
245. Kotiranta, A. et al., Surface structure, hydrophobicity, phagocytosis, and adherence to matrix proteins of *Bacillus cereus* cells with and without the crystalline surface protein layer, *Infect. Immun.*, 66, 4895, 1998.
246. Tresse, O. et al., Comparative evaluation of adhesion, surface properties, and surface protein composition of *Listeria monocytogenes* strains after cultivation at constant pH of 5 and 7, *J. Appl. Microbiol.*, 101, 53, 2006.
247. Calvo, E. et al., Analysis of the *Listeria* cell wall proteome by two-dimensional nanoliquid chromatography coupled to mass spectrometry, *Proteomics*, 5, 433, 2005.
248. Ren, Q., Kang, K.H., and Paulsen, I.T., TransportDB: A relational database of cellular membrane transport systems, *Nucl. Acids Res.*, 32, D284, 2004.
249. Hofmann, K. and Stoffel, W., PROFILEGRAPH: An interactive graphical tool for protein sequence analysis, *Comput. Appl. Biosci.*, 8, 331, 1992.

250. Claros, M.G. and von Heijne, G., TopPred II: An improved software for membrane protein structure predictions, *Comput. Appl. Biosci.,* 10, 685, 1994.
251. Rost, B. et al., Transmembrane helices predicted at 95% accuracy, *Protein Sci.,* 4, 521, 1995.
252. Csero, M. et al., Prediction of transmembrane alpha-helices in prokaryotic membrane proteins: The dense alignment surface method, *Protein Eng.,* 10, 673, 1997.
253. Persson, B. and Argos, P., Prediction of membrane protein topology utilizing multiple sequence alignments, *J. Protein Chem.,* 16, 453, 1997.
254. Kihara, D., Shimizu, T., and Kanehisa, M., Prediction of membrane proteins based on classification of transmembrane segments, *Protein Eng.,* 11, 961, 1998.
255. Sonnhammer, E.L., von Heijne, G., and Krogh, A., A hidden Markov model for predicting transmembrane helices in protein sequences. *Proc. Int. Conf. Intell. Syst. Mol. Biol.,* 6, 175, 1998.
256. Hirokawa, T., Boon-chieng, S., and Mitaku, S., SOSUI: Classification and secondary structure prediction system for membrane proteins, *Bioinformatics,* 14, 378, 1998.
257. Pasquier, C. and Hamodrakas, S.J., An hierarchical artificial neural network system for the classification of transmembrane proteins, *Protein Eng.,* 12, 631, 1999.
258. Pilpel, Y., Ben-Tal, N., and Lancet, D., kPROT: A knowledge-based scale for the propensity of residue orientation in transmembrane segments. Application to membrane protein structure prediction, *J. Mol. Biol.,* 294, 921, 1999.
259. Senes, A., Gerstein, M., and Engelman, D.M., Statistical analysis of amino acid patterns in transmembrane helices: The GxxxG motif occurs frequently and in association with beta-branched residues at neighboring positions, *J. Mol. Biol.,* 296, 921, 2000.
260. Tusnady, G.E. and Simon, I., The HMMTOP transmembrane topology prediction server, *Bioinformatics,* 17, 849, 2001.
261. Deber, C.M. et al., TM Finder: A prediction program for transmembrane protein segments using a combination of hydrophobicity and nonpolar phase helicity scales, *Protein Sci.,* 10, 212, 2001.
262. Cserzo, M. et al., On filtering false positive transmembrane protein predictions, *Protein Eng.,* 15, 745, 2002.
263. Juretic, D., Zoranic, L., and Zucic, D., Basic charge clusters and predictions of membrane protein topology, *J. Chem. Inf. Comput. Sci.,* 42, 620, 2002.
264. Zhou, H. et al., Web-based toolkits for topology prediction of transmembrane helical proteins, fold recognition, structure and binding scoring, folding-kinetics analysis and comparative analysis of domain combinations, *Nucleic Acids Res.,* 33, W193, 2005.
265. Taylor, P.D., Attwood, T.K., and Flower, D.R., BPROMPT: A consensus server for membrane protein prediction, *Nucleic Acids Res.,* 31, 3698, 2003.
266. Yuan, Z., Mattick, J.S., and Teasdale, R.D., SVMtm: Support vector machines to predict transmembrane segments, *J. Comput. Chem.,* 25, 632, 2004.
267. Arai, M. et al., ConPred II: A consensus prediction method for obtaining transmembrane topology models with high reliability, *Nucleic Acids Res.,* 32, W390, 2004.
268. Bagos, P.G., Liakopoulos, T.D., and Hamodrakas, S.J., Algorithms for incorporating prior topological information in HMMs: Application to transmembrane proteins, *BMC Bioinform.,* 7, 189, 2006.
269. Cao, B. et al., Enhanced recognition of protein transmembrane domains with prediction-based structural profiles, *Bioinformatics,* 22, 303, 2006.
270. Jones, T.D., Improving the accuracy of transmembrane protein topology prediction using evolutionary information, *Bioinformatics,* 23, 538, 2007.

Section IV

Immunity and Vaccines

13 Innate Immunity

Gernot Geginat and Silke Grauling-Halama

CONTENTS

13.1 INTRODUCTION

Innate immunity is a phylogenetically ancient defense system that provides a first barrier against a broad range of pathogens through nonspecific defense mechanisms. In the absence of immune defense, a single bacterium replicating every 60 min could have, within 3 days, a theoretical progeny of approximately 2^{72} (i.e., more than 10^{21} bacteria). The innate, nonadaptive immune defense of the host has a formidable task to keep the multiplication and the spread of pathogens under control until a specific adaptive immune response has developed. Here, we outline the numerous lines of innate immediate defense mechanisms that protect hosts from infection with the facultative intracellular bacterium *Listeria monocytogenes* (Figure 13.1).

Natural infection with *L. monocytogenes* occurs in general after oral uptake of contaminated food. The first barrier that a pathogen entering the host via the oral route has to overcome is the bactericidal action of lysozyme in mucosal secretions, gastric acid in the stomach, and antibacterial lectins in the intestine. In the intestine, *L. monocytogenes* competes also with numerous symbiotic bacteria for nutrients and space and survives only temporarily. *L. monocytogenes* is able to invade the mucosa of the gastrointestinal tract and disseminates via the blood stream into the spleen, liver, and, sometimes, the central nervous system. Dissemination may in part occur as blind passengers of macrophages and dendritic cells (DCs), a strategy that avoids the direct contact of bacteria with the complement system. The innate immune system is activated by pattern recognition receptors (PRRs) recognizing pathogen-associated molecular patterns (PAMPs) from intra- and extracellular bacteria (Table 13.1, Figure 13.2). Infection of macrophages results in the secretion of various cytokines such as interleukin (IL)-1, IL-6, IL-12, and IL-18, and the chemokine monocyte chemo-attractant protein-1 (MCP-1), which activate and recruit polymorphonuclear neutrophils, monocytes, T cells, natural killer (NK) cells, and DCs. IL-12 and IL-18 trigger the production of gamma interferon (IFN-γ) and tumor necrosis factor (TNF) by T cells and NK cells. The secreted cytokines, IFN-γ and TNF, act back on macrophages and strongly enhance their bactericidal activity (Figure 13.3). Apart from macrophages and neutrophils, only intestinal Paneth cells and a specialized subpopulation of TNF- and nitric oxide (NO)-producing DCs possess direct bactericidal activity. The mechanisms that result in the killing of bacteria are not completely understood, but they may include the attack of bacteria by the main bactericidal effector molecules lysozyme, antimicrobial peptides, reactive oxygen radicals, and reactive nitrogen compounds (Table 13.2).

The innate, nonadaptive immune system is only able to control *L. monocytogenes* infection over a limited period of time. This has been shown by studies in SCID (severe combined immunodeficiency) mice that exhibit an inborn deficiency of both the cellular T cell–mediated and the humoral antibody-mediated immunity. Sterilizing immunity after *L. monocytogenes* infection always requires a specific $CD4^+$ and $CD8^+$ T cell response, which is the focus of chapter 14.

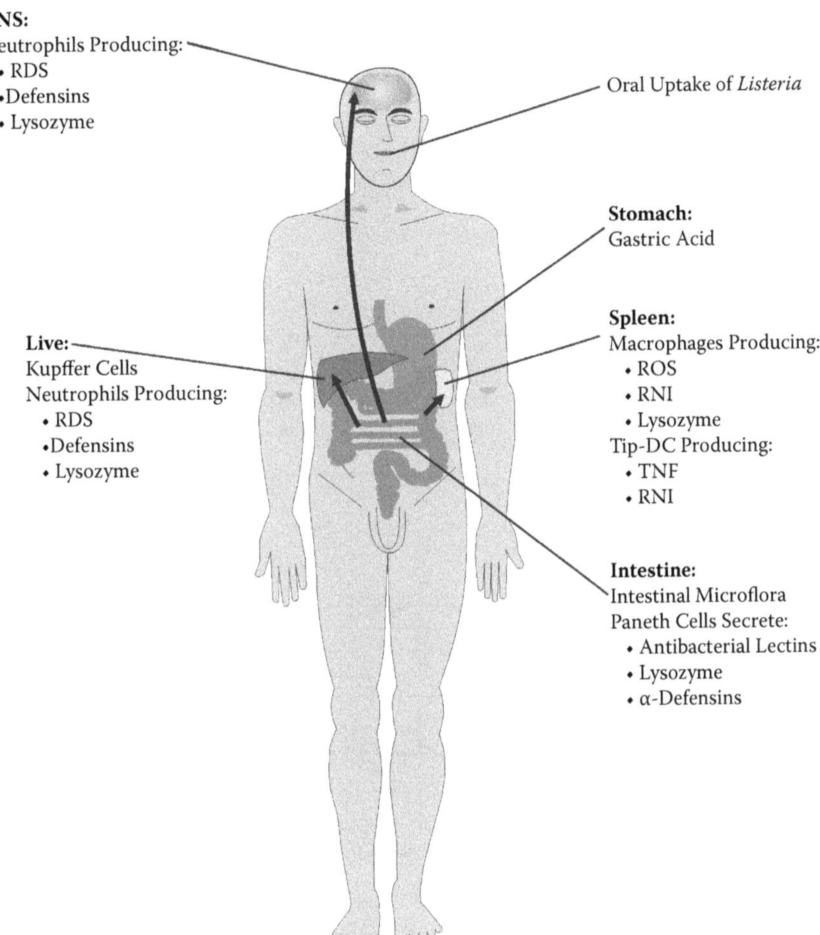

FIGURE 13.1 Schematic overview of the main target organs of *L. monocytogenes* and the principal cells and effector molecules mediating organ-specific anti-*Listeria* immunity. Arrows indicate the spreading of *L. monocytogenes* after oral infection. Abbreviations: RNI, reactive nitrogen intermediate; ROS, reactive oxygen species; Tip-DCs, TNF- and iNOS-producing dendritic cells.

TABLE 13.1
Principal Molecules from *L. monocytogenes* that Trigger Innate Immune Responses

Listeria-Derived Ligand	Cellular Sensor Molecule[a]
Diaminopimelic acid-containing dipeptide or tripeptide motifs	NOD1
Muramyl dipeptide	NOD2
Lipoteichoic acid	TLR2, scavenger receptor A
Flagellin	TLR5
Bacterial DNA	Receptor not yet defined, TLR9-independent

[a] Refer to the main text for abbreviations.

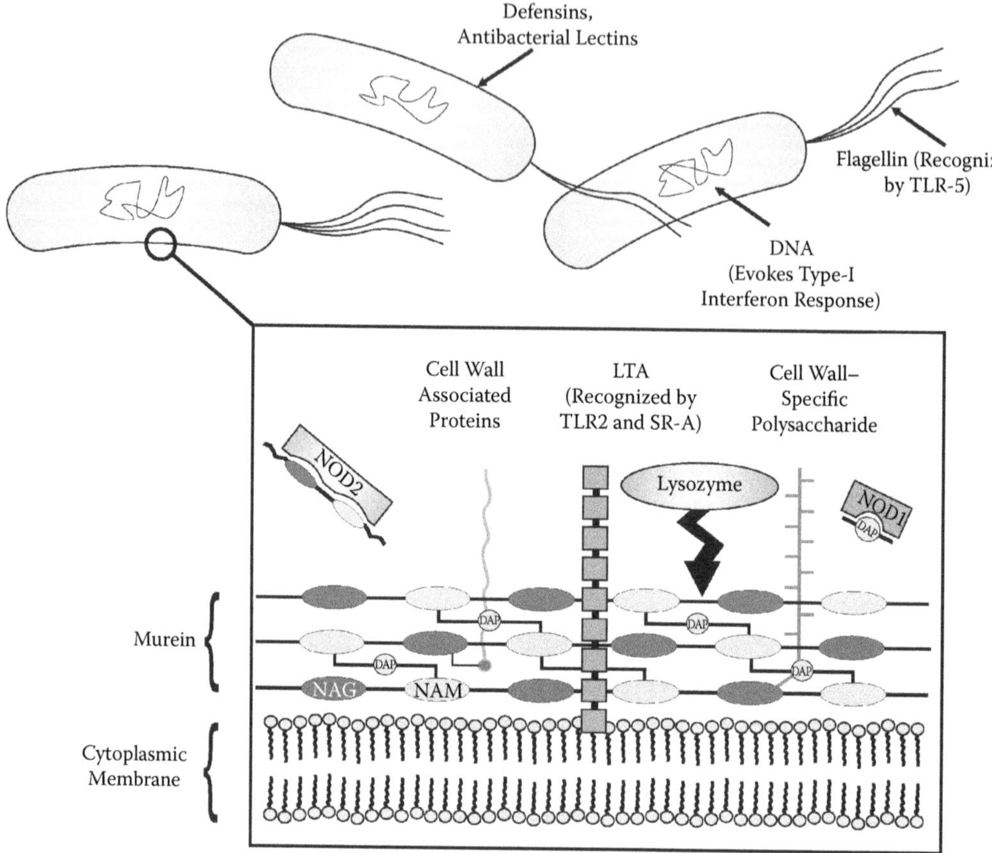

FIGURE 13.2 Illustration of the main structures of *L. monocytogenes* that are sensed by pattern recognition receptors of the innate immune system. The accessibility of murein (peptidoglycan)-derived ligands is enhanced by lysozyme-mediated hydrolysis of the bacterial cell wall. Abbreviations: DAP, diaminopimelic acid; LTA, lipoteichoic acid; NAG, N-acetylglucosamine; NAM, N-acetylmuramic acid; NOD, nucleotide-binding oligomerization domain; SR-A, group A scavenger receptor; TLR, Toll-like receptor.

13.2 SENSOR MOLECULES

13.2.1 *LISTERIA*-SPECIFIC SENSOR MOLECULES

The innate immune system recognizes pathogen-associated molecular patterns (PAMPs) that come from infectious agents, but not from the host.[1] The typical bacterial PAMPs include bacterial carbohydrates (e.g., lipopolysaccharide [LPS] and mannose), nucleic acids (including both DNA and RNA), peptidoglycans, lipoteichoic acids, etc. For that purpose the immune system makes use of a number of pattern recognition receptor molecules (PRRs) that are proteins expressed by immune cells, enabling the innate immune defense system to distinguish microbes from self without the time-consuming clonal selection process that forms the basis of adaptive immunity. Depending on their locations, PRRs are divided into membrane-bound (e.g., Toll-like receptors [TLRs] and mannose receptor), cytoplasmic (e.g., NOD-like receptors [NLRS]), and secreted (e.g., complement receptors and collectins).

PAMPs are expressed by all bacteria, invasive pathogens as well as noninvasive commensals. The maintenance of a proper function of the innate immune system requires that the sensor molecules of the innate immune system are activated only by invasive bacteria entering the host by

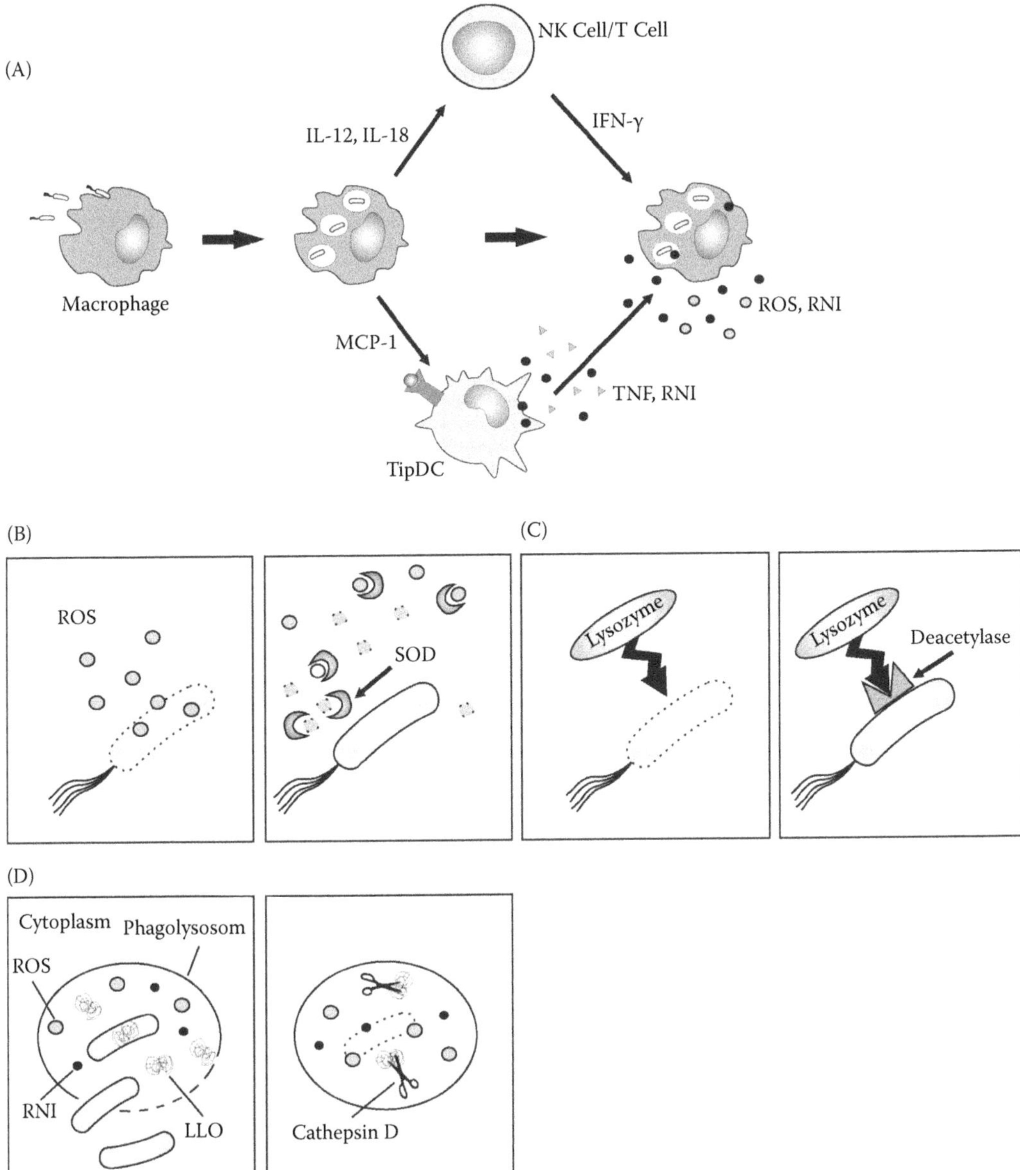

FIGURE 13.3 Schematic illustration of the bactericidal attack of *L. monocytogenes* by macrophages and counteracting bacterial immune evasion mechanisms. (A) Macrophages infected with *L. monocytogenes* secrete interleukin (IL)-12 and IL-18, which trigger interferon (IFN)-γ and tumor necrosis factor (TNF) release by natural killer (NK) cells and T cells. IFN-γ and TNF elicit and enhance the bactericidal activity of infected macrophages. Release of monocyte chemoattractant protein-1 (MCP-1) triggers the release of precursor cells from the bone marrow that differentiate into TNF- and iNOS-producing dendritic cells (Tip-DCs). Tip-DCs enhance the bactericidal activity of macrophages by the secretion of TNF and also act directly bactericidal by the release of reactive nitrogen intermediates (RNI). (B) By the release of superoxide dismutase (SOD) *L. monocytogenes* is able to partially neutralize the attack by reactive oxygen species (ROS). (C) *L. monocytogenes* protects itself from lysozyme attack by deacetylation of cell wall peptidoglycan. (D) *L. monocytogenes* avoids the bactericidal attack of ROS and RNI by listeriolysin O (LLO)-mediated escape into the host cell cytoplasm. The endosomal protease cathepsin D counteracts this bacterial immune evasion strategy by cleavage and inactivation of LLO.

TABLE 13.2
Bactericidal Effector Molecules that Target *L. monocytogenes*

Type	Molecule[a]	Target and Action	Cell Types
Enzymes	Lysozyme (muramidase)	Hydrolyzes cell wall peptidoglycan	Paneth cells, neutrophils, macrophages
	Cathepsin D (protease)	Prevents phagolysosomal escape of *L. monocytogenes* by inactivation of LLO	Macrophages
Antimicrobial peptides	α-defensin	Bacterial cell wall, pore formation	Paneth cells, neutrophils
	Ubiquicidin	Probably same as α-defensins	Macrophages
	IFN-γ-inducible tripeptide motif CXC chemokines	Probably same as α-defensins	Monocytes
Lectin	Mouse: RegIIIγ human: HIP/PAP	Antibacterial lectin targeting bacterial peptidoglycan, bactericidal mechanism unknown	Paneth cells
Reactive small molecules	RNI (iNOS)	Unspecific cell damage, possible inactivation of virulence factors	Macrophages
	RNI (eNOS)		Neutrophils
	ROS	Unspecific cell damage, possible inactivation of virulence factors	Neutrophils, macrophages
Gastric acid	Hydrochloric acid	Low pH	Stomach parietal cells

[a] Refer to the main text for abbreviations.

breaching the mucous membranes or the surface of the skin. For the innate immunity to *L. monocytogenes*, the importance of two types of pattern recognition receptors, the TLRs and the NLRs, is well documented. In addition, some experimental evidence suggests the involvement of scavenger receptors and a TLR-9-independent cytosolic sensor system for bacterial DNA, which still awaits detailed characterization (Table 13.1, Figure 13.2).

13.2.1.1 Toll-like Receptors

Toll-like receptors (TLRs) are transmembrane proteins that are present in both vertebrates and invertebrates. They function as pattern recognition receptors that detect PAMPs once pathogens have crossed a host's physical barriers (e.g., the skin or the mucosa of the intestinal tract).[2] They activate the production of cytokines for the subsequent development of an effective immune response. TLRs are expressed by macrophages, DCs, neutrophils, and other cell types involved in innate immune responses. They are named for their similarity to Toll, a receptor with developmental function first identified in the fruit fly *Drosophila melanogaster*. Up to the time of writing, 13 different TLRs (termed TLR1–TLR13) have been identified, but only two TLRs (TLR2 and TLR5) play a role in the recognition and detection of *Listeria*. TLR2 and TLR5 are able to sense microbe-derived ligands on the cell surface as well as within phagosomes of phagocytic cells. TLRs launch an intracellular signaling cascade by the recruitment of cytoplasmic adapter molecules. The main adaptor molecule for TLR2 and TLR5 is the myeloid differentiation primary response gene 88 (MyD88). Disruption of the gene coding for MyD88 results in dramatically reduced resistance of mice against *L. monocytogenes* infection.[3,4]

Toll-like receptor 2. TLR2 is specific for lipoteichoic acids of Gram-positive bacteria, including *L. monocytogenes*.[5] TLR2 forms heterodimers with TLR1 and TLR6, which improve the recognition of bacterial lipoteichoic acids.[5] Because peptidoglycan preparations are often contaminated with lipoteichoic acid, earlier studies concluded that peptidoglycans are TLR2 ligands. TLR2 is expressed on the cell surface as well as within phagolysosomes and therefore is not able to sense bacteria residing within the host cell cytoplasm. The activation of TLR2 by bacterial lipoteichoic acid requires the release of lipoteichoic acid by the lysozyme-mediated hydrolysis of the bacterial cell wall peptidoglycan.[6] *L. monocytogenes*, as well as some other Gram-positive bacteria, protect themselves from lysozyme attack by deacetylation of N-acetylglucosamine residues.[6,7] Deacetylase deletion mutants of *L. monocytogenes* induce a massive IL-6 and IFN-β response in a mainly TLR2-dependent manner and are almost avirulent *in vivo* since they are rapidly destroyed in macrophage vacuoles.[6] Thus, the inhibition of lysozyme attack by deacetylation of N-acetylglucosamine residues prevents TLR2-mediated signaling and the activation of innate immune responses.

The effect of disruption of the gene coding for TLR2 on the early protection against *L. monocytogenes* seems to be variable. In the original study, *L. monocytogenes*-infected TLR2-deficient mice were as resistant as wild-type mice.[4] However, a later analysis, using a slightly different experimental setting, showed a protective effect of TLR2 during early *Listeria* infection.[8]

Toll-like receptor 5. TLR5 recognizes the flagellin from flagella of *L. monocytogenes* as well as from numerous other bacteria by binding to a region common to the flagellin of Gram-positive as well as Gram-negative bacteria.[9,10] Since activation of TLR5 mobilizes nuclear transcription factor κB (NF-κB) and stimulates TNF production, TLR5 may have evolved to allow mammals to recognize flagellated bacterial pathogens and may serve as a general alarm system for a broad group of motile pathogens. However, flagellin-deficient *L. monocytogenes* deletion mutants show no significant alteration in virulence during the course of infection, suggesting that TLR5-mediated signaling is not essential for pathogenesis and immunity after *L. monocytogenes* infection.[11]

13.2.1.2 Nucleotide-Binding Oligomerization Domain (NOD)-like Receptors

The NOD-like receptors (NLRs) are a group of cytosolic pattern recognition receptors that comprises NOD1, NOD2, NALP1 (NACHT-, LRR-, and PYRIN domain-containing protein 1), NALP3, neuronal apoptosis inhibitory protein-5, and ICE protease activating factor.[12,13] NLRs have important roles in innate immunity as sensors of microbial components and cell injury in the cytosol. The presence of a mutation in the human gene coding for NOD2 has been associated with a dysregulated mucosal immune response against a luminal antigen, possibly a bacterium in patients suffering from idiopathic inflammatory bowel disease. NLRs mediate proinflammatory signals through NF-κB and activation of caspase-1. Mice with a targeted deletion of the genes coding for NF-κB[14] or Rip-2, which acts as a main activator of NF-κB[15] as well as of caspase-1,[16] demonstrate significantly reduced anti-*Listeria* protection in the early phase of *L. monocytogenes* infection. It was proposed that NLR and caspase-1 form a proinflammatory multiprotein complex termed inflammasome. Activation of caspase-1 leads to the cleavage and activation of proinflammatory cytokines such as interleukin-1β and IL-18, as well as apoptotic cell death. Both NOD1 and NOD2 play an important role in the innate immune response against *L. monocytogenes,* whereas the possible involvement of other NLRs has yet to be established.

Nucleotide-binding oligomerization domain 1. NOD1 senses diaminopimelic acid-containing dipeptide or tripeptide motifs generated by the lysozyme-mediated cleavage of peptidoglycans from most Gram-negative and some Gram-positive bacteria, including the genus *Listeria* (Figure 13.2).[17–19] As described before for TLR2, the inhibition of lysozyme-mediated hydrolysis of cell wall peptidoglycan by deacetylation prevents also the release of NOD1 ligands and NOD1-mediated activation of the innate immune system.[6]

Nucleotide-binding oligomerization domain 2. Muramyl dipeptides are final degradation products of peptidoglycans and represent the main ligand of NOD2 (Figure 13.2).[20,21] Efficient detection of bacteria by NOD2 similar to NOD1 and TLR2 probably also requires lysozyme-mediated hydrolysis of the cell wall peptidoglycan. The expression of cryptidins by intestinal Paneth cells requires NOD2-mediated signaling and NOD2-deficient mice are highly susceptible to *L. monocytogenes* infection via the oral route, but normally susceptible to infection via intravenous or peritoneal inoculation.[22]

13.2.1.3 Scavenger Receptors

Scavenger receptors are a group of receptors that engage in the removal of foreign and waste substances in the body through recognition of low-density lipoproteins that are modified by oxidation or acetylation. Scavenger receptors are grouped into classes A, B, and C, according to their structural characteristics, of which the class A scavenger receptor is an 80-kDa protein composed of six domains, including a single transmembrane domain, and is expressed mainly in the macrophages. The class B scavenger receptor differs from class A scavenger receptor by possessing two transmembrane domains instead of one; the class C scavenger receptor is a transmembrane protein with an extracellularly located N-terminus. In turn, the class A scavenger receptor is separated into two types (type I and type II scavenger receptors), which recognize a variety of polyanions including lipoteichoic acid. After *L. monocytogenes* infection, uptake and killing of bacteria by Kupffer cells in wild-type mice are more efficient than in type I and type II scavenger receptor-deficient mice.[23] These scavenger receptor–deficient mice also demonstrate significantly reduced resistance in the early phase of *L. monocytogenes* infection. *L. monocytogenes* rapidly lyses the phagosomal membrane and escapes to the cytosol in macrophages of type I and type II macrophage scavenger receptor-negative mice, suggesting that scavenger receptors not only act as a receptor but also mediate a mechanism that prevents the listeriolysin O-dependent escape from the phagolysosome.[24]

13.2.1.4 Sensor Molecules of *Listeria*-Derived Nucleic Acids

During the early stage of infection, invading *L. monocytogenes* is attacked by macrophages and other immune cells. As a result, a majority of *L. monocytogenes* is destroyed and bacterial DNA is released. In addition, as a fundamental process in all prokaryotes, live *L. monocytogenes* may secrete DNA actively. The availability of this bacterial DNA initiates a cytosolic recognition process that constitutes part of host response for eventual elimination of the invading pathogen. Currently, TLR9 is the only known sensor of foreign DNA, which detects viral DNA via both endosomal, TLR-dependent and cytosolic, RIG-I/MDA5-dependent pathways. It appears that cytosolic DNA from *L. monocytogenes* activates a type I IFN response in macrophages independent of TLR signaling, which requires IRF3 but occurs without detectable activation of NF-κB and MAP kinases. This unique bacterial DNA recognition in the cytosol may account for the potent induction of type I IFN that occurs in *L. monocytogenes* infection. Live *Listeria* activate an RIP2-dependent pathway in macrophages that is separable from the IFN-β and IL-6 response to bacterial DNA.[25]

13.2.2 Nonspecific Sensor Molecules

13.2.2.1 Ubiquitin System

Ubiquitin is a conserved small regulatory protein (made up of 76 amino acids with a molecular mass of 8.5 kDa) that is present in all eukaryotic cells. The main function of ubiquitin is to mark other proteins for proteolytic destruction by the proteasome. Proteolysis by the proteasome generates short peptides that, via the transporter associated with antigen processing, are transported into the endoplasmic reticulum where peptides together with β_2-microglobulin and major histocompatibility complex (MHC) class I heavy chain form a ternary complex. The peptide/MHC complex is subsequently transported to the cell surface, where it is presented to CD8 T cells. During ubiquitination, several ubiquitin molecules bind to a target protein, and then move the protein to a proteasome where the proteolysis occurs. Once the condemned protein is destroyed, the ubiquitin moiety is cleaved off the protein by deubiquitinating enzymes, and ubiquitin is recycled for future use. The presence of bacteria in the cytoplasm can be sensed by the host ubiquitin system.[26] In macrophages, this triggers subcellular relocalization of the proteasome, which associates with the surface of bacteria in the cytosol and inhibits bacterial replication. Intracellular *L. monocytogenes* is able to avoid the recognition by the ubiquitin system by using actin-based motility.[26] Nevertheless, in infected cells, secreted proteins of intracellular *L. monocytogenes* are efficiently processed and presented by MHC class I molecules on the surface of infected cells via the cytosolic antigen presentation pathway.

13.3 BACTERICIDAL EFFECTOR MOLECULES

Hosts employ a large arsenal of soluble effector molecules to combat invasive pathogens (Table 13.2). These effector molecules have a direct bacteriostatic or even bactericidal effect and play an essential role in the innate immune defense against *L. monocytogenes.* Only recently have we begun to understand how *L. monocytogenes* is able to evade at least some of these multiple lines of ingenious defense against pathogens (Figure 13.3).

13.3.1 Gastric Acid and Intestinal Microflora

Mice and rats are highly resistant to *L. monocytogenes* infection via the gastrointestinal tract. The resistance of mice to intestinal infection is mainly the consequence of a single mutation in the mouse epithelial cadherin (E-cadherin), which functions as receptor for internalin A of *L. monocytogenes* on the surface of epithelial cells.[27] In addition, the secretion of gastric acid[28] and also the intestinal microflora[29–31] seem to enhance the resistance of mice to intestinal infection with *L. monocytogenes.* Vice-versa adaptation of *L. monocytogenes* to acidic pH increases bacterial invasion and also enhances virulence *in vivo.*[32,33] Enhanced bacterial glutamate decarboxylase activity increases acid resistance of *L. monocytogenes.*[34]

13.3.2 Acute Phase Proteins

Acute phase proteins are produced and secreted into the bloodstream by the liver in response to local or systemic inflammatory stimuli. In humans, acute phase proteins comprise, among others, complement factors, C-reactive protein (CRP), coagulation factors, and serum amyloids, of which CRP is a useful marker of inflammation, as its level increases notably during the inflammatory process that occurs in the body. The main function of CRP is to enhance complement binding to foreign and damaged cells for their eventual removal.

Acute phase serum of humans is more bactericidal to *L. monocytogenes* than normal serum. The Group IIA phospholipase A2, an acute phase protein that increases up to 500-fold in the blood plasma of patients with severe acute diseases, is capable of killing 90% of *L. monocytogenes in vitro* at concentrations even found in normal human serum.[35] Purified serum amyloid P component, the major acute-phase reactant of mice, augments the *in vitro* listericidal activity of activated macrophages.[36] *L. monocytogenes* activates complement via the alternative complement activation pathway, resulting in enhanced phagocytosis and killing of bacteria.[37–39]

13.3.3 Transferrin

Transferrin is a glycoprotein of 80 kDa that is present predominantly in blood plasma. This protein contains two specific high-affinity Fe(III) binding sites, whose chief function is for iron ion delivery. Transferrin is also found in the mucosa. Through its binding with iron, it creates an environment low in free iron, which is unfavorable to the survival of most bacteria. Thus, transferrin forms part of the host innate immune system as an anti-acute phase protein, which decreases in case of systemic inflammation. Uptake of iron bound to transferrin increases the ability of activated macrophages to kill *L. monocytogenes*[40] and also reduces the availability of iron, which is required for bacterial growth.[41]

13.3.4 Nitric Oxide

Nitric oxide (NO) is a lipid- and water-soluble, highly reactive gas that reacts in water with molecular oxygen and other reactive oxygen compounds to yield a diverse array of reactive nitrogen intermediates (RNI), which exhibit direct bactericidal activity. NO is generated enzymatically by NO synthases (NOS), which oxidize L-arginine to L-citrulline. There are three isoforms of NOS: the neuronal form nNOS (= NOS1), the inducible NOS (iNOS = NOS2), and eNOS (= NOS3), which was primarily discovered in epithelial cells.[42] The kinetics of NO production by NOS isoforms differ. Inducible iNOS generates large, toxic amounts of NO, whereas NO synthesis by nNOS and eNOS is much weaker, has a shorter duration, and also acts over a very short distance only. In mice, the main producers of RNI are macrophages, monocytes, and TNF/iNOS-producing (Tip) DCs,[43] which represent a specialized subset of DCs capable of NO synthesis (see later discussion). In contrast, neutrophils express eNOS only. Induction of NO synthesis can be initiated by TLR ligands as well as by inflammatory cytokines like IFN-γ, TNF-α, or IL-1. The inhibition of iNOS by aminoguanidine,[44,45] as well as the disruption of the gene coding for iNOS,[46] significantly reduces the resistance of mice during early *L. monocytogenes* infection.

13.3.5 Reactive Oxygen Species

Reactive oxygen species (ROS) are important bactericidal effector molecules of epithelial cells, macrophages, and, in particular, neutrophils. The extracellular release of large amounts of ROS, produced by macrophages and polymorphonuclear neutrophils during the respiratory burst, is an important mechanism of pathogen killing. The major ROS generated by cells are superoxide anions, hydrogen peroxide, and hydroxyl radicals. They are produced by members of the NADPH oxidase (NOX) family (reviewed in reference 47). ROS also play a role in a large number of reversible regulatory processes in virtually all cells and tissues. Being small molecules with important roles in cell signaling, the ROS are implicated in cellular activity associated with a variety of inflammatory responses.

In macrophages and neutrophils, the phagocyte oxidase $gp91^{phox}$ (NOX2) is highly expressed. The active NOX2 complex comprises NOX2 itself, GTP-Rac, the membrane subunit $p22^{phox}$, and the organizer subunit $p47^{phox}$, which subsequently binds the cytosolic subunits $p40^{phox}$ and $p67^{phox}$. The disruption of the genes coding for $p47^{phox}$ or $gp91^{phox}$ results in abolished NOX2 activity by macrophages and neutrophils and a significantly enhanced susceptibility of mice during early *L. monocytogenes* infection.[46,48] *L. monocytogenes* counteracts ROS by the enzyme superoxide dismutase. Deletion of the gene coding for the bacterial superoxide dismutase drastically reduces the virulence of *L. monocytogenes* in mice.[49] The partial neutralization of ROS by superoxide dismutase may explain why the virulence of *L. monocytogenes* in iNOS-deficient mice is higher than in NOX2-deficient mice.

Mice producing neither ROS nor RNI due to a targeted deletion of both enzymes NOX2 and iNOS are much more susceptible to *L. monocytogenes* than mice with a single deficiency of either NOX2 or iNOS alone.[46] Macrophages from mice with a combined NOX2/iNOS deficiency are not at all able to kill virulent *L. monocytogenes*, indicating that NOX2 and iNOS may be able to compensate at least partially for each other's deficiency.

13.3.6 Antibacterial Lectins

Lectins are ubiquitous carbohydrate-binding proteins or glycoproteins with specific affinity for sugar moieties. Apart from serving a number of biological functions such as in the regulation of cell adhesion to glycoprotein synthesis and control of protein levels in the blood, lectins also play a critical part in the immune system by recognizing carbohydrates that are present exclusively on pathogens, or that are not ordinarily accessible on host cells. Lectins are regarded as predecessors to the immune system; NK cell receptors are a typical example of lectins that contribute to host innate immune response against pathogens.

Numerous beneficial symbiotic bacteria populate the mammalian intestine without penetrating the mucosal surface. The mechanisms that prevent mucosal invasion by commensal bacteria include secretory immunoglobulin A[50] as well as antibacterial lectins.[51] Microbial colonization of germ-free mice triggers epithelial expression of RegIIIγ, a secreted C-type lectin.[51] The murine RegIIIγ and its human counterpart HIP/PAP bind to the peptidoglycan carbohydrates of *L. monocytogenes* as well as of *L. innocua* and act directly bactericidal in a complement-independent manner. Expression of intestinal bactericidal lectins is up-regulated only by bacterial invasion of the mucosal surface. Thus, by their expression scheme, intestinal bactericidal lectins are preferentially directed against invasive pathogens, whereas noninvasive commensals of the intestinal microflora are spared.

13.3.7 Lysozyme

The human lysozyme (also known as 1,4-beta-N-acetylmuramidase C) is an enzyme (EC 3.2.1.17) with the ability to kill bacteria. The molecule consists of a single chain of 130 amino acid residues. It is abundantly present in mucosal secretions and is also contained in cytoplasmic granules of polymorphonuclear neutrophils. In the intestine large amounts of lysozyme are produced by Paneth cells.[52] The enzyme attacks peptidoglycans and hydrolyzes the bond that connects N-acetyl muramic acid with the fourth carbon atom of N-acetylglucosamine.[53] *L. monocytogenes*[6] and other Gram-positive bacteria[7] protect themselves from lysozyme attack by deacetylation of N-acetylglucosamine residues. *L. monocytogenes* mutants without a functional peptidoglycan deacetylase gene are extremely sensitive to the bacteriolytic activity of lysozyme.[6] Deacylation of N-acetylglucosamine residues not only inhibits direct lysozyme activity but also prevents the stimulation of the innate immune system via release of muropeptides and lipoteichoic acids (see sections 13.2.1.1 and 13.2.1.2). As *L. monocytogenes* actually is quite well protected from lysozyme, it is currently not known whether the remaining activity of lysozyme against this bacterium plays any role in anti-*Listeria* protection.

13.3.8 Antimicrobial Peptides

Antimicrobial peptides (also known as host defense peptides) are evolutionarily conserved components of the innate immune response with antibiotic property and are found in both vertebrates and invertebrates. They are short proteins that are produced by cells of the immune system as well as by epithelial cells. They kill bacteria by disrupting bacterial membrane, interfering with metabolism, and targeting cytoplasmic components. Furthermore, they also demonstrate a number of immunomodulatory functions to help clear infection, by acting as chemokines and/or inducing chemokine production, inhibiting LPS-induced proinflammatory cytokine production, and modulating the responses of DCs and other cells of the immune system.

Two main groups of antimicrobial peptides are distinguished: the defensins and the cathelicidin-derived peptides.[54] The largest and probably most important group, the defensins, are oligopeptides that consist of 28–42 amino acids with three intramolecular disulfide bonds. Most defensins function by attaching to microbial cell walls by electrostatic binding and subsequently form ion-permeable channels. Intestinal Paneth cells respond to bacterial components as lipoteichoic acid or muropeptides with the production of α-defensins, also known as crypt defensins or cryptidins.[55] Defensins are also a major constituent of the granula of neutrophils. However, the precise role of the defensins of neutrophils in the innate immune defense against *L. monocytogenes* is still unknown.

Some chemokines in structure and size resemble defensins. Interestingly, some of the chemokines produced by IFN-γ-stimulated human peripheral blood mononuclear cells (recombinant monokine induced by IFN-γ [CXCL9], IFN-γ-inducible protein of 10 kDa [CXCL10], and IFN-inducible T cell alpha chemoattractant [CXCL11]), which all belong to the IFN-γ-inducible tripeptide motif Glu-Leu-Arg CXC chemokines, have a direct antimicrobial activity against *L. monocytogenes*.[56]

Ubiquicidin is another antimicrobial peptide that was isolated from an IFN-γ activated murine macrophage cell line and has a direct bactericidal effect on *L. monocytogenes*.[57] Although these macrophage-derived antimicrobial peptides unequivocally possess some bactericidal activity, their role in the early immune defense against *L. monocytogenes in vivo* cannot yet be judged conclusively.

13.3.9 Phagolysosomal Protease

Phagolysosomes are membrane-bound organelles formed in phagocytic cells by the fusion of lysosomes with phagosomes. Lysosomes contain various enzymes such as proteases, lipases, carbohydrases, nucleases, and phosphatases. Phagolysosomal fusion is required for the expression of the antibacterial activity of macrophages. Within the phagolysosome, the TNF- and IFN-γ-responsive acid sphingomyelinase generates the signaling molecule ceramide, resulting in the activation of proteases like cathepsin D. Mice lacking a functional sphingomyelinase demonstrate a dramatically increased susceptibility to *L. monocytogenes*.[58] Although macrophages and neutrophils from sphingomyelinase-deficient mice are able to produce NO and ROS, sphingomyelinase-deficient macrophages are completely incapable of controlling the growth of *L. monocytogenes in vitro*. Also, cathepsin D-deficient mice show decreased resistance against *L. monocytogenes* infection.[49] In the phagolysosome of infected cells cathepsin D prevents the escape of *L. monocytogenes* into the cytoplasm by destroying listeriolysin O (LLO). Bacteria retained in the phagolysosome are subsequently killed by bactericidal effector molecules like ROS and RNI (Figure 13.3D).

13.4 CELLS INVOLVED IN INNATE IMMUNITY

Cells represent the integrating compartments of the innate immune system. On the one hand, cells directly sense the presence of microbes by pattern recognition receptors, or are indirectly alerted and attracted to areas of bacterial infection by inflammatory signals. On the other hand, cells produce and release antibacterial effector molecules as well as cytokines and chemokines in order to directly kill pathogens and to alert and recruit other cells. The molecular sensors that detect bacteria, the effector molecules that attack bacteria, and also the cytokines, chemokines, and their respective receptors that orchestrate the innate immune defense are distributed among various cell types. Thus, the selective depletion of an individual factor in most instances does not allow one to directly infer the role of a specific cell type. Only in some instances cell populations (e.g., NK cells or neutrophils) can effectively be depleted *in vivo*, which is required in order to evaluate the contribution of a specific cell type to antibacterial resistance in infected mice. In all other cases, the role of an individual cell population has to be inferred indirectly by monitoring the phenotype, abundance, and functional state of infiltrating cells at sites of bacterial replication in infected organs.

13.4.1 Paneth Cells

The gastrointestinal barrier represents the first line of defense against invasive pathogens in the small intestine. The function of the gastrointestinal barrier is far from passive. A specialized subset of secretory cells, Paneth cells in intestinal crypts, with functional similarity to neutrophils, provide active host defense against microbes in the small intestine. Paneth cells are named after the Austrian physiologist Joseph Paneth who, in 1888, described this secretory cell type at the base of intestinal Lieberkühn glands. The functional role of Paneth cells has remained unknown for more than 100 years and only recently has their role in innate mucosal immunity been discovered. Upon exposure to bacteria, Paneth cells secrete several antimicrobial molecules such as α-defensins (or cryptidin), lysozyme, and phospholipase A2 into the lumen of the intestinal crypts. The peptides of α-defensins possess hydrophobic and positively charged domains that interact with phospholipids in bacterial cell membranes (which tend to be negatively charged) and cause cell lysis. In addition, both lysozyme and phospholipase A2 molecules are active against microbial pathogens. It seems that activation of NOD2 is required for the expression of cryptidins and NOD2-deficient mice are susceptible to bacterial infection via the oral route but not through intravenous or peritoneal delivery.[22] By the expression of membrane-bound antibacterial lectins, Paneth cells selectively attack invasive bacterial pathogens such as *L. monocytogenes* while beneficial bacteria of the intestinal microflora are spared.[51]

13.4.2 Polymorphonuclear Neutrophils

Polymorphonuclear neutrophil granulocytes, generally referred to as neutrophils, are the most abundant type of white blood cells. Neutrophils are short-lived—living just a few days—and are among the first cell types arriving at sites of infection. In particular, in response to bacterial infection, neutrophils leave the blood stream and migrate toward the site of inflammation. Chemotaxis of neutrophils is elicited by certain chemokines like the CXC chemokine IL-8, complement factors C3a and C5a, and formyl peptides. Formyl peptides are short di-, tri-, or tetrapeptides of bacterial origin with an N-formylated methionin at the N-terminus. A typical formyl peptide is N-formyl methyonyl-leucyl-phenylalanine (fMLF), which acts on cell surface fMLF receptors expressed on neutrophils as well as on macrophages. Neutrophils are active phagocytes. Ingestion of microorganisms triggers the production of large quantities of ROS in a single respiratory burst. Neutrophils also trap and kill extracellular bacteria by so-called neutrophil extracellular traps. They catch bacteria in a fine mesh formed of chromatin and serine proteases. It is suspected that this mesh acts as a physical barrier that hinders the spreading of bacteria and also increases the local concentration of bactericidal effector molecules by reducing diffusion.

Considering the intracellular lifestyle of *L. monocytogenes,* polymorphonuclear neutrophils play an unexpectedly important role in the protection against *L. monocytogenes.* Depletion of neutrophils *in vivo* with a specific monoclonal antibody results in significantly reduced anti-*Listeria* resistance.[59–61] In particular, the control of bacterial replication in hepatocytes requires neutrophils. The mechanisms by which neutrophils recognize and kill intracellular *L. monocytogenes* are not well understood. Neutrophils express only eNOS and no iNOS and thus are probably not able to produce enough NO for a far-reaching NO-mediated bactericidal effect.[42] However, neutrophils produce defensins and ROS, which both act bactericidally on *L. monocytogenes*. Granulocyte-mediated killing of *L. monocytogenes* in the liver is accompanied by apoptosis of hepatocytes and a strong inflammatory tissue reaction.

13.4.3 Macrophages

Macrophages are tissue cells that differentiate from blood monocytes. Depending on the tissue, monocytes differentiate into Kupffer cells in the liver, histiocytes in connective tissue, microglia in neural tissue, and sinusoidal lining cell in the spleen. Both macrophages and monocytes are phagocytic cells that play an important role in innate immunity. Their function is to phagocytose and kill pathogens and to stimulate other immune cells to respond to the pathogen. Macrophages are attracted by certain chemokines, complement factors, and N-formylated peptides; through chemotaxis they actively migrate toward foci of infection. The life span of macrophage ranges from months to years. During listeriosis macrophages are both targets for infection with *L. monocytogenes* as well as effector cells that are believed to play an important role in the innate immune response against *L. monocytogenes*. While macrophages are easily infected by *L. monocytogenes* and play an important role in the spreading of bacteria,[62] they are also required for the detection and killing of *L. monocytogenes*. Pathogen detection by macrophages involves TLR2, TLR5, NOD1, and NOD2 as well as type 1 and type 2 scavenger receptors. During *L. monocytogenes* infection, macrophages are important sources of cytokines and chemokines. Upon infection, macrophages rapidly produce TNF, IL-1, and IL-6, as well as IL-12, IL-18, and MCP-1.[63–66] The main activators of macrophages are IFN-γ and TNF produced mainly by T cells and NK cells (Figure 13.3A).

Direct evidence that macrophages *summa summarum* play a beneficial role during early listeriosis comes from macrophage depletion studies with the drug dichloromethylene diphosphonate encapsulated in liposomes, which significantly reduces early natural resistance to *L. monocytogenes* in mice.[67] The main effector molecules of activated human and murine macrophages are NO, ROS, lysozyme, and antimicrobial peptides such as ubiquicidin (Table 13.2). Human and murine macrophages do not produce defensins. *L. monocytogenes* avoids being killed after phagocytosis by macrophages by escaping from phagocytotic vacuoles into the host cell cytoplasm. However, activated macrophages are listericidal, in part because they can retain *L. monocytogenes* in phagocytotic vacuoles, where they are exposed to high doses of antibacterial effector molecules. Vacuolar retention and killing of *L. monocytogenes* require cathepsin D[49] as well as ROS and is augmented by RNI.[68] As the study of mice with a targeted deletion of the genes coding for both NOX2 and iNOS has shown, ROS and RNI can compensate to some degree for each other's deficiency.[46] Apoptosis of *L. monocytogenes*-infected macrophages is prevented by the anti-apoptotic factor SPα triggered by liver X receptors, which are nuclear receptors with established roles in the regulation of lipid metabolism.[69]

13.4.4 Monocytes and TNF/iNOS-Producing DCs

Monocytes represent a heterogeneous leucocyte population produced by the bone marrow from hematopoietic stem cell precursors. Monocytes circulate in the bloodstream for about 1–3 days before moving into tissues. In the tissue monocytes differentiate into various types of macrophages, depending on the anatomical location. A subset of monocytes expressing chemokine receptor 2 (CCR2) plays an important role in the murine host defense against *L. monocytogenes* infections. After *L. monocytogenes* infection of mice, signals mediated by CCR2 are required for monocyte emigration of CCR2-positive monocytes from the bone marrow into the blood.[70] The major ligand of CCR2 is MCP-1 produced by infected macrophages.[66] Once monocytes enter infected tissues, they differentiate into a special subset of TNF/iNOS-producing (Tip) DCs.[43] This differentiation step is MyD88-dependent but IL-1 and IL-18-independent, suggesting that it is probably TLR mediated.[66] Tip-DCs are the major producers of TNF-α and iNOS during the early phase of *L. monocytogenes* infection and are able to exert direct bactericidal activity against *L. monocytogenes.* In contrast to conventional DCs, Tip-DCs seem to play no major role as antigen presenting cells, as efficient priming of *L. monocytogenes*-specific CD8 T cells occurs also in the absence of Tip-DCs.

13.4.5 "Bystander" $CD8^+$ T Cells

$CD8^+$ T cells are also referred to as cytotoxic T cells. They are able to destroy infected cells and tumor cells. On the cells' surfaces they express the CD8 glycoprotein and a highly variable T cell receptor (TCR). Activation of $CD8^+$ T cells requires the presentation of short antigenic peptides in the context of MHC class I molecules on the cell surface of target cells. $CD8^+$ T cells play a pivotal role during the later adaptive phase of anti-*Listeria* immunity (see chapter 14). However, in the absence of cognate antigen, both effector and memory $CD8^+$ T cells still have the potential to induce an early IFN-γ secretion in an antigen-independent manner in response to IL-12 and IL-18 produced mainly by activated *Listeria*-infected macrophages (Figure 13.3A).[71–73] By contrast, infection with other intracellular pathogens does not trigger such an early IFN-γ response. Interestingly, $CD8^+$ T cells from C57BL/6 mice, which were infected with an *Escherichia coli* strain expressing a pore-forming toxin LLO from *L. monocytogenes,* also displayed a rapid production of IFN-γ, suggesting the importance of LLO for stimulating this early IFN-γ production. By this mechanism, "bystander" $CD8^+$ T cells, which have not been previously exposed to *L. monocytogenes*-derived antigens, provide an innate host resistance to listeriosis in mice.[74]

13.4.6 Natural Killer Cells

Natural killer (NK) cells are a population of cytotoxic lymphocytes that play a major role in the host rejection of tumors and virally infected cells. Release of perforin and granzymes from NK cells induces apoptosis of target cells. NK cells do not express TCRs or the pan T cell marker CD3, but most NK cells express the surface markers CD6 (FcγRIII) and CD56 in humans, and NK1.1/NK1.2 in mice. NK cells are activated in response to type I interferons and macrophage-derived cytokines such as IL-12 and IL-18. The exact mechanisms by which NK cells recognize target cells are still unclear. NK cells express a number of activating as well as inhibitory receptors (NKG2, Ly49, and others), which are able to sense altered self on virally infected cells or tumor cells. Cells expressing low level of self MHC class I cell surface molecules are often killed by NK cells due to the lack of inhibitory signaling by MHC-specific NK cell receptors. NK cells belong to the earliest producers of IFN-γ after infection of mice with *L. monocytogenes.* IFN-γ production by NK cells is triggered by IL-12 and IL-18, which are produced by activated macrophages.[65] Macrophages stimulated with TLR ligands themselves express ligands for the NKG2D receptor, which is found on NK cells and activated CD8 T cells, providing an additional cell contact–dependent pathway of macrophage/NK cell interaction.[75] However, the precise role of NK cells during *L. monocytogenes* infection is controversial. Depending on experimental conditions, depletion of NK cells reduces[76] or enhances[77,78] early resistance of mice to *L. monocytogenes.* The direct comparison of nonspecific CD8 T cells and NK cells adoptively transferred into *L. monocytogenes*-infected IFN-γ-deficient mice has shown that bystander CD8 T cells are even more protective than NK cells.[74] These observations suggest that NK cells in the presence of other IFN-γ producers (e.g., CD8 T cells) probably play only a minor role in the protection during early *L. monocytogenes* infection.

13.4.7 Natural Killer T Cells

Natural killer T (NKT) cells are a subset of T cells that coexpress the αβ T cell receptor (TCR) with a variety of molecular markers that are typically associated with NK cells, such as NK1.1. Unlike NK cells, NKT cells develop in the thymus and express a rearranged TCR. In contrast to typical T cells, NKT cells respond to antigen presented by the atypical MHC class I molecule, CDld. NKT cells are either $CD4^+$ or $CD4^-CD8^-$, in contrast to typical $CD8^+$ MHC class I-restricted T cells. NKT cells express an extremely limited T cell repertoire. The best-known subset of CDld-dependent NKT cells expresses an invariant T cell receptor α (TCR-α) chain. These cells are conserved between humans and mice and are called type I or invariant NKT cells (iNKT) cells. Splenic and hepatic Vα14(+) NKT cells in C57BL/6 mice are early producers of IFN-γ after *L. monocytogenes* infection, and adoptively transferred Vα14(+) NKT cells provide early protection against enteric *L. monocytogenes* infection in these mice.[79] Also, the absence of CDld results in increased susceptibility toward *L. monocytogenes* in the early phase of infection.[80]

13.5 REGULATION

The innate immune response is regulated by cytokines in an intricate intercellular communication network. Cytokines are short proteins or glycoproteins with a mass of 8–30 kDa or even smaller peptides that play a particularly important role in both innate and adaptive immune responses. While hormones are released from specific organs into the blood and neurotransmitters are released by nerves, cytokines are released by many types of cells, including both hematopoietic and nonhematopoietic cells. The function of cytokines is mediated by specific cytokine receptors expressed on the cell surface of various cell types. Activation of these receptors often triggers the release of other cytokines, resulting in intricate cytokine cascades and regulatory networks. The large group of cytokines was divided into interleukins, interferons, and chemokines. However, as the functions of these different groups of cytokines overlap, today this division is mostly of historical interest. Interleukins were first seen to be expressed by blood leukocytes; the chemokines, which also include IL-8, are a subclass of cytokines that generally play a role in chemotactic attraction of cells, while interferons were first described as cytokines that interfere with the replication of intracellular, in particular, viral, pathogens.

One of the earliest events in the proinflammatory cytokine cascade is the formation of the inflammasome, a cytosolic multiprotein complex that allows activation of precursors of proinflammatory caspases, which then cleave the precursors of IL-1β and IL-18 into the active forms. The secretion of both cytokines leads to a potent inflammatory response. The central component of the inflammasome is a member of the NLR family (see section 13.2.1.2), which associates with the adaptor protein apoptosis–associated speck-like protein, which in turn recruits proinflammatory caspase precursors.

In response to *L. monocytogenes* macrophages secrete the proinflammatory cytokines IL-1, IL-6, IL-12, IL-18, and TNF as well as the chemokine MCP-1 and IL-8. In response to IL-12 and IL-18, T cells and NK cells secrete IFN-γ and lymphotoxins. Subsequently, IFN-γ further activates macrophages in a positive feedback loop. On the other hand, anti-inflammatory cytokines, such as IL-10 and possibly IL-4, prevent the host from damage due to an inflammatory overreaction.

13.5.1 Proinflammatory Cytokines

13.5.1.1 Tumor Necrosis Factor

Tumor necrosis factor (TNF) was previously termed TNF-α to differentiate it from TNF-β. Because the latter is now named lymphotoxin (LT), the term TNF-α is now synonymous with TNF. Tumor necrosis factor is an endogenous pyrogen and triggers together with IL-1 and IL-6 the acute phase response that, among other effects, includes fever, release of acute phase proteins into the circulation, increase of vascular permeability, and leucocytosis. The main cellular sources of TNF during the nonadaptive innate immune response are monocytes, macrophages, Tip-DCs, and NK cells.

TNF belongs to the earliest cytokines produced by *L. monocytogenes*-infected macrophages, but is also produced by neutrophils, T cells, and B cells. TNF alone or in synergism with IFN-γ activates macrophages[81,82] and also enhances IFN-γ synthesis by NK cells.[83] The biological activities of TNF are mediated by two distinct receptors, TNFR1 (p55) and TNFR2 (p75), which are coexpressed by most cell types. Mice with a genetic disruption of the gene coding for TNFR1[84,85] or TNF[86] demonstrate a strongly reduced innate resistance against early *L. monocytogenes* infection. In contrast, the engagement of TNFR2 seems to play no major role in innate immunity against *L. monocytogenes*.[87,88] For protection against *L. monocytogenes* in particular, TNF production by macrophages and neutrophils is required.[89]

13.5.1.2 Lymphotoxins

Lymphotoxins (LTs) belong to the TNF family. They are mainly secreted by activated T cells, as well as by fibroblasts and endothelial and epithelial cells. The two forms of lymphotoxin, LT-α and LT-β, modulate diverse immune functions. LT-β and the LT-β receptors are required for the formation of lymphoid tissues during embryogenesis. Although the protective role of TNF during the early innate immune response against *L. monocytogenes* is well established, the anti-*Listeria* functions of LT-α and LT-β are still a matter of debate. A principal problem of the study of LT-α- and LT-β-deficient mice is their lack of peripheral lymphoid tissues. Therefore, infection studies have to be performed in bone marrow chimeras reconstituted with bone marrow from gene knockout mice. Chimeric mice reconstituted with bone marrow from LT-β receptor–deficient mice demonstrate a moderately decreased resistance to early *L. monocytogenes* infection,[90] but mice reconstituted with bone marrow from LT-β-deficient mice are almost as resistant as wild-type mice.[91] During early listeriosis LT-α seems to play a more important role than LT-β, as mice reconstituted with bone marrow from LT-α-deficient mice are almost as susceptible to early listeriosis as TNF-deficient mice.[91] Although these LT-α-deficient mice are able to control bacterial replication as well as wild-type mice during the first 3 days after infection, they are not able to control bacterial replication within infectious foci later. LT-α-deficient mice die around day 5 postinfection, probably due to inefficient recruitment of lymphocytes and monocytes to sites of infection.

13.5.1.3 Interferon-γ

Interferons in humans are divided into three types in accordance with the type of receptor with which they interact. Type I interferons bind to a specific cell surface receptor complex known as the IFN-α receptor (IFNAR), consisting of IFNAR1 and IFNAR2 chains. IFN-γ is the single representative of the type II interferon family and signals through the IFNGR (interferon gamma receptor). The recently described type III interferons recognize a receptor complex consisting of IL-10 receptor 2 and IFN-lambda receptor 1 chains. IFN-γ is also referred to as immune interferon because it is mostly produced by T cells and NK cells in contrast to type I interferons, which are secreted by a diverse array of cells including fibroblasts and endothelial cells. The biologically active IFN-γ dimer is formed by antiparallel interlocking of two IFN-γ monomers. IFN-γ alters transcription in numerous genes producing antiviral, immunoregulatory, and antitumor responses. In particular, IFN-γ enhances the antiviral and bactericidal activity of macrophages, up-regulates antigen processing and antigen presentation to T cells by MHC class I and II molecules on the surface of infected cells, and promotes Th_1 differentiation through up-regulation of the transcription factor T-bet.

Secretion of IFN-γ by CD8 T cells and NK cells in the early phase of the murine *L. monocytogenes* infection is triggered by IL-12 and IL-18, which are mainly produced by *L. monocytogenes*-infected macrophages. IFN-γ triggers activation of cells through interaction with a heterodimeric receptor consisting of IFNGR1 and IFNGR2 chains. The role of IFN-γ in the early phase of the host defense against *L. monocytogenes* is demonstrated by the high susceptibility of mice with a disruption of the genes coding for IFN-γ[92] or the IFNGR.[93] IFN-γ triggers bactericidal activities by macrophages as the oxidative burst, production of RNI, and secretion of antimicrobial peptides.

13.5.1.4 Type I Interferons

The type I interferons IFN-α and IFN-β play a major role in the antiviral immune defense. IFN-β expression by macrophages can also be triggered by *L. monocytogenes.* A known activator of IFN-β expression is cytosolic bacterial DNA, which triggers IFN-β expression by a TLR9-independent mechanism.[25] Mice lacking the type I interferon receptor[94,95] or the interferon regulatory factor 3, which plays an important role in type I interferon signaling,[95] are, compared to wild-type mice, significantly more resistant to early *L. monocytogenes* infection. Enhanced resistance in mice lacking the type I interferon receptor correlates with increased IL-12 levels in the blood, increased numbers of TNF-producing macrophages, and a block of *L. monocytogenes*-induced splenic apoptosis.[94,95]

13.5.1.5 Interleukin-1

The interleukin-1 (IL-1) family includes the agonistic cytokines IL-1α and IL-1β, and the antagonistic IL-1 receptor (IL-1R) inhibitor. IL-1β is processed from an inactive precursor by the IL-1β-converting enzyme caspase-1. IL-1, together with TNF, is responsible for the general host inflammatory response to infection. As an endogenous pyrogen, IL-1 induces fever by direct action on the central nervous system and also triggers the release of IL-6 and acute phase proteins. The blockade of the IL-1/IL-1R interaction with monoclonal antibodies prevents macrophage activation in infected mice.[96,97] Two types of IL-1R are known. For the mediation of proinflammatory signals during *L. monocytogenes* infection, IL1R1 is required. Variable degrees of early anti-*Listeria* resistance have been reported for mice with a genetic disruption of the gene coding for IL-1R1.[98,99]

13.5.1.6 Interleukin-6

Interleukin-6 (IL-6) is a pleiotropic cytokine secreted mainly by T cells, macrophages, and endothelial cells in response to PAMPs of diverse microbial pathogens. Upon binding with the detection molecules of the innate immune system that are present on the cell surface (or in intracellular compartments), the PAMPs induce intracellular signaling cascades leading to inflammatory cytokine production. After *L. monocytogenes* infection IL-6 is rapidly produced by Kupffer cells, which probably are the major source of IL-6 during listeriosis.[100] Among other effects, IL-6 induces peripheral blood neutrophilia and synthesis and release of acute phase proteins by hepatocytes. During early *L. monocytogenes* infection IL-6-deficient mice show a significantly reduced resistance compared to control mice.[101,102] IL-6 signals through a cell-surface type I cytokine receptor complex consisting of the ligand-binding IL-6Rα chain (CD126), and the signal-transducing component gp130 (also called CD130). Type I cytokine receptors share a common amino acid motif in the extracellular portion adjacent to the cell membrane. Members of the type I cytokine receptor family include the receptors for IL-2, IL-4, IL-6, and IL-12.

13.5.1.7 Interleukin-12

Interleukin-12 (IL-12) is a heterodimeric cytokine encoded by two separate genes: IL-12A (p35) and IL-12B (p40). IL-12 is a key macrophage product that is a T cell stimulating factor, whose type I cytokine receptor is a heterodimer formed by IL-12R-β1 and IL-12R-β2. In response to antigenic stimulation, IL-12 activates the production of IFN-γ and TNF from T and NK cells and reduces IL-4 mediated suppression of IFN-γ. In addition, IL-12 enhances the cytotoxic activity of NK cells and CD8 T cells, and fosters the differentiation of naïve T cells into Th1 cells. There is strong evidence that IL-12 provides protection during early *Listeria* infection by inducing IFN-γ production by NK cells and T cells.[103]

13.5.1.8 Interleukin-18

Interleukin-18 (IL-18) belongs to the IL-1 superfamily of cytokines. It is produced by macrophages upon exposure to microbial PAMPs such as bacterial lipopolysaccharides (LPS). During listeriosis IL-18 synergizes with IL-12 to induce IFN-γ production by natural killer and T cells and also directly triggers NO and TNF release by macrophages.[104] Caspase-1 is required to process IL-1β and IL-18 precursor molecules to yield the biologically active forms of IL-1β and IL-18 and also plays a role in the induction of apoptosis. Caspase-1-deficient mice exhibit a decreased resistance to early *L. monocytogenes* infection, which can be reconstituted to normal levels by injection of exogenous IL-18 and IFN-γ into mice.[16] Similar effects are achieved in mice lacking a functional apoptosis-associated speck-like protein, which plays an important role in caspase-1 activation by its C-terminal caspase recruitment domain.[105] Interestingly, mice with a targeted deletion of caspase-12, which also plays a role in proinflammatory cytokine processing, show significantly increased resistance to peritonitis and sepsis after *L. monocytogenes* infection by dampening the production of the proinflammatory cytokines IL-1β, IL-18, and IFN-γ, but not of TNF and IL-6.[106]

13.5.1.9 Chemokines

Chemokines are a family of small secreted chemotactic cytokines. All chemokines share structural characteristics such as small size (8–10 kDa) and the presence of four cysteine residues in conserved locations. Chemokines are grouped according to the position of the two cysteine residues near the amino terminus. Near the N-terminus, C chemokines have only one cysteine residue, CC chemokines have two N-terminal cysteines without additional intervening residues, CXC chemokines have one additional amino acid residue between the two N-terminal cysteine residues, and CX3C chemokines have three additional residues between the two N-terminal cysteine residues. Up to now, the possible role in anti-*Listeria* protection of only a few selected chemokines has been studied in detail. Some chemokines also exhibit a directly bactericidal effect on *L. monocytogenes* (see section 13.3).

13.5.1.10 Interleukin-8

Also known as CXC chemokine ligand 8 (CXCL8), IL-8 is a chemokine that is secreted by macrophages and other cells (e.g., intestinal and cervical epithelial cells[107] and endothelial cells[108]). When macrophages phagocytose an antigen (or invading bacteria such as *L. monocytogenes* and other intracellular bacteria), they release chemokines (e.g., IL-8) to attract other immune cells (e.g., neutrophils) to the site of inflammation. Thus, IL-8 acts as a potent chemoattractant and activator of polymorphonuclear neutrophils. Triggering of IL-8 secretion is NOD1-dependent.[108]

13.5.1.11 Monocyte Chemoattractant Protein-1

MCP-1, also known as CC chemokine ligand 2 (CCL2), is the main ligand of CC chemokine receptor 2 (CCR2). Mice lacking CCR2[109] or MCP-1[66] show significantly reduced protection against *L. monocytogenes* during the early phase of infection. MCP-1 is secreted by *L. monocytogenes*-infected macrophages. Triggering of MCP-1 secretion requires cytosolic localization of bacteria.[66] *L. monocytogenes* infection of CCR2-deficient mice results in accumulation of activated monocytes in the bone marrow but reduced numbers of peripheral circulating monocytes, indicating that signaling via CCR2 is required for monocyte emigration from the bone marrow.[70]

No major influence during early *L. monocytogenes* infection has been reported for CCR5, which acts as the main receptor for the chemokines RANTES (regulated upon activation, normal T cell expressed and secreted), macrophage inflammatory protein1α (MIP-1α = CCL3), and MIP-1β (= CCL4).[110]

13.5.2 Anti-inflammatory Cytokines

13.5.2.1 Interleukin-4

Interleukin-4 (IL-4) is a cytokine produced by Th_2 cells whose major biological functions include the stimulation of activated B cell and T cell proliferation, and the differentiation of CD4 T cells into Th_2 cells. As a key regulator in humoral immunity, IL-4 induces B cell class switching to IgE and directs the T cell response toward a TH_2 type response. After infection of mice with *L. monocytogenes,* a rapid transient IL-4 burst has been demonstrated.[111] As *L. monocytogenes*, however, induces a TH_1 type T cell response, the role of IL-4 during the early phase of listeriosis is still unclear.

13.5.2.2 Interleukin-10

Previously also known as human cytokine synthesis inhibitory factor (CSIF), IL-10 is a homodimer with each subunit containing 178 amino acid residues. It is an anti-inflammatory cytokine that is mostly expressed in monocytes, Th_2 cells, mast cells, and some activated T and B cells. Capable of inhibiting synthesis of proinflammatory cytokines (e.g., IFN-γ, IL-2, IL-3, TNF, and GM-CSF), IL-10 is also stimulatory to certain T cells, mast cells, and B cells. Mice with a targeted disruption of the gene coding for IL-10 compared to wild-type mice show significantly reduced bacterial burdens at day 3 after infection with *L. monocytogenes.*[112] *In vitro* increased resistance of IL-10-deficient mice correlates with an enhanced proinflammatory IL-1, IL-6, IL-12, TNF, and IFN-γ response.

13.6 ORGAN-SPECIFIC INNATE IMMUNITY

The intestinal mucosa is the point of bacterial entry into the host. Extended bacterial replication in the host, however, only occurs after bacterial spread. The main target organs in which *L. monocytogenes* replicates are the spleen, liver, and central nervous system. Spread of bacteria requires infected cells that are exploited as Trojan horses to enter their target organs without being exposed to the immune system.[62,113,114] Once bacterial infection of target organs occurs, the control of bacterial replication requires an effective immediate immune defense in order to prevent overwhelming bacterial replication and sepsis. Some important organ-specific differences exist in regard to the innate immune defense mechanisms that control bacterial replication in the early phase of infection.

13.6.1 INTESTINE

The epithelial surface of the intestine provides a first line of defense against bacterial invasion. Due to a mutation in the E-cadherin, which is exploited by *L. monocytogenes* as receptor for invasion of epithelia, mice and rats are highly resistant to *L. monocytogenes* via the intragastral route.[27] This inherent resistance long has hampered the development of a reliable model for gastrointestinal *L. monocytogenes* infection. After high-dose oral infection of mice, there are few signs of local replication and inflammation of the intestinal mucosa.[115] The early events of bacterial invasion have been studied in detail in a rat-ligated ileal loop system.[116] In this system dendritic cells in Peyer's patches are the first target cells for *L. monocytogenes* after intestinal inoculation. As a blind passenger of DCs, *L. monocytogenes* subsequently enters draining mesenterial lymph nodes, from which further spread occurs. Survival of *L. monocytogenes* in the lumen of the intestine is self-limited. The high colonization resistance of the intestine depends on the normal microflora as well as on the innate mucosal immune defense mechanisms. The Paneth cell has been identified as the main effector cell in mucosal immunity against *L. monocytogenes.* Paneth cells express antibacterial lectins as well as bactericidal defensins, which protect the host from bacterial invasion while preserving the beneficial microflora of the intestine. After Paneth cells, the second line of early innate intestinal immune defense against enteroinvasive *L. monocytogenes* is dependent on neutrophils and TNF.[117]

13.6.2 SPLEEN

In listeriosis the spleen has a dual role by being a target organ for bacteria in which extensive replication occurs and also being a secondary lymphoid organ actively involved in the anti-*Listeria* immune defense. However, splenectomized[118] and T cell-depleted mice,[119] as well as mice genetically deficient in lymphocytes, such as nude mice,[120] SCID mice,[121] and rag-deficient mice,[122] are more resistant to early *L. monocytogenes* infection, indicating that the large number of lymphocytes in the spleen is more a burden than an asset in this early phase of infection.

For the spread of *L. monocytogenes* into the spleen and also for bacterial replication within the spleen, the presence of CD8α-positive DCs is required.[114] Later in the course of infection viable bacteria can also be recovered from other cell types, including macrophages and neutrophils. Early after infection of the spleen, *L. monocytogenes* is preferentially found in the T cell–independent marginal zone, from where it subsequently translocates rapidly into the T cell–dependent white pulp.[123] Translocation into the white pulp is associated with influx of neutrophils into this compartment. In contrast to the liver, however, depletion of neutrophils during the first day of *L. monocytogenes* infection does not impair the control of bacterial replication in the spleen, indicating that neutrophils do not play a major role for infection control in the spleen at this stage of infection.[59] In the white pulp, lymphocytopenia occurs independently of, but in coincidence with the influx of neutrophils.[59] Lymphocytopenia during the acute phase of *L. monocytogenes* infection in T cell–dependent areas of spleen and lymph nodes is due to apoptosis of lymphocytes,[124] which affects predominantly unspecific "bystander" CD4 and CD8 T cells.[125]

Apoptosis of numerous lymphocytes triggers the production and release by macrophages of large amounts of the anti-inflammatory cytokine IL-10, which results in the attenuation of early innate immunity in the spleen.[122] As susceptibility to *L. monocytogenes*-induced apoptosis depends on IFN-β,[94,95] the prevention of lymphocyte apoptosis may be the reason for the enhanced anti-*Listeria* resistance of mice lacking a functional type I IFN receptor. Subsequent control of bacterial replication in the spleen and sterile immunity require CD4 and CD8 T cells. Adoptive transfer studies in SCID mice showed that both CD4 and CD8 T cells contribute to clearance of bacteria, granuloma formation, and disappearance of microabscesses.[126]

13.6.3 Liver

The liver is one of the largest organs of the body and plays a vital part in the clearing of systemic bacterial infections (Figure 13.4). The liver is a main target organ for *L. monocytogenes* not only after i.v. infection of mice but also in naturally occurring infections of humans. Already 10 minutes after i.v. infection of mice, more than 60% of injected bacteria are taken up by the liver and subsequently replicate mainly in hepatocytes.[127] Uptake of bacteria from the circulation depends on hepatic Kupffer cells, which represent the largest compartment of tissue macrophages present in the body. Depletion of Kupffer cells by a liposome-encapsulated toxin results in enhanced blood bacteremia and significantly reduced hepatic bacterial load in *L. monocytogenes*-infected mice.[128] After infection with *L. monocytogenes*, Kupffer cells produce proinflammatory cytokines and are the main source of IL-1, IL-6, IL-12, and TNF in the liver.[100] Bacteria attach externally to the surface of Kupffer cells without being efficiently killed. Killing of bacteria requires infiltrating bactericidal neutrophils. Depletion of neutrophils by treatment with a monoclonal antibody significantly increases the bacterial load in the liver.[59] Influx of neutrophils is mediated by the expression of integrins by Kupffer cells. The blockade of the intercellular adhesion molecule 1 (ICAM-1), as well as of its ligand CD11b/CD18, blocks the influx of neutrophils and enhances the bacterial burden in the liver.[128] In contrast to CD11b/CD18, the leukocyte function-associated antigen 1 (LFA-1, CD11a/CD18), which also binds ICAM-1, seems not to be required for liver infiltration by neutrophils.[129]

13.6.4 Central Nervous System

In the mouse model of *L. monocytogenes* infection, the primary target organs after oral, intraperitoneal, or intravenous infection are the liver and spleen. After high-dose intravenous inoculation with *L. monocytogenes*, mice succumb to infection before bacteria establish central nervous system (CNS) infection. The repeated oral administration of a sublethal dose of *L. monocytogenes* for up to 10 consecutive days leads to the development of CNS infection in only 25% of mice.[130] In order to overcome this experimental limitation most researchers perform direct intracranial infections for the study of CNS listeriosis. Target cells of *L. monocytogenes* are choroid plexus cells, epithelial cells, ependymal cells, macrophages, microglia, and, to some degree, also neurons.[131] Neutrophils are the main inflammatory cells present in CNS lesions of listeriosis. In the CNS as in the liver, the influx of neutrophils plays an important role in early anti-*Listeria* protection.[132] The early control of CNS listeriosis also requires TNF, which, despite its strong proinflammatory activity, is not responsible for brain edema and the destruction of brain tissue.[133]

13.7 CONCLUSIONS AND PERSPECTIVES

The infection of mice with *L. monocytogenes* is one of the paradigmatic models for the investigation of all facets of antibacterial protection. Probably no other immune response has been scrutinized in as much detail as the innate and adaptive immune defense of mice against *L. monocytogenes*. A significant part of our current knowledge of innate antibacterial immunity in general originates from the study of *L. monocytogenes*-infected mice, which has become a quasi-standard test in the

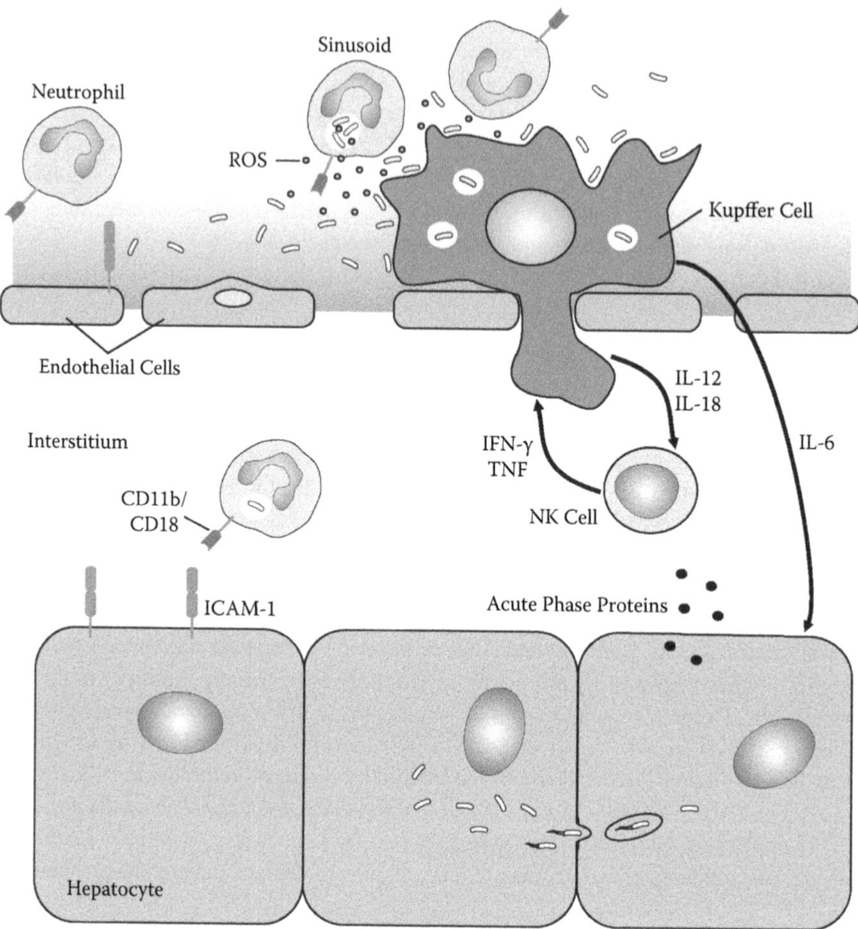

FIGURE 13.4 Innate immune response against *L. monocytogenes* in the liver. In humans and mice the liver is the largest target organ of *L. monocytogenes*. Once infection is established, *L. monocytogenes* rapidly multiplies within hepatocytes. Kupffer cells in liver sinusoids act as filters that remove corpusculate particles like bacteria from the blood stream. Killing of *L. monocytogenes* attached to the surface of Kupffer cells requires the influx of neutrophils into the liver that take up and kill bacteria. Influx of neutrophils into the liver is regulated by expression of the intercellular adhesion molecule (ICAM)-1 on the surface of endothelial cells and hepatocytes, which interacts with CD11b/CD18 on neutrophils. In *L. monocytogenes*-infected mice, Kupffer cells are the main producers of IL-6, which triggers the release of acute phase proteins from hepatocytes. Abbreviations: interferon, IFN; natural killer cell, NK cell; reactive oxygen species, ROS; tumor necrosis factor, TNF.

workup of mice with a targeted gene disruption (Table 13.3). A general conclusion that emerges from the study of these numerous mutant mouse strains is that the immune system employs redundant mechanisms to achieve a single goal. This redundancy of innate immune mechanisms makes it often difficult to properly estimate the physiological relevance of individual immune defense mechanisms. For example, the significance of ROS and RNI for the protection against *L. monocytogenes* was only fully realized after double knockout mice lacking NOS as well as PHOX became available.[46]

Probably, the full appreciation of the role of other bactericidal effector mechanisms such as lysozyme and defensins, as well as of the various sensor molecules of *L. monocytogenes*-associated PAMPs, will require a similar approach. An inherent problem of mice, however, with multiple targeted gene disruptions is their high susceptibility even to nonpathogenic bacteria of the normal microflora,[46] rendering the breeding of such mouse strains a very demanding task. A possible way

TABLE 13.3
Modulation of Early Innate Anti-*Listeria* Resistance in Mice with Spontaneous or Targeted Gene Disruptions

Target[a]	Function	Resistance[b]	Ref.
	General Adaptive Immunity		
Spontaneous mutation of protein kinase Prkdc (SCID mouse)	B and T cell deficiency	→	121
Spontaneous mutation of Foxn1 (nude mouse)	Thymic dysgenesis, T cell deficiency	→	120
Recombination activating gene 1	B and T cell deficiency	→	122
	Sensor Molecules		
TLR2	Activated by lipoteichoic acid	→↓	3, 4, 8
TLR4	Activated by lipopolysaccharide	→	3
NOD2	Activated by muramyl dipeptides	p.o. ↓↓ i.v. →	22
Mouse scavenger receptor A	Activated by lipoteichoic acid	↓	23, 24
Liver X receptor	Protection from apoptosis	↓↓	69
	Effector Molecules		
p47phox Subunit of NOX2	ROS production	↓	48
gp91phox Subunit of NOX2	ROS production		66
iNOS	RNI production	↓	46
NOX2 and iNOS	ROS and RNI production	↓↓↓	46
Cathepsin-D	LLO inactivation	↓	49
Sphingomyelinase	Activation of cathepsin D	↓↓	58
	Integrins		
LFA-1	Systemic neutrophilia	↑	129
	Cytokines		
IFN-γ	Macrophage activation	↓↓↓	92
IFN-γ receptor	Macrophage activation	↓↓↓	93
TNF	Endogenous pyrogen, macrophage activation	↓↓↓	86
Cell type–specific conditional TNF inactivation	Macrophages, neutrophils	↓	89
	T cells	↓	89
	B cells	→	89
TNF receptor p55	TNF-mediated signaling	↓↓↓	84, 85
TNF receptor p75	TNF-mediated signaling	→	88
TNF apoptosis related ligand	Apoptose	↑	135
LT-β	Recruitment of effector cells	↓	91
LT-β	Organization of lymphoid organs	→	91
LT-β receptor	LT-β-mediated signaling	↓	90
IL-1 receptor 1	IL-1 signaling	→↓	98, 99
IFN-α receptor	Type I IFN signaling	↑	94, 95, 136
IL-6	Synthesis and release of acute phase proteins	↓	101, 102
IL-10	Anti-inflammatory activity	↑	112
IL-12	NK and T cell activation	↓	136, 137
MCP-1 (= CCL2)	Monocyte recruitment	↓↓	66
CCR2	Main MCP-1 receptor	↓↓↓	66, 109
CCR5	Main receptor for RANTES, MIP-1α, and MIP-1β	→	110

(continued on next page)

TABLE 13.3 (continued)
Modulation of Early Innate Anti-*Listeria* Resistance in Mice with Spontaneous or Targeted Gene Disruptions

Target[a]	Function	Resistance[b]	Ref.
	Caspases		
Caspase-1	IL-1β and IL-18 maturation, apoptose	↓	16
Caspase-11	Caspase-1 activation (mice only)	→	138
Caspase-12	Caspase-1 activation	↑	106
Apoptosis-associated speck-like protein	Caspase-1 activation	↓	105
	Signal Transduction		
NFκB	Main transcription factor for various immune functions	↓↓	14
MyD88	TLR signaling	↓↓↓	3, 4
Receptor interacting protein 2	TLR and NOD signaling	↓	15
IFN regulatory factor 3	Type 1 IFN signaling	↑	95
IL-1 receptor associated kinase	IL-1 and IL-18 signal transduction	→	139

[a] Refer to the main text for abbreviations.

[b] Arrows denote suppression (↓), enhancement (↑), or no effect (→) on early anti-*Listeria* resistance. The strength of the effect is indicated by the number of arrows.

to overcome these limitations could be the application of soluble inhibitors in knockout mice. Also, the advancement of small inhibitory RNA (siRNA) technology might in the future open new ways for the synchronous inhibition of multiple genes *in vivo*. Another general problem of knockout mice is the possible development of compensatory mechanisms in these mice. The use of inducible knockout strains is a possible approach to avoid some of the typical problems encountered when working with mice with targeted gene disruptions.[134]

Scientists prefer experimental systems in which early innate and later adaptive immune responses can be separated neatly and studied independently. An important goal for the future will be to understand how the innate immune response directs the evolving adaptive immunity in the early phase of an infection and, conversely, how in the later phases of infection the adaptive immune response directs the still active primary innate immune defense system. As far as we currently know, CD4 and CD8 T cells—the main cell populations involved in the adaptive anti-*Listeria* defense—are not directly bactericidal. Thus, the killing and removal of bacteria in the adaptive phase of the antibacterial immune response finally also depend on the various bactericidal effector mechanisms provided by cells of the innate immune system. In principle, the use of inducible knockout mice would allow one to dissect the role of antibacterial effector molecules like lysozyme or ROS in the innate and adaptive phases of the antibacterial immune response.

The very successful application of the murine *L. monocytogenes* infection model over the past decades lets us forget that the murine listeriosis is not a perfect model for human *L. monocytogenes* infection. In contrast to humans, mice are highly resistant to oral infection and only rarely develop infection of the central nervous system, because wild-type mice generally lack a human-like E-cadherin receptor for the *L. monocytogenes* InlA, which is vital for the bacterium's successful passage through the host's intestinal barrier. The recent development of a transgenic mouse model that incorporates a human-like E-cadherin receptor in the intestine[27] makes it possible to correlate the principal findings from the murine infection model with the situation in human infection. Listeriosis is still a relatively frequent and often fatal infection of humans. Thus, it will be important to apply our knowledge of innate and adaptive immunity to provide better protection against *L. monocytogenes* infection for individuals who are—for example, during pregnancy or immunosuppression—at an increased risk of listeriosis-related morbidity and mortality.

REFERENCES

1. Medzhitov, R., and Janeway, C., Jr., Innate immune recognition: Mechanisms and pathways, *Immunol. Rev.*, 173, 89, 2000.
2. Kawai, T., and Akira, S., Pathogen recognition with Toll-like receptors, *Curr. Opin. Immunol.*, 17, 338, 2005.
3. Seki, E. et al., Critical roles of myeloid differentiation factor 88-dependent proinflammatory cytokine release in early phase clearance of *Listeria monocytogenes* in mice, *J. Immunol.*, 169, 3863, 2002.
4. Edelson, B.T., and Unanue, E.R., MyD88-dependent but Toll-like receptor 2-independent innate immunity to *Listeria*: No role for either in macrophage listericidal activity, *J. Immunol.*, 169, 3869, 2002.
5. Travassos, L.H. et al., Toll-like receptor 2-dependent bacterial sensing does not occur via peptidoglycan recognition, *EMBO Rep.*, 5, 1000, 2004.
6. Boneca, I.G. et al., A critical role for peptidoglycan N-deacetylation in *Listeria* evasion from the host innate immune system, *Proc. Natl. Acad. Sci. USA*, 104, 997, 2007.
7. Clarke, A.J., and Dupont, C., O-acetylated peptidoglycan: Its occurrence, pathobiological significance, and biosynthesis, *Can. J. Microbiol.*, 38, 85, 1992.
8. Torres, D. et al., Toll-like receptor 2 is required for optimal control of *Listeria monocytogenes* infection, *Infect. Immun.*, 72, 2131, 2004.
9. Hayashi, F. et al., The innate immune response to bacterial flagellin is mediated by Toll-like receptor 5, *Nature*, 410, 1099, 2001.
10. Jacchieri, S.G., Torquato, R., and Brentani, R.R., Structural study of binding of flagellin by Toll-like receptor 5, *J. Bacteriol.*, 185, 4243, 2003.
11. Way, S.S. et al., Characterization of flagellin expression and its role in *Listeria monocytogenes* infection and immunity, *Cell. Microbiol.*, 6, 235, 2004.
12. Delbridge, L.M., and O'Riordan M, X., Innate recognition of intracellular bacteria, *Curr. Opin. Immunol.*, 19, 10, 2007.
13. Mariathasan, S., and Monack, D.M., Inflammasome adaptors and sensors: Intracellular regulators of infection and inflammation, *Nat. Rev. Immunol.*, 7, 31, 2007.
14. Sha, W.C. et al., Targeted disruption of the p50 subunit of NF-kappa B leads to multifocal defects in immune responses, *Cell*, 80, 321, 1995.
15. Kobayashi, K. et al., RICK/Rip2/CARDIAK mediates signaling for receptors of the innate and adaptive immune systems, *Nature*, 416, 194, 2002.
16. Tsuji, N.M. et al., Roles of caspase-1 in *Listeria* infection in mice, *Int. Immunol.*, 16, 335, 2004.
17. Girardin, S.E. et al., Nod1 detects a unique muropeptide from Gram-negative bacterial peptidoglycan, *Science*, 300, 1584, 2003.
18. Girardin, S.E. et al., Peptidoglycan molecular requirements allowing detection by Nod1 and Nod2, *J. Biol. Chem.*, 278, 41702, 2003.
19. Chamaillard, M. et al., An essential role for NOD1 in host recognition of bacterial peptidoglycan containing diaminopimelic acid, *Nat. Immunol.*, 4, 702, 2003.
20. Girardin, S.E. et al., Nod2 is a general sensor of peptidoglycan through muramyl dipeptide (MDP) detection, *J. Biol. Chem.*, 278, 8869, 2003.
21. Inohara, N. et al., Host recognition of bacterial muramyl dipeptide mediated through NOD2. Implications for Crohn's disease, *J. Biol. Chem.*, 278, 5509, 2003.
22. Kobayashi, K.S. et al., Nod2-dependent regulation of innate and adaptive immunity in the intestinal tract, *Science*, 307, 731, 2005.
23. Suzuki, H. et al., A role for macrophage scavenger receptors in atherosclerosis and susceptibility to infection, *Nature*, 386, 292, 1997.
24. Ishiguro, T. et al., Role of macrophage scavenger receptors in response to *Listeria monocytogenes* infection in mice, *Am. J. Pathol.*, 158, 179, 2001.
25. Stetson, D.B., and Medzhitov, R., Recognition of cytosolic DNA activates an IRF3-dependent innate immune response, *Immunity*, 24, 93, 2006.
26. Perrin, A.J. et al., Recognition of bacteria in the cytosol of mammalian cells by the ubiquitin system, *Curr. Biol.*, 14, 806, 2004.

27. Lecuit, M. et al., A transgenic model for listeriosis: Role of internalin in crossing the intestinal barrier, *Science*, 292, 1722, 2001.
28. Schlech, W.F., III, Chase, D.P., and Badley, A., A model of food-borne *Listeria monocytogenes* infection in the Sprague–Dawley rat using gastric inoculation: Development and effect of gastric acidity on infective dose, *Int. J. Food Microbiol.*, 18, 15, 1993.
29. Zachar, Z., and Savage, D.C., Microbial interference and colonization of the murine gastrointestinal tract by *Listeria monocytogenes*, *Infect. Immun.*, 23, 168, 1979.
30. Czuprynski, C.J., and Balish, E., Pathogenesis of *Listeria monocytogenes* for gnotobiotic rats, *Infect. Immun.*, 32, 323, 1981.
31. Okamoto, M., Nakane, A., and Minagawa, T., Host resistance to an intragastric infection with *Listeria monocytogenes* in mice depends on cellular immunity and intestinal bacterial flora, *Infect. Immun.*, 62, 3080, 1994.
32. Saklani-Jusforgues, H., Fontan, E., and Goossens, P.L., Effect of acid-adaptation on *Listeria monocytogenes* survival and translocation in a murine intragastric infection model, *FEMS Microbiol. Lett.*, 193, 155, 2000.
33. O'Driscoll, B., Gahan, C.G., and Hill, C., Adaptive acid tolerance response in *Listeria monocytogenes*: Isolation of an acid-tolerant mutant which demonstrates increased virulence, *Appl. Environ. Microbiol.*, 62, 1693, 1996.
34. Cotter, P.D., Gahan, C.G., and Hill, C., A glutamate decarboxylase system protects *Listeria monocytogenes* in gastric fluid, *Mol. Microbiol.*, 40, 465, 2001.
35. Gronroos, J.O., Laine, V.J., and Nevalainen, T.J., Bactericidal group IIA phospholipase A2 in serum of patients with bacterial infections, *J. Infect. Dis.*, 185, 1767, 2002.
36. Singh, P.P. et al., Serum amyloid P-component-induced enhancement of macrophage listericidal activity, *Infect. Immun.*, 52, 688, 1986.
37. Drevets, D.A., and Campbell, P.A., Roles of complement and complement receptor type 3 in phagocytosis of *Listeria monocytogenes* by inflammatory mouse peritoneal macrophages, *Infect. Immun.*, 59, 2645, 1991.
38. Croize, J. et al., Activation of the human complement alternative pathway by *Listeria monocytogenes*: Evidence for direct binding and proteolysis of the C3 component on bacteria, *Infect. Immun.*, 61, 5134, 1993.
39. Drevets, D.A., Leenen, P.J., and Campbell, P.A., Complement receptor type 3 (CD11b/CD18) involvement is essential for killing of *Listeria monocytogenes* by mouse macrophages, *J. Immunol.*, 151, 5431, 1993.
40. Alford, C.E., King, T.E., Jr., and Campbell, P.A., Role of transferrin, transferrin receptors, and iron in macrophage listericidal activity, *J. Exp. Med.*, 174, 459, 1991.
41. Cowart, R.E., and Foster, B.G., Differential effects of iron on the growth of *Listeria monocytogenes*: Minimum requirements and mechanism of acquisition, *J. Infect. Dis.*, 151, 721, 1985.
42. Ignarro, L.J., Nitric oxide as a unique signaling molecule in the vascular system: A historical overview, *J. Physiol. Pharmacol.*, 53, 503, 2002.
43. Serbina, N.V. et al., TNF/iNOS-producing dendritic cells mediate innate immune defense against bacterial infection, *Immunity*, 19, 59, 2003.
44. Beckerman, K.P. et al., Release of nitric oxide during the T cell-independent pathway of macrophage activation. Its role in resistance to *Listeria monocytogenes*, *J. Immunol.*, 150, 888, 1993.
45. Boockvar, K.S. et al., Nitric oxide produced during murine listeriosis is protective, *Infect. Immun.*, 62, 1089, 1994.
46. Shiloh, M.U. et al., Phenotype of mice and macrophages deficient in both phagocyte oxidase and inducible nitric oxide synthase, *Immunity*, 10, 29, 1999.
47. Bedard, K., and Krause, K.H., The NOX family of ROS-generating NADPH oxidases: Physiology and pathophysiology, *Physiol. Rev.*, 87, 245, 2007.
48. Endres, R. et al., Listeriosis in p47(phox-/-) and TRp55-/- mice: Protection despite absence of ROI and susceptibility despite presence of RNI, *Immunity*, 7, 419, 1997.
49. del Cerro-Vadillo, E. et al., Cutting edge: A novel nonoxidative phagosomal mechanism exerted by cathepsin-D controls *Listeria monocytogenes* intracellular growth, *J. Immunol.*, 176, 1321, 2006.
50. Macpherson, A.J. et al., A primitive T cell-independent mechanism of intestinal mucosal IgA responses to commensal bacteria, *Science*, 288, 2222, 2000.
51. Cash, H.L. et al., Symbiotic bacteria direct expression of an intestinal bactericidal lectin, *Science*, 313, 1126, 2006.

52. Dommett, R. et al., Innate immune defense in the human gastrointestinal tract, *Mol. Immunol.*, 42, 903, 2005.
53. Johnson, L.N. et al., Protein-oligosaccharide interactions: Lysozyme, phosphorylase, amylases, *Curr. Top. Microbiol. Immunol.*, 139, 81, 1988.
54. Wehkamp, J., Schauber, J., and Stange, E.F., Defensins and cathelicidins in gastrointestinal infections, *Curr. Opin. Gastroenterol.*, 23, 32, 2007.
55. Ayabe, T. et al., Secretion of microbicidal alpha-defensins by intestinal Paneth cells in response to bacteria, *Nat. Immunol.*, 1, 113, 2000.
56. Cole, A.M. et al., Cutting edge: IFN-inducible ELR- CXC chemokines display defensin-like antimicrobial activity, *J. Immunol.*, 167, 623, 2001.
57. Hiemstra, P.S. et al., Ubiquicidin, a novel murine microbicidal protein present in the cytosolic fraction of macrophages, *J. Leukoc. Biol.*, 66, 423, 1999.
58. Utermohlen, O. et al., Severe impairment in early host defense against *Listeria monocytogenes* in mice deficient in acid sphingomyelinase, *J. Immunol.*, 170, 2621, 2003.
59. Conlan, J.W., and North, R.J., Neutrophils are essential for early anti-*Listeria* defense in the liver, but not in the spleen or peritoneal cavity, as revealed by a granulocyte-depleting monoclonal antibody, *J. Exp. Med.*, 179, 259, 1994.
60. Czuprynski, C.J. et al., Administration of antigranulocyte mAb RB6-8C5 impairs the resistance of mice to *Listeria monocytogenes* infection, *J. Immunol.*, 152, 1836, 1994.
61. Conlan, J.W., Critical roles of neutrophils in host defense against experimental systemic infections of mice by *Listeria monocytogenes*, *Salmonella typhimurium*, and *Yersinia enterocolitica*, *Infect. Immun.*, 65, 630, 1997.
62. Drevets, D.A., Dissemination of *Listeria monocytogenes* by infected phagocytes, *Infect. Immun.*, 67, 3512, 1999.
63. Kuhn, M., and Goebel, W., Induction of cytokines in phagocytic mammalian cells infected with virulent and avirulent *Listeria* strains, *Infect. Immun.*, 62, 348, 1994.
64. Demuth, A. et al., Differential regulation of cytokine and cytokine receptor mRNA expression upon infection of bone marrow-derived macrophages with *Listeria monocytogenes*, *Infect. Immun.*, 64, 3475, 1996.
65. Tripp, C.S., Wolf, S.F., and Unanue, E.R., Interleukin 12 and tumor necrosis factor alpha are costimulators of interferon gamma production by natural killer cells in severe combined immunodeficiency mice with listeriosis, and interleukin 10 is a physiologic antagonist, *Proc. Natl. Acad. Sci. USA*, 90, 3725, 1993.
66. Serbina, N.V. et al., Sequential MyD88-independent and -dependent activation of innate immune responses to intracellular bacterial infection, *Immunity*, 19, 891, 2003.
67. Pinto, A.J. et al., Selective depletion of liver and splenic macrophages using liposomes encapsulating the drug dichloromethylene diphosphonate: Effects on antimicrobial resistance, *J. Leukoc. Biol.*, 49, 579, 1991.
68. Myers, J.T., Tsang, A.W., and Swanson, J.A., Localized reactive oxygen and nitrogen intermediates inhibit escape of *Listeria monocytogenes* from vacuoles in activated macrophages, *J. Immunol.*, 171, 5447, 2003.
69. Joseph, S.B. et al., LXR-dependent gene expression is important for macrophage survival and the innate immune response, *Cell*, 119, 299, 2004.
70. Serbina, N.V., and Pamer, E.G., Monocyte emigration from bone marrow during bacterial infection requires signals mediated by chemokine receptor CCR2, *Nat. Immunol.*, 7, 311, 2006.
71. Berg, R.E. et al., Memory $CD8^+$ T cells provide innate immune protection against *Listeria monocytogenes* in the absence of cognate antigen, *J. Exp. Med.*, 198, 1583, 2003.
72. Lertmemongkolchai, G. et al., Bystander activation of $CD8^+$ T cells contributes to the rapid production of IFN-gamma in response to bacterial pathogens, *J. Immunol.*, 166, 1097, 2001.
73. D'Orazio, S.E., Troese, M.J., and Starnbach, M.N., Cytosolic localization of *Listeria monocytogenes* triggers an early IFN-gamma response by $CD8^+$ T cells that correlates with innate resistance to infection, *J. Immunol.*, 177, 7146, 2006.
74. Berg, R.E. et al., Relative contributions of NK and CD8 T cells to IFN-gamma mediated innate immune protection against *Listeria monocytogenes*, *J. Immunol.*, 175, 1751, 2005.
75. Hamerman, J.A., Ogasawara, K., and Lanier, L.L., Cutting edge: Toll-like receptor signaling in macrophages induces ligands for the NKG2D receptor, *J. Immunol.*, 172, 2001, 2004.
76. Dunn, P.L., and North, R.J., Early gamma interferon production by natural killer cells is important in defense against murine listeriosis, *Infect. Immun.*, 59, 2892, 1991.

77. Teixeira, H.C., and Kaufmann, S.H., Role of NK1.1+ cells in experimental listeriosis. NK1+ cells are early IFN-gamma producers but impair resistance to *Listeria monocytogenes* infection, *J. Immunol.*, 152, 1873, 1994.
78. Takada, H. et al., Analysis of the role of natural killer cells in *Listeria monocytogenes* infection: Relation between natural killer cells and T-cell receptor gamma delta T cells in the host defense mechanism at the early stage of infection, *Immunology*, 82, 106, 1994.
79. Ranson, T. et al., Invariant V alpha 14+ NKT cells participate in the early response to enteric *Listeria monocytogenes* infection, *J. Immunol.*, 175, 1137, 2005.
80. Arrunategui-Correa, V. and Kim, H.S., The role of CDld in the immune response against *Listeria* infection, *Cell. Immunol.*, 227, 109, 2004.
81. Havell, E.A., *Listeria monocytogenes*-induced interferon-gamma primes the host for production of tumor necrosis factor and interferon-alpha/beta, *J. Infect. Dis.*, 167, 1364, 1993.
82. Leenen, P.J. et al., TNF-alpha and IFN-gamma stimulate a macrophage precursor cell line to kill *Listeria monocytogenes* in a nitric oxide-independent manner, *J. Immunol.*, 153, 5141, 1994.
83. Wherry, J.C., Schreiber, R.D., and Unanue, E.R., Regulation of gamma interferon production by natural killer cells in SCID mice: Roles of tumor necrosis factor and bacterial stimuli, *Infect. Immun.*, 59, 1709, 1991.
84. Pfeffer, K. et al., Mice deficient for the 55 kDa tumor necrosis factor receptor are resistant to endotoxic shock, yet succumb to *L. monocytogenes* infection, *Cell*, 73, 457, 1993.
85. Rothe, J. et al., Mice lacking the tumor necrosis factor receptor 1 are resistant to TNF-mediated toxicity but highly susceptible to infection by *Listeria monocytogenes*, *Nature*, 364, 798, 1993.
86. Pasparakis, M. et al., Immune and inflammatory responses in TNF alpha-deficient mice: A critical requirement for TNF alpha in the formation of primary B cell follicles, follicular dendritic cell networks and germinal centers, and in the maturation of the humoral immune response, *J. Exp. Med.*, 184, 1397, 1996.
87. Sheehan, K.C. et al., Monoclonal antibodies specific for murine p55 and p75 tumor necrosis factor receptors: Identification of a novel *in vivo* role for p75, *J. Exp. Med.*, 181, 607, 1995.
88. Peschon, J.J. et al., TNF receptor-deficient mice reveal divergent roles for p55 and p75 in several models of inflammation, *J. Immunol.*, 160, 943, 1998.
89. Grivennikov, S.I. et al., Distinct and nonredundant *in vivo* functions of TNF produced by T cells and macrophages/neutrophils: Protective and deleterious effects, *Immunity*, 22, 93, 2005.
90. Ehlers, S. et al., The lymphotoxin beta receptor is critically involved in controlling infections with the intracellular pathogens *Mycobacterium tuberculosis* and *Listeria monocytogenes*, *J. Immunol.*, 170, 5210, 2003.
91. Roach, D.R. et al., Independent protective effects for tumor necrosis factor and lymphotoxin alpha in the host response to *Listeria monocytogenes* infection, *Infect. Immun.*, 73, 4787, 2005.
92. Harty, J.T., and Bevan, M.J., Specific immunity to *Listeria monocytogenes* in the absence of IFN gamma, *Immunity*, 3, 109, 1995.
93. Huang, S. et al., Immune response in mice that lack the interferon-gamma receptor, *Science*, 259, 1742, 1993.
94. Auerbuch, V. et al., Mice lacking the type I interferon receptor are resistant to *Listeria monocytogenes*, *J. Exp. Med.*, 200, 527, 2004.
95. O'Connell, R.M. et al., Type I interferon production enhances susceptibility to *Listeria monocytogenes* infection, *J. Exp. Med.*, 200, 437, 2004.
96. Rogers, H.W. et al., Interleukin 1 participates in the development of anti-*Listeria* responses in normal and SCID mice, *Proc. Natl. Acad. Sci. USA*, 89, 1011, 1992.
97. Havell, E.A. et al., Type I IL-1 receptor blockade exacerbates murine listeriosis, *J. Immunol.*, 148, 1486, 1992.
98. Glaccum, M.B. et al., Phenotypic and functional characterization of mice that lack the type I receptor for IL-1, *J. Immunol.*, 159, 3364, 1997.
99. Labow, M. et al., Absence of IL-1 signaling and reduced inflammatory response in IL-1 type I receptor-deficient mice, *J. Immunol.*, 159, 2452, 1997.
100. Gregory, S.H. et al., IL-6 produced by Kupffer cells induces STAT protein activation in hepatocytes early during the course of systemic *Listeria* infections, *J. Immunol.*, 160, 6056, 1998.
101. Dalrymple, S.A. et al., Interleukin-6-deficient mice are highly susceptible to *Listeria monocytogenes* infection: Correlation with inefficient neutrophilia, *Infect. Immun.*, 63, 2262, 1995.
102. Kopf, M. et al., Impaired immune and acute-phase responses in interleukin-6-deficient mice, *Nature*, 368, 339, 1994.

103. Tripp, C.S. et al., Neutralization of IL-12 decreases resistance to *Listeria* in SCID and C.B-17 mice. Reversal by IFN-gamma, *J. Immunol.*, 152, 1883, 1994.
104. Neighbors, M. et al., A critical role for interleukin 18 in primary and memory effector responses to *Listeria monocytogenes* that extends beyond its effects on interferon gamma production, *J. Exp. Med.*, 194, 343, 2001.
105. Ozoren, N. et al., Distinct roles of TLR2 and the adaptor ASC in IL-1beta/IL-18 secretion in response to *Listeria monocytogenes*, *J. Immunol.*, 176, 4337, 2006.
106. Saleh, M. et al., Enhanced bacterial clearance and sepsis resistance in caspase-12-deficient mice, *Nature*, 440, 1064, 2006.
107. Eckmann, L., Kagnoff, M.F., and Fierer, J., Epithelial cells secrete the chemokine interleukin-8 in response to bacterial entry, *Infect. Immun.*, 61, 4569, 1993.
108. Opitz, B. et al., *Listeria monocytogenes* activated p38 MAPK and induced IL-8 secretion in a nucleotide-binding oligomerization domain 1-dependent manner in endothelial cells, *J. Immunol.*, 176, 484, 2006.
109. Kurihara, T. et al., Defects in macrophage recruitment and host defense in mice lacking the CCR2 chemokine receptor, *J. Exp. Med.*, 186, 1757, 1997.
110. Zhong, M.X. et al., Chemokine receptor 5 is dispensable for innate and adaptive immune responses to *Listeria monocytogenes* infection, *Infect. Immun.*, 72, 1057, 2004.
111. Kaufmann, S.H. et al., Interleukin-4 and listeriosis, *Immunol. Rev.*, 158, 95, 1997.
112. Dai, W.J., Kohler, G., and Brombacher, F., Both innate and acquired immunity to *Listeria monocytogenes* infection are increased in IL-10-deficient mice, *J. Immunol.*, 158, 2259, 1997.
113. Join-Lambert, O.F. et al., *Listeria monocytogenes*-infected bone marrow myeloid cells promote bacterial invasion of the central nervous system, *Cell. Microbiol.*, 7, 167, 2005.
114. Neuenhahn, M. et al., CD8alpha+ dendritic cells are required for efficient entry of *Listeria monocytogenes* into the spleen, *Immunity*, 25, 619, 2006.
115. Marco, A.J. et al., A microbiological, histopathological and immunohistological study of the intragastric inoculation of *Listeria monocytogenes* in mice, *J. Comp. Pathol.*, 107, 1, 1992.
116. Pron, B. et al., Dendritic cells are early cellular targets of *Listeria monocytogenes* after intestinal delivery and are involved in bacterial spread in the host, *Cell. Microbiol.*, 3, 331, 2001.
117. Conlan, J.W., Neutrophils and tumor necrosis factor-alpha are important for controlling early gastrointestinal stages of experimental murine listeriosis, *J. Med. Microbiol.*, 46, 239, 1997.
118. Skamene, E., and Chayasirisobhon, W., Enhanced resistance to *Listeria monocytogenes* in splenectomized mice, *Immunology*, 33, 851, 1977.
119. Chan, C., Kongshavn, P.A., and Skamene, E., Enhanced primary resistance to *Listeria monocytogenes* in T cell-deprived mice, *Immunology*, 32, 529, 1977.
120. Emmerling, P., Finger, H., and Bockemuhl, J., *Listeria monocytogenes* infection in nude mice, *Infect. Immun.*, 12, 437, 1975.
121. Bancroft, G.J. et al., Regulation of macrophage Ia expression in mice with severe combined immunodeficiency: Induction of Ia expression by a T cell-independent mechanism, *J. Immunol.*, 137, 4, 1986.
122. Carrero, J.A., Calderon, B., and Unanue, E.R., Lymphocytes are detrimental during the early innate immune response against *Listeria monocytogenes*, *J. Exp. Med.*, 203, 933, 2006.
123. Conlan, J.W., Early pathogenesis of *Listeria monocytogenes* infection in the mouse spleen, *J. Med. Microbiol.*, 44, 295, 1996.
124. Merrick, J.C. et al., Lymphocyte apoptosis during early phase of *Listeria* infection in mice, *Am. J. Pathol.*, 151, 785, 1997.
125. Jiang, J., Lau, L.L., and Shen, H., Selective depletion of nonspecific T cells during the early stage of immune responses to infection, *J. Immunol.*, 171, 4352, 2003.
126. Bhardwaj, V. et al., Chronic *Listeria* infection in SCID mice: Requirements for the carrier state and the dual role of T cells in transferring protection or suppression, *J. Immunol.*, 160, 376, 1998.
127. Gregory, S.H., Sagnimeni, A.J., and Wing, E.J., Bacteria in the bloodstream are trapped in the liver and killed by immigrating neutrophils, *J. Immunol.*, 157, 2514, 1996.
128. Gregory, S.H. et al., Complementary adhesion molecules promote neutrophil–Kupffer cell interaction and the elimination of bacteria taken up by the liver, *J. Immunol.*, 168, 308, 2002.
129. Miyamoto, M. et al., Neutrophilia in LFA-1-deficient mice confers resistance to listeriosis: Possible contribution of granulocyte-colony-stimulating factor and IL-17, *J. Immunol.*, 170, 5228, 2003.
130. Altimira, J. et al., Repeated oral dosing with *Listeria monocytogenes* in mice as a model of central nervous system listeriosis in man, *J. Comp. Pathol.*, 121, 117, 1999.

131. Schlüter, D. et al., Intracerebral targets and immunomodulation of murine *Listeria monocytogenes* meningoencephalitis, *J. Neuropathol. Exp. Neurol.*, 55, 14, 1996.
132. Lopez, S. et al., Critical role of neutrophils in eliminating *Listeria monocytogenes* from the central nervous system during experimental murine listeriosis, *Infect. Immun.*, 68, 4789, 2000.
133. Virna, S. et al., TNF is important for pathogen control and limits brain damage in murine cerebral listeriosis, *J. Immunol.*, 177, 3972, 2006.
134. Jung, S. et al., *In vivo* depletion of CD11c(+) dendritic cells abrogates priming of CD8(+) T cells by exogenous cell-associated antigens, *Immunity*, 17, 211, 2002.
135. Zheng, S.J. et al., Reduced apoptosis and ameliorated listeriosis in TRAIL-null mice, *J. Immunol.*, 173, 5652, 2004.
136. Carrero, J.A., Calderon, B., and Unanue, E.R., Type I interferon sensitizes lymphocytes to apoptosis and reduces resistance to *Listeria* infection, *J. Exp. Med.*, 200, 535, 2004.
137. Oxenius, A. et al., IL-12 is not required for induction of type 1 cytokine responses in viral infections, *J. Immunol.*, 162, 965, 1999.
138. Mueller, N.J., Wilkinson, R.A., and Fishman, J.A., *Listeria monocytogenes* infection in caspase-11-deficient mice, *Infect. Immun.*, 70, 2657, 2002.
139. Thomas, J.A. et al., Impaired cytokine signaling in mice lacking the IL-1 receptor-associated kinase, *J. Immunol.*, 163, 978, 1999.

14 Adaptive Immunity

Masao Mitsuyama

CONTENTS

14.1 INTRODUCTION

Until the end of the 1960s, humoral immunity had been thought to be the primary mechanism of immune defense against bacterial infections by means of complement-dependent bacterial lysis and the opsonizing activity of antibodies. The alternative, nonhumoral type of acquired resistance mediated by cellular components was observed only in tuberculosis and brucellosis. However, due to technical difficulties in animal experimentation using *Mycobacterium tuberculosis* and the susceptibility of virulent strains of *Brucella* to both opsonization and lysis by complement in the presence of specific antibodies, the precise mechanism of cell-mediated acquired immunity had not been characterized.

Listeria monocytogenes was first identified by Murray et al. and reported under the name of *Bacterium monocytogenes* in 1926.[1] Osebold and Sawyer were probably the first to introduce experimental infection in mice with this bacterium. They showed that the spleen and liver were the primary target organs in the infected mice and the passive transfer of anti-*Listeria* antisera from goat or rabbit was ineffective. However, active immunization with a sublethal dose of viable bacteria provided a high level of protective immunity in response to the challenge infection.[2] At that time, the concept of cell-mediated protective immunity was demonstrated by Mackaness using an elegant series of experiments based on both *in vivo* and *in vitro* infections with *L. monocytogenes*. In his first and the most significant paper published in 1962, he observed that macrophages (mononuclear phagocytes) were responsible for the acquired protection and that this acquired immunity persisted as long as the infected host was hypersensitive.[3] Since then, this Gram-positive bacterium of a relatively low impact as human pathogen has been employed extensively for the study of cell-mediated acquired immunity in rodents and other animal models all over the world.

Being a facultative intracellular bacterium, *L. monocytogenes* spends the bulk of its time within a host's intracellular environment during its infection cycle. After attachment and entry into a host's phagocytic cells passively through phagocytosis and nonphagocytic cells actively through actions of internalins, *L. monocytogenes* soon moves into the vacuole, where it stays for a brief period. It then escapes to the cytosol and undergoes a rapid multiplication. Bacterial spread to neighboring cells is achieved via direct cell-to-cell contact. By undertaking its infection processes intracellularly, *L. monocytogenes* avoids detection and recognition by the host's humoral immunity. Thus, the host's innate and specific immunity to listeriosis is dominated by cell-mediated responses.

Innate immunity represents an early and nonspecific host response to restrict *L. monocytogenes* expansion at an early stage of infection, and it is exemplified by actions of natural killer (NK) cells (see chapter 13). NK cells, together with dendritic cell (DC)-primed CD4 T helper cells that develop at the later stage of infection, are potent producers of gamma interferon (IFN-γ). Among a number of cytokines produced during the initial stage of infection, IFN-γ is highly potent in activating antimicrobial macrophages. Despite its rapid response upon the encounter with initial *L. monocytogenes* invasion, innate immunity does not usually result in a total decimation of the bacterium, and the development of antigen-specific protective immunity is required.

Specific immunity (also known as acquired or adaptive immunity) is crucial for complete elimination of *L. monocytogenes* and for buildup of resistance to subsequent infection. It is characterized by expansion of antigen-specific, cytotoxic $CD8^+$ T cells via major histocompatibility complex (MHC) class I processing and presentation passage, further expression of IFN-γ by NK and DC-primed $CD4^+$ T cells, and up-regulation of molecules in antigen presentation and activation of antimicrobial macrophages. Importantly, the formation of a strong and enduring $CD8^+$ T cell response against *L. monocytogenes* and other intracellular pathogens is dependent on the generation of a robust and durable $CD4^+$ T cell response, which is principally primed by DCs. Thus, early production of interleukin-12 (IL-12) by macrophages and of IFN-γ by NK cells promotes the development of specific immunity that begins with activation of $CD4^+$ T cells, followed by expansion of $CD8^+$ T cells. The host T cell response reaches its peak by 4–5 days after primary infection. As bacterial antigens are cleared, the T cells contract and differentiate into memory type $CD8^+$ and $CD4^+$ T cells that can last for several weeks in mice.

In this chapter, we first review key historical events that led to the unraveling of adaptive immunity to *L. monocytogenes* in mice. Then we examine in detail the roles of $CD8^+$ and $CD4^+$ T cells in the initiation and maintenance of *L. monocytogenes*-specific adaptive immunity. Finally, we discuss the aspects of adaptive immunity that require additional research in order to fully understand the host defense mechanisms against the *L. monocytogenes* pathogen.

14.2 ADAPTIVE IMMUNITY AS AN EVOLVING CONCEPT

Using a mouse infection model, Mackaness clearly demonstrated the cell-mediated nature of the immune response to infection with *L. monocytogenes*.[3] The most critical observation in this study was that functional changes in macrophages were responsible for the acquired protection observed in convalescent mice, which is consistent with the current model for macrophage activation. Subsequently, Mackaness and his colleagues showed that lymphocytes, not humoral antibodies, were essential for the expression of protective immunity by using passive transfer of spleen cells from infected mice[4] and by observing the effect of antilymphocyte globulin.[5] Based on the antigen specificity in comparative transfer experiments employing the spleen cells from mice immunized with either BCG or *L. monocytogenes*, Mackaness proposed that the immune lymphocytes expressed the antigen specificity, but the antibacterial activity of macrophages activated by lymphocytes was nonspecific.[6] Although Mackaness and his colleagues were unable to pinpoint the molecular mechanism of adaptive immunity at that time, they postulated that some form of specific interaction between the immune lymphocytes and the infecting organism resulted in the nonspecific macrophage activation.

Using the passive transfer of thoracic duct cells from infected rats that had been injected with ^{3}H-thymidine, Mackaness's group observed that the effector cells in the passive protection were small lymphocytes with a short life span. They proposed that this type of effector cell played a crucial role in the activation of the host macrophages.[7] With further characterization of thymus-dependent lymphocytes in the 1970s, researchers were able to demonstrate that the cell-mediated protective immunity was thymus dependent; in other words, T cells were the effector lymphocytes that mediated acquired immunity.[8]

Several subsequent experiments suggested a close relationship between delayed-type hypersensitivity and the acquired cellular resistance in animals infected with facultative intracellular parasitic bacteria, including *L. monocytogenes, Brucella abortus,* and *Salmonella typhimurium.*[9] This led to the biochemical characterization of the host factor from T cells mediating macrophage activation and delayed-type hypersensitivity. One of the T cell–derived factors studied extensively was macrophage migration inhibitory factor (MIF). MIF was identified as a lymphocyte-derived factor that was produced in an antigen-specific manner and that suppressed the *in vitro* migration of macrophages. It was also a reliable marker of the antigen-specific T cell response for an extended period of time.[10,11] The precise identity of MIF and macrophage activating factor (MAF) was unclear for many years, but this issue was resolved by the development of monoclonal antibody (mAb)-producing hybridomas. Schreiber et al. established a hybridoma that produced mAb against IFN-γ and demonstrated that macrophage activation by T cell–derived activating factors was completely neutralized by that antibody.[12] Since then, IFN-γ has been recognized as the major T cell–derived factor that activates macrophages and contributes to the resistance to intracellular parasitic bacteria. Several years later, the human MIF gene was cloned,[13] and MIF is now believed to play important roles not only in protective immunity but also in inflammatory and autoimmune diseases.[14]

Regarding the T cells that mediate macrophage activation, a subset of cells expressing the Lyt1 (L3T4 later) marker was identified using negative selection with antibody and complement, which are identical to those known now as $CD4^+$ T cells. Production of cytokines was observed in *Listeria*-specific Lyt1+ T cells after antigen stimulation *in vitro*,[15] and these helper T cells were believed to be the major effector T cells for both acquired protective immunity and delayed-type hypersensitivity. In 1982, Kaufmann and Hahn established a number of *Listeria*-specific T cell lines using T lymphocytes from the peritoneal exudate cells of mice immunized with heat-killed *L. monocytogenes.* Each of the T cell lines propagated from colonies in a double-layer soft agar showed an antigen-specific proliferation, helper activity for B cell maturation, and adoptive transfer of both delayed-type hypersensitivity and passive protection against a challenge infection.[16] A representative clone clearly exhibited an MHC class II restriction in the proliferative response and interleukin production, consistent with previous findings by Zinkernagel's group.[17] It was concluded that the diverse biological functions of acquired antilisterial immunity could be mediated by a single helper T cell population.

Until 1985, it was assumed that the protective immunity against intracellular bacteria including *L. monocytogenes* was mediated by antigen-specific, $L3T4^+$, $Lyt\text{-}2^-$ (currently, $CD4^+$, CD8-, respectively), MHC class II-restricted helper/inducer T cells that secrete multiple lymphokines (cytokines) that induce macrophage activation. This model was contradicted by a study that employed IL-2 and accessory cells pulsed with viable (not killed) *L. monocytogenes*, in an attempt to establish *Listeria*-specific $Lyt2^+$ (currently $CD8^+$) T cell clones of the cytotoxic T lymphocyte (CTL) type. A majority of the established CTL clones were able to lyse the peritoneal exudate macrophages infected with *L. monocytogenes* when determined by ^{51}Cr-release assay, and some clones produced IFN-γ in an MHC class I–restricted manner.[18] Later, the same group reported the isolation of *Listeria*-specific but non-MHC-restricted T cell clones that were cytolytic *in vitro* against infected target cells and protective upon passive transfer *in vivo.*[19] Both helper type $CD4^+$ and CTL type $CD8^+$ T cells were therefore identified as mediators of the adaptive protective immunity to *L. monocytogenes.*

The importance of MHC class I–restricted $CD8^+$ T cells was also demonstrated by several experiments using *in vivo* administration of antibodies specific for each T cell subset. Mice treated with purified anti-Lyt-2 mAb displayed a delayed elimination of *L. monocytogenes* in the primary infection from their spleens. Treatment with anti-L3T4 mAb, alone or in combination with anti-Lyt-2 mAb, resulted in a similar increase in the numbers of bacteria recovered from the spleens on day 7 after the challenge. Treatment with anti-L3T4 mAb before immunization did not prevent mice from developing an increased anti-*Listeria* resistance if they were then immunized with a sublethal dose of *L. monocytogenes.* Treatment of mice with anti-Lyt-2 mAb or anti-L3T4 mAb before immunization, however, reduced the ability of their spleen cells to transfer anti-*Listeria* resistance to recipient mice.[20] This study indicated that Lyt-2$^+$ cells make substantial contributions to the resistance of mice to primary *L. monocytogenes* infection and to the ability of spleen cells from *Listeria*-immune mice to transfer resistance to naïve recipients.

In the late 1980s, the development of transgenic mice allowed the production of β2-microglobulin-deficient mice. Because β2-microglobulin is an essential component of the intact major histocompatibility complex (MHC) class I molecule, these mice completely lacked the functional MHC class I–restricted CD8$^+$ T cells.[21] The mice were immunized with *L. monocytogenes,* and a one-spleen equivalent of T cells was adoptively transferred to naïve mice. The recipient mice were then challenged with a lethal dose. No protection was afforded by transfer of CD4$^+$ or CD8$^+$ cells from immune knockout mice, in comparison to more than 4 log units of protection provided by transfer of cells from the immune wild-type mice.[22]

MHC is a gene-dense region in the mammalian genome that is vital to the normal functionality of the immune system, as the proteins encoded by the MHC are capable of displaying the fragments of molecules (antigenic epitopes) from invading microbes to T cells for specific recognition. The MHC complex can be further separated into three subgroups: MHC class I, MHC class II, and MHC class III. The MHC classes I and II encode heterodimeric peptide-binding proteins, whereas the MHC class III region encodes other immune components (e.g., complement components and cytokines including tumor necrosis factor-alpha, or TNF-α). As heterodimers, MHC class I molecules comprise a single transmembrane polypeptide chain (the α-chain) and a β_2 microglobulin (which is encoded elsewhere, not in the MHC), whereas MHC class II molecules contain two homologous peptides, an α and β chain, both of which are encoded in the MHC. MHC class I is present in almost every nucleated cell of the body, and it can be further divided into classical MHC class Ia (K, D, L) and nonclassical MHC class Ib (Q, T, M) molecules. If a host cell is infected by an intracellularly parasitic microbe such as *L. monocytogenes*, it may display the antigenic epitope on its surface in the context of class I MHC molecules. MHC class II molecules are found only on a few specialized cell types that form so-called professional antigen-presenting cells (APCs) (e.g., macrophages, dendritic cells, and activated B cells). The antigenic epitopes presented by class II molecules are derived from extracellular proteins (not cytosolic as in class I). Class II molecules interact exclusively with CD4$^+$ ("helper") T cells.

Based on the results from these representative studies, the essential contribution of CD8$^+$ T cells to acquired protective immunity was accepted. The CD8$^+$ T cells could thus play several potentially important roles in this process, including (1) contribution to the production of IFN-γ by themselves, (2) reduction of bacterial multiplication by eliminating the host cells refractory to further activation, and (3) facilitation of intracellular killing by an altered internal mechanism triggered by CTL.[23] Based on current knowledge, both possibilities 1 and 2 are now generally accepted, but possibility 3 may need further study.

Whereas copious data confirm the role of the cell-mediated response in the adaptive immunity to *L. monocytogenes,* several recent reports indicate that antibody response may also contribute partially to the control of listeriosis. Specifically, although serum antibodies produced by the infected hosts provide no protection against *L. monocytogenes,* murine mAb to listeriolysin O (LLO) does help lower bacterial titers in the spleen and liver and increase animal survival after lethal challenge. LLO is a pore-forming protein toxin that demonstrates multiple functions under *in vitro* conditions. Apart from having the capacity to cause apoptosis, it can also activate MAP kinase and NF-κB, induce phosphatidylinositol metabolism and cytokines, and activate neutrophils. However, the significance of these properties *in vivo* remains to be elucidated.

14.3 CD8⁺ PROTECTIVE T CELLS

During their development in the thymus, T lymphocytes evolve into either a $CD4^+$ or $CD8^+$ subset, depending on whether their T cell receptors (TcRs) recognize an MHC class I presented antigen (CD8) or an MHC class II presented antigen (CD4). $CD8^+$ T cells belong to a subset of T lymphocytes that are capable of killing pathogen-infected cells, and hence they are also referred to as cytotoxic T cells (CTL) or killer T cells. $CD8^+$ T cells often express TcRs that recognize a specific peptide bound to Class I MHC molecules, and the affinity between CD8 and the MHC molecule facilitates the binding of the CTL to the target cells during antigen-specific activation. The activation of cytotoxic T cells relies on interactions between molecules expressed on the surface of the T cells and molecules on the surface of the antigen presenting cells (APCs). Once activated, $CD8^+$ T cells are considered to have a predefined cytotoxic role within the immune system. Upon exposure to infected somatic cells, $CD8^+$ T cells release cytotoxins perforin and granulysin that form pores in the target cell membrane, causing ions and water to flow into the target cell that expand and lyse eventually.

14.3.1 INDUCTION OF *LISTERIA*-SPECIFIC CD8⁺ T CELL RESPONSE

The kinetics of bacterial growth in the primary infection of mice with *L. monocytogenes* show three distinct phases, consisting of initial decline in the liver during the first 6 h of infection, continuous multiplication for 3–4 days, and then a decline phase leading to complete elimination from the liver and spleen (Figure 14.1). The convalescent mice become extremely resistant to a subsequent challenge infection.[24] As discussed previously, $CD8^+$ T cells play a critical role in the antigen-specific acquired protective immunity in mice immunized with a nonlethal dose of viable *L. monocytogenes*. The reason why *L. monocytogenes* so efficiently induces class I MHC-restricted $CD8^+$ T cells is ascribed to the intracellular fate of bacteria inside host cells, especially the cells capable of anti-

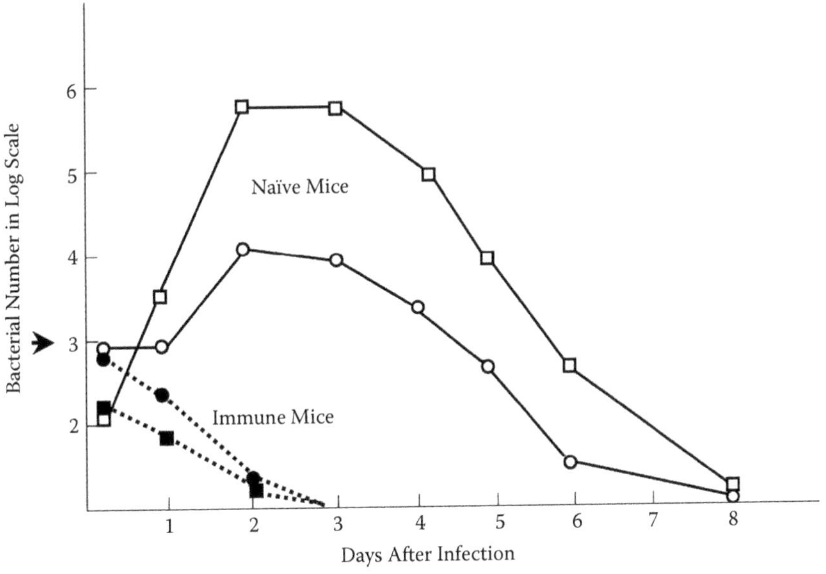

FIGURE 14.1 The kinetics of *L. monocytogenes* growth in naïve and immune mice. Each symbol indicates Log CFU in the spleen (open and closed square, i.e., □ and ■) and liver (open and closed circle, i.e., ○ and ●) of naïve mice (—) and mice immunized with a non-lethal dose of *L. monocytogenes* 10 days before the challenge.

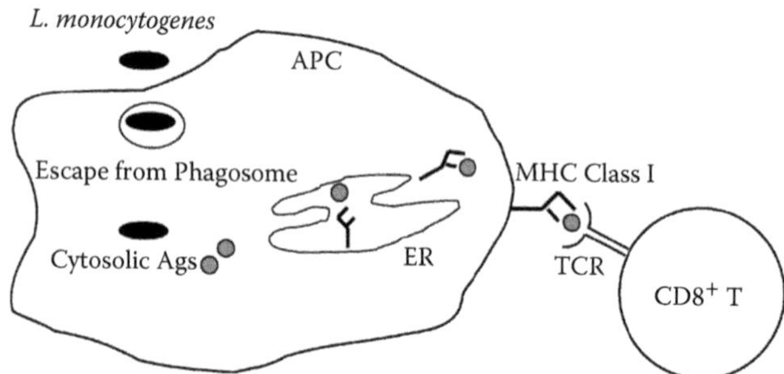

FIGURE 14.2 Priming of CD8+ T cells by *L. monocytogenes* antigens in the context of class I MHC. Cytosolic antigens produced after bacterial escape from the phagosome are degraded by proteasome into antigenic epitopes and then presented by MHC class I molecules in ER.

gen presentation. Among various intracellular parasitic bacteria, *L. monocytogenes* is representative of those that escape from the phagosome into the cytosolic space, by means of LLO encoded by *hly* located in the major pathogenicity island (which was previously known as the PrfA-regulated virulence gene cluster and later renamed as LIPI-1).[25] In the cytosolic space, this bacterium can multiply and disseminate to the neighboring cells with the help of several factors encoded by LIPI-1.[26] Bacterial products released in the cytosolic space are processed and presented in the context of MHC class I molecule and prime specific CD8+ T cells (Figure 14.2).

In natural infection as food-borne listeriosis, *L. monocytogenes* first enters the intestinal cells with assistance from a surface protein InlA (internalin A) that is expressed on the bacterial surface targeting E-cadherin of the host cells.[27] Nonhematopoietic cells are believed to be incompetent to initiate the MHC class I–restricted T cell response.[28] With an experimental infection via systemic injection of bacterial suspension, however, *L. monocytogenes* mainly invades mononuclear phagocytic cells in the liver and spleen. Among mononuclear phagocytes, dendritic cells (DCs) are highly effective in the priming of CD8+ T cells for initiating the cytotoxic T lymphocyte (CTL) response. Mice carrying a transgene encoding a diphtheria toxin (DT)-sensitive receptor were established so that the expression of DT receptor (DTR) would be under the control of the promoter of murine CD11c that is exclusively expressed in DC compartment. A single injection of DT into these transgenic mice resulted in a rapid depletion of CD11c+ DCs persisting for 2 days. In the DT-treated DTR transgenic mice, the *in vivo* proliferative response of CTL precursors was completely eliminated.[29] Maturation of DCs was induced by IFN-γ produced after an intracellular sensing of cytosolic *L. monocytogenes*.[30]

A novel type of DC has been recently identified as Tip-DCs, which are TNF-α and inducible nitric oxide synthase (iNOS)-producing DCs. Tip-DCs are recruited to the spleen of mice infected with *L. monocytogenes* in a CCR2-dependent manner. CCR2 is a murine chemokine receptor that mediates *in vitro* response to a monocyte chemoattractant protein JE, which has been shown to be important for monocyte/macrophage recruitment *in vivo*. Therefore, CCR2 acts as a key mediator for the host's inflammatory and immune responses. In CCR2-deficient mice, the recruitment of Tip-DCs was completely absent and uncontrolled, and a lethal bacterial replication was observed.[31] Therefore, Tip-DCs play an essential role during the acute phase of infection. However, it appears that Tip-DCs are not required for mounting an effective CD8+ T cell response, since there was no significant level of reduction in LLO-specific CD8+ T cell as well as in specific CD4+ T cell response in CCR2-deficient mice.[31]

MHC class I molecules are critically important in CD8+ T cell response. *Listeria*-specific T cells restricted by both classical MHC class Ia (K, D, L) and nonclassical MHC class Ib (Q, T, M) molecules are activated during infection. In an experiment using MHC class Ia–deficient mice that showed a severe reduction in the number of circulating CD8+ T cells, a strong protective immunity reaction was induced by immunization with a sublethal dose of *L. monocytogenes*, and a passive transfer revealed the generation of protective CD8+ T cells exhibiting IFN–γ production and cytotoxicity to infected cells. Even a limited number of MHC class Ib–restricted T cells were sufficient to generate the rapid recall response required for protection against the secondary infection.[32] MHC class Ib response is well characterized in H2-M3-restricted $CD8^+$ T cells. Compared to the MHC class Ia–restricted T cell response that expands later during the primary infection and forms a long-lasting memory T cell population, the H2-M3-restricted T cell response occurs earlier, but the memory response upon re-infection is limited.[33] Although H2-M3-restricted T cells and classical MHC class Ia–restricted T cells are indistinguishable by the expression of cell surface memory markers and longevity, the different response appears to reflect some intrinsic differences between these T cell populations.[34]

14.3.2 Antigens Recognized by $CD8^+$ T Cells

L. monocytogenes multiplies after evasion from the phagosome and produces a number of proteins, including virulence factors as well as the products of housekeeping genes. Therefore, any antigenic epitopes derived from components or products of bacteria growing inside the cytoplasm may gain access to the MHC class I pathway, leading to the priming of specific $CD8^+$ T cell clones. Among such T cell epitopes, LLO and p60 have been analyzed extensively and identified as the major antigens detected by $CD8^+$ T cells.

LLO is a major virulence-associated protein synthesized as a precursor consisting of 529 amino acid residues, with a typical signal sequence at its N-terminus. After cleavage of the signal sequence, a mature form of LLO is secreted as a monomeric protein. The essential function of this protein is to form multiple membrane pores after binding to cholesterol-containing membrane[35] through domain 4 of LLO,[36] thus enabling the evasion of bacteria from the phagosomal compartment. Since LLO is a secretory protein produced in high levels in culture and not produced by an LLO-negative mutant that is unable to induce protective T cell response *in vivo*, LLO was the first target of interest as a putative T cell antigen. By using a panel of $CD8^+$ T cell clones and synthesized overlapping oligopeptides, two major LLO epitopes, LLO91-99 and LLO215-234, have been identified (Table 14.1). CTL clones derived from H-2^d × H-2^b F1 mice immunized with viable *L. monocytogenes* were tested for their ability to lyse the target cells pulsed with monomer peptides of H-$2K^d$ peptide-binding motif. LLO91–99 was shown to be recognized efficiently by one clone.[37] LLO 215–234 was found to be another dominant CTL epitope recognized in the context of H-$2E^k$ by screening of IL-2 response of immune T cells against a panel of overlapping LLO peptides.[38] Compared with extracellular *L. monocytogenes,* which secretes LLO at a much higher rate, cytosolic *L. monocytogenes* produces LLO at a rate equal to about 2% of that by extracellular bacteria. However, the accumulation of H-$2K^d$-associated LLO 91–99 in the cytosol is quite effective, which suggests the rapid degradation of the LLO molecule and an effective processing into an MHC class I–associated epitope.[39]

As a major extracellular component that is encoded by the *L. monocytogenes iap* gene (for invasion-associated protein), p60 is a 60-kDa protein with murein hydrolase activity. This protein is

TABLE 14.1
$CD8^+$ T Cell Epitopes

Epitope	MHC Restriction	Amino Acid Sequence
LLO 91–99	H-2Kd	GYKDGNEYI
LLO 215–234	H-2Ek	SQLIAKFGTAFKAVNNSLNV
LLO 296–304	H-2Kb	VAYGRQVYL
p60 217–225	H-2Kd	KYGVSVQDI
p60 449–457	H-2Kd	IYVGNGQMI
p60 476–484	H-2Kd	KYLVGFGRV
Mpl 84–92	H-2Kd	GYLTDNDEI

essential for the bacterial division at a later stage and is possibly involved in the invasion of mammalian cells[40] p60 is clearly expressed during *L. monocytogenes* growth in human hosts, as antibodies to p60 are frequently detected in sera of humans with listeriosis. *L. monocytogenes* mutants with impaired synthesis of p60 often display a rough-colony morphology (R mutants) and notable virulence attenuation in mice. Structurally, p60 possesses two highly conserved regions at the N and C termini (with about 100 and 120 amino acids, respectively); its central portion is a highly variable region dominated by the threonine-asparagine repeats. The C-terminal region of p60 shows homology to several hydrolytic enzymes, which may be responsible for the hydrolytic activity that is detected in p60.

Two major H-2K^d-restricted p60 epitopes, p60 217–225[41] and p60 449–457, have been identified. A quantitative analysis revealed the amount of p60 449–457 in infected cells to be approximately 10-fold greater than the amount of p60 217–225.[42] Based on the results obtained by using p60-deletion mutants, p60 protein is believed to be an N-end rule substrate because the N-terminal amino acid and multiple internal regions of p60 influenced its stability in the cytosol of infected cells. Since there is a direct correlation between the degradation rates and epitope production, recognition and destruction of cytosolic protein derived from bacteria by the host cell, proteolytic mechanisms appear to be the limiting step in CTL epitope generation.[43] The importance of p60 217–225 processing in the appropriate T cell response is demonstrated by the absence of p60 217–225-specific T cell responses in mice infected with an *L. monocytogenes* mutant secreting altered p60 with a significantly reduced processing efficiency.[44]

Other CD8$^+$ T cell epitopes so far reported include metalloprotease (Mpl84–92, encoded by *mpl*) that binds H-2K^d with a high affinity,[45] LLO296–304, and p60 476–484.[46] Since Mpl84–92 but not other secreted Mpl, is presented to CTL by the H-2K^d molecule during *L. monocytogenes* infection, it underscores the significance of high-affinity binding between antigenic peptides and class I MHC molecules for CTL priming in the host immune response against invading bacterial pathogens. In addition, ActA, a 639-amino-acid surface protein encoded by *actA* gene, is required for bacterial movement inside host cell cytoplasm, propelled by continuous actin assembly at one pole of the bacterium.[47] Even though ActA-specific CD8$^+$ T cells were generated after immunization with *L. monocytogenes*, CTL against the membrane protein ActA did not protect mice against the challenge infection.[48] The lack of protection by ActA-specific CTL was probably due to a deficient presentation of this membrane-bound protein.

After bacterial evasion into the cytosolic compartment, not only secreted protein antigens but also nonsecreted, somatic antigens may gain access to the MHC class I pathway. There is some evidence that only the secreted proteins are effective in the generation of protective CD8$^+$ T cells, in spite of the fact that secreted proteins account for only 1–2% of total bacterial protein synthesis.[49] However, it is not conclusive that only the secreted antigens are immunogenic for CD8$^+$ T cells, because a detailed subsequent study indicated that both *L. monocytogenes* isogenic strains expressing either nonsecreted LCMV antigen or secreted LCMV antigen were able to induce protective immunity to LCMV infection involving CD8$^+$ T cells only.[50] In any case, the immunodominance of secreted proteins in the induction of protective CD8$^+$ T cells in natural infection with *L. monocytogenes* may therefore have critical implications in the design of novel bacterial vaccines.[51]

14.3.3 Mechanism of Protection by CD8$^+$ T Cells

The precise mechanisms by which *Listeria*-specific CD8$^+$ T cells protect mice against secondary infection are still unclear. However, there are two major possibilities, including the activity of effector cytokines produced by CD8$^+$ T cells and the direct cytolysis of infected cells through CTL activity (Figure 14.3). CD8$^+$ T cells can produce both IFN-γ and TNF-α, which are highly capable of activating the microbicidal activity of macrophages nonspecifically. CTL activity is unique in its effect only on the infected host cells expressing antigenic epitopes in the context of MHC class I molecules. Although CTL is unable to kill *L. monocytogenes* directly, it is plausible that the lysis of

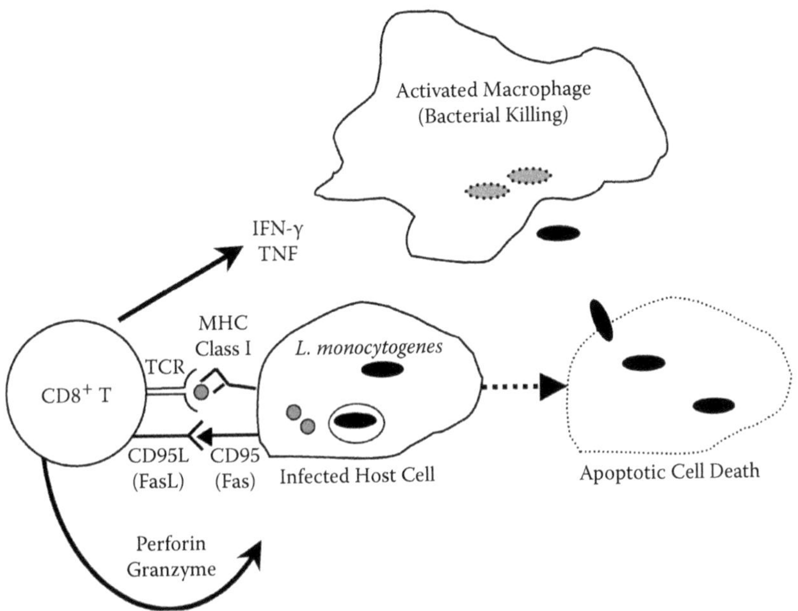

FIGURE 14.3 Possible mechanism of CD8+ T cell-mediated protection against *L. monocytogenes.* Upon the interaction of CD8+ T cell with infected host cell, CD8+ T cell produces IFN-γ and TNF that activate macrophage listericidal activity. T cell-derived perforins form membrane pores through which granzyme enters the cell to cause cell death. Interaction of CD95L and CD95 results in the apoptotic cell death. Bacteria released from the dead host cell may be captured by activated macrophages.

infected host cells by CTLs, particularly when the cells are not of phagocytic lineage, results in the release of growing bacteria so that activated macrophages can capture and kill the target bacteria effectively.

Evidence on the contribution of cytokines to the protection mechanism was obtained by observing the *in vivo* effect of neutralizing antibodies directed against candidate cytokines. Nakane et al. observed an enhanced bacterial multiplication and a high mortality in immunized mice after the secondary infection following simultaneous *in vivo* administration of anti-IFN-γ and anti-TNF antibodies.[52] In an experiment using different antibodies and an adoptive cell transfer system, however, the adoptive protection afforded with transfer of LLO-specific CD8+ T cells was not affected by anti-IFN-γ antibody.[53] The effect of IFN-γ neutralization by injection of antibody-producing cells was also marginal, while TNF neutralization provided a significant level of inhibition of an established protective immunity.[54] The absence of specific dependence on IFN-γ was supported by findings obtained in IFN-γ gene knockout (GKO) mice. Immunization of GKO mice with an attenuated strain of *L. monocytogenes* induced an antigen-specific CD8+ T cell response that could transfer immunity to naïve mice, and immune GKO mice themselves exhibited a highly increased resistance to challenge with a virulent strain, which appeared to be mediated by CD8+ T cells.[55] Specific anti-*Listeria* immunity could be observed by adoptive transfer of CD8+ T cells even in the IFN-γ-depleted recipient.[56] Therefore, in contrast to the established importance of IFN-γ in the primary defense,[57] the contribution of IFN-γ to the acquired immunity mediated by CD8+ T cells may apparently be marginal or nil. Indeed, a more recent report confirmed that, despite having a lower hemolytic activity and generating a smaller (or negligible) amount of IFN-γ than virulent strain EGD (of serotype 1/2a), naturally avirulent strain HCC23 (of serotype 4a) was capable of eliciting a durable protective immunity against *L. monocytogenes* virulent strain challenge (Liu et al., *Arch. Microbiol.* in press).

TNF-α is a pleiotrophic cytokine that mediates a broad range of proinflammatory activities (see chapter 13). Its essential role in host defense against primary infection was documented using

TNF-deficient mice[59] as well as mice deficient for the TNF receptor 1.[60] In contrast, the role of TNF in the mediation of secondary infection in immune animals is not fully understood. In a recent study, immune T cells obtained from mice lacking soluble TNF and expressing only membrane TNF conferred a significant level of protection against a high-dose infection, even in the recipients deficient for TNF.[61] Therefore, the release of soluble TNF may not be essentially required for the expression of acquired protective immunity; however, the precise role remains to be further elucidated.

CD8+ T cells, following ligation of TCR by the complex of specific antigenic epitope and MHC class I molecule, induce their target to undergo programmed cell death. One pathway for this process is dependent on perforin and granzyme. The other is the CD95L (FasL) and CD95 pathway. Perforin is expressed uniquely in CTL granules and is required for the function of the granule exocytosis pathway of cytotoxicity. This 555-amino-acid glycoprotein is inserted into the target cell membrane, resulting in the formation of structural and functional pores. These pores allow granzymes, which are exogenous serine proteases also present in CTL granules, to gain access to the cytoplasm of the target cells and induce activation of the caspase cascade, resulting in cell death. CD95L (FasL) is a ligand expressed on the activated CD8+ T cells and, upon ligation of this protein on target cells, initiates intracellular signaling events that activate the caspase cascade and cause apoptosis of target cells. The contribution of these molecules in the target cell death has been examined extensively, but many contradictory results have been reported.

In mice deficient for either perforin or Fas (CD95), Fas was shown to be more important in the defense against primary infection and double knockout resulted in up to a 3-log increase in bacterial burden. Acquired protection against secondary infection was significantly reduced in perforin/Fas double knockout mice. The lytic activity of immune CD8+ T cells against infected hepatocytes was severely affected in the absence of Fas.[62] The relative absence of perforin involvement in CD8+ T cell-mediate protective immunity was demonstrated, but the immunity provided by perforin-deficient CD8+ T cells was independent of CD95.[63] White et al. analyzed the CD8+ T cells specific for three identified *L. monocytogenes*-derived epitopes in perforin-deficient mice and found that all the CD8+ T cell populations specific for LLO91–99 and p60 217–225 were perforin independent, though the level of involvement appeared to be different in the liver and spleen.[64] Badovinac and Harty also observed that perforin was not required for the development or expression of adaptive immunity mediated by CD8+ T cells specific for immunodominant LLO91–99.[65] Therefore, perforin, which is believed to be critically important for the lysis of tumor target cells, is apparently of insignificant impact in the expression of adaptive immunity by CD8+ CTL against intracellular bacteria. The perforin-mediated cytolysis was highlighted again by a study conducted under the assumption that the lysis of infected cells may play a pivotal role in the inhibition of cell-to-cell spread of *L. monocytogenes*. Perforin-deficient antigen-specific CD8+ T cells did not confer significant protective immunity against a recombinant strain of *L. monocytogenes* capable of cell-to-cell spread, but protection against challenge with the *actA* mutant strain that is defective in cell-to-cell spread was not dependent on perforin.[66]

Jiang et al. constructed two isogenic recombinant *L. monocytogenes* (rLM) strains that secrete a fusion protein with or without an MHC class Ia–restricted CD8 T cell epitope (NP118–126) derived from the nuclear protein of lymphocytic choriomeningitis virus (LCMV). LCMV-immune mice were highly resistant to the challenge with rLM-NP(+) but not with the rLM-NP(–) strain. Against a mixed infection of the two rLM strains, the protection afforded by immunization with LCMV was effective only in the elimination of the rLM-NP(+) strain, thus suggesting the absence of any bystander effect in the antigen-specific CD8+ T cells.[67]

The experimental findings so far may not be sufficient to establish a comprehensive model of the effector mechanism of *L. monocytogenes*-specific CD8+ CTLs. However, it is apparent that the significant level of protective immunity afforded by CD8+ T cells against virulent *L. monocytogenes* is a combination of several processes, including (1) lysis of infected host cells resulting in the interruption of intracellular multiplication or cell-to-cell spread, (2) delivery to activated phagocytes by releasing the intracellular bacteria, and (3) activation of effector macrophages by production of IFN-γ and TNF.

Granulysin is a 9-kDa protein located in the granules of human NK cells and CTLs[68] and is implicated in the killing of bacteria, including intracellular parasitic *Mycobacterium tuberculosis*[69]—probably via its saponin-like activity against the host membrane. There is a recent report showing that recombinant human granulysin exhibited a rapid and direct lethal activity against *L. innocua* in culture in a dose-dependent manner, and mediated the lysis of *L. innocua* inside dendritic cells after binding the lipid raft on the cell surface.[70] While this observation may be interesting, it is questionable whether granulysin is one of the effector molecules of CTL against *L. monocytogenes* in mice, as *L. innocua* is a nonvirulent saprophytic bacterium and granulysin has been studied only in human NK or T cells.

14.3.4 Maintenance of *Listeria*-Specific CD8$^+$ T Cells

The mechanism of maintenance of *L. monocytogenes*-specific CD8$^+$ T cells has been studied extensively. *In vivo* priming of antigen-specific CD8$^+$ T cells results in their proliferation/differentiation into effector CTLs and commitment into the stable memory T cells. Even though CD8$^+$ T cell clones are different in antigenic specificity and bacterial MHC class I–restricted antigens may differ in the stability, they undergo a synchronous *in vivo* expansion. Administration of antibiotics at different times to limit *L. monocytogenes in vivo* growth in experimental mice did not alter the duration and kinetics of CD8$^+$ T cell expansion.[71] In an *in vitro* experiment carried out by the same research group, a transient *in vitro* stimulation for periods as short as 2.5 h was sufficient to program CD8$^+$ T cells to undergo further cycles of division even in the absence of continuous antigenic stimulation. Such an expansion of antigen-specific CD8$^+$ T cells was probably driven by IL-2 and further augmented by IL-7 and IL-15.[72] A similar finding was obtained in a system employing transgenic CD8$^+$ T cells that express a TCR specific for MHC class I–restricted epitopes of LCMV glycoprotein and recombinant *L. monocytogenes* that expresses the specific epitope.[73] The results of a study using varying doses of *actA*-deficient strain for immunization indicated that the magnitude of CD8$^+$ T cell expansion was dependent on the level of antigens, but, again, the onset and kinetics of CD8$^+$ T cell commitment were independent of the dose and duration of antigenic stimulation.[74] This confirmed that parental naïve CD8$^+$ T cells can, once activated by antigenic epitope in the context of the MHC class I molecule, undergo an instructive developmental program that does not cease until the formation of memory CD8$^+$ T cells.

A suppressive effect of CTLs on CD8$^+$ T cell priming has been observed. The effective expansion of adoptively transferred CD8$^+$ naïve T cells was limited to the initial stage of immunization in spite of the continued presence of infection, and was inversely correlated with the development of antigen-specific CTL.[75] It was thus proposed that bacterial growth occurs in cells incapable of promoting CD8$^+$ T cell proliferation, while APC capable of priming naïve T cells are the first to be eliminated by the nascent CTL response. A previous observation that the expansion of LLO91–99-specific CD8$^+$ T cells was enhanced in perforin-deficient mice, suggesting the suppressive effect of perforin on CD8$^+$ T cell expansion,[76] may therefore be consistent with the preceding interpretation (Figure 14.4).

Following the initial phase of proliferation and differentiation, most of the activated CD8$^+$ T cells die by apoptosis during the contraction phase. The few that survive become memory CD8$^+$ T cells and persist for a long period of time.[77] In various experimental systems, several mechanisms have been implicated in the contraction of activated T cells, including activation-induced cell death and activated T cell autonomous cell death. In contrast to IL-2 that appears to be involved in the process of apoptotic cell death, IL-15, which shares many properties of IL-2, is required for the maintenance of memory type of CD8$^+$ T cells. In IL-15 transgenic mice, Yajima et al. observed a significant increase in the number of CD8$^+$ T cells with the memory phenotype at 6 weeks after

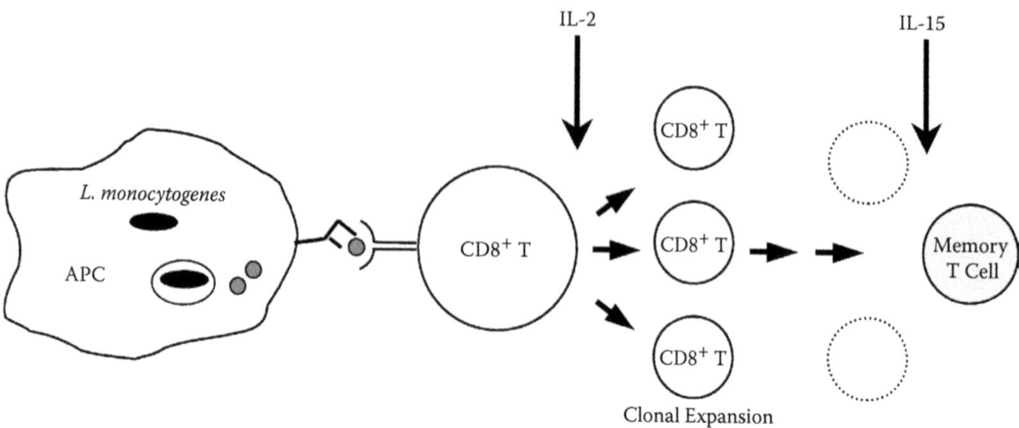

FIGURE 14.4 Emergence of protective CD8+ T cells and generation of memory. CD8+ T cell precursor shows a clonal expansion, but major portion of such activated T cells undergoes apoptosis during the contraction phase. Only a small number of cells survive as memory T cells.

infection with *L. monocytogenes*.[78] They have also found that an up-regulation of Bcl-2 expression, a process that resists cell death, is involved in the IL-15-dependent survival of CD8+ T cells.[79] In addition, IL-7 and IL-7 receptors may be involved in the contraction phase and the maintenance of memory CD8+ T cells specific for *L. monocytogenes,* as observed in response to viral infection.[80]

14.4 CD4+ T CELLS

CD4+ T cells are a subset of T lymphocytes that express surface protein CD4, which has an affinity for class II MHC. CD4+ T cells do not have cytotoxic or phagocytic activity, and they are considered as having a predefined role as helper T cells within the immune system. CD4+ T cells are essential in the activation and growth of cytotoxic T cells, in the maximization of bactericidal activity of phagocytes such as macrophages, and in the determination of B cell antibody class switch through their influence on cytokine secretion. Thus, CD4+ T cells are also referred to as T helper (Th) cells. During their activation and maturation processes, T helper cells release a potent T cell growth factor interleukin-2 (IL-2), which activates the T cell's proliferation pathways (a process called autoregulation). Eventually, T helper cells may evolve into effector Th cells, memory Th cells, and suppressor Th cells after many cell generations after undergoing activation and maturation processes. In turn, effector T cells may develop into Type 1 (Th1) and Type 2 (Th2) helper T cells.

Helper T cells influence other immune cells through generation of extracellular signals such as cytokines, which help determine the outcome of infection. For instance, Th1 helper T cells typically promote the secretion of IFN-γ and TNF-α, maximize the killing efficacy of the macrophages, and in the proliferation of cytotoxic CD8+ T cells. In particular, IFN-γ increases the production of interleukin-12 (IL-12) by dendritic cells and macrophages, and the resulting IL-12 has a positive impact on the production of IFN-γ in helper T cells, thereby strengthening the Th1 profile. In addition, IFN-γ also inhibits the production of cytokines such as IL-4, which is associated with the Th2 response. In doing so, Th1 helper cells act to maintain their own response.

On the other hand, Th2 helper T cells tend to promote the production of cytokines such as IL-4, IL-5, IL-6, IL-10, and IL-13, which help stimulate B-cells into proliferation, induce B-cell antibody class switching, and increase antibody production. In turn, IL-4 acts upon helper T cells to increase the production of the preceding Th2 cytokines including itself, while IL-10 inhibits IL-2 and IFN-γ in helper T cells and IL-12 in dendritic cells and macrophages. As a consequence, the Type 2 response is upheld.

14.4.1 Priming of CD4⁺ T Cells and T Cell Epitopes

CD4$^+$ T cells recognize specific antigens in the context of MHC class II molecules and are highly capable of producing various cytokines. After infection, *L. monocytogenes* stays in the endosomal vacuole until successful escape into the cytosolic space and some bacteria are killed before evading from the phagosome. Therefore, both somatic and secreted antigens may be processed by MHC class II pathway as conventional exogenous antigens by antigen-presenting cells, including macrophages and dendritic cells in which bacteria reside (Figure 14.5). As CD4$^+$ T cells show a robust response to heat-killed bacteria that are neither capable of secreting protein antigens nor capable of escaping from the endosomal compartment, various bacterial antigens other than secreted proteins are likely to be presented to CD4$^+$ T cells. However, further analyses of CD4$^+$ T cell epitopes of *L. monocytogenes* have shown that protein antigens identified as CD8$^+$ T cell antigens are also recognized by CD4$^+$ T cells (Table 14.2). H-2E^k-restricted LLO215–234[38] and H-2A^d-restricted p60 301–302[81] are known as the major epitopes stimulating CD4$^+$ T cells in the context of MHC class II molecule. Another study using a method of direct *ex vivo* epitope mapping of LLO and p60 identified a total of 10 additional CD4 T cell epitopes as well as two new CD8 T cell epitopes.[46]

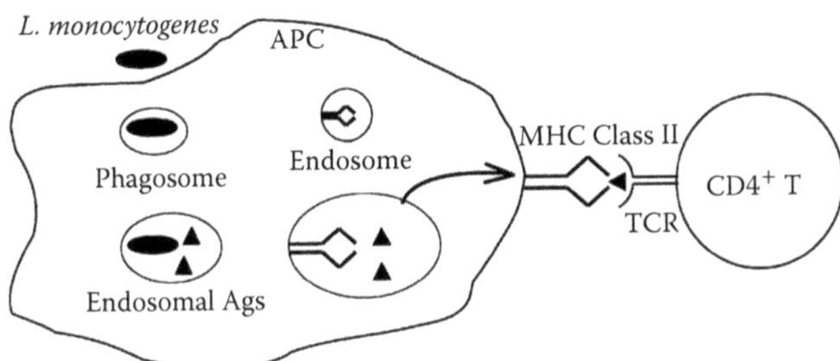

FIGURE 14.5 Priming of CD4$^+$ T cells by *L. monocytogenes* antigens in the context of the MHC class II molecule. Bacterial antigens inside the phagosome are processed as exogenous antigen and, during the endosomal maturation steps, appropriate antigenic epitopes are presented on the membrane surface in the context of MHC class II molecules.

TABLE 14.2
CD4$^+$ T Cell Epitopes

Epitope	MHC Restriction	Amino Acid Sequence
LLO 189–200	H-2b,d	WNEKYAQAYPNV
LLO 190–201	H-2b	NEKYAQAYPNVS
LLO 215–234	H-2Ek	SQLIAKFGTAFKAVNNSLNV
LLO 216–227	H-2d	QLIAKFGTAFKA
LLO 211–222	H-2d	AYSESQLIAKFG
LLO 318–329	H-2b	AFDAAVSGKSVS
p60 177–188	H-2b	TTQATTPAPKVA
p60 300–311	H-2d	TEAAKPAPAPST
p60 301–312	H-2Ad	EAAKPAPAPSTN
p60 367–378	H-2d	SSASAIIAEAQK
p60 401–412	H-2b	KYVFAKAGISLP
p60 418–429	H-2d	QYASTTRISESQ

Cross-presentation (or cross-priming), a system that originally explains MHC class I–dependent response against intact antigens added to the immune system from outside the antigen-presenting cells, may also account for the redundancy of antigen specificity between MHC class I–restricted CD8+ T cells and MHC class II–restricted CD4+ T cells. Cross-presentation denotes the ability of some APCs to take up, process, and present extracellular antigens with class I MHC molecules to CD8 cells, facilitating a host's defense against invading microbes and tumors that do not infect APCs. To achieve this, protein antigens are first transported into cytosol, where they are processed by proteosomes into peptides. These peptides then form complexes with class I MHC molecules, which are detectable CD8+ T cells. Cross-presentation of *L. monocytogenes*-derived CD4+ T cell epitopes is indicated by an *in vitro* observation that, in spite of *L. monocytogenes*-infected macrophages and dendritic cells being poor presenters of CD4+ T cell epitopes, more efficient presentation occurred after cocultivation of noninfected DCs or macrophages with infected cells.[82] It is shown also that, in contrast to cross-presentation of CD4 T cell epitopes, MHC class I–restricted cross-presentation of CD8+ T cell epitopes required viable antigen donor cells and continuous bacterial protein biosynthesis during cross-presentation.[83] However, it is unlikely that such a cross-presentation of nonsecreted protein to CD8+ T cells induces a protective immunity.[84]

The precise analysis of CD4+ T cell epitopes will be feasible only when the sequence of candidate antigen is completely defined. Therefore, the current information on CD4+ T cell epitopes that are rather limited to a few antigenic proteins may not rule out the presence of a panel of antigenic epitopes derived from nonsecreted proteins that await further investigation.

14.4.2 Role of CD4+ T Cells

After intravenous infection, activation and proliferation of CD4+ T cells occur in the spleen in the first few days in a manner synchronous with the expansion of the CD8+ T cell responses. Similar to the expansion and contraction of CD8+ T cells, antibiotic treatment of mice as early as 24 h after infection had minimal impact on the expansion and contraction of CD4+ T cells.[85] Enzyme-linked immunospot (ELISPOT) assay revealed an increase of CD4+, IFN-γ-producing T cells from 10 in 10^6 cells in uninfected mice to 500 in 10^6 cells on day 7 of immunizing infection.[86] Using an MHC class II–presented epitope LLO189–201, the organ-specific CD4+ T cell response during infection was examined. A strong peptide-specific CD4+ T cell response with frequency of 1/100 was observed in the spleen and liver, the major sites of bacterial multiplication. At the peak of secondary response, the frequency reached as high as 1/30 CD4+ spleen cells and a high frequency of epitope-specific CD4+ memory was detected after 5–8 weeks.[87] Upon passive transfer of T cell populations from *Listeria*-immune mice into naïve recipient and subsequent restimulation *in vivo*, transferred T cells represented more than one third of the total CD4+ and CD8+ T cell populations. After stimulation *in vitro*, 40% of these CD4+ T cells responded to heat-killed bacteria with the production of IFN-γ.[88] Thus, despite some experimental results indicating a considerable level of CD4+ T cell expansion and maintenance in mice infected with *L. monocytogenes*, the precise role of CD4+ T cells displaying such a robust expansion during the course of primary and secondary infection is still poorly understood. The mechanism of contraction of *Listeria*-specific CD4+ T cells is not known well, but a recent study has implicated a crucial role of IFN-γ not only in the timing of the contraction but also in the determination of the phenotype of CD4+ T cells.[89]

According to the profile of major cytokines produced, CD4+ T cells are divided into two functionally distinct subsets, Th1 and Th2,[90] although no surface marker that distinguishes these two subsets has yet become available. Th1 cells play an essential role in the immune response to *L. monocytogenes*. Compared to the period when Lyt1+ (CD4+) T cells were believed to be the main and sole effector T cells in adaptive immunity as mentioned before, the general interest in this subset at present appears to be limited. However, the role of CD4+ T cells in the overall host immune response in mice infected with *L. monocytogenes* should not be underestimated. CD4+ T cell help is essentially required for the development of effective CD8+ T cell recall response. Two different but similar experiments clearly indicated that the generation of functional CD8+ T cell memory was severely affected in the absence of CD4+ T cells during the priming with *L. monocytogenes*.[91,92] It was also indicated that once CD8+ T cells were programmed in the presence of CD4+ T cell help, CD4+ T cells were no longer required for the function of CD8+ T cells. IL-2 produced by Th1 type of cells may also play a major role in the propagation of CD8+ T cells as the essential growth factor.

An important functional aspect of Th1 cells is to mediate delayed-type hypersensitivity (DTH). DTH is a T cell–mediated, inflammatory skin reaction that develops 24–72 h after the administration of antigen into the skin of animals immunized with protein antigen plus adjuvant or with intracellular bacteria; it is characterized by a massive infiltration of mononuclear cells at the site of antigen administration. A typical example of infection-induced DTH is the PPD skin reaction in animals immune to *M. tuberculosis. L. monocytogenes* infection also induces a strong DTH reaction as determined by the delayed footpad reaction in mice.[93] DTH is thought to be mediated exclusively by Th1-type CD4+ T cells and to contribute to the granuloma formation.[94] The appearance of CD4+ T cells expressing both TNF-α and IFN-γ or those expressing TNF-α, IFN-γ, and IL-2 was observed in both primary and secondary response to *Listeria* infection.[95] These cytokines appeared to be involved in the T cell–dependent granuloma formation.[96]

The T-box transcription factor (or T-bet), which was originally identified in Th1-type CD4+ T cells, is considered to be essential for the generation of IFN-γ-producing Th1 CD4+ T cells, in addition to its role in the regulation of IFN-γ production in CD8+ T cells, NK cells, and DCs. By stimulating an increased IFN-γ production, T-bet contributes to the suppression of Th2 cytokine production by CD4 T cells and orients the host's immune response toward a Th1 bias. There is evidence that T-bet is more important for sustained secretion of IFN-γ than the induction of IFN-γ-producing CD4 T cells. A recent study indicated that only a modest level of IFN-γ reduction was observed in infection with *L. monocytogenes* in T-bet-deficient mice, which makes a significant contrast to almost a complete loss of this type of cell in the response to *Leishmania major* infection.[97] Apparently, in the *L. monocytogenes* infection model, host resistance and generation of IFN-γ-producing cells are not drastically affected without T-bet. Therefore, the regulatory mechanism in the generation of Th1 cells might be different among the types of intracellular pathogens. Considering that STAT-4 signaling is capable of inducing IFN-γ in CD4 T cells *in vitro* in the absence of T-bet, there is a distinct possibility that STAT-4 signaling in response to IL-12, which is generated in *L. monocytogenes* infection, may permit the induction of IFN-γ from T-bet-deficient CD4 T cells.

A novel subset of CD4+ T cells expressing CD25 is now believed to regulate various immune responses and contribute to autoimmune disease as the regulatory T cells (Tregs).[98] It is most notable that CD25+CD4+ Tregs are largely produced under developmental control by the normal thymus, and persist in the periphery with stable function. Following depletion of CD25+CD4+ Tregs, experimental animals often develop autoimmune disease, display heightened immune responses to autologous tumor cells and invading microbes, and show magnified allergic reactions to otherwise innocuous environmental substances. It is not yet clear whether Tregs are involved in the regulation of immune response to *L. monocytogenes*, but experimental data available suggest the effect of Tregs on CD8+ T cell response. Kursar et al. observed a marked enlargement of the size of antigen-specific CD8+ T cell populations when mice were depleted for CD4+ T cells using anti-CD4 antibody in a secondary response and the CD4-dependent suppression was likely due to CD4+ CD25+ Tregs.[99] The depletion of CD4+ T cells during boost immunization with killed bacteria enhanced the frequencies of LLO91–99-specific CD8+ T cells and the level of bacterial elimination was enhanced.[100] Thus, CD4+CD25+ regulatory T cells may play a potential role in the down-regulation of possibly harmful CD8+ T cells after eradication of this pathogen (Figure 14.6).

14.4.3 Role of LLO in the Development of T1 Cells

As mentioned earlier, the role for LLO, the primary virulence factor of *L. monocytogenes*, in the host's immune response has been analyzed mainly as a specific antigen recognized by CD4+ T and

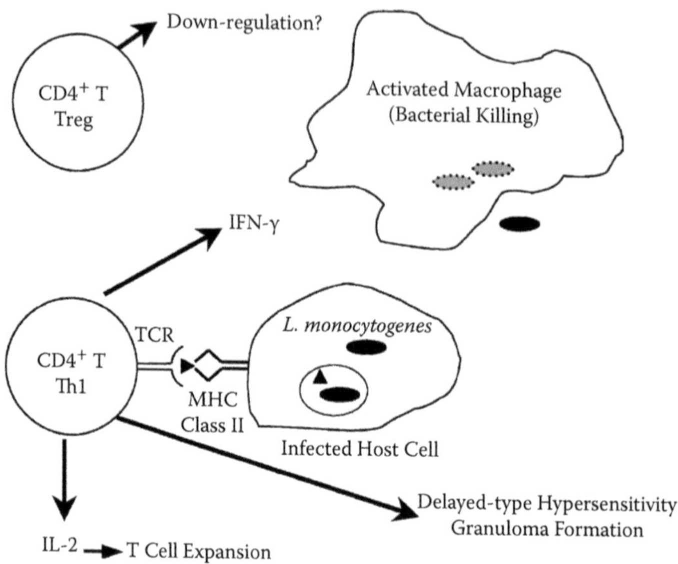

FIGURE 14.6 Possible function of *L. monocytogenes*-specific CD4+ T cells. The major function of CD4+ T cells is production of various cytokines resulting in macrophage activation, delayed-type hypersensitive reaction, and granuloma formation. The possible role of *Listeria*-specific Treg is not clearly indicated.

CD8+ T cell clones. A series of recent studies has identified another role of LLO as a strong T cell adjuvant that supports Th1 development. Infection-induced early IFN-γ production is dependent on NK1.1+ CD11c+ cells,[101] and IFN-γ plays an important part in nonspecific macrophage activation for an enhanced listericidal activity[102] and Th1 development. In an *in vitro* study that employed various recombinant LLO preparations, the N-terminus portion of LLO was shown to display a strong activity to induce IFN-γ production from naïve spleen cells of mice, even in the absence of the C-terminus portion essential for pore formation in the host cell membrane.[36] Induction of IFN-γ *in vitro* by LLO was highly dependent on both IL-12 and IL-18 presumably derived from macrophages or DCs.[103] A similar activity was observed also in seeligeriolysin O (LSO) of *L. seeligeri*[104] via a Toll-like receptor,[105] but such an activity was not observed in ivanolysin O (ILO) of *L. ivanovii*,[106] despite a high sequence homology among LLO, LSO, and ILO. Recombinant LLO was highly effective in skewing the Th2-dependent response induced by ovalbumin,[107,108] suggesting its adjuvant-like activity for Th1 dominance *in vivo*. While suspension of killed *L. monocytogenes* is notoriously ineffective in the induction of T cell–dependent adaptive immune response that can be elicited easily by a small number of viable bacteria, an immune response including protective immunity and a massive expansion of CD4+ Th1 cells could be induced in mice immunized with killed bacteria and purified LLO[109] or with killed bacteria plus recombinant LLO (unpublished observation). These findings suggest that an adjuvant-like activity appears to be involved in the robust Th1 response in mice infected with viable, LLO-producing strains of *L. monocytogenes* (Figure 14.7).

14.5 CONCLUSIONS AND PERSPECTIVES

The infection of mice with *L. monocytogenes* has been regarded as a prototype model of cell-mediated immunity to bacterial infection, and has also been increasingly utilized in the studies of other immunological responses. This may be due to the relative stability and reproducibility of *L. monocytogenes* infection, a short infection period (as compared to the animal model of tuberculosis that employs a slowly growing *M. tuberculosis*), a dynamic response, and the contribution and involvement of a large number of cell types. Indeed, a whole spectrum of host cells (e.g., macrophages,

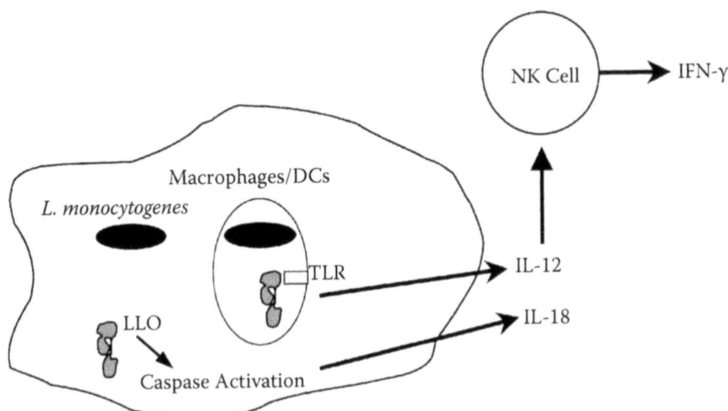

FIGURE 14.7 Possible involvement of listeriolysin O in the generation of IFN-γ-dependent T cell response. LLO produced inside the phagosome may stimulate TLRs for NF-κB-dependent cytokine production and LLO released after evasion from the phagosome may activate caspases, resulting in the secretion of mature IL-18. These two major cytokines, in turn, stimulate NK cells for the production of IFN-γ.

DCs, NK cells, γδT cells, $CD4^+$ T cells, and $CD8^+$ T cells) are activated after a systemic infection with this bacterium.

Not surprisingly, with voluminous experimental data being generated using murine models over the years, the fundamental aspects of host innate and adaptive responses to *L. monocytogenes* as well as host–parasite interplay have been delineated. In contrast, there is limited information on the T cell response in human listeriosis, since clinical cases are rather sporadic and it is not clear whether any immune status develops after the recovery from an acute stage of human listeriosis. In one study, it was shown that most of the T cells enriched from the peripheral blood mononuclear cells from 30 healthy individuals displayed a significant level of proliferative response when stimulated *in vitro* with *Listeria* organisms of not only pathogenic species but also nonpathogenic species.[110] As the direct mitogenic activity of *Listeria* organisms could be ruled out, this study suggested the possibility that a subclinical infection results in the T cell response among healthy individuals, yet the contribution of such *Listeria*-specific T cell response to the host defense is still obscure. As described elsewhere, the clinical feature of human listeriosis is different from that of systemic infection in mice. It is likely that there is a limited contribution of T cell-dependent immunity in human listeriosis. In this respect, there is scope for continued application of murine models that will reveal new details on host responses to *L. monocytogenes* with relevance to human listeriosis and other intracellular parasitic microbes.

Nonetheless, despite the accumulation of a large amount of experimental findings related to *L. monocytogenes* infection, there are still numerous questions that need to be fully addressed. For example, what is the ultimate effector mechanism for the complete resolution of *L. monocytogenes* infection? How do the $CD8^+$ T cells contribute to the clearance of bacterial burden? What are the precise roles of $CD4^+$ T cells that are activated robustly during infection? Why did most of the previous studies demonstrate the exclusive contribution of Lyt1+ ($CD4^+$) T cells? Why is this pathogen capable of inducing such a strong cell-mediated immune response in mice? To provide comprehensive answers to all these questions, further investigations using innovative experimental approaches are definitely warranted.

Study of *L. monocytogenes* infection represents a typical example of the recent advances in the understanding of the genetic and molecular mechanisms of bacterial virulence expression. It is remarkable that *L. monocytogenes* has the capability to adhere and invade the host cells, escape from the phagosome, multiply and migrate inside the cytosol, and spread to neighboring cells. It is even more impressive that during the course of infection, *L. monocytogenes* activates a large number of host immune cells, but manages to avoid total decimation by the host. To fully appreciate the complexity and ingenuity of this bacterium, a combination of bacteriological, immunological, genetic, proteomic, and other innovative approaches is undoubtedly required.

REFERENCES

1. Murray, E.G.D., Webb, R.A., and Swann, M.B.R., A disease of rabbits characterized by a large mononuclear monocytosis, caused by a hitherto undescribed bacillus *Bacterium monocytogenes*, *J. Pathol. Bacteriol.*, 29, 407, 1926.
2. Osebold, J.W., and Sawyer, M.T., Immunization studies on listeriosis in mice, *J. Immunol.*, 78, 262, 1957.
3. Mackaness, G.B., Cellular resistance to infection, *J. Exp. Med.*, 116, 381, 1962.
4. Miki, K., and Mackaness, G.B., The passive transfer of acquired resistance to *Listeria monocytogenes*, *J. Exp. Med.*, 120, 93, 1964.
5. Mackaness, G.B., and Hill, W.C., The effect of antilymphocyte globulin on cell-mediated resistance to infection, *J. Exp. Med.*, 129, 993, 1969.
6. Mackaness, G.B., The influence of immunologically committed lymphoid cells on macrophage activity *in vivo*, *J. Exp. Med.*, 129, 973, 1969.
7. McGregor, D.D., Koster, F.T., and Mackaness, G.B., The short-lived lymphocytes as a mediator of cellular immunity, *Nature*, 228, 855, 1970.
8. McGregor, D.D., Hahn, H.H., and Mackaness, G.B., The mediator of cellular immunity. V. Development of cellular resistance to infection in thymectomized irradiated rats, *Cell Immunol.*, 6, 186, 1973.
9. Mackaness, G.B., The relationship of delayed hypersensitivity to acquired cellular resistance, *Brit. Med. Bull.*, 23, 52, 1967.
10. Remold, H.G., and David, J.R., Further studies on migration inhibitory factor (MIF): Evidence for its glycoprotein nature, *J. Immunol.*, 107, 1090, 1971.
11. Rocklin, R.E., Remold, H.G., and David, J.R., Characterization of human migration inhibitory factor (MIF) from antigen-stimulated lymphocytes, *Cell. Immunol.*, 5, 436, 1972.
12. Schreiber, R.D. et al., Monoclonal antibodies to murine gamma-interferon which differentially modulate macrophage activation and antiviral activity, *J. Immunol.*, 134, 1609, 1985.
13. Weiser, W.Y. et al., Molecular cloning of a cDNA encoding a human macrophage migration inhibitory factor, *Proc. Natl. Acad. Sci. USA*, 86, 7522, 1989.
14. Calandra, T., and Roger, T., Macrophage migration inhibitory factor: A regulator of innate immunity, *Nat. Rev. Immunol.*, 3, 791, 2003.
15. Kaufmann, S.H. et al., Interleukin 2 induction in Lyt1^{+}23^{-} T cells from *Listeria monocytogenes*-immune mice, *Infect. Immun.*, 37, 1292, 1982.
16. Kaufmann, S.H.E., and Hahn, H., Biological functions of T cell lines with specificity for the intracellular bacterium *Listeria monocytogenes in vitro* and *in vivo*, *J. Exp. Med.*, 155, 1754, 1982.
17. Zinkernagel, R.M. et al., *H-2* restriction of cell-mediated immunity to an intracellular bacterium. Effector T cells are specific for *Listeria* antigen in association with *H-2I* region-coded self-markers, *J. Exp. Med.*, 145, 1353, 1977.
18. Kaufmann, S.H., Hug, E., and De Libero, G., *Listeria monocytogenes*-reactive T lymphocyte clones with cytolytic activity against infected target cells, *J. Exp. Med.*, 164, 363, 1986.
19. Kaufmann, S.H. et al., Cloned *Listeria monocytogenes* specific non-MHC-restricted Lyt-2^{+} T cells with cytolytic and protective activity, *J. Immunol.*, 140, 3173, 1988.
20. Czuprynski, C.J., and Brown, J.F., Effects of purified anti-Lyt-2 mAb treatment on murine listeriosis: Comparative roles of Lyt-2^{+} and L3T4^{+} cells in resistance to primary and secondary infection, delayed-type hypersensitivity and adoptive transfer of resistance, *Immunology*, 71, 107, 1990.
21. Zijlstra, M. et al., Beta2-microglobulin deficient mice lack CD4-8+ cytolytic T cells, *Nature*, 344, 742, 1990.
22. Roberts, A.D., Ordway, D.J., and Orme, I.M., *Listeria monocytogenes* infection in beta 2 microglobulin-deficient mice, *Infect. Immun.*, 61, 1113, 1993.
23. Kaufmann, S.H., *Listeria monocytogenes*-specific T-cell lines and clones, *Infection*, 16, S128, 1988.
24. Mitsuyama, M. et al., Three phases of phagocyte contribution to resistance against *Listeria monocytogenes*, *J. Gen. Microbiol.*, 106, 165, 1978.
25. Kayal, S., and Charbit, A., Listeriolysin O: A key protein of *Listeria monocytogenes* with multiple functions, *FEMS Microbiol. Rev.*, 30, 514, 2006.

26. Hamon, M., Bierne, H., and Cossart, P., *Listeria monocytogenes*: A multifaceted model, *Nat. Rev. Microbiol.*, 4, 423, 2006.
27. Mengaud, J. et al., E-cadherin is the receptor for internalin, a surface protein required for entry of *L. monocytogenes* into epithelial cells, *Cell,* 84, 923, 1996.
28. Lenz, L.L., Butz, E.A., and Bevan, M.J., Requirements for bone marrow-derived antigen-presenting cells in priming cytotoxic T cell responses to intracellular pathogens, *J. Exp. Med.*, 192, 1135, 2000.
29. Jung, S. et al., *In vivo* depletion of CD11c(+) dendritic cells abrogates priming of CD8(+) T cells by exogenous cell-associated antigens, *Immunity*, 17, 211, 2002.
30. O'Riordan, M. et al., Innate recognition of bacteria by a macrophage cytosolic surveillance pathway, *Proc. Natl. Acad. Sci. USA*, 99, 13861, 2002.
31. Serbina, N.V. et al., TNF/iNOS-producing dendritic cells mediate innate immune defense against bacterial infection, *Immunity*, 19, 59, 2003.
32. D'Orazio, S.E.F. et al., Class Ia MHC-deficient BALB/c mice generate $CD8^+$ T cell-mediated protective immunity against *Listeria monocytogenes* infection, *J. Immunol.*, 171, 291, 2003.
33. Kerksiek, K.M. et al., H2-M3–restricted T cells in bacterial infection: Rapid primary but diminished memory responses, *J. Exp. Med.*, 190, 195, 1999.
34. Kerksiek, K.M. et al., H2-M3-restricted memory T cells: Persistence and activation without expansion, *J. Immunol.*, 170, 1862, 2003.
35. Jacobs, T. et al., Listeriolysin O: Cholesterol inhibits cytolysis but not binding to cellular membranes, *Mol. Microbiol.*, 28, 1081, 1998.
36. Kohda, C. et al., Dissociated linkage of cytokine-inducing activity and cytotoxicity to different domains of listeriolysin O from *Listeria monocytogenes, Infect. Immun.*, 70, 1334, 2002.
37. Pamer, E.G. et al., Precise prediction of a dominant class I MHC-restricted epitope of *Listeria monocytogenes, Nature*, 353, 852, 1991.
38. Safley, S.A. et al., Role of listeriolysin-O (LLO) in the lymphocyte response to infection with *Listeria monocytogenes*. Identification of T cell epitopes of LLO, *J. Immunol.*, 146, 3604, 1991.
39. Villanueva, M.S., Sijts, A.J., and Pamer, E.G., Listeriolysin is processed efficiently into an MHC class I-associated epitope in *Listeria monocytogenes*-infected cells, *J. Immunol.*, 155, 5227, 1995.
40. Wuenscher, M.D. et al., The *iap* gene of *Listeria monocytogenes* is essential for cell viability, and its gene product, p60, has bacteriolytic activity, *J. Bacteriol.*, 175, 3491, 1993.
41. Pamer, E.G., Direct sequence identification and kinetic analysis of an MHC class I-restricted *Listeria monocytogenes* CTL epitope, *J. Immunol.*, 152, 686, 1994.
42. Sijts, A.J. et al., Two *Listeria monocytogenes* CTL epitopes are processed from the same antigen with different efficiencies, *J. Immunol.*, 156, 683, 1996.
43. Sijts, A.J., Pilip, I., and Pamer, E.G., The *Listeria monocytogenes*-secreted p60 protein is an N-end rule substrate in the cytosol of infected cells. Implications for major histocompatibility complex class I antigen processing of bacterial proteins, *J. Biol. Chem.*, 272, 19261, 1997.
44. Vijh, S., Pilip, I.M., and Pamer, E.G., Effect of antigen-processing efficiency on *in vivo* T cell response magnitudes, *J. Immunol.*, 160, 3971, 1998.
45. Busch, D.H. et al., A nonamer peptide derived from *Listeria monocytogenes* metalloprotease is presented to cytolytic T lymphocytes, *Infect. Immun.*, 65, 5326, 1997.
46. Geginat, G. et al., A novel approach of direct ex vivo epitope mapping identifies dominant and subdominant CD4 and CD8 T cell epitopes from *Listeria monocytogenes, J. Immunol.*, 166, 1877, 2001.
47. Lasa, I. et al., The amino-terminal part of ActA is critical for the actin-based motility of *Listeria monocytogenes*; the central proline-rich region acts as a stimulator, *Mol. Microbiol.*, 18, 425, 1995.
48. Darji, A. et al., The role of the bacterial membrane protein ActA in immunity and protection against *Listeria monocytogenes, J. Immunol.*, 161, 2414, 1998.
49. Villanueva, M.S. et al., Efficiency of antigen processing: A quantitative analysis, *Immunity*, 1, 479, 1994.
50. Shen, H. et al., Compartmentalization of bacterial antigens: Differential effects on priming of CD8 T cells and protective immunity, *Cell,* 92, 535, 1998.
51. Pamer, E.G., Cell-mediated immunity: The role of bacterial protein secretion, *Curr. Biol.*, 8, R457, 1998.
52. Nakane, A. et al., Interactions between endogenous gamma interferon and tumor necrosis factor in host. Resistance against primary and secondary *Listeria monocytogenes* infections, *Infect. Immun.*, 57, 3331, 1989.
53. Harty, J.T., Schreiber, R.D., and Bevan, M.J., CD8 T cells can protect against an intracellular bacterium in an interferon γ-independent fashion, *Proc. Natl. Acad. Sci. USA*, 89, 11612, 1992.

54. Samsom, J.N. et al., Tumor necrosis factor, but not interferon-gamma, is essential for acquired resistance to *Listeria monocytogenes* during a secondary infection in mice, *Immunology*, 86, 256, 1995.
55. Harty, J.T., and Bevan, M.J., Specific immunity to *Listeria monocytogenes* in the absence of IFN-γ, *Immunity*, 3, 109, 1995.
56. White, D.W., and Harty, J.T., Perforin-deficient CD8+ T cells provide immunity to *Listeria monocytogenes* by a mechanism that is independent of CD95 and IFN-γ but requires TNF-α, *J. Immunol.*, 160, 898, 1998.
57. Mielke, M.E., Peters, C., and Hahn, H., Cytokines in the induction and expression of T-cell-mediated granuloma formation and protection in the murine model of listeriosis, *Immunol. Rev.*, 158, 79, 1997.
58. Liu, D. et al., Characteristics of cell-mediated, anti-listerial immunity induced by a naturally avirulent *Listeria monocytogenes* serotype 4a strain HCG 23, *Arch. Microbiol.*, 188, 251, 2007.
59. Roach, D.R. et al., Independent protective effects for tumor necrosis factor and lymphotoxin alpha in the host response to *Listeria monocytogenes* infection, *Infect. Immun.*, 73, 4787, 2005.
60. Rothe, J. et al., Mice lacking the tumor necrosis factor receptor 1 are resistant to IMF-mediated toxicity but highly susceptible to infection by *Listeria monocytogenes*, *Nature*, 364, 798, 1993.
61. Musicki, K. et al., Differential requirements for soluble and transmembrane tumor necrosis factor in the immunological control of primary and secondary *Listeria monocytogenes* infection, *Infect. Immun.*, 74, 3180, 2006.
62. Jensen, E.R. et al., Fas (CD95)-dependent cell-mediated immunity to *Listeria monocytogenes*, *Infect. Immun.*, 66, 4143, 1998.
63. Harty, J.T., and White, D.W., A knockout approach to understanding CD8+ cell effector mechanisms in adaptive immunity to *Listeria monocytogenes*, *Immunobiology*, 201, 196, 1999.
64. White, D.W. et al., Perforin-deficient CD8+ T cells: *In vivo* priming and antigen-specific immunity against *Listeria monocytogenes*, *J. Immunol.*, 162, 980, 1999.
65. Badovinac, V.P., and Harty, J.H., Adaptive immunity and enhanced CD8+ T cell response to *Listeria monocytogenes* in the absence of perforin and IFN-γ, *J. Immunol.*, 164, 6444, 2000.
66. Mateo, L.R.S. et al., Perforin-mediated CTL cytolysis counteracts direct cell-cell spread of *Listeria monocytogenes*, *J. Immunol.*, 169, 5202, 2002.
67. Jiang, J. et al., Activation of antigen-specific CD8 T cells results in minimal killing of bystander bacteria, *J. Immunol.*, 171, 6032, 2003.
68. Stenger, S. et al., Granulysin: A lethal weapon of cytolytic T cells, *Immunol. Today*, 20, 390, 999.
69. Ernst, W.A. et al., Granulysin, a T cell product, kills bacteria by altering membrane permeability, *J. Immunol.*, 165, 7102, 2000.
70. Walch, M. et al., Uptake of granulysin via lipid rafts leads to lysis of intracellular *Listeria innocua*, *J. Immunol.*, 174, 4220, 2005.
71. Mercado, R. et al., Early programming of T cell populations responding to bacterial infection, *J. Immunol.*, 165, 6833, 2000.
72. Wong, P., and Pamer, E.G., Cutting edge: Antigen-independent CD8+ T cell proliferation, *J. Immunol.*, 166, 5864, 2001.
73. Kaech, S.M., and Ahmed, R., Memory CD8+ T cell differentiation: Initial antigen encounter triggers a developmental program in naïve cells, *Nat. Immunol.*, 2, 415, 2001.
74. Badovinac, V.P., Porter, B.B., and Harty, J.T., Programmed contraction of CD8+ T cells after infection, *Nat. Immunol.*, 3, 619, 2002.
75. Wong, P., and Pamer, E.G., Feedback regulation of pathogen-specific T cell priming, *Immunity*, 18, 499, 2003.
76. Badovinac, V.P., Tvinnereim, A.R., and Harty, J.T., Regulation of antigen-specific CD8+ T cell homeostasis by perforin and interferon-gamma, *Science*, 290, 1354, 2000.
77. Wong, P., and Pamer, E.G., CD8 T cell responses to infectious pathogens, *Annu. Rev. Immunol.*, 21, 29, 2003.
78. Yajima, T. et al., Overexpression of IL-15 *in vivo* increases antigen-driven memory CD8+ T cells following a microbe exposure, *J. Immunol.*, 168, 1198, 2002.
79. Yajima, T. et al., IL-15 regulates CD8+ T cell contraction during primary infection, *J. Immunol.*, 176, 507, 2006.
80. Kaech, S.M. et al., Selective expression of the interleukin 7 receptor identifies effector CD8 T cells that give rise to long-lived memory cells, *Nat. Immunol.*, 4, 1191, 2003.
81. Geginat, G. et al., Th1 cells specific for a secreted protein of *Listeria monocytogenes* are protective *in vivo*, *J. Immunol.*, 160, 6046, 1998.

82. Skoberne, M. et al., Cross-presentation of *Listeria monocytogenes*-derived CD4 T cell epitopes, *J. Immunol.*, 169, 1410, 2002.
83. Janda, J. et al., Cross-presentation of *Listeria*-derived CD8 T cell epitopes requires unstable bacterial translation products, *J. Immunol.*, 173, 5644, 2004.
84. Zenewicz, L.A., Nonsecreted bacterial proteins induce recall CD8 T cell responses but do not serve as protective antigens, *J. Immunol.*, 169, 5805, 2002.
85. Corbin, G.A., and Harty, J.T., Duration of infection and antigen display have minimal influence on the kinetics of the CD4+ T cell response *to Listeria monocytogenes* infection, *J. Immunol.*, 173, 5679, 2004.
86. Mannering, S.I., Zhong, J., and Cheers, C., T cell activation, proliferation and apoptosis in primary *Listeria monocytogenes* infection, *Immunology,* 106, 87, 2002.
87. Kursar, M. et al., Organ-specific CD4+ T cell response during *Listeria monocytogenes* infection, *J. Immunol.*, 168, 6382, 2002.
88. Mittrucker, H.W., Kohler, A., and Kaufmann, S.H., Substantial *in vivo* proliferation of CD4(+) and CD8(+) T lymphocytes during secondary *Listeria monocytogenes* infection, *Eur. J. Immunol.* 30, 1053, 2000.
89. Haring, J.S., and Harty, J.T., Aberrant contraction of antigen-specific CD4 T cells after infection in the absence of gamma interferon or its receptor, *Infect. Immun.*, 74, 6252, 2006.
90. Mosmann, T.R. et al., Two types of murine helper T cell clone. 1. Definition according to profiles of lymphokine activities and secreted proteins, *J. Immunol.*, 136, 2348, 1996.
91. Shedlock, D.J., and Shen, H., Requirement for CD4 T cell help in generating functional CD8 T cell memory, *Science*, 300, 337, 2003.
92. Sun, J.C., and Bevan, M.J., Defective CD8 T cell memory following acute infection without CD4 T cell help, *Science,* 300, 339, 2003.
93. Mitsuyama, M. et al., Generation of *Listeria monocytogenes*-specific T cells mediating delayed footpad reaction and protection in neonatally thymectomized mice but not in nude mice, *Med. Microbiol. Immunol*, 177, 207, 1988.
94. Ehlers, S., Mielke, M.E., and Hahn, H., CD4+ T cell associated cytokine gene expression during experimental infection with *Listeria monocytogenes*: The mRNA phenotype of granuloma formation, *Int. Immunol.*, 6, 1727, 1994.
95. Freeman, M.M., and Ziegler, H.K., Simultaneous Th-1-type cytokine expression is a signature of peritoneal CD4+ T lymphocytes responding to infection with *Listeria monocytogenes, J. Immunol.*, 175, 394, 2005.
96. Mielke, M.E., Peters, C., and Hahn, H., Cytokines in the induction and expression of T-cell-mediated granuloma formation and protection in the murine model of listeriosis, *Immunol. Rev.*, 158, 79, 1997.
97. Way, S.S., and Wilson, C.B., Cutting edge: Immunity and IFN-γ-production during *Listeria monocytogenes* infection in the absence of T-bet, *J. Immunol.*, 173, 5918, 2004.
98. Sakaguchi, S. et al., Foxp3+CD25+CD4+ natural regulatory T cells in dominant self-tolerance and autoimmune disease, *Immunol. Rev.*, 212, 9, 2006.
99. Kursar, M. et al., Regulatory CD4+CD25+ T cells restrict memory CD8+ T cell responses, *J. Exp. Med.*, 196, 1585, 2002.
100. Kursar, M. et al., Depletion of CD4+ T cells during immunization with nonviable *Listeria monocytogenes* causes enhanced CD8+ T cell-mediated protection against listeriosis, *J. Immunol.*, 172, 3167, 2004.
101. Chang, S.R. et al., Characterization of early gamma interferon (IFN-γ) expression during murine listeriosis: Identification of NK1.1+ CD11c+ cells as the primary IFN-γ-expressing cells, *Infect. Immun.*, 75, 1167, 2007.
102. Ohya, S. et al., The contribution of reactive oxygen intermediates and reactive nitrogen intermediates to listericidal mechanisms differ in macrophages activated pre- and postinfection, *Infect. Immun.*, 66, 4043, 1998.
103. Nomura, T. et al., Essential role of interleukin 12 (IL-12) and IL-18 for gamma interferon production induced by listeriolysin O in the spleen cells of mice, *Infect. Immun.*, 70, 1049, 2002.
104. Ito, Y. et al., Seeligeriolysin O, a cholesterol-dependent cytolysin of *Listeria seeligeri*, induces gamma interferon from spleen cells of mice, *Infect. Immun.*, 71, 234, 2003.
105. Ito, Y. et al., Seeligeliolysin O, a protein toxin of *Listeria seeligeri*, stimulates macrophage cytokine production via Toll-like receptors in a profile different from that induced by other bacterial ligands, *Int. Immunol.*, 17, 1597, 2005.
106. Kimoto, T. et al., Differences in gamma interferon production induced by listeriolysin O and ivanolysin O result in different levels of protective immunity in mice infected with *Listeria monocytogenes* and *Listeria ivanovii, Infect. Immun.*, 71, 2447, 2003.

107. Yamamoto, K. et al., Listeriolysin O, a cytolysin derived from *Listeria monocytogenes,* inhibits generation of ovalbumin-specific Th2 immune response by skewing maturation of antigen-specific T cells into Th1 cells, *Clin. Exp. Immunol.*, 142, 2005.
108. Yamamoto, K. et al., Listeriolysin O derived from *Listeria monocytogenes* inhibits the effector phase of an experimental allergic rhinitis induced by ovalbumin in mice, *Clin. Exp. Immunol.*, 144, 475, 2006.
109. Xiong, H. et al., Administration of killed bacteria together with listeriolysin O induces protective immunity against *Listeria monocytogenes* in mice, *Immunology*, 94, 14, 1998.
110. Munk, M.E. et al., *Listeria monocytogenes* reactive T lymphocytes in healthy individuals, *Microb. Pathog.*, 5, 49, 1988.

15 Anti-Infective Vaccine Strategies

Toshi Nagata and Yukio Koide

CONTENTS

15.1 INTRODUCTION

Listeria monocytogenes is a Gram-positive, opportunistic intracellular bacterium that can cause serious diseases (e.g., meningitis and death) in immunocompromised hosts, pregnant women, and neonates.[1–3] During the course of its infection, *L. monocytogenes* employs a collection of purposefully made molecules that facilitate its efficient adhesion and entry to host cells, escape from vacuoles, replication in the cytoplasm, and spread to neighboring cells—provoking a cascade of innate and adaptive immune reactions in its wake without being completely eliminated by the host. The host responses to *L. monocytogenes* have been investigated extensively during the past 50 years and have contributed to the understanding of the key concepts in innate and adaptive immunity. Indeed, the *L. monocytogenes* infection model is considered indisputably a paradigm for the study of cell-mediated immunity[4–6] (see chapter 14).

As a food-borne pathogen, the natural route of infection with *L. monocytogenes* is oral ingestion. After crossing the mucosal membrane (i.e., intestine), *L. monocytogenes* eventually settles in a host's dendritic cells and macrophages. Upon escape from the phagolysosome of the host cells by virtue of listeriolysin O (LLO), it replicates rapidly in the cytoplasm, where *L. monocytogenes* antigens are delivered to related immune cells via the major histocompatibility complex (MHC) class I presentation pathway as well as the MHC class II presentation pathway of host cells. Both $CD8^+$ cytotoxic T-lymphocytes (CTL) and type 1 $CD4^+$ helper T-lymphocytes (Th1) have been shown to be amplified during *L. monocytogenes* infection (see chapter 14), and these lymphocytes play a pivotal role in a host's buildup of protective immunity, as confirmed by experimental depletion and adoptive transfer studies[7–9] or by analysis of mutant mice with a defect in a gene for β2-microglobulin or H2-Aβ.[10,11] Several T cell epitopes (MHC-binding antigenic peptides) in *L. monocytogenes* antigens have been reported.[6,12]

While prior exposures to *L. monocytogenes* virulent strains contribute to the development of an enduring protective immunity in the host against listeriosis, it is not practical to apply intact virulent strains as a vaccine due to obvious health considerations. Hence, a variety of approaches has been taken to attenuate *L. monocytogenes* virulent strains in order for them to be useful in vaccine applications. These include chemical/physical inactivation (or killing), live attenuation, and identification of protective subunits from the bacterium. Although immunization with heat-killed *L. monocytogenes* is in general not sufficiently protective, a combined injection of killed bacteria and LLO leads to a protective immunity against listeriosis.[13] Attenuation of *L. monocytogenes* virulent strains by deletion or replacement of certain virulence-associated genes such as *actA, hly,* and *inlB* or other essential genes has underscored the development of viable, safe, and effective vaccines that mimic natural infection without causing unintended pathogenic consequences. Given that *L. monocytogenes* comprises strains of negligible virulence within the species (e.g., serotype 4a strains), the potential of using these naturally nonpathogenic strains as candidate antilisterial vaccines can be also explored.

Furthermore, vaccine preparations containing defined and subcellular components (e.g., LLO) have been experimented with for protection against listeriosis. Besides its value as a vaccine candidate, attenuated live *L. monocytogenes* has also proven useful as an excellent antigen carrier that delivers target molecules into intracellular environs to strengthen a host's $CD8^+$ and $CD4^+$ T cell responses as well as innate immune responses against other infective agents (e.g., virus and parasite) and cancers. Indeed, compared with other bacterial vectors (e.g., *Mycobacterium bovis* Bacille de Calmette et Guérin (or BCG), *Salmonella, Shigella,* and *Escherichia coli*), *L. monocytogenes* is not only easier to maintain due to its tolerance of extreme stresses, but also has a much higher safety threshold than other bacterial vectors (with 10^{11} organisms being the lethal dose for *L. monocytogenes* wild-type strain vs. 20 organisms for *Salmonella*). Thus, *L. monocytogenes* has increasingly been recognized as a preferred vehicle for delivering protective molecules from various bacterial, viral, and parasitic pathogens.

In this chapter, attenuation of *L. monocytogenes* and the antigen delivery systems based on this bacterium for vaccine development are reviewed. The trials to induce protective immunity against listeriosis with various killed and live *L. monocytogenes* preparations are also examined. Next, the application of attenuated live *L. monocytogenes* for delivering vaccine molecules from other intracellularly localized pathogens including viruses, bacteria, and parasites—protection against which requires involvement of specific cell-mediated immunity—is summarized. Finally, perspectives on the future research that may lead to further improvement in the efficiency and applicability of *L. monocytogenes*-based vaccine strategies against various infective agents are discussed.

15.2 ATTENUATION OF *LISTERIA* FOR VACCINE PURPOSES

Microbial pathogens may lose part or all of their pathogenicity after going through various chemical, physical, biological, and/or genetic treatments. The resulting bacterial preparations can then be applied for protection of hosts against subsequent infections with related bacteria. In addition, identification of defined, protective subcellular components offers another avenue for vaccine development.

15.2.1 Killed Vaccines

Many chemical (e.g., formalin, ether or beta-propiolactone) and/or physical (e.g., heat, ultraviolet, or γ-irradiation) measures are lethal to microbial organisms, and can lead to the killing or inactivation of the bacteria concerned. While these processes may destroy certain antigenic structures and formations in some bacteria, which are essential for introduction of protective immunity, they seem to have minimal influence on others. This is reflected by the extraordinary success with killed vaccine preparations for prevention of several high-profile bacterial infections in humans (e.g., typhoid fever, cholera, plague, lyme disease, and pertussis). In general, production of the killed or inactivated vaccines is technically simple, with some well-defined and quantifiable manufacturing processes. However, care should be taken to ensure that bacterial aggregates are completely inactivated to prevent some unwanted clinical complication or disease transmission. Additionally, since the killed or inactivated vaccines do not replicate inside the host, they often require multiple injections of relatively high doses to be effective, or occasionally they rely on adjuvants to exert their full immunogenic potential.

The attempts to use killed *L. monocytogenes* preparations for protecting mice against listeriosis have been well documented.[14–17] It is apparent that some of the key antigenic components in *L. monocytogenes* are prone to disruption by chemical or physical treatments, as application of formalin- or heat-killed *L. monocytogenes* has often resulted in minimal immunity against listeriosis in experimental animal models. In Mackaness's landmark paper,[18] in which the importance of cell-mediated immunity for protection against *L. monocytogenes* was highlighted, inactivated *Listeria* failed to protect mice against lethal *Listeria* challenge, to influence the growth curve of *Listeria* in the spleen, and to induce delayed-type hypersensitivity. Many subsequent investigations have since confirmed that heat-killed *L. monocytogenes* (HKLM) does not confer efficient protection[19–23] (see Table 15.1), although some early reports found that HKLM can elicit protective immune responses with the help of polyanions[24] or in C3H/HeJ mouse strain.[25] In the latter case, ineffective killing of *Listeria* at 56°C was suspected for stimulation of such an immunity. It appears that a host's immunity against listeriosis is mediated by lymphoid cells, not by antibodies (Abs),[15,26–30] and that induction of protective immunity by HKLM requires additional manipulations to the host, such as supplemental LLO[13] or IL-12[31,32] injections, anti-CD40 monoclonal antibody (mAb) treatment,[33] or anti-CD4 mAb treatment.[34]

For example, depletion of regulatory $CD4^+$ T cells with anti-CD4 mAb during boost immunization with HKLM dramatically increased the number of *L. monocytogenes*-specific $CD8^+$ T cells, which in turn expressed effector functions such as IFN-γ production.[34] While administration of HKLM alone generally stimulated a T cell response that was insufficiently protective, combining HKLM with purified LLO offered protection to mice against *L. monocytogenes* challenge.[13] Similarly, injection of HKLM together with IL-12 elicited vigorous Th1-type T cell responses and cytokine profiles in mice, resulting in a durable immunity against listeriosis.[31,32] These results suggest that LLO is liable to heat or formalin destruction, and that LLO is an essential component for stimulating production of endogenous IFN-γ at the early stage of immunization and for activation of protective $CD8^+$ T cell against *L. monocytogenes in vivo*. The role of LLO in the generation of endogenous IFN-γ can be overtaken or compensated by IL-12, which induces alternative IFN-γ and IL-2 production through activation of $CD4^+$ Th1 cells, leading to activation of $CD8^+$ CTL. The regulatory $CD4^+$ T cells appear to have a negative impact on $CD8^+$ T cells, as removal of the $CD4^+$ T cells during immunization leads to the expansion of *L. monocytogenes*-specific $CD8^+$ T cells.

On the other hand, Szalay and colleagues[35] reported recently that the protective antigens of HKLM can be introduced into the MHC class I pathway through cross-presentation, facilitating their recognition by host $CD8^+$ T cells and leading to the introduction of a protective immunity. The difference between the early results by others and the study of Szalay and colleagues[35] may be due to the mode of application. In the early reports, injection of HKLM was done only once and not repeatedly as described by Szalay and colleagues,[35] who injected HKLM three times at 5-day intervals. After an examination of immune responses with HKLM immunization, Lauvau and colleagues[36] showed that HKLM immunization primed memory $CD8^+$ T cell populations, which were ineffective at providing protection from subsequent virulent *Listeria* challenge despite their substantial size. This can be attributable to a qualitative difference of memory $CD8^+$ T cells between live *Listeria* infection and HKLM immunization. In addition, HKLM immunization primed $CD8^+$

TABLE 15.1
***L. monocytogenes* Vaccines against Virulent Strain Challenge**

Strain	Attenuation	Mouse Strain	Immunization Route[a]	Immune Responses	Protection	Ref.
		Killed *L. monocytogenes*				
10403	Formalin treatment	Swiss–Webster	i.v.	ND[b]	–	19
Not described	Heat-killed + dextran sulphate (DS 500)	BALB/c	i.p.	ND	+	24
Not described	Heat-killed	C3H/HeJ, C3HeB/FeJ, BALB/c	i.v.	CD8	+	25
Serotype 4b	Heat-killed	C3H/HeJ	i.v.	ND	–	20
EGD	Heat-killed	C57BL/6	i.v., s.c.	CD4	–	21
EGD	Heat-killed	C3H/He	i.v.	CD4	–	22
EGD	Heat-killed	C3H/He	i.v.	CD4	–	23
EGD	Heat-killed	C57BL/6	i.v.	CD8	+	35
ATCC 43251	Heat-killed + IL-12	C3HeB/FeJ, C3H/HeJ	i.p.	CD4	+	31, 32
EGD	Heat-killed + LLO	C3H/He	i.v.	CD4	+	13
EGD	Heat-killed + anti-CD40 mab	C57BL/6	i.v.	CD4, CD8	+	33
EGD	Heat-killed + anti-CD4 mab	BALB/c	i.v.	CD8	+	34
DP-L4056	Δ*uvrAB*	BALB/c, C57BL/6	i.v.	CD8	+	72
10403S	γ-irradiation	C57BL/6	i.v.	CD8	+	38
		Auxotroph Strains				
1070138	Δ*pheA*	MF1	i.v.	ND	+	41
10403S	Δ*dal*, Δ*dat*	BALB/c	i.v.	CD8	+	42
EGD	Δ*aroA*, Δ*aroB*, Δ*aroA*/*B*, Δ*aroE*	BALB/c	i.v.	CD8	+	46
10403S	Δ*dal*, Δ*dat*, SPAC/pL*idal*	BALB/c	i.v.	CD8	+	44
10403S	Δ*dal*, Δ*dat*, pRRR	BALB/c	i.v.	CD8	+	45
		Virulence Gene-Deficient Strains				
EGD	Δ*hly*	BALB/c	i.v.	CD4	–	56
10403S	Δ*hly*	BALB/c	i.v.	ND	–	57
10403S	Δ*hly*	BALB/c	i.v.	CD8	+	58
10403S	Δ*hly*, CytoLLO	BALB/c, C57BL/6	i.v.	CD8	+	60
LO28	Δ*actA*	C3H	i.v.	CD8	+	62
10403S	Δ*actA*	BALB/c	oral	IgA	ND	64
LO28	Δ*actA*Δ*plcB*, Δ*plcA*	BALB/c	i.p.	CD4	ND	65
10403S	Δ*actA*Δ*plcB*	Human volunteers[c]	oral	IgG, T cells	ND	66
EGD	Δ*actA*, Δ*actA*Δ*plcB*	C57BL/6	i.v.	CD8	+	67
EGD	Δ*actA*, Δ*actA*Δ*plcB*	BALB/c	i.v.	CD8	+	68
EGD	Δ*mpl2*	BALB/c	i.p., oral	CD8, Ab	ND	69

[a] i.p.: intraperitoneal; i.v.: intravenous; s.c.: subcutaneous.
[b] ND: not determined.
[c] This study was carried out in human volunteers, not in mice.

T cell populations that did not acquire effector functions. Upon further investigation on the difference of CD8+ T cell responses between HKLM and live *L. monocytogenes* infection, it is clear that HKLM immunization did not induce as high levels of CD40, CD80, and CD86 expression on CD11c^{high} dendritic cells as live *L. monocytogenes* infection, leading to the difference of CD8+ T cell priming between them.[37]

Datta and colleagues[38] evaluated γ-irradiation as a means to prepare *Listeria* vaccine. Previously, radiation-attenuated *Plasmodium* sporozoite vaccines have been applied against malaria parasites in mice and humans, and protection of humans by immunization with these vaccines has been documented.[39] Whereas the irradiated malaria vaccines were attenuated "live" vaccines, the γ-irradiated *Listeria* vaccine of Datta and colleagues[38] was "killed" *Listeria*. This irradiated *L. monocytogenes* efficiently activated dendritic cells via Toll-like receptors (TLRs) and stimulated protective T cell responses in mice. Immunization with the irradiated *Listeria* strain facilitated cross-presentation of listerial antigens to CD8+ T cells in TAP- and proteasome-dependent manners. Thus, it is possible that γ-irradiation did not alter antigenic and adjuvant structures, which can be destroyed by traditional heat or chemical inactivation. These results indicate that γ-irradiation is applicable to bacterial vaccine candidates including *L. monocytogenes* to induce cell-mediated as well as humoral immunity.

15.2.2 Attenuated Live Vaccine

Bacteria undergoing repeated *in vitro* cultures or passages through non-natural hosts may demonstrate lowered pathogenicity, but their capability to mimic the natural infection and their desired immunogenicity are unaltered. Given that these live attenuated bacteria maintain some self-limiting capacity to infect and replicate in the hosts, show negligible pathogenicity, and yet are highly immunogenic, they have the potential to confer protection against respective infections at relative low doses in the absence of adjuvants. The usefulness of the live attenuated bacteria as vaccines against related bacterial infections is exemplified by attenuated *M. bovis* Bacillus de Calmette et Guérin (i.e., BCG) and *Salmonella typhi* TY21a against human tuberculosis and typhoid fever, respectively. However, a well-known attenuated *L. monocytogenes* strain, ATCC 15313, whose attenuation resulted from successive laboratory subculturing, does not offer protection against listeriosis. The inability of ATCC 15313 to elicit a protective immunity is most likely due to a mutation in its *hly* gene leading to a truncated LLO protein (GenBank Accession No. AY750900), since LLO plays a vital part in *L. monocytogenes*' escape from vacuoles and subsequent stimulation of endogenous cytokine production.[4–6,40] More recently, genetic manipulation techniques have been used to create attenuated live vaccines that lack certain essential genes or key virulence-associated genes, which are the focus of the discussion that follows.

15.2.2.1 Attenuation Strategies

Auxotroph strains. An ideal vaccine strain of *L. monocytogenes* would be absolutely avirulent but fully immunogenic. One well-known strategy to make attenuated bacterial strain is to develop auxotroph strains—namely, strains that require specific nutrients for growth (Table 15.1). Alexander and colleagues[41] described a transposon-insertion mutant of *L. monocytogenes* deficient in prephenate dehydratase gene (Δ*pheA*). The enzyme catalyzes the conversion of prephenic acid to phenylpyruvic acid, which is involved in the biosynthesis pathway of phenylalanine. Mice vaccinated with this mutant *L. monocytogenes* were protected significantly from subsequent challenge with virulent *L. monocytogenes*. However, the mutant persisted in the tissues of infected mice and elicited splenomegaly.

D-alanine is required for the synthesis of the mucopeptide component of the cell walls of virtually all bacteria and is found almost exclusively in the microbial world. Thompson and colleagues[42] constructed an attenuated *Listeria* strain deficient in two genes—that is, an alanine racemase gene (*dal*), which catalyzes the reaction L-alanine to D-alanine or vice versa, and a D-amino acid aminotransferase gene (*dat*), which catalyzes the reaction D-glutamic acid + pyruvate to α-ketoglutaric acid + D-alanine, or vice versa. The mutant strain can be grown in the laboratory with D-alanine supplement, but is unable to grow without such supplement. Immunization with the strain induced $CD8^+$ T cell responses in mice and generated a protective immunity against lethal challenge by wild-type *L. monocytogenes*, which equaled that elicited by infection with a sublethal dose of wild-type *L. monocytogenes.*

The D-alanine-deficient *Listeria* has also been examined for plasmid DNA delivery.[43] The D-alanine-deficient *L. monocytogenes* mutant (Δ*dal*Δ*dat*) was transformed with a plasmid encoding green fluorescent protein (GFP) under control of the cytomegalovirus immediate early (CMV-IE) promoter/enhancer. No GFP-positive cells in the infected cells were observed when D-alanine was omitted. On the other hand, if D-alanine was supplied at the high concentration or for a lengthy period to allow for sustained bacterial growth, the infected host cells were often killed by the bacteria, resulting in fewer GFP-expressing cells. These results suggest that efficient DNA delivery by transformed *Listeria* must balance invasion and survival of *Listeria* with health and survival of target host cells.[43]

By developing additional systems that obviate the need for exogenous D-alanine administration, the utility of a mutant *Listeria* strain as a vaccine vector can be further improved. Li and colleagues[44] developed a recombinant *Listeria* strain that expresses a copy of the *Bacillus subtilis* racemase gene (*dal*) under the control of a tightly regulated isopropyl-β-D-thiogalactopyranoside (IPTG)-inducible promoter in a multicopy plasmid. The recombinant *L. monocytogenes* demonstrated the strict dose-dependent growth in the presence of IPTG. After removal of IPTG, bacterial growth ceased within two replication cycles. Nevertheless, a single immunization evoked a state of long-lasting protective immunity against wild-type *L. monocytogenes.*

Zhao and colleagues[45] devised a new version of the Δ*dal*Δ*dat* mutant *Listeria* strain, which expresses a *dal* gene and synthesizes D-alanine under highly selective conditions. The suicide plasmid designated pRRR carries a *dal* gene surrounded by two *res1* sites and a resolvase gene, *tnpR*, which acts at the *res1* sites. The resolvase gene is regulated by a promoter activated upon exposure to host cell cytoplasm (*actA* promoter). Therefore, the recombinant *Listeria* (*L. monocytogenes* Δ*dal*Δ*dat*/pRRR strain) is able to grow in liquid culture and to infect host cells without D-alanine supplementation. This system allows only transient growth of the recombinant *Listeria* in infected cells and survival in animals for only 2–3 days. Mice immunized with the mutant *Listeria* produced specific effector and memory $CD8^+$ T cells, and they were protected against lethal challenge by wild-type *L. monocytogenes.*

Stritzker and colleagues[46] constructed a series of *L. monocytogenes* mutants with deletions in genes of the common branch of the biosynthesis pathway leading to aromatic compounds. *aroA* (encoding the first enzyme in aromatic amino acid biosynthesis, 3-deoxy-D-arabino-heptulosonate-7-phosphate synthase), *aroB* (encoding the second enzyme, 3-dehydroquinate synthase), and *aroE* (encoding 5-enolpyruvylshikimate-3-phosphate synthase) mutants showed greatly reduced growth rates in epithelial cells and in rich culture media. All *aro* mutants displayed significant virulence attenuation in mice. That is, the 50% lethal dose (LD_{50}) in BALB/c mice was increased at least 10^4-fold for the *aroA*, *aroB*, and *aroA/B* mutants and >10^5-fold for the *aroE* mutant compared to the parent strain. Mice immunized with *aro* mutant bacteria developed strong T cell responses and full protection against subsequent virulent *L. monocytogenes* challenge.

Strains lacking essential virulence genes. Another approach to obtain attenuated strains of *L. monocytogenes* is to create bacteria that lack essential virulence genes (Table 15.1). Two of the major virulence factors of *L. monocytogenes* are the pore-forming cytolysin, LLO, and the actin-recruiting and -organizing protein, ActA. LLO is a 58-kDa protein, with essential functions in vacuolar lysis and escape into the cytoplasm of host cells[40] and ActA is necessary for *L. monocytogenes* to spread to adjacent host cells.[47–49] Besides acting as a major virulence factor, LLO is also an immunodominant antigen.[50–52] Previous studies indicate that *L. monocytogenes* with a deficiency in *hly* gene encoding LLO was highly attenuated (10^5-fold less virulent[40,53–56]) and that immunization with such a strain generally failed to elicit protective immunity.[56,57]

Recently, Hamilton and colleagues[58] reexamined the role of LLO in listerial immunity by using a recombinant *Listeria* strain that is deficient in most of the *hly* structural gene but expresses an additional $CD8^+$ T cell epitope derived from lymphocytic choriomeningitis virus (LCMV). Injection of the *Listeria* mutant-evoked sizable priming of the epitope-specific $CD8^+$ T cells and development of a functional memory cell population. Further, mice immunized with the *Listeria* mutant were resistant against high-dose challenge of virulent *L. monocytogenes* and also against heterologous challenge with LCMV. It is possible that priming with a low dose of *hly*-deficient *Listeria*, which occurred in an environment with a reduced level of IFN-γ, allowed rapid amplification of antigen-specific $CD8^+$ T cells by booster immunization, despite an undetectable primary response. Thus, the generation of protective immunity by the *hly*-deficient *Listeria* strain may represent a useful platform for vaccine delivery. Harty's group also showed that using dendritic cell vaccine for priming naïve $CD8^+$ T cells in an environment of low inflammation (IFN-γ) accelerated the generation of memory $CD8^+$ T cells.[59] These results indicate that immunization protocols, in addition to selection of attenuated *Listeria* strains, are critical for induction of specific protective immunity.

Bouwer and colleagues[60] described a "cytoLLO" *Listeria* strain, DH-L1233, which harbors a deletion in its *hly* gene (ΔLLO), resulting in a truncated LLO lacking its N-terminal secretion signal sequence (cytoLLO). This strain has an additional mutation in gene encoding PrfA transcriptional activator, leading to constitutive overexpression of the integrated cytoLLO construct and all other PrfA-controlled virulence genes.[61] Although it failed to escape the phagosome and did not replicate within bone marrow-derived macrophages, this strain showed the capacity to deliver native antigens to the MHC class I antigen presentation pathway. Being attenuated 10^5-fold in BALB/c mice compared with wild-type *L. monocytogenes* strain 10403S, it was cleared rapidly in normal and immunocompromised mice. However, antigen-specific $CD8^+$ effector T cells were stimulated after immunization, and mice immunized with the mutant *Listeria* were protected against lethal challenge with a virulent wild-type *L. monocytogenes* strain. In this system, the cytoLLO protein is used for phagosomal lysis after the bacteria die in the phagosome and release antigens into the cytoplasm of host cells, facilitating antigen presentation through the MHC class I pathway.

ActA is an *L. monocytogenes* surface protein involved in the actin nucleation at the bacterial surface and cell-to-cell spread. The *L. monocytogenes actA* mutant could enter and multiply in the cytoplasm of the infected host cell, but had deficiency in actin-dependent cell-to-cell spread. Goossens and Milon[62] showed that the LD_{50} of an attenuated *actA* mutant was 10^3-fold higher than that of the wild-type strain. A single intravenous infection with the maximum sublethal dose of the mutant induced a long-lasting immunity and that protection was mainly conferred by $CD8^+$ T cells, as depletion of $CD4^+$ T cells had no significant effect on the level of protection.[63] Manohar and colleagues[64] applied the *actA* mutant for induction of gut mucosal immune responses. Through oral immunization, the *actA* mutant was capable of colonizing the intestinal mucosa of formerly germ-free mice for a long period without causing disease while eliciting secretory immunoglobulin (Ig) A response. The IgA antibodies recognized the 96-, 60-, 40-, and 14-kDa proteins of *L. monocytogenes.*

Although disruption of the *actA* gene does not interfere with the ability of *Listeria* to replicate in the cytoplasm and subsequent presentation of listerial antigens through the MHC class I pathway, the *actA* deletion may not be sufficient for attenuation of *L. monocytogenes* since these bacteria persist in mice in the liver for up to 7 days.[64] Thus, *L. monocytogenes* mutants containing additional deletions apart from *actA* locus have been examined.[65–67] The *actA/plcB* double mutant was clinically evaluated for safety. In a study involving human volunteers, Angelakopoulos and colleagues[66] showed that 20 healthy volunteers, each receiving a single oral dose (10^6–10^9 CFUs) of *actA/plcB*-deficient *L. monocytogenes* orally, developed no serious long-term health side effects, despite displaying humoral, mucosal, and cellular immune responses to the inoculated *L. monocytogenes* strain. The *actA/plcB* double mutant was also evaluated using the mouse model.[67,68] The double mutant strain was extremely low in virulence and was rapidly eliminated by the murine host during the first days of infection. Interestingly, the mutant strain exhibited a significantly reduced ability to induce CD4$^+$ T cell-mediated inflammation and a single immunization with the strain efficiently induced and maintained effector memory CD8$^+$ T cells, which protected the immunized mice against wild-type *L. monocytogenes.*

Another virulence gene–deficient attenuated *L. monocytogenes* strain, Δ*mpl2*, which results in the secretion of nonfunctional metalloproteinase (Mpl), has been evaluated as an oral vaccine carrier for a mouse fibrosarcoma.[69] The Mpl proteinase is a zinc-dependent protein, which is required for the production of biologically active phospholipase B for lysis of the double membrane vacuole, and hence for permitting *Listeria* to spread to adjacent cells. Oral and intraperitoneal immunization of mice with this strain stimulated a long-lasting CD8$^+$ T cell response against fibrosarcoma expressing β-galactosidase.

Further, the entire lecithinase operon (consisting of the genes *mpl*, *actA*, and *plcB*)-deficient *L. monocytogenes* strain, designated Δ2, has been constructed[70] and examined as a vaccine carrier.[71,72] Infection of BALB/c mice with the Δ2 strain led to a 10^3–fold higher intravenous LD_{50} value compared with wild-type *L. monocytogenes* EGD strain.[71] Immunization with the Δ2 strain harboring DNA vaccines for *Mycobacterium tuberculosis* Ag85 family antigens has been shown to induce specific T cell responses and protective immunity in mice.[72]

Brockstedt and colleagues[73] reported a metabolically active yet replication-defective *L. monocytogenes* strain that was developed with DNA cross-linking procedure. This *Listeria* strain takes advantage of the potency of live vaccine and the safety of killed vaccine simultaneously as the gene required for nucleotide excision repair, *uvrAB*, is removed, rendering it exquisitely sensitive to photochemical inactivation with psoralen and long-wave-length ultraviolet light. Psoralen treatment of this strain leaves intact the ability of a population of inactivated bacteria to express its genes, but prevents its productive growth and its disease-causing ability in the immunized host. The psoralen-inactivated *L. monocytogenes* Δ*actA*/Δ*uvrAB*-OVA, which has ovalbumin (OVA) gene as a model antigen gene, was capable of inducing OVA-specific CD8$^+$ T cells and protective immunity in mice against OVA-expressing vaccinia virus and also virulent wild-type *L. monocytogenes* challenge.

Besides these virulence factor–deficient mutants, newly identified virulent factors can be also targeted for development of new attenuated *Listeria* strains. A new approach to identify genes involved in the virulency of *Listeria* using signature-tagged transposon mutagenesis has been reported.[74]

Suicide gene. In addition to the aforementioned attenuation strategies for *Listeria*, a "suicide" gene has been used to ease the destruction of the bacterium. A typical suicide gene that has proven useful in the preparation of attenuated *Listeria* vaccines is an endolysin gene, *ply118,* derived from *Listeria* temperate phage A118 attacking *L. monocytogenes* serovars 1/2.[75] This gene encodes for a cell-wall lytic enzyme that cleaves between L-alanine and D-glutamate residues of listerial peptidoglycan, but does not cleave the peptidoglycan of other bacteria in infected host cells.

Antibiotics. Antibiotics treatment (e.g., penicillin, streptomycin, and tetracycline) is very efficient in killing *L. monocytogenes in vivo.*[71,76] Ampicillin administration with the drinking water appears to influence the duration of *L. monocytogenes* infection in mice and on the magnitude of T cell responses[77–79] and has been shown to eliminate viable bacteria from spleens of *L. monocytogenes*-infected mice within 12 hours after administration of the antibiotic.[80] Mice undergoing repeated *L. monocytogenes* infection followed by antibiotic treatment may acquire an increased level of immunity against listeriosis.

15.2.2.2 Pros and Cons of Attenuated Live Vaccines

Schafer and colleagues[81] were the first to demonstrate the feasibility of using *L. monocytogenes* as an efficient attenuated live vaccine for induction of cell-mediated immune responses, especially CD8$^+$ T cell responses. Since then, many live attenuated *L. monocytogenes* vaccines have been developed. These live attenuated vaccines possess some clear advantages and potential drawbacks in comparison with other vaccine strategies, which are discussed next.

The advantages of these vaccines include:

- Propensity to infect antigen-presenting cells (APCs). *L. monocytogenes* specifically infects professional APCs, which include monocytes, macrophages,[82,83] and dendritic cells (DCs).[84,85] After being taken up by these cells, some bacteria escape from the phagocytic vacuoles into the cytoplasm and grow there. *L. monocytogenes* protein antigens are then presented to the MHC class I pathway to CD8$^+$ T cells. In addition, bacterial protein antigens are also delivered to CD4$^+$ T cells via the MHC class II pathway during listeriosis (see chapter 14). Thus, *L. monocytogenes* represents an ideal carrier for antigen-delivery to elicit a cell-mediated immune response.
- Ease of genetic manipulation. *L. monocytogenes* can be transformed relatively easily with plasmid vaccine vectors using electroporation, although the transformation efficiency of *L. monocytogenes* is lower than that of *Escherichia coli*.[86] Methods for growing and processing *L. monocytogenes* have been established,[87,88] along with an increasing number of genetic tools available for its genetic manipulation.[89] For example, shuttle vectors have been developed to facilitate easy insertion of foreign genes stably into the *Listeria* chromosome[90] or to allow conjugative transfer of plasmid DNA from *E. coli* to *L. monocytogenes*[91,92] (see chapter 9).
- Adjuvanticity. Murine immune responses to *L. monocytogenes in vivo* are essentially of Th1-type, in which interferon (IFN)-γ and interleukin (IL)-12 productions are predominant.[93,94] Cell-surface moieties of *L. monocytogenes* are recognized by host cells as "danger" signals, against which innate immune responses are provoked via the TLRs, the pattern recognition receptors on host cells. Cell activation by lipoteichoic acid and soluble peptidoglycan, the main stimulatory components of Gram-positive bacteria, has been shown to be mediated by TLR2.[95] Since lipopolysaccharide (LPS) in the cell wall of Gram-negative bacteria can cause severe side effects such as septic shock *in vivo*, Gram-positive bacteria constitute a safe and preferred option for vaccine delivery.

  Lipoteichoic acid, a predominant surface glycolipid of *L. monocytogenes* potently induces IL-12 p40 gene expression similar to LPS.[96] IL-12 has been shown to augment IFN-γ production by natural killer and T cells. Lipoteichoic acid derived from *L. monocytogenes* has been also shown to induce a transient IκBα degradation. After *L. monocytogenes* enters the cytoplasm of the host cell, its virulence genes, *plcA* and *plcB*, express two phospholipases, which lead to persistent induction of NF-κB DNA-binding activity.[70] In turn, this activity induces expression of a series of target genes including IL-12 p40 gene.

  Furthermore, the unmethylated cytidine–phosphate–guanosine (CpG) sequence abundant in *Listeria* genomic DNA may stimulate innate immunity through TLR9-mediated signal. Therefore, live attenuated *Listeria* acts as a "natural adjuvant," thereby attracting cells involved in innate immunity and promoting APC maturation and activation. Thus, dendritic cells, by far the most potent APCs, are activated to express the necessary costimulatory molecules that enhance subsequent antigen presentation activity, and bias the ensuing immune response in the direction of cell-mediated immunity.
- Simplicity of handling and storage. *L. monocytogenes* is a robust bacterium that can be stored as lyophilized powder, similar to the current procedure for *Mycobacterium bovis* BCG vaccine. In addition, it is easily controlled by common antibiotics such as penicillin and tetracycline, an option that is usually unavailable for viral vectors. In the case of a *Listeria*-carrying DNA vaccine system, plasmids for DNA vaccine do not require purification, which is an indispensable step for naked DNA vaccination.
- Mucosal route of immunization. Many human infectious diseases are initiated at mucosal surfaces. Attenuated live *Listeria*-carrying vaccine can be administered effectively by any one of several mucosal routes (e.g., oral, intranasal, intragastric, intravaginal, or rectal).[69,97] Administration of *L. monocytogenes* vaccines through these routes often elicits the immune

responses in the host that emulate those by natural infections and can lead to long-lasting protective mucosal and systemic immunity. Moreover, vaccination via a mucosal route is associated with fewer side effects and in many cases lower delivery costs. One caveat of the murine model of the oral *Listeria* vaccine system would be that murine E-cadherin, a *Listeria* surface protein that serves as a receptor for internalin A, was reported to have a glutamine to a proline substitution.[98,99] As *L. monocytogenes* surface proteins InlA and InlB are responsible for invasion of nonphagocytic cells and seem to determine the specificity of the cell type to be infected, wild-type murine models of *Listeria* infection are not appropriate for addressing internalin function *in vivo.*

- Amplification of DNA vaccine plasmids *in vivo.* The low efficiency of traditional naked DNA vaccination can be due to the limited amounts of DNA vaccine plasmids available *in vivo.* As *L. monocytogenes* harboring DNA vaccine plasmid allows the plasmid replication inside, the amounts of *Listeria*-carrying DNA vaccines will increase *in vivo.*

Like other attenuated bacterial vaccines and naked DNA vaccines, *Listeria*-carrying vaccines have several biosafety risks. Potential drawbacks include:

- potential risk of integration of *Listeria*-carrying genes to the host cell chromosome
- the spread to, and long-term persistence of, the plasmid in multiple tissues
- the induction of tolerance to the immunization antigens
- the risk of autoimmune disease by elicitation of anti-DNA Abs
- possibility of reversion to toxic/virulent phenotype

A live vaccine strain should contain appropriate attenuation mutations in genes that are essential for survival in the host cells. Strains containing more than two genetically unlinked attenuating mutations are infinitely better than those that have one mutation. However, too much attenuation may reduce immunogenicity of the vaccine. Therefore, the balance between attenuation (safety) and maintaining immunogenicity of vaccines is one of the important issues for clinical adaptation of vaccine candidates.

The bacterial restriction and modification system may sometimes hamper the maintenance of plasmids introduced into *L. monocytogenes*. Host specificity in a bacterial strain is the result of the action of particular enzymes that impose a "modification" pattern on DNA. The pattern identifies the source of the DNA. Modification allows the bacterium to distinguish between its own DNA and any "foreign" DNA, which lacks the characteristic host modification pattern. This difference renders an invading foreign DNA susceptible to attack by restriction enzymes that recognize the absence of methyl groups at the appropriate sites. Such "modification and restriction" systems are widespread in bacteria, although some bacterial strains lack any restriction system. *E. coli* strains widely used in laboratories all over the world have mutations in these systems by genetic manipulation. However, live *Listeria* strains used as vaccine carriers may still have these systems and interfere with maintenance of exogenous plasmids in them after introduction of plasmids by electroporation. These "modification and restriction" systems will not be a problem after introduction of plasmids by conjugation.

Intrinsic immunogenicity of viable bacteria itself is another issue. Live attenuated *L. monocytogenes* is composed of a variety of proteins, lipid, sugar, and so on, which are themselves immunogenic in host cells. So, repeated immunization of live attenuated *Listeria* vaccine is possible to cause rapid elimination of live *Listeria* vaccine. A similar situation has been noticed with adenovirus-based vaccines. One useful strategy to overcome this possibility is to combine different vaccination methods (so-called prime-boost strategy). This is achieved by immunization with naked DNA vaccination first, and followed by boost immunization with live *Listeria* vaccine.[100–102] Alternatively, it can be done by injection with different carrier bacteria for each immunization step.

Preexisting immunity to *Listeria* is also one of the issues concerning the use of attenuated *Listeria* carrier vaccines. *L. monocytogenes* is ubiquitously distributed in nature, being found commonly in soil, decaying vegetation, water, and as part of the normal flora of mammalian intestines[2] (see chapter 2). However, the incidence of listeriosis is low. Therefore, preexisting immunity to the vector may not be widespread in humans. After evaluating whether existing antilisterial immunity limits or alters efficacy of *Listeria* vaccine, several investigators confirmed that antilisterial immunity does not inhibit the development of recall responses as well as primary immune responses to antigens delivered by the *Listeria* vaccine system and that preexisting immunity to *L. monocytogenes* does not prevent induction of immune responses by *Listeria* vaccines.[103–105]

15.2.3 Naturally Avirulent Live Vaccines

Although *L. monocytogenes* has been regarded as being pathogenic at the species level, it encompasses a diversity of strains with varied pathogenic potential. Of the 11 common serotypes (i.e., 1/2a, 1/2b, 1/2c, 3a, 3b, 3c, 4a, 4b, 4c, 4d, and 4e), three (1/2a, 1/2b, and 4b) have accounted for over 98% of clinical cases of human listeriosis. Through analysis in mouse models, it is clear that except for serotype 4a, all other *L. monocytogenes* serotypes are potentially virulent and able to cause mortality in the A/J mouse strain via an intraperitoneal route. Given that naturally avirulent serotype 4a strains are not isolated from human patients and are readily eliminated by mammalian hosts, they can be useful as candidate vaccine against virulent strain challenge,[106–109] and also as a vaccine vehicle for delivery of protective molecules against other microbial pathogens.

The concept of using naturally avirulent bacterial strains as vaccines against virulent strains has been explored successfully in several bacterial systems. For instance, an avirulent live *Lawsonia intracellularis* strain administered orally to pigs facilitated the development of protective immunity against challenge with virulent *L. intracellularis.*[110] A *Mycoplasma gallisepticum* (MG) isolate (K5054) causing very mild lesions prevented challenge infection with a virulent MG strain in turkeys.[111] Immunization of chicken with avirulent *Salmonella* carrying *Campylobacter cjaA* gene hindered the heterologous wild-type *C. jejuni* strain to colonize the bird cecum.[112] Furthermore, mice injected with naturally avirulent *L. monocytogenes* serotype 4a strains ATCC 19114, L99, and HCC23 developed a strong immunity (both short and long term) against virulent strains' challenges.[106–109]

Originally isolated from catfish, serotype 4a strain HCC23 is nonpathogenic in mouse virulence assay, and it lacks several virulence-specific genes such as those encoding internalins (*lmo2821* and *lmo2470*) and transcriptional regulators (*lmo0833, lmo1116, lmo1134,* and *lmo2672*). Recent DNA sequencing analysis revealed that, apart from a deletion of 105 nucleotides in its *actA* gene leading to a reduction of 35 amino acids in the encoded ActA protein, HCC23 possesses largely intact *prfA, plcA, hly, mpl,* and *plcB* genes in the *prfA* virulence gene cluster in comparison with EGD (GenBank accession nos. AY878649 and DQ118415). Whereas EGD ActA protein possesses four copies of proline-rich repeats (DFPPPPTDEEL) that control *L. monocytogenes* movement between cells (with each repeat contributing about 2 μm/min), the deletion of 35 amino acids in HCC23 ActA protein effectively removes two copies of the proline-rich repeats (Genbank accession no. DQ118415). Thus, it is no surprise that while serotype 4a strains readily replicate within fibroblast L929 and macrophage J774 cells, they are unable to undergo cell-to-cell spread. Apparently, serotype 4a strains produce a nonfunctional ActA protein that is 5 kDa smaller than that from virulent strain EGD.[106–109]

IFN-γ is a vital component in the cell-mediated immune response, which activates antimicrobial macrophages and NK cells for subsequent development of host natural resistance and adaptive immunity to listeriosis. In contrast to mice injected with virulent strain EGD that produced a large amount of IFN-γ (about 580 pg/ml of serum on day 2 after exposure), mice injected with avirulent strains HCC23 and ATCC 19114 generated a much lower level of IFN-γ production (about 50 pg/ml of serum on day 2 after exposure). However, similar to mice immunized with virulent strain EGD, mice immunized with avirulent strains HCC23 and ATCC 19114 all developed a strong immunity against *L. monocytogenes* challenge.[108,109] These results suggest that in spite of generating a relatively smaller (or negligible) amount of IFN-γ than virulent strain EGD, avirulent strains HCC23 and ATCC 19114 are equally efficient in initiating a protective immunity against listeriosis. It is thus apparent that induction of excessive amounts of IFN-γ, such as in the case with EGD, may not only be unnecessary, but actually detrimental to the host. As IFN-γ enhances secretion of TNF-α, it may bring such a potent immune response that the host may succumb as a consequence, and naturally avirulent serotype 4a strains may cause much less potential harm to the vaccine recipient through reduced IFN-γ and TNF-α production.

Besides being useful as a candidate vaccine against listeriosis, *L. monocytogenes* avirulent serotype 4a strains also hold promise as a potential vaccine carrier for other bacterial, viral, fungal, and parasitic diseases as well as cancers. The application of a naturally avirulent strain not only negates the necessity to attenuate *L. monocytogenes* virulent strains through deletion or replacement of certain virulence-associated genes such as *actA*, *hly*, and *inlB*, but provides a much safer approach for effective delivery of vaccine molecules. Further study is required to insert a foreign gene into a site (e.g., *actA*) in an avirulent *L. monocytogenes* serotype 4a strain followed by assessment of its vaccine efficacy. This will help verify the potential of using naturally avirulent strains as an effective alternative to live attenuated *L. monocytogenes* strains for vaccine delivery.

15.2.4 Subunit Vaccines

Built on a detailed knowledge of the molecular and immune mechanisms of bacterial pathogens, it becomes possible to utilize more defined, protective subcellular components for enhanced vaccine efficiency and safety. This not only helps eliminate some unquantifiable and unintended side effects associated with the use of whole cell vaccines, but also gives batch-wise reliability and consistency. Similarly, utilizing plasmids, which are replicable in *E. coli* but not in hosts, to express protective proteins *in situ* creates so-called DNA vaccines that further improve the vaccine efficacy and streamline the vaccine production processes. The subcellular bacterial vaccines against diphtheria/tetanus/pertussis (acellular), *Hemophilus influenza* b, *Streptococcus pneumoniae,* and meningococci are some of the excellent examples of the types of vaccines that have been successfully developed.

As the 58-kDa LLO of *L. monocytogenes* is required for the initiation of specific immunity, many innovative designs incorporating this protein have been described to enhance vaccine potency.[113–121] *L. monocytogenes* LLO is a pore-forming, cholesterol-dependent cytolysin belonging to a family of homologous proteins present in several pathogenic bacteria, such as perfringolysin O derived from *Clostridium perfringens* and streptolysin O from *Streptococcus pyogenes*. Using plasmid DNA constructs encoding recombinant forms of LLO, peptide-specific $CD8^+$ immune T cells exhibiting *in vitro* cytotoxic activity were activated in immunized mice, and provided protection against a subsequent *L. monocytogenes* challenge.[113–115] Injection of IFN-γ or IFN-γ receptor knockout BALB/c mice with plasmid DNA constructs encoding recombinant forms of the *L. monocytogenes* LLO or a minigene encoding LLO 91–99 (GYKDGNEYI) elicited the peptide-specific $CD8^+$ T cell response and also resulted in protection against listeriosis.[116,117] A recombinant LLO 91–99 minigene vaccinia virus infection provided partial protection against listeriosis.[118] Injection with LLO 91–99-coated DCs[119] or DCs retrovirally transduced with LLO 91–99 minigene[120] protected mice against high-dose challenge with virulent *L. monocytogenes*. Mice immunized with plasmid DNA constructs for an Ii-LLO 215–226 fusion protein in which LLO 215–226 is a dominant Th epitope also developed significant protective immunity against *L. monocytogenes*.[121]

15.3 *LISTERIA*-BASED ANTIGEN DELIVERY SYSTEMS

15.3.1 Delivery of Antigenic Protein

Listeria-carrying vaccines that produce antigenic proteins of interest have been used to induce cell-mediated immunity, especially specific CD8+ T cell responses.[122–124] Essentially, there are two ways to deliver antigens using the attenuated *Listeria* vaccine system. One is antigenic protein delivery by *Listeria* that itself produces antigens, and the other is to use attenuated *Listeria* as a DNA vaccine carrier (see section 15.3.2). Secreted proteins of intracellular bacteria such as *Listeria* and *Salmonella* have been shown to be good inducers of cell-mediated immune responses because bacterial secreted proteins are vulnerable to the protein degradation system of host cells and the antigen processing and presenting systems. Only secreted (not nonsecreted) proteins of *Salmonella* have been shown to be effectively presented by the MHC class I molecules, followed by induction of CD8+ T cell responses.[125,126] In contrast, infection of mice with recombinant *L. monocytogenes* expressing a model epitope in either secreted or nonsecreted form resulted in similar CD8+ T cell priming efficacy by a cross-presentation mechanism.[127–129] Interestingly, Shen and colleagues[127] reported that antigen-specific CD8+ T cells conferred protection only against *Listeria* expressing the secreted antigen, but not against *Listeria* expressing the nonsecreted form of the same antigen, thus indicating that nonsecreted antigens in *L. monocytogenes* have immunogenicity for CD8+ T cells, but do not necessarily become targets of the CD8+ T cells.

Foreign genes encoding antigens of interest may be carried on plasmid or integrated in the *Listeria* genome. Plasmids harboring foreign genes are maintained episomally in *Listeria*. A number of plasmid vectors developed for use in other Gram-positive bacteria have been applied successfully in *L. monocytogenes* and several of these have been improved or modified. A typical plasmid for antigen production in *Listeria* is shown in Figure 15.1A. The plasmid should contain replication origins of both *L. monocytogenes* and *E. coli*. A replication origin of pAMβ1 derived from *Streptococcus faecalis*[130]

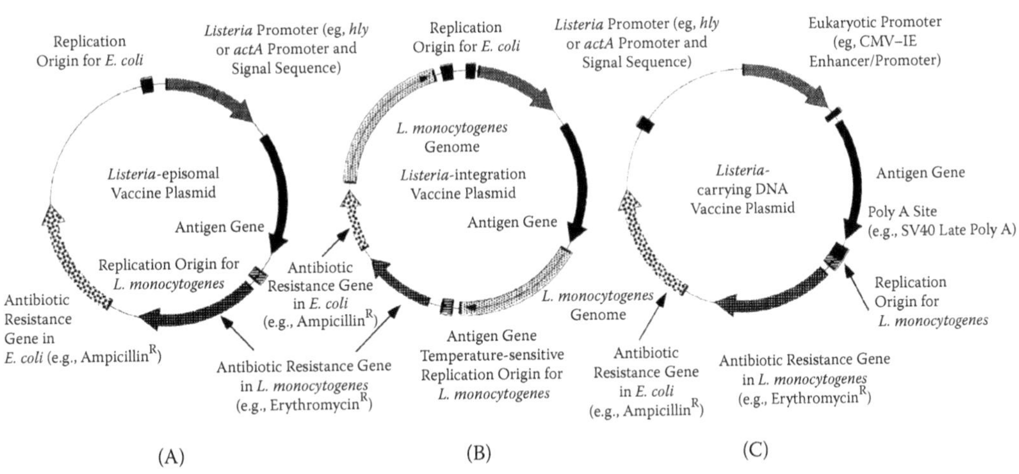

FIGURE 15.1 Schematic drawings of typical plasmids for antigen expression in *Listeria*-carrying vaccine system. (A) A typical plasmid for antigen expression in *Listeria* contains replication origins for *L. monocytogenes* and *E. coli*, antibiotic resistance genes for *L. monocytogenes* and *E. coli*, and a gene that encodes the target antigen, which is driven by a strong *Listeria* promoter. In the case of genome integration plasmid (B), upstream region and downstream region of the antigen gene should be *Listeria* genome sequence for homologous recombination with *Listeria* genome and a temperature-sensitive origin of replication of *Listeria* facilitates integration of the plasmid into *Listeria* genome. (C) A plasmid for antigen expression in infected host cells contains replication origins for *L. monocytogenes* and *E. coli*, antibiotic resistance genes for *L. monocytogenes* and *E. coli*, and a gene that encodes the target antigen, which is driven by a strong eukaryotic promoter such as CMV-IE enhancer/promoter.

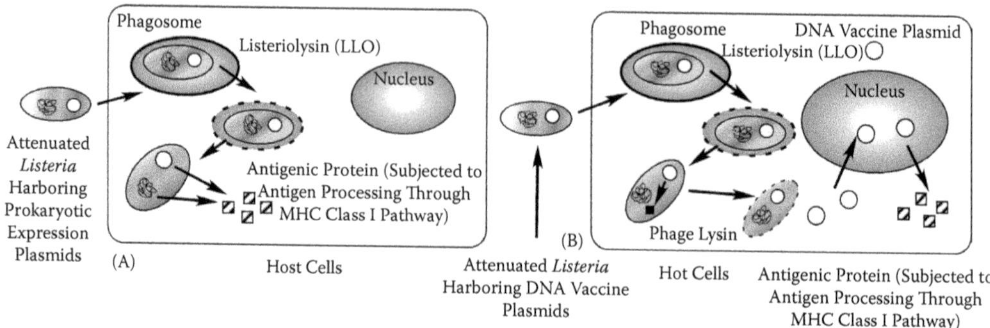

FIGURE 15.2 Schematic drawings of antigen delivery system by *Listeria*-carrying vaccines, with antigens being produced either in *Listeria* (A) or in infected host cells (B). (A) An attenuated *Listeria* strain harboring vaccine plasmid or an integrated antigen gene in the genome infects macrophages and is released into the cytoplasm of infected host cells by virtue of LLO of *L. monocytogenes*. The bacterium produces the antigenic proteins (usually secreted proteins) *in situ* in the cytoplasm of the infected host cells by using *Listeria* transcription and translation machinery. Produced antigenic proteins are then subjected to antigen processing and presentation, mainly through the MHC class I pathway, leading to specific CD8$^+$ T cell responses. (B) An attenuated *Listeria* strain harboring DNA vaccine plasmids infects macrophages and is released into the cytoplasm of infected host cells as in (A). Then, if plasmids contain a suicide gene such as listeriophage gene, *ply118*, the gene product lyses *Listeria* from inside and releases DNA vaccine plasmids into the cytoplasm of the infected host cells, allowing expression of the antigen by using host cell transcription and translation machinery. Produced antigenic proteins are subjected to antigen processing and presentation systems of host cells, mainly through the MHC class I pathway, leading to specific CD8$^+$ T cell responses.

has been the most adequate for stable replication in *L. monocytogenes*.[131] Antibiotic resistance genes in *L. monocytogenes* and also *E. coli* have been employed to facilitate selection of *L. monocytogenes* and *E. coli* possessing the plasmid. A heterologous antigen gene is driven by *Listeria* promoter, usually the *hly* (encoding LLO) or *actA* promoter. Codon usage change may be necessary for high expression of the heterologous antigen gene, as optimal codons for expression of proteins in a variety of organisms substantially differ. For example, *L. monocytogenes* usually used A+T-rich codons and *M. tuberculosis* tends to use G+C-rich codons. Introduction of plasmid DNA into *L. monocytogenes* can be accomplished by transformation of *L. monocytogenes* protoplasts[132] or by electroporation.[86]

The antigen delivery mechanism by *Listeria* is shown in Figure 15.2(A). *L. monocytogenes* is released to the cytoplasm of host cells from phagosomes with assistance from LLO, and antigenic proteins produced in recombinant *L. monocytogenes* are secreted to the cytoplasm of host cells, which are then subjected to antigen processing and presenting machinery of host cells through the MHC class I and also class II pathways. Usage of high-copy-number plasmids would lead to a high yield of foreign antigens in *Listeria,* although the forced expression of foreign genes might be deleterious in terms of growth of the bacteria. One of the issues pertaining to this episomal system is that the plasmid would be cured easily *in vivo* without selection pressure. However, as elimination of *Listeria* just 1 day after infection has been reported to have minimal impact on CD8$^+$ and CD4$^+$ T cell responses, short-time antigen display *in vivo* with recombinant *Listeria* vaccines may be sufficient for induction of immune responses.

Ikonomidis and colleagues[133] reported that infection of mice with the mutant *L. monocytogenes* secreting the LLO-influenza virus nucleoprotein (NP) fusion protein (Lm-NP) led to *in vivo* priming of CTLs in BALB/c mice, which recognized and lysed influenza virus–infected syngeneic target cells. The gene coding for the transcriptional activator prfA was introduced on the plasmid, as PrfA is a key transcription factor for a series of virulent factor genes including LLO and necessary for maintaining virulence of the bacterium. A replication-defective *prfA* (–) *L. monocytogenes* strain was transformed with the plasmid, and the resulting transformants retained the plasmid stably in the episomal state *in vivo* and efficiently secreted the fusion protein.

Verch and colleagues[134] developed a new system in which D-alanine racemase-deficient *Listeria* mutant (Δ*dal*Δ*dat*; *Lmdd*)[42] carrying vaccine plasmids was utilized. The vaccine plasmid incorporated *Listeria dal* gene driven by *Listeria* p60 promoter, whose expression supplements the requirement of exogenous D-alanine for the bacteria growth. Retention of the antibiotic-resistant gene-free plasmid in the *Listeria* carrier in the system resulted in the development of *Listeria* episomal vaccines without using the antibiotic-resistant gene, which complies with Food and Drug Administration regulations.

Alternatively, delivery of antigenic protein by *Listeria* can be achieved through the genome integration system, in which foreign antigen genes are integrated in *Listeria* genome, which allows stable retention of foreign genes. Several different strategies for integration of a heterologous antigen gene into *Listeria* genome have been reported. The plasmids can be grouped according to their usages: plasmids used for the delivery of transposons into *L. monocytogenes*[135] and plasmids used for homologous recombination with *Listeria* chromosomal sequences. Gram-positive, temperature-sensitive origins of replication have been utilized to facilitate integration and excision of plasmid vectors. One representative scheme is shown in Figure 15.1B. In this case, integration of the antigen gene is accomplished after two successive homologous recombination events in two *Listeria* genome homologous regions in the plasmid. The temperature-sensitive replication origin of *L. monocytogenes* facilitates the events as mentioned before. After introduction of the plasmid into *L. monocytogenes* by electroporation, the plasmid-carrying bacteria are selected by antibiotic addition. Then the antibiotic-resistant bacteria are subjected to growth at nonpermitted temperature (e.g., 40°C) to prevent replication of the plasmid. Thus, only bacteria that integrate the plasmid into the genome are positively selected. Once the plasmid is integrated into *Listeria* genome, the system does not require the presence of an antibiotic to select for the retention. It is possible for the expression level to be low due to low copy number (usually one copy) of a foreign gene in *Listeria* genome, but selection of an appropriate promoter for expression of a foreign gene will help to enhance the antigen expression.

Shen and colleagues[136] used the genome integration system for attenuated *Listeria* vaccines against LCMV. Several plasmids for stable, site-specific integration of antigen expression cassettes into the *Listeria* chromosome were constructed by sequential homologous recombination events. The plasmids were designed to drive an antigen gene, LCMV-NP, by the promoter of the *hly* gene and to integrate the antigen gene into the *Listeria* chromosome region just downstream from a transcriptional terminator between the lecithinase (*plcB*) and lactate dehydrogenase (*ldh*) operons. LCMV antigen expression was regulated *in vitro* in a manner identical to the chromosomal *hly* locus and expression of LCMV-NP was enough to induce perfect protection against LCMV challenge infection.

Lauer and colleagues[137] developed a *Listeria* site-specific phage integration vector system (i.e., pPL2 plasmid), which carries listeriophage PSA integrase and PSA *attPP′* sites to integrate a foreign gene into the 3′ end of an arginine tRNA gene of the *L. monocytogenes* genome. As commonly used *L. monocytogenes* strains 10403S and EGDe do not have a prophage at the $tRNA^{Arg}$ attachment site, pPL2 plasmid is readily utilized in these backgrounds for integration-type attenuated *Listeria* vaccine construction. In addition, the pPL2 plasmid can integrate in an *L. monocytogenes* strain that already carries a prophage at the $tRNA^{Arg}$ attachment site, such as LO28, which further increases the usefulness of pPL2 as a genetic tool for integration of antigen genes into the *L. monocytogenes* genome.

15.3.2 Delivery of DNA

Another way to deliver antigens is to use attenuated *Listeria* as a DNA vaccine carrier—in other words, as a vehicle of genes encoding target antigens. This bacteria-mediated DNA delivery to host mammalian cells has been designated as bactofection.[138,139] Bacteria as DNA vaccine carriers for genetic immunization have been reviewed by several investigators.[123,140–151] Usually, the genes are driven by mammalian promoter (Figure 15.1C). *Listeria* is used as only a vehicle/vector to transport the gene into the eukaryotic cells. The released plasmids get into the nucleus and the transgene is expressed by the eukaryotic transcription and translation machinery (Figure 15.2B). The ability of *L. monocytogenes* to enter the cytoplasm of host cells after phagocytosis and deliver plasmid DNA directly to the cytoplasm makes it an attractive DNA delivery candidate to induce cellular immune responses.

RNA interference has been used for post-transcriptional gene silencing, which has been observed in many eukaryotic cell types. Genetic interference with specific double-stranded RNA has been reported via bactofection.[152] By feeding recombinant *E. coli* to *Caenorhabiditis elegans*, a free-living nematode, it may offer a new approach for construction of attenuated *Listeria* strains for antigenic delivery.

Naked DNA vaccine strategy has a variety of advantages over other vaccine methods,[153,154] but its chief weaknesses include low immunogenicity, a need for adjuvant, and the high cost of purifying injection-grade plasmid DNA. Use of *Listeria* to deliver plasmid DNA negates the need to purify plasmid prior to injection, and thus it provides a more efficient and low-cost alternative to the naked DNA vaccine strategy.

Figure 15.1C illustrates a typical plasmid for a *Listeria*-carrying DNA vaccine system. Structure of plasmid DNA for a *Listeria*-carrying DNA vaccine system is essentially the same as that of the plasmid DNA for naked DNA vaccination. DNA vaccines are composed of an antigen-encoding gene whose expression is driven by a strong eukaryotic promoter such as CMV I.E. enhancer/promoter. The plasmid possesses a polyadenylation termination sequence such as the sequence derived from simian virus 40 or bovine growth hormone gene and a selective marker such as ampicillin resistance gene to facilitate selection of *E. coli* containing the plasmid. Usually, plasmids for naked DNA vaccines include special nucleotide sequences for enhancing the immunogenicity and an unmethylated CpG dinucleotide with appropriate flanking regions. The CpG motif stimulates the innate immune system through TLR9 to produce a series of immunomodulatory cytokines such as IL-12 and IFN-γ, which promote the development of Th1 cells.[155–157] However, in the case of *Listeria*-carrying DNA vaccines, components of *Listeria* itself, such as lipoteichoic acids, stimulate the innate immune system. Therefore, the contribution of the CpG sequence in *Listeria*-carrying DNA vaccines as an immune adjuvant is low. Plasmid DNA for *Listeria*-carrying DNA vaccines should contain an origin of replication for *Listeria*. Introduction of plasmid DNA into *L. monocytogenes* is again accomplished by transformation of *L. monocytogenes* protoplasts[132] or by electroporation.[86]

Hense and colleagues[76] evaluated *Listeria* as a vehicle for gene transfer using a variety of cell lines. Gene transfer to host cells was achieved after treating host cells infected with plasmid-carrying *Listeria* with tetracycline, an antibiotic that is only bacteriostatic, as tetracycline treatment makes the bacteria susceptible to cellular defense and degradation mechanisms. It appeared that bacterial properties required for delivery of the eukaryotic expression plasmids were strictly dependent on the ability of *Listeria* to both invade eukaryotic cells and egress from the vacuole into the cytoplasm of the infected host cells, and that macrophage-like cells or primary, peritoneal macrophages were almost refractory to *Listeria*-mediated gene transfer.

Dietrich and colleagues[71] demonstrated the feasibility of a DNA vaccination system of an attenuated self-destructing *L. monocytogenes* strain in a cell culture system. A deletion mutant strain of *L. monocytogenes*, Δ2, that lacks the entire lecithinase operon, including the virulence-associated genes *actA*, *mpl*, and *plcB*, can infect macrophages and replicate in the cytoplasm of host cells, but cannot spread to adjacent cells.[70,71] This attenuated mutant was introduced with a plasmid containing the gene for lysis protein PLY118 of the listerial bacteriophage A118. The *ply118* gene expression is controlled by the *actA* promoter, which is active when *L. monocytogenes* is in the host cell cytoplasm. Thus, this *L. monocytogenes* mutant escapes from the phagosome and then lyses when the *ply118* gene is expressed in the cytoplasm of host cells. Autolysis of the *L. monocytogenes* mutant apparently releases the plasmid DNA into the host cell cytoplasm, allowing expression of the transgene in the host cells (Figure 15.2B). The DNA vaccine transfer system was first assessed with mouse macrophage cell line, and further evaluation was conducted *in vivo* by infection of

BALB/c mice and cotton rats with the attenuated suicide Δ2 mutant *Listeria* strain carrying GFP-expressing DNA vaccine.[143] Peritoneal macrophage cells isolated from cotton rats infected with the *Listeria*-carrying vaccine exhibited highly efficient expression of GFP, demonstrating that DNA vaccine delivery by the *Listeria* strain is functional *in vivo*. The same system has been also evaluated with primary human DCs and up to 1% of the DCs were shown to express GFP derived from plasmid DNA transferred from the *Listeria* strain.[144,145] Miki and colleagues[72] applied this system for DNA vaccines against *M. tuberculosis*, through construction of the self-destructing attenuated *L. monocytogenes* Δ2 strains carrying eukaryotic expression plasmids for mycobacterial antigen 85 complex molecules (Ag85A and Ag85B) and for the MPB/MPT51 molecule. Immunization of these *Listeria*-carrying DNA vaccines to BALB/c mice intravenously elicited significant protective responses against virulent *M. tuberculosis* challenge.

Although these plasmids were capable of inducing immune responses against *M. tuberculosis in vivo*, they tended to show *in vivo* instability within the carrier *Listeria*. Pilgrim and colleagues[131] modified the *Listeria* system in order to stabilize the plasmid in the *L. monocytogenes* carrier strain. They constructed an *L. monocytogenes* strain that has the chromosomal deletion region encompassing *trpS* gene (encoding tryptophanyl-tRNA synthetase) and also *actA* gene. As *trpS* gene is essential for viability of the bacterium, *trpS*-deleted *Listeria* can maintain only in the presence of a plasmid carrying *trpS* gene. DNA vaccine plasmids were constructed that have *trpS* gene in addition to a listerial autolysis cassette consisting of the *ply118* lysis gene under the control of the *actA* promoter, which is activated in the cytoplasm of infected host cells. No plasmid loss for more than 50 generations of the *Listeria* strain was noted. This new *Listeria*-carrying DNA vaccine permits more efficient cell-to-cell spread than the nonspreading counterparts such as Δ2 mutant *Listeria* strain. However, DNA uptake into the nucleus of host cells is still a major limiting step for optimal gene expression.

Schoen and colleagues[158] described a further improved *Listeria*-carrying vaccine system in which translation-competent mRNA is released directly into the cytoplasm of host cells, leading to immediate translation and production of antigen molecules in host cells. They noticed a much earlier expression of a model protein, enhanced GFP (EGFP), which was detectable within 4 hours of infection, with a much higher number of EGFP-expressing mammalian cells being generated with this novel mRNA delivery system compared to the plasmid DNA delivery system. Loeffler and colleagues[159] also assessed the *in vivo* activation of antigen-specific $CD8^+$ and $CD4^+$ T cells by *L. monocytogenes*-secreted antigen, as well as by *L. monocytogenes* self-destructing, *aroA/B* mutant strain–harboring antigen-encoding plasmid DNA or mRNA. It seems that secretion of a model antigen (OVA) by *L. monocytogenes* yielded the strongest immune responses involving OVA-specific $CD8^+$ and $CD4^+$ T cells. Infection of mice with self-destructing *L. monocytogenes* delivering mRNA for OVA resulted in a significant OVA-specific $CD8^+$ T cell response. However, infection with *L. monocytogenes* delivering OVA-encoding DNA failed to generate specific T cells.

15.4 ANTIVIRAL VACCINES

Recombinant *L. monocytogenes* have been successfully used as biologic vaccine vectors against LCMV, murine influenza virus, and human immunodeficiency virus (HIV) (Tables 15.2 and 15.3). In murine virus infection systems of LCMV and influenza virus, $CD8^+$ T cell responses have been shown to be pivotal for induction of protective immunity through adoptive transfer of virus-specific $CD8^+$ T cells or $CD8^+$ T cell-oriented vaccination studies. In addition to $CD8^+$ T cell responses, *L. monocytogenes* has been shown to induce $CD4^+$ T cell responses. Adjuvanticity of the bacterium directs the $CD4^+$ T cell responses toward Th1-type responses, which are important for protective immunity against viruses.

Murine LCMV infection has been employed as an excellent prototype model of immune responses, especially $CD8^+$ T cell responses, against viruses. Study of murine LCMV infection led to the discovery of MHC restriction.[160] Recombinant *Listeria* vaccines have been developed for

TABLE 15.2
Recombinant *L. monocytogenes* Vaccines for Lymphocytic Choriomeningitis Virus (LCMV) and Influenza Virus

Attenuation	Antigen	Promoter	Immunization Route	Immune Responses	Protection	Ref.
			LCMV			
actA	NP	*hly* (episomal)	i.v.	CD8	+	161
Wild type	NP	*hly* (chromosomal); *actA* (chromosomal)	i.v.	CD8	+	136, 163
Wild type	NP 118–126 (secreted and nonsecreted forms)	*hly* (chromosomal)	i.v.	CD8	ND[a]	127
Wild type	NP	*actA* (chromosomal)	i.v.	CD8	+	162
hly	NP 118–126 (secreted form)	*hly* (chromosomal)	i.v.	CD8	+	58
			Influenza Virus			
hly	LLO-flu NP fusion	*hly* (episomal)	i.v.	CD8 (LLO$^+$ Listeria); CD4 (LLO$^-$ Listeria)	+	133
hly	LLO-flu NP fusion	*hly* (episomal)	i.v.	CD8	+	164

[a] ND: not determined.

induction of specific CD8$^+$ T cell responses and protective immunity against LCMV. Goossens and colleagues[161] reported an episomal *Listeria* vaccine for LCMV. Immunization of *L. monocytogenes* secreting LLO-NP fusion protein induced LCMV-specific CD8$^+$ T cells and protective immunity in mice against an otherwise lethal intracerebral LCMV challenge. The plasmid was maintained in *L. monocytogenes* for a short time, which appeared adequate for provoking effective antigen-specific CD8$^+$ T cell responses.

Shen and colleagues[136] used a CD8$^+$ T cell epitope in NP of LCMV as a model antigen to examine T cell responses to the secreted or nonsecreted form of the antigen produced by *L. monocytogenes*. Induction of CD8$^+$ T cell responses by both secreted and nonsecreted forms of the antigen was observed. However, only the secreted antigen was capable of being a target of CD8$^+$ T cells. Further, the LCMV-NP-producing recombinant *Listeria* vaccines were protective against subsequent LCMV challenge. Analysis of kinetics of challenged LCMV titer levels in LCMV-NP-producing *Listeria*-immune mice indicates that the LCMV was cleared at 28 days after LCMV challenge, which was later than the virus-cleared timing in LCMV-immunized mice.[162] Duration of LCMV antigen existing *in vivo* was paralleled with the strength of the protective immunity induced by LCMV or by the recombinant *Listeria* vaccine.[163]

Influenza virus NP-expressing *Listeria* vaccine has been also examined for the efficacy to elicit protective immunity against flu virus infection. Ikonomidis and colleagues[133] prepared *L. monocytogenes* secreting a fusion protein consisting of LLO and influenza virus NP, a major protective antigen of influenza virus. Infection of wild-type *L. monocytogenes* strain secreting the LLO-NP fusion protein was capable of processing and presenting the NP antigen to CD8$^+$ T cells through the MHC class I pathway. In contrast, infection of a *hly*-deficient *L. monocytogenes* strain secreting

TABLE 15.3
Recombinant *L. monocytogenes* Vaccines for Human Immunodeficiency Virus (HIV)

Attenuation	Antigen	Promoter	Immunization Route[a]	Immune Responses	Protection	Ref.
Wild type	HIV-1 Gag	*hly* (chromosomal)	i.v., i.p.	CD8	ND[b]	169
hly, actA	HIV-1 gp120	*actA* (episomal)		CD4	ND	173
Wild type	HIV-1 Gag	*hly* (chromosomal)	i.p.	CD4	ND	171
dal, dat	HIV-1 Gag	*hly* (chromosomal)		CD8	ND	174
Wild type	HIV-1 Gag	*hly* (chromosomal)	i.p.	CD8, CD4	+[c]	172
dal, dat	HIV-1 Gag	*hly* (chromosomal)	i.p., i.g.	CD8	+[c]	97
dal, dat	HIV-1 Gag	*hly* (chromosomal)	i.p.	CD8	+[c]	175
dal, dat	HIV-1 Gag	*hly* (chromosomal)	i.p.	CD8	+[c]	176
Wild type	HIV-1 Gag	*hly* (chromosomal)	i.v., i.g., Intrarectal	CD8	ND	178
Wild type	FIV Gag; FIV Env	*hly* (chromosomal) + *CMV-IE* (episomal)	Oral	IgA, IgG	+[d]	179, 180
Wild type	SIV Gag; SIV Env	*hly* (chromosomal)	Oral	CD8	+[e]	100–102[f]
dal, dat (pARS)	HIV-1 Gag	*hly* (chromosomal)	i.v., i.g.	CD8	+[c]	177

[a] i.g., intragastric.
[b] ND: not determined.
[c] Protection against HIV-1 Gag-expressing vaccinia virus in mice.
[d] Protection against FIV in cats.
[e] Protection against SIV in Chinese rhesus macaques.
[f] These three studies used prime-boost strategy (i.e., DNA vaccination followed by *Listeria* vaccination).

the LLO-NP fusion protein was subjected to processing and presenting of the NP antigen to CD4+ T cells, not to CD8+ T cells. As the study involved the episomal *Listeria* vaccine, almost all *Listeria* colonies recovered from organs of mice 48 hours after immunization had lost the plasmid. To avoid the loss of plasmid in recombinant *Listeria*, a *prfA*-deficient *L. monocytogenes* mutant was used as the carrier of a plasmid containing *prfA* and the LLO-NP fusion gene, as mentioned before.[133] Two recombinant *Listeria* vaccines—one strain expressing only an H2-K^d-restricted NP epitope (NP 147-155) and the other strain expressing the full-length NP sequence fused with LLO—helped clear influenza virus infection in mice rapidly, suggesting the feasibility of recombinant *Listeria* vaccines for future influenza virus vaccine.[164]

L. monocytogenes has been well studied as a vaccine platform for HIV.[165,166] A series of studies on *Listeria* vaccines against HIV have used HIV-1 Gag as the target protein. HIV-1 Gag is a viral core protein highly conserved among different virus clades. This protein has been shown to be one of the major CTL antigens in HIV-positive individuals.[167,168] Frankel and colleagues[169] examined the first-generation recombinant *Listeria* carrying HIV-1 Gag or Nef. These genes were inserted into the nonessential *sepA* region of the *Listeria* chromosome, which resulted in three logs of attenuation and induced potent CD8+ T cell responses to the HIV gene products. The HIV-1 Gag-expressing *L. monocytogenes* vaccine was useful for identification of murine CTL epitope of HIV-1 Gag.[170] Following these studies, Mata and Paterson[171] examined CD4+ T cell responses with the same HIV-1 Gag-expressing *L. monocytogenes* immunization. BALB/c and C57BL/6 mice immunized with HIV-1 Gag-expressing *L. monocytogenes* produced CD4+ T cells specific for HIV Gag that secreted high amounts of IFN-γ, indicating Th1 phenotype. Mata and colleagues[172] showed that immunization of HIV-1 Gag-expressing *L. monocytogenes* protected mice against challenge infection by a recombinant vaccinia virus expressing HIV-1 Gag. This protective immunity was mainly attributable to a specific CD4+ T cell subset with a minor contribution of a specific CD8+ T cell subset.

Human CD4+ T cell responses have been also evaluated with recombinant *Listeria* vaccines. Guzmán and colleagues[173] noted that infection of *L. monocytogenes* expressing a helper T cell epitope (pep24) of HIV gp120 envelope glycoprotein to human macrophages and dendritic cells was capable of processing and presenting the epitope to a specific CD4+ T cell line in the context of the MHC class II molecules. These results were obtained using *hly*-deficient and *actA*-deficient carrier *L. monocytogenes* strains. The epitope was designed to be secreted, to be associated with the bacterial surface, or to be restricted to the bacterial body. Recombinant *Listeria* delivered antigens to the MHC class II pathway irrespective of *Listeria* localization in host cells (namely, phagosomes in *hly*-deficient strain and the cytoplasm in *actA*-deficient strain) or the mode of antigen display (secreted or nonsecreted).

Friedman and colleagues[174] introduced the HIV-1 *gag* gene into the highly attenuated *L. monocytogenes ΔdalΔdat* strain (*Lmdd*). The ability of the *Listeria* (*Lmdd-gag*) to infect human monocytes and to present HIV-1 Gag antigen to CD8+ T cells of HIV-infected donors for induction of a secondary T cell immune response was evaluated. The *Lmdd-gag* provided a strong stimulus for Gag-specific CTL in HIV-infected donor peripheral blood mononuclear cells as Gag-expressing wild-type *L. monocytogenes* did. Immunization of mice with this *Lmdd-gag* strain elicited strong and prolonged CD8+ T cell responses both systematically and in Peyer's patches and mesenteric lymph nodes of immunized mice. Furthermore, Rayevskaya and colleagues[175] evaluated *Lmdd-gag* immunization in neonatal mice as well as adult mice against the challenge of vaccinia virus expressing HIV-1 Gag. Immunization of *Lmdd-gag* was protective against the challenge both in adult and neonatal mice. Upon analysis of spleen cells of mice immunized with *Lmdd*-gag, Rayevskaya and colleagues[176] found that the antigen-specific CTL activity reached a maximum at about 9 days after immunization of *Lmdd-gag*. Concomitant with the fall of CTL activity of immune spleen cells, the number of CD11a^{high} antigen-specific CD8+ T cells increased. The cells showed little or no 4-hour CTL activity, but had high delayed (18-hour) CTL activity. Zhao and colleagues[177] evaluated a modified *Lmdd* system as HIV vaccine. The authors developed a new *Lmdd-gag*, in which the transient supply of D-alanine is provided by a *dal* gene–carrying plasmid from which the *dal* gene is excised during cytosolic growth of the bacterium through the action of the *res*/resolvase recombination system.[45] Oral immunization of the recombinant attenuated *Listeria* resulted in strong dose-dependent, Gag-specific CTL in the mucosal-associated lymphoid tissues and protective immunity against vaginal challenge of HIV-1 Gag-expressing vaccinia virus.[177]

Natural transmission of HIV occurs at the mucosa and, hence, mucosa-associated lymphoid tissues may be the earliest site for virus replication. Therefore, successful induction of robust mucosal immunity may require vaccination by a mucosal route. Rayevskaya and Frankel[97] reported that parenteral immunization with *Lmdd-gag* strain provided complete protection against systemic and mucosal challenges with a recombinant vaccinia virus expressing HIV-1 Gag and that oral immunization with the attenuated strain also produced complete, long-lasting protection against the recombinant virus, but only against mucosal virus challenge. Peters and colleagues[178] showed that oral immunization of mice with HIV-1-Gag-expressing recombinant *L. monocytogenes* has led to Gag-specific responses in about 35% of lamina propria CD8+ T cells and that significant levels of Gag- and LLO-specific CD8+ T cells were observed in mucosal lymphoid tissues only after two immunizations. Immunization of recombinant *Listeria* vaccines twice enhanced T cell responses against HIV-1 Gag protein, confirming that the immunity induced against *Listeria* vector after the priming immunization does not limit immune responses induced by boosting immunization of the same *Listeria* vaccines.

To evaluate the potential of *L. monocytogenes* as a biologic vaccine vector against HIV, recombinant *L. monocytogenes* needs to be investigated in an *in vivo* challenge system. In the mouse system, a challenge study has been performed using vaccinia viruses expressing an HIV antigen (often HIV-1 Gag). Stevens and colleagues[179] used the feline immunodeficiency virus (FIV) system to evaluate recombinant *L. monocytogenes* vaccine in cats, in which FIV infection has been widespread. FIV is a pathogen of cats that induces a disease syndrome similar to that of HIV infection in humans. Like HIV, FIV infection leads to chronic immune dysfunctions, including depletion of $CD4^+$ T cells, inversion of CD4/CD8 T cell ratios, decreased lymphocyte proliferation, and increased susceptibility to opportunistic infections. The feline model of infection and disease progression is uniquely relevant for the evaluation of vaccine design and immune response upon challenge. Cats are the natural hosts for FIV and can be infected by the vaginal route with either cell-free or cell-associated virus, thereby mimicking the natural route of infection by HIV. A single oral immunization with a novel recombinant *L. monocytogenes* strain conferred some control of viral load after vaginal challenge with FIV. Preexisting immunity to *L. monocytogenes* did not prevent induction of immune responses to FIV by a recombinant *Listeria* vaccine secreting HIV-1 Gag protein and also harboring DNA vaccine plasmid for FIV Env protein.[180]

The prime-boost vaccine strategy has been examined to improve vaccine efficacy. The combinations of naked DNA vaccination and viral vector, such as the vaccinia virus Ankara strain, have been examined.[181] The combinations of naked DNA vaccination and recombinant *Listeria* vaccination have been examined against simian immunodeficiency virus (SIV), a counterpart of HIV in nonhuman primates.[100–102] Boyer and colleagues[100] reported a prime-boost study using a nonhuman primate (rhesus monkeys) with a DNA vaccine and a recombinant *Listeria* vaccine that expresses and secretes SIV Gag and Env antigens followed by a challenge with SIV239. A recombinant DNA vaccine delivered intramuscularly and a recombinant *L. monocytogenes* delivered orally have the ability to induce $CD8^+$ and $CD4^+$ T cell immune responses in a nonhuman primate. The combined vaccine was able to induce cellular immune responses in the nonhuman primate. Thus, this strategy enhanced the efficacy of a DNA vaccine. Further, the same group reported that the DNA prime-oral *Listeria* boost strategy induced mucosal SIV-Gag-specific $CD8^+$ T cell responses characterized by expression of the $\alpha4\beta7$ integrin gut-homing receptor.[101]

15.5 ANTIBACTERIAL VACCINES

With over 8 million new cases and 2 million deaths each year and the appearance of multi-drug-resistant *M. tuberculosis* strains, tuberculosis still remains an urgent public health problem worldwide.[182] Since the protective efficacy of the currently used *M. bovis* BCG vaccine strain ranges from 0 to 85% in different controlled studies,[183] there is a great need for an improved vaccine. DNA vaccines have been shown to be one of the most promising new approaches to this end. Miki and colleagues[72] reported on the induction of specific protective cellular immunity against *M. tuberculosis* employing vaccination with recombinant attenuated *L. monocytogenes* strains (Table 15.4). C57BL/6 mice immunized intraperitoneally with the attenuated self-destructing *L. monocytogenes* $\Delta2$ strain carrying plasmids for eukaryotic expression of *M. tuberculosis* Ag85 complex molecules

TABLE 15.4
Recombinant *L. monocytogenes* Vaccines for Bacteria and Parasites

Attenuation	Antigen	Promoter	Immunization Route	Immune Responses	Protection	Ref.
		Mycobacterium tuberculosis				
mpl, actA, plcB	Ag85A, Ag85B MPT51	CMV-IE (episomal)	i.v., i.p.	T cells	+	72
		Leishmania major				
actA	LACK	*hly* (episomal)	i.v.	CD4	+	193
actA	LACK	*hly* (chromosomal)	i.g., i.p.	CD4	+	192
actA	LACK	*hly* (chromosomal)	i.g.	CD4	+	194
Wild type	IL-12	*actA* (episomal)	i.p.	CD4	+	196

(Ag85A and Ag85B) and the MPB/MPT51 molecule showed specific Th1-type cellular immune responses. Furthermore, BALB/c mice immunized intravenously with these recombinant strains mounted protective cellular immunity against intravenous challenge with *M. tuberculosis* H37Rv comparable to that evoked by the conventional live *M. bovis* BCG vaccine strain.

Because *L. monocytogenes* membrane-perforating protein LLO plays a key role in activating CD4$^+$ and CD8$^+$ T cell responses, it has been applied in other bacterial systems to enhance immune response and protection. For example, the tuberculosis vaccine *M. bovis* BCG was combined with *L. monocytogenes hly* gene (encoding LLO) to form a recombinant BCG (*hly*+ rBCG), which improved specific protection against *M. tuberculosis*. By creating a urease C-deficient *hly*+ rBCG (i.e., Δ*ureC hly* + rBCG) to provide an intraphagosomal pH closer to the acidic pH optimum for *hly* activity, a much higher vaccine efficacy and cross-priming were obtained. This recombinant strain (i.e., Δ*ureC hly* + rBCG) also conferred profound protection against the *M. tuberculosis* Beijing/W genotype, for which the parental BCG is only marginally effective.[184] Recombinant *M. bovis* strains secreting *L. monocytogenes* LLO improved MHC class I presentation of cophagocytosed soluble protein, leading to enhanced capacity to stimulate CD8$^+$ T cells and protection against tuberculosis.[185]

15.6 ANTIPARASITIC VACCINES

Pathogenic parasites such as malaria parasites and *Leishmania* have their peculiar complex life cycles, which are divided into multiple stages. The appropriate types of immune responses have been required for protective immunity for these parasitic diseases. Cell-mediated immunity has been shown to be indispensable to contain intracellular protozoa such as *Leishmania*, *Toxoplasma*, and *Trypanosoma*.[186] Therefore, intracellular protozoa have been major target organisms of *Listeria*-mediated vaccines. So far, *Listeria*-mediated vaccines have been examined for the murine infection model of *Leishmania major*, an obligate intracellular protozoan (Table 15.4).

Leishmania exists in two major forms: the promastigote, which is a flagellated organism found in the sand fly, and the amastigote, which multiplies within macrophages of the host animals. Once inoculated into the mammalian host, the promastigotes rapidly invade cells such as macrophages and are transformed to amastigotes. Different from *Trypanosoma* and *Toxoplasma*, *Leishmania* can survive after the phagosomes have fused with the lysosomes. Therefore, antigens are degraded in the phagosomes and tend to be presented by the MHC class II pathway, leading to inducing specific CD4$^+$ T cells. *Leishmania*-specific CD8$^+$ T cells may also play a role in protective immunity against *Leishmania* infection, but the extent to which the T cell subset contributes remains to be examined.

Especially, Th1-type responses are required for effective protection against *L. major* infection. The murine infection model of *L. major* has been considered as one of the best models of Th1/Th2 dichotomy; the disease resistance correlates with the appearance of Th1-type immune responses to *L. major* and the susceptibility with that of Th2-type immune responses to the parasite.[187,188] IL-12 plays a central role in initiating protective immune responses against *L. major* infection by promoting IFN-γ production and the development of Th1-type immune responses.[189] A protein designated as a *Leishmania* homolog of receptors for activated C kinase (LACK) has been shown to be the major protective antigen and used as a target gene with *Listeria* carrier vaccine studies.[190] LACK is a 36-kDa protein that is highly conserved among various *Leishmania* species and expressed at both promastigote and amastigote stages.

Using naked DNA vaccine, Gurunathan and colleagues[191] showed that vaccination with DNA encoding LACK conferred protective immunity to mice infected with *L. major*. Soussi and colleagues[192] established the efficacy of an attenuated Δ*actA* recombinant *L. monocytogenes* strain expressing the heterologous LACK protein of *L. major* to induce protective Th1 CD4$^+$ T cell responses. In the study, they adopted a chromosomal integration-type plasmid system for expression of the LACK antigen, in comparison with a previous report in which an episomal plasmid system for LACK antigen expression was used.[193] The multicopy plasmid in the episomal plasmid system was not retained for more than 48 hours and the presence of readily detectable Th1-type LACK-reactive T cells was ineffective to limit *L. major* lesion progression in BALB/c mice. In contrast, when a chromosomal integration-type plasmid system was used for the antigen expression, significant protection against *L. major* infection was observed in BALB/c mice, ranging from delay in the lesion onset to full protection in 80% of the challenged mice, depending on the size of the parasite inoculum for challenge. They also compared the intragastric route and intraperitoneal route for immunization of the attenuated *Listeria* vaccine. It appeared that the intragastric route led to a higher protection level than did the intraperitoneal one.

Saklani-Jusforgues and colleagues[194] described the local and extraintestinal dynamics of the CD4$^+$ T cell populations primed after enteric delivery of the attenuated recombinant *L. monocytogenes*. They examined the timing, magnitude, and persistence of the LACK-reactive IFN-γ- and IL-4-secreting CD4$^+$ T cell immune responses generated during enteric immunization with the *Listeria* vaccine in the Peyer's patches, mesenteric lymph nodes, spleen, liver, and blood. Efficient priming of IFN-γ-secreting LACK-specific CD4$^+$ T cells was detected in all tested tissues. These results indicate that enteral immunization of the Δ*actA L. monocytogenes* mutant efficiently induced CD4$^+$ T cells. They used intragastric delivery with an acid-adapted *L. monocytogenes* strain that survived the low gastric pH and spread physiologically along the whole gut.[195] In terms of induction of protective immunity, intragastric injections of *actA*-deficient *L. monocytogenes* expressing LACK antigen led to slower lesion progression and more chronic process without ulceration compared with the unimmunized group after a subcutaneous high-dose *L. major* challenge (2×10^5 stationary-phase promastigotes). Even though no recall responses were observed at the systemic level after intragastric reinoculations, both the delay in the onset of the lesion and the attenuation of clinical signs in the *L. major*-infected cutaneous site suggested the existence of local immune responses that extend the functions of these locally primed CD4$^+$ T cells to the skin-draining lymph nodes.

L. monocytogenes is also useful for the delivery of expression plasmids encoding immunomodulatory cytokines in addition to genes for the target antigen molecules. Shen and colleagues[196] showed that *Listeria* delivered *IL-10* and *IL-12* genes into mammalian cells, which led to regression of the progress of *Leishmania* infection. While injection of wild-type *L. monocytogenes* into hind legs of mice limited local *L. major* expansion, inoculation of IL-12-expressing *L. monocytogenes* resulted in an enhanced protective effect. It was possible that the injected *Listeria* modulated the murine immune responses to Th1-type responses, which are favorable to the inhibition of *L. major* infection. Co-delivery of genes for cytokines and antigenic proteins by using one or two separate plasmids of *L. monocytogenes* may lead to a much higher vaccine efficacy in the future.

15.7 CONCLUSIONS AND PERSPECTIVES

L. monocytogenes is an extraordinary intracellular bacterium that can be exploited as a vaccine vehicle for initiating both CD4$^+$ and CD8$^+$ T cell responses, promoting IFN-γ and IL-12 production, and delivering protective molecules into the cytoplasm. In comparison with other microbial vector systems, live *Listeria*-carrying vaccine platforms have several distinct advantages, such as mucosal route of immunization, propensity to infect professional APCs, ease of genetic manipulation, adjuvanticity, amplification of DNA vaccine plasmids *in vivo*, and simplicity of handling and storage. At present, live attenuated *L. monocytogenes* vaccines mostly belong to either auxotrophic mutants or virulence gene-deficient mutants. Auxotrophic mutants typically include a double mutant of the alanine racemase gene (*dal*) and the D-amino acid aminotransferase gene (*dat*), which mutants require D-alanine supplementation for growth, and a series of *aro* mutants such as *aroA* and *aroB* with deletions in genes of the common branch of the biosynthesis pathway leading to aromatic compounds. Virulence gene–deficient mutants frequently comprise mutants of the *hly* gene encoding LLO and *actA* gene encoding ActA. Use of these live attenuated *Listeria*-carrying vaccines has contributed to the development of enhanced protective immunity against other bacterial, viral, and parasitic infections.

Expression of genes encoding target antigens can be driven by the *Listeria* promoter or eukaryotic promoter (bactofection). A variety of episomal plasmids in *Listeria* and genome-integration plasmids (controlled by *Listeria* promoter) and episomal DNA vaccine plasmids (controlled by eukaryotic promoter) has been developed. Genome-integration plasmids enable maintenance of heterologous antigen genes in *Listeria* and hence long-time production of antigens *in situ* compared with episomal vaccine plasmids. Further, many systems have been developed for retention of plasmids episomally in *L. monocytogenes.* Target pathogens of *Listeria*-carrying vaccines reported so far contain viruses (e.g., HIV, LCMV, and influenza virus), bacteria (e.g., *M. tuberculosis*), and parasites (e.g., *L. major*). Since *Listeria*-carrying vaccines are capable of inducing specific $CD8^+$ and $CD4^+$ T cell responses and effective protective immunity, they show enormous promise for developing more effective vaccine against intracellular pathogens.

To date, the live attenuated *L. monocytogenes* strains have been derived from parent strains of high virulence and pathogenicity, which often provoke a potent immune response in the host through generations of large amounts of IFN-γ and other molecules, contributing to the host's illness. While deletion or modification of one or two key genes may reduce the pathogenicity of these strains, it may not change many other aspects of *L. monocytogenes* strains, such as the level of IFN-γ production. It is possible that these attenuated strains may maintain their ability to induce a strong immune response, which may be unnecessary and potentially detrimental to the vaccine recipients. The fact that *L. monocytogenes* consists of naturally avirulent strains (of serotype 4a) with the capability of stimulating a durable immunity against listeriosis highlights the potential benefit of using the naturally avirulent strains as vaccine carriers. Given that avirulent serotype 4a strains stimulate relatively small amounts of IFN-γ in comparison with EGD, they may cause fewer side effects without sacrificing their ability to induce an effective T cell immune response. Therefore, once the ability of recombinant naturally avirulent stains to initiate specific immunity to pathogens of interest is confirmed, it can be envisaged that naturally avirulent strains will offer a useful alternative to the live attenuated strains for delivering protective molecules against infective agents of medical and veterinary importance.

REFERENCES

1. Gellin, B.G. and Broome, C.V., Listeriosis, *J. Am. Med. Assoc.*, 261, 1313, 1989.
2. Lorber, B., *Listeria monocytogenes*, in *Principles and practice of infectious diseases,* 6th ed., Mandell, G.L., Bennett, J.E., and Dolin, R., eds., Elsevier, Inc., Philadelphia, 2000, p. 2478.
3. Vazquez-Boland, J.A. et al., *Listeria* pathogenesis and molecular virulence determinants, *Clin. Microbiol. Rev.*, 14, 584, 2001.
4. Cossart, P. and Mengaud, J., *Listeria monocytogenes:* A model system for the molecular study of intracellular parasitism, *Mol. Biol. Med.*, 6, 463, 1989.
5. Portnoy, D.A. et al., Molecular determinants of *Listeria monocytogenes* pathogenesis, *Infect. Immun.*, 60, 1263, 1992.
6. Pamer, E.G., Immune responses to *Listeria monocytogenes*, *Nat. Rev. Immunol.*, 4, 812, 2004.
7. Kaufmann, S.H.E. et al., Effective protection against *Listeria monocytogenes* and delayed-type hypersensitivity to listerial antigens depend on cooperation between specific $L3T4^+$ and $Lyt2^+$ T cells, *Infect. Immun.*, 48, 263, 1985.
8. Czuprynski, C.J. and Brown, J.F., Effects of purified anti-Lyt-2 mAb treatment on murine listeriosis: Comparative roles of $Lyt\text{-}2^+$ and $L3T4^+$ cells in resistance to primary and secondary infection, delayed-type hypersensitivity and adoptive transfer of resistance, *Immunology*, 71, 107, 1990.
9. Sasaki, T. et al., Roles of $CD4^+$ and $CD8^+$ cells, and the effect of administration of recombinant murine interferon γ in listerial infection, *J. Exp. Med.*, 171, 1141, 1990.
10. Roberts, A.D., Ordway, D.J., and Orme, I.M., *Listeria monocytogenes* infection in β2 microglobulin-deficient mice, *Infect. Immun.*, 61, 1113, 1993.
11. Ladel, C.H. et al., Studies with MHC-deficient knock-out mice reveal impact of both MHC I- and MHC II-dependent T cell responses on *Listeria monocytogenes* infection, *J. Immunol.*, 153, 3116, 1994.

12. Geginat, G. et al., A novel approach of direct ex vivo epitope mapping identifies dominant and subdominant CD4 and CD8 T cell epitopes from *Listeria monocytogenes*, *J. Immunol.*, 166, 1877, 2001.
13. Xiong, H. et al. Administration of killed bacteria together with listeriolysin O induces protective immunity against *Listeria monocytogenes* in mice, *Immunology*, 94, 14, 1998.
14. Graham, R., Morrill, C.C., and Levine, N.D., Studies on *Listerella*. IV. Unsuccessful attempts at immunization with living and dead *Listerella* cultures, *Cornell Vet.*, 30, 291, 1940.
15. Osebold, J.W. and Sawyer, M.T., Immunization studies on listeriosis in mice, *J. Immunol.*, 78, 262, 1957.
16. Hasenclever, H.F. and Karakawa, W.W., Immunization of mice against *Listeria monocytogenes*, *J. Bacteriol.*, 74, 584, 1957.
17. Armstrong, A.S. and Sword, C.P., Cellular resistance in listeriosis, *J. Infect. Dis.*, 114, 258, 1964.
18. Mackaness, G.B., Cellular resistance to infection, *J. Exp. Med.*, 116, 381, 1962.
19. Kearns, R.J. and Hinrichs, D.J., Kinetics and maintenance of acquired resistance in mice to *Listeria monocytogenes*, *Infect. Immun.*, 16, 923, 1977.
20. Wirsing von Koenig, C.H. and Finger, H., Failure of killed *Listeria monocytogenes* vaccine to produce protective immunity, *Nature*, 297, 233, 1982.
21. Yamamoto, K., Kato, K., and Kimura, T., Killed *Listeria*-induced suppressor T cells involved in suppression of delayed-type hypersensitivity and protection against *Listeria* infection, *Immunology*, 55, 609, 1985.
22. Koga, T. et al., Induction by killed *Listeria monocytogenes* of effector T cells mediating delayed-type hypersensitivity but not protection in mice, *Immunology*, 62, 241, 1987.
23. Mitsuyama, M. et al., Difference in the induction of macrophage interleukin-1 production between viable and killed cells of *Listeria monocytogenes*, *Infect. Immun.*, 58, 1254, 1990.
24. van der Meer, C., Hofhuis F.M., and Willers, J.M., Killed *Listeria monocytogenes* vaccine becomes protective on addition of polyanions, *Nature*, 269, 594, 1977.
25. van Dijk, H. et al., Killed *Listeria monocytogenes* vaccine is protective in C3H/H3eJ mice without addition of adjuvants, *Nature*, 286, 713, 1980.
26. Miki, K. and Mackaness, G.B., The passive transfer of acquired resistance to *Listeria monocytogenes*, *J. Exp. Med.*, 120, 93, 1964.
27. Mackaness, G.B., The influence of immunologically committed lymphoid cells on macrophage activity *in vivo*, *J. Exp. Med.*, 129, 973, 1969.
28. Mackaness, G.B. and Hill, W.C., The effect of antilymphocyte globulin on cell-mediated resistance to infection, *J. Exp. Med.*, 129, 993, 1969.
29. North, R.J., Cellular mediators of anti-*Listeria* immunity as an enlarged population of short-lived, replicating T cells, *J. Exp. Med.*, 138, 342, 1973.
30. Cheers, C. et al., Resistance and susceptibility of mice to bacterial infection: Course of listeriosis in resistant or susceptible mice, *Infect. Immun.*, 19, 763, 1978.
31. Miller, M.A., Skeen, M.J., and Ziegler, H.K., Nonviable bacterial antigens administered with IL-12 generate antigen-specific T cell responses and protective immunity against *Listeria monocytogenes*, *J. Immunol.*, 155, 4817, 1995.
32. Miller, M.A., Skeen, M.J., and Ziegler, H.K., Protective immunity to *Listeria monocytogenes* elicited by immunization with heat-killed *Listeria* and IL-12, *Ann. NY Acad. Sci.*, 797, 207, 1996.
33. Rolph, M.S. and Kaufmann, S.H.E., CD40 signaling converts a minimally immunogenic antigen into a potent vaccine against the intracellular pathogen *Listeria monocytogenes*, *J. Immunol.*, 166, 5115, 2001.
34. Kursar, M. et al., Depletion of $CD4^+$ T cells during immunization with nonviable *Listeria monocytogenes* causes enhanced $CD8^+$ T cell-mediated protection against listeriosis, *J. Immunol.*, 172, 3167, 2004.
35. Szalay, G., Ladel, C.H., and Kaufmann, S. H. E., Stimulation of protective $CD8^+$ T lymphocytes by vaccination with nonliving bacteria, *Proc. Natl. Acad. Sci. USA*, 92, 12389, 1995.
36. Lauvau, G. et al., Priming of memory but not effector CD8 T cells by a killed bacterial vaccine, *Science*, 294, 1735, 2001.
37. Muraille, E. et al., Distinct *in vivo* dendritic cell activation by live versus killed *Listeria monocytogenes*, *Eur. J .Immunol.*, 35, 1463, 2005.
38. Datta, S.K. et al., Vaccination with irradiated *Listeria* induces protective T cell immunity, *Immunity*, 25, 143, 2006.
39. Hoffman, S.L. et al., Protection of humans against malaria by immunization with radiation-attenuated *Plasmodium falciparum* sporozoites, *J. Infect. Dis.*, 185, 1155, 2002.
40. Portnoy, S., Jacks, P.S., and Hinrichs, D.J., Role of hemolysin for the intracellular growth of *Listeria monocytogenes*, *J. Exp. Med.*, 167, 1459, 1988.

41. Alexander, J.E. et al., Characterization of an aromatic amino acid-dependent *Listeria monocytogenes* mutant: Attenuation persistence, and ability to induce protective immunity in mice, *Infect. Immun.*, 61, 2245, 1993.
42. Thompson, R.J. et al., Pathogenicity and immunogenicity of a *Listeria monocytogenes* strain that requires D-alanine for growth, *Infect. Immun.*, 66, 3552, 1998.
43. Simon, B.E. et al., Plasmid DNA delivery by D-alanine-deficient *Listeria monocytogenes*, *Biotechnol. Prog.*, 22, 1394, 2006.
44. Li, Z. et al., Conditional lethality yields a new vaccine strain of *Listeria monocytogenes* for the induction of cell-mediated immunity, *Infect. Immun.*, 73, 5065, 2005.
45. Zhao, X. et al., Pathogenicity and immunogenicity of a vaccine strain of *Listeria monocytogenes* that relies on a suicide plasmid to supply an essential gene product, *Infect. Immun.*, 73, 5789, 2005.
46. Stritzker, J. et al., Growth, virulence, and immunogenicity of *Listeria monocytogenes aro* mutants, *Infect. Immun.*, 72, 5622, 2004.
47. Domann, E. et al., A novel bacterial virulence gene in *Listeria monocytogenes* required for host cell microfilament interaction with homology to the proline-rich region of vinculin, *EMBO J.*, 11, 1981, 1992.
48. Kocks, C. et al., *L. monocytogenes*-induced actin assembly requires the *actA* gene product, a surface protein, *Cell*, 68, 521, 1992.
49. Portnoy, D.A., Auerbuch, V., and Glomiski, I.J., The cell biology of *Listeria monocytogenes* infection: The intersection of bacterial pathogenesis and cell-mediated immunity, *J. Cell Biol.*, 158, 409, 2002.
50. Berche, P. et al., T cell recognition of listeriolysin O is induced during infection with *Listeria monocytogenes*, *J. Immunol.*, 139, 3813, 1987.
51. Cossart, P. et al., Listeriolysin O is essential for virulence of *Listeria monocytogenes*: Direct evidence obtained by gene complementation, *Infect. Immun.*, 57, 3629, 1989.
52. Hiltbold, E.M., Safley, S.A., and Ziegler, H.K., The presentation of class I and class II epitopes of listeriolysin O is regulated by intracellular localization and by intracellular spread of *Listeria monocytogenes*, *J. Immunol.*, 157, 1163, 1996.
53. Kathariou, S. et al., Tn916-induced mutations in the hemolysin determinant affecting virulence of *Listeria monocytogenes*, *J. Bacteriol.*, 169, 1291, 1987.
54. Brunt, L.M., Portnoy, D.A., and Unanue, E.R., Presentation of *Listeria monocytogenes* to CD8+ T cells requires secretion of hemolysin and intracellular bacterial growth, *J. Immunol.*, 145, 3540, 1990.
55. Michel, E. et al., Attenuated mutants of the intracellular bacterium *Listeria monocytogenes* obtained by single amino acid substitutions in listeriolysin O, *Mol. Microbiol.*, 4, 2167, 1990.
56. Berche, P., Gaillard, J.L., and Sansonetti, P.J., Intracellular growth of *Listeria monocytogenes* as a prerequisite for *in vivo* induction of T cell-mediated immunity, *J. Immunol.*, 138, 2266, 1987.
57. Barry, R.A. et al., Pathogenicity and immunology of *Listeria monocytogenes* small-plaque mutants defective for intracellular growth and cell-to-cell spread, *Infect. Immun.*, 60, 1625, 1992.
58. Hamilton, S.E. et al., Listeriolysin O-deficient *Listeria monocytogenes* as a vaccine delivery vehicle: Antigen-specific CD8 T cell priming and protective immunity, *J. Immunol.*, 177, 4012, 2006.
59. Badovinac, V.P. et al., Accelerated CD8+ T-cell memory and prime-boost response after dendritic-cell vaccination, *Nat. Med.*, 11, 748, 2005.
60. Bouwer, H.G.A. et al., Directed antigen delivery as a vaccine strategy for an intracellular bacterial pathogen, *Proc. Nat. Acad. Sci. USA*, 103, 5102, 2006.
61. Ripio, M.-T. et al., A Gly145Ser substitution in the transcriptional activator PrfA causes constitutive overexpression of virulence factors in *Listeria monocytogenes*, *J. Bacteriol.*, 179, 1533, 1997.
62. Goossens, P.L. and Milon, G., Induction of protective CD8+ T lymphocytes by an attenuated *Listeria monocytogenes actA* mutant, *Int. Immunol.*, 4, 1413, 1992.
63. Brundage, R.A. et al., Expression and phosphorylation of the *Listeria monocytogenes* ActA protein in mammalian cells, *Proc. Natl. Acad. Sci. USA*, 90, 11890, 1993.
64. Manohar, M. et al., Gut colonization of mice with *actA*-negative mutant of *Listeria monocytogenes* can stimulate a humoral mucosal immune response, *Infect. Immun.*, 69, 3542, 2001.
65. Rudnicka, W. et al., The host response to *Listeria monocytogenes* mutants defective in genes encoding phospholipases C (*plcA*, *plcB*) and actin assembly (*actA*), *Microbiol. Immunol.*, 41, 847, 1997.
66. Angelakopoulos, H. et al., Safety and shedding of an attenuated strain of *Listeria monocytogenes* with a deletion of *actA/plcB* in adult volunteers: A dose escalation study of oral inoculation, *Infect. Immun.*, 70, 3592, 2002.
67. Peters, C. et al., Tailoring host immune responses to *Listeria* by manipulation of virulence genes—The interface between innate and acquired immunity, *FEMS Immunol. Med. Microbiol.*, 35, 243, 2003.

68. Darji, A. et al., Induction of immune responses by attenuated isogenic mutant strains of *Listeria monocytogenes*, *Vaccine*, 21, S2/102, 2003.
69. Paglia, P. et al., The defined attenuated *Listeria monocytogenes* Δ*mpl2* mutant is an effective oral vaccine carrier to trigger a long-lasting immune response against a mouse fibrosarcoma, *Eur. J. Immunol.*, 27, 1570, 1997.
70. Hauf, N. et al., *Listeria monocytogenes* infection of P388D_1 macrophages results in a biphasic NF-κB (RelA/p50) activation induced by lipoteichoic acid and bacterial phospholipases and mediated by IκBα and IκBβ degradation, *Proc. Natl. Acad. Sci. USA*, 94, 9394, 1997.
71. Dietrich, G. et al., Delivery of antigen-encoding plasmid DNA into the cytosol of macrophages by attenuated suicide *Listeria monocytogenes*, *Nat. Biotechnol.*, 16, 181, 1998.
72. Miki, K. et al., Induction of protective cellular immunity against *Mycobacterium tuberculosis* by recombinant attenuated self-destructing *Listeria monocytogenes* strains harboring eukaryotic expression plasmids for antigen 85 complex and MPB/MPT51, *Infect. Immun.*, 72, 2014, 2004.
73. Brockstedt, D.G. et al., Killed but metabolically active microbes: A new vaccine paradigm for eliciting effector T-cell responses and protective immunity, *Nat. Med.*, 11, 853, 2005.
74. Autret, N. et al., Identification of new genes involved in the virulence of *Listeria monocytogenes* by signature-tagged transposon mutagenesis, *Infect. Immun.*, 69, 2054, 2001.
75. Loessner, M.J., Wendlinger, G., and Scherer, S., Heterologous endolysins in *Listeria monocytogenes* bacteriophages: A new class of enzymes and evidence for conserved holing genes within the siphoviral lysis cassettes. *Mol. Microbiol.*, 16, 1231, 1995.
76. Hense, M. et al., Eukaryotic expression plasmid transfer from the intracellular bacterium *Listeria monocytogenes* to host cells, *Cell. Microbiol.*, 3, 599, 2001.
77. Wong, P. and Pamer, E.G., Feedback regulation of pathogen-specific T cell priming, *Immunity*, 18, 499, 2003.
78. Williams, M.A. and Bevan, M.J., Shortening the infectious period does not alter expansion of CD8 T cells but diminishes their capacity to differentiate into memory cells, *J. Immunol.*, 173, 6694, 2004.
79. Corbin, G.A. and Harty, J.T., Duration of infection and antigen display have minimal influence on the kinetics of the $CD4^+$ T cell response to *Listeria monocytogenes* infection, *J. Immunol.*, 173, 5679, 2004.
80. Mercado, R. et al., Early programming of T cell populations responding to bacterial infection, *J. Immunol.*, 165, 6833, 2000.
81. Schafer, R. et al., Induction of a cellular immune response to a foreign antigen by a recombinant *Listeria monocytogenes* vaccine, *J. Immunol.*, 149, 53, 1992.
82. de Chastellier, C. and Berche, P., Fate of *Listeria monocytogenes* in murine macrophages: Evidence for simultaneous killing and survival of intracellular bacteria, *Infect. Immun.*, 62, 543, 1994.
83. Barsig, J. and Kaufmann, S.H.E., The mechanism of cell death in *Listeria monocytogenes*-infected murine macrophages is distinct from apoptosis, *Infect. Immun.*, 65, 4075, 1997.
84. Guzman, C.A. et al., Interaction of *Listeria monocytogenes* with mouse dendritic cells, *Infect. Immun.*, 63, 3665, 1995.
85. Kolb-Mäurer, A. et al., *Listeria monocytogenes*-infected human dendritic cells: Uptake and host cell response, *Infect. Immun.*, 68, 3680, 2000.
86. Park, S.F. and Stewart, G.S., High-efficiency transformation of *Listeria monocytogenes* by electroporation of penicillin-treated cells, *Gene*, 94, 129, 1988.
87. Schäferkordt, S., Domann, E., and Chakraborty, T., Molecular approaches for the study of *Listeria*, *Methods Microbiol.*, 27, 421, 1998.
88. Busch, D.H., Vijh, S., and Pamer, E.G., Animal model for infection with *Listeria monocytogenes*, in *Current protocols in immunology*, Coligan, J. E., ed., John Wiley & Sons, Inc., Unit 19.9, 2002.
89. Freitag, N.E., Genetic tools for use with *Listeria monocytogenes*, in *Gram-positive pathogens*, Fischetti, V. A., Novick, R. P., Ferretti, J. J., Portnoy, D. A., and Rood, J. I., eds., American Society for Microbiology, Washington, D.C., 2000, 488.
90. Schäferkordt, S. and Chakraborty, T., Vector plasmid for insertional mutagenesis and directional cloning in *Listeria* spp., *BioTechniques*, 19, 720, 1995.
91. Trieu-Cuot, P. et al., A pair of mobilizable shuttle vectors conferring resistance to spectinomycin for molecular cloning in *Escherichia coli* and in Gram-positive bacteria, *Nucleic Acids Res.*, 18, 4296, 1990.
92. Trieu-Cuot, P. et al., Enhanced conjugative transfer of plasmid DNA from *Escherichia coli* to *Staphylococcus aureus* and *Listeria monocytogenes*, *FEMS Microbiol. Lett.*, 109, 19, 1993.
93. Hsieh, C.S. et al., Development of Th1 $CD4^+$ T cells through IL-12 produced by *Listeria*-induced macrophages, *Science*, 260, 547, 1993.

94. Mielke, M.E.A., Peters, C., and Hahn, H., Cytokines in the induction and expression of T-cell-mediated granuloma formation and protection in the murine model of listeriosis, *Immunol. Rev.*, 158, 79, 1997.
95. Schwandner, R. et al., Peptidoglycan- and lipoteichoic acid-induced cell activation is mediated by Toll-like receptor 2, *J. Biol. Chem.*, 274, 17406, 1999.
96. Cleveland, M.G. et al., Lipoteichoic acid preparations of Gram-positive bacteria induce interleukin-12 through a CD14-dependent pathway, *Infect. Immun.*, 64, 1906, 1996.
97. Rayevskaya, M.V. and Frankel, F.R., Systemic immunity and mucosal immunity are induced against human immunodeficiency virus Gag protein in mice by a new hyperattenuated strain of *Listeria monocytogenes*, *J. Virol.*, 75, 2786, 2001.
98. Lecuit, M. et al., A single amino acid in E-cadherin responsible for host specificity towards the human pathogen *Listeria monocytogenes*, *EMBO J.*, 18, 3956, 1999.
99. Lecuit, M. et al., A transgenic model for listeriosis: Role of internalin in crossing the intestinal barrier, *Science*, 292, 1722, 2001.
100. Boyer, J.D. et al., DNA prime *Listeria* boost induces a cellular immune response to SIV antigens in the rhesus macaque model that is capable of limited suppression of SIV 239 viral replication, *Virology*, 333, 88, 2005.
101. Neeson, P. et al., A DNA prime-oral *Listeria* boost vaccine in rhesus macaques induces an SIV-specific CD8 T cell mucosal response characterized by high levels of α4β7 integrin and an effector memory phenotype, *Virology*, 354, 299, 2006.
102. Boyer, J.D. et al., Rhesus macaques with high levels of vaccine induced IFN-gamma producing cells better control viral set-point following challenge with SIV239, *Vaccine*, 24, 4498, 2006.
103. Bouwer H.G.A. et al., Existing antilisterial immunity does not inhibit the development of a *Listeria monocytogenes*-specific primary cytotoxic T-lymphocyte response, *Infect. Immun.*, 67, 253, 1999.
104. Starks, H. et al., *Listeria monocytogenes* as a vaccine vector: Virulence attenuation or existing anti-vector immunity does not diminish therapeutic efficacy, *J. Immunol.*, 173, 420, 2004.
105. Stevens, R. et al., Pre-existing immunity to pathogenic *Listeria monocytogenes* does not prevent induction of immune responses to feline immunodeficiency virus by a novel recombinant *Listeria monocytogenes* vaccine, *Vaccine*, 23, 1479, 2005.
106. Kaufmann, S.H., Acquired resistance to facultative intracellular bacteria: Relationship between persistence, cross-reactivity at the T-cell level, and capacity to stimulate cellular immunity of different *Listeria* strains, *Infect. Immun.*, 45, 34, 1984.
107. Chakraborty, T. et al., Naturally occurring virulence-attenuated isolates of *Listeria monocytogenes* capable of inducing long-term protection against infection by virulent strains of homologous and heterologous serotypes, *FEMS Immunol. Med. Microbiol.*, 10, 1, 1994.
108. Liu, D., *Listeria*-based anti-infective vaccine strategies, *Recent Patents Anti-infect. Drug Disc.*, 1, 281, 2006.
109. Liu, D. et al., Characteristics of cell-mediated, antilisterial immunity induced by a naturally avirulent *Listeria monocytogenes* serotype 4a strain HCC23, *Arch. Microbiol.*, 188, 251, 2007.
110. Kroll, J.J., Roof, M.B., and McOrist, S., Evaluation of protective immunity in pigs following oral administration of an avirulent live vaccine of *Lawsonia intracellularis*, *Am. J. Vet. Res.*, 65, 559, 2004.
111. Ferguson, N.M., Leiting, V.A., and Klevena, S.H., Safety and efficacy of the avirulent *Mycoplasma gallisepticum* strain K5054 as a live vaccine in poultry, *Avian Dis.*, 48, 91, 2004.
112. Wyszynska, A. et al., Oral immunization of chickens with avirulent *Salmonella* vaccine strain carrying *C. jejuni* 72Dz/92 *cjaA* gene elicits specific humoral immune response associated with protection against challenge with wild-type *Campylocacter*, *Vaccine*, 22, 1379, 2004.
113. Uchijima, M. et al., Optimization of codon usage of plasmid DNA vaccine is required for the effective MHC class I-restricted T cell responses against an intracellular bacterium, *J. Immunol.*, 161, 5594, 1998.
114. Cornell, K.A. et al., Genetic immunization of mice against *Listeria monocytogenes* using plasmid DNA encoding listeriolysin O, *J. Immunol.*,163, 322, 1999.
115. Fensterle, J. et al., Effective DNA vaccination against listeriosis by prime/boost inoculation with the gene gun, *J. Immunol.*, 163, 4510, 1999.
116. Yoshida, A. et al., Protective CTL response is induced in the absence of $CD4^+$ T cells and IFN-γ by gene gun DNA vaccination with a minigene encoding a CTL epitope of *Listeria monocytogenes*, *Vaccine*, 19, 4297, 2001.
117. Barry, R.A. et al., Protection of interferon-γ knockout mice against *Listeria monocytogenes* challenge following intramuscular immunization with DNA vaccines encoding listeriolysin O, *Vaccine*, 21, 2122, 2003.

118. An, L.-L., Pamer, E., and Whitton, J.L., A recombinant minigene vaccine containing a nonameric cytotoxic-T-lymphocyte epitope confers limited protection against *Listeria monocytogenes* infection, *Infect. Immun.*, 64, 1685, 1996.
119. Hamilton, S.E., and Harty, J.T., Quantitation of CD8+ T cell expansion, memory, and protective immunity after immunization with peptide-coated dendritic cells, *J. Immunol.*, 169, 4936, 2002.
120. Nakamura, Y. et al., Induction of protective immunity to *Listeria monocytogenes* with dendritic cells retrovirally transduced with a cytotoxic T lymphocyte epitope minigene, *Infect. Immun.*, 71, 1748, 2003.
121. Nagata, T. et al., Induction of protective immunity to *Listeria monocytogenes* by immunization with plasmid DNA expressing a helper T-cell epitope that replaces the class II-associated invariant chain peptide of the invariant chain, *Infect. Immun.*, 70, 2676, 2002.
122. Shen, H., Tato, C.M., and Fan, X., *Listeria monocytogenes* as a probe to study cell-mediated immunity, *Curr. Opin. Immunol.*, 10, 450, 1998.
123. Weiskirch, L. and Paterson, Y., The use of *Listeria monocytogenes* recombinants as vaccine vectors in infectious and neoplastic disease, in *Intracellular bacterial vaccine vectors*, Paterson, Y., ed., John Wiley & Sons, Inc., New York, 1999, 223.
124. Dietrich, G., Gentschev, I., and Goebel, W., Delivery of protein antigens and DNA vaccines by *Listeria monocytogenes*, in *Vaccine delivery strategies*, Dietrich, G. and Goebel, W., eds., Horizon Scientific Press, Wymondham, U.K., 2002, 263.
125. Hess, J. et al., Superior efficacy of secreted over somatic antigen display in recombinant *Salmonella* vaccine induced protection against listeriosis, *Proc. Natl. Acad. Sci. USA*, 93, 1458, 1996.
126. Darji, A. et al., Oral somatic transgene vaccination using attenuated *S. typhimurium, Cell*, 91, 765, 1997.
127. Shen, H. et al., Compartmentalization of bacterial antigens: Differential effects on priming of CD8 T cells and protective immunity, *Cell*, 92, 535, 1998.
128. Tvinnereim, A.R. and Harty, J.T., CD8+ T-cell priming against a nonsecreted *Listeria monocytogenes* antigen is independent of the antimicrobial activities of gamma interferon, *Infect. Immun.*, 68, 2196, 2000.
129. Tvinnereim, A.R., Hamilton, S.E., and Harty, J.T., CD8+ T-cell response to secreted and nonsecreted antigens delivered by recombinant *Listeria monocytogenes* during secondary infection, *Infect. Immun.*, 70, 153, 2002.
130. Flamm, R.K., Hinrichs, D.J., and Thomashow, M.F., Introduction of pAMβ1 into *Listeria monocytogenes* by conjugation and homology between native *L. monocytogenes* plasmids, *Infect. Immun.*, 44, 157, 1984.
131. Pilgrim, S. et al., Bactofection of mammalian cells by *Listeria monocytogenes*: Improvement and mechanism of DNA delivery, *Gene Ther.*, 10, 2036, 2003.
132. Camilli, A., Portnoy, D.A., and Youngman, P., Insertional mutagenesis of *Listeria monocytogenes* with a novel Tn917 derivative that allows direct cloning of DNA flanking transposon insertions, *J. Bacteriol.*, 172, 3738, 1990.
133. Ikonomidis, G. et al., Delivery of a viral antigen to the class I processing and presentation pathway by *Listeria monocytogenes, J. Exp. Med.*, 180, 2209, 1994.
134. Verch, T., Pan, Z.-K., and Paterson, Y., *Listeria monocytogenes*-based antibiotic resistance gene-free antigen delivery system applicable to other bacterial vectors and DNA vaccines, *Infect. Immun.*, 72, 6418, 2004.
135. Gaillard, J.L., Berche, P., and Sansonetti, P., Transposon mutagenesis as a tool to study the role of hemolysin in the virulence of *Listeria monocytogenes, Infect. Immun.*, 52, 50, 1986.
136. Shen, H. et al., Recombinant *Listeria monocytogenes* as a live vaccine vehicle for the induction of protective antiviral cell-mediated immunity, *Proc. Natl. Acad. Sci. USA*, 92, 3987, 1995.
137. Lauer, P. et al., Construction, characterization, and use of two *Listeria monocytogenes* site-specific phage integration vectors, *J. Bacteriol.*, 184, 4177, 2002.
138. Pálffy, R. et al., Bacteria in gene therapy: Bactofection versus alternative gene therapy, *Gene Ther.*, 13, 101, 2006.
139. Souders, N.C., Verch, T., and Paterson, Y., *In vivo* bactofection: *Listeria* can function as a DNA-cancer vaccine, *DNA Cell Biol.*, 25, 142, 2006.
140. Dietrich, G. et al., Delivery of DNA vaccines by attenuated intracellular bacteria, *Immunol. Today*, 20, 251, 1999.
141. Grillot-Courvalin, C., Goussard, S., and Courvalin, P., Bacteria as gene delivery vectors for mammalian cells, *Curr. Opin. Biotechnol.*, 10, 477, 1999.

142. Dietrich, G. et al., Bacterial systems for the delivery of eukaryotic antigen expression vectors, *Antisense Nucleic Acid Drug Dev.*, 10, 391, 2000.
143. Spreng, S. et al., Novel bacterial systems for the delivery of recombinant protein or DNA, *FEMS Immunol. Med. Microbiol.*, 27, 299, 2000.
144. Gentschev, I. et al., Delivery of protein antigens and DNA by virulence-attenuated strains of *Salmonella typhimurium* and *Listeria monocytogenes*, *J. Biotechnol.*, 83, 19, 2000.
145. Gentschev, I. et al., Recombinant attenuated bacteria for the delivery of subunit vaccines, *Vaccine*, 19, 2621, 2001.
146. Weiss, S. and Chakraborty, T., Transfer of eukaryotic expression plasmids to mammalian host cells by bacterial carriers, *Curr. Opin. Biotechnol.*, 12, 467, 2001.
147. Drabner, B. and Guzmán, C.A., Elicitation of predictable immune responses by using live bacterial vectors, *Biomol. Eng.*, 17, 75, 2001.
148. Mollenkopf, H., Dietrich, G., and Kaufmann, S.H.E., Intracellular bacteria as targets and carriers for vaccination, *Biol. Chem.*, 382, 521, 2001.
149. Gentschev, I. et al., Delivery of protein antigens and DNA by attenuated intracellular bacteria, *Int. J. Med. Microbiol.*, 291, 577, 2002.
150. Dietrich, G. et al., Live attenuated bacteria as vectors to deliver plasmid DNA vaccines, *Curr. Opin. Mol. Ther.*, 5, 10, 2003.
151. Schoen, C. et al., Bacteria as DNA vaccine carriers for genetic immunization, *Int. J. Med. Microbiol.*, 294, 319, 2004.
152. Timmons, L., Court, D.L., and Fire, A., Ingestion of bacterially expressed dsRNAs can produce specific and potent genetic interference in *Caenorhabditis elegans*, *Gene*, 263, 103, 2001.
153. Sharma, A.K. and Khuller, G.K., DNA vaccines: Future strategies and relevance to intracellular pathogens, *Immunol. Cell Biol.*, 79, 537, 2001.
154. Nagata, T. et al., Cytotoxic T-lymphocyte-, and helper T-lymphocyte-oriented DNA vaccination, *DNA Cell Biol.*, 23, 93, 2004.
155. Krieg, A.M. et al., CpG motifs in bacterial DNA trigger direct B-cell activation, *Nature*, 374, 546, 1995.
156. Klinman, D.M. et al., CpG motifs present in bacterial DNA rapidly induce lymphocytes to secrete interleukin 6, interleukin 12, and interferon γ, *Proc. Natl. Acad. Sci. USA*, 93, 2879, 1996.
157. Roman, M. et al., Immunostimulatory DNA sequences function as T helper-1-promoting adjuvants, *Nat. Med.*, 3, 849, 1997.
158. Schoen, C. et al., Bacterial delivery of functional messenger RNA to mammalian cells, *Cell. Microbiol.*, 7, 709, 2005.
159. Loeffler, D.I.M. et al., Comparison of different live vaccine strategies *in vivo* for delivery of protein antigen or antigen-encoding DNA and mRNA by virulence-attenuated *Listeria monocytogenes*, *Infect. Immun.*, 74, 3946, 2006.
160. Zinkernagel, R.M. and Doherty, P.C., Restriction of *in vivo* T cell-mediated cytotoxicity lymphocytic choriomeningitis within a syngeneic and semiallogeneic system, *Nature*, 248, 701, 1974.
161. Goossens, P.L. et al., Attenuated *Listeria monocytogenes* as a live vector for induction of CD8$^+$ T cells *in vivo*: A study with the nucleoprotein of the lymphocytic choriomeningitis virus, *Int. Immunol.*, 7, 797, 1995.
162. Slifka, M.K. et al., Antiviral cytotoxic T-cell memory by vaccination with recombinant *Listeria monocytogenes*, *J. Virol.*, 70, 2902, 1996.
163. Ochsenbein, A.F. et al., A comparison of T cell memory against the same antigen induced by virus versus intracellular bacteria, *Proc. Natl. Acad. Sci. USA*, 96, 9293, 1999.
164. Ikonomidis, G. et al., Influenza-specific immunity induced by recombinant *Listeria monocytogenes* vaccines, *Vaccine*, 15, 433, 1997.
165. Lieberman, J. and Frankel, F.R., Engineered *Listeria monocytogenes* as an AIDS vaccine, *Vaccine*, 20, 2007, 2002.
166. Paterson, Y. and Johnson, R. S., Progress towards the use of *Listeria monocytogenes* as a live bacterial vaccine vector for the delivery of HIV antigens, *Expert Rev. Vaccines*, 3, S119, 2004.
167. Johnson, R.P. et al., HIV-1 gag-specific cytotoxic T lymphocytes recognize multiple highly conserved epitopes. Fine specificity of the gag-specific response defined by using unstimulated peripheral blood mononuclear cells and cloned effector cells, *J. Immunol.*, 147, 1512, 1991.
168. Littaua, R.A. et al., An HLA-C-restricted CD8$^+$ cytotoxic T-lymphocyte clone recognizes a highly conserved epitope on human immunodeficiency virus type 1 gag, *J. Virol.*, 65, 4051, 1991.
169. Frankel, F.R. et al., Induction of cell-mediated immune responses to human immunodeficiency virus type 1 Gag protein by using *Listeria monocytogenes* as a live vaccine vector, *J. Immunol.*, 155, 4775, 1995.

170. Mata, M. et al., The MHC class I-restricted immune response to HIV-gag in BALB/c mice selects a single epitope that does not have a predictable MHC-binding motif and binds to K^d through interactions between a glutamine at P3 and pocket D, *J. Immunol.*, 161, 2985, 1998.
171. Mata, M. and Paterson, Y., Th1 T-cell responses to HIV-1 Gag protein delivered by a *Listeria monocytogenes* vaccine are similar to those induced by endogenous listerial antigens, *J. Immunol.*, 163, 1449, 1999.
172. Mata, M. et al., Evaluation of a recombinant *Listeria monocytogenes* expressing an HIV protein that protects mice against viral challenge, *Vaccine*, 19, 1435, 2001.
173. Guzmán, C.A. et al., Attenuated *Listeria monocytogenes* carrier strains can deliver an HIV-1 gp120 T helper epitope to MHC class II-restricted human $CD4^+$ T cells, *Eur. J. Immunol.*, 28, 1807, 1998.
174. Friedman, R.S. et al., Induction of human immunodeficiency virus (HIV)-specific CD8 T-cell responses by *Listeria monocytogenes* and a hyperattenuated *Listeria* strain engineered to express HIV antigens, *J. Virol.*, 74, 9987, 2000.
175. Rayevskaya, M., Kushnir, N., and Frankel, F.R., Safety and immunogenicity in neonatal mice of a hyperattenuated *Listeria* vaccine directed against human immunodeficiency virus, *J. Virol.*, 76, 918, 2002.
176. Rayevskaya, M., Kushnir, N., and Frankel, F.R., Antihuman immunodeficiency virus-gag $CD8^+$ memory T cells generated *in vitro* from *Listeria*-immunized mice, *Immunology*, 109, 450, 2003.
177. Zhao, X. et al., Vaginal protection and immunity after oral immunization of mice with a novel vaccine strain of *Listeria monocytogenes* expressing human immunodeficiency virus type 1 gag, *J. Virol.*, 80, 8880, 2006.
178. Peters, C. et al., The induction of HIV gag-specific $CD8^+$ T cells in the spleen and gut-associated lymphoid tissue by parenteral or mucosal immunization with recombinant *Listeria monocytogenes* HIV Gag, *J. Immunol.*, 170, 5176, 2003.
179. Stevens, R. et al., Oral immunization with recombinant *Listeria monocytogenes* controls virus load after vaginal challenge with feline immunodeficiency virus, *J. Virol.*, 78, 8210, 2004.
180. Stevens, R. et al., Pre-existing immunity to pathogenic *Listeria monocytogenes* does not prevent induction of immune responses to feline immunodeficiency virus by a novel recombinant *Listeria monocytogenes* vaccine, *Vaccine*, 23, 1479, 2005.
181. Ramshaw, I.A. and Ramsay, A.J., The prime-boost strategy: Exciting prospects for improved vaccination, *Immunol. Today*, 21, 163, 2000.
182. Dye, C. et al., Global burden of tuberculosis: Estimated incidence, prevalence, and mortality by country, *J. Am. Med. Assoc.*, 282, 677, 1999.
183. Colditz, G.A. et al., Efficacy of BCG vaccine in the prevention of tuberculosis: Meta-analysis of the published literature, *J. Am. Med. Assoc.*, 271, 698, 1994.
184. Grode, L. et al., Increased vaccine efficacy against tuberculosis of recombinant *Mycobacterium bovis* bacilli Calmette-Guérin mutants that secrete listeriolysin, *J. Clin. Invest.*, 115, 2472, 2005.
185. Hess, J. et al., *Mycobacterium bovis* bacilli Calmette-Guérin strains secreting listeriolysin of *Listeria monocytogenes*, *Proc. Natl. Acad. Sci. USA*, 95, 5299, 1998.
186. Kima, P.E., Ruddle, N.H., and McMahon-Pratt, D., Presentation via the class I pathway by *Leishmania amazonensis*-infected macrophages of an endogenous leishmanial antigen to $CD8^+$ T cells, *J. Immunol.*, 159, 1828, 1997.
187. Heinzel, F.P. et al., Reciprocal expression of interferon γ or interleukin 4 during the resolution or progression of murine leishmaniasis, *J. Exp. Med.*, 169, 59, 1989.
188. Gumy, A., Louis, J.A., and Launois, P., The murine model of infection with *Leishmania major* and its importance for the deciphering of mechanisms underlying differences in Th cell differentiation in mice from different genetic backgrounds, *Int. J. Parasitol.*, 34, 433, 2004.
189. Afonso, L.C. et al., The adjuvant effect of interleukin-12 in a vaccine against *Leishmania major*, *Science*, 263, 235, 1994.
190. Julia, V., Rassoulzadegan, M., and Glaichenhaus, N., Resistance to *Leishmania major* induced by tolerance to a single antigen, *Science*, 274, 421, 1996.
191. Gurunathan, S. et al., Vaccination with DNA encoding the immunodominant LACK parasite antigen confers protective immunity to mice infected with *Leishmania major*, *J. Exp. Med.*, 186, 1137, 1997.
192. Soussi, N. et al., Effect of intragastric and intraperitoneal immunization with attenuated and wild-type LACK-expressing *Listeria monocytogenes* on control of murine *Leishmania major* injection, *Vaccine*, 20, 2702, 2002.
193. Soussi, N. et al., *Listeria monocytogenes* as a short-lived delivery system for the induction of type 1 cell-mediated immunity against the p36/LACK antigen of *Leishmania major*, *Infect. Immun.*, 68, 1498, 2000.

194. Saklani-Jusforgues, H. et al., Enteral immunization with attenuated recombinant *Listeria monocytogenes* as a live vaccine vector: Organ-dependent dynamics of CD4 T lymphocytes reactive to a *Leishmania major* tracer epitope, *Infect. Immun.*, 71, 1083, 2003.
195. Saklani-Jusforgues, H., Fontan, E., and Goossens, P.L., Effect of acid-adaptation on *Listeria monocytogenes* survival and translocation in a murine intragastric infection model, *FEMS Microbiol. Lett.*, 193, 155, 2000.
196. Shen, H. et al., Modulation of the immune system by *Listeria monocytogenes*-mediated gene transfer into mammalian cells, *Microbiol. Immunol.*, 48, 329, 2004.

16 Anticancer Vaccine Strategies

Matthew M. Seavey, Thorsten Verch, and Yvonne Paterson

CONTENTS

16.1 INTRODUCTION

16.1.1 CANCER VACCINES

The race to cure communicable diseases is at a feverish pace with pending flu and AIDS epidemics; however, humankind's greatest threat of all may reside in its own DNA. According to the World Health Organization, in 2005, cancer was only second to cardiovascular disease as the highest killer among the noncommunicable diseases, responsible for 7.6 million deaths worldwide. Until only recently has the idea of using the immune system to fight cancer become so apparent.[1–4] For decades, the use of highly toxic drugs that target rapidly dividing cells has been the conventional cancer therapy. However, targeting cancer cells using the immune system has several advantages that far outweigh the use of chemotherapeutics, which carry many side effects.

The earliest attempt to harness the immune system to target cancer was carried out, unwittingly, by William B. Coley over 100 years ago. Coley observed that the tumors of several sarcoma patients who developed erysipelas (a streptococcal infection) shrank and in some cases disappeared.[5] He developed this into a cancer treatment using heat-killed, Gram-positive streptococci combined with Gram-negative *Serratia marcescens* directly injected into tumors.[6] These bacterial mixtures, known as "Coley's toxins," were also injected systemically in the gluteus maximus or pectoral muscles in addition to the intratumoral injections.[6] Coley managed to achieve a cure rate in patients treated with his bacterial toxins of about 10%, which is remarkable considering that these patients had advanced cancer and massive tumor load.[7] While little was known about the immune system at that time, we now recognize that Coley and his contemporaries were pioneers in the field of tumor immunotherapy.

Even though surgery is effective at removing primary tumors, it is metastatic disease that is the primary killer of patients inflicted with cancer. The ultimate goal of tumor immunotherapy is to utilize the power of the immune system to target and destroy tumor cells at these distant metastatic sites. Effective tumor immunotherapy holds several potential advantages over traditional radiotherapy or chemotherapeutic approaches to treating metastatic disease, including tumor specificity and a reduction in toxicity. Modern tumor immunotherapy includes both passive and adoptive techniques. Cancer vaccines, which aim to stimulate the patient's own immune system to respond to tumor cells, belong to the latter approach. However, in order for cancer vaccines to be effective, tumors must express antigens that are in some way specific to the tumor itself and can be recognized by the immune system.

Early immunologists speculated on the existence of tumor antigens in the 1900s,[8] but none were actually identified until the 1990s, when they were first isolated and cloned. Tumor antigens are a heterogeneous group of proteins that can be loosely classified into four groups: (1) mutated self-antigens, (2) cancer-testis tumor antigens, (3) tissue-specific antigens, and (4) viral antigens.[9] Few of these are truly tumor specific but may be tumor associated either by overexpression in the tumor—for example, receptors involved in cell signaling—or because they are shared by a limited number of normal cells—for example, tissue-specific antigens shared between the tumor and cells of its tissue origin. Mutated self-antigens arise when normal genes undergo mutation as part of the neoplastic process and are commonly the result of mutations in oncogenes such as p53, retinoblastoma, or members of the epidermal growth factor (EGF) receptor family. However, they can also arise from genes that encode proteins that have no direct role in the transformed phenotype. Often the mutations are unique to the tumor arising in a single patient and may not be shared by other tumors of the same type. Cancer-testis tumor antigens are not specific to one tumor, but are expressed by multiple tumors that arise independently. These are more useful for cancer vaccines because the same vaccine can target tumors that arise in a population of patients that have the same type of cancer. Cancer-testis tumor antigens include the MAGE, BAGE, and GAGE families of antigens originally isolated from human melanoma but now known to be present in other tumors.[10] Their name is derived from the fact that they are expressed at low levels in normal tissues considered to be immune privileged, such as the testis or the placenta. They are thought to be the products of genes expressed during embryonic development that have been silenced by the time of birth.

Tissue-specific tumor antigens are also expressed by the normal tissue from which the tumor has arisen. They are useful targets of the immune response only if the tissue is not required to maintain life. Many melanoma-associated tumor antigens such as tyrosinase and tyrosinase-related proteins 1 and 2 (TRP-1, TRP-2), which are involved in the biosynthesis of melanin from tyrosine, belong to this category since they are also expressed by normal melanocytes. Immune responses that target these antigens can kill normal melanocytes leading to the autoimmune disease vitiligo, in which white patches appear in the skin.[11] However, although vitiligo can be mildly disfiguring, it is considered a small price to pay for eradicating a potentially fatal disease such as melanoma. Finally, 20–30% of all human cancer is caused by transforming viruses that express viral-specific oncogenes. Epstein–Barr virus and human papillomavirus are examples of transforming viruses commonly associated with specific human tumors.[9] An advantage of targeting virally derived proteins with cancer vaccines is that these antigens are immunologically "foreign" and should not have induced central tolerance in the host.

16.1.2 The Immune Response to Tumor Antigens

Although early attempts at harnessing the immune response to fight cancer focused on the induction of humoral immunity, it is now clear that tumor rejection is largely mediated by a strong cellular immune response,[12] because cellular immune responses are capable of targeting antigens regardless of their compartmentalization within the cell. Tumor antigens can have one of five general locations or destinations within the tumor cell. They can be localized in the nucleus, which is especially true of oncogenes, the mitochondria, or the cytoplasm; they can be expressed on the surface of the cell, such as constitutively active EGF receptors or they can be secreted. Typically, secreted antigens or cell surface antigens are the only types accessible to a humoral immune response. In contrast, peptides derived from proteins, regardless of their cellular or extracellular destination, can be presented by major histocompatibility complex (MHC) class I molecules to responding $CD8^+$ T cells.[13]

MHC class I molecule/peptide complexes give the immune system a means of sampling the autologous proteins expressed by a cell. Tumor antigens must be able to access the MHC class I antigen processing pathway for presentation to responding T cells because presentation of antigenic peptides by the tumor cells themselves is required for their recognition and destruction by responding T cells. However, the initial priming of the antitumor response must be mediated by antigen presenting cells (APCs) such as macrophages and dendritic cells (DCs). Indeed, it has been shown in transplantable tumor models in the mouse that bone marrow–derived APCs are responsible for priming a tumor rejection response.[14] The presentation of tumor antigens by "professional" antigen presenting cells is commonly called cross-priming.[15] Cancer vaccines must therefore target tumor antigens to professional APCs in order to prime a potent tumor-specific immune response. This ensures not only the appropriate presentation of tumor antigens to responding lymphocytes but also appropriate costimulation through cognate interactions with costimulatory molecules on APCs, as well as growth factors and cytokine stimulation.

Cytokines play a major role in tumor rejection responses. Cytokines of the Th1-type produced by tumor-infiltrating $CD4^+$ T cells are particularly effective for inducing inflammatory responses within the tumor, which can have both direct cytotoxic[16] as well as angiostatic effects by blocking the formation of new blood vessels within tumors.[17] The hallmark of a Th1-type cytokine profile includes the cytokines IL-2, IL-12, tumor necrosis factor (TNF)-α, and interferon (IFN)-γ. Production of the cytokine IL-12 early in the immune response by innate immune cells is known to be the driving force behind the shift to a Th1-type cytokine profile.[18,19] IL-12 secreted by macrophages and DCs stimulates the secretion of IFN-γ by activated and resting natural killer (NK) cells. Further, IL-12 release by phagocytic cells is boosted by IFN-γ.[20,21] Later in the immune response, IFN-γ is released from activated T cells.[20,21] Activated T cells have also been shown to induce IL-12 secretion by macrophages and dendritic cells through the interaction of CD40L on the surface of activated T cells with CD40 on the surface of IL-12 producing cells.[22,23] IFN-γ up-regulates the expression of MHC class I and II molecules as well as costimulatory molecules of the B7 family on APCs, rendering them more effective at stimulating responding T cells.[20] In mouse models both, TNF-α[24,25] and IL-2[26–28] have been shown to significantly impact tumor growth.

The cytokine granulocyte macrophage-colony stimulating factor (GM-CSF) has also been shown to be a useful adjuvant for tumor immunotherapy.[29] GM-CSF expressed in the tumor microenvironment promotes the maturation of DCs, which are required APCs for the presentation of antigen to, and subsequent activation of, naive T cells.[15] Even the Th-2 type cytokine, IL-4, has been shown to assist in the eradication of tumors in some mouse tumor models. IL-4-induced killing is largely mediated by tumor–infiltrating eosinophils, which create an inflammatory environment within the tumor.[30] Antigen presenting cells and cytokines cooperate to promote cognate stimulation through the B7/CD28 costimulatory pathway in order to activate tumor-specific T cells.[31] It is the activation of these cells that will ultimately determine the outcome of the antitumor response. Activated T cells must be able to overcome any tolerance to tumor antigens as well as possible immunosuppressive activities mediated by the tumor itself. Additionally, these T cells must be able to mediate tumor destruction more rapidly than the tumor can replicate. The outcome of this race determines the success of the immunotherapy.

It is well known that the immune system does not respond well to tumor-associated antigens. Resident T cells with receptors specific for tumor antigens tend to remain ignorant rather than become activated. There are a number of reasons for this. One is certainly that tumor-associated antigens are often homologous with self-proteins, or may indeed be self-proteins, and thus the immune system has been rendered tolerant of their presence. However, even tumor antigens of "foreign" origin are ignored by the immune system. One reason may be due to a lack of costimulation in the tumor site.[32] It has also been proposed that tumors do not provide the "danger signal" that infectious pathogens do, which provides inflammation and an influx of antigen presenting cells.[33] Aside from failing to produce a danger signal, tumors have been shown to actively evade the immune response. Tumors can often down-regulate expression of tumor antigens,[34–36] MHC molecules,[37–40] and the antigen processing machinery.[41]

Recently, we have shown that, similar to chronic viral infections, CTL escape mutants can arise in response to immune pressure on tumors.[42] Some tumors secrete immunosuppressive cytokines such as TGF-β[43] and others have been shown to express Fas ligand, which can eliminate Fas expressing tumor–infiltrating T cells by inducing apoptosis.[44] Thus, designing effective cancer vaccines poses problems usually not encountered in conventional prophylactic vaccine development. To cross these hurdles, tumor immunologists attempt to develop therapeutics that overcome natural ignorance to tumor antigens and provide the necessary costimulatory signals, antigen presentation, and cytokines to ensure recognition and lysis of tumor cells in an antigen-specific manner. Theoretically, the best immunotherapeutic approach would be to induce a cellular immune response with a Th1 type cytokine profile targeting a tumor antigen that is effectively presented by both MHC class I and II molecules in a cytokine environment that stimulates the development of long-lived adaptive immunity. Clearly, this is precisely the kind of immune response induced by live infectious organisms such as intracellular bacteria and some viruses. Intracellular bacteria are particularly attractive vehicles to induce such immunity. Not only has bacterial vaccination been shown to mitigate against growing tumors,[45] but it has also been shown to be very effective in preventing the reestablishment of new tumors.[46]

The use of intracellular bacteria as an anticancer vaccine vector has several advantages over other live carriers such as viral vector systems. First of all, intracellular bacterial vectors can be easily controlled using antibiotics. This gives the clinician precise control over bacterial load and thus vaccine exposure. Second, one of the major problems with viral vectors as vaccine agents is the generation of neutralizing antibodies that are usually induced after initial priming. This may become a problem during future rounds of boosting. However, clearance of intracellular bacteria, such as *Listeria monocytogenes*, does not require neutralizing antibodies and humoral responses to *Listeria* are weak.[46–48] Finally, bacteria contain several pathogen-associated molecular patterns (PAMPS) (e.g., unmethylated CpG DNA, flagellin, LPS, and other cell wall components) responsible for activating the innate immune system and shaping the following adaptive immune response.

16.1.3 *L. monocytogenes* and the Cellular Immune Response

L. monocytogenes is a Gram-positive rod that can infect humans and animals. Soon after entering the blood stream, these bacteria are phagocytosed by macrophages, Kupffer cells, neutrophils, and

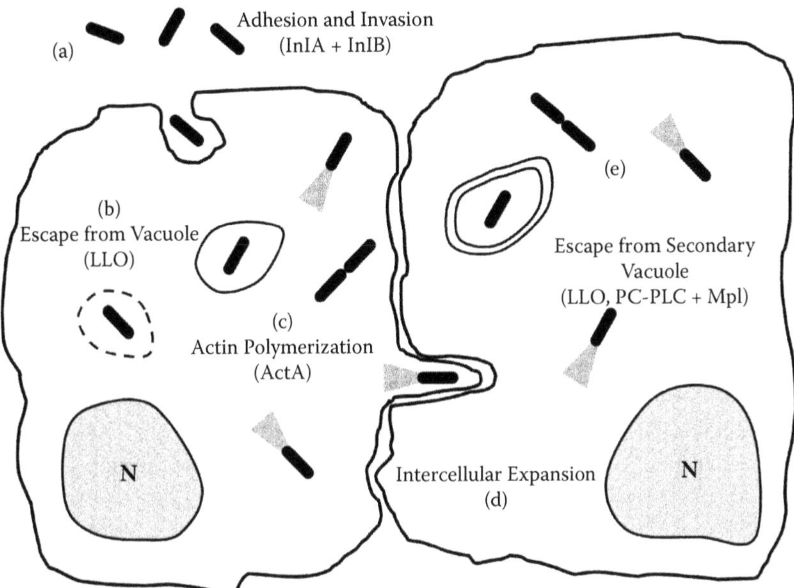

FIGURE 16.1 Interaction of *Listeria monocytogenes* with host phagocytes. Cellular infection with *Listeria* involves several steps that eventually lead to intracellular transmission and bacterial propagation. (a) Adhesion and invasion are mediated by molecules internalin A (InlA) and internalin B (InlB). (b) After cellular entry, *L. monocytogenes* escapes the phagosome by secreting listeriolysin-O (LLO), which lyses the vacuolar membrane, releasing the bacteria into the host's cytosol. (c) *Listeria* polymerizes host actin molecules mediated by ActA; this serves to move the bacterium throughout the cytosol and eventually into neighboring cells. (d) Intense actin polymerization at the tail of the bacterium can literally shoot the organism into adjacent cells, generating a double membrane vacuole. (e) Escape from this secondary vacuole requires three secreted bacterial proteins: LLO, a broad-range phosholipase C (PC-PLC), and a metalloprotease (Mpl). Escape from this secondary vacuole allows safe delivery to a second host cell without further alert of the immune system or interaction with the outside environment.

monocytes. Intracellular bacteria differ on where they replicate in the host cell. Both *Salmonella* and *Mycobacteria* escape host lysis by remodeling and remaining in the phagosome.[45] *Listeria* and *Shigella* avoid the antimicrobial environment by escaping into the cytosol (*Listeria*, Figure 16.1). *L. monocytogenes* escapes from the phagocytic vacuole, largely by the action of the thiol-activated, pore-forming hemolysin, listeriolysin-O (LLO), and enters the cytoplasm of the cell.[49,50] The gene *plcA* encodes a phosphatidylinositol-specific phospholipase C (PI-PLC) that improves the efficiency of this vacuole escape.[51] The cytoplasm of a cell serves as a rich growth medium for the bacteria, allowing *L. monocytogenes* to undergo replication rapidly.[52]

The listerial virulence gene *actA* encodes a protein, ActA, which is expressed in a polar manner on the outer surface of the bacterium. ActA polymerizes the host cell actin, forming a tail of actin that pushes the listeriae around the cytoplasm of the cell.[50,53] This intracellular motility allows *L. monocytogenes* to form pseudopods that extend from the infected cell and are taken up by neighboring cells. In this manner, *L. monocytogenes* can spread from one cell to another without leaving the cytoplasm or contacting the extracellular milieu. Thus, *L. monocytogenes* remains largely sequestered from the humoral arm of the immune system. Through the action of LLO and a second, broad-range phospholipase C (PLC), *L. monocytogenes* escapes from the double–walled phagosome and begins the intracellular growth and replication cycle again.[54,55] The virulence genes encoding the PLCs, *plcA* and *plcB*, and the other virulence genes, *hly* and *actA*, are regulated by the pluripotential transcription factor PrfA, which is encoded by the *prfA* gene.[56]

This escape mechanism allows the bacteria to avoid death; however, proteins expressed and secreted in the cytosol are targeted for degradation by the proteosome and the resulting peptides can be transported to the endoplasmic reticulum (ER) for loading onto intact host MHC I. This targeting of bacterial proteins to the MHC I pathway will result in cytotoxic T lymphocyte (CTL) priming or activation, and ultimately death, of the infected cell. If tumor antigens are fused to bacterial virulence factors normally secreted by the invading bacterium, the tumor antigen will also be targeted to the MHC I pathway. If this tumor antigen is combined with a molecular adjuvant, priming of potent, tumor-specific CTLs will result (Figure 16.2). Further, bacteria remaining in the lysosomal compartment will be destroyed and their proteins shuttled into the MHC II pathway for $CD4^+$ T cell activation (Figure 16.2).

L. monocytogenes infection in the mouse has served as a classic model of cellular immunity for decades.[57] Control of the infection is mediated by a complex interaction of macrophages, NK cells, neutrophils, and T cells.[58] Macrophages and neutrophils are the first line of defense against listerial infection. These phagocytic cells secrete IL-12 on infection and TNF-α, which stimulate NK cells to secrete IFN-γ.[58] IL-12 influences the development of the T cell response, driving it to a Th1 type.[59] The responding T cells secrete IFN-γ, which in turn stimulates macrophages and dendritic cells to become activated, up-regulate class I and class II MHC molecules and molecules of the costimulatory B7 family, and also to secrete more IL-12. This inflammatory cytokine cascade early in infection is particularly useful in jump-starting the immune response against poorly immunogenic molecules such as tumor antigens and breaking tolerance against them.

The use of *Listeria*, as opposed to other intracellular bacteria such as *Shigella* and *Salmonella*, combines the powerful feature of MHC I pathway targeting, with the safety of using a low pathogenicity strain only known to cause disease in immunocompromised individuals.[60] Infection with *Listeria* generates a very powerful Th1-mediated immune response, needed for macrophage activation and *Listeria* elimination.[61] In addition, Th1 cells help $CD8^+$ T cell maturation and memory recall.[62] Both cell types contribute to effective antitumor therapy.[63] Furthermore, *Listeria* infection can activate γδ T cells and NK cells, both of which play a role in the recognition and elimination of neoplastic cell types.[64,65]

This chapter will focus on the general principles and practicalities needed for the efficient use of *Listeria* as a bacterial vector for cancer immunotherapy. The basic principles of the methods described for *Listeria* in this chapter can be easily adapted to other bacterial species with some modifications that can be obtained from the literature. We will also discuss the construction of *Listeria* vectors, examples of current experimental vaccines, and vaccines in trial. Furthermore, we will provide a current review of the literature of using *Listeria* as an anticancer vaccine. The interested reader may also consider several review articles on bacterial vector systems as a guide through the literature.[45,66–69]

16.2 GENERAL PRINCIPLES FOR CONSTRUCTING *L. MONOCYTOGENES* CANCER VACCINES

16.2.1 MATERIALS

Bacterial strains. Most vector strains have been constructed from *L. monocytogenes* strain 10403S, a natural streptomycin–resistant strain derived from 10403.[70] The sequence of 10403S is very close to the EGD-e strain whose complete genome sequences are available; thus, EGD-e can be used as a basis for primer and cloning design. A mutant strain, *Listeria* XFL7, in which the *prfA* virulence gene has been inactivated, has been described previously.[71]

Plasmids. All genetic engineering steps need to be performed in *E. coli* since the transformation efficiency of *Listeria* is very low and requires relatively large amounts of purified plasmid DNA (approximately 1 μg). As an example of the vector systems used, a schematic for the expression of human papilloma virus (HPV) E7 in a *Listeria* shuttle plasmid, capable of replicating in Gram-positive and Gram-negative bacteria, can be found in Figure 16.3B. Several shuttle plasmids have been published, of which we suggest the use of two different types:

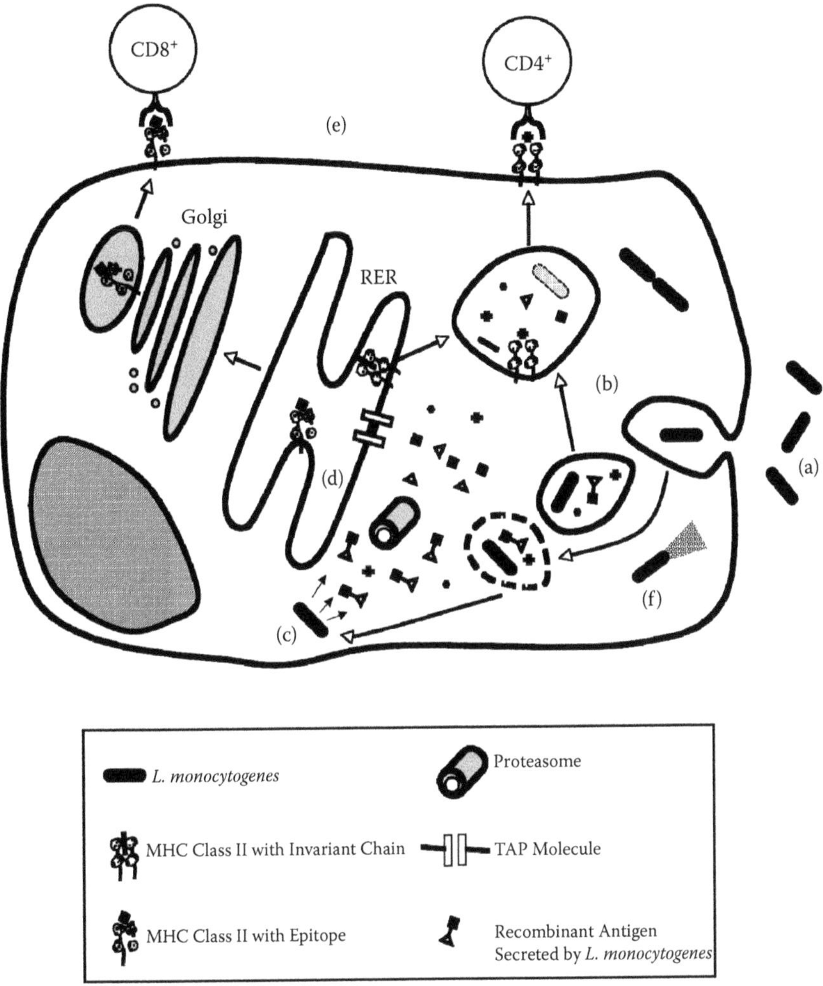

FIGURE 16.2 *L. monocytogenes* vectors can deliver a secreted tumor antigen to both cytotoxic CD8+ T cells and helper CD4+ T cells. (a) *Listeria* attaches and invades a phagocytic cell. (b) Once inside the phagosome prfA up-regulates virulence genes, releasing the recombinant antigen. Only about 10% of the total engulfed *Listeria* will escape the vacuole and gain entry to the host cytosol. The other 90% will be eliminated by phagolysosomal formation. Epitopes generated will be loaded onto MHC II molecules shuttling to the surface for presentation to CD4+ T cells. (c) *Listeria* that has escaped death now secretes the recombinant protein into the host cytosol. These proteins are destined to be cleaved by the host proteosome. (d) Epitopes generated by this ubiquitin-dependent process will be selected by the TAP transporter to gain access to the rough endoplasmic reticulum (RER). Cognate peptides will be loaded onto stable MHC I molecules destined for the Golgi and eventually the surface for presentation to CD8+ T cells. (e) Since secreted proteins from *Listeria* gain entry to both intracellular and extracellular compartments, recombinant proteins expressed will gain entry to MHC I and MHC II pathways for both CD8+ and CD4+ T cell activation. (f) Following replication in cytosol, *Listeria* migrates toward cell membrane for spread to neighboring cells.

- Integration of recombinant genes into the *Listeria* genome is achieved by using pKSV7, which contains a temperature-sensitive origin of replication.[72] This plasmid was modified by introducing approximately 500 bp of *Listeria* open reading frame (ORF) Y and ORF Z, interrupted by a multiple cloning site for the recombinant gene in order to allow for integration specifically into the ORF XYZ genome region.[72]
- Plasmid-based gene expression can be achieved using shuttle plasmid pAM401,[73] which contains replication and antibiotic resistance selection genes for both *E. coli* and *Listeria*.

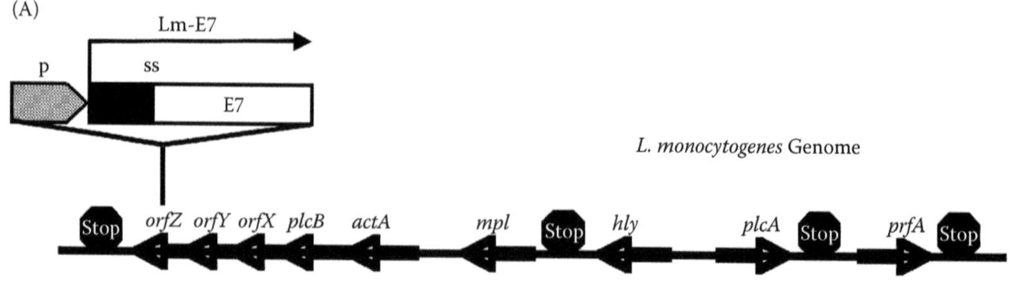

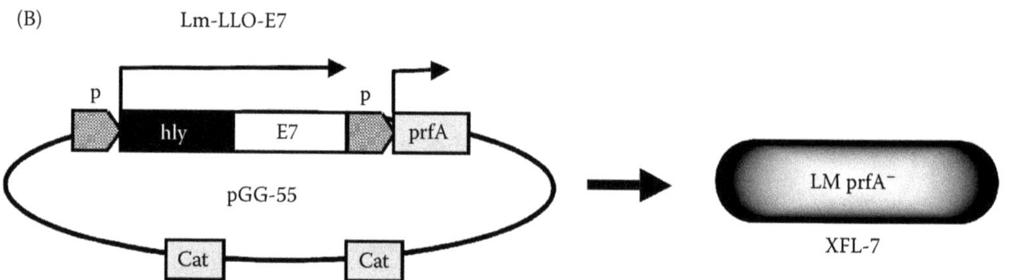

FIGURE 16.3 Expression of foreign genes in *Listeria* (here as an example: HPV E7). A. Schematic map of genomic integration. The Lm-E7 expression cassette on a plasmid containing flanking homologous sequences is introduced by homologous recombination into the *orfZ* domain of the *L. monocytogenes* genome. (The *L. monocytogenes* genome diagram was adapted from Paterson et al.)[45] Expression of the virulence genes *actA*, *hly*, *mpl*, and *plcB*, along with *plcA* and open reading frames *X*, *Y*, and *Z* are up-regulated by *prfA*. Transcription termination signals (STOP symbols) follow *hly*, *orfZ*, *plcA*, and *prfA*. B. The Lm-LLO-E7 strain was generated by transforming the *prfA*⁻ *Listeria* strain XFL-7 with the plasmid pGG-55. HPV E7 was genetically fused to the LLO encoding gene, *hly*, under control of the virulence transcription factor, *prfA*. This plasmid also contains CAT(–) and CAT(+) chloramphenicol transferase genes for *in vitro* selection in Gram-negative and Gram-positive bacteria, along with replication genes for both bacterial types.

> Recombinant gene expression was first made possible by introducing the *Listeria hly* promoter and the *prfA* virulence gene resulting in pDP2028.[74] We modified this plasmid to facilitate the insertion of new antigens.[71] The resulting plasmid, pGG55 (Figure 16.3B), is the basis of most of our *Listeria*-based cancer vaccine strains. Also, since the shuttle plasmids are relatively large, all initial PCR cloning steps are performed in TA-cloning vector pCR2.1 (Invitrogen) prior to moving the recombinant genes into pKSV7 or pAM401.

Media. L. monocytogenes is best cultured in brain–heart infusion (BHI) media, in which it doubles approximately every 40 minutes at 37°C. The transcription factor gene *prfA* is not expressed when bacteria are cultured in iron-rich media such as BHI,[75] so the expression of the *prfA*-regulated virulence genes—including *hly,* the gene that encodes LLO—is also suppressed. To test for recombinant antigen expression when using the *hly* promoter, the strains must be cultured in either plain Luria–Miller (LB) broth or alternatively in BHI treated with charcoal to remove iron-containing compounds.

Mouse strains. Mice can be readily infected by *Listeria* when injected i.p. or s.c. Oral infection is possible usually only for strains with an $LD_{50} \leq 10^7$ colony forming units (CFUs), due to inefficient translocation of the bacteria through the gut barrier. The virulence of *L. monocytogenes* strains varies between different mouse species.[76] For example, Balb/c and A/J mice are more susceptible than C57BL/6 mice. Further, the virulence of *Listeria* expressing different passenger antigens will vary, depending on the protein. In general, our recombinant strains are highly attenuated, with an LD_{50} of about 10^9 CFUs and are cleared by innate immunity in SCID mice within 5 days postinjection.

16.2.2 Antigen Expression in *Listeria*

Subcellular targeting. Usually, it is preferable for the cancer antigen to be secreted outside the bacterial cell into the host cell cytosol in order to improve efficacy of antigen presentation by the host cell. Placing the LLO leader sequence (amino acids [aa] 1–25) in fusion with the recombinant protein sequence will ensure targeting of the protein to the secretory pathway.[71] Due care needs to be taken to exclude hydrophobic sequences in the recombinant antigen that might interfere with secretion. In our experience, transmembrane domains in particular tend to cause problems, but other highly hydrophobic regions should preferably be removed also. In addition, many proteins larger than 400 aa are poorly expressed. Better expression can be achieved by dividing the protein between separate constructs with overlapping sequences.[77,78] Individual strains will need to be established for each construct and immunization can be performed either individually[77] or as a cocktail.[78]

Virulence-regulated promoters. To achieve high antigen expression levels, a strong promoter is necessary such as the *hly* promoter (Genebank accession No. M24199, bp 1256–1488). Expression of *hly* is regulated by the virulence factor PrfA and is highly expressed in the phagolysosome. Although it is down-regulated to a certain degree when *Listeria* reaches the host cytosol,[79] it is still active. In our hands, translation levels from the *hly* promoter were always sufficient to express the recombinant antigen in high enough levels to evoke a strong immune response. The *actA* promoter (Genebank accession No. AL591974, bp 9270–9470) may represent a useful alternative, as it is up-regulated 30 times compared to LLO in the cytosol.[79] Both *hly* and *actA*, which are *prfA* regulated, have the advantage that transcription can be suppressed *in vitro* by culturing in BHI so that proteins that are potentially problematic for *Listeria* growth could still be delivered *in vivo* without interfering with growth outside the host and the production of bacterial stocks.

Constitutive promoters. Alternatively to *hly* or *actA*, a constitutive promoter could be chosen in specific cases if desired. For example, the promoter for the invasion associated protein (*iap*) gene, p60, is well characterized.[80,81] We have successfully used the *iap* promoter (Genebank accession No. X52268, bp 287–443) on a recombinant plasmid in *Listeria* to drive the expression of the D-alanine racemase gene.[82]

Introduction of recombinant genes. Recombinant antigen expression can be achieved using different strategies and the choice of the expression system depends on the demands of the vaccine and its efficacy. The recombinant expression cassette can be introduced into *Listeria* either as a single copy into the genome or episomally on a multicopy plasmid.[71] Since both methods initially require transformation with a plasmid, it should be noted that the bacteria have to be competent for the uptake of plasmid DNA. As a Gram-positive bacterium, *Listeria* requires the partial degradation of the cell wall prior to electroporation.[83] This is usually achieved by culturing the bacteria during log phase in the presence of penicillin G (10–12.5 mg/ml) or lysozyme (2–5 kUnits). In case of the penicillin G method, the antibiotic is added to the culture when it reaches $OD_{600\ nm}$ of 0.2–0.4 followed by additional 2–2.5 h of culturing at 37°C and shaking at 250 rpm. The addition of 0.5 *M* sucrose to the culture medium as an osmotic buffer can improve results further. In the second method, bacteria are washed with 1 m*M* HEPES and 0.5 *M* sucrose prior to the addition of lysozyme and incubation at 37°C for 30 min with little shaking. After further washes in 0.5 *M* sucrose in both protocols, the bacteria are ready for electroporation (0.1-mm path length, V = 1.2 kV, R = 200 Ω, C = 25 μF).

Genomic integration. Genomic integration by homologous recombination can potentially circumvent the use of antibiotic resistance genes in the final product and results in a stable recombinant strain. Briefly, bacteria are transformed with temperature-sensitive pKSV7 containing the homologous recombination sites and the recombinant gene. Serial culture under chloramphenicol selective pressure at 30°C, which is permissive for plasmid replication and during which time random DNA crossover events occur, allows the integration of the shuttle plasmid into the genome, thus initially transferring chloramphenicol resistance. Bacteria containing a chromosomally integrated plasmid copy are selected by growth under chloramphenicol-selected pressure during a temperature shift to 40°C, which is nonpermissive for plasmid replication. After verification of the integration event by colony PCR, the growth temperature is shifted again to 30°C to allow for a second DNA crossover to occur at the homologous sites, thus excising unwanted plasmid sequences and leaving only the recombinant gene copy behind in the *Listeria* chromosome. The excised plasmid is prohibited from replication during another temperature shift to 40°C so that it will be diluted out during expansion of the bacterial culture. Bacteria that are chloramphenicol sensitive are selected by subsequent replica plating. Colony PCR using primers that bind genomic sequences on either side of the recombinant gene then verify the integration of the foreign gene into the desired chromosomal region. Specificity of chromosomal targeting is achieved with this method by exact design of the homologous recombination sequences. A detailed protocol has been described.[51]

In our[71,78] and others' hands,[84] the ORF XYZ locus of *Listeria* is a particularly useful integration site since its interruption does not seem to affect essential *Listeria* functions. *cis*-acting effects through proximity to transcription factor prfA and a cluster of important virulence genes including *hly* and *actA*,[51,84] (see Figure 16.3A) may also potentially enhance expression levels of the recombinant gene in this position. Other sites in the *Listeria* genome may also be suitable for integration of foreign genes, but we have not tested the effects on bacterial metabolism and virulence. As the genome of *L. monocytogenes* EGD strain has been sequenced (Genebank accession No. NC_003210), rational integration design can also be performed for alternative positions.

Episomal transformation. Episomal transformation with multicopy plasmids has the advantage of increasing the expression levels of the recombinant antigen that can be obtained in bacteria containing only a single integrated gene copy. However, if using this approach, the stability of the plasmid *in vitro* as well as *in vivo* must be carefully determined. Since *Listeria* has a low rate of transfection, all genetic manipulations and amplification of plasmid material are less time consuming if they are performed in *E. coli* prior to electroporation of the final plasmid into *Listeria*. Convenient shuttle plasmids that are capable of replication and *in vitro* antibiotic selection in both *E. coli* and *Listeria* have long been available.[73] However, plasmids can be rapidly lost *in vivo* where no antibiotic pressure is present and if the expression of the passenger antigen places a metabolic load on the bacterium. To overcome this problem we have used complementation systems for *in vivo* stabilization. The addition of the *prfA* gene into the plasmid in combination with a *Listeria* strain lacking this gene helped to dramatically improve plasmid stability *in vivo*.[71,74] Similar complementation systems using other genes have been described by our group as well as others.[82,85]

Enhancement of the immunogenicity of recombinant antigens. While the vector features described before are quite predictable, we found additional factors that influence vaccine efficacy. We routinely fuse cancer antigens to the virulence factor LLO to improve antigen presentation *in vivo*. A putative PEST (proline, glutamic acid, serine, threonine)-like domain at the N-terminus of LLO (amino acids: NKENS ISSMA PPASP PASPK TPIEK KHADE ID)[86,87] is suspected to direct fusion proteins to the proteosome and thus enhance loading of peptides onto MHC I complexes. Although there has been recent evidence of involvement of other mechanisms, the fusion of antigens to LLO or the PEST-like sequence of LLO or ActA, which also contains PEST domains, does seem to enhance the immunogenicity of the fused antigen.[71,88,89] In order to avoid overexpression of toxic LLO, we truncated the protein at aa 441, thereby deleting the hemolytic domain. It should be noted that after cleavage of the signal sequence, the secreted fusion protein contains the first 417 aa of the mature LLO. Other virulence factors containing putative PEST-like domains such as ActA were also found to enhance vaccine efficacy when fused to the recombinant antigen.[89]

Restoration of vector virulence. During the extended *in vitro* culture and genetic manipulation required to generate a *Listeria*-based vaccine, we have found that *Listeria* can down-regulate its virulence factors required for survival *in vivo*. Virulence can be restored by passaging vaccine strains in mice.[90,91] Besides up-regulation of bacterial virulence, *in vivo* passaging also enhances the cell-mediated immune response against both bacterial and passenger antigens.[90] Usually only two passaging steps are required to restore virulence. After passaging a new vector, we establish a cell bank from which we produce fresh immunization stocks as needed. As most vaccine strains are attenuated, either by deletion of required *Listeria* genes for safety compliance or due to the metabolic load imposed by expression of a recombinant antigen, larger inoculums of bacteria, which would be lethal for wild-type *Listeria,* can be used for immunization. We routinely start with 10^7–10^8 CFUs injected i.p. On day 3, the spleen is isolated and homogenized in 5–10 ml of PBS and 50–200 µl are plated in order to recover single colonies. Four colonies are pooled and expanded in liquid culture. The bacteria are frozen in 20% glycerol and a dilution series is prepared to count CFUs ranging from $1{:}10^1$–$1{:}10^{10}$ (plate 5–10 µl). An injection stock is prepared from the liquid culture and a second passage is then performed.[90] The bacteria isolated on day 3 are counted. If the second passage does not greatly increase the colonies per spleen, we prepare vaccine stocks and determine the virulence of this final stock. However, although the greatest benefits are usually observed after two passages, it may be necessary to increase vaccine efficacy by passaging up to four times.[90]

16.2.3 Immunization of Mice

Boosting protocol. Humoral immunity to intracellular bacteria such as *Listeria* is weak. Thus, unlike the case with viruses that induce neutralizing antibodies in the host, multiple vaccination boosts with *Listeria* are possible.[47,48] We have observed increasing antitumor immunity with up to four immunizations of $0.1 \times LD_{50}$ in mice for some tumor models.[47] However, immunity against the bacterial vector does increase with each boost, resulting in more rapid clearance.[92] To overcome this, higher booster doses can be used than those used for primary immunization without increasing the toxicity (i.e., $0.2 \times LD_{50}$ or $0.3 \times LD_{50}$). For most tumor models that we have worked with, we find that two immunizations at 1-week intervals with $0.1 \times LD_{50}$ (prime plus one boost) can be sufficient to eliminate established tumors. However, on occasion, optimal efficacy has required a more extensive immunization protocol.[47]

Immunization route. The route of immunization depends on the type of immune response that needs to be achieved and on practical considerations. As most tumors do not require a mucosal immune response, we usually inject the vaccine i.p. or i.v, which we found to be equally efficient in the E7 tumor system (unpublished data). Either route of injection resulted in similar antitumor immunity. Although s.c. immunization may appear to be a logical, convenient, and safe route of immunization, for an s.c tumor we found decreased efficacy using this route in tumor regression studies. The oral route is difficult to test since *Listeria* does not readily cross the gut barrier in mice. Guinea pigs are a more suitable study subject,[93] but cancer models have not been well established in these animals. Using a less attenuated vaccine strain (LD_{50}: 1×10^7) than usual (LD_{50}: 1×10^9), we were able to immunize orally against challenge with a tumor cell line expressing influenza nucleoprotein.[94] Other groups have also been successful with the oral route.[95] Additional attempts with more attenuated strains, however, have not yet been successful in our hands.

16.3 PRACTICAL EXAMPLES OF *LISTERIA*-BASED CANCER VACCINES

In the past decade, since we first demonstrated that *Listeria* could induce strong antitumor immunity,[46] many examples of the power and utility of this immunotherapeutic approach against cancer have been demonstrated by us and others. Here we will briefly summarize the published examples of *Listeria* as a cancer vaccine vector and then focus on two examples from our own laboratory because they provide insights in the ability of *Listeria* to "break" immune tolerance.

One of the greatest challenges in developing anticancer vaccines is overcoming self-tolerance. Unlike adoptive cell therapy (ACT) or other immunotherapies such as monoclonal antibody therapy that rely on external generation and manipulation, anticancer vaccines try to utilize the body's own immune system to fight disease. Many tumor antigens are homologous to self-proteins; thus, tolerance to self-tissues becomes a major obstacle in trying to target and amplify self-reactive T cells. In order for a cancer immunotherapy to really be effective, it will have to overcome several tolerance mechanisms. The ability of *Listeria*-based vaccines to break tolerance has not been extensively tested, but immune tolerance pertains to most cancers; the possible exceptions are tumors caused by transforming viruses, where the viral oncogenes may be considered to be foreign antigens.

16.3.1 Examples of *Listeria*-Based Cancer Vaccines

Listeria has been tested as a cancer vaccine using a variety of proteins. These include model tumor antigens such as β-galactosidase, ovalbumin, influenza nucleoprotein (NP), and lymphocytic choriomeningitis virus (LCMV) NP. Moreover, mouse tumor antigens such as Moloney murine leukemia virus gp70 and tryrosinase-related protein-2 (TRP-2), and a rabbit antigen, cottontail rabbit papillomavirus (CRPV) antigen E1, have also been explored (see Table 16.1). Our laboratory has focused on the human tumor antigens HPV-16 E7 oncogene and the epidermal growth factor-like receptor HER-2/neu in recent years.

Jensen et al. generated a recombinant *L. monocytogenes* that expressed the CRPV antigen E1 and tested its ability to protect outbred rabbits against challenge with CRPV- and CRPV–DNA-induced tumors.[96] Since HPV does not infect other species, CRPV has been used as a convenient animal model for oncogenic papillomavirus transformation and to test tumor vaccine approaches that could be applied to HPV.[97] CRPV infection results in papillomas that progress to malignant carcinoma with high frequency. Several CRPV antigens have been targeted for vaccine therapy,

TABLE 16.1
Summary of *Listeria* Strains Used as Live Replicating Cancer Vaccines

Vaccine Name[a]	Strain Modification[a]	Design	Antigen	Ref.
Lm-LLO-NP	*prfA*⁻	Plasmid	NP, influenza	46
Lm-LLO-E7	*prfA*⁻	Plasmid	E7, HPV-16	71
Lm-E7	WT	Targeted recombination into the chromosome	E7, HPV-16	71
Lm-LLO-EC1, EC2, EC3, IC1, IC2	*prfA*⁻	Plasmid	HER-2/neu, breast cancer	77
Lm^{DD}-LLO-E7	*dal*⁻, *dat*⁻	Plasmid	E7, HPV-16	82
Lm-PEST-E7	*prfA*	Plasmid	E7, HPV-16	88
Lm-ActA-E7	*prfA*	Plasmid	E7, HPV-16	89
rLm-E7	WT	Targeted recombination into the chromosome	E7, HPV	95
E1-rLm	WT	Targeted recombination into chromosome	E1, CRPV	96
rLm-NP	WT	Targeted recombination into chromosome	NP, LCMV	98
actA⁻/*inlB*⁻AH1-A5	*actA*⁻, *inlB*⁻	Inserted into a phage integration site in the genome	Gp70 epitope	99
actA⁻/*inlB*⁻-OVA	*actA*⁻, *inlB*⁻	Inserted into a phage integration site in the genome	OVA	99
Lm-TRP2	WT	Targeted recombination into chromosome	TRP-2, melanoma	100
Lm-v1/Lm-v2 DNA vaccines	*dal*⁻, *dat*⁻	Plasmid	E7, HPV	182
Dmpl2GK20	*mpl2*⁻	Plasmid	*E. coli* beta-galactosidase epitope	187

[a] All *Listeria* strains were initially based on the wild-type (WT) strain 10403S.

including the early proteins E1 and E2, which are required for viral DNA replication, maintenance, and papilloma formation. E1 and E2 proteins are believed to be required to maintain the viral DNA as an episome and to ensure the correct segregation of these episomes during cell division. In addition, the capsid proteins L1 and L2 have been intensely studied as targets for prophylactic vaccines for HPV. L1 self-assembles into virus-like particles (VLPs), which have been shown to generate neutralizing antibodies in rabbits that protect against CRPV infection.[97]

Jensen and associates vaccinated rabbits with wild-type *L. monocytogenes* or the recombinant expressing CRPV E1 and then challenged with CRPV or viral DNA. Although the rabbits developed papillomas after CRPV challenge, the tumors were of smaller size than those from control vaccinated rabbits and many lesions regressed. In fact, more than two thirds of the vaccinated rabbits generated protective immunity that controlled and induced complete regression of papillomas induced by CRPV. In contrast none of the control vaccinated rabbits showed papilloma regression. Several months after the tumors had regressed they tested the sites where the papillomas had developed and could not detect any latent viral DNA. Lm-E1-vaccinated rabbits were completely resistant to papilloma formation from challenge with viral DNA.[96]

A number of investigators have focused on viral antigens as model tumor antigens. We have used the influenza antigen, NP, as a model antigen to test the ability of *Listeria* expressing NP to protect against challenge by NP expressing tumor lines and to eradicate established tumors from several different tissues.[46,47,94] We constructed an expression system in which the NP antigen was expressed as a fusion protein consisting of a truncated and nonfunctional virulence factor, LLO, joined to the influenza NP and showed that this was processed and presented to NP-specific T cells.[74] To ensure retention of the plasmid *in vivo*, a copy of the *prfA* gene was included on the plasmid, which was then used to transform a *prfA* negative mutant of *L. monocytogenes* that is incapable of *in vivo* replication in the absence of the episomal expression of *prfA*. This Lm-NP recombinant efficiently induced NP-specific $CD8^+$ CTLs in mice following infection. As expected, escape of Lm-NP from the phagosome was necessary for recognition by $CD8^+$ T cells, but not MHC class II restricted $CD4^+$ T cells.

We then used Lm-NP to test the ability of recombinant *L. monocytogenes* as a tumor immunotherapy using the BALB/c mouse-derived tumors CT26-NP and Renca-NP.[46] Not only did Lm-NP treatment protect mice in an NP-specific manner from subsequent tumor challenge, but it also induced the complete regression of established tumors.[46] This study also verified the importance of $CD8^+$ T cells to the efficacy of the Lm-NP vaccine in that Lm-NP treatment induced CTLs that lysed CT26-NP and Renca-NP *in vitro* and depletion of $CD8^+$ T cells *in vivo* completely abrogated the antitumor efficacy of Lm-NP.[46] $CD4^+$ T cells were also shown to be important for protection from tumor challenge since Lm-NP immunized mice that were depleted of $CD4^+$ T cells prior to tumor challenge were only able to slow the growth of the CT26-NP tumors, not to reject them.

Liau et al. also used a recombinant *L. monocytogenes* that expressed a viral antigen, LCMV-NP, as a model tumor antigen to induce a response to a rat glioma transduced with LCMV-NP. They demonstrated that vaccination with rLm-LCMV-NP protected against subcutaneous, but not intracerebral, tumor challenge.[98] However, mice that survived subcutaneous challenge were then protected against intracerebral challenge with even the parental tumor not expressing the NP antigen, presumably by a mechanism of epitope spreading. Brockstedt et al. were able to induce immunity to AH1-A5, a peptide analog with 80% homology to the gp70 antigen, endogenous to the murine colon tumor line, CT26.[99] They showed that vaccination with a *Listeria* expressing this AH1-A5 epitope could significantly reduce the number of CT26 lung metastases in a tumor challenge model.[99] Bruhn et al. demonstrated that it was possible to even drive protective responses against the self-antigen murine tyrosinase–related protein-2 (TRP2), a nonmutated melanocyte-derived differentiation antigen highly expressed in melanomas.[100] They used a recombinant strain of *L. monocytogenes* expressing the TRP-2 antigen and were able to protect mice from B16 melanoma challenge.[100] A complete list of published studies using *Listeria* as a vaccine vector for tumor antigens is shown in Table 16.1.

The studies briefly described here tested the efficacy of the *Listeria* cancer vaccines in transplantable mouse tumor models. This is a reasonable first approach to evaluating the effectiveness of a therapy; however, a more physiological system would be to use a mouse that endogenously expresses the tumor antigen and is tolerant. Using mice transgenic for the target tumor antigen better tests vaccines to overcome self-tolerance to endogenous protein. Next, we describe our studies using the human tumor antigens, HPV-16 E7 and HER-2/neu, in conjunction with our *Listeria* vectors, to break tolerance in preclinical transgenic systems.

16.3.2 *Listeria* Targeting HPV-16 E7 as a Vaccine for Cervical Cancer

16.3.2.1 HPV-16 and Cancer

Human papilloma viruses are a heterogeneous group of viruses with over 100 different subtypes that inhabit the squamous epithelium of the mucocutaneous surface. Acute infections cause either minor subclinical disease or chronic infections such as warts. Severe life-threatening disease occurs when latent infection leads to malignant disease caused by a small subset of papilloma viruses, predominantly strains 16, 18, 31, and 45.[101] Cervical HPV infections are largely immunologically quiescent.[102] However, from the studies of HPV-related diseases in immunocompromised HIV patients, both men[103,104] and women,[105–108] it is clear that the immune system is effective in eliminating HPV-infected cells in healthy people. Nevertheless, normal immune responses do not protect a significant percentage of women against incidence of cervical cancer that occurs after papilloma infection with a number of different serotypes, possibly due to a lack of local APCs[102] or from loss or mutation of immunologically important target molecules such as MHC class I.[109] The goal for immunologists interested in the disease is to change the immune response in individuals with inadequate immune responses to HPV, either qualitatively or quantitatively, so as to treat HPV-associated disease. Treatment requires the development of an effective cytotoxic immune response against viral antigens expressed during the latent phase.

Cervical cancer is the second most common female malignancy, inducing disease in over 400,000 women per year worldwide, and is a leading cause of death when left untreated. Human papilloma virus serotypes 16 and 18 are the primary cause of anogenital cancer and are strongly associated with certain forms of head and neck cancer. Both cervical and squamous cell carcinomas of the tonsil show correlation with HPV-16.[110,111] HPV is the predominant etiologic agent of cervical cancer and encodes three transforming oncogenes: E5, E6, and E7. The early transforming proteins, E6 and E7, are almost ideal candidates for vaccine approaches against HPV neoplasia because they are the only open reading frames from the HPV genome that are constitutively expressed in HPV transformed tissues and are thought to be necessary to maintain the transformed state of these cells.[112] In addition, there is a wealth of evidence that these two proteins are immunogenic in humans with the production of both humoral and cell-mediated responses.[113] Both E6 and E7 are responsible for the induction of cancer by binding to and inhibiting the function of two major tumor suppressor genes, pRb and p53.[114] Both pRb and p53 play major roles in cellular replication and DNA damage/repair. Due to the actions of E6 and E7, the cervix epithelium will over time become immortalized, ultimately leading to neoplasia and cervical cancer.

16.3.2.2 *Listeria* Vaccines Targeting HPV-16 E7

We have used a mouse model utilizing the HPV-16 E6 and E7 immortalized mouse tumor line TC-1 to test the antitumor effectiveness of the E7-specific immune responses induced by E7 secreting *Listeria* strains. TC-1 is a lung epithelial cell immortalized by HPV-16 E6 and E7 and transformed by pVEJB expressing activated human c-Ha-*ras*.[115] It is an aggressive tumor, syngeneic with the C57Bl/6 mouse. Like human HPV-associated tumors, TC-1 constitutively expresses E6 and E7. We constructed two *Listeria* strains that either secreted E7 alone or a fusion protein consisting of a truncated LLO joined at the C-terminus to E7.[71,88] E7 is a small protein (98 residues) and is readily expressed and secreted by *Listeria*. Although both *L. monocytogenes* recombinants secreted the E7 tumor antigen, they induced very different antitumor therapeutic outcomes. Lm-LLO-E7 treatment effectively eradicated established TC-1 in the majority of tumor-bearing mice. *In vivo* antibody depletion studies demonstrated that the antitumor response required CD4+ and CD8+ T cells as well as IFN-γ.[71] However, Lm-E7 treatment of tumor-bearing mice had little impact on the growth of the tumor. Unexpectedly, the depletion of CD4+ T cells greatly improved the effectiveness of the Lm-E7 treatment and induced the regression of TC-1 in about 25% of treated mice. This suggested that CD4+ T cells induced by Lm-E7 were immunosuppressive. To test this hypothesis we adoptively transferred CD4+ T cells from mice immunized with Lm-E7 or a control *Listeria* vector. We found that CD4+ T cells induced by Lm-E7, but not the control vector, could negate the anti-TC1 immune response induced by Lm-LLO-E7-treated recipient mice.[71]

We next determined whether the Lm-E7-induced immunosuppression was mediated by the CD4+CD25+ regulatory T cell subset (Tregs) important for the maintenance of self-tolerance[116] and that have been shown to suppress both proliferation and IFN-γ production by CD8+ T cells.[117] CD4+CD25+ Treg cells have also been demonstrated to aid tumor growth by suppressing antitumor immune responses[118] Furthermore, transforming growth factor-β (TGF-β) has been reported to be secreted by Treg cells[119] and to protect tumors from immune responses in immune competent hosts.[43,120] The depletion of both CD25+ cells and TGF-β in Lm-E7-treated mice greatly improved the antitumor response.[71] In addition, we were able to demonstrate the presence of CD4+CD25+ T cells in the tumors of Lm-E7 immunized mice that secreted both TGF-β and IL-10.[121] These studies demonstrate that recombinant listerial vectors have the potential to induce both effective and suppressive immunity.

In addition to the *Listeria*-based vaccine Lm-LLO-E7, we have also created a recombinant *Listeria* that expresses E7 as a fusion protein with the virulence factor ActA, Lm-ActA-E7, as an alternative vaccine for HPV-16-induced cancers.[89] As for Lm-LLO-E7, we truncated the ActA virulence factor to remove possible cell cytotoxicity. However, our truncations retained the PEST-like domains, in ActA and LLO, which in the case of LLO may be involved in the targeting of the LLO degradation and cross-presentation.[122,123] Fusing our target protein to the PEST region of LLO or ActA allowed us to utilize these virulence factors as adjuvants, rapidly targeting the protein for host degradation, but without the possibility of toxic side effects. Fusion of our tumor antigens to LLO or the PEST-like sequence significantly enhanced the immunogenicity of the fused antigen.[71,88] Both Lm-LLO-E7 and Lm-ActA-E7 vaccines induce CTLs that can penetrate and kill the solid tumor in the transplantable TC-1 tumor model syngeneic with C57Bl/6 mice.[3,71,88,89]

16.3.2.3 The E6/E7 Thyroid Cancer Mouse

As discussed previously, studying cancer vaccines in a tolerance model better mimics the human scenario. We thus generated a mouse transgenic for HPV-16 E6 and E7 under the thyroglobulin promoter in order to test the ability of the vaccines to break self-tolerance.[124] The genetic background of the E6/E7 transgenic mouse is identical to C57BL/6 except for the tissue-specific expression of the E6/E7 transgenes. E7 was expressed in the thyroid and in mouse thymic epithelial cells (mTECs), which are thought to present self-antigens to T cells in the thymus to ensure negative selection of high avidity T cells responsive to tissue-specific antigens.[125] mTECs can delete autoreactive T cells either directly or indirectly by thymic DC cross-presentation of the peripheral antigen to the autoreactive T cell.[126] mTECs express the autoimmune regulator (AIRE) gene, which is a transcription factor critical for maintaining tolerance to self.[127,128] AIRE controls the expression of peripheral antigens and the processing and presentation of those antigens.[129] Mice in which the AIRE gene has been deleted rapidly succumb to a variety of autoimmune disorders directed toward peripheral tissue, including the thyroid. mTECs thus play a critical role in deleting T cells reactive to thyroid antigens during repertoire selection in the thymus in an AIRE-dependent manner.[130]

We found that the E7 specific CD8+ T cell response induced by the *Listeria* vaccines, Lm-LLO-E7 and Lm-ActA-E7, were of lower avidity in the E6/E7 trangenic mouse compared to the wild-type mouse consistent with the expression of the E7 transgene in mTECs. The expansion of E7 specific T cells in the transgenic mouse is probably due to the activation of low avidity T cells that are not deleted during thymic selection. However, Lm-LLO-E7 and Lm-ActA-E7 still showed therapeutic efficacy in the E6/E7 transgenic mouse as measured by the regression of implanted TC-1 tumors, although the wild-type mice mounted a more robust immune response, presumably due to a lack of tolerance to the E7 antigen, thus allowing for a high-avidity anti-E7 CTL response. It was recently shown in a transgenic mouse model of breast cancer that the CD4+ CD25+ regulatory T cell subset controls the outcome of the immune response observed.[131] However, when we examined whether depleting this specific T cell subset *in vivo* improved the antitumor efficacy of Lm-LLO-E7 or Lm-ActA-E7 in the E6/E7 transgenic mouse, we concluded that CD4+ CD25+ regulatory T cells do not play a significant role in maintaining tolerance to E7 in young transgenic mice.[124]

16.3.2.4 E6/E7 Transgenic Mouse Models

There are a number of transgenic mice that express the E6 and E7 oncogenes of HPV-16 from various promoters. In one model, E6 and E7 are expressed in the suprabasal layers of the epidermis under the control of the keratin 10 promoter.[132,133] The α-crystallin promoter has also been used to target HPV-16 E6 and E7 to the ocular lens.[134] Interestingly, expression of E6 and E7 occurred in a number of nonlens tissue in some lines of these mice. In particular a line (line 19) was developed that expressed E6/E7 in the skin. Skin dysplasia occurred in these mice and about 20% progressed to squamous cell tumors. However, these transgenic mice did not appear to be tolerant to E7 and both humoral and cell-mediated immunity could be readily induced to E7 in the line 19 mice at levels comparable to normal mice.[135] This is curious as it was clearly shown by Derbinski et al. that crystallin was expressed in the thymus specifically, but not limited to mTECs.[125] When E6 and E7 oncoproteins were expressed under the control of the keratin 14 promoter, E7 was shown to be expressed in the thymus in addition to epithelia.[136,137] These mice developed progressive papillomatosis but practically no mice developed tumors spontaneously in the absence of the application of carcinogens to the skin. Further, there is some controversy as to the level of tolerance of T cells in K14-Tg mice to E7.[135,137,138]

However, despite the earlier studies, immune tolerance has been observed in this mouse and was presumed to be due to mostly peripheral mechanisms.[139] In the K14-Tg mouse, the E7 transgene was highly expressed in the cortex of the thymus in addition to the epithelium.[136] Moreover, this mouse showed a down-regulation of all T cell–mediated immune responses to a variety of antigens, including E7, and the avidities of E7 specific CD8+ cells in these mice were similar to those in wild-type mice.[139] This differs from the E6/E7 transgenic mice we have made, which appear to be competent to mount T cell responses to the foreign antigen chicken ovalbumin.[124] In addition, low-avidity E7-specific CD8+ T cells can be expanded in this transgenic mouse and are capable of killing solid tumors, which differs from the observations in the other E7 transgenic mice.

The ability of *Listeria*-based vaccines to break tolerance has not been extensively examined. As we described earlier, one study used *Listeria* that secretes the self-melanoma antigen, TRP-2, treated in conjunction with imiquimod.[100] However, a drawback to this model is that the level of tolerance to this antigen could not be examined because there is no syngeneic mouse that lacks TRP-2. Another group has claimed to break tolerance to a self-antigen using *Listeria* targeting the gp70 retroviral gene in the CT26 mouse tumor model; however, the epitope that their strain secreted did not have complete homology with the target TAA sequence and therefore cannot stringently be considered a self-antigen.[99]

In contrast, we observed clear tolerance to the E6/E7 proteins (e.g., poorer tumor regression and less potent CTLs in transgenic vs. wild-type mice). However, our *Listeria* vaccines were still able to eradicate established solid TC-1 tumors in some mice.[124] We have also recently shown that both Lm-LLO-E7 and Lm-ActA-E7 can impact the growth of autochthonous tumors that arise in the E6/E7 transgenic mouse (Sewell, Pan, and Paterson, unpublished observations).

16.3.3 *Listeria* Expressing HER-2/neu as a Vaccine for Breast Cancer

16.3.3.1 The Potential of Immunotherapy for HER-2/neu Associated Malignancies

The term "breast cancer" usually refers to the most common malignant neoplasia of mammary tissue, breast adenocarcinoma. This is a highly heterogeneous group of tumors that are not all causally or genetically related. Nevertheless, tumor-associated antigens have been identified for certain of these tumors. Of these, the oncogene product, HER-2/neu, is of great interest because it is overexpressed in 25–40% of breast adenocarcinoma and over half of ductal carcinomas *in situ*.[140] The presence of HER-2/neu on breast adenocarcinomas is associated with metastatic disease and poor prognosis and survival.[141] HER-2/neu is a 185-kD transmembrane glycoprotein and a member of the epidermal growth factor receptor family of tyrosine kinases that appear to become constitutively activated in certain malignancies. A ligand specific for HER-2/neu has still not been discovered but may not be required for its activation. Thus, activation of neu can occur through the formation of constitutive homodimers, which can be brought about by a variety of mutations in the transmembrane[142] or extracellular domains.[143] HER-2/neu can also form heterodimers with other members of the EGF family and extend and amplify the signaling that results from their interaction with their ligands.[144] Thus, the oncogenic potential of HER-2/neu results from its overexpression on malignant cells, which would result in sustained signaling of a wide number of EGF receptors by multiple stroma-derived growth factors.

HER-2/neu containing heterodimers have been shown to generate more potent signals and enhanced activation of MAPK, c-src, PK-Akt signaling pathways, gene transcription, and cellular proliferation.[145] In normal tissue the formation of HER-2/neu heterodimers is strictly controlled. Although HER-2/neu is expressed on normal tissue, there is evidence that it is not immunologically silent when overexpressed on malignant cells. Humoral,[146] CD8$^+$ T cell,[140] and CD4$^+$ T cell[147] immune responses directed at HER-2/neu have been observed in patients with breast cancer. Thus, the human immune repertoire of lymphocytes reactive to HER-2/neu appears not to have been deleted by central tolerance, allowing the exploitation of this tumor-associated antigen for cancer immunotherapy.[140,148–151]

Breast cancer is the most common form of cancer in females. Approximately one out of nine women will be affected by breast cancer sometime in her life. It is the second most fatal cancer in women (after lung cancer). The incidence of breast cancer in males is significantly lower than in females. Targeting breast cancer with a *Listeria*-based vaccine has the benefit that a strong CTL response will be generated. Since many of breast cancer patients contain mammary tumors that overexpress HER-2/neu and because HER-2/neu contributes to the tumor phenotype, many investigators[140,150–154] have found that this is a good target antigen for breast cancer. HER-2/neu contains three domains: an extracellular domain, a hydrophobic transmembrane domain, and an intracellular domain that contains a kinase domain and carboxyl terminal tail for intracellular signaling. Patients with cancers that overexpress HER-2/neu exhibit extreme immune tolerance regardless of the fact that they can generate CD4$^+$ and CD8$^+$ T cell responses and detectable humoral responses.[146,150,152,153]

The standard approved therapies for breast cancer are surgery, radiation, or chemotherapy, or some combination. Passive immunization with a humanized monoclonal antibody against HER-2/neu, trastuzumab or Herceptin®, has also recently entered the clinic. With the approval and commercialization of Herceptin, the treatment of both early- and late-stage patients by passive immunization looks promising. The costs of life-long antibody therapy, however, are very high and their affordability for the healthcare system has been the focus of recent discussion. In addition, antibody therapy requires repeated infusions, whereas a vaccine would provide long-lasting immunity and may only require boosting steps.

16.3.3.2 *Listeria* as a Vector for HER-2/neu

Because the size of the HER-2/neu molecule exceeded that which *Listeria* could readily secrete and also contained a hydrophobic transmembrane domain, we made five different *Listeria*-based vaccines, each expressing some overlapping region of the HER-2/neu molecule[77]: extracellular region-1 (EC1; HER-2/neu aa 20–326), extracellular region-2 (EC2; HER-2/neu aa 303–501), extracellular region-3 (EC3; aa 479–655), intracellular region-1 (IC1; HER-2/neu aa 690–1091), and intracellular region-2 (IC2; HER-2/neu aa 1020–1260. All of the HER-2/neu fragments were fused to truncated, nonhemolytic LLO. The models we used to test these vaccines were a transplantable NT-2 model in the FVB/N mouse[154] and a tolerance model using a transgenic FVB/N mouse overexpressing the rat HER-2/neu gene under the control of the MMTV promoter, which spontaneously develops tumors between 4 and 9 months of age.[155,156] All five vaccines were equally able to induce regression of the implantable, HER-2/neu-expressing tumor line (NT-2) in the wild-type FVB/N mouse.[77]

Previous studies in the FVB/N mouse described only one CTL epitope.[154] In this study Ercolini et al. cloned overlapping regions of the HER-2/neu molecule into 3T3 fibroblasts that were used as targets in GM-CSF assays that contained neu-specific, FVB/N-derived T cell clones.[154] They found that the MHC I-restricted rat HER-2/neu peptide$_{420-429}$ (rNEU$_{420-429}$) could protect mice from mammary tumor outgrowth.[154] The rNEU$_{420-429}$ peptide is derived from the extracellular portion of the HER-2/neu molecule and is recognized by all neu-specific CD8$^+$ T cell lines from FVB/N mice vaccinated with 3T3 cells expressing the rat HER-2/neu protein.[154] In contrast, we have found not 1, but at least 13, different CTL epitopes.[42,77,157,158] Thus, our method of vaccination helped to reveal cryptic epitopes that were previously overlooked, implying that the vaccination strategy can determine the immunodominance of epitopes. We hypothesize that the strong immune response induced by *Listeria* infection helps to establish CD8$^+$ T cell effectors against epitopes in different regions of the HER-2/neu molecule.

16.3.3.3 Transgenic Mouse Models for Breast Cancer

The oncogenic properties of HER-2/neu have been harnessed to develop transgenic mouse models of breast cancer.[155] A transgenic mouse that overexpresses activated rat HER-2/neu under the transcriptional control of the mouse mammary tumor virus promoter and enhancer has been developed as a model for human breast cancer in the laboratory of Dr. William Muller.[155] This particular mouse model of breast cancer has proved to be extremely useful for immunotherapeutic studies because it closely resembles the ontogeny and progression of human breast cancer. Focal mammary tumors develop stochastically in the female transgenic mice after a latency of about 7 months and metastatic disease in the lung occurs in many of the tumor-bearing mice. Escape tumor variants that have low MHC class I expression occur on about half of the tumors.[159] This mouse model thus provides a test for cancer immunotherapies *in vivo* where tumors have arisen spontaneously and that may, therefore, have induced immune unresponsiveness in the host. Although the mouse HER-2/neu gene differs from the rat neu by <6% of amino acid residues, there is evidence that rat neu is immunogenic in the mouse,[160] implying that the rat HER-2/neu-expressing tumors commonly used to test immune strategies may be immunogenic in wild-type mice. Thus, rat HER-2/neu transgenic mice are a more stringent model to test the ability of immunotherapy to break tolerance against tumor antigens, such as HER-2/neu, that are expressed at low levels in normal tissue.

16.3.3.4 The Ability of *Listeria*-Based HER-2/neu Vaccines to Overcome Tolerance

We tested our five Lm-LLO-HER-2/neu vaccines for their ability to eradicate NT-2 in the transgenic mouse.[158] We found that they were all equally capable of slowing or halting tumor growth despite the fact that the CD8+ cytotoxic T cells that we isolated from immunized mice were of lower avidity than those from wild-type mice.[158] We also tested the ability of Lm-LLO-HER-2/neu vaccines to delay the appearance of spontaneous tumors in the transgenic HER-2/neu mouse.[42] Interestingly, the vaccines differed in their ability in this regard. When we tracked the tumors that grew over time, we found that the tumors that emerged had mutated residues within CTL epitopes of the HER-2/neu molecule. These mutations resided in the exact regions that were targeted by the *Listeria* vaccines, which suggested that the rate of generation of escape mutants was a significant factor in the efficacy of each vaccine. Thus, mutations in the HER-2/neu regions allowed for escape mutants from immunotherapy to form, leading to the outgrowth of spontaneous tumors in the transgenic HER-2/neu mouse model.[42]

A longer delay in the onset of tumors after immunotherapy occurred with the vaccine that targeted the kinase domain, which is contained in the fragment IC1. We verified that the mutations in this domain occurred within novel CD8+ T cell epitopes and that the mutation of these residues abrogated cytotoxic T cell lymphocyte responses to these epitopes.[42] The long delay in the onset of tumors after immunotherapy targeting the kinase domain may be because this region of HER-2/neu cannot undergo extensive mutations without impairing the ability of the molecule to signal cell growth. Thus, tumors targeted by Lm-LLO-IC1 may grow after a delay due to a decrease in signaling and a subsequent decrease in the proliferative capacity of the tumor cells.

Our findings have implications for improving cancer immunotherapy and preclinical testing strategies. For an autochthonously arising tumor in an animal for which the target antigen is a self-protein, similar to human cancers, there is a clear difference in vaccine efficacy among the five Lm-LLO-HER-2/neu vaccines. In a transplantable tumor model in which rat HER-2/neu is a foreign antigen, all five vaccines are equally efficacious, suggesting that transplantable tumor models, which are frequently used for preclinical testing, may not be a reliable indicator of vaccine efficacy. Further, a cancer immunotherapy that shifts immune pressure to epitopes in a different region of the molecule could provide a more effective therapeutic outcome. Finally, this study demonstrates that cancer immunotherapies can force the editing of oncogenes, even in critical domains necessary for signaling. Beyond basic immunosurveillance, this is the first proof that therapeutic vaccines can cause these specific changes in tumors.

Our studies, taken together, provide evidence that *Listeria* can overcome tolerance to a self-antigen and expand autoreactive T cells normally too low in number and avidity to drive antitumor responses. These data provide an optimistic view for the use of *Listeria* in the treatment of cancer.

16.3.4 Alternative Approaches

In addition to using *Listeria* as a vaccine vector, several other immunization strategies have arisen in our and others' attempts to optimize the immunological impact of our vaccines while retaining a margin of safety. For example, the virulence factor LLO appears to have adjuvant-like properties. LLO is a 529-amino-acid protein that is secreted by *L. monocytogenes* and mediates escape from the phagosome. We have found that fusion of a tumor antigen to the first 420 aa or so of LLO, which excludes the hemolytic domain, may facilitate secretion of the antigen,[71,74] increase antigen presentation,[88] and help to stimulate the maturation of dendritic cells.[161] The augmentation of antigen presentation seems to reside in an N-terminal PEST-sequence in LLO, which is a region rich in the following amino acids: proline (P), glutamic acid (E), serine (S), and threonine (T).[88] Fusion of target tumor antigens directly to LLO and use in boosting steps may overcome the need for using whole live bacteria. Current experiments strongly suggest this may be the case when tested in conventional models for non-Hodgkin's lymphoma (NHL)[162] (Neeson and Paterson, unpublished data).

In general, it is believed that live replicating vectors induce more powerful immunity.[163,164] In addition, suitable attenuation of the wild-type strains of *Listeria* should address safety concerns in using a live *Listeria* vector. Immunization, with heat-killed *L. monocytogenes* (HKLM), has long been known not to induce protective immunity against live *Listeria* infection.[164] Pamer and colleagues have also shown that although HKLM immunization primes memory CD8+ T lymphocyte populations they are ineffective at providing protection from subsequent *L. monocytogenes* infection.[164] In contrast to live infection, which elicits large numbers of effector CD8+ T cells, HKLM immunization primes T lymphocytes that are not cytolytic and produce lower levels of IFN-γ compared to CD8+ T cells that arise from live infection.[164] Nevertheless, recent studies suggest that *Listeria* incapable of replicating may be effective as vectors for foreign antigens.[166–168]

As *Listeria* targets APCs for infection it is also a potential delivery vehicle for DNA vaccines while providing at the same time the necessary "danger signals" to the immune system. Its ability to enter the cytosol and replicate there ensures that the DNA plasmid vaccine is also delivered to that compartment. DNA vaccines delivered by *Listeria* may be more appropriate than secreted protein delivery for some antigens. Empirically, we have found that *Listeria* cannot secrete highly hydrophobic proteins, probably due to difficulties in translocating across the membrane or cell wall. This immediately limits the type of proteins available for vaccination using *Listeria*. In addition, *Listeria* often cannot secrete a protein bigger than 50 kDa, so for proteins like HER-2/neu, multiple strains must be generated. In these cases bactofection, or bacterial mediated intradomain gene transfer,[169,170] is a potential means to develop novel therapeutics for these cancer antigens. In this section we will also describe the use of *Listeria* as a DNA vaccine delivery vehicle.

16.3.4.1 LLO-Fusion Proteins as Immunizing Agents

Besides the use of live bacterial vectors as antigen delivery systems, the fusion of *Listeria* virulence factors LLO or ActA to cancer antigens can also increase the immunogenicity when delivered by other means, including viral vectors[171] and DNA vaccines.[172] Using vaccinia virus as a live vector we showed that the expression of HPV-16 E7 fused to nonhemolytic LLO induces an immune response that eradicates established HPV-16 immortalized tumors in C57BL/6 mice.[171] The vaccinia construct expressing LLO-E7 (VacLLO-E7) was compared with two previously described vaccinia virus constructs: one that expresses unmodified E7 (VacE7) and another that expresses E7 fused between the signal sequence and the cytoplasmic tail of the lysosome-associated membrane protein-1 (LAMP-1), which directs it to intracellular lysosomal compartments and improves MHC class II restricted responses (VacSigE7LAMP-1).

C57BL/6 mice bearing established HPV-16 immortalized tumors of 5 or 8 mm were treated with each of these vaccines. Of the mice treated with VacLLO-E7, 50% remained tumor free 2 months after tumor inoculation, whereas 12–25% of the mice were tumor free following treatment with VacSigE7LAMP-1 (depending on the size of the tumor). No mice were tumor free in the group given VacE7. Compared to VacE7, mice that received VacSigE7LAMP-1 and VacLLO-E7 had increased numbers of E7-specific CD8+ T cells in the spleen, which produced IFN-γ and TNF-α. E7 specific T cells were also found in high numbers in the tumor sites of mice treated with VacLLO-E7. An increased efficiency of E7-specific lysis by splenocytes from mice immunized with VacLLO-E7, compared to Vac-E7 and VACSigE7LAMP-1, was also observed. This study demonstrated that the delivery of the fusion protein LLO-E7 by a nonlisterial vector not only enhances antitumor therapy by improving the tumoricidal function of E7-specific CD8+ T cells but also increases the number of antigen specific CD8+ T cells in the tumor, the principal site of antigen expression.[171]

To address the question of how fusion with LLO improves vaccine efficacy, we turned to using DNA plasmid vaccines as a tool to isolate the adjuvant effect of LLO in the absence of other endogenous *Listeria* effects.[172] We constructed DNA vaccines bearing the tumor antigen HPV-16 E7 either alone or in combination with LLO, as a chimera or in a bicistronic construct, to address the importance of fusion between these elements. Interestingly, LLO and E7 as both chimeric and bicistronic vaccines were effective against HPV-16 immortalized tumors transplanted into mice, suggesting that LLO can act as an adjuvant that does not require fusion with the tumor antigen to mediate its effect. However, fusion of LLO to E7 was required to induce E7-specific CD8+ T cell responses and mice immunized with LLO-E7 DNA showed the greatest antitumor response *in vivo*. We also showed that LLO acts as an adjuvant to augment E7-specific CD4+ T cell responses either in the form of bicistronic message or as a mixture of E7 and LLO or as an LLO-E7 fused gene. These data,[172] together with our previous observations of DC maturation in response to vaccines that include fusion with LLO,[161] suggest this may be a valuable component of an effective DNA vaccine.

Thus, we have repeatedly found that the best antitumor response is achieved when the tumor antigen is fused to a nonhemolytic version of the virulence factor LLO. (The use of hemolytic LLO to enhance the immunogenicity of vaccines has been discussed elsewhere.[173,174]) We hypothesized that LLO may have adjuvant properties when used in the form of a recombinant protein. We thus constructed expression systems where the idiotype (i.e., B cell receptor) of a B cell lymphoma clone was chemically fused to the full-length LLO protein and used in challenge studies to see if anti-idiotype responses could be generated for the potential treatment of NHL. Besides passive antibody therapy with Rituximab,[175,176] the currently active immunotherapy being tested in clinical trials for NHL is similar to our LLO-fusion approach: keyhole limpet hemocyanin (KLH) is fused to the patient's cloned B cell idiotype and the patient is re-immunized in order to generate an anti-lymphoma response. We found that conjugation of the mouse B cell lymphoma 38C13 idiotype[177] to LLO was better able to control tumor growth compared to 38C13 conjugated to KLH (the conventional method) for up to 62 days after tumor establishment.[162] Anti-38C13 antibodies were present in both immunization groups, but higher titers were achieved with the 38C13 idiotype conjugated to LLO, compared to KLH, after a single immunization.

16.3.4.2 Nonreplicating *Listeria*-Based Vaccines

Several recent studies have revisited the use of bacteria that are unable to replicate *in vivo* as carriers for foreign antigens.[166–168] Brockstedt et al. generated a *Listeria* strain that was very sensitive to photochemical inactivation through psoralens by removing the genes required for nucleotide excision repair (*uvrAB*). Psoralens intercalcate in DNA as it is dividing and cross-link with pyrimidine bases of DNA and RNA upon illumination with long-wavelength ultraviolet light. Since the primary repair pathway for psoralen cross-links is nucleotide excision repair mediated by the ultraviolet light response (*uvr*) genes, deletion of any one of these genes renders bacteria highly susceptible to ultraviolet light–induced DNA damage and to psoralen-mediated cross-linking. However, although psoralen-treated bacteria are incompetent to replicate, they are still able to express their genes and synthesize proteins.

Brockstedt and colleagues designate these psoralen-treated bacteria as killed but metabolically active (KBMA). To test the potential of KBMA bacteria as vaccine strains they used *L. monocytogenes* as a model intracellular pathogen and showed that, upon treatment with psoralen and long-wavelength UV-light, *Listeria* became inactivated, but could still escape into the cytosol of the cell. They also demonstrated that a psoralen-inactivated *Listeria* that expressed OVA could induce CD4+ and CD8+ T cells against this model antigen and could protect mice from viral infection with a vaccinia virus expressing OVA.[166] To test the potential of KBMA *Listeria* as a cancer vaccine, they tested mice with psoralen-inactivated *Listeria* expressing the AH1–A5 peptide analog, from the CT26 tumor line gp70 antigen, 3 days before IV challenge with CT26 and showed a significant reduction in lung tumor burden compared to control vaccinated mice.

Despite the lack of potency of heat-killed bacteria,[164,165] Datta and colleagues revisited the idea of using killed bacteria as vaccines.[167] They reasoned that γ-irradiation could preserve antigenic and adjuvant structures destroyed by heat inactivation. They found that irradiated *L. monocytogenes*, unlike heat-killed *Listeria*, efficiently activated dendritic cells through TLRs. Irradiation preserved the ability of the bacteria to synthesize new proteins and induced protective T cell responses against challenge by live *L. monocytogenes* in mice. Like live *Listeria*, irradiated *Listeria* also induced TLR-independent T cell priming. Using OVA as a model antigen, they showed that cross-presentation of irradiated listerial antigens to CD8+ T cells involved TAP- and proteasome-dependent cytosolic antigen processing. Thus, killed *Listeria*, if appropriately inactivated, can induce protective T cell responses, previously thought to require live infection.

Surprisingly, John Harty's group has recently shown that *Listeria* deficient in LLO, and thus unable to escape into the host's cytosol, could still be used as a vaccine to protect against viral infection.[168] This LLO-negative *Listeria* strain was engineered to express a CD8+ T cell epitope derived from the LCMV virus.[168] It could prime epitope-specific CD8+ T cells against both LCMV and an endogenous listerial antigen, p60, detectable by tetramer binding and intracellular cytokine staining for IFN-γ. If 3×10^9 CFUs of the nonhemolytic *Listeria* and 3×10^3 wild-type (hemolytic) *Listeria*—both expressing NP—were used to immunize, similar numbers of CD8+ T cells were induced. The CD8+ T cells isolated 30 days after infection with the LLO-negative *Listeria* had an effector-memory phenotype and were functional in that they secreted IFN-γ and TNF-α with a smaller subset also making IL-2. However, lowering the dose of LLO-deficient *Listeria* infection by 100- to 1000-fold did not result in priming of detectable numbers of antigen-specific CD8+ T cells, although the vaccinated mice were able to mount an enhanced response to an early (day 7 after primary infection) booster immunization with wild-type *Listeria*. Vaccination could partially protect mice from high-dose *Listeria* challenge, 65 days later, as well as from heterologous challenge with lymphocytic choriomeningitis virus, 30 days later.[168] However, whether long-term memory was induced by these vaccines was not examined.

These findings challenge the paradigm that CD8+ T cell responses to antigens delivered by *Listeria* have an absolute requirement for escape from the phagosome[61] and are consistent with a role for cross-priming in generating cell-mediated immunity against *Listeria*.[178] In that regard it is interesting that the LCMV antigen delivered by nonhemolytic *Listeria* was equally effective at stimulating CD8+ T cell responses even if it was not secreted outside the bacterial cell wall but retained inside the cell.[167] This scenario, since antigen could only be released after cell death in the phagolysosome, also argues for a mechanism of cross-priming.

16.3.4.3 *Listeria* as a DNA Cancer Vaccine

Dietrich and Goebel pioneered the use of *Listeria* as a DNA vaccine delivery vehicle. To ensure plasmid release after the bacterium had escaped into the cytosol, they engineered *Listeria* to lyse upon expression of a phage lysin under the control of the *actA* promoter.[179] Ply118 is a highly specific L-alanyl-D-glutamate peptidase that can only access the peptidoglycan wall if a pore is present in the cell membrane, as this gene product has no known way to spontaneously translocate through the cell membrane in a healthy bacterium.[180] Under normal circumstances during a bacteriophage infection, Hol118 creates a pore by inserting into the listerial cell membrane, which allows Ply118 to reach the outer surface of the peptidoglycan cell wall.[179] Additionally, since phage lysins are rapid killers of bacteria in a matter of seconds, some other mechanism of plasmid release must be taking place.[181] It is possible that death of some of the bacteria from natural causes releases Ply118, which binds to the peptidoglycan wall of bystander *Listeria*, lysing those cells and thus repeating the cycle until all bacteria in the infected cell are dead, releasing the plasmid into host cytosol.[85] How the plasmid enters the macrophage nucleus is largely unknown. Presumably, the plasmid DNA binds to some host cell factor, which would facilitate entry into the host cell nucleus.

DNA vaccines encode for a gene of interest to be expressed in host cells, but, despite attempts to improve methods of administration and the inclusions of various adjuvants in the vaccine protocol, generally, they are poorly immunogenic and incapable of eliminating macroscopic tumors. Hypothetically, using *Listeria* to deliver a plasmid should be a better vaccine than naked DNA alone because *Listeria* would deliver the plasmid to an APC, whereas DNA, injected intramuscularly, is thought to transfect myocytes. Additionally, by administering the *Listeria*-based DNA vaccine using the *actA–ply118* suicide release system, the complication of having a prokaryote secreting a eukaryotic protein can be avoided by having a eukaryotic host cell, presumably a macrophage, properly produce the protein of interest. In contrast to viral DNA vectors, bacteria also offer additional safety features since they can be readily controlled *in vivo* by antibiotics, thus reducing the risk of unforeseen side effects. By encoding for the programmed lysis of *Listeria* at a certain point in its intracellular life cycle, an inherent safety feature is incorporated to both allow for the delivery of the plasmid vaccine and elimination of the bacteria.[179]

A minimal DNA vaccine would comprise sequences necessary for plasmid replication in *E. coli* and *Listeria*, plasmid sequences for *in vitro* and *in vivo* retention in both bacterial species, the antigen expression cassette under the control of a promoter active in mammalian cells, and a lysis gene that releases the plasmid from the bacteria into the host cytosol for better transfection efficiency. We generated a *Listeria* vector via this bactofection method, containing a plasmid encoding for an HPV-16 E7 under the control of a CMV promoter in an attempt to induce an antitumor response. Utilizing a plasmid release mechanism involving a suicide cassette, we were able to engineer *Listeria* to release the plasmid encoding our tumor antigen of interest (in this case E7) into the cell and induce antitumor immunity against TC-1 solid tumors.[182] This method was more effective than delivering plasmid alone, but not as effective as the protein delivery system Lm-LLO-E7.[182]

We believe that this is the first series of experiments that directly compares a protein-secreted bacterial vaccine and a DNA-delivered vaccine in which both use *Listeria* as the vector in a physiologically relevant model. A number of elegant studies have been done previously using *Salmonella*, establishing that both respective technologies could potentially serve a protective purpose, but, generally, neither protein nor DNA delivery systems were compared directly to measure therapeutic efficacy. *Salmonella* has shown a plethora of instances that demonstrate the protective effect and potential relevance of these vaccines for a variety of cancers or infectious diseases. Reisfeld and colleagues have shown that a DNA-based *Salmonella* vaccine could demonstrate protective immunity in a mouse model of melanoma,[183] and Paglia et al. were able to utilize *Salmonella* to treat a murine fibrosarcoma model.[183,184] In both cases, their vaccines were able to deliver their plasmid to APCs and induce a downstream T cell response. Lewis and Hone were also able to elicit an immune response to HIV gag using *Salmonella* to deliver a DNA plasmid.[185] It would be intriguing to determine whether the protein or DNA-based vaccines would work optimally or perhaps even equally in the *Salmonella* system in a therapeutic model. Additionally, the mechanism of plasmid release is unclear in the *Salmonella* studies, unlike the *pactA–plyl18* release system that we studied, and this also would be another interesting area of investigation. Perhaps the intrinsic characteristics of the bacteria used would determine which type of vaccine is more effective.

16.4 CONCLUSIONS AND PERSPECTIVES

16.4.1 Moving Bacterial Vectors into the Clinic

Regulatory bodies, such as the FDA, are likely to inspect biologicals, and especially live vectors, for patient and environmental safety with high scrutiny. Some live attenuated bacteria, such as *Salmonella* and BCG, have been used as vaccines against the pathogenic version of the bacterial strain for many years, but the use of bacteria as antigen vectors is relatively recent. Because of this, bacterial vectors are often regulated together with DNA vaccines or viral vectors, for which there are extensive guidelines.

The ideal vaccine vector should display no pathogenicity and should be sufficiently attenuated without the presence of the recombinant antigen. Although expression of a foreign antigen can result in serious attenuation of recombinant bacteria,[186] this is not considered sufficient for safety evaluation. A difference of at least two to three log units between the virulence of the vaccine strain and wild-type strain is preferable. In addition, the attenuation must remain stable with no possibility of reversion to wild type during the time of clinical therapy. For each vaccine strain, the mechanism of attenuation as well as the general genotype and phenotype must be well documented, which is often not the case for laboratory strains that have been transferred between different laboratories over several years. To avoid future problems with regulatory bodies, it is best to start with a strain obtained from a tissue type culture collection such as ATCC.

Any live vector has the potential to cause serious adverse events in patients, particularly cancer patients who have often undergone prior therapy that is immunosuppressive. To avoid adverse events, suitably attenuated strains should be used and patients should be screened for immune deficiencies prior to treatment. However, bacteria have an advantage over viral vectors because of the availability of a wide range of antibiotics that can be used to curtail infection. Thus, potential adverse events can be avoided by the application of a suitable antibiotic before the bacterial infection is beyond control. Most of the safety concerns raised for *Listeria*-based vectors, such as the possibility of bacterial meningitis, can easily be addressed this way. Additionally, an antibiotic regimen should be included in clinical protocols with live bacterial vectors at the assumed end of the vector infection for safety reasons.

Concerns are also often raised about the possibility of *Listeria* contagion from the patient to healthcare workers or the patient's family. However, although this may be a cause for concern with *Shigella* or *Salmonella* vectors based on typhoid strains when not sufficiently attenuated, person-to-person transmission of *Listeria* has not been documented even in clinically confirmed cases of listeriosis (Bennet Lorber, personal communication). A final concern is the possibility of environmental spread of the vector, particularly one that harbors antibiotic resistance genes. The FDA strongly discourages the presence of any antibiotic resistance genes in the final vector. Thus, it is preferable if the recombinant antigen is maintained stably, either on a plasmid or integrated into the bacterial chromosome, without the use of antibiotic selection. Whenever a bacterial vector is moved into clinical applications, these concerns are likely to arise. Thus, to avoid problems later they should be addressed early in the preclinical phase of vector design.

16.4.2 Future Considerations

We and others have demonstrated that the generation of *Listeria*-based vaccines for cancer therapy is relatively easy and the efficacy of these vaccine vectors in the treatment of cancer patients is likely to be determined in the not too distant future. Indeed, a Phase I trial for the E7 HPV-16 vaccine described in this chapter (Lm-LLO-E7) has been mounted by Advaxis, Inc. (North Brunswick, New Jersey) and is currently in progress. Advaxis is finishing some preclinical testing for the human versions of the rat HER-2/neu vaccines described in this chapter. The company hopes to have the HER-2/neu vaccines ready for Phase I trials as early as 2008. From the overwhelming data showing that *Listeria*-based vaccines can elicit strong immune responses against solid, well vascularized tumors with a low level of pathogenicity, we are confident that *Listeria* will move into the spotlight as a new weapon in the war on cancer.

ACKNOWLEDGMENTS

We would like to thank the present and past members of the Paterson laboratory for their contributions to the studies described here. The authors wish to disclose that they have a financial interest in Advaxis, Inc., a vaccine and therapeutic company that has licensed or has an option to license all patents from the University of Pennsylvania concerning the use of *Listeria* or listerial products as vaccines.

REFERENCES

1. Leen, A.M., Rooney, C.M., and Foster, A.E., Improving T cell therapy for cancer, *Annu. Rev. Immunol.*, 25, 243, 2007.
2. Buchsel, P.C., and DeMeyer, E.S., Dendritic cells: Emerging roles in tumor immunotherapy, *Clin. J. Oncol. Nurs.* 10, 629, 2006.

3. Hussain, S. F., and Paterson, Y., What is needed for effective antitumor immunotherapy? Lessons learned using *Listeria monocytogenes* as a live vector for HPV-associated tumors, *Cancer Immunol. Immunother.* 54, 577, 2005.
4. Banchereau, J., and Palucka, A.K., Dendritic cells as therapeutic vaccines against cancer, *Nat. Rev. Immunol.* 5, 296, 2005.
5. Coley, W.B., The treatment of malignant tumors by repeated inoculations of erysipelas: With a report of 10 original cases, *Am. J. Med. Sci.* 105, 487, 1893.
6. Coley, W.B., The treatment of inoperable sarcoma by bacterial toxins (the mixed toxins of the streptococcus of erysipelas and the *Bacillus prodigiosus*), *Practitioner* 83, 589, 1909.
7. Wiemann, B., and Starnes, C.O., Coley's toxins, tumor necrosis factor and cancer research: A historical perspective, *Pharmacol. Ther.* 64, 529, 1994.
8. Schreiber, H. et al., Unique tumor-specific antigens, *Annu. Rev. Immunol.* 6, 465, 1988.
9. Boon, T. et al., Tumor antigens recognized by T lymphocytes, *Annu. Rev. Immunol.* 12, 337, 1994.
10. Chomez, P. et al., An overview of the MAGE gene family with the identification of all human members of the family, *Cancer Res.* 61, 5544, 2001.
11. Overwijk, W.W. et al., Vaccination with a recombinant vaccinia virus encoding a "self" antigen induces autoimmune vitiligo and tumor cell destruction in mice: Requirement for CD4(+) T lymphocytes, *Proc. Natl. Acad. Sci. USA* 96, 2982, 1999.
12. Berd, D., Cancer vaccines: Reborn or just recycled? *Semin. Oncol.* 25, 605, 1998.
13. York, I.A., and Rock, K.L., Antigen processing and presentation by the class I major histocompatibility complex, *Annu. Rev. Immunol.* 14, 369, 1996.
14. Huang, A.Y. et al., Role of bone marrow-derived cells in presenting MHC class I-restricted tumor antigens, *Science* 264, 961, 1994.
15. Pardoll, D.M., Cancer vaccines, *Nat. Med.* 4, 525, 1998.
16. Pace, J.L. et al., Comparative effects of various classes of mouse interferons on macrophage activation for tumor cell killing, *J. Immunol.* 134, 977, 1985.
17. Beatty, G., and Paterson, Y., IFN-gamma-dependent inhibition of tumor angiogenesis by tumor-infiltrating CD4+ T cells requires tumor responsiveness to IFN-gamma, *J. Immunol.* 166, 2276, 2001.
18. Hsieh, C.S. et al., Development of TH1 CD4+ T cells through IL-12 produced by *Listeria*-induced macrophages, *Science* 260, 547, 1993.
19. Manetti, R. et al., Natural killer cell stimulatory factor (interleukin 12 [IL-12]) induces T helper type 1 (Th1)-specific immune responses and inhibits the development of IL-4-producing Th cells, *J. Exp. Med.* 177, 1199, 1993.
20. Boehm, U. et al., Cellular responses to interferon-gamma, *Annu. Rev. Immunol.* 15, 749, 1997.
21. Trinchieri, G., and Scott, P., Interleukin-12: Basic principles and clinical applications, *Curr. Topics Microbiol. Immunol.* 238, 57, 1999.
22. Shu, U. et al., Activated T cells induce interleukin-12 production by monocytes via CD40–CD40 ligand interaction, *Eur. J. Immunol.* 25, 1125, 1995.
23. Cella, M. et al., Ligation of CD40 on dendritic cells triggers production of high levels of interleukin-12 and enhances T cell stimulatory capacity: T-T help via APC activation, *J. Exp. Med.* 184, 747, 1996.
24. Creasey, A.A., Reynolds, M.T., and Laird, W., Cures and partial regression of murine and human tumors by recombinant human tumor necrosis factor, *Cancer Res.* 46, 5687, 1986
25. Haranaka, K., Satomi, N., and Sakurai, A., Antitumor activity of murine tumor necrosis factor (TNF) against transplanted murine tumors and heterotransplanted human tumors in nude mice, *Int. J. Cancer* 34, 263, 1984.
26. Rosenberg, S.A., Immunotherapy of cancer by systemic administration of lymphoid cells plus interleukin-2, *J. Biol. Respon. Modif.* 3, 501, 1984.
27. Rosenberg, S.A. et al., Regression of established pulmonary metastases and subcutaneous tumor mediated by the systemic administration of high-dose recombinant interleukin 2, *J. Exp. Med.* 161, 1169, 1985.
28. Winkelhake, J.L., Stampfl, S., and Zimmerman, R.J., Synergistic effects of combination therapy with human recombinant interleukin-2 and tumor necrosis factor in murine tumor models, *Cancer Res.* 47, 3948, 1987.
29. Dranoff, G. et al., Vaccination with irradiated tumor cells engineered to secrete murine granulocyte-macrophage colony-stimulating factor stimulates potent, specific, and long-lasting anti-tumor immunity, *Proc. Natl. Acad. Sci. USA* 90, 3539, 1993.
30. Tepper, R.I., Coffman, R.L., and Leder, P., An eosinophil-dependent mechanism for the antitumor effect of interleukin-4, *Science* 257, 548, 1992.

31. June, C.H. et al., Role of the CD28 receptor in T-cell activation, *Immunol. Today* 11, 211, 1990.
32. Melero, I., Bach, N., and Chen, L., Costimulation, tolerance and ignorance of cytolytic T lymphocytes in immune responses to tumor antigens, *Life Sci.* 60, 2035, 1997.
33. Fuchs, E.J., and Matzinger, P., Is cancer dangerous to the immune system? *Semin. Immunol.* 8, 271, 1996.
34. Wortzel, R.D., Philipps, C., and Schreiber, H., Multiple tumor-specific antigens expressed on a single tumor cell, *Nature* 304, 165, 1983.
35. Urban, J.L. et al., Mechanisms of syngeneic tumor rejection. Susceptibility of host-selected progressor variants to various immunological effector cells, *J. Exp. Med.* 155, 557, 1982.
36. Uyttenhove, C., Maryanski, J., and Boon, T., Escape of mouse mastocytoma P815 after nearly complete rejection is due to antigen-loss variants rather than immunosuppression, *J. Exp. Med.* 157, 1040, 1983.
37. Hui, K., Grosveld, F., and Festenstein, H., Rejection of transplantable AKR leukemia cells following MHC DNA-mediated cell transformation, *Nature* 311, 750, 1984.
38. Haywood, G.R., and McKhann, C.F., Antigenic specificities on murine sarcoma cells. Reciprocal relationship between normal transplantation antigens (H-2) and tumor-specific immunogenicity, *J. Exp. Med.* 133, 1171, 1971.
39. Trowsdale, J. et al., Expression of HLA-A, -B, and -C and beta 2-microglobulin antigens in human choriocarcinoma cell lines, *J. Exp. Med.* 152, 11s, 1980.
40. Wallich, R. et al., Abrogation of metastatic properties of tumor cells by de novo expression of H-2K antigens following H-2 gene transfection, *Nature* 315, 301, 1985.
41. Restifo, N.P. et al., Defective presentation of endogenous antigens by a murine sarcoma. Implications for the failure of an antitumor immune response, *J. Immunol.* 147, 1453, 1991.
42. Singh, R., and Paterson, Y., Immunoediting sculpts tumor epitopes during immunotherapy, *Cancer Res.* 67, 1887, 2007.
43. Torre-Amione, G. et al., A highly immunogenic tumor transfected with a murine transforming growth factor type beta 1 cDNA escapes immune surveillance, *Proc. Natl. Acad. Sci. USA* 87, 1486, 1990.
44. O'Connell, J. et al., The Fas counterattack: Fas-mediated T cell killing by colon cancer cells expressing Fas ligand, *J. Exp. Med.* 184, 1075, 1996.
45. Paterson, Y., The relationship between intra-cellular bacterial life-styles and immune responsiveness to bacterial delivered antigens. In *Intracellular bacterial vaccine vectors.* Paterson, Y., ed. John Wiley & Sons, Inc., New York, 1–24, 1999.
46. Pan, Z.K. et al., A recombinant *Listeria monocytogenes* vaccine expressing a model tumor antigen protects mice against lethal tumor cell challenge and causes regression of established tumors, *Nat. Med.* 1, 471, 1995.
47. Pan, Z.K., Weiskirch, L.M., and Paterson, Y., Regression of established B16F10 melanoma with a recombinant *Listeria monocytogenes* vaccine, *Cancer Res.* 59, 5264, 1999.
48. Starks, H. et al., *Listeria monocytogenes* as a vaccine vector: Virulence attenuation or existing antivector immunity does not diminish therapeutic efficacy, *J. Immunol.* 173, 420, 2004.
49. Gaillard, J.L. et al., *In vitro* model of penetration and intracellular growth of *Listeria monocytogenes* in the human enterocyte-like cell line Caco-2, *Infect. Immun.* 55, 2822, 1987.
50. Tilney, L.G., and Portnoy, D.A., Actin filaments and the growth, movement, and spread of the intracellular bacterial parasite, *Listeria monocytogenes*, *J. Cell Biol.* 109, 1597, 1989.
51. Camilli, A., Tilney, L.G., and Portnoy, D.A., Dual roles of *plcA* in *Listeria monocytogenes* pathogenesis, *Mol. Microbiol.* 8, 143, 1993.
52. Marquis, H. et al., Intracytoplasmic growth and virulence of *Listeria monocytogenes* auxotrophic mutants, *Infect. Immun.* 61, 3756, 1993.
53. Kocks, C. et al., *L. monocytogenes*-induced actin assembly requires the *actA* gene product, a surface protein, *Cell* 68, 521, 1992.
54. Cossart P. et al., Listeriolysin O is essential for virulence of *Listeria monocytogenes*: Direct evidence obtained by gene complementation, *Infect. Immun.* 57, 3629, 1989.
55. Vazquez-Boland, J.A. et al., Nucleotide sequence of the lecithinase operon of *Listeria monocytogenes* and possible role of lecithinase in cell-to-cell spread, *Infect. Immun.* 60, 219, 1992.
56. Portnoy, D.A. et al., Molecular determinants of *Listeria monocytogenes* pathogenesis, *Infect. Immun.* 60, 1263, 1992.
57. Mackaness, G.B., The influence of immunologically committed lymphoid cells on macrophage activity *in vivo*, *J. Exp. Med.* 129, 973, 1969.
58. Unanue, E.R., Studies in listeriosis show the strong symbiosis between the innate cellular system and the T-cell response, *Immunol. Rev.* 158, 11, 1997.

59. Mata, M., and Paterson, Y., Th1 T cell responses to HIV-1 Gag protein delivered by a *Listeria monocytogenes* vaccine are similar to those induced by endogenous listerial antigens, *J. Immunol.* 163, 1449, 1999.
60. Southwick, F.S., and Purich, D.L., Intracellular pathogenesis of listeriosis, *New Engl. J. Med.* 334, 770, 1996.
61. Pamer, E.G., Immune responses to *Listeria monocytogenes*, *Nat. Rev. Immunol.* 4, 812, 2004.
62. Sun, J.C., Williams, M.A., and Bevan, M.J., CD4+ T cells are required for the maintenance, not programming, of memory CD8+ T cells after acute infection, *Nat. Immunol.* 5, 927, 2004.
63. Cerundolo, V., T cells work together to fight cancer, *Curr. Biol.* 9, R695, 1999.
64. Bottino, C., Moretta, L., and Moretta, A., NK cell activating receptors and tumor recognition in humans, *Curr. Topics Microbiol.Immunol.* 298, 175, 2006.
65. Ferrarini, M. et al., Human γδ T cells: A nonredundant system in the immune-surveillance against cancer, *Trends Immunol.* 23, 14, 2002.
66. Weiss, S., and Chakraborty, T., Transfer of eukaryotic expression plasmids to mammalian host cells by bacterial carriers, *Curr. Opin. Biotechnol.* 12, 467, 2001.
67. Chabalgoity, J.A. et al., Live bacteria as the basis for immunotherapies against cancer, *Expert Rev. Vaccines* 1, 495, 2002.
68. Pawelek, J.M., Low, K.B., and Bermudes, D., Bacteria as tumor-targeting vectors, *Lancet Oncol.* 4, 548, 2003.
69. Gunn, G.R., Zubair, A., and Paterson, Y., Harnessing bacteria for cancer immunotherapy. In *Vaccines: Delivery systems,* Goebel, W., and Dietrich, G., eds. Scientific Horizon Press, Norwich, U.K., 315–348, 2002.
70. Bishop, D.K., and Hinrichs, D.J., Adoptive transfer of immunity to *Listeria monocytogenes.* The influence of *in vitro* stimulation on lymphocyte subset requirements, *J. Immunol.* 139, 2005, 1987.
71. Gunn, G.R. et al., Two *Listeria monocytogenes* vaccine vectors that express different molecular forms of human papilloma virus-16 (HPV-16) E7 induce qualitatively different T cell immunity that correlates with their ability to induce regression of established tumors immortalized by HPV-16, *J. Immunol.* 167, 6471, 2001.
72. Smith, K., and Youngman, P., Use of a new integrational vector to investigate compartment-specific expression of the *Bacillus subtilis spoIIM* gene, *Biochimie* 74, 705, 1992.
73. Wirth, R., An, F.Y., and Clewell, D.B., Highly efficient protoplast transformation system for *Streptococcus faecalis* and a new *Escherichia coli-S. faecalis* shuttle vector, *J. Bacteriol.* 165, 831, 1986.
74. Ikonomidis, G. et al., Delivery of a viral antigen to the class I processing and presentation pathway by *Listeria monocytogenes*, *J. Exp. Med.* 180, 2209, 1994.
75. Ripio, M. T. et al., Transcriptional activation of virulence genes in wild-type strains of *Listeria monocytogenes* in response to a change in the extracellular medium composition, *Res. Microbiol.* 147, 371, 1996.
76. Cheers, C. et al., Resistance and susceptibility of mice to bacterial infection: Course of listeriosis in resistant or susceptible mice, *Infect. Immun.* 19, 763, 1978.
77. Singh, R. et al., Fusion to listeriolysin O and delivery by *Listeria monocytogenes* enhances the immunogenicity of HER-2/neu and reveals subdominant epitopes in the FVB/N mouse, *J. Immunol.* 175, 3663, 2005.
78. Boyer, J. D. et al., DNA prime *Listeria* boost induces a cellular immune response to SIV antigens in the rhesus macaque model that is capable of limited suppression of SIV239 viral replication, *Virology* 333, 88, 2005.
79. Milenbachs Lukowiak, A. et al., Deregulation of *Listeria monocytogenes* virulence gene expression by two distinct and semi-independent pathways, *Microbiology* 150, 321, 2004.
80. Kohler, S. et al., The gene coding for protein p60 of *Listeria monocytogenes* and its use as a specific probe for *Listeria monocytogenes*, *Infect. Immun.* 58, 1943, 1990.
81. Kohler, S. et al., Expression of the iap gene coding for protein p60 of *Listeria monocytogenes* is controlled on the posttranscriptional level, *J. Bacteriol.* 173, 4668, 1991.
82. Verch, T., Pan, Z.K., and Paterson, Y., *Listeria monocytogenes*-based antibiotic resistance gene-free antigen delivery system applicable to other bacterial vectors and DNA vaccines, *Infect. Immun.* 72, 6418, 2004.
83. Park, S.F., and Stewart, G.S., High-efficiency transformation of *Listeria monocytogenes* by electroporation of penicillin-treated cells, *Gene* 94, 129, 1990.
84. Shen, H. et al., Recombinant *Listeria monocytogenes* as a live vaccine vehicle for the induction of protective antiviral cell-mediated immunity, *Proc. Natl. Acad. Sci. USA* 92, 3987, 1995.
85. Pilgrim, S. et al., Bactofection of mammalian cells by *Listeria monocytogenes:* Improvement and mechanism of DNA delivery, *Gene Ther.* 10, 2036, 2003.

86. Decatur, A.L., and Portnoy, D.A., A PEST-like sequence in listeriolysin O essential for *Listeria monocytogenes* pathogenicity, *Science* 290, 992, 2000.
87. Dubail, I., Berche, P., and Charbit, A., Listeriolysin O as a reporter to identify constitutive and *in vivo*-inducible promoters in the pathogen *Listeria monocytogenes*, *Infect. Immun.* 68, 3242, 2000.
88. Sewell, D.A. et al., Recombinant *Listeria* vaccines containing PEST sequences are potent immune adjuvants for the tumor-associated antigen human papillomavirus-16 E7, *Cancer Res.* 64, 8821, 2004.
89. Sewell, D.A. et al., Regression of HPV-positive tumors treated with a new *Listeria monocytogenes* vaccine, *Arch. Otolaryngol. Head Neck Surg.* 130, 92, 2004.
90. Peters, C., and Paterson, Y., Enhancing the immunogenicity of bioengineered *Listeria monocytogenes* by passaging through live animal hosts, *Vaccine* 21, 1187, 2003.
91. Vahidy, R., Waseem, M., and Khalid, S.M., A comparative study of unpassaged and animal passaged cultures of *Listeria monocytogenes* in rabbits, *Ann. Acad. Med. Singapore* 25, 139, 1996.
92. Peters, C. et al., The induction of HIV-Gag specific $CD8^+$ T cells in the spleen and GALT by parenteral or mucosal immunization using recombinant *Listeria monocytogenes*-HIV-Gag, *J. Immunol.* 170, 5176, 2003.
93. Lecuit, M. et al., A transgenic model for listeriosis: Role of internalin in crossing the intestinal barrier, *Science* 292, 1722, 2001.
94. Pan, Z.K. et al., Regression of established tumors in mice mediated by the oral administration of a recombinant *Listeria monocytogenes* vaccine, *Cancer Res.* 55, 4776, 1995.
95. Lin, C.W. et al., Oral vaccination with recombinant *Listeria monocytogenes* expressing human papillomavirus type 16 E7 can cause tumor growth in mice to regress, *Int. J. Cancer* 102, 629, 2002.
96. Jensen, E.R. et al., Recombinant *Listeria monocytogenes* vaccination eliminates papillomavirus-induced tumors and prevents papilloma formation from viral DNA, *J. Virol.* 71, 8467, 1997.
97. Christensen, N.D., Cottontail rabbit papillomavirus (CRPV) model system to test antiviral and immunotherapeutic strategies, *Antiviral Chem. Chemother.* 16, 355, 2005.
98. Liau, L.M. et al., Tumor immunity within the central nervous system stimulated by recombinant *Listeria monocytogenes* vaccination, *Cancer Res.* 62, 2287, 2002.
99. Brockstedt, D.G. et al., *Listeria*-based cancer vaccines that segregate immunogenicity from toxicity, *Proc. Natl. Acad. Sci. USA* 101, 13832, 2004.
100. Bruhn, K. W. et al., Characterization of anti-self CD8 T-cell responses stimulated by recombinant *Listeria monocytogenes* expressing the melanoma antigen TRP-2, *Vaccine* 23, 4263, 2005.
101. Tindle, R.W., Human papillomavirus vaccines for cervical cancer, *Curr. Opin. Immunol.* 8, 643, 1996.
102. Stanley, M., Coleman, N., and Chambers, M., The host response to lesions induced by human papillomavirus, *Ciba Foundation Symposium* 187, 21, 1994.
103. Palefsky, J.M., and Barrasso, R., HPV infection and disease in men, *Obst. Gynecol. Clin. N. Am.* 23, 895, 1996.
104. Unger, E.R. et al., Human papillomavirus type in anal epithelial lesions is influenced by human immunodeficiency virus, *Arch. Pathol. Lab. Med.* 121, 820, 1997.
105. Braun, L., Role of human immunodeficiency virus infection in the pathogenesis of human papillomavirus-associated cervical neoplasia, *Am. J. Pathol.* 144, 209, 1994.
106. Petry, K.U. et al., Cellular immunodeficiency enhances the progression of human papillomavirus-associated cervical lesions, *Int. J. Cancer* 57, 836, 1994.
107. Ho, G.Y. et al., Risk of genital human papillomavirus infection in women with human immunodeficiency virus-induced immunosuppression, *Int. J. Cancer* 56, 788, 1994.
108. Sun, X.W. et al., Human papillomavirus infection in women infected with the human immunodeficiency virus, *New Engl. J. Med.* 337, 1343, 1997.
109. Ellis, J.R. et al., The association of an HPV16 oncogene variant with HLA-B7 has implications for vaccine design in cervical cancer, *Nat. Med.* 1, 464, 1995.
110. Gillison, M.L. et al., Evidence for a causal association between human papillomavirus and a subset of head and neck cancers, *J. Natl. Cancer Inst.* 92, 709, 2000.
111. Lowy, D.R., Kirnbauer, R., and Schiller, J.T., Genital human papillomavirus infection, *Proc. Natl. Acad. Sci. USA* 91, 2436, 1994.
112. Seedorf, K. et al., Identification of early proteins of the human papilloma viruses type 16 (HPV 16) and type 18 (HPV 18) in cervical carcinoma cells, *EMBO J.* 6, 139, 1987.
113. Tindle, R.W., and Frazer, I.H., Immune response to human papillomaviruses and the prospects for human papillomavirus-specific immunisation, *Curr. Topics Microbiol. Immunol.* 186, 217, 1994.
114. Scheffner, M. et al., The E6 oncoprotein encoded by human papillomavirus types 16 and 18 promotes the degradation of p53, *Cell* 63, 1129, 1990.

115. Lin, K.Y. et al., Treatment of established tumors with a novel vaccine that enhances major histocompatibility class II presentation of tumor antigen, *Cancer Res.* 56, 21, 1996.
116. Sakaguchi, S. et al., Immunologic self-tolerance maintained by activated T cells expressing IL-2 receptor alpha-chains (CD25). Breakdown of a single mechanism of self-tolerance causes various autoimmune diseases, *J. Immunol.* 155, 1151, 1995.
117. Piccirillo, C.A., and Shavach, E.M., Cutting edge: control of CD8(+) T cell activation by CD4(+)CD25(+) immunoregulatory cells, *J. Immunol.* 167, 1137, 2001.
118. Onizuka, S. et al., Tumor rejection by *in vivo* administration of anti-CD25 (interleukin-2 receptor alpha) monoclonal antibody, *Cancer Res.* 59, 3128, 1999.
119. Read, S., Malmstrom, V., and Powrie, F., Cytotoxic T lymphocyte-associated antigen 4 plays an essential role in the function of CD25(+)CD4(+) regulatory cells that control intestinal inflammation, *J. Exp. Med.* 192, 295, 2000.
120. Chang, H.L. et al., Increased transforming growth factor beta expression inhibits cell proliferation *in vitro*, yet increases tumorigenicity and tumor growth of Meth A sarcoma cells, *Cancer Res.* 53, 4391, 1993
121. Hussain, S.F., and Paterson, Y., $CD4^+CD25^+$ Suppressor T cells that secrete inhibitory cytokines TGF-beta and IL-10 are preferentially induced by a vaccine vector, *J. Immunother.* 27, 339, 2004.
122. Rechsteiner, M., and Rogers, S.W., PEST sequences and regulation by proteolysis, *Trends Biochem. Sci.* 21, 267, 1996.
123. Rogers, S., Wells, R., and Rechsteiner, M., Amino acid sequences common to rapidly degraded proteins: The PEST hypothesis, *Science* 234, 364, 1986.
124. Souders, N.C. et al., *Listeria*-based vaccines can partially overcome tolerance by expanding low avidity $CD8^+$ T cells capable of eradicating a solid tumor in a transgenic mouse model of cancer, *Cancer Immun.* 7, 2, 2007.
125. Derbinski, J. et al., Promiscuous gene expression in medullary thymic epithelial cells mirrors the peripheral self, *Nat. Immunol.* 2, 1032, 2001.
126. Gallegos, A.M., and Bevan, M.J., Central tolerance to tissue-specific antigens mediated by direct and indirect antigen presentation, *J. Exp. Med.* 200, 1039, 2004.
127. Anderson, M. S. et al., Projection of an immunological self shadow within the thymus by the Aire protein, *Science* 298, 1395, 2002.
128. Liston, A. et al., Aire regulates negative selection of organ-specific T cells, *Nat. Immunol.* 4, 350, 2003.
129. Anderson, M.S. et al., The cellular mechanism of Aire control of T cell tolerance, *Immunity* 23, 227, 2005.
130. Liston, A. et al., Gene dosage—Limiting role of Aire in thymic expression, clonal deletion, and organ-specific autoimmunity, *J. Exp. Med.* 200, 1015, 2004.
131. Ercolini, A.M. et al., Recruitment of latent pools of high-avidity CD8(+) T cells to the antitumor immune response, *J. Exp. Med.* 201, 1591, 2005.
132. Auewarakul, P., Gissmann, L., and Cid-Arregui, A., Targeted expression of the E6 and E7 oncogenes of human papillomavirus type 16 in the epidermis of transgenic mice elicits generalized epidermal hyperplasia involving autocrine factors, *Mol. Cell Biol.* 14, 8250, 1994.
133. Riezebos-Brilman, A. et al., Induction of human papilloma virus E6/E7-specific cytotoxic T-lymphocyte activity in immune-tolerant, E6/E7-transgenic mice, *Gene Ther.* 12, 1410, 2005.
134. Lambert, P.F. et al., Epidermal cancer associated with expression of human papillomavirus type 16 E6 and E7 oncogenes in the skin of transgenic mice, *Proc. Natl. Acad. Sci. USA* 90, 5583, 1993.
135. Herd, K. et al., E7 oncoprotein of human papillomavirus type 16 expressed constitutively in the epidermis has no effect on E7-specific B- or Th-repertoires or on the immune response induced or sustained after immunization with E7 protein, *Virology* 231, 155, 1997.
136. Doan, T. et al., Mice expressing the E7 oncogene of HPV16 in epithelium show central tolerance, and evidence of peripheral energizing tolerance, to E7-encoded cytotoxic T-lymphocyte epitopes, *Virology* 244, 352, 1998.
137. Melero, I. et al., Immunological ignorance of an E7-encoded cytolytic T-lymphocyte epitope in transgenic mice expressing the E7 and E6 oncogenes of human papillomavirus type 16, *J. Virol.* 71, 3998, 1997
138. Frazer, I.H. et al., Split tolerance to a viral antigen expressed in thymic epithelium and keratinocytes, *Eur. J. Immunol.* 28, 2791, 1998.
139. Tindle, R.W. et al., Nonspecific down-regulation of $CD8^+$ T-cell responses in mice expressing human papillomavirus type 16 E7 oncoprotein from the keratin-14 promoter, *J. Virol.* 75, 5985, 2001.
140. Kiessling, R. et al., Cellular immunity to the Her-2/neu protooncogene, *Adv. Cancer Res.* 85, 101, 2002.
141. Treurniet, H.F. et al., Differences in breast cancer risk factors to neu (c-erbB-2) protein overexpression of the breast tumor, *Cancer Res.* 52, 2344, 1992.

142. Weiner, D.B. et al., A point mutation in the neu oncogene mimics ligand induction of receptor aggregation, *Nature* 339, 230, 1989.
143. Siegel, P.M., and Muller, W.J., Mutations affecting conserved cysteine residues within the extracellular domain of Neu promote receptor dimerization and activation, *Proc. Natl. Acad. Sci. USA* 93, 8878, 1996.
144. Klapper, L.N. et al., The ErbB-2/HER2 oncoprotein of human carcinomas may function solely as a shared coreceptor for multiple stroma-derived growth factors, *Proc. Natl. Acad. Sci. USA* 96, 4995, 1999.
145. Prenzel, N. et al., The epidermal growth factor receptor family as a central element for cellular signal transduction and diversification, *Endocrine-Related Cancer* 8, 11, 2001.
146. Disis, M.L. et al., Existent T-cell and antibody immunity to HER-2/neu protein in patients with breast cancer, *Cancer Res.* 54, 16, 1994.
147. Tuttle, T.M. et al., Proliferative and cytokine responses to class II HER-2/neu-associated peptides in breast cancer patients, *Clin. Cancer Res.* 4, 2015, 1998.
148. Brugger, W. et al., Approaches to dendritic cell-based immunotherapy after peripheral blood stem cell transplantation, *Ann. NY Acad. Sci.* 872, 363, 1999.
149. Disis, M.L. et al., Generation of immunity to the HER-2/neu oncogenic protein in patients with breast and ovarian cancer using a peptide-based vaccine, *Clin. Cancer Res.* 5, 1289, 1999.
150. Disis, M.L. et al., Pre-existent immunity to the HER-2/neu oncogenic protein in patients with HER-2/neu overexpressing breast and ovarian cancer, *Breast Cancer Res. Treat.* 62, 245, 2000.
151. Disis, M.L., and Cheever, M.A., HER-2/neu protein: A target for antigen-specific immunotherapy of human cancer, *Adv. Cancer Res.* 71, 343, 1997.
152. Coronella, J.A. et al., Evidence for an antigen-driven humoral immune response in medullary ductal breast cancer, *Cancer Res.* 61, 7889, 2001.
153. Disis, M.L. et al., High-titer HER-2/neu protein-specific antibody can be detected in patients with early-stage breast cancer, *J. Clin. Oncol.* 15, 3363, 1997.
154. Ercolini, A.M. et al., Identification and characterization of the immunodominant rat HER-2/neu MHC class I epitope presented by spontaneous mammary tumors from HER-2/neu-transgenic mice, *J. Immunol.* 170, 4273, 2003.
155. Muller, W.J., Expression of activated oncogenes in the murine mammary gland: Transgenic models for human breast cancer, *Cancer Metastasis Rev.* 10, 217, 1991.
156. Guy, C.T. et al., Expression of the neu protooncogene in the mammary epithelium of transgenic mice induces metastatic disease, *Proc. Natl. Acad. Sci. USA* 89, 10578, 1992.
157. Singh, R., and Paterson, Y., Vaccination strategy determines the emergence and dominance of $CD8^+$ T-cell epitopes in a FVB/N rat HER-2/neu mouse model of breast cancer, *Cancer Res.* 66, 7748, 2006.
158. Singh, R., and Paterson, Y., In the FVB/N HER-2/neu transgenic mouse both peripheral and central tolerance limit the immune response targeting HER-2/neu induced by *Listeria monocytogenes*-based vaccines, *Cancer Immunol. Immunother.* 56, 927, 2007.
159. Lollini, P.L. et al., Down-regulation of major histocompatibility complex class I expression in mammary carcinoma of HER-2/neu transgenic mice, *Int. J. Cancer* 77, 937, 1998.
160. Nagata, Y. et al., Peptides derived from a wild-type murine proto-oncogene c-erbB-2/HER2/neu can induce CTL and tumor suppression in syngeneic hosts, *J. Immunol.* 159, 1336, 1997.
161. Peng, X., Hussain, S.F., and Paterson, Y., The ability of two *Listeria monocytogenes* vaccines targeting human papillomavirus-16 E7 to induce an antitumor response correlates with myeloid dendritic cell function, *J. Immunol.* 172, 6030, 2004.
162. Neeson, P., Pan, Z.-K., and Paterson, Y., Listeriolysin O is an improved protein carrier for lymphoma immunoglobulin idiotype and provides protection against 38C13 lymphoma, *Cancer Immunol. Immunother.*, 2007 Sep 18; [Epub ahead of print].
163. Schafer, R. et al., Induction of a cellular immune response to a foreign antigen by a recombinant *Listeria monocytogenes* vaccine, *J. Immunol.* 149, 53, 1992.
164. Lauvau, G. et al., Priming of memory but not effector CD8 T cells by a killed bacterial vaccine, *Science* 294, 1735, 2001.
165. von Koenig, C.H., Finger, H., and Hof, H., Failure of killed *Listeria monocytogenes* vaccine to produce protective immunity, *Nature* 297, 233, 1982.
166. Brockstedt, D.G. et al., Killed but metabolically active microbes: A new vaccine paradigm for eliciting effector T-cell responses and protective immunity, *Nat. Med.* 11, 853, 2005.
167. Datta, S.K. et al., Vaccination with irradiated *Listeria* induces protective T cell immunity, *Immunity* 25, 143, 2006.

168. Hamilton, S.E. et al., Listeriolysin O-deficient *Listeria monocytogenes* as a vaccine delivery vehicle: Antigen-specific CD8 T cell priming and protective immunity, *J. Immunol.* 177, 4012, 2006.
169. Dietrich, G. et al., Live attenuated bacteria as vectors to deliver plasmid DNA vaccines, *Curr. Opin. Mol. Ther.* 5, 10, 2003.
170. Dietrich, G. et al., Delivery of DNA vaccines by attenuated intracellular bacteria, *Immunol. Today* 20, 251–253, 1999.
171. Lamikanra, A. et al., Regression of established human papillomavirus type 16 (HPV-16) immortalized tumors *in vivo* by vaccinia viruses expressing different forms of HPV-16 E7 correlates with enhanced CD8(+) T-cell responses that home to the tumor site, *J. Virol.* 75, 9654, 2001.
172. Peng, X., Treml, J., and Paterson, Y., Plasmid listeriolysin O (LLO) acts as an adjuvant to enhance the potency of an HPV-16 E7 DNA vaccine by improving both $CD4^+$ and $CD8^+$ T cell responses, *Cancer Immunol. Immunother.* 56, 797, 2007.
173. Dietrich, G. et al., From evil to good: A cytolysin in vaccine development, *Trends Microbiol.* 9, 23, 2001.
174. Gunn, G.R., Peters, C., and Paterson, Y., Listeriolysin—A useful cytolysin, *Trends Microbiol.* 9, 161, 2001.
175. Cvetkovic, R.S., and Perry, C.M., Rituximab: A review of its use in non-Hodgkin's lymphoma and chronic lymphocytic leukemia, *Drugs* 66, 791, 2006.
176. Marcus, R., Use of 90Y-ibritumomab tiuxetan in non-Hodgkin's lymphoma, *Semin. Oncol.* 32, S36, 2005.
177. Bergman, Y., and Haimovich, J., Characterization of a carcinogen-induced murine B lymphocyte cell line of C3H/eB origin, *Eur. J. Immunol.* 7, 413, 1977.
178. Janda, J. et al., Cross-presentation of *Listeria*-derived CD8 T cell epitopes requires unstable bacterial translation products, *J. Immunol.* 173, 5644, 2004.
179. Dietrich, G. et al., Delivery of antigen-encoding plasmid DNA into the cytosol of macrophages by attenuated suicide *Listeria monocytogenes*, *Nat. Biotechnol.* 16, 181, 1998.
180. Loessner, M.J., Wendlinger, G., and Scherer, S., Heterogeneous endolysins in *Listeria monocytogenes* bacteriophages: A new class of enzymes and evidence for conserved holin genes within the siphoviral lysis cassettes, *Mol. Microbiol.* 16, 1231, 1995.
181. Loeffler, J.M., Nelson, D., and Fischetti, V.A., Rapid killing of *Streptococcus pneumoniae* with a bacteriophage cell wall hydrolase, *Science* 294, 2170, 2001.
182. Souders, N.C., Verch, T., and Paterson, Y., *In vivo* bactofection: *Listeria* can function as a DNA-cancer vaccine, *DNA Cell Biol.* 25, 142, 2006.
183. Xiang, R. et al., An autologous oral DNA vaccine protects against murine melanoma, *Proc. Natl. Acad. Sci. USA* 97, 5492, 2000.
184. Paglia, P. et al., Gene transfer in dendritic cells, induced by oral DNA vaccination with *Salmonella typhimurium*, results in protective immunity against a murine fibrosarcoma, *Blood* 92, 3172, 1998.
185. Shata, M.T. et al., Mucosal and systemic HIV-1 Env-specific CD8(+) T-cells develop after intragastric vaccination with a *Salmonella* Env DNA vaccine vector, *Vaccine* 20, 623, 2001.
186. Galen, J.E., and Levine, M.M., Can a "flawless" live vector vaccine strain be engineered? *Trends Microbiol.* 9, 372, 2001.
187. Paglia, P. et al., The defined attenuated *Listeria monocytogenes* delta mp12 mutant is an effective oral vaccine carrier to trigger a long-lasting immune response against a mouse fibrosarcoma, *Eur. J. Immunol.* 27, 1570, 1997.

Index

B

C

D

E

F

G

H

I

J

K

L

M

N

O

P

Q

R

S

T

U

V

W

X

Y

Z

Milton Keynes UK
Ingram Content Group UK Ltd.
UKHW052026071024
449327UK00027B/2449